Advances in Intelligent Systems and Computing

Volume 845

Series editor

Janusz Kacprzyk, Polish Academy of Sciences, Warsaw, Poland
e-mail: kacprzyk@ibspan.waw.pl

The series "Advances in Intelligent Systems and Computing" contains publications on theory, applications, and design methods of Intelligent Systems and Intelligent Computing. Virtually all disciplines such as engineering, natural sciences, computer and information science, ICT, economics, business, e-commerce, environment, healthcare, life science are covered. The list of topics spans all the areas of modern intelligent systems and computing such as: computational intelligence, soft computing including neural networks, fuzzy systems, evolutionary computing and the fusion of these paradigms, social intelligence, ambient intelligence, computational neuroscience, artificial life, virtual worlds and society, cognitive science and systems, Perception and Vision, DNA and immune based systems, self-organizing and adaptive systems, e-Learning and teaching, human-centered and human-centric computing, recommender systems, intelligent control, robotics and mechatronics including human-machine teaming, knowledge-based paradigms, learning paradigms, machine ethics, intelligent data analysis, knowledge management, intelligent agents, intelligent decision making and support, intelligent network security, trust management, interactive entertainment, Web intelligence and multimedia.

The publications within "Advances in Intelligent Systems and Computing" are primarily proceedings of important conferences, symposia and congresses. They cover significant recent developments in the field, both of a foundational and applicable character. An important characteristic feature of the series is the short publication time and world-wide distribution. This permits a rapid and broad dissemination of research results.

More information about this series at http://www.springer.com/series/11156

Aboul Ella Hassanien · Mohamed F. Tolba
Khaled Shaalan · Ahmad Taher Azar
Editors

Proceedings of the International Conference on Advanced Intelligent Systems and Informatics 2018

Springer

Editors
Aboul Ella Hassanien
Information Technology Department,
Faculty of Computers and Information
Cairo University
Giza, Egypt

Mohamed F. Tolba
Ain Shams University
Cairo, Egypt

Khaled Shaalan
Dubai International Academic City
The British University in Dubai
Dubai, United Arab Emirates

Ahmad Taher Azar
Faculty of Computers and Information
Benha University
Benha, Egypt

ISSN 2194-5357 ISSN 2194-5365 (electronic)
Advances in Intelligent Systems and Computing
ISBN 978-3-319-99009-5 ISBN 978-3-319-99010-1 (eBook)
https://doi.org/10.1007/978-3-319-99010-1

Library of Congress Control Number: 2018951098

This Springer imprint is published by the registered company Springer Nature Switzerland AG
The registered company address is: Gewerbestrasse 11, 6330 Cham, Switzerland

Preface

This volume constitutes the refereed proceedings of the 4th International Conference on Advanced Intelligent Systems and Informatics (AISI 2018), which took place in Cairo, Egypt, during September 3–5, 2018, and is an international interdisciplinary conference covering research and development in the field of informatics and intelligent systems.

In response to the call for papers for AISI2018, 142 papers were submitted for presentation and inclusion in the proceedings of the conference. After a careful blind refereeing process, 61 papers were selected for inclusion in the conference proceedings. The papers were evaluated and ranked on the basis of their significance, novelty, and technical quality by at least two reviewers per paper. After a careful blind refereeing process, 61 papers were selected for inclusion in the conference proceedings. The papers cover current research in machine learning, robot modeling and control systems, sentiment analysis and Arabic text mining, Part (V) deep learning and cloud computing, Part (VII) data mining, visualization and E-learning, and intelligence swarms and optimization. In addition to these papers, the program included one keynote talk by Professor Zbigniew Suraj, University of Rzeszów, Poland, on linking fuzzy Petri nets with interval analysis.

We express our sincere thanks to the plenary speakers, workshop chairs, and international program committee members for helping us to formulate a rich technical program. We would like to extend our sincere appreciation for the outstanding work contributed over many months by the organizing committee: Local Organization Chair and Publicity Chair. We also wish to express our appreciation to the SRGE members for their assistance. We would like to emphasize that the success of AISI2018 would not have been possible without the support of many committed volunteers who generously contributed their time, expertise, and resources toward making the conference an unqualified success. Finally, thanks to

Springer team for their supporting in all stages of the production of the proceedings. We hope that you will enjoy the conference program.

Aboul Ella Hassanien
Mohamed F. Tolba
Ahmad Taher Azar
Khaled Shaalan

Organization

Honorary Chair

Fahmy Tolba, Egypt

General Chairs

Aboul Ella Hassanien	Egypt
Khaled Shaalan	British University in Dubai

Program Chairs

Siddhartha Bhattacharyya, India
Mohamed Elhoseny, Egypt
Ashraf Darwish, Egypt

International Advisory Board

Neetu Agarwal, India
Swagatam Das, India
Xiao-Zhi Gao, Finland
Hesham Hefny, Egypt
Fahmy Tolba, Egypt
Fatos Xhafa, Spain
Vaclav Snasel, Czech Republic
Janusz Kacprzyk, Poland
Tai-hoon Kim, Korea
Qing Tan, Canada
Hiroshi Sakai, Japan
Khaled Shaalan, Egypt

Publicity Chairs

Saurav Karmakar, India
Deepak Gupta, India
Abdelhameed Ibrahim, Egypt
Mohamed Abd Elfattah, Egypt

Technical Program Committee

Said Broumi	Morocco
Nahla Zaaboub Haddar	University of Sfax, Tunisia
Maytham Alabbas	UK
Essam Hamed	Egypt
Sherif Abuelenin	Egypt
Ashraf AbdelRaouf	Egypt
Aarti Singh	India
Tahani Alsubait	UK
Evgenia Theodotou	Greece
Pavel Kromer	Czech Republic
Irma Aslanishvili	Czech Republic
Jan Platos	Czech Republic
Ivan Zelinka	Czech Republic
Sebastian Tiscordio	Czech Republic
Natalia Spyropoulou	Hellenic Open University, Greece
Dimitris Sedaris	Hellenic Open University, Greece
Vassiliki Pliogou	Metropolitan College, Greece
Pilios Stavrou	Metropolitan College, Greece
Eleni Seralidou	University of Piraeus, Greece
Stelios Kavalaris	Metropolitan College, Greece
Litsa Charitaki	University of Athens, Greece
Elena Amaricai	University of Timisoara, Greece
Qing Tan	Athabasca University, Greece
Pascal Roubides	Broward College, Greece
Manal Abdullah	King Abdulaziz University, KSA
Mandia Athanasopoulou	Metropolitan College, Greece
Vicky Goltsi	Metropolitan College, Greece
Mohammad Reza Noruzi	Tarbiat Modares University, Iran
Abdelhameed Ibrahim	Egypt
Ahmad Taher Azar	Egypt
Ahmed Elhayek	Germany
Alaa Tharwat	Belgium
Amira S. Ashour	KSA
Boyang Albert Li	.

Edgard Marx	Germany
Fatma Helmy	Egypt
Hany Alnashar	Egypt
Islam Amin	Egypt
Ivan Ermilov	Germany
Mahmoud Awadallah	USA
Maryam Hazman	Egypt
Minu Kesheri	India
Mohamed Hamed	Germany
Mohamed Shrif	Germany
Mohamed Abdelfatah	Egypt
Mohammed Abdel-Megeed	Egypt
Mona Solyman	Egypt
Muhammad Saleem	Germany
Nabiha Azizi	Algeria
Namshik Han	UK
Noreen Kausar	KSA
Rania Hodhod	Georgia
Reham Ahmed	Egypt
Sara Abdelkader	Canada
Sayan Chakraborty	India
Shoji Tominaga	Japan
Siva Ganesh Malla	India
Soumya Banerjee	India
Sourav Samanta	India
Suvojit Acharjee	India
Swarna Kanchan	India
Takahiko Horiuchi	Japan
Tommaso Soru	Germany
Wahiba Ben Abdessalem	KSA
Zeineb Chelly	Tunis

Local Arrangement Chairs

Mohamed Abd Elfattah (Chair), Egypt
Hassan Aboul Ella Hassanien
Heba Aboul Ella Hassanien
Taha Aboul Ella Hassanien

Linking Fuzzy Petri Nets with Interval Analysis (Invited Speaker)

Zbigniew Suraj

University of Rzeszów in Poland

Abstract. Fuzzy Petri nets (FPNs) are a modification of the classic Petri nets that enable modeling of knowledge-based systems (KBSs) with uncertain information. FPNs have been widely used in knowledge representation and approximate reasoning in KBSs. Previous FPN models, as indicated in the literature on this subject, are plagued by a number of shortcomings and are not suitable for increasingly complex KBSs. Therefore, many alternative models have been proposed in the literature to increase the power of FPN knowledge representation and to apply rule-based reasoning more intelligently and more efficiently. It is almost universally known that interval analysis has been developed to make it easier to deal with the processing of uncertain information. In this lecture, we present an approach based on the use of positive aspects of interval analysis to strengthen the FPN into a more realistic model.

Brief Biography:

Zbigniew Suraj received a PhD from the University of Warsaw in mathematics and D.Sc. (habilitation) in the field of computer science at the Institute of Computer Science of the Polish Academy of Sciences in Warsaw. He is a titular professor in Poland. He is a Full Professor at the University of Rzeszów, Poland. He is also the Head of the Department of Computer Science at the University. He visited many universities in Europe (Amsterdam, Bratislava, Kiev, Sofia, Uppsala), West Asia (Thuwal), and Canada (Winnipeg). He is a Member of the Editorial Board of the Transactions on Rough Sets (Springer) and International Journal of Rough Sets and Data Analysis (IGI Global). He was an editor and co-editor of many special issues of Fundamenta Informaticae (IOS Press) and edited books, as well as two monographs “Rough Computing: theories, technologies and applications” (IGI Global, 2008) and “Rough Sets and Intelligent Systems - Professor Zdzisław Pawlak in Memoriam” (Springer, 2013). He is the co-author of the monograph “Inhibitory Rules in Data Analysis” (Springer, 2009), as well as a co-author of several books in Polish. He has published over 200 scientific papers in prestigious journals and conference proceedings. He is the leader of Rough Set and Petri Net Research Group. His main research interests are focused on the modeling and analysis of intelligent systems, knowledge representation, approximate reasoning, fuzzy sets, rough sets, Petri nets, and advanced data mining techniques including process mining. He has organized many domestic and international conferences. He is a member of the steering/program committee of many international conferences concerning rough sets, concurrency, data mining including CS&P, DATA, IJCRS, ISDA, ISMIS, PReMI, RSFDGrC, RSKT. He is a senior Member of the International Rough Set Society and a Member of the Geselschaft für Informatik Special Interest Group FG 0.0.1 Petri Nets and Related System Models (Germany), and others. He received a number of awards for scientific achievements in the field of computer science, including the individual award of the Minister of Science, Higher Education and Technology, and the team award of the Ministry of National Education in Poland.

Contents

Sentiment Analysis and Arabic Text Mining

Swarm Optimizations and Applications

Deep Learning and Cloud Computing

Data Mining, Visualization and E-learning

Intelligent Systems

Robot Modeling and Control Systems

Study of the Effect of Magnetic Field and Pulsating Flow on the Thermoelectric Cooler Performance Using Fuzzy Logic Control

M. Sh. Nassar(✉), A. A. Hegazi, and M. G. Mousa

Mechanical Power Engineering, Engineering Faculty, Mansoura University, Mansoura, Egypt
d_mahmoud2006@yahoo.com, ahmedabd_elsallam@yahoo.com, mgmousa@mans.edu.eg

Abstract. A forced convective pulsating flow of the air and magnetic field effect on the thermoelectric cooler performance is experimentally investigated. The experimental tests are conducted at a constant flow rate with Reynolds number of 5871. The applied electric power, magnetic field, pulsation rate and the output temperatures for both TEC sides are recorded and manipulated to evaluate the coefficient of the performance (COP) of the TEC. The obtained temperature differences are between 5.56 °C and 19.8 °C. TEC's COPc ranges from 0.122 to 0.277. The optimum COP of 0.277 during these experimental tests occurred at 52.247 W input power, magnetic field of 1.25 T, and 41.28 Womersley number. Comparing with Fuzzy logic results it shows a good agreement (T; Tesla).

Keywords: Thermoelectric cooler · Magnetic field · Pulsating flow Coefficient of performance · Channel flow

1 Introduction

Thermoelectric devices are considered the future devices because of its advantages and possibility to be used in many applications. These types of devices are considered a promising eco-friendly [1, 2], small sizes, silent and safe in operation. It can be used for controlling hot spots, accurate temperature control, cooling of infrared detectors. Also, it can be used in medical applications like brain local cooling to avoid the brain cell stokes. Achieving the local cooling of the brain by using shielded helmet including a set of thermoelectric coolers, using an effective controller with Fuzzy logic to investigate this helmet effectiveness [3]. TEC works on the Peltier effect which is discovered by Jean Peltier in 1834 [4]. Pushkarny et al. [5] experimentally studied the performance of a refrigerator under different parameters of AC and DC power supply. They found that the temperature difference with respect to time came out more in case of DC supply compared with AC supply. Moreover, while observing the graph of time versus actual COP they had found that the COP rises with positive slope till it reached to its peak point and then it started declining in the negative slope. The works of [6, 7] were an experimental and theoretical investigation held on a Peltier module with heat sink assembly.

A. E. Hassanien et al. (Eds.): AISI 2018, AISC 845, pp. 3–12, 2019.
https://doi.org/10.1007/978-3-319-99010-1_1

Results obtained showed a good reduction with respect to heat sink fins of air temperatures, moreover an increase of the module COP. The study of the effect of the pulsating flow done by many researchers aimed to clarify whether this type of flow would affect the convective heat transfer process or not. Depending on their studies results, they concluded that the pulsating flow could enhance the heat transfer process or deteriorate it depending on the parameters of flow [8, 9]. Young et al. [10] numerically investigated the periodically pulsating flow in a compound channel. An enhancement of heat transfer of about 35% was noticed which was mainly due to the lateral pulsating flow, gap width, and depth. Magnetic field effect is considered one of the interesting research subjects due to the continuous demand of enhancing the forced convection mechanism in the practical life application. The magnetic field affects both the semiconductors performance and the fluid flow behavior. Lukasz et al. [11] observed that the magnetic field could accelerate the fluid flow in the heating area, while it suppresses the flow of the fluid in the cooling area. Karapetyan et al. [12] reported that the magnetic field affects the semiconductors' material properties during the manufacturing process and Peltier effect as well. Also, they concluded that the effect of magnetic field enhances the performance of the semiconductors. Kiema et al. [13] numerically studied the effect of the magnetic field and the angle of inclination on the fluid flow velocity profile for a laminar flow between two parallel plates. Their results showed that the fluid flow velocity profile changed with the change of the inclination angle and magnetic field induction value. Rashidi and Esfahani [14] numerically studied the effect of magnetic field on the heat transfer by forced convection in a built-in a square obstacle in a channel. They concluded that the existence of the square obstacles decreased the effect of magnetic field on the convective heat transfer coefficient and Nusselt number. The aim of this study is to evaluate experimentally the thermoelectric cooler performance exposed to both effects of the magnetic field and the air convective pulsating flow at different input electric current values using fuzzy logic control and comparing it with the experimental results. The main idea behind a fuzzy system is the usage of the concept of linguistic variables to make decisions based on fuzzy rules and thereby getting a better response compared with a system using crisp values. The main components in the fuzzy system are fuzzification interface, knowledge base, decision making logic and defuzzification interface [15–17].

2 Proposed Experimental Method

The experimental test rig consists of four thermoelectric coolers [TEC-12706] attached with eight heat sinks are distributed on the hot and cold sides of the four TEC's (1), and two suction fans for the upper and lower sections (2) of the test channel. A pulsating mechanism at the inlet of the test channel (3). Magnetic field induction source (4) for both sides of the TEC modules of maximum capacity of 2 T. Temperature sensors LM35 are used to detect and measure the temperatures at a different points on both sides of the TEC's with measurement accuracy of ± 0.75 °C at temperature range of -55 – 150 °C. Also, Hall sensors S49E are used to detect the presence of the magnetic field of typical accuracy of ± 1.8 mV/G. All enclosed in an acrylic enclosure of rectangular cross-section as shown in Fig. 1. And a set of power supplies with the desired power output.

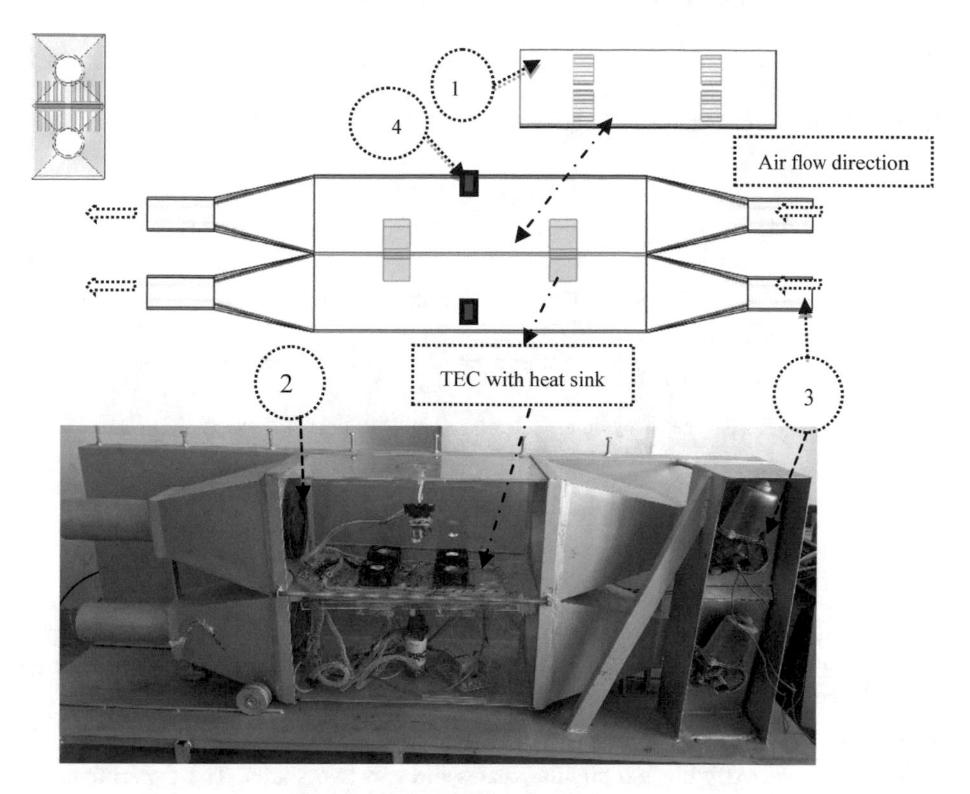

Fig. 1. Shows the experimental test rig

The measurable physical quantities are controlled using microcontroller (Arduino Mega) for recoding and manipulation. The experimental tests are conducted at different input power values 60%, 80%, 90%, and 100% of maximum value 90.871 W, and magnetic field values of 70%, 80%, 90%, and 100% of the maximum value of 1.25 T. Also, the pulsating rates are 0, 30, 40, and 50 pulses/min, which are equivalent to Womersley number values of 0, 31.96, 36.93, and 41.28. Each test period was 30 min to reach approximately a constant temperature, and 3 h minimum to be able to conduct another experiment in order to be able to meet initial conditions.

3 Calculation Method

The heat flow inside the TEC is generated in terms of three different types, Peltier cooling, Joule heating, and heat transfer by conduction respectively, as shown in the energy balance Eq. (1).

$$Q_c = \alpha I T_c - 0.5 I^2 R - K(T_h - T_c) \quad (1)$$

The total consumed electrical of the module is given by;

$$P = \alpha I \Delta T + I^2 R_m \quad (2)$$

The COP of the TEC module is given by;

$$COP = \frac{Q_c}{P} \quad (3)$$

Table 1. TEC module material properties

Thermoelectric module properties	Relation
Seebeck coefficient, α	$\alpha = \frac{V_{max}}{T_h}$
Resistance of module, R_m	$R_m = \frac{[(T_h - \Delta T_{max}) \times V_{max}]}{I T_h}$
Thermal conductivity, K	$K = \frac{[(T_h - \Delta T_{max}) \times I_{max} V_{max}]}{2(\Delta T_{max} \times T_h)}$

where; α: is the average Seebeck coefficient; R_m is the electric, K is the thermal conductivity which can be calculated as listed in Table 1. And I is the input current while T_h and T_c are hot and cold sides' temperature. ΔT_{max} is the maximum temperature difference, and V_{max} is the applied voltage.

Pulsating Mechanism. It consists of two DC motors responsible for the air stream pulsation connected with two circular plates through two shafts. The experiments as held with constant Reynolds number of 5781. Moreover, the pulsating rates can be introduced by Womersley number (W_O) where it can be expressed as following;

$$W_o = \frac{D_h}{2} \sqrt{{}^{w}\!/_{v}} \quad (4)$$

where, ω is the angular frequency = $2\pi f$ (rad/s), while f is the pulsating frequency.

4 Experimental Uncertainty Analysis

The reliability of the experimental results is determined by the uncertainty analysis as shown in Eq. (5) [18], where the values of the accuracy and the uncertainty of the measuring devices are known. The calculated uncertainty in the experimentally obtained COP is ±3.208%.

$$\omega_Z = \left[\sum_i^N \left(\frac{\partial Z}{\partial x_i}\omega_i\right)^2\right]^{0.5} \tag{5}$$

where i, N, ω, and z are the certain parameter under the study, the total number of independent parameters, uncertainty of the variables, and the independent parameters.

5 Design Fuzzy Logic Model of Thermoelectric Cooling System

The proposed fuzzy model has three input variables (electrical power, affecting magnetic field and air pulsating rate) used in evaluating the possibility of the output grades in order to determine the final temperature's change as shown in Fig. 2.

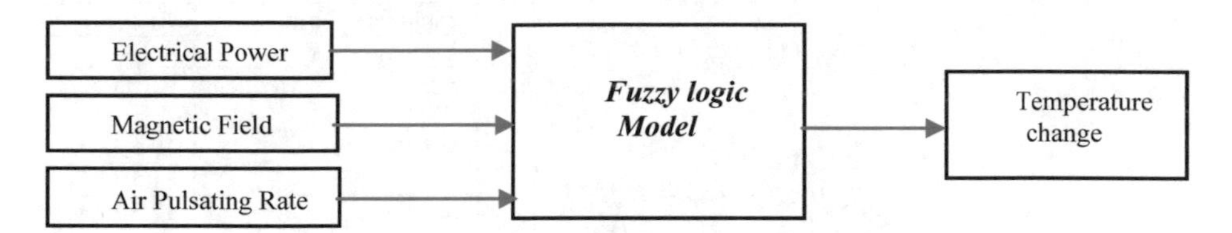

Fig. 2. Fuzzy logic model block diagram.

5.1 Fuzzification of Input and Output

The input variables distribution of each linguistic values are shown in Figs. 3, 4 and 5 and their ranges.

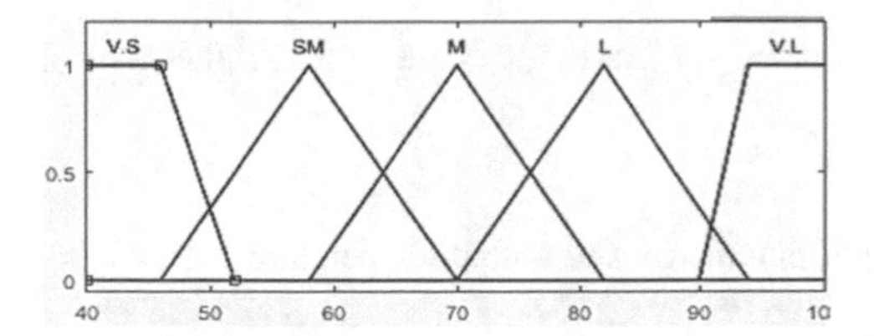

Linguistic Value	Range
Very Small (V.S)	40 :52
Small (SM)	48 : 70
Medium (M)	52 : 82
Large (L)	70 : 94
Very large (V.L)	90 : 100

Fig. 3. Membership's linguistic values and ranges of the electrical input power

Linguistic Value	Range
Small (SM)	70 : 85
Medium (M)	85 : 92
Large (L)	92: 100

Fig. 4. Membership's linguistic values and ranges of the affecting the pulsating air flow rate

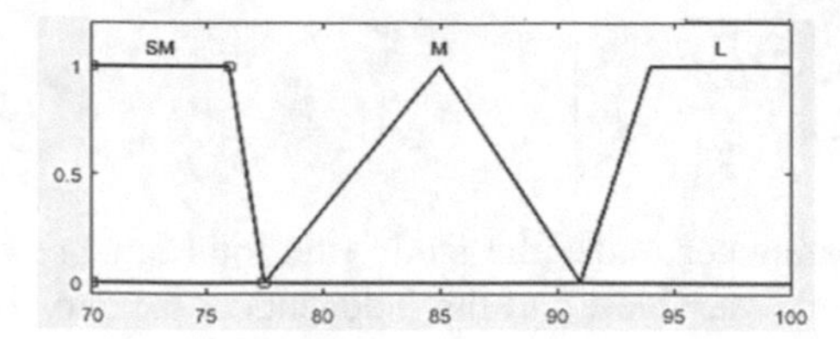

Linguistic Value	Range
Small Large (S.L)	70 : 85
Medium (M)	85 : 92
Full	92: 100

Fig. 5. Membership's linguistic values and ranges of affecting magnetic field

The proposed output linguistic variables were taken from the experimental study which is the change in temperature as represented in Fig. 6.

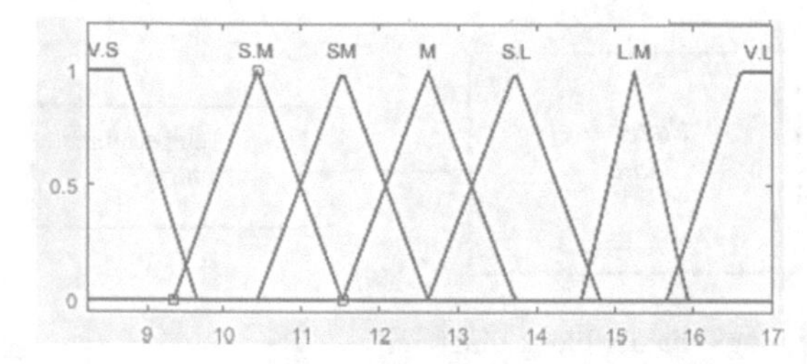

Linguistic Value	Range
Very Small (V.S)	8:9.8
Small Medium (S.M)	9.2:11.5
Small (SM)	10.6:12.5
Medium (M)	11.5:13.7
Small Large (S.L)	12.5:14.8
Large Medium (S.L)	14.6:15.9
Very large (V.L)	15.7:16.8

Fig. 6. Membership's linguistic values and ranges of the output temperature

5.2 Rules Base of the Model

Forty-five rules were modeled and inferred from the three input variables. These rules are based on the fuzzified inputs. The rules are arranged as shown in Table 2. The intersected cells between all input variables of linguistic values form the overall modeling results. For instance:

Table 2. Rule base of the input parameters and output temperature

Elec. Power	V.S			S.M			M			L			V.L		
Pulsating flow	Magnetic field														
	S.L	M	F	S.L	M	F	S.L	M	F	S.L	M	F	S.L	M	F
SM	*V.S*	*V.S*	*V.S*	*V.S*	*S.M*	*M*	*S.M*	*S.M*	*S.M*	*S.L*	*S.L*	*S.L*	*L.M*	*L.M*	*L.M*
M	*S.M*	*S.M*	*S.M*	*S.M*	*M*	*M*	*S.L*	*S.L*	*S.L*	*M*	*M*	*M*	*M*	*M*	*M*
L	*S.M*	*V.S*	*S.M*	*S.L*	*S.L*	*V.L*	*S.L*	*S.L*	*M*	*M*	*S.L*	*S.L*	*V.L*	*L.M*	*V.L*

If (Elect.Power is M) and (Magnatic.Field is SM) and (Pulseting.Rate is SL) then (Temp.Out is S.M)
If (Elect.Power is L) and (Magnatic.Field is L) and (Pulseting.Rate is SL) then (Temp.Out is S.L)

The trial and error for selecting the appropriate parameters of the fuzzy model, the AND operation in applying rules is carried out using the input variable.

5.3 Proposed Defuzzification Method

Some of the input is discretized into five or three possibilities are given after the process of defuzzification (Very Small, Small, Medium, Large, and Very large), the final output temperature is brought in by finding the maximum among these crisp values. The center of Maxima method proved to yield better results during the tuning stage of the fuzzy model [15, 17].

6 Results and Discussion

The proposed fuzzy classifier Mamdani's inference engine is chosen to be the core interference engine is designed for controlling the system and membership functions. This model has three input variables used in evaluating the possibility of the output grades to figure out the final temperature difference between the hot and cold surfaces, then the COP of the thermoelectric cooler as explained by the below figures. It is observed that the outputs of the prepared software and the outputs of experimental results good matched up. The temperature difference variation along with the time at 100% magnetic field of 1.25 T at different input power values at W_O of value 41.28. A 15.57 °C maximum temperature difference is attained at both 100% magnetic field and input power as shown in Fig. 7.

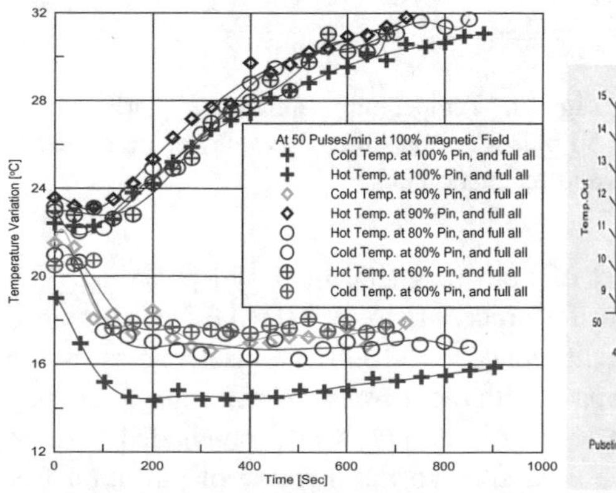

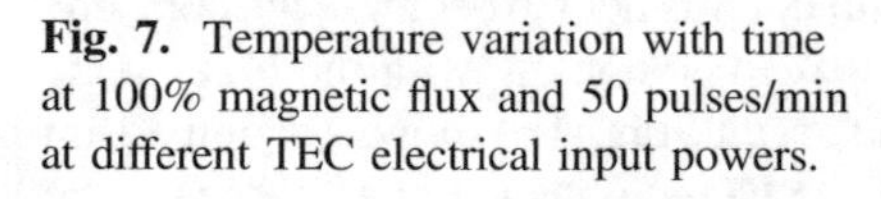
Fig. 7. Temperature variation with time at 100% magnetic flux and 50 pulses/min at different TEC electrical input powers.

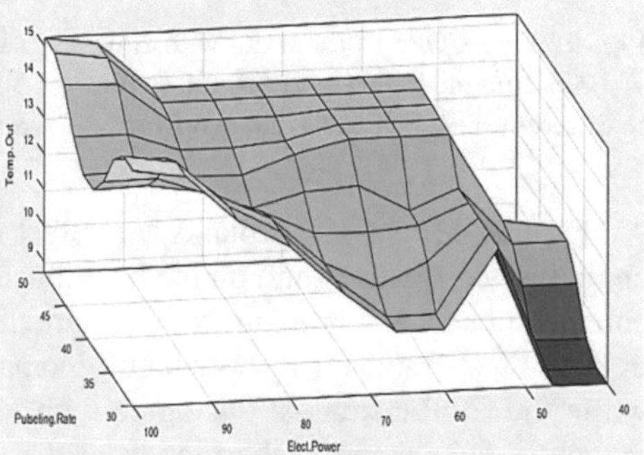

Fig. 8. Temperature difference variation at 60% magnetic field at different values of input power and pulsating flow rate.

While Fig. 8, nearly matches with the experimental results at 50 pulses/min and 100% input power value. This enhancement occurs due to the effect of magnetic field as it excites the charge carriers in the semiconductor material which generate an extra electromotive force then generating an electric current that enhance the performance of the thermoelectric cooler. Moreover it increase the heat transfer by convection due to the deceleration and acceleration of air motion at the cold and hot sides of the TEC respectively. Moreover the pulsating flow effect deteriorates the effect of air flow deceleration on the cold side, which leads to an increase in the heat transfer process as a whole. Figure 9, shows the variation of the TEC COP with time, according to the definition of the coefficient of the performance increases with the decrease of the input electric power. Moreover, the COP decreases with time as the temperature difference between both sides of the TEC increase. Figure 10 shows the variation of the temperature difference along with the different values of the input power and magnetic field at 50 pulses/min. the maximum temperature of 16.4 °C occurs at 60% input power and 100% magnetic field, also approaches same range at 100% input power but at 70% magnetic field which nearly agrees with the obtained experimental results.

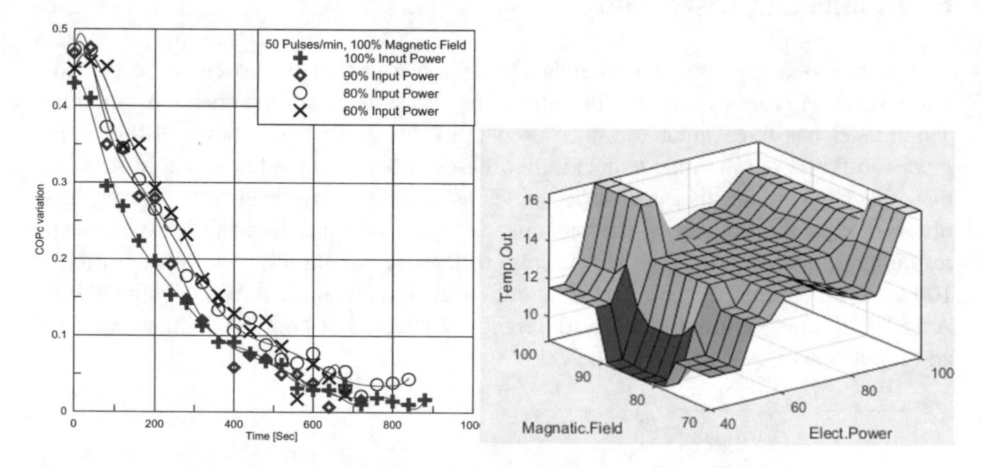

Fig. 9. The COPc of the TEC with time at 100% magnetic flux and 50 pulses/min at different TEC electrical input powers.

Fig. 10. Temperature difference variation at 50 pulses/min at different values of input power and magnetic field.

Figures 11 and 12, explains the effect of both magnetic field and pulsating flow changing to 30 pulses/min on the heat transfer process moreover the TEC performance enhancement. Comparing the temperature difference for 50 and 30 pulses/min as shown in both Figs. 7 and 11, it show that it increase with the increase of pulsation frequency while the COP increases, this agree with Lukasz et al. [11] which concluded that the magnetic field suppress the motion at the cold side, so the increase of pulsation frequency deteriorate the effect of magnetic field with in turns help to decrease the cold side temperature more than in case of 30 pulses/min. Also comparing with the fuzzy logic software results it agrees with the experimental results Fig. 13, shows a comparison between samples of the experimental results of the COPc obtained at 100% magnetic

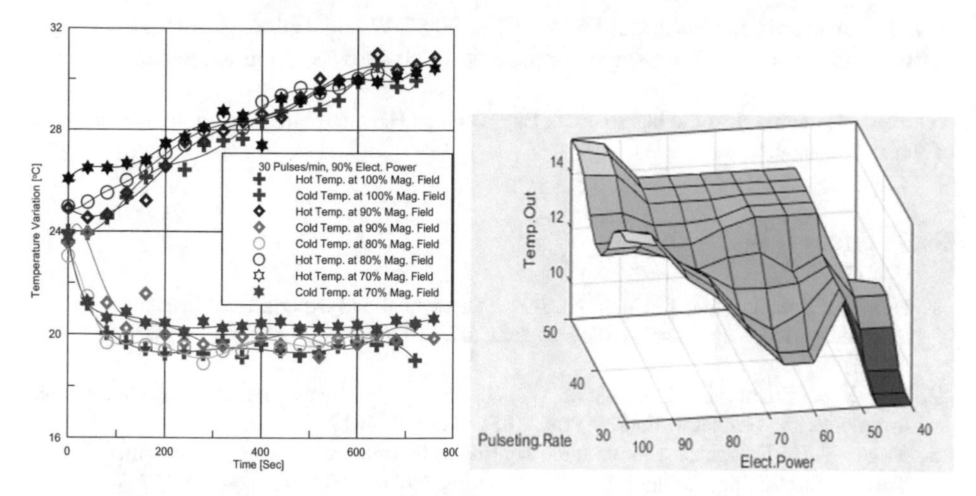

Fig. 11. Shows temperature change with time at different magnetic field intensity and 30 pulses/min at 90% TEC powers

Fig. 12. Variation of temperature with the input power at different pulsating flow rate at 100% magnetic field.

field intensity that is about 1.25 T at different pulsating flow rate and input power. The COP is enhanced by the combined effect of the magnetic and pulsating flow.

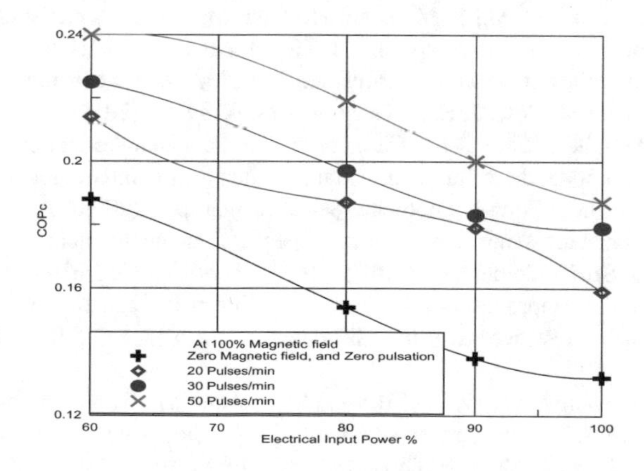

Fig. 13. Shows the COPc at different values of input power at 100% magnetic field at different pulsating rate.

7 Conclusion

Thermoelectric cooler performance is strongly affected by both magnetic field and pulsating rate effects, also the input power (electric current). It can be concluded that the maximum COP of the TEC module is obtained at 1.25 T and 54.52 W at 50 pulses/min of value 0.277. The maximum temperature difference in case of introducing pulsating flow and magnetic field is about 15.98 °C, at 90.87 W and 1 T. Also from fuzzy logic software is about 16.1 °C at same conditions which shows a good agreement.

Acknowledgement. I would like to thank Eng. Hussam Hassan for his support in understanding of the Fuzzy logic.

References

1. He, W., Zhou, J., Hou, J., Chen, C., Ji, J.: Theoretical and experimental investigation on a thermoelectric cooling and heating system driven by solar. Appl. Energy **107**(C), 89–97 (2013)
2. Liu, Z.B., Zhang, L., Luo, Y.Q., Wu, J., Meng, F.: Investigation on a photovoltaic thermoelectric ventilator. Energy Proc. **105**, 511–517 (2017)
3. Yavuz, A.H.: Design of a fuzzy logic controlled thermoelectric brain hypothermia system. Turk. J. Electr. Eng. Comput. Sci. (2016). https://doi.org/10.3906/elk-1405-137
4. http://www.thermoelectrics.caltech.edu/. Accessed on 10 July 2016
5. Pushkarny, B.H., Khan, N., Parulkar, A., Rai, H., Patel, D.: Solar operated cooling using thermo-electric module. Int. J. Sci. Res. Dev. **3**, 2321–2613 (2016)
6. Dhumal, G.S., Deshmukh, P.A., Kulkarni, M.L.: Experimental investigation of thermoelectric refrigeration system running on solar energy and development of mathematical model. Int. Eng. Res. J. (IERJ) **1**, 232–238 (2015)
7. Chavhan, D.K., Mahajan, S.D.: Testing and validation of thermoelectric coolers. Int. J. Eng. Res. Appl. **5**, 2248–9622 (2015)
8. Elshafei, E.A.M., Safwat, M.M., Mansour, H., Sakr, M.: Experimental study of heat transfer in pulsating turbulent flow in a pipe. Int. J. Heat Fluid Flow **29**, 1029–1038 (2008)
9. Elshafei, E.A.M., Safwat, M.M., Mansour, H., Sakr, M.: Numerical study of heat transfer in pulsating turbulent air flow. J. Eng. Technol. Res. **4**, 89–97 (2012)
10. Hong, S.H., Seo, J.S., Shin, J.K., Choi, Y.D.: Numerical investigation of turbulent flow pulsation in compound rectangular channel. In: Sixth International Symposium on Turbulence and Shear Flow Phenomena, Seoul, Korea, p. 6 (2009)
11. Pleskacz, L., Fornalik-Wajs, E.: Various aspects of magnetic field influence on forced convection. In: SEED Conference (2016). https://doi.org/10.1051/e3sconf/20161000120.
12. Karapetyan, G.Y., Dneproveski, V.G., Wu, P.C.: Effect of magnetic field on thermoelectric coefficient value and Peltier factor in InSb, chap. 11, pp. 127–133 (2014). ISBN 978-3-319-03748-6
13. Kiema, D.W., Manyonge, W.A., Bitok, J.K., Adenyah, R.K., Barasa, J.S.: On the Steady MDH Couette flow between two infinite parallel plates in an uniform transverse magnetic field. J. Appl. Math. Bioinform. **5**(1), 87–99 (2015)
14. Rashidi, S., Esfahani, J.A.: The effect of magnetic field on instabilities of heat transfer from an obstacle in a channel. J. Magn. Magn. Mater. **391**, 5–11 (2015)
15. Rani, C., Deepa, S.N.: Design of optimal fuzzy classifier system using particle swarm optimization. In: International Conference of Innovative Computing Technologies (ICICT) (2010)
16. Zadeh, L.A.: Outline of a new approach to the analysis of complex systems and decision processes. IEEE Trans. Syst. Man Cyber. SMC **3**, 28–44 (1973)
17. Von Altrock, C.: Fuzzy logic technologies in automotive engineering. In: IEEE Proceedings of WESCON, Anaheim, CA, 27–29 September 1994 (1994)
18. Holman, J.P.: Experimental Method for Engineers, 6th edn. McGraw-Hill, Singapore (1994)

Two-Degree of Freedom Proportional Integral Derivative (2-DOF PID) Controller for Robotic Infusion Stand

Ahmad Taher Azar[1,2](✉), Hossam Hassan[2], Mohd Saiful Akmal Bin Razali[2], Gabriel de Brito Silva[2], and Hedaya Rafat Ali[2]

[1] Faculty of Computers and Information, Benha University, Benha, Egypt
ahmad_t_azar@ieee.org, ahmad.azar@fci.bu.edu.eg
[2] School of Engineering and Applied Sciences, Nile University, Sheikh Zayed District, 6th of October City, Giza, Egypt
hhassan@nu.edu.eg, {saifulakmal.razali,gabriel.silva, hedaya.ali}@eu4m.eu

Abstract. Infusion Stand is one of the medical supportive tools in the field of biomedical that assist in holding and carrying medications to patients via intravenous injections. Mobilization of Infusion Stand from a place to another place is necessary not only for the patients itself but also for the nurses. Therefore, this leads to not only uneasiness but also inconvenience for both parties. Therefore, to improve the existing situation and current Infusion Stand in the market, a proposal to design and implement a prototypic Robotic Infusion Stand is submitted. In this paper, 2-Degree of Freedom Proportional Integral Derivative (2-DOF PID) controller is proposed for Robotic Infusion Stand after comparison between 1-Degree of Freedom Proportional Integral Derivative (1-DOF PID) to find the most suitable controller. Analysis of reference tracking, disturbance rejection and controller effort are performed which demonstrate the ability of the proposed approach within the system parameters.

Keywords: Proportional-Integral-Derivative (PID) controller
2-Degree of Freedom PID · Robotic infusion stand

1 Introduction

Infusion Stand is one of the medical supportive tools that not only used extensively in hospitals, clinics, physical practices but also in the supportive care provided in the home [1, 2]. Normally, a conventional Infusion Stand consists of a stand-alone metal structure which includes a rod, a chassis which typically equipped with 3, 4 or 5 legs with and a wheels & a hanger with one or more hooks at the upper part of the Infusion Stand [3]. There are a lot of accessories that will be attached to the Infusion Stand such as bag of fluids like a water, medication & blood, urinary hooks, temporary pacemakers, patient handles & support trays [4].

Research studies from the hospitals, clinics, physical practices and in the supportive care provided in the home found that one of the main drawback of Infusion Stand that

A. E. Hassanien et al. (Eds.): AISI 2018, AISC 845, pp. 13–25, 2019.
https://doi.org/10.1007/978-3-319-99010-1_2

currently available in the market is their difficulty to maneuver freely [5]. When a patient wants to move away from a confined area, he or she need to drag their own Infusion Stand or in worse case, to get an assistance from the nurse or any other person nearby him or her [6]. Consequently, the nurse will use up his or her time accompanying the patient itself and serving the patient by dragging the Infusion Stand which ends up give inconvenience to the patient and unproductive day to the nurse itself [7]. Besides, patient and nurse complaint about the massive size of Infusion Stand that may causes issues while maneuvering the Infusion Stand [8].

Wiederhold et al. proposed a new approach to enhance the acceptability of robotic systems, based on the introduction of affective dimensions in human-robot interaction. By implementing this proposed strategy, they can enhance the user's acceptance and adoption of the assistive robotic system [9]. Besides, Qureshi et al. come out with a proposal by introducing a robot in daily operations. With the advancement of the healthcare sector, Qureshi et al. introduced robots in the day-to-day operations. Not only that, the sector should provide a proper training and development for their human resources to keep their and knowledge up to date, which at the ends would stimulate the health care professionals to work in parallel with robots [10, 11]. Moreover, Binger et al. have suggested an autonomous Infusion Stand. Nevertheless, the proposed system was not being mounted with an alarm system that capable to detect the IV bag medication level. Therefore, the proposed design was not able to handle common medical equipment such as infusion pumps and oxygen tanks [7].

Meanwhile, Robotic Infusion Stand is an improved version of existing Infusion Stand whereby it will moving with an input given due to the new elongation of the joystick, which influence by the strip reaction. Our proposed design will implement a Robotic Infusion Stand that tether to the patient and able to maneuver with a given direction. This article presents a simulation of our designed Infusion Stand by using 1DOF. Then, the results from this 1DOF will be compared with 2DOF. The method discussed in this article try to provide best outputs performances. The controller proposed results with a fast time response and high stability. The obtained results are very promising.

The rest of the paper is organized as follows: The mathematical model is presented in Sect. 2. While in Sect. 3, the controller design is introduced and explained. In Sect. 4, all results are shown under different tests and inputs. In addition, the discussion section and a conclusion are written to conclude this article.

2 Mathematical Model

In the mathematical model, the kinematics and dynamic of the Robotic Infusion robot as Nonholonomic Wheeled Mobile Robot are being modeled to get the electrical model of the DC Motor.

2.1 Kinematic Model of Robotic Infusion Stand

From the kinematic model of wheeled mobile robots in Fig. 1, the following equations is generated:

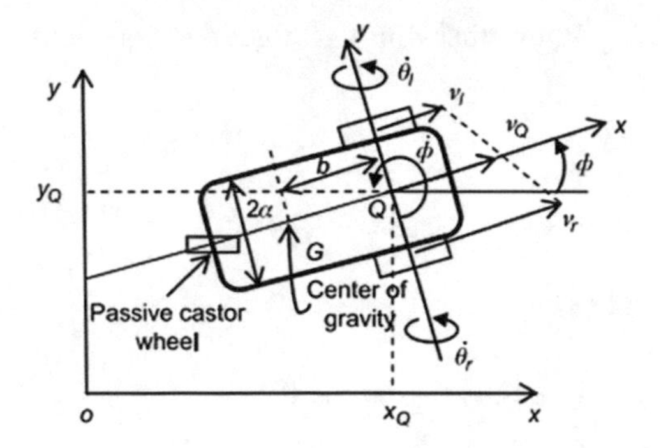

Fig. 1. Kinematic model of wheeled mobile robots

$$\dot{X} = v \cos \theta; \tag{1}$$

$$\dot{Y} = v \sin \theta \tag{2}$$

Eliminating v from Eqs. (1) and (2) to get:

$$-\dot{X} \sin \theta + \dot{Y} \cos \theta = 0 \tag{3}$$

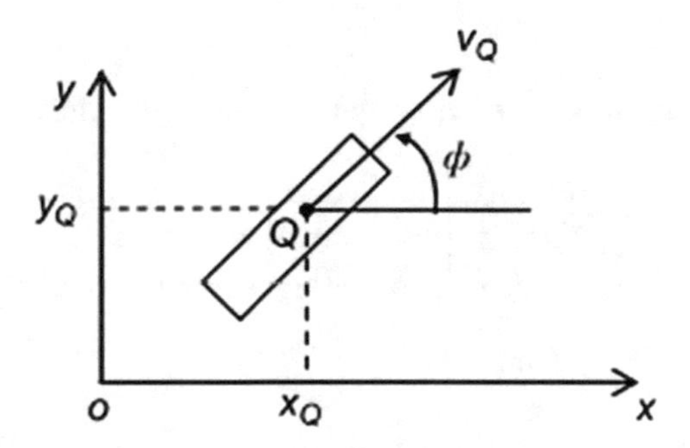

Fig. 2. Kinematic structure of unicycle

From Fig. 2,

$$V_r = v + a\ \dot{\theta}; \quad V_l = v - a\dot{\theta} \tag{4}$$

Whereby, v is the linear velocity of the wheels, V_r and V_1 are the linear velocity of the left and right motors.

As

$$V_r = r\,\omega_r; \tag{5}$$

$$V_l = r\,\omega_l \tag{6}$$

Adding and subtracting V_r together with V_l, to get two equations

$$v = \frac{1}{2}(V_r + V_l) \tag{7}$$

$$2a\dot{\theta} = V_r - V_l \tag{8}$$

From Eqs. (1), (7) and (8);

$$\begin{aligned} \dot{X} &= \frac{r}{2}(\omega_r \cos\theta + \omega_l \cos\theta); \\ \dot{Y} &= \frac{r}{2}(\omega_r \sin\theta + \omega_l \sin\theta); \quad \dot{\theta} = \frac{r}{2a}(\omega_r - \omega_l) \end{aligned} \tag{9}$$

Using p and q vectors to describe the position and velocity vectors respectively as:

$$\begin{bmatrix} & X \\ p = & Y \\ & \theta \end{bmatrix}; \tag{10}$$

$$\begin{bmatrix} \dot{q} = & \omega_r \\ & \omega_l \end{bmatrix} \tag{11}$$

where by ω_r is the angular right velocity and ω_l is the angular left velocity.

$$\begin{bmatrix} & \dot{X} \\ \dot{p} = & \dot{Y} \\ & \dot{\theta} \end{bmatrix} = \begin{bmatrix} \cos\theta \\ \sin\theta \\ 0 \end{bmatrix} v + \begin{bmatrix} 0 \\ 0 \\ 1 \end{bmatrix} \dot{\theta} \tag{12}$$

And the Jacobian of the system,

$$J = \begin{bmatrix} \cos\theta & 0 \\ \sin\theta & 0 \\ 0 & 1 \end{bmatrix} \tag{13}$$

From Eq. (9), (12) and (13),

$$\dot{p} = \begin{bmatrix} \frac{r}{2}\cos\theta \\ \frac{r}{2}\sin\theta \\ \frac{r}{2a} \end{bmatrix} \omega_r + \begin{bmatrix} \frac{r}{2}\cos\theta \\ \frac{r}{2}\sin\theta \\ -\frac{r}{2a} \end{bmatrix} \omega_l = J\dot{q} \tag{14}$$

Hence, since

$$v = \dot{X}\cos\theta + \dot{Y}\sin\theta$$

$$r\omega_r = \dot{X}\cos\theta + \dot{Y}\sin\theta + a\dot{\theta} \tag{15}$$

$$r\omega_l = \dot{X}\cos\theta + \dot{Y}\sin\theta - a\dot{\theta} \quad (16)$$

which equal to

$$\begin{bmatrix} \omega_r \\ \omega_l \end{bmatrix} = \frac{1}{r}\begin{bmatrix} \cos\theta & \sin\theta & a \\ \cos\theta & \sin\theta & -a \end{bmatrix}\begin{bmatrix} \dot{X} \\ \dot{Y} \\ \dot{\theta} \end{bmatrix} \quad (17)$$

2.2 Dynamic Modeling of Robotic Infusion Stand

Using Lagrange Dynamic model of nonholonomic wheeled mobile robot. The non-holonomic constraint matrix is:

$$M(q) = [-\sin\theta \quad \cos\theta \quad 0] \quad (18)$$

To get a constrain free model, use a $n(n - m)B(q)$ matrix defined as:

$$B^T(q)\,M^T(q) = 0 \quad (19)$$

$$B(q) = \begin{bmatrix} \cos\theta & 0 \\ \sin\theta & 0 \\ 0 & 1 \end{bmatrix} \quad (20)$$

From Eqs. (18) and (19), there is a (n–m) vector in term of v(t) which is:

$$\dot{q}(t) = B(q)v(t) \quad (21)$$

Therefore, from Eqs. (20) and (21), the dynamic model of wheeled mobile robot is:

$$\begin{bmatrix} \dot{X} \\ \dot{Y} \\ \dot{\theta} \end{bmatrix} = \begin{bmatrix} \cos\theta & 0 \\ \sin\theta & 0 \\ 0 & 1 \end{bmatrix}; \quad (22)$$

$$\begin{bmatrix} v \\ w \end{bmatrix} = \begin{bmatrix} v\cos\theta \\ v\sin\theta \\ \omega \end{bmatrix} \quad (23)$$

2.3 Mathematical Modelling of a DC Motor

$$T = K_t i \quad (24)$$

$$e = K_e\dot{\theta} \quad (25)$$

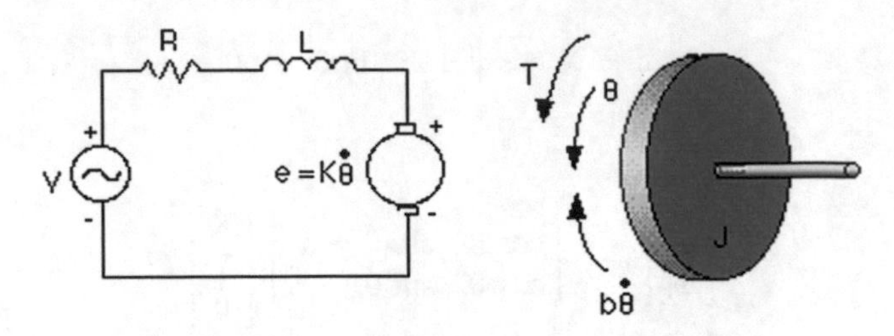

Fig. 3. Model representation of DC motor.

Table 1. Physical parameter of DC motor

J	Moment of inertia of the rotor (kg m^2)
b	Motor viscous friction constant (N m s)
K_t	Electromotive force constant $\left(\frac{\mathrm{V\,s}}{\mathrm{rad}}\right)$
K_e	Motor torque constant $\left(\frac{\mathrm{N\,m}}{\mathrm{A}}\right)$
R	Electric resistance (Ω)
L	Electric inductance (H)
$\dot{\theta}$	Output rotating speed $\left(\frac{\mathrm{m}}{\mathrm{s}}\right)$
V	Voltage source (V)

In SI units, $K_t = K_e$; therefore, K is used to represent both the motor torque constant and the electromotive force (Fig. 3; Table 1).

$$J\ddot{\theta} + b\,\dot{\theta} = ki \tag{26}$$

$$L\frac{di}{dt} + Ri = V - K\dot{\theta} \tag{27}$$

From Eqs. (26) and (27), the Laplace transform is applied. The results are shown by the modeling equations:

$$s(Js + b)\Omega(s) = kI(s) \tag{28}$$

$$s(Ls + R)I(s) = V(s) - Ks\Omega(s) \tag{29}$$

Then, the system model can be concluded as Fig. 4.

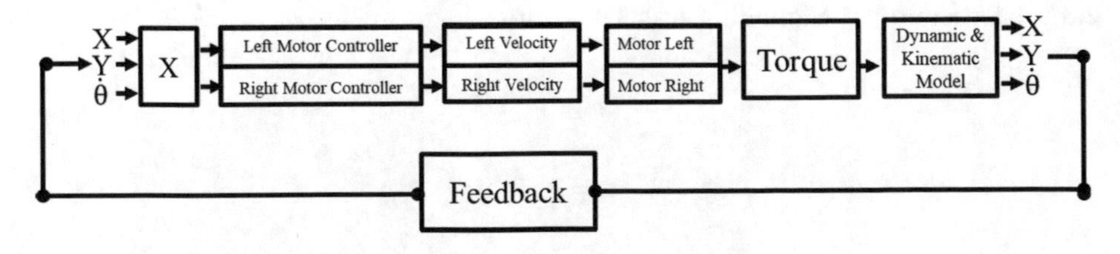

Fig. 4. Model of transfer function for DC motor.

2.4 Two-Degree-of-Freedom Control

To develop the Two-Degree-of-Freedom from the time domain, the Eqs. (30–35) are followed like showed [12–16]:

$$u(t) = K_p\left\{e_p(t) + \frac{1}{T_i}\int_0^1 e_i(\xi)d\xi + T_d\frac{de_d(t)}{dt}\right\} \tag{30}$$

Applying the Laplace Transformation, the control transfer function in the s domain is reached.

$$u(t) = K_p\left\{e_p(s) + \frac{1}{T_i s}e_i(s) + \frac{T_d s}{\alpha T_d s + 1}e_d(s)\right\} \tag{31}$$

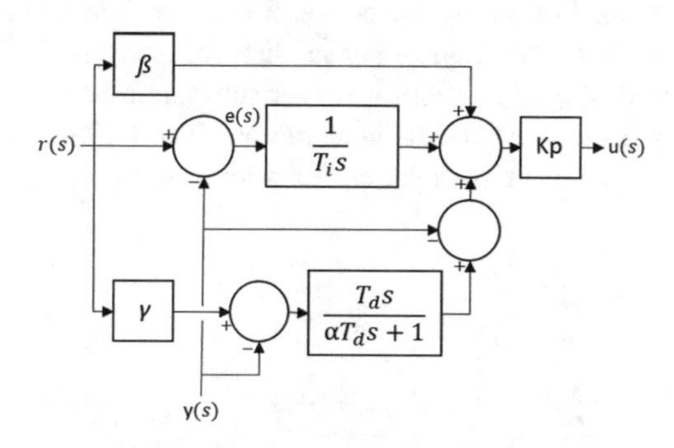

Fig. 5. Two-degree-of-freedom control diagram

With all the parameters presented on Fig. 5 placed in the transfer function, the following transfer function shows off.

$$\mathrm{u(t) = K_p\left\{\beta r(s) - y'(s) + \frac{1}{T_i s}[r(s) - y'(s)] - \left(\frac{T_d s}{\alpha T_d s + 1}\right)y'(s)\right\}} \tag{32}$$

Once it is assumed that the measurement noise is filtered, it is possible to deduce that $y'(s) \approx y(s)$. Rearranged variables and coefficients, other equation is attained, as follows:

$$\mathrm{u(s) = K_p\left\{\beta + \frac{1}{T_i s}r(s) - K_p\left(1 + \frac{1}{T_i s} + \frac{T_d s}{\alpha T_d s + 1}\right)y(s)\right\}} \tag{33}$$

where the $C_r(s)$ and $C_y(s)$ controller aspects read as:

$$\mathrm{C_r(s) = K_p\left(\beta + \frac{1}{T_i s}\right)} \tag{34}$$

$$C_y(s) = K_p\left(1 + \frac{1}{T_i s} + \frac{T_d s}{\alpha T_d s + 1}\right), \tag{35}$$

being the controller parameters $\theta_c = \{K_p, T_i, T_d, \alpha, \beta, \gamma = 0\}$.

3 Schematic and Hardware Design

3.1 Schematic Design

Figure 6 shows a schematic diagram of Robotic Infusion Stand. In general, the system can be divided into 3 main sub-systems which are input, microcontroller and motor driver sub-system. For this project, joystick is used to provide an analog input to the microcontroller which is Arduino Mega. Arduino Mega will reads raw data from the measurement sub-system and sends PWM signals to the motor driver to be converted to mechanical motion [17]. For safety purposes, 3 distance sensors are installed on the body of the Robotic Infusion Stand. These distance sensors will send an Infrared (IR) output to Arduino Mega. The motor driver subsystem consists of a motor driver, L298N H Bridge connected to two DC gear motors. The L298N H Bridge is powered by 3.5 Li-Ion batteries connected in series for a total of 14 V.

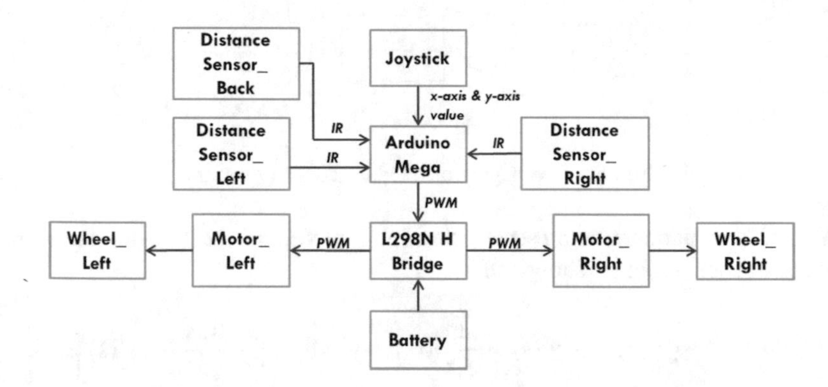

Fig. 6. Schematic diagram of robotic infusion stand.

3.2 Hardware Design

The overall design of this Robotic Infusion Stand consists of 2 layers of base, wheels, a stand with a joystick. This Robotic Infusion Stand is designed in a triangle shape to ease the placement of distance sensors. Three distance sensors will be located at 3 different sides to provide an accurate motion sensing. To maximize the availability of the space, the Arduino Mega, L298N H bridge, breadboard is installed on upper side of the first base. Below side of the first base is attached with a L-shape motor hub, motors, 65 mm wheels & caster ball wheel. On the second base, 4 3.5 V Li-Ion batteries is installed to make sure that the center of mass is located above the pivot point. Then, a stand is placed on the first base connected through the second base with a hole in the middle of the base itself. Joystick is placed on the highest point of the Robotic Infusion Stand itself which is on the stand with a strip attached to the joystick (Fig. 7).

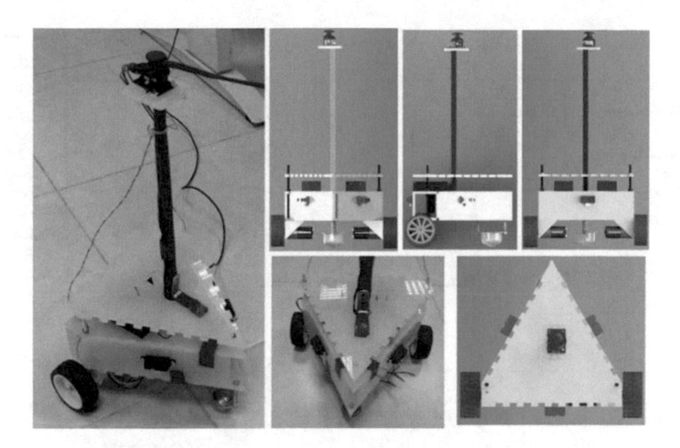

Fig. 7. Hardware design of robotic infusion stand

4 Controller Design

4.1 Concept of Joystick

Joystick is theoretically made up of two potentiometers which are connected to the analog inputs of the microcontroller [18]. This joystick will have values from 0 to 1023 [19]. Value of the joystick on both axes (x-axis & y-axis) is 512 when the joystick stays in the center position. The minimum and maximum value of the joystick are from 0 to 1023 (Fig. 7).

4.2 Concept of Controller

The design target of the system is to control the Robotic Infusion Stand with a tuned Proportional-Integral-Derivative (PID) controller. Figure 8 shows an overall block diagram of this Robotic Infusion Stand while Figs. 9 and 10 show a block diagram for motor subsystem Left and Right. The controllers are divided by two which are Controller 1 and Controller 2. Controller 1 is a translational movement in a y-axis which have values from 0 (downward) to 1023 (upward). Controller 2 is a rotation movement in x-axis which have values from 0 (left) to 1023 (right). Two set point speed for the motor left and motor right are recorded, and this value will be used in the Subsystem Motor Left Speed Control and Subsystem Motor Right Speed Control. −1 values are needed to turn the Robotic infusion stand in the opposite direction. Then, the current set point speed for the motor is subtracted with the current motor speed determined by encoder of the motor to get the error of motor speed. Using a PID Controller, a modification technique for getting analog results with digital means called PWM is established to control the PWM output and get the motor transfer function [20]. Then,

Robotic Infusion Stand will be at it physical state which means that, robot moves to new position and the strip reaction influences the new elongation of the joystick (Table 2).

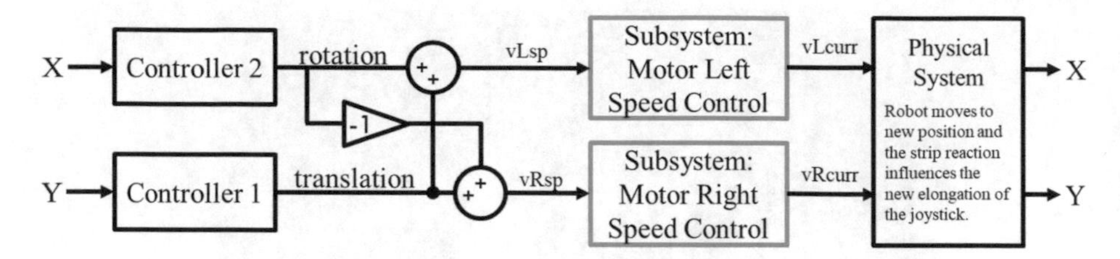

Fig. 8. Block diagram for robotic infusion stand

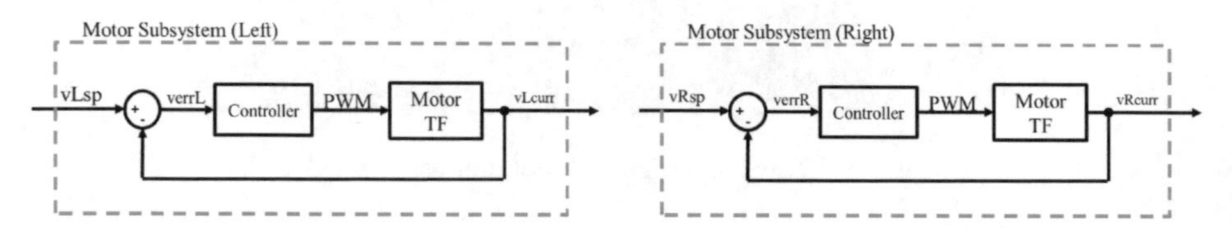

Fig. 9. Block diagram for motor subsystem (left)

Fig. 10. Block diagram for motor subsystem (right)

Table 2. Physical parameter for motor speed controller.

Abbreviations	Description
vLsp	Set point speed for motor left
vRsp	Set point speed for motor right
vLcurr	Current motor left speed determined by an encoder
vRcurr	Current motor right speed determined by an encoder
verrL	Error of motor left speed
verrR	Error of motor right speed
PWM	Pulse width modulation
TF	Transfer function

5 Results and Discussion

After the identification of the transfer function of the DC motors, the transfer function recorded is used to complete the actual transfer function of the whole system with a kinematic model. The focus is around two types of compensators, the complete PID with One Degree of Freedom (1-DOF) and the PID with Two Degrees of Freedom (2-DOF). With the estimation of each one separately, comparisons are made between both, to understand the benefits of each method. The transfer function of the whole system is:

$$\frac{3.022}{s^4 + 24.21s^3 + 174.21Ss^2 + 324.1s + 30.99}$$

Using the step input over this transfer function, it is possible to obtain the characteristic behavior of the system, and the factors such as overshoot, settling time, rise time, steady state error. The first try was the 1-DOF PID controller. After getting satisfactory outcomes in the factors to a good compensator, the values of K_P, K_I, K_d and the compensator transfer function itself are exported to the MATLAB workspace. Since is not possible to generate graph for the Controller Effort due to number of zeros is higher than number of poles for the transfer function, the only criterion worked to define a suitable controller was the best fitting. Is it being possible to see, when both controllers are compared in Figs. 11 and 12, indeed 1-DOF controller is faster than the 2-DOF controller. However, the values acquired to K_P, K_I and K_d for 2-DOF are lower and better applicable to the real hardware. While the difference between the time of

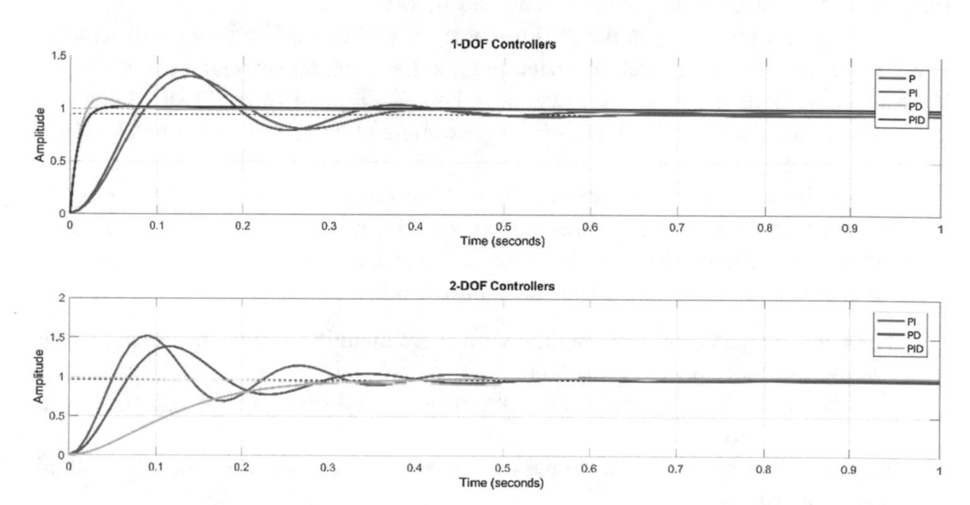

Fig. 11. Comparison between 1-DOF controllers and 2-DOF controllers

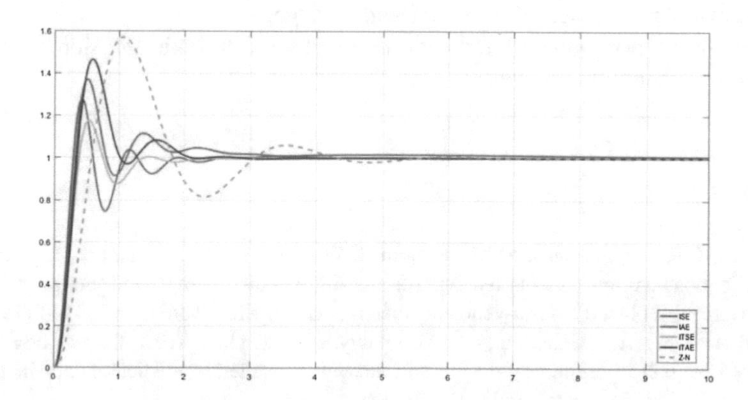

Fig. 12. Performance criteria of the whole system

both controllers is not much expressive, difference between the K_P, K_I and K_d can be a considerable distinction of performance. In addition, the 2-DOF controller has no overshoot while the 1-DOF has an overshoot of 1.7%. The 2-DOF PID, as a modified form of PID control, proposed to overcome the limitations of PID controllers [21], has the ability of a fast disturbance rejection without any significant increment on the overshoot factor. Moreover, this controller can also be used to relieve changes influence in the input signal.

6 Conclusion

By referring to Fig. 11, the chosen controller for this Robotic Infusion Stand which is PID Controller is the best in comparison with other available in the sense of Proportional-Integral-Derivative controllers. Therefore, this Fig. 11 validates the Results and Discussion that being considered before.

Figure 12 shows some of the performance criteria of the whole system which is ways in the modern optimization technique such as Integral of Squared Error (ISE), Integral of Absolute Error (IAE), Integral of Time Multiplied Error (ITSE), Integral of Time Multiplied Absolute Error (ITAE) and example of controller performed by Zigler Nichols (ZN) methods.

Robotic Infusion Stand can maneuver with an input given on the joystick by the strip reaction and try to follow the person who is in control of the strip. Besides, Robotic Infusion Stand also able to avoid the obstacles. Moreover, below are some recommendations that we can do for the future work:

i. Improving current joystick module with more quantity of potentiometers, so that we can get more accurate location.
ii. Transforming the current joystick into an advanced drive though automating the Robotic Infusion Stand.
iii. Implement an alarm system to alert the patients & nurses when the bag of fluids was emptied.
iv. Improving the drive of motors between the transmitting side & receiving side of Robotic Infusion Stand.
v. Optimize obstacle detection with a wider range.
vi. Implement a non-human intervention whereby the Robotic Infusion Stand is not directly tether to the patient.

References

1. Graham, D.R., Keldermans, M.M., Klemm, L.W., Semenza, N.J., Shafer, M.L.: Infectious complications among patients receiving home intravenous therapy with peripheral, central, or peripherally placed central venous catheters. Am. J. Med. **91**(3), S95–S100 (1991)
2. Sayed-Kassem, A., Ghandour, A., Hamawy, L., Zaylaa, A.J.: Cutting-edge robotic intravenous pole: preliminary design and survey in academic medical center in Lebanon. J. Biomed. Eng. Med. Dev. **2**(1), 1–9 (2017)

3. Simons, S.H., Van Dijk, M., van Lingen, R.A., Roofthooft, D., Duivenvoorden, H.J., Jongeneel, N., Tibboel, D.: Routine morphine infusion in preterm newborns who received ventilatory support: a randomized controlled trial. JAMA **290**(18), 2419–2427 (2003)
4. Vignali, M.: Intravenous Stand Design. Project Partners from Huazhong University of Science and Technology, China (2006)
5. Wilt Jr., C.F.: St. Joseph's Hospital, Medical Center, assignee. Apparatus for facilitating intravenous feeding during transportation of patient. United States Patent US 4,511,157 (1985)
6. Qureshi, M.O., Syed, R.S.: The impact of robotics on employment and motivation of employees in the service sector, with special reference to health care. Saf. Health Work **5**, 198–202 (2014)
7. Binger, M., Conway, C., Goddard, N., Jacobs, N., Pysher, C., et al.: Project Number: P15073 Autonomous IV Stand
8. Westbrook, J.I., Rob, M.I., Woods, A., Parry, D.: Errors in the administration of intravenous medications in hospital and the role of correct procedures and nurse experience. BMJ Quality Saf. **20**(12), 1027–1034 (2011)
9. Carelli, L., Gaggioli, A., Pioggia, G., De Rossi, F., Riva, G.: Affective robot for elderly assistance. Stud. Health Technol. Inform. **144**(2009), 44–49 (2009)
10. Qureshi, M.O., Syed, R.S.: The impact of robotics on employment and motivation of employees in the service sector, with special reference to health care. Saf. Health Work **5**(4), 198–202 (2014)
11. Azar, A.T., Eljamel, M.S.: Medical robotics. In: Sobh, T., Xiong, X. (eds.) Prototyping of Robotic Systems: Applications of Design and Implementation. IGI Global, USA (2012). ISBN 978-1466601765
12. Åström, K.J., Hägglund, T.: PID Controllers: Theory, Design and Tuning. Instrument Society of America, Research Triangle Park (1995)
13. Azar, A.T., Vaidyanathan, S.: Computational Intelligence Applications in Modeling and Control: Studies in Computational Intelligence, vol. 575. Springer, Germany (2015). ISBN 978-3-319-11016-5
14. Azar, A.T., Vaidyanathan, S.: Handbook of research on advanced intelligent control engineering and automation. In: Advances in Computational Intelligence and Robotics (ACIR) Book Series. IGI Global, USA (2015). ISBN 9781466672482
15. Sung, S.W., Lee, I.B.: Process Identification and PID Control. Wiley, Singapore (2009)
16. Azar, A.T., Serrano, F.E.: Robust IMC-PID tuning for cascade control systems with gain and phase margin specifications. Neural Comput. Appl. **25**(5), 983–995 (2014). https://doi.org/10.1007/s00521-014-1560-x
17. Blum, J.: Exploring Arduino: Tools and Techniques for Engineering Wizardry. Wiley, London (2013)
18. Immega, G., Antonelli, K.: The KSI tentacle manipulator. In: 1995 IEEE International Conference on Robotics and Automation, vol. 3, pp. 3149–3154. IEEE (1995)
19. Ritchie, G.J., Turner, J.A.: Input devices for interactive graphics. Int. J. Man Mach. Stud. **7** (5), 639–660 (1975)
20. Rech, C., Pinheiro, H., Grundling, H.A., Hey, H.L., Pinheiro, J.R.: Analysis and design of a repetitive predictive-PID controller for PWM inverters. In: 2001 IEEE 32nd Annual Power Electronics Specialists Conference, PESC, vol. 2, pp. 986–991, IEEE (2001)
21. Adar, N.G., Kozan, R.: Comparison between real time PID and 2-DOF PID controller for 6-DOF robot arm. Acta Phys. Pol. A **130**(1), 269–271 (2016)

Ultrasound Transducer Quality Control and Performance Evaluation Using Image Metrics

Amr A. Sharawy[1(✉)], Kamel K. Mohammed[2], Mohamed Aouf[3], and Mohammed A.-M. Salem[4,5]

[1] Biomedical Engineering and System Department, Cairo University, Cairo, Egypt
amrarsh@gmail.com
[2] Center for Virus Research and Studies, Al-Azhar University, Cairo, Egypt
k.eel@hotmail.com
[3] Biomedical Engineering Department, HTI, Tenth of Ramadan City, Egypt
maoufmedical@yahoo.com
[4] Faculty of Computer and Information Sciences, Ain Shams University, Cairo, Egypt
salem@cis.asu.edu.eg
[5] German University in Cairo, Cairo, Egypt
mohammed.salem@guc.edu.eg

Abstract. This paper aims to two main goals, first goal is to achieve the characterization of quality control of ultrasound scanners based on the potential image metrics. On the other hand, the most effective goal is how to classify ultrasound scanners based on image metrics to evaluate performance of ultrasound transducer. The authors utilize the metrics to give information about the spatial arrangement of the gray levels in the specific interest region. The execution of ultrasound images metric based on a set of 19 metrics (i.e. contrast, gradient and Laplacian). This set reflects quality control of ultrasound scanners. The wok of this paper based on the best 6 metrics from 19 metrics which extracted from linear discriminative analysis (LDA). The classification methods used for minimum numbers of metrics are fused using support vector machine (SVM) and the highest classification method is back propagation neural network (BPNN) classifiers to get the main target of paper. Finally, the results show that objective performance evaluation of ultrasound transducer accuracy was 100% by using back propagation neural network classifier.

Keywords: Focal lesion · Contrast · Resolution · Speckle noise Phantom

1 Introduction

Quality control is most invaluable to estimate image quality and machine precision both during the presentation of a modern technology and with regard to performance constancy of over time. Quality control of various ultrasound units is performed on the basis of detection of focal lesions against background tissue because of high spatial resolution and contrast sensitivity requirements [1]. Spatial resolution is defined in

A. E. Hassanien et al. (Eds.): AISI 2018, AISC 845, pp. 26–39, 2019.
https://doi.org/10.1007/978-3-319-99010-1_3

terms of axial resolution which will lies between 0.5 mm and 0.1 mm) and lateral resolution which will lies between 5 mm and 0.5 mm [2, 3]. Contrast sensitivity is defined as the lesion's signal to noise ratio (LSNR), which usually lies between 2 (i.e. 3 dB) and 5 (i.e. 7 dB) [4]. When a focal lesion against background tissue is scanned ultrasonically (e.g. in cysts or tumors in the liver or breast), there is generally a background speckle pattern on the B-mode image, due to small objects of the size compared with the wavelength of the sound and due to lesions having usually low contrast sensitivity relative to the surrounding tissue. If the difference in average brightness level of two areas in the image is small compared to average fluctuations in brightness due to speckle, the tissue will not appear to be different. Rownd [5] evolved tissue-mimicking phantoms and an automated system to test the combined effects of spatial resolution. Unluckily, the position of the spheres were exactly coplanar, the industrialization process did not confirm the exact position to closer than 3 mm. This doubt prevented precise defining of sphere positions in images of the phantom; thus, the computation of (LSNR) values could not be done. This work had been extended by Kofler and Madsen [6] and Kofler et al. [7] to design phantoms containing low echo spheres with coplanar centers forming a regular array alongside a strong software for calculating detectability (i.e. LSNR) and for quantifying the human observer's ability to detect these spheres. This software completely permitted the determination of all object spheres in the image, thus permitting the calculation of LSNR values for target lesions. An important restriction is the utilization of the gray scale instead of the relative echo scale that is combined as well in the International Electrotechnical Commission (IEC) [8]. So, the echo scale (in dB) is utilized in our work for the assessment of most of the quality measures: overall dynamic range, contrast resolution, contrast sensitivity, and spatial resolution. Our aim was to develop an algorithm consisting of 19 metrics to perform two tasks, firstly to analyze and assess ultrasound images, and secondly for the objective performance evaluation of ultrasound transducer. The metrics are related to the spatial and contrast resolution, and the contrast sensitivity of ultrasound images, which are specifically used for quality assessment and performance testing of medical ultrasound equipment [2, 3, 8]. Additionally, this assessment of ultrasound images used for quality control analyzes the images at different depths (specifically at 6 cm and 8 cm) to determine the depth ranges over which lesions of given diameters (6 mm and 8 mm in our method) and phantom material contrast (15 dB & −15 dB in our case) are detected. However, the selected metrics are fused using support vector machine and back propagation neural network classifiers to get an objective performance evaluation of ultrasound transducer and to increase the evaluation accuracy and speed, in order to ensure that the ultrasound transducer is operating consistently before it affects patient diagnosis or image quality, thus preserving the diagnostic power of the technique [9–11]. Additionally, objective performance measures of ultrasound transducers are a required element of the main international scientific organizations programs, including the American Institute of Ultrasound in Medicine (AIUM) and the (IEC) [12]. These international scientific organizations, in addition to comparable requirements imposed by private insurance providers, have increased the focus on accreditation and performance evaluation of medical imaging equipment, which is a common component of these programs [13]. However, as with all technical systems, the utilization of the equipment over time may lead to malfunction, which may affect the quality of the diagnosis. Ultrasound transducers have demonstrated to be the most failure-prone piece of the signal series [9–11, 13]. Our approach focuses on one repeatedly arising transducer defect, the local damage of the array, e.g. due to the failure of individual piezoelectric elements, or element groups in the transducer array, which is hardly observed by the user [9, 10] but can be detected with the support of the gray value objects of the phantom, however only in those situations in which the fault affected sound field portion really interferes with the object under consideration [9, 14]. The faulty transducers were simulated by blurred images as lose in image clarity.

2 Materials and Methods

We implemented the Solo Compact Ultrasound Imaging System (International Biomedical Engineering Technologies, Egypt), a Fukuda, an Esaote, and a GE (General Electric) scanner because they are the most popular scanners in the Egypt. The quality control and quality assurance of ultrasound transducer was performed by using a tissue mimicking "phantom" Model 48-340 General Purpose Multi-Tissue Ultrasound [15]. Its properties are critical for imaging because they simulate focal lesions against background tissue used for quality control and testing of ultrasound transducer. The algorithm is shown in Fig. 1.

In the preprocessing stage, the region of interest (ROI) is selected from digital ultrasound images. Then we delineate a speculated lesion (i.e. with minimum echo level) (e.g. cysts or tumors) against back-ground tissue that can be detected by the human observer anywhere in the image. Based on Weber's law, as shown in Fig. 2 the human's ability to resolve two visual stimuli with different intensities, I (e.g. background tissue) and I + ΔI (e.g. focal lesion), is determined by the ratio ΔI/I over a wide intensity range.

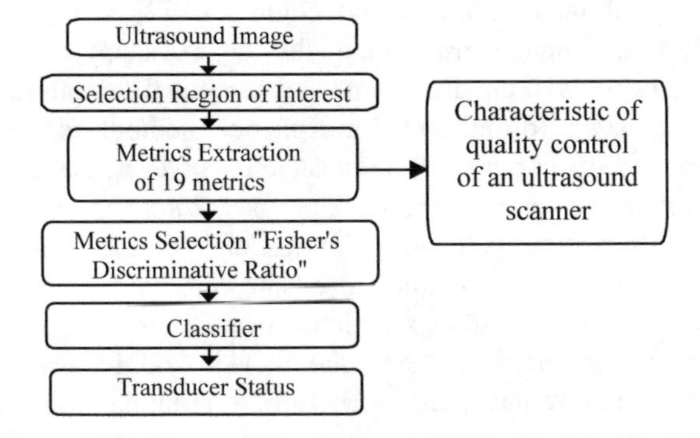

Fig. 1. Block diagram of the proposed system.

In other words, a larger intensity difference (ΔI) is needed for a target to be detected in a brighter background of intensity [16] and therefore we deliberately extracted an

ROI of size 45 × 45 pixels with the object centered in the window as in Fig. 3 according to the shape of the objects in the phantom. Additional to our research, the local failures have coincided with the region of interest which is a particular cyst or gray target in the phantom [11, 16].

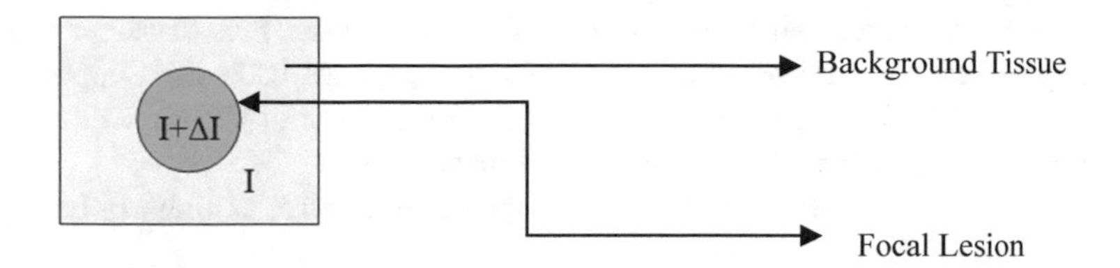

Fig. 2. Basic experimental setup used to characterize brightness discrimination.

Fig. 3. (a) Original image. (b) Region of interest 45 × 45

19 metrics were extracted from the ROI itself. The actual reason for using these metrics, are related to the maximization of contrast sensitivity and spatial resolution, besides their direct relationship to contrast resolution, which altogether reflects on maximizing contrast, variance, cluster shade, and the range of gray levels. We also used two Laplacian metrics, three percentiles, two gradients, the standard deviation of Laplacian, and two autocorrelations for the measured contrast resolution. The metrics also produce minimum values in blurred images and minimum values at greater depth in ultrasound images. The metrics were used for the analysis and assessment of ultrasound images from different aspects. A metrics reduction algorithm is necessary to select few of those extracted features which are most significant, and which best describe the image characteristics. Linear discriminative analysis is used because only two classes were used (maximum quality image, related to intact transducers and minimum quality image, related to fault transducers) [17]. We used Fisher's Discrimination Ratio (FDR) for the highly discriminative metrics [18]. A maximum value of FDR implies that the metrics have more power to differentiate between both classes. The Fisher's Discrimination Ratio is used to select the six metrics that give best results (Eq. 1), where μ_{ai} and σ_{ai} are the mean and the standard deviation of metrics M_j in the minimum quality images (class 'a'), and μ_{bi} and σ are the mean and standard deviation of the same metrics (M_i) in the maximum quality (class 'b'). FDRn is the rank of metrics n; the bigger FDR_n, the bigger the difference between the metrics values of blurred image and maximum quality image relative to metrics n. It turns out that contrast1, gradient and Inverse difference moment are the best metrics for that purpose.

$$\mathrm{FDR_n} = \left| \frac{(\mu_{an} - \mu_{bn})}{\sqrt{(\sigma_{an}^2 + \sigma_{bn}^2)}} \right| \quad (1)$$

Classification: The classification step is the final step where the reduced metrics set is the input to the classifier, while the output is the image class. This classifier uses only six metrics selected on the basis of FDR. In this study, we used the Support Vector Machine (SVM) classifier [27] and the BPNN Neural Network (NN) classifier, in order to determine the classifier with the best performance.

Before running an algorithm, there are appropriate control settings to be adjusted and some issues to be considered:

(a) Scan the phantom as if it were a patient and adjust the control settings (e.g. maximum depth, single focus, maximum gain, and time gain compensation (TGC)) to produce the best possible clinical image and be sure to read and write the final settings on the data sheet and use them every time the tests are performed.
(b) Next degrade the image or reduce image clarity to produce a blurred image (simulating a malfunction in ultrasound transducer) [19]. We actually used the function "imfilter" with a 5 × 5 mask in Matlab. However, a change in resolution (i.e. axial or lateral resolution) or contrast would indicate that there is serious malfunction with the transducer and problems would be noticed [10, 11, 20, 21]. A tissue phantom image of a good array versus a defective array (6 dead elements) is shown in Fig. 4 [11]. Reasons for degraded axial resolution include damaged transducers (broken crystals, loose facing or backing material, or broken electrical connections) and changes in the pulser and/or receiver characteristics. Lateral resolution is typically affected by the loss of transducer elements or by problems in the system's beam-forming and receiving circuits [11, 22]. From Table 1 we notice that axial or lateral resolution are degraded to lower values after blurring images of the GE scanners, and therefore we can interpret degraded images as blurred images.

Table 1. Spatial resolution for original and blurred images of GE imaging system

	Image 1	Degraded image 1	Image 2	Degraded image 2	Image 3	Degraded image 3	Image 4	Degraded image 4
Lateral resolution at 6 cm for GE	0.06	0.37	0.06	0.41	0.07	0.41	0.06	0.41
Axial resolution at 6 cm for GE	0.04	0.44	0.03	0.45	0.01	0.48	0.03	0.47
Lateral resolution at 8 cm for GE	0.38	0.62	0.09	0.42	0.13	0.36	0.12	0.40
Axial resolution at 8 cm for GE	0.24	0.81	0.07	0.27	0.03	0.52	0.03	0.25

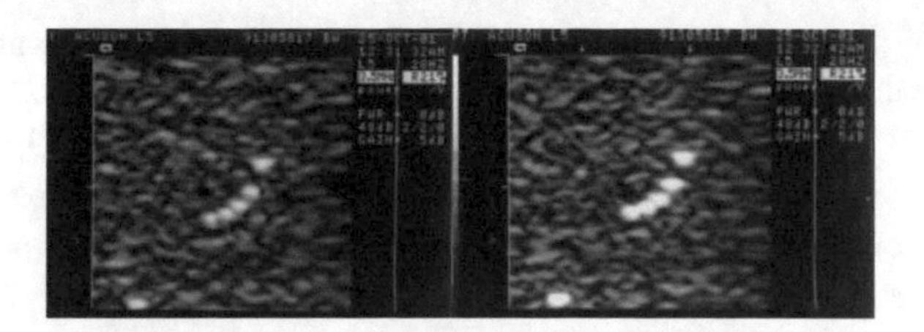

Fig. 4. Intact transducer (left), faulty transducer (right)

2.1 Contrast Resolution

The contrast resolution is represented by the gamma of the system [8].

2.2 Contrast Sensitivity

The contrast sensitivity is to estimate the lesion signal-to-noise ratio, SNRL (as in Eq. 2):

$$SNR_L = \frac{|\langle \mu_2 \rangle - \langle \mu_1 \rangle|}{\sqrt{\sigma^2_{\mu_2} + \sigma^2_{\mu_1}}} \tag{2}$$

where $\langle \mu_2 \rangle$ and $\langle \mu_1 \rangle$ are mean grey level within a circular area of surrounding (background) tissue (2), and of lesion (1), averaged over the ensemble of images from independent scans. $\sigma^2_{\mu_2}$ and $\sigma^2_{\mu_1}$ are variance of mean grey level of area of background (1) and of lesion (2) [8].

2.3 Spatial Resolution (2D In-Plane Point-Spread-Function, IP-PSF)

The spatial resolution is defined as the axial, or lateral, the full width at half maximum (FWHM), i.e. −6 dB width, of the image of a small object at the focus (i.e. both elevation and azimuth foci) [8].

2.4 Statistical Metrics

2.4.1 First Order Statistics

In our study, seven first order statistics are used, as defined below.

Contrast 2

Minimum quality images tend to exhibit a minimum difference between light and dark regions, and therefore it is difficult to observe objects as compared with maximum contrast images [23, 24].

$$\sigma = \sqrt{\frac{1}{n(n-1)}\left(n\sum g_{ij}^2 - \left(\sum g_{ij}\right)^2\right)} \tag{3}$$

where σ is the (S.D.) of gray level values in the ROI image, n is the number of gray level values in the ROI image and g is the grey level of the pixels. Maximum contrast is related to intact transducers and minimum contrast is related to fault transducers.

Variance

The variance is formed and used to measure the degree of variation or dispersion of data around the mean spread or variability.

$$\mathrm{v} = \frac{1}{m}\sum_{i=1}^{m}(\mathrm{i} - \mu)^2 \tag{4}$$

where m is the number of gray level values in the image and μ is the mean of the image pixels of individual gray levels, i.

Maximum Grey Level and Range

Maximum grey level of pixel is related to maximum echo amplitudes [25]. It is the difference between the maximum and minimum values in the sample [25].

$$\mathrm{R} = g_{max} - g_{min} \tag{5}$$

where g is grey level value of the image pixel.

Cumulative Frequency Graph and Percentiles

Cumulative relative frequency, gives the percentage of pixels having a measurement less than or equal to the upper boundary of the class interval:

$$\mathrm{P_n} = \frac{100}{\mathrm{N}}\left(\mathrm{n} - \frac{1}{2}\right) \tag{6}$$

where N is the number of elements in the sample, n is the rank of the percentile. Nine percentile features were used ranking from (10, 20, …, 90) [25].

2.4.2 Second Order Statistics

We have also extracted four second order statistics (derived from gray-level co-occurrence matrix) as defined below.

A gray-level co-occurrence matrix (GLCM) is the method of computing the frequency of pixel pairs having the same gray level in the image [26]. In this paper the matrix is calculated for only one direction (θ = 0) and one distance (d = 1).

Contrast 1

Contrast measures the difference between the maximum and minimum value of a contiguous set of pixels.

$$\text{Contrast 1} = \sum_{\mathrm{i}=1}^{\mathrm{k}}\sum_{\mathrm{j}=1}^{\mathrm{k}}(\mathrm{i} - \mathrm{j})^2 \times \mathrm{P(i,j)} \tag{7}$$

The probability p_{ij} is the ij-th element of G/n, where n is equal to the sum of the elements of G, G is referred to as a gray-level (or intensity) co-occurrence matrix, k is the row (or column) dimension of square matrix G.

Correlation
Returns a measure of a how correlated a pixel is to its neighbor over the entire image. High correlated pixels result in minimum contrast image and therefore it is a minimum quality image and vice versa [27].

$$\text{Correlation} = \sum_{i=0}^{K-1}\sum_{j=0}^{K-1}\frac{\{i \times j\} \times P(i,j) - \{\mu_x \times \mu_y\}}{\sigma_x \times \sigma_y}$$

where $\mu_x, \mu_y, \sigma_x,$ and σ_y are the means and standard deviations of P_x and P_y.

Inverse Difference Moment (IDM)
IDM or homogeneity measures the closeness of distribution of the gray level co-occurrence-matrix (GLCM) elements to main diagonal. The more concentration along main diagonal in GLCM leads to more homogeneous area and minimum contrast [27].

$$\text{Homogeneity} = \sum_{i=0}^{K-1}\sum_{j=0}^{K-1}\frac{P(i,j)}{1+(i-j)^2} \tag{9}$$

Cluster Shade
Cluster shade is a measure of skewness of the matrix [27].

$$\text{Cluster shade} = \sum_{i,j=0}^{K-1}(i+j-\mu_i-\mu_j)^4 P(i,j) \tag{10}$$

2.4.3 Higher Order Features

Complexity
Complexity refers to the visual information content of the image [28]. The complexity is

$$\text{Complexity} = \sum_{c=1}^{C}\sum_{b=1}^{C}\frac{|c-b|}{(M-2d)(N-2d)(h_b+h_c)}(h_c t_c + h_b t_b) \tag{11}$$

2.5 Resolution Metrics

Five resolution metrics were used in our study, namely two gradient features, two Laplacian features, and standard deviation of the Laplacian.

2.5.1 Gradient

If the object in the phantom has thicker edges or boundaries, the gradient of the image becomes maximum [16, 29]. Given an image of intensity g(i, j) at pixel (i, j) we define:

$$\text{Gradient 1} = \sum_{ij} \left| \begin{bmatrix} 1 \\ 0 \\ -1 \end{bmatrix} * g_{ij} \right| \tag{12}$$

$$\text{Gradient 2} = \sum_{ij} \left| \begin{bmatrix} 1 \\ -1 \end{bmatrix} * g_{ij} \right| \tag{13}$$

2.5.2 Laplacian

The Laplacian is a derivative operator (namely a second-order derivative) [16, 29].

Given an image of intensity g(i, j) at pixel (i, j) we define:

$$\text{Laplacian 1} = \sum_{ij} \left| \begin{bmatrix} -1 & -2 & -1 \\ -2 & 12 & -2 \\ -1 & -2 & -1 \end{bmatrix} * g_{ij} \right| \tag{14}$$

$$\text{Laplacian 2} = \sum_{ij} \left| \begin{bmatrix} 0 & -1 & 0 \\ -1 & 4 & -1 \\ 0 & -1 & 0 \end{bmatrix} * g_{ij} \right| \tag{15}$$

2.5.3 Standard Deviation of the Laplacian (SDL)

The equation of standard deviation above "Contrast 2" was proposed by the Kernforschungszentrum Karlsruhe GmbH [24]. Standard Deviation of the Laplacian (SDL) combines "Contrast 2" and a 3 × 3 sharpening filter. The frequency spectrum is independent of the contrast [24, 30]. The converse is not true: filtering the image can change the contrast.

$$\text{SDL} = \sqrt{\frac{1}{n(n-1)} \left(n \sum_{ij} p_{ij}^2 - \left(\sum_{ij} p_{ij} \right)^2 \right)} \tag{16}$$

$$p_{ij} = \begin{bmatrix} -1 & -1 & -1 \\ -1 & 9 & -1 \\ -1 & -1 & -1 \end{bmatrix} * g_{ij} \tag{17}$$

$$g_{ij} = \text{pixel value at coordinate } i, j$$

2.6 Autocorrelation

A maximum in the correlation function should occur for the best image contrast [24].

$$\text{Autocorrelation 1} = \sum_{ij} g_{ij}^2 - \sum_{ij} g_{ij} g_{i,j+1} \tag{18}$$

$$\text{Autocorrelation 2} = \sum_{ij} g_{ij} g_{i,j+1} - \sum_{ij} g_{ij} g_{i,j+2} \tag{19}$$

3 Performance Measures

Usually, an image region can be called a maximum quality image (negative) if it is related to intact transducers or a minimum quality image (positive) if it is related to faulty transducers, and a decision for a detection result can be either correct (true) or incorrect (false). The correctness of the results using True positive (TP), False positive (FP), True negative (TN), False negative (FN), Sensitivity, Specificity, and Accuracy are measured in terms of the metrics. The performance of the designed neural network classifier is measured in terms of accuracy and depends on the ultrasound control settings that are carefully selected, and it is independent of a specific scanner.

4 Results and Discussion

A group of 32 ultrasound images and a group of 32 blurred ultrasound images at different depths were used to evaluate the metrics. After selecting instrument settings, all 19 metrics were calculated from the selected regions in ultrasound images or regions of interest (ROI) as in Table 3. We used a 45 $\times$ 45-pixel ROI. When we compared the results for the original images and degraded images (i.e. simulating a malfunction in ultrasound transducer), we note that the metrics provide significant differences amongst two classes of ultrasound images. Original images provide maximum values of metrics (e.g. Laplacian, Gradient, etc.) as compared to degraded images. Therefore, when ultrasound transducer malfunction is suspected or when a degradation in image quality or a loss in image clarity exists due to probe performance degradation, the metrics can be employed to detect the malfunction or degradation before they affect the patient scan and thus before the diagnosis may become incorrect. Our algorithm can verify that the equipment is operating correctly and that repairs are done properly, and it can be used to compare image quality between different ultrasound machines (i.e. quality control of various ultrasound units). Therefore, as in Table 2 the GE imaging system will detect focal lesions better than the Esaote imaging system because the value of contrast sensitivity for GE is better than the value of contrast sensitivity for Esaote. We also observe from Table 3 how metrics change to low values at a greater depth because of the increase in attenuation of ultrasound with increasing depth of penetration of the beam into the medium. Additionally, we also observe from Table 2 that high contrast sensitivity and high spatial resolution are related to maximum values of contrast 2, variance, cluster shade, maximum gray level or the range of gray levels. We also observe from results that the two Laplacian metrics, three percentiles, two gradients, the standard deviation of the Laplacian and two autocorrelations can be used to measure contrast resolution. Performance results of the proposed neural network classifier for the Esaote and GE imaging data show that the best selected metrics yield an accuracy of 100%, a sensitivity of 100%, a specificity of 100%, a false positive rate computed of 0%, a false negative rate computed of 0%, and a misclassification rate of 0%. The size of the input dataset loaded in the network was of 24 samples, out of which 24 samples were correctly classified, and 0 samples were misclassified by this network as shown in Fig. 5a.

Table 2. Contrast sensitivity values of GE and Esaote imaging systems.

Contrast resolution for Esaote	Gamma = 2.8 [echo levels/dB]
LSNR	Signal to noise ratio = 1.6
Contrast Resolution for GE	Gamma = 1.1 [echo levels/dB]
LSNR	Signal to noise ratio = 2.5

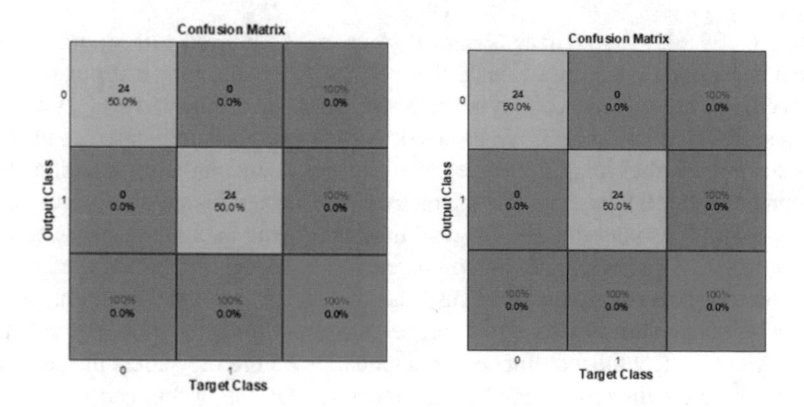

Fig. 5. Left: Confusion matrix for back propagation neural network used for classification between original images and blurred images for GE Ultrasound Imaging System. Right: Confusion matrix for back propagation neural network used for classification between original images and blurred images for Esaote.

Table 3. Metrics for ROI = 45 × 45 at a depth of 6 cm & 8 cm for a GE scanner (one sample normal and degraded image shown)

	Depth = 6 cm		Depth = 8 cm	
	Normal image	Degraded	Normal image	Degraded
Laplacian 1	5609692	2740880	2851520	1443510
Laplacian 2	1517774	707332	752898	376670
Gradient 1	1025674	558710	539066	331796
Autocorrelation 1	157341	133109	68271	60661
Autocorrelation 2	184136	144809	81956	67993
Contrast 1	29.4	6.03	11.4	4.15
Correlation	0.9432	0.9862	0.9639	0.9842
Cluster shade	2.44×10^4	9.85×10^3	1.66×10^4	1.12×10^4
Contrast 2	16.0073	14.149	12.5107	11.412
Variance	256.23365	200.19420	156.51761	130.23374
Max. gray level	146	106	88	78
Gray level range	121	80	69	58
7th percentile	0.2471	0.251	0.1686	0.1725

5 Conclusion

This study aims to confirm the feasibility of using image metrics for the characteristic of quality control of ultrasound scanners from ultrasound images, and to determine the optimal metrics. This method may also be applicable to other types (i.e., other than ultrasound) of imaging systems which produce noisy images. It is anticipated that in the near future, most ultrasound instruments will incorporate facilities for performing such image analysis. Our study results show 100% of accuracy, 100% of sensitivity, and 100% of specificity, by using 6 metrics extracted from ultrasound images for the evaluation of ultrasound transducer. We proved that our method covered the following characteristics: (1) It is quantitative (i.e., it yields numerical results), (2) It is reproducible, and (3) It speeds up the evaluation of ultrasound transducer, given the right conditions and equipment.

References

1. Hall, T.J., Insana, M.F., Harrison, L.A., et al.: Ultrasound contrast detail analysis: a comparison of low contrast detectability among scanhead designs. J. Med. Phys. **22**, 1117–1125 (1995)
2. Hoskins, P., Martin, K., Thrush, A.: Diagnostic Ultrasound: Physics and Equipment, p. 147. Cambridge University Press, Cambridge (2010)
3. Zdero, R., Fenton, P.V., Bryant, J.T.: A digital image analysis method for diagnostic ultrasound calibration. Ultrasonics **39**, 695–702 (2002)
4. Lopez, H., Loew, M.H., Goodenough, D.J.: Objective analysis of ultrasound images by use of a computational observer. IEEE Trans. Med. Imaging **11**, 496–506 (1992)

5. Rownd, J.J., Madsen, E.L., Zagzebski, J.A., et al.: Phantoms and automated system for testing the resolution of ultrasound scanners. J. Ultrasound Med. Biol. **23**, 245–260 (1997)
6. Kofler, J.M., Madsen, E.L.: Improved method for determining resolution zones in ultrasound phantoms with spherical simulated lesions. Ultrasound Med. Biol. **27**, 1667–1676 (2001)
7. Kofler, J.M., Lindstrom, M.J., Kelcz, F., et al.: Association of automated and human observer lesion detecting ability using phantoms. J. Ultrasound Med. Biol. **31**, 351–359 (2005)
8. Thijssen, J.M., Weijers, G., De Korte, C.L.: Objective performance testing and quality assurance of medical ultrasound equipment. J. Ultrasound Med. Biol. **33**, 460–471 (2007)
9. Rosenfeld, E., Wolter, S., Kopp, A., et al.: Investigation of the suitability of tissue phantoms for testing the constancy of ultrasonic transducer arrays in quality assurance. Ultraschall Med. **33**, 289–294 (2012)
10. Rosenfeld, E., Jenderka, K.V., Kopp, A., et al.: How perfect are you with defective probes? Information on the results of the mini-trial on technical quality assurance during the "Ultraschall 2012" conference in Davos. Ultraschall Med. **34**, 185–188 (2013)
11. Weigang, B., Moore, G.W., Gessert, J., et al.: The method and effects of transducer degradation on image quality and the clinical efficacy of diagnostic sonography. J. Diagn. Med. Sonogr. **19**, 3–13 (2003)
12. Cozzolino, P., Stramare, R., Udilano, A., et al.: Quality control of ultrasound transducers: analysis of evaluation parameters and results of a survey of 116 transducers in a single hospital. Radiol. Med. (Torino) **115**, 668–677 (2010)
13. Hangiandreou, N.J., Stekel, S.F., Tradup, D.J., et al.: Four-year experience with a clinical ultrasound quality control program. Ultrasound Med. Biol. **37**, 1350–1357 (2011)
14. Wolter, S., Kopp, A., Liebscher, E., et al.: Consistency check of diagnostic ultrasound transducer arrays using tissue equivalent Phantoms. AIP Conf. Proc. **1433**, 644–647 (2012)
15. Filho, A.C., Rodrigues, E.P., Junior, J.E., et al.: A computational tool as support in B-mode ultrasound diagnostic quality control, Master dissertation. Braz. J. Biomed. Eng. **30**, 402–405 (2014)
16. Gonzalez, R.C.: Digital Image Processing, pp. 125–128. Prentice Hall, Prentice (2002)
17. Satonkar, S.S., Kurhe, A.B., Khanale, P.B.: Face recognition using principal component analysis and linear discriminant analysis on holistic approach in facial images database. J. Eng. **2**, 15–23 (2012)
18. Singh, M., Singh, S., Gupta, S.: An information fusion-based method for liver classification using image metrics of ultrasound images. Inf. Fusion **19**, 91–96 (2014)
19. Jaffe, C.C., Harris, D.J., Taylor, K.J.W., Viscomi, G., Mannes, E.: Sonographic transducer performance cannot be evaluated with clinical images. Am. J. Roentgenol. **137**, 1239–1243 (1981)
20. Lualdi, M., Gamberale, L., Pignoli, E.: A novel computerized method for quality assurance of medical ultrasound probes. Phys. Med. **32**, 81 (2016)
21. Montani, L., Paoli, M., Camarda, M., et al.: Implementation of a quality assurance program for ultrasound transducers. Phys. Med. **32**, 136 (2016)
22. Goodsitt, M.M., Carson, P.L., Witt, S., et al.: Real-time B-mode ultrasound quality control test procedures. Report of AAPM Ultrasound Task Group No. 1. J Med. Phys. **25**, 1385–1406 (1998)
23. Kalyan, K., Lele, R.D., Jakhia, B., et al.: Artificial neural network application in the diagnosis of disease conditions with liver ultrasound images. Adv. Bioinform. **2014**, 1–14 (2014)
24. Price, J.H., Gough, D.A.: Comparison of phase-contrast and fluorescence digital autofocus for scanning microscopy. J. Cytom. **16**, 283–297 (1994)

25. Mabrouk, M., Karrar, A., Sharawy, A.: Computer aided detection of large lung nodules using chest computer tomography image. Int. J. Appl. Inf. Syst. **3**, 12–18 (2012)
26. Gonzalez, R.C.: Digital Image Processing Using Matlab, pp. 601–607. Mc Graw Hill Education, London (2009)
27. Abduh, Z., Abdel Wahed, M.A., Kadah, Y.M.: Robust computer-aided detection of pulmonary nodules from chest computed tomography. J. Med. Imaging Health Inf. **6**, 1–7 (2016)
28. Javed, U., Riaz, M.M., Cheema, T.A.: MRI brain classification using texture features, fuzzy weighting and support vector machine. J. Progr. Electromagn. Res. B **53**, 73–88 (2013)
29. McGee, K.P., Manduca, A., Felmlee, J.P., et al.: Image metric-based correction (autocorrection) of motion effects: analysis of image metrics. J. Magn. Resonan. Imaging **11**, 174–181 (2000)
30. Vollath, D.: The influence of the scene parameters and of noise on the behavior of automatic focusing algorithms. J. Microsc. **151**, 133–146 (1988)

Comparative Study of Two Level and Three Level PWM-Rectifier with Voltage Oriented Control

Arezki Fekik[1], Hakim Denoun[1], Ahmad Taher Azar[2,3](✉), Mohamed Lamine Hamida[1], Mustapha Zaouia[1], and Nabil Benyahia[1]

[1] LATAGE Research Laboratory, University Mouloud Mammeri, Tizi-Ouzou, Algeria
arezkitdk@yahoo.fr, akim_danoun2002dz@yahoo.fr, ml_hamida@yahoo.com, zbmust@yahoo.fr, benyahia.ummto@yahoo.fr

[2] Faculty of Computers and Information, Benha University, Benha, Egypt
ahmad_t_azar@ieee.org

[3] School of Engineering and Applied Sciences, Nile University Campus, Sheikh Zayed District, Juhayna Square, 6th of October City, Giza 12588, Egypt

Abstract. This article presents performance evaluation and comparison between Voltage Oriented Control (VOC) methods for PWM-rectifiers, two levels and three levels, in order to demonstrate the great advantages of using a three-level Neutral Point Clamped (NPC). The control of the DC bus voltage is carried out using the PI controller. The effectiveness of this approach is illustrated by simulation results using MATLAB/Simulink.

Keywords: Neutral point clamped (NPC) · PWM-rectifier
Total harmonic distortion (THD) · Voltage oriented control (VOC)

1 Introduction

Nowadays, the use of PWM-converters in the field of renewable energy such as solar and wind energy is in progress [1–8]. The tendency to use the AC/DC converter know a considerable increase seen to the advantages that it offers: (1) Ensure that the direct current of the THD alternating current is less than 5% of the total fluctuation to reduce the adverse effects on the grid; (2) Guarantee the power factor close to it and consider the rectifier as a "pure resistive load" in terms of grid and (3) Improve the dynamic characteristic of the DC bus voltage regulation, and reduce the dynamic response time, etc. [9–12].

In [13], the various current control techniques applied to the MLI rectifier are classified into two classes, linear current controller (PI-Stationary, PI-rotating reference, etc.) and non-linear current controller (fuzzy logic control, neural networks, hysteresis current control, etc.). The high-performance control strategies of PWM rectifiers are mainly Voltage oriented control (VOC) [11, 12, 14] and direct power control (DPC) [15, 16], which are similar to vector control (VC) and control direct

A. E. Hassanien et al. (Eds.): AISI 2018, AISC 845, pp. 40–51, 2019.
https://doi.org/10.1007/978-3-319-99010-1_4

current torque (DTC) [17] for ac machines. The VOC control technique is based on transforms in two coordinate systems. The first is the fixed coordinate system ($\alpha - \beta$), and the second is the rotating coordinate system (d − q). The measured values of three phases are converted into an equivalent system of two phases ($\alpha - \beta$) and are then transformed to the rotating coordinate system. By means of this type of transformation, the control variables are continuous signals, an inverse transformation (d − q)/($\alpha - \beta$) is performed on the output of the control system which gives the reference signals of the rectifier in the fixed coordinates.

On the other hand, multi-level inverters have become a very interesting solution for high-power applications [18, 19]. The three-stage neutral point clamped (NPC) inverter is one of the most widely used multi-level inverters in high-power AC drives. By comparing the standard level of two inverter levels, the three-level inverter has its superiority in terms of lower semiconductor stresses, lower voltage distortion, less harmonic content and lower switching frequency [20]. Three-level inverters are of great interest in the field of high voltages and high powers because they introduce less distortion and low losses with relatively low switching frequency [21].

This article presents a brief description and comparison of VOC methods to control the PWM rectifier at two and three levels and demonstrate the brilliant advantages of using three levels NPC type converters. The dc-bus controller output is provided by the conventional PI controller.

This paper is organized as follows: Sect. 2 presents the modelling of PWM Rectifier two level. Section 3 modelling of PWM Rectifier two level Sect. 4 gives an overview about the VOC algorithm used in this study. The simulation results and discussion are described in Sect. 5. Finally, conclusion is presented in Sect. 6.

2 The Principle and Modeling of the Two Level PWM-Rectifier

The structure of the three-phase PWM-rectifier with two voltage levels is illustrated in Fig. 1. The PWM-rectifier is connected to the three phases of the source via the smoothing L and the internal resistance R. The inductance acts as a line filter to smooth the line currents with minimal ripples.

Isolated gate bipolar transistors (IGBTs) are used as rectifier supply switches because IGBTs have high power characteristics, simple door control requirements and are suitable for high frequency switching applications.

The resistive load is assumed to be pure R_d and in parallel with DC capacitor C.

The logic states impose the input voltages of the PWM-rectifier with two levels voltage are given as follows

$$\begin{cases} u_{ea} = S_a.V_{dc} \\ u_{eb} = S_b.V_{dc} \\ u_{ec} = S_c.V_{dc} \end{cases} \quad (1)$$

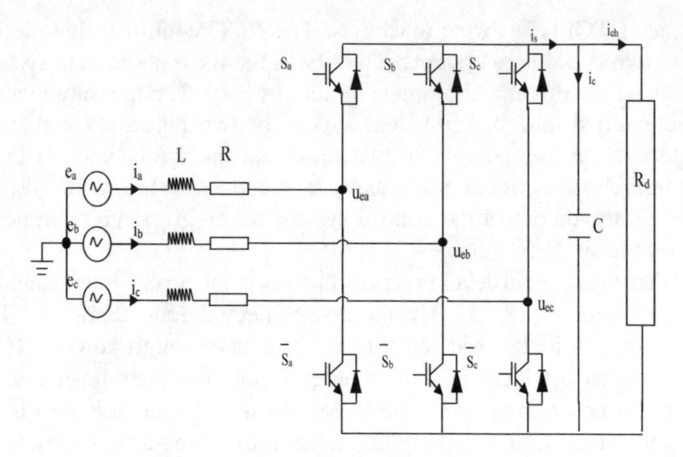

Fig. 1. PWM rectifier two level voltage

S_a, S_b, S_c are the switching states of the PWM-rectifier, the input voltages of the PWM-converter are equal to:

$$\begin{bmatrix} u_{ea} \\ u_{eb} \\ u_{ec} \end{bmatrix} = V_{dc} \begin{pmatrix} \frac{2}{3} & \frac{-1}{3} & \frac{-1}{3} \\ \frac{-1}{3} & \frac{2}{3} & \frac{-1}{3} \\ \frac{-1}{3} & \frac{-1}{3} & \frac{2}{3} \end{pmatrix} \begin{pmatrix} S_a \\ S_b \\ S_c \end{pmatrix} \tag{2}$$

The alternative side equation can be modelled as follows:

$$\begin{cases} u_{ea} = e_a - Ri_a - L\dfrac{di_a}{dt} \\ u_{eb} = e_b - Ri_b - L\dfrac{di_b}{dt} \\ u_{ec} = e_c - Ri_c - L\dfrac{di_c}{dt} \end{cases} \tag{3}$$

The direct current as a function of the switching states can be given by the following equation

$$i_s = S_a i_a + S_b i_b + S_c i_c \tag{4}$$

3 The Principle and Modeling of the Three Level PWM-Rectifier

In the neutral point rectifier illustrated in Fig. 2, the converter is built around twelve switching cells (based on IGBT) and six clamping diodes; each phase can produce three distinct levels by connecting the output to the positive (Vdc/2), negative (−Vdc/2) or zero (0) potential. In a system three, there are $3^3 = 27$ output voltage vectors linked to 19 possible voltage vectors at the output of the converter (see Fig. 3).

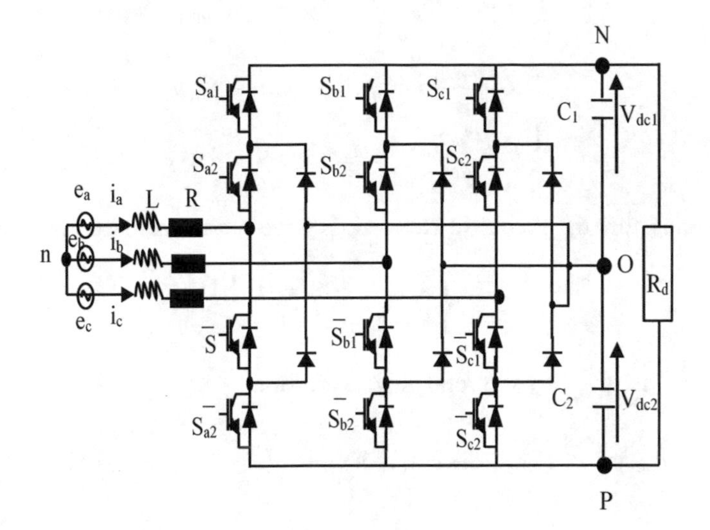

Fig. 2. PWM rectifier three level voltage

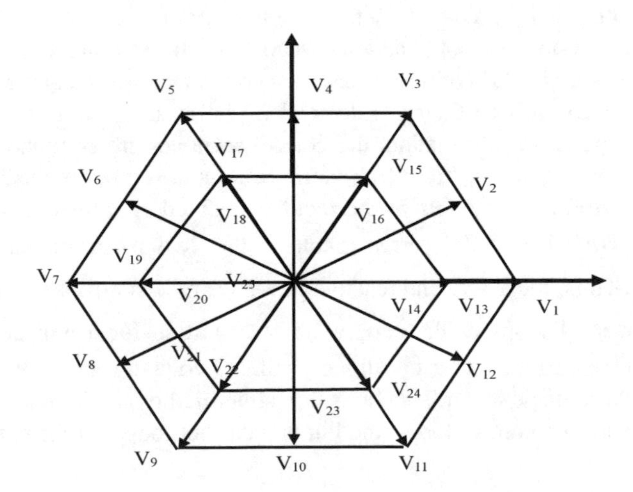

Fig. 3. Space vector diagram of three-level inverter.

According to the definition of the switching functions, the equations of the converter can be developed in the three-phase stationary coordinate system abc.

In addition to the line currents, the dynamics of the capacitors is taken into account and is selected as the state variable. However, the phase voltage of the gate and the load current are considered disturbances.

$$\begin{cases} L\frac{di_a}{dt} = e_a - Ri_a - S_{a1}.V_{dc1} + S_{a2}.V_{dc2} - u_{on} \\ L\frac{di_b}{dt} = e_b - Ri_b - S_{b1}.V_{dc1} + S_{b2}.V_{dc2} - u_{on} \\ L\frac{di_c}{dt} = e_c - Ri_c - S_{c1}.V_{dc1} + S_{c2}.V_{dc2} - u_{on} \end{cases} \quad (5)$$

$$\begin{cases} C_1\dfrac{dv_{dc1}}{dt} = S_{a1}.i_a + S_{b1}.i_b + S_{c1}.i_c \\ C_2\dfrac{dv_{dc2}}{dt} = S_{a2}.i_a - S_{b2}.i_b - S_{c1}.i_c \end{cases} \quad (6)$$

Considering that the Electrical Network is three-phase balanced one can write

$$\begin{cases} e_a + e_b + e_c = 0 \\ i_a + i_b + i_c = 0 \end{cases} \quad (7)$$

the voltage u_{on} is given by the following expression:

$$u_{on} = -\frac{1}{3}(S_{a1} + S_{b1} + S_{c1}).V_{dc1} + \frac{1}{3}(S_{a2} + S_{b2} + S_{c2}).V_{dc2} \quad (8)$$

4 Voltage Oriented Control Strategy

This control technique is based on transforms in two coordinate systems. The first is the fixed coordinate system ($\alpha - \beta$), and the second is the rotating coordinate system ($d - q$). The measured values of three phases are converted into an equivalent system of two phases ($\alpha - \beta$) and are then transformed to the rotating coordinate system. By means of this type of transformation, the control variables are continuous signals, an inverse transformation ($d - q$)/($\alpha - \beta$) is carried out on the output of the control system which gives the reference signals of the rectifier in fixed coordinates. In the rotating coordinate system ($d - q$) the current vector $\vec{i}$ has two perpendicular components $\vec{i} = \begin{bmatrix} i_d & i_q \end{bmatrix}$. Thus, the active and reactive powers can be controlled indirectly by the intermediate internal loops of the currents. The condition for a unit power factor is obtained when the current vector $\vec{i}$ is aligned with the voltage vector $\vec{e}$ by choosing the orientation of the voltage towards the axis d, a simplified dynamic model is obtained.

The VOC had two control loops, the internal current loop and the external voltage loop.

4.1 The Internal Current Loop

The voltage equations in the synchronous frame ($d - q$) are

$$\begin{aligned} e_d &= Ri_d + L\frac{di_d}{dt} + v_d + wLi_q \\ e_q &= Ri_q + L\frac{di_q}{dt} + v_q - wLi_d \end{aligned} \quad (9)$$

Decoupling between the axes d and q is carried out by the variable h_p and h_q:

$$\begin{aligned} h_d &= e_d - v_d - wLi_q = Ri_d + L\frac{di_d}{dt} \\ h_q &= e_q - v_q + wLi_d = Ri_q + L\frac{di_q}{dt} \end{aligned} \tag{10}$$

The system of uncoupled state presented by:

$$\begin{bmatrix} \frac{di_d}{dt} \\ \frac{di_q}{dt} \end{bmatrix} = \begin{bmatrix} -\frac{R}{L} & 0 \\ 0 & -\frac{R}{L} \end{bmatrix} \begin{bmatrix} i_d \\ i_q \end{bmatrix} + \begin{bmatrix} \frac{1}{L} & 0 \\ 0 & \frac{1}{L} \end{bmatrix} \begin{bmatrix} h_d \\ h_q \end{bmatrix} \tag{11}$$

4.2 The External Voltage Loop

The controller in the external control loop of the PWM rectifier is used to regulate the DC voltage side and to generate the amplitude of the reference line current which will be multiplied by the DC voltage to obtain the reference of the instantaneous active power to have the current I_{dref} reference. In this work the regulator used is the conventional PI illustrated in Fig. 4, and to have a unit power factor it is necessary that i_{qref} equal to zero.

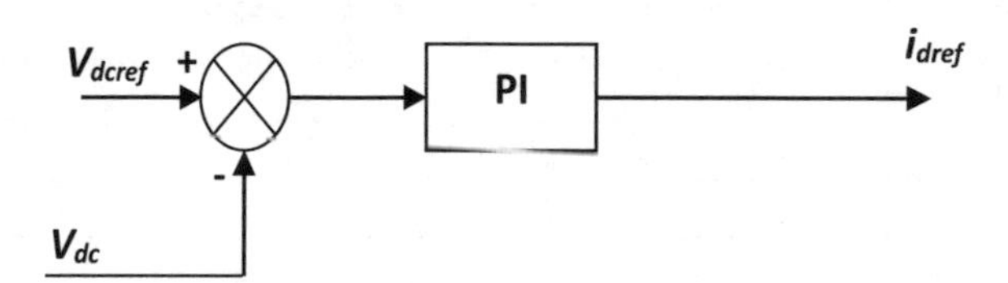

Fig. 4. DC voltage regulation

The transfer function of the studied system is given by:

$$v_{dc}^2 = P_{ref} \cdot \frac{2}{C.s} \tag{12}$$

The transfer function of the PI controller can be expressed by:

$$K_p + \frac{K_i}{s} = \frac{1 + \tau s}{s.T_i} \tag{13}$$

The transfer function of the closed loop system is given by:

$$F(s) = \frac{\omega_0^2.(1 + \tau s)}{s^2 + 2\varepsilon_0 \omega_0 s + \omega_0^2} \tag{14}$$

With:

$$\omega_0 = \sqrt{\frac{2}{CT_i}} \text{ and } \varepsilon_0 = \frac{\tau}{\sqrt{2CT_i}}.$$

After calculation; we find

$$K_p = \frac{\tau}{T_i} \text{ and } K_i = \frac{1}{T_i}.$$

5 Simulation Results

In order to validate the effectiveness of the control strategy developed in this paper for the control of the PWM-rectifier either at two levels or at three levels a numerical simulation was carried out under MATLAB/SIMULINK. The system parameters are summarized in Table 1.

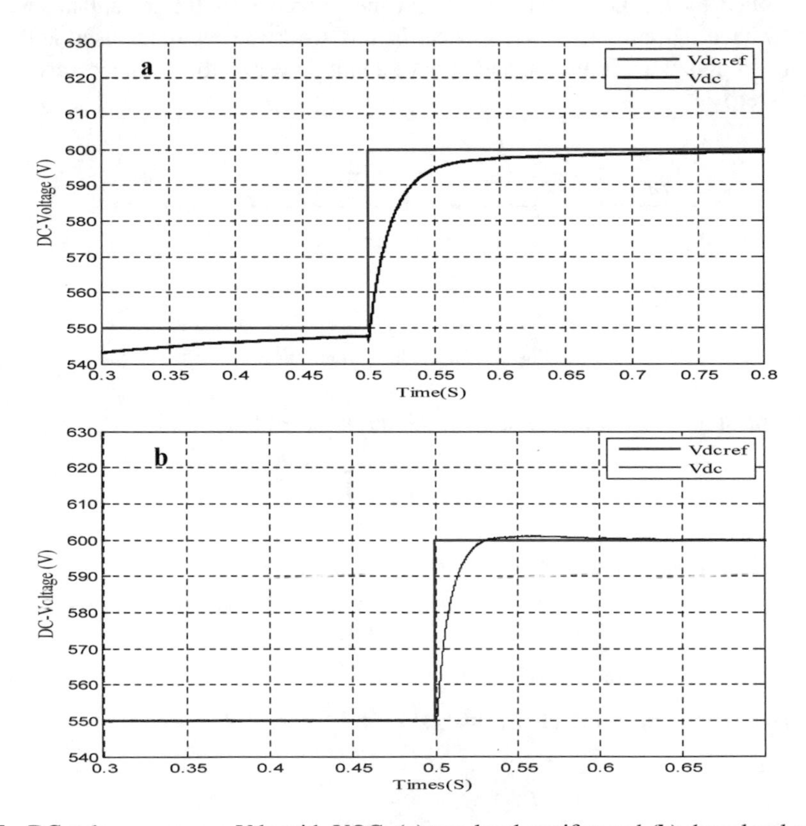

Fig. 5. DC voltage response Vdc with VOC: (**a**) two-level rectifier and (**b**) three-level rectifier

The DC voltage control system for both PWM-converter control by Voltage Oriented Control is tested following a DC-voltage variation occurred at t = 0.5 s from 550 V to 600 V.

Figure 5a and b shows the DC voltage for the two and three levels of PWM rectifiers. The DC voltage measurement follows its new reference when applying a V_{dcref} to the time (t = 0.5 S). It's noted that the response of the DC voltage is faster for a three-level than two-level PWM rectifier.

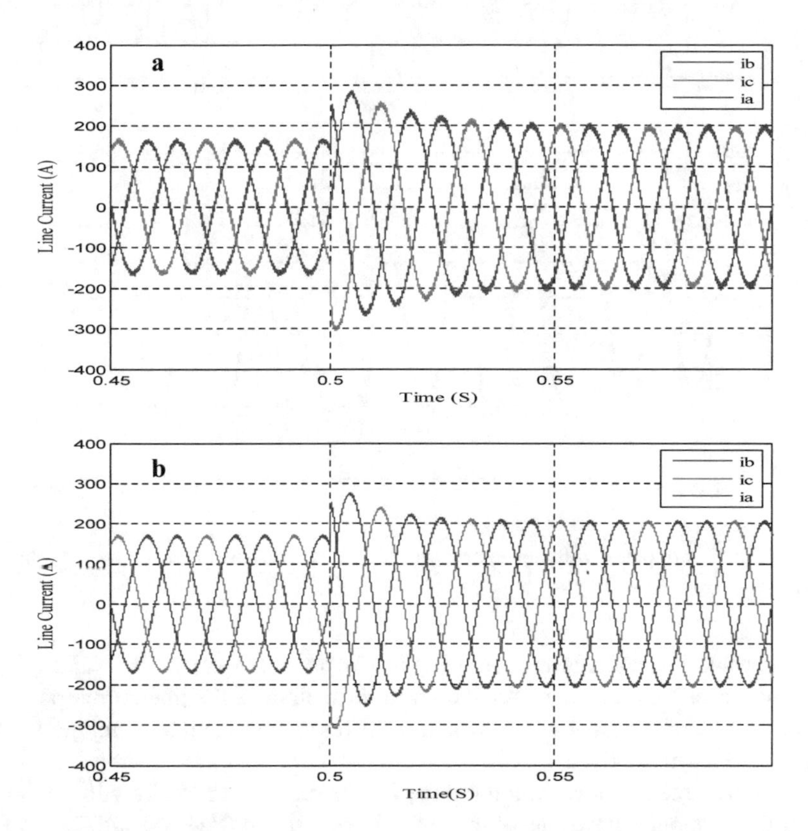

Fig. 6. Line currents with VOC: (**a**) two-level rectifier and (**b**) three-level rectifier.

Figure 6a and b shows that the line currents of the two PWM-rectifier structures are substantially sinusoidal. When changing the DC voltage reference, the current maintains its new value with acceptable response time and a good signal quality of the current for the three-level structure.

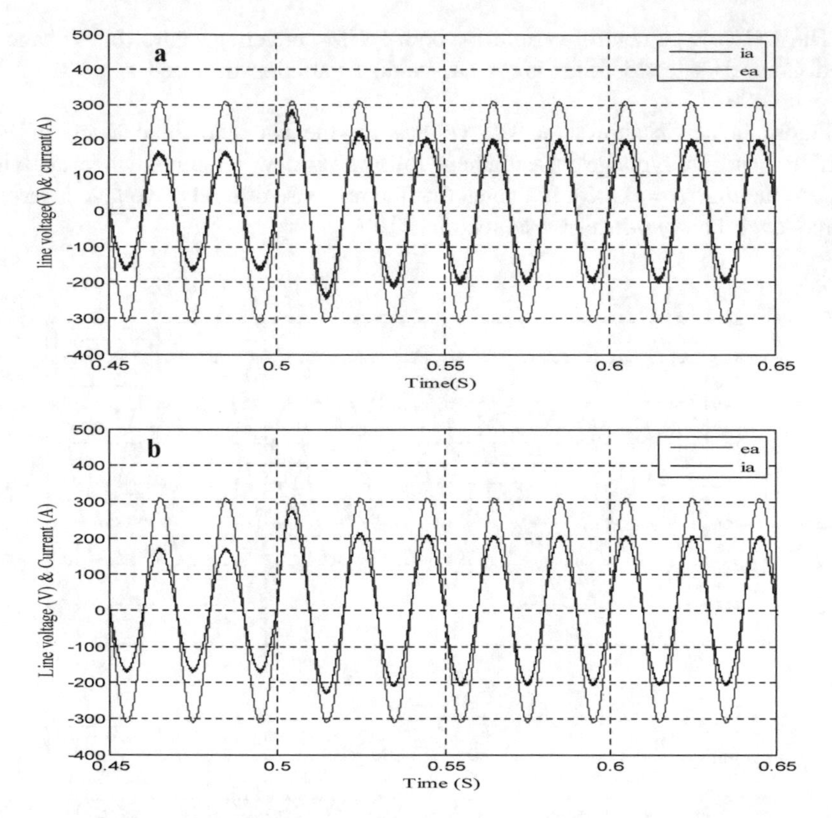

Fig. 7. Voltage is in phase with current ia with VOC: (**a**) two-level rectifier and (**b**) three-level rectifier.

Current and line voltage are shown (Fig. 7a) for a two-level VOC and (Fig. 7b). For a three-level VOC rectifier. As shown in these figures, the line current is in phase with the voltage of the same phase, which confirms operation under a unit power factor and has zero reactive power.

To compare the two level and the three level structure of PWM rectifier with VOC strategy the harmonic spectrums of the current are given in (Fig. 8a) and (Fig. 4b). It is shown that the three level structure (THD = 1.16%) is better than the two level structure (THD = 3.06%).

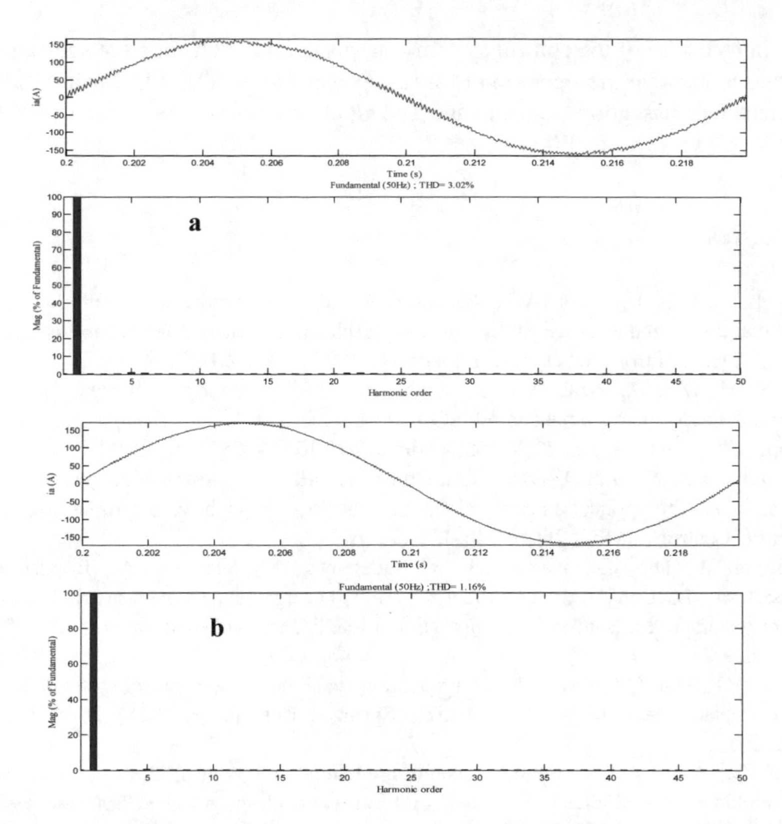

Fig. 8. Harmonic spectrum of the line current with VOC: (**a**) two-level rectifier (THD = 3.06%) and (**b**) three-level (THD = 1.16%) rectifier.

Table 1. System parameters

R	Line resistance
L	Line inductance
C	DC-capacitor
R_d	Load resistance
e_{abc}	Peak amplitude of line voltage
f	Source voltage frequency
f_c	Switching frequency
V_{dcref}	DC-Voltage reference
R	Line resistance
L	Line inductance

6 Conclusion

In this work, the VOC control strategy for a two-level and three-level PWM rectifier is presented. To predict the behavior of the PWM-three-phase rectifier VOC under different load and feed conditions, the dynamic model is implemented in SIMULINK/MATLAB.

The main objective of the control system is to maintain the DC bus voltage at a desired value and to achieve the operation of a unit power factor. The VOC using three-level PWM-rectifiers has good performance and ripple reductions, compared to the VOC using PWM-two-level rectifier.

References

1. Billel, M., Dib, D., Azar, A.T., Saadoun, A.: Effective supervisory controller to extend optimal energy management in hybrid wind turbine under energy and reliability constraints. Int. J. Dyn. Control (2017). https://doi.org/10.1007/s40435-016-0296-0
2. Billel, M., Dib, D., Azar, A.T.: A second-order sliding mode and fuzzy logic control to optimal energy management in PMSG wind turbine with battery storage. Neural Comput. Appl. **28**(6), 1417–1434 (2017). https://doi.org/10.1007/s00521-015-2161-z
3. Ghoudelbourk, S., Dib, D., Azar, A.T., Omeiri, A.: MPPT control in wind energy conversion systems and the application of fractional control (PIα) in pitch wind turbine. Int. J. Model. Identif. Control (IJMIC) **26**(2), 140–151 (2016)
4. Meghni, B., Dib, D., Azar, A.T., Ghoudelbourk, S., Saadoun, A.: Robust adaptive supervisory fractional order controller for optimal energy management in wind turbine with battery storage. In: Studies in Computational Intelligence, vol. 688, pp. 165–202. Springer, Heidelberg (2017)
5. Azar, A.T., Zhu, Q.: Advances and applications in sliding mode control systems. In: Studies in Computational Intelligence, vol. 576. Springer, Heidelberg (2015). ISBN 978-3-319-11172-8
6. Zhu, Q., Azar, A.T.: Complex system modelling and control through intelligent soft computations. In: Studies in Fuzziness and Soft Computing, vol. 319. Springer, Heidelberg (2015). ISBN 978-3-319-12882-5
7. Azar, A.T., Vaidyanathan, S.: Handbook of research on advanced intelligent control engineering and automation. In: Advances in Computational Intelligence and Robotics (ACIR) Book Series. IGI Global, USA (2015). ISBN 9781466672482
8. Azar A.T., Vaidyanathan, S.: Computational intelligence applications in modeling and control. In: Studies in Computational Intelligence, vol. 575. Springer, Heidelberg (2015). ISBN 978-3-319-11016-5
9. Huang, J., Zhang, A., Zhang, H., Wang, J.: A novel fuzzy-based and voltage-oriented direct power control strategy for rectifier. In: IECON 2011 - 37th Annual Conference on IEEE Industrial Electronics Society, Melbourne, VIC, Australia, 7–10 November 2011
10. Hu, J.B., et al.: Direct active and reactive power regulation of grid-connected DC/AC converters using sliding mode control approach. IEEE Trans. Power Electron. **26**(1), 210–222 (2011)
11. Fekik, A., Denoun, H., Benamrouche, N., Benyahia, N., Zaouia, M., Haddad, S.: Comparative study of PI and fuzzy DC voltage control for voltage oriented control-PWM rectifier. In: WSEAS 14th International Conference on Circuits, Systems, Electronics, Control and Signal Processing, Konya, Turkey, pp. 103–110. CSECS (2015)
12. Fekik, A., Denoun, H., Benamrouche, N., Benyahia, N., Zaouia, M.: A fuzzy logic based controller for three phase PWM rectifier with voltage oriented control strategy. Int. J. Circuits Syst. Signal Process. **9**, 412–419 (2015)
13. Kazmierkowski, M.P., Malesani, L.: Current control techniques for three-phase voltage-source PWM converter: a survey. IEEE Trans. Ind. Electron. **45**(5), 691–703 (1998)

14. Song, W.S., Feng, X.Y., Xiong, C.L.: A neutral point voltage regulation method with SVPWM control for single-phase three-level NPC converters. In: IEEE Vehicle Power and Propulsion Conference, pp. 1–4 (2008)
15. Fekik, A., Denoun, H., Benamrouche, N., Benyahia, N., Badji, A., Zaouia, M.: Comparative analysis of direct power control and direct power control with space vector modulation of PWM rectifier. In: 4th IEEE International Conference on Control Engineering and Information Technology (CEIT-2016) Tunisia, Hammamet, 16–18 December 2016
16. Fekik, A., Denoun, H., Zaouia, M., Benyahia, N., Benamrouche, N., Badji, A., Vaidyanathan, S.: Improvement of the performances of the direct power control using space vector modulation of three phases PWM-rectifier. Int. J. Control Theory Appl. **10**(30), 133–145 (2017)
17. Ortega, C., Arias, A., Caruana, C., Balcells, J., Asher, G.: Improved waveform quality in the direct torque control of matrix-converter-fed PMSM drives. IEEE Trans. Ind. Electron. **57** (6), 2101–2110 (2010)
18. Rodriguez, J., Bernet, S., Wu, B., Pontt, J.O., Kouro, S.: Multilevel voltage-source-converter topologies for industrial medium-voltage drives. IEEE Trans. Ind. Electron. **54**(6), 2930–2945 (2007)
19. Zhang, Y., Zhao, Z.: Study on capacitor voltage balance for multilevel inverter based on a fast SVM algorithm. Proc. CSEE **26**(18), 71–76 (2006). (in Chinese)
20. Dalessandro, L., Round, S.D., Kolar, J.W.: Center-point voltage balancing of hysteresis current controlled three-level PWM rectifiers. IEEE Trans. Power Electron. **23**(5), 2477–2488 (2008)
21. Ouboubker, L., Khafallah, M., Lamterkati, J., Chikh, K.: Comparison between DTC using a two-level inverters and DTC using a three level inverters of induction motor. In: 2014 International Conference Multimedia Computing and Systems (ICMCS), pp. 1051–1058 (2014)

New Control Schemes for Fractional Chaos Synchronization

Adel Ouannas[1], Giuseppe Grassi[2], Ahmad Taher Azar[3,4](✉), and Shikha Singh[5]

[1] Laboratory of Mathematics, Informatics and Systems (LAMIS), University of Larbi Tebessi, 12002 Tebessa, Algeria
ouannas.a@yahoo.com

[2] Dipartimento Ingegneria Innovazione, Università del Salento, 73100 Lecce, Italy
Giuseppe.grassi@unisalento.it

[3] Faculty of Computers and Information, Benha University, Benha, Egypt
ahmad.azar@fci.bu.edu.eg

[4] School of Engineering and Applied Sciences, Nile University Campus, 6th of October City, Giza, Egypt
ahmad_t_azar@ieee.org

[5] Faculty of Natural Sciences, Jamia Millia Islamia, New Delhi, India
sshikha7014@gmail.com

Abstract. Chaos theory deals with the behavior of dynamical systems that are highly sensitive to initial conditions. Chaotic systems are characterized by the property that small changes in the initial conditions result in widely diverging responses. In this paper, new control schemes of synchronization for different arbitrary incommensurate and commensurate fractional order chaotic systems are presented. Synchronization stability, based on stability of linear fractional-order systems and fractional Lyapunov stability, is proved theoretically. Numerical examples are given to show the effectiveness of the proposed method.

Keywords: Fractional order chao · Chaos control · Synchronization
Incommensurate system · Commensurate system · Fractional stability

1 Introduction

Recently, fractional calculus has became a great tool in the modeling of many physical phenomena and engineering problems [1,4,8]. Onc of the very important areas of application of fractional calculus is chaos theory [2,17]. In recent years, researcher's interest has been shifted from integer order chaotic systems to fractional order chaotic systems. In the control literature, many research studies have been carried out for the control and synchronization of integer order chaotic systems [3,21–25]. Moreover, recent studies show that chaotic fractional order systems can also be synchronized [18,26].

Recently, studying the synchronization of fractional order chaotic systems has become an active research area. So far, a wide variety of approaches and

A. E. Hassanien et al. (Eds.): AISI 2018, AISC 845, pp. 52–63, 2019.
https://doi.org/10.1007/978-3-319-99010-1_5

techniques have been proposed for the synchronization of the fractional-order chaotic [12–16].

The main aim of this paper is to present new constructive schemes to investigate synchronization between n-dimensional different master and slave fractional-order chaotic systems. Firstly, we propose a new general scheme, using some properties of the Caputo fractional derivative and stability theory of fractional linear systems, to achieve synchronization between incommensurate fractional chaotic systems with respect to the fractional derivative orders of the master system. Secondly, Secondly, to study synchronization between commensurate fractional chaotic systems with respect to the fractional derivative order of the master system, based on Laplace transform theory and fractional Lyapunov stability theory, universal control law is presented. The rest of this paper is organized as follows. Section 2 provides some preliminaries about fractional systems. Our main results are presented in Sects. 3 and 4. In Sect. 5, numerical experiments are given. Finally, Sect. 6 is the brief conclusion.

2 Preliminaries

Caputo fractional derivative [5] is defined as follows

$$D_t^p x(t) = J^{m-p} x^m(t) \quad \text{with} \quad 0 < p \leq 1 \tag{1}$$

where $m = [p]$, i.e., m is the first integer which is not less than p, x^m is the m-order derivative in the usual sense, and $J^q\,(q > 0)$ is the q-order Reimann-Liouville integral operator with expression:

$$J^q y(t) = \frac{1}{\Gamma(q)} \int_0^t (t-\tau)^{q-1} y(\tau)\, d\tau \tag{2}$$

where Γ denotes Gamma function.

Lemma 1 [19]. *Suppose $f(t)$ has a continuous kth derivative on $[0, t]$ ($k \in N$, $t > 0$), and let $p, q > 0$ be such that there exists some $\ell \in N$ with $\ell \leq k$ and p, $p + q \in [\ell - 1, \ell]$. Then*

$$D_t^p D_t^q f(t) = D_t^{p+q} f(t) \tag{3}$$

Remark 1. *Note that the condition requiring the existence of the number ℓ with the above restrictions in the property is essential. In this paper, we consider the case that p, $q \in (0, 1]$ and $p + q \in (0, 1]$. Apparently, under such conditions this property holds.*

Lemma 2 [20]. *The Laplace transform of the Caputo fractional derivative rule reads*

$$L\{D_t^p f(t)\} = s^p F(s) - \sum_{k=0}^{n-1} s^{p-k-1} f^{(k)}(0), \qquad (p > 0, \quad n-1 < p \leq n). \tag{4}$$

Particularly, when $p \in (0, 1]$, we have $L\{D_t^p f(t)\} = s^p F(s) - s^{p-1} f(0)$.

Lemma 3 [20]. *The Laplace transform of the Riemann-Liouville fractional integral rule satisfies*

$$L\left\{J^{p}f(t)\right\}=s^{-p}F\left(s\right),\quad(p>0).\tag{5}$$

Lemma 4 [9]. *The n-dimensional fractional order linear system:* $D_t^{p_i}x_i\left(t\right)=\sum_{j=1}^{n}a_{ij}x_j\left(t\right),\ 1\leq i\leq n$, *where* $0<p_i\leq1$, *is asymptotically stable if all roots* λ *of the equation*

$$det\left(diag\left(\lambda^{Mp_1},\lambda^{Mp_2},...,\lambda^{Mp_n}\right)-A\right)=0\tag{6}$$

satisfy $\left|\arg\left(\lambda\right)\right|>\frac{\pi}{2M}$, *where* M *is the least common multiple of the denominators of* p_i*'s and* $A=\left(a_{ij}\right)_{n\times n}$.

Lemma 5 [11]. *The n-dimensional fractional order system:* $D_t^{p}X\left(t\right)=AX\left(t\right)$, *where* $0<p\leq1$ *and* $A\in\mathbf{R}^{n\times n}$, *is asymptotically stable if* A *is a negative definite matrix.*

3 Synchronization of Incommensurate Fractional Chaotic Systems

We consider the master system as

$$D_t^{p_i}x_i\left(t\right)=f_i\left(X\left(t\right)\right),\quad i=1,2,...,n,\tag{7}$$

where $X\left(t\right)=\left(x_1\left(t\right),x_2\left(t\right),...,x_n\left(t\right)\right)^T$ is the state vector of the master system (7), $f_i:\mathbb{R}^n\rightarrow\mathbb{R}$, p_i is a rational number between 0 and 1 and $D_t^{p_i}$ is the Caputo fractional derivative of order p_i. Also, consider the slave system as

$$D_t^{q_i}y_i\left(t\right)=\sum_{j=1}^{n}b_{ij}y_j\left(t\right)+g_i\left(Y\left(t\right)\right)+u_i,\quad i=1,2,...,n,\tag{8}$$

where $Y(t)=\left(y_1\left(t\right),y_2\left(t\right),...,y_n\left(t\right)\right)^T$ is the state vector of the slave system (8), $\left(b_{ij}\right)\in\mathbb{R}^{n\times n}$, $g_i:\mathbb{R}^n\rightarrow\mathbb{R}$, are nonlinear functions, q_i is a rational number between 0 and 1, $D_t^{q_i}$ is the Caputo fractional derivative of order q_i and u_i, $1\leq i\leq n$, are controllers to be designed.

Let us define the synchronization errors, between the master system (7) and the slave system (8), as

$$e_i\left(t\right)=y_i\left(t\right)-x_i\left(t\right),\quad i=1,2,...,n.\tag{9}$$

In this case, we assume that $0<q_i<p_i\leq1$.

Theorem 1. *The master system (7) and the slave (8) are globally synchronized under the following controllers*

$$\begin{aligned}u_i=&-\sum_{j=1}^{n}b_{ij}y_j\left(t\right)-g_i\left(Y\left(t\right)\right)\\&+J^{p_i-q_i}\left(-\left|b_{ii}\right|e_i\left(t\right)+f_i\left(X\left(t\right)\right)\right),\quad i=1,2,...,n.\end{aligned}\tag{10}$$

Proof. The Caputo fractional derivative of order p_i of the error system of (9) can be described as

$$D_t^{p_i} e_i(t) = D_t^{p_i} y_i(t) - f_i(X(t)), \quad i = 1, 2, ..., n. \tag{11}$$

By substituting Eq. (10) into Eq. (8), we can rewrite the slave system as

$$D_t^{q_i} y_i(t) = J^{p_i - q_i}\left(-|b_{ii}| e_i(t) + f_i(X(t))\right), \quad i = 1, 2, ..., n. \tag{12}$$

By applying the Caputo fractional derivative of order $p_i - q_i$, to both the left and right sides of Eq. (12), we obtain

$$\begin{aligned} D_t^{p_i} y_i(t) &= D_t^{p_i - q_i}\left(D_t^{q_i} y_i(t)\right) \\ &= D_t^{p_i - q_i} J^{p_i - q_i}\left(-|b_{ii}| e_i(t) + f_i(X(t))\right) \\ &= -|b_{ii}| e_i(t) + f_i(X(t)) \end{aligned} \tag{13}$$

Note that $p_i - q_i$ satisfies $p_i - q_i \in (0, 1]$ for $i = 1, 2, ..., n$. According to Lemma 1 the above statement holds. By substituting Eq. (13) into Eq. (11), the synchronization errors can be written as

$$D_t^{q_i} e_i(t) = -|b_{ii}| e_i(t), \quad i = 1, 2, ..., n, \tag{14}$$

Rewriting the error system (14) in the compact form

$$D_t^q e(t) = Be(t), \tag{15}$$

where $e(t) = (e_1(t), e_2(t), ..., e_n(t))^T$, $D_t^q = [D_t^{q_1}, D_t^{q_2}, ..., D_t^{q_n}]$ and

$$B = \begin{pmatrix} -|b_{11}| & 0 & \cdots & 0 \\ 0 & -|b_{22}| & \cdots & 0 \\ \vdots & \vdots & \ddots & \vdots \\ 0 & 0 & \cdots & -|b_{nn}| \end{pmatrix}. \tag{16}$$

Then, we can show that all roots λ of $det\left(\text{diag}\left(\lambda^{Mq_1}, ..., \lambda^{Mq_{nq}}\right) - B\right) = 0$, where M is the least common multiple of the denominators of q_i's, can be described as $\lambda^{Mq_i} = -|b_{ii}|$, $i = 1, 2, ..., n$, this implies that $\arg(\lambda) = \frac{\pi}{q_i M} > \frac{\pi}{2M}$, $i = 1, 2, ..., n$. Then, according to Lemma 4, the error system (15) is asymptotically stable. Therefore, the systems (7) and (8) are globally synchronized.

4 Synchronization of Commensurate Fractional Chaotic Systems

Now, the master and the slave chaotic systems can be considered in the following forms

$$D_t^p X(t) = AX(t) + f(X(t)) \tag{17}$$

$$D_t^q Y(t) = g(Y(t)) + U \tag{18}$$

where $X(t) = (x_1(t), x_2(t), ..., x_n(t))^T$, $Y(t) = (y_1(t), y_2(t), ..., y_n(t))^T$ are the states of the master system (17) and the slave system (18), $A \in \mathbb{R}^{n \times n}$, $f : \mathbb{R}^n \to \mathbb{R}^n$, is a nonlinear function, $g : \mathbb{R}^n \to \mathbb{R}^n$, $0 < q < p \leq 1$ and $U = (u_i)_{1 \leq i \leq n}$ is a vector controller to be designed.

In this case, the error system between the master system (17) and the slave system (18), can be defined as

$$e(t) = X(t) - Y(t) \tag{19}$$

Theorem 2. *The master system (17) and the slave (18) are globally synchronized under the following controller*

$$U = -g(Y(t)) + J^{p-q}[CX(t) + (A - C)Y(t) + f(X(t))] \tag{20}$$

where $C = (c_{ij})_{n \times n}$

$$c_{ij} = \begin{cases} a_{ij} & \text{if } i \neq j \\ a_{ii} + |a_{ii}| & \text{if } i = j \end{cases}$$

Proof. By inserting Eq. (20) into Eq. (18), we can rewrite the slave system as follows:

$$D_t^q Y(t) = J^{p-q}[CX(t) + (A - C)Y(t) + f(X(t))] \tag{21}$$

Applying the Laplace transform to Eq. (21) and letting

$$\mathbf{F}(s) = \mathbf{L}(Y(t)) \tag{22}$$

we obtain,

$$\begin{gathered} s^q \mathbf{F}(s) - s^{q-1} Y(0) = \\ s^{q-p} \mathbf{L}(CX(t) + (A - C)Y(t) + f(X(t))) \end{gathered} \tag{23}$$

multiplying both the left-hand and right-hand sides of Eq. (23) by s^{p-q} and applying the inverse Laplace transform to the result, we obtain a new equation for the slave system

$$D_t^p Y(t) = CX(t) + (A - C)Y(t) + f(X(t)) \tag{24}$$

The Caputo fractional derivative of order p of the error system of (19) can be written as

$$D_t^p e(t) = AX(t) + f(X(t)) - D_t^p Y(t) \tag{25}$$

and by substituting Eq. (24) into Eq. (25), the error system can be described as

$$D_t^p e(t) = (A - C)e(t) \tag{26}$$

It is easy to see that $A - C$ is a negative definite matrix. Then, according to Lemma 5, the error system (26) is asymptotically stable. Therefore, the systems (17) and (18) are globally synchronized.

5 Numerical Examples

5.1 Synchronization of Fractional Modified Coupled Dynamos System and Fractional Unified System

In this example, we consider the incommensurate fractional order modified coupled dynamos system as the master system and the controlled incommensurate fractional order unified system as the slave system. The master system is defined as

$$\begin{cases} D^{p_1}x_1 = -\alpha x_1 + (x_3 + \beta)\, x_2 \\ D^{p_2}x_2 = -\alpha x_2 + (x_3 - \beta)\, x_1 \\ D^{p_3}x_3 = x_3 - x_1 x_2 \end{cases} \tag{27}$$

This system, as shown in [10], exhibits chaotic behaviors when $(p_1, p_2, p_3) = (0.9, 0.93, 0.96)$ and $(\alpha, \beta) = (2, 1)$. The chaotic attractors of the incommensurate fractional order modified coupled dynamos system (27) are shown in Fig. 1.

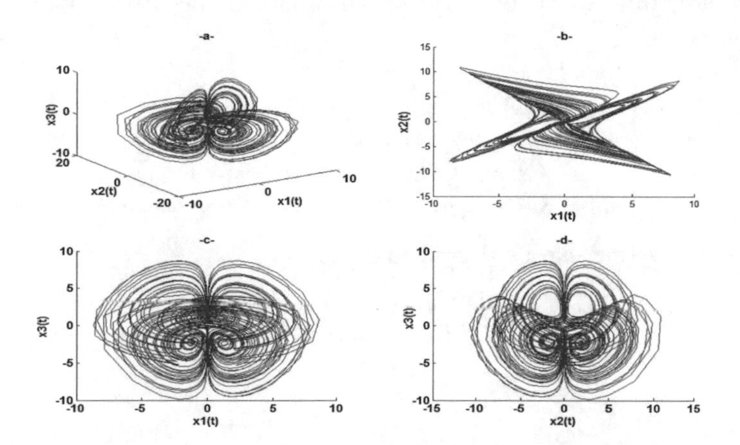

Fig. 1. Chaotic attractors of the incommensurate fractional order modified coupled dynamos system when $(p_1, p_2, p_3) = (0.9, 0.93, 0.96)$ and $(\alpha, \beta) = (2, 1)$.

The slave system is described by

$$\begin{cases} D^{q_1}y_1 = (25a + 10)\,(y_2 - y_1) + u_1 \\ D^{q_2}y_2 = (28 - 35a)y_1 + (29a - 1)y_2 + y_1 y_3 + u_2 \\ D^{q_3}y_3 = \frac{-(a+8)}{3} y_3 + y_1 y_2 + u_3 \end{cases} \tag{28}$$

This system, as shown in [7], still exhibits chaotic behaviors when $(q_1, q_2, q_3) = (0.85, 0.9, 0.95)$ and $a = 1$. The chaotic attractors of the incommensurate fractional order unified system (ie, the system (28) with $u_1 = u_1 = u_1 = 0$) are shown in Fig. 2.

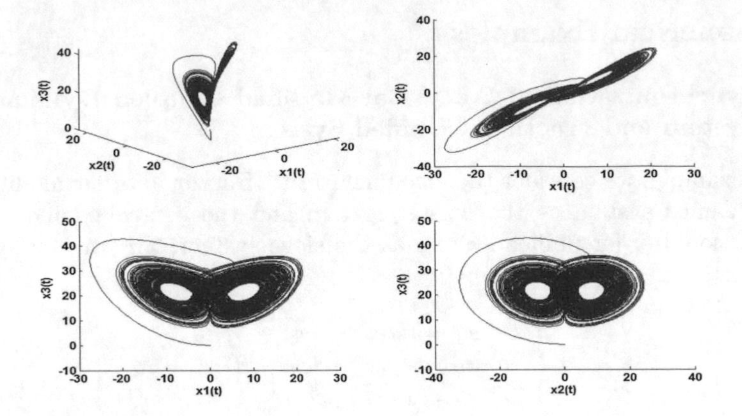

Fig. 2. Chaotic attractors of the incommensurate fractional order unified when $(q_1, q_2, q_3) = (0.85, 0.9, 0.95)$ and $a = 1$.

Then, according to Theorem 1, the synchronization controllers u_1, u_2 and u_3 can be chosen as

$$\begin{cases} u_1 = -35\left(y_2 - y_1\right) \\ \qquad +J^{0.05}\left(-35y_1 + 33x_1 + \left(x_3 + 1\right)x_2\right) \\ u_2 = 7y_1 - 28y_2 - y_1y_3 \\ \qquad +J^{0.07}\left(-28y_2 + 26x_2 + \left(x_3 - 1\right)x_1\right) \\ u_3 = 3y_3 - y_1y_2 + J^{0.01}\left(-3y_3 + 4x_3 - x_1x_2\right) \end{cases} \tag{29}$$

and the error system can be described as

$$\begin{pmatrix} D^{0.9}e_1 \\ D^{0.93}e_2 \\ D^{0.96}e_3 \end{pmatrix} = B \times \begin{pmatrix} e_1 \\ e_2 \\ e_3 \end{pmatrix} \tag{30}$$

where

$$B = \begin{pmatrix} -35 & 0 & 0 \\ 0 & -28 & 0 \\ 0 & 0 & -3 \end{pmatrix}$$

Then, the roots of

$$\det\left(\mathrm{diag}\left(\lambda^{M0.9}, \lambda^{M0.93}, \lambda^{M0.96}\right) - B\right) = 0,$$

where M is the least common multiple of the denominators of 0.85, 0.9, and 0.95, are obtained as follows:

$$\begin{aligned} \lambda_1 &= 35^{\frac{1}{M0.9}}\left[\cos\left(\tfrac{\pi}{M0.9}\right) + \mathbf{i}\sin\left(\tfrac{\pi}{M0.9}\right)\right] \\ \lambda_2 &= 28^{\frac{1}{M0.93}}\left[\cos\left(\tfrac{\pi}{M0.93}\right) + \mathbf{i}\sin\left(\tfrac{\pi}{M0.93}\right)\right] \\ \lambda_3 &= 3^{\frac{1}{M0.96}}\left[\cos\left(\tfrac{\pi}{M0.96}\right) + \mathbf{i}\sin\left(\tfrac{\pi}{M0.96}\right)\right] \end{aligned} \tag{31}$$

It is easy to see that $\arg\left(\lambda_i\right) > \frac{\pi}{2M}$, $i = 1, 2, 3$. Therefore, the systems (27) and (28) are globally synchronized. The numerical simulations of the error functions evolution are shown in Fig. 3.

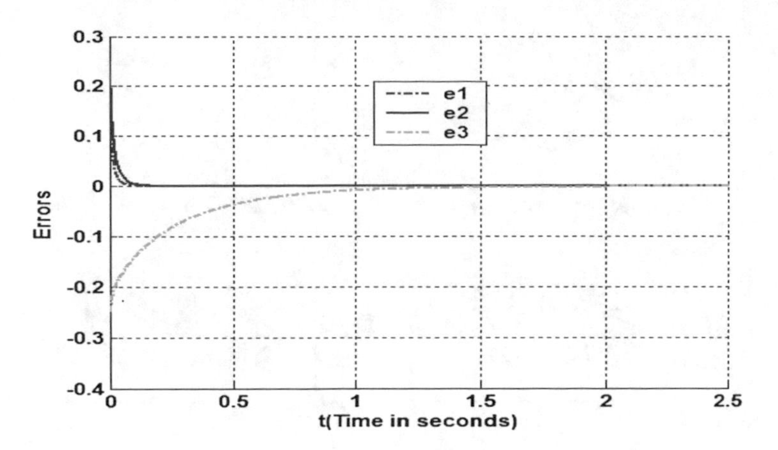

Fig. 3. Time evolution of the synchronization errors between systems (27) and (28).

5.2 Synchronization of Fractional Generalized Volta's System and Fractional Lu System

In this example, we take as the master the commensurate fractional generalized Volta's system and as the slave system the controlled commensurate fractional Lu system. The master system can described by

$$\begin{cases} D^p x_1 = -x_1 - \alpha x_2 - x_3 x_2 \\ D^p x_2 = -x_2 - \beta x_1 - x_3 x_1 \\ D^p x_3 = \gamma x_3 + x_1 x_2 + 1 \end{cases} \tag{32}$$

When $p = 0.98$ and $(\alpha, \beta, \gamma) = (19, 11, 0.73)$, system (32) exhibits chaotic behaviors [17]. Chaotic attractors of the fractional generalized Volta's system are shown in Fig. 4.

Also, the slave system can described by

$$\begin{cases} D^q y_1 = a\,(y_2 - y_1) + u_1 \\ D^q y_2 = b y_1 - y_1 y_3 + u_2 \\ D^q y_3 = -c y_3 + y_1 y_2 + u_3 \end{cases} \tag{33}$$

where $(u_1, u_2, u_3)^T$ is a control law. And when $u_1 = u_2 = u_3 = 0$, $(q_1, q_2, q_3) = (0.85, 0.9, 0.95)$ and $a = 1$, system (33), as shown in [6], still exhibits chaotic behaviors. Chaotic attractors of the commensurate fractional Lu system are given in Fig. 5.

Then, according to Theorem 2, the controllers u_1, u_2 and u_3 can be chosen as follows:

$$\begin{cases} u_1 = -35\,(y_2 - y_1) + J^{0.08}\,[-19 x_2 - y_1 - x_3 x_2] \\ u_2 = -28 y_1 + y_1 y_3 + J^{0.08}\,[-11 x_1 - y_2 - x_3 x_1] \\ u_3 = 3 y_3 - y_1 y_2 + J^{0.08}\,[1.46 x_3 - 0.73 y_3 + x_1 x_2 + 1] \end{cases} \tag{34}$$

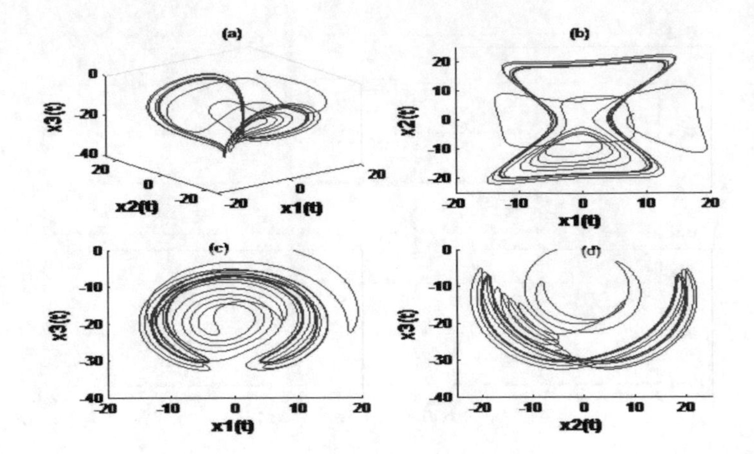

Fig. 4. Chaotic attractors of the fractional generalized Volta's system when $p = 0.98$ and $(\alpha, \beta, \gamma) = (19, 11, 0.73)$.

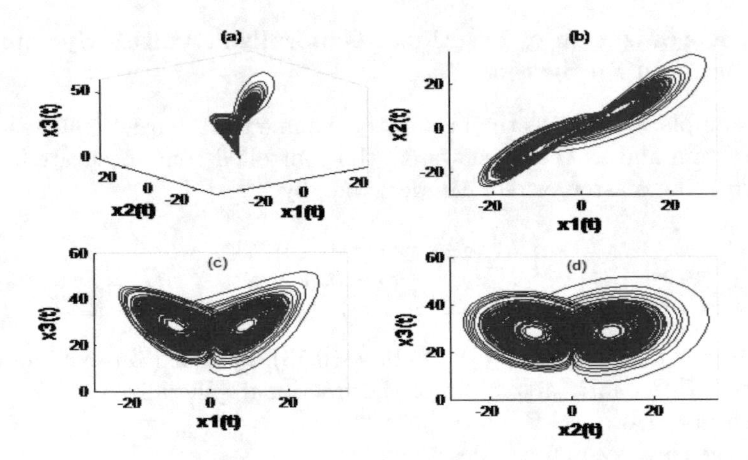

Fig. 5. Chaotic attractors of the fractional Lu system when $p = 0.98$ and $(\alpha, \beta, \gamma) = (19, 11, 0.73)$.

and the errors system can be written as

$$\begin{pmatrix} D^{0.98}e_1 \\ D^{0.98}e_2 \\ D^{0.98}e_3 \end{pmatrix} = L \times \begin{pmatrix} e_1 \\ e_2 \\ e_3 \end{pmatrix} \tag{35}$$

where

$$L = \begin{pmatrix} -1 & -19 & 0 \\ -11 & -1 & 0 \\ 0 & 0 & -0.73 \end{pmatrix}$$

In this case, L is a negative definite matrix. Then, the systems (32) and (33) are globally synchronized. The numerical simulations of the error functions evolution are shown in Fig. 6.

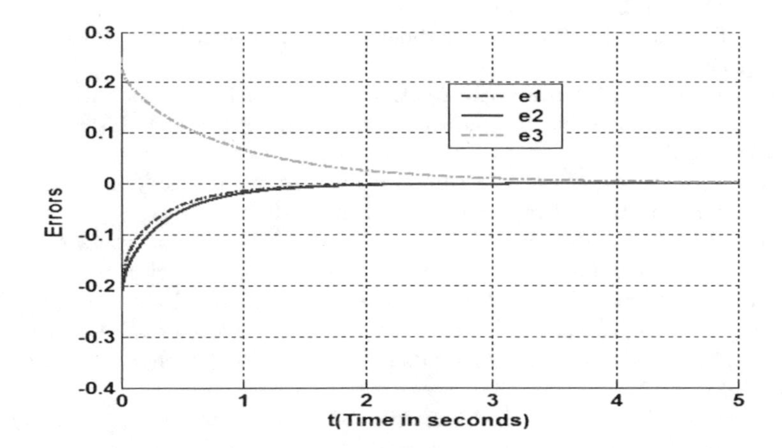

Fig. 6. Time evolution of the error system between systems (32) and (33)

6 Conclusion

In this study, new complex control schemes were designed to achieve synchronization between coupled of $n-$dimensional different fractional-order chaotic systems. The first scheme was obtained via controlling the linear part of the slave system with respect to fractional derivative order of the master system and the second one is constructed based on the control of the linear part of the master system. Numerical simulations are given to verify the effectiveness of the proposed synchronization schemes.

References

1. Azar, A.T., Vaidyanathan, S.: Chaos Modeling and Control Systems Design. Studies in Computational Intelligence, vol. 581. Springer, Berlin (2015)
2. Azar, A.T., Vaidyanathan, S., Ouannas, A.: Fractional Order Control and Synchronization of Chaotic Systems. Studies in Computational Intelligence, vol. 688. Springer, Berlin (2017)
3. Azar, A.T., Volos, C., Gerodimos, N.A., Tombras, G.S., Pham, V.T., Radwan, A.G., Vaidyanathan, S., Ouannas, A., Munoz-Pacheco, J.M.: A novel chaotic system without equilibrium: dynamics, synchronization, and circuit realization. Complexity 2017, 11 p. (2017). Article ID: 7871467
4. Azar, A.T., Vaidyanathan, S.: Advances in Chaos Theory and Intelligent Control. Studies in Fuzziness and Soft Computing, vol. 337. Springer, Berlin (2016)
5. Caputo, M.: Linear models of dissipation whose Q is almost frequency independent-II. Geophys. J. Roy. Astron. Soc. **13**(5), 529–539 (1967)

6. Dastranj, M., Moghaddas, M., Rad, P., Ebrahimi, H.: Synchronization of chaotic fractional-order Lu-Lu systems with active sliding mode control. J. Artif. Intell. Electr. Eng. **2**(8), 59–67 (2014)
7. Deng, W., Li, C.: The evolution of chaotic dynamics for fractional unified system. Phys. Lett. A **372**(4), 401–407 (2008)
8. Magin, R.: Fractional Calculus in Bioengineering. Begell House Publishers, Redding (2006)
9. Matignon, D.: Stability results for fractional differential equations with applications to control processing. In: Computational Engineering in Systems Applications, pp. 963–968 (1996)
10. Ming-Jun, W., Xing-Yuan, W.: Dynamic analysis of the fractional order Newton-Leipnik system. Acta Phys. Sinica **59**(3), 1583 (2010)
11. Ouannas, A., Al-sawalha, M.M., Ziar, T.: Fractional chaos synchronization schemes for different dimensional systems with non-identical fractional-orders via two scaling matrices. Optik - Int. J. Light Electron Opt. **127**(20), 8410–8418 (2016)
12. Ouannas, A., Abdelmalek, S., Bendoukha, S.: Coexistence of some chaos synchronization types in fractional-order differential equations. Electron. J. Differ. Eqn. **128**, 1–15 (2017)
13. Ouannas, A., Azar, A.T., Vaidyanathan, S.: A new fractional hybrid chaos synchronisation. Int. J. Model. Ident. Control **27**(4), 314–322 (2017)
14. Ouannas, A., Azar, A.T., Vaidyanathan, S.: A robust method for new fractional hybrid chaos synchronization. Math. Methods Appl. Sci. **40**(5), 1804–1812 (2017). mma.4099
15. Ouannas, A., Grassi, G., Ziar, T., Odibat, Z.: On a function projective synchronization scheme for non-identical fractional-order chaotic (hyperchaotic) systems with different dimensions and orders. Optik - Int. J. Light Electron Opt. **136**, 513–523 (2017)
16. Ouannas, A., Odibat, Z., Hayat, T.: Fractional analysis of co-existence of some types of chaos synchronization. Chaos, Solitons Fractals **105**, 215–223 (2017)
17. Petras, I.: Fractional-Order Nonlinear Systems: Modeling, Analysis and Simulation. Higher Education Press/Springer, Beijing/Heidelberg (2011)
18. Pham, V.T., Ouannas, A., Volos, C., Kapitaniak, T.: A simple fractional-order chaotic system without equilibrium and its synchronization. AEU - Int. J. Electr. Commun. **86**, 69–76 (2018)
19. Podlubny, I.: Fractional Differential Equations. Academic Press, New York (1999)
20. Samko, S.G., Klibas, A.A., Marichev, O.I.: Fractional Integrals and Derivatives: Theory and Applications. Gordan and Breach, Amsterdam (1993)
21. Vaidyanathan, S., Azar, A.T.: A novel 4-D four-wing chaotic system with four quadratic nonlinearities and its synchronization via adaptive control method. In: Azar, A.T., Vaidyanathan, S. (eds.) Advances in Chaos Theory and Intelligent Control. Studies in Fuzziness and Soft Computing, vol. 337, pp. 203–224. Springer, Berlin (2016)
22. Vaidyanathan, S., Azar, A.T.: Adaptive backstepping control and synchronization of a novel 3-D jerk system with an exponential nonlinearity. In: Azar, A.T., Vaidyanathan, S. (eds.) Advances in Chaos Theory and Intelligent Control. Studies in Fuzziness and Soft Computing, vol. 337, pp. 249–274. Springer, Berlin (2016)
23. Vaidyanathan, S., Azar, A.T.: Adaptive control and synchronization of Halvorsen circulant chaotic systems. In: Azar, A.T., Vaidyanathan, S. (eds.) Advances in Chaos Theory and Intelligent Control. Studies in Fuzziness and Soft Computing, vol. 337, pp. 225–247. Springer, Berlin (2016)

24. Vaidyanathan, S., Azar, A.T.: Dynamic analysis, adaptive feedback control and synchronization of an eight-term 3-D novel chaotic system with three quadratic nonlinearities. In: Azar, A.T., Vaidyanathan, S. (eds.) Advances in Chaos Theory and Intelligent Control. Studies in Fuzziness and Soft Computing, vol. 337, pp. 155–178. Springer, Berlin (2016)
25. Vaidyanathan, S., Azar, A.T.: Generlized projective synchronization of a novel hyperchaotic four-wing system via adaptive control method. In: Azar, A.T., Vaidyanathan, S. (eds.) Advances in Chaos Theory and Intelligent Control. Studies in Fuzziness and Soft Computing, vol. 337, pp. 275–296. Springer, Berlin (2016)
26. Wang, X., Ouannas, A., Pham, V.T., Abdolmohammadi, H.R.: A fractional-order form of a system with stable equilibria and its synchronization. Adv. Differ. Eqn. **1**, 20 (2018)

Self-balancing Robot Modeling and Control Using Two Degree of Freedom PID Controller

Ahmad Taher Azar[1,2](✉), Hossam Hassan Ammar[2], Mohamed Hesham Barakat[2], Mahmood Abdallah Saleh[2], and Mohamed Abdallah Abdelwahed[2]

[1] Faculty of Computers and Information, Benha University, Benha, Egypt
ahmad_t_azar@ieee.org
[2] School of Engineering and Applied Sciences, Nile University Campus, Sheikh Zayed District, Juhayna Square, 6th of October City, Giza 12588, Egypt
{hhassan,mhesham,mabdullah,moabdullah}@nu.edu.eg

Abstract. This paper represents the control of a two-wheel self-balancing robot based on the theory of controlling the inverted pendulum. This paper dividing the system modeling into two main parts. The first part is the dc motor and the second part are the whole mechanical design and its characteristics as a function in the motor speed and the torque depending on the system, creating two control closed loops inner and outer. The study uses conventional proportional–integral–derivative (PID) and two degree of freedom PID controllers to obtain a robust controller for the system. The inner loop controls the motor speed use the motor speed feedback signal from the encoder. The outer loop keeps the robot always in the accepted vertical angle boundary, using a six-degree of freedom gyroscope and accelerometer as a feedback signal. A state space model is obtained considering some assumptions and simplifications. The results are verified through simulations and experiments. Numerical simulation results indicate that the 2-DOF PID controller is superior to the traditional PID controller.

Keywords: Self-balancing robot · Segway · Inverted pendulum
Two-degree of freedom PID controller

1 Introduction and the Related Work

The self-balancing robot is an important device as a base to many applications in practical life like some of mobile robots. This type of robots has an enormous diversity in shape and applications. This paper represents the concept of controlling this unstable, nonlinear system. The inverted pendulum is an interesting point to have a good understanding of this problem [1, 2].

The robot structure is consisting of three horizontal layers are spaced in the vertical direction for mounting the control components [3]. In the lower face of the lower layer the motors are mounted at the upper face the inertial measurement unit (IMU), the component which is responsible for measuring the acceleration and the inclination in the three directions. At the second layer the controller and at the upper layer the driving batteries are mounted respectively [4]. The used controller is Arduino mega which is

A. E. Hassanien et al. (Eds.): AISI 2018, AISC 845, pp. 64–76, 2019.
https://doi.org/10.1007/978-3-319-99010-1_6

cheaper and easy to control driving a dc geared motors with built in encoders and motor drive unit as an interface between the two voltage levels control and drive motors.

One of the most important problems is to obtain a good transfer function for the motor. Obtaining the proper PID parameters and set them into the controller with fine tuning [4]. The second approach is using the two degree of freedom PID controller which could attain a higher performance than the one degree of freedom.

This paper is organized as follows: In Sect. 2, dynamical modeling of self-balancing robot is introduced. Section 3 describes the design and structure of two degree of freedom PID controllers. In Sect. 4, simulation results of control methods are presented and discussed. Finally, in Sect. 5, concluding remarks are given.

2 Dynamical Modeling of Self-balancing Robot

The objective of a self-balancing robot is to maintain it always stand on two wheels with changing in the horizontal position and oscillations keeping it always vertically [5].

The robot could be modeled through two sub systems first of them is inverted pendulum system and the other one is DC motor that supply the inverted pendulum with the sufficient force/torque to maintain its position as shown in Fig. 1.

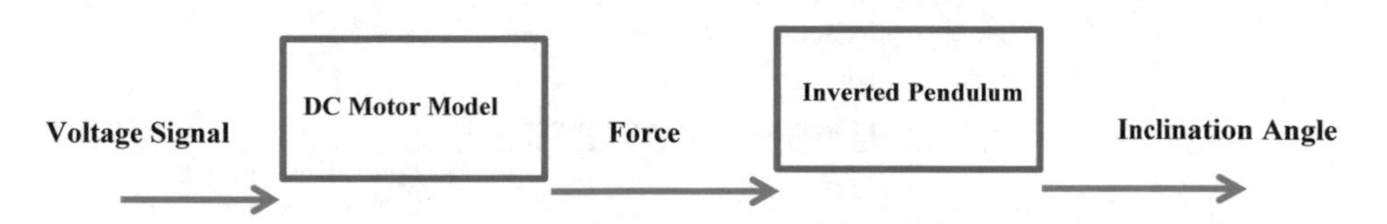

Fig. 1. Self-balancing robot model including the two subsystem needed

2.1 Dynamical Modeling of Inverted Pendulum

Figure 2 illustrates the theoretical mathematical model for an inverted pendulum on a moving carriage. Free body diagram is used here to obtain the dynamic equations of motion [6]. The self-balancing robot can be represented as an inverted pendulum on a moving carriage where the pendulum has a mass of m at a distance L from the pivot point, and an inclination angle θ from the vertical, an acceleration of $\ddot{x}$.The carriage has a mass M and external applied force of F [7]. Inverted pendulum physical parameters are summarized in Table 1.

The free-body diagrams of the two elements of the inverted pendulum system [8] is shown in Fig. 3. The forces in the free-body diagram of the carriage in the horizontal direction generates the equation of motion (1) and the forces of the pendulum in the horizontal direction generates the following equation of motion (2).

$$M\ddot{x} + b\dot{x} + N = F \tag{1}$$

$$N = m\ddot{x} + ml\ddot{\theta}\cos\theta - ml\dot{\theta}^2\sin\theta \tag{2}$$

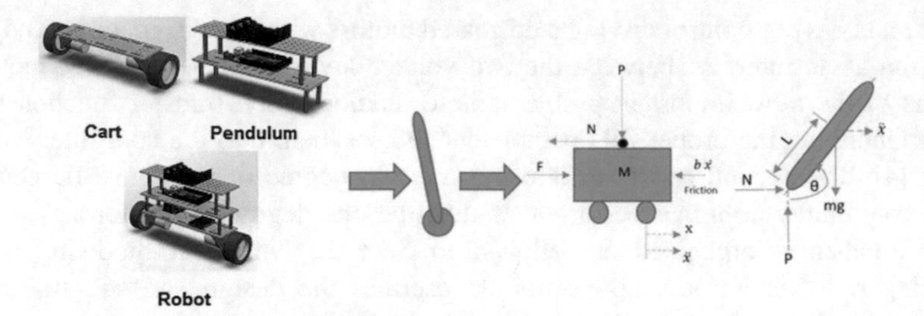

Fig. 2. The two-wheel self-balancing robot model as an inverted pendulum on a moving carriage

Table 1. Inverted pendulum physical parameters

Symbol	Parameter
M	Cart mass
m	Pendulum mass
b	Cart Friction
l	Length to pendulum center of mass
I	Pendulum Inertia
F	Force applied to the cart
N, P	Reaction forces
x	Cart position coordinate
θ	Pendulum angle from vertical

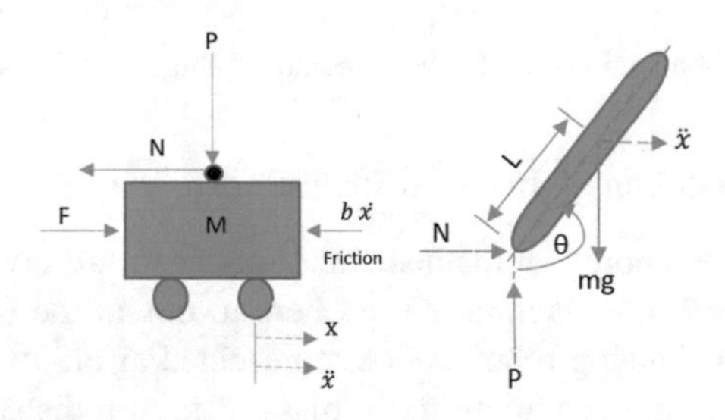

Fig. 3. Inverted pendulum free body diagram

Substituting (1) in (2) gives one of the two governing equations for this system:

$$(M+m)\ddot{x}+b\dot{x}+ml\ddot{\theta}\cos\theta-ml\dot{\theta}^{2}\sin\theta=F \tag{3}$$

Summing the forces along the axis perpendicular to the pendulum generates the Eq. (4)

$$P\,sin\,\theta+N\cos\theta-mg\sin\theta=ml\ddot{\theta}+m\ddot{x}\ddot{\theta}\cos\theta \tag{4}$$

Summing the moments about the centroid of the pendulum generates Eq. (5), combining (4) and (5) getting (6)

$$-Pl\,sin\theta - Nl\cos\theta = I\ddot{\theta} \quad (5)$$

$$\left(I + ml^2\right)\ddot{\theta} + mgl\sin\theta = -ml\ddot{x}\cos\theta \quad (6)$$

This set of equations needs to be linearized so as to be used in the linear control system techniques, so the equations will be linearized around the vertical upward equilibrium position, where $\theta = \pi$, assuming that the system stays within a small neighborhood of this equilibrium. Let ø represent the deviation of the pendulum's position from equilibrium, where $\theta = \pi + ø$, small angle approximation can be used in the nonlinear functions in the system equations:

$$cos\,\theta = \cos(\pi + \varnothing) \approx -1 \quad (7)$$

$$sin\,\theta = \sin\,(\pi + \varnothing) \approx -\varnothing \quad (8)$$

$$\dot{\theta}^2 = \dot{\varnothing}^2 \approx 0 \quad (9)$$

Substituting (7), (8) and (9) into (3) and (6) getting Eqs. (10) and (11) noting that u substituted for the input F.

$$\left(I + ml^2\right)\ddot{\varnothing} - mgl\,\varnothing = ml\ddot{x} \quad (10)$$

$$(M + m)\ddot{x} + b\,\dot{x} - ml\ddot{\varnothing} = u \quad (11)$$

Transfer Function generation

Taking the Laplace transform of the system equations assuming zero initial conditions to obtain the transfer functions of the linearized system equations, resulting Eqs. (12) and (13) [9].

$$\left(I + ml^2\right)\varnothing(\mathrm{s})s^2 - mgl\,\varnothing(\mathrm{s}) = ml\mathrm{X}(\mathrm{s})s^2 \quad (12)$$

$$(M + m)X(\mathrm{s})s^2 + b\,\mathrm{X}(\mathrm{s})s^2 - ml\varnothing(\mathrm{s})s^2 = U(s) \quad (13)$$

Obtaining a transfer function between the output ø and the input U(s) by eliminating the X(s) from Eqs. (12) and (13) as the transfer function only represents the relationship between a single input and a single output at a time, this yields Eq. (14).

$$X(s) = \left[\frac{I + ml^2}{ml} - \frac{g}{s^2}\right]\varnothing(s) \quad (14)$$

Substituting (14) into (13) gives (15).

$$(M+m)\left[\frac{I+ml^2}{ml}-\frac{g}{s^2}\right]\varnothing(s)s^2+b\left[\frac{I+ml^2}{ml}-\frac{g}{s^2}\right]\varnothing(s)s-ml\,\varnothing(s)s^2=U(s) \quad (15)$$

Rearranging (15) yields (16).

$$\frac{\varnothing(s)}{U(s)}=\frac{\frac{ml}{q}s^2}{s^4+\frac{b(I+ml^2)}{q}s^3-\frac{(M+m)mgl}{q}s^2-\frac{bmgl}{q}s} \quad (16)$$

Whereas,

$$q=\left[(M+m)\left(I+ml^2\right)-(ml)^2\right] \quad (17)$$

From (16) it's observed that there is a pole and zero located at the origin, so these can be canceled, and the transfer function will be:

$$P_{pend}(s)=\frac{\varnothing(s)}{U(s)}=\frac{\frac{ml}{q}s}{s^4+\frac{b(I+ml^2)}{q}s^2-\frac{(M+m)mgl}{q}s-\frac{bmgl}{q}}\left[\frac{rad}{N}\right] \quad (18)$$

Obtaining a transfer function between the output X(s) and the input U(s) by eliminating ø from Eqs. (12) and (13) yields Eq. (19).

$$P_{cart}(s)=\frac{X(s)}{U(s)}=\frac{\frac{\left(I+ml^2\right)s^2-gml}{q}}{s^4+\frac{b(I+ml^2)}{q}s^3-\frac{(M+m)mgl}{q}s^2-\frac{bmgl}{q}s}\left[\frac{rad}{N}\right] \quad (19)$$

State Space Model

Representing the two transfer functions (18) and (19) in the standard matrix form since the two equations are now linear after linearizing in State-Space form after rearranging to a series of first order differential equations as follows [9]:

$$\begin{bmatrix}\dot{x}\\ \ddot{x}\\ \dot{\theta}\\ \ddot{\theta}\end{bmatrix}=\begin{bmatrix}0 & 1 & 0 & 0\\ 0 & \frac{-(I+ml^2)b}{I(M+m)+Mml^2} & \frac{m^2gl^2}{I(M+m)+Mml^2} & 0\\ 0 & 0 & 0 & 1\\ 0 & \frac{-mlb}{I(M+m)+Mml^2} & \frac{mgl(M+m)}{I(M+m)+Mml^2} & 0\end{bmatrix}\begin{bmatrix}x\\ \dot{x}\\ \theta\\ \dot{\theta}\end{bmatrix}+\begin{bmatrix}0\\ \frac{I+ml^2}{I(M+m)+Mml^2}\\ \theta\\ \frac{ml}{I(M+m)+Mml^2}\end{bmatrix}U \quad (20)$$

$$y = \begin{bmatrix} 1 & 0 & 0 & 0 \\ 0 & 0 & 1 & 0 \end{bmatrix} \begin{bmatrix} x \\ \dot{x} \\ \theta \\ \dot{\theta} \end{bmatrix} + \begin{bmatrix} 0 \\ 0 \end{bmatrix} u \tag{21}$$

As due to the cart position and the pendulum's position are part of the output, so the C matrix has 2 rows.

Using *SolidWorks™* CAD program to estimate the parameters of inverted pendulum model as shown in Table 2.

Table 2. Inverted model estimated from CAD Model

Symbol	Quantity
M	0.3 kg
m	0.4 kg
b	0.1 N/m/sec
I	0.0004 $Kg.m^2$
g	9.8 N/m^2
L	0.084 m

Substituting with the robot measured quantities in the State-Space model gives the two governing Eqs. (22) and (23) as follows:

$$\frac{X(s)}{U(s)} = \frac{2.86s^2 + 5.08e^{-15}s - 292.2}{s^4 + 0.286s^3 - 204.6s^2 - 29.22s} \tag{22}$$

$$\frac{\varnothing(s)}{U(s)} = \frac{29.82s + 1.665e^{-15}}{s^3 + 0.286s^2 - 204.6s - 29.22} \tag{23}$$

Using Eq. (23) which describes the system transfer function where the force is the input and inclination angle is the output.

2.2 DC Motor Model and Parameters Estimation

One of the most commonly actuator used in the control system is DC motor. It directly generates rotary motion which is coupled to wheels that could provide translational motion for the robot [10, 11]. The equivalent system for DC motor which consist of equivalent electric circuit of the armature and the free-body diagram or the rotor is shown in Fig. 4. The torque generated by a DC motor is proportional to the armature current and the strength of the magnetic field. In this case the used motor is a permanent magnet DC motor. So, magnetic field is assumed to be constant and therefore the motor torque is proportional to only the armature current I by a constant factor K_t as shown in the equations below.

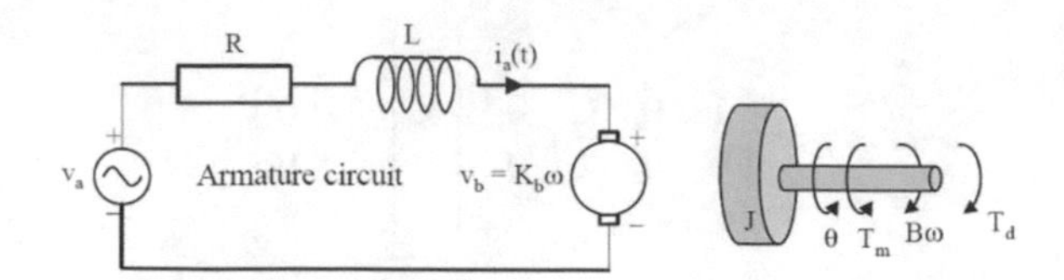

Fig. 4. DC motor free body diagram

The Dc Motor parameters are summarized as follows:

(J) Moment of inertia of the rotor = 0.02 kg.m^2
(b) Motor viscous friction constant = 0.1 N.m.s
(K) Motor torque constant and the back emf constant = 0.2
(R) Electric resistance = 2.111 Ohm
(L) Electric inductance = 0.002 H

$$T = K_t i \tag{24}$$

$$e = K_e \dot{\theta} \tag{25}$$

In SI units, Kt = Ke; therefore, K is used to represent both the motor torque constant and the electromotive force.

$$J\ddot{\theta} + b\dot{\theta} = k\,i \tag{26}$$

$$L\frac{di}{dt} + R\,i = V - K\dot{\theta} \tag{27}$$

From Eqs. (13) and (14), the Laplace transform is applied and the results are shown by the modeling equations:

$$s(Js + b)\dot{\theta}(s) = k\,I(s) \tag{28}$$

$$s(Ls + R)I(s) = V(s) - Ks\dot{\theta}(s) \tag{29}$$

$$\frac{\dot{\theta}(s)}{V(s)} = \frac{K}{(Js + b)(Ls + R) + K^2} \tag{30}$$

Similarly, for torque relation vs input voltage

$$\frac{T(s)}{V(s)} = \frac{K(s + Jb)}{(sL + R)(sJ + b) + K^2} \tag{31}$$

For the DC motor used to drive the robot, National Instrument Elvis 2 Board is used to measure the electric resistance and inductance for used motor and MATLAB system identification tool using the given data for the DC Motor. Figure 5 illustrates the estimated DC motor rtep response versus the measured response.

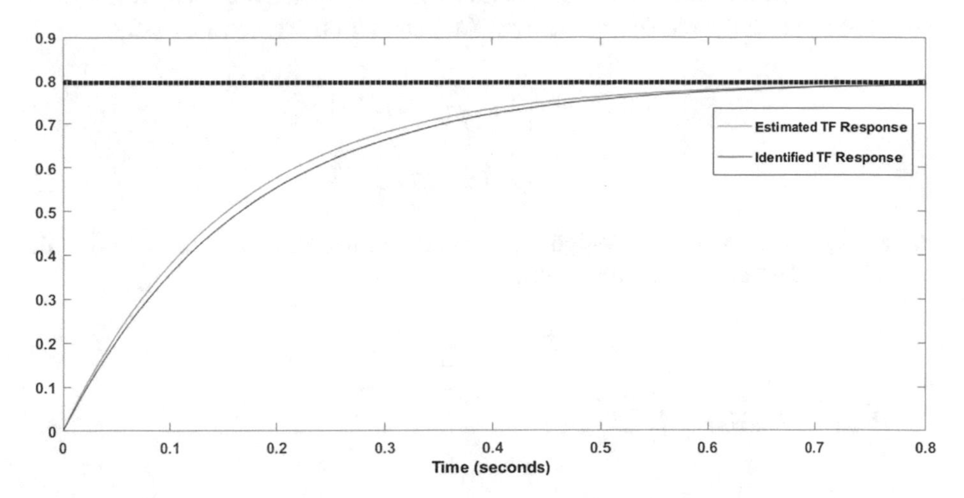

Fig. 5. Estimated DC motor step response vs measured response. voltage signal as an input and motor speed as an output

Finally, with substitution in Eqs. (30) and (31) to get two transfer functions one of them Speed with input voltage ad other one is torque with input voltage as following

$$G(s) - \frac{\dot{\theta}(s)}{V(s)} = \frac{0.2}{0.0422s + 0.2511} \tag{32}$$

$$P(s) = \frac{T(s)}{V(s)} = \frac{0.2s + 0.0004}{0.04222s + 0.2511} \tag{33}$$

3 Design of Two Degree of Freedom PID Controller

The two-degree-of-freedom (2-DOF) based controller has recently attracted the attention in different areas of control engineering community because of their better control quality for both smooth set point variable tracking and good disturbance rejection. The 2-DOF controller generates an output signal based on the difference between a reference signal and a measured system output. It computes a weighted difference signal for each of the proportional, integral, and derivative actions according to the set-point weights. The controller output is the sum of the proportional, integral, and derivative actions on the respective difference signals, where each action is weighted according to the chosen gain parameters. In general, 2DOF PID structure improves the overall closed loop performance of the process. A detailed study on various 2DOF structures are clearly presented by Araki and Taguchi [12]. In this work, the 2DOF PID

Feedforward type is used as shown in Fig. 6 where P(s) is the controlled process transfer function, C_f (s) is the feedforward compensator transfer function, C(s) is the serial (or main) compensator transfer function, *r* is the set-point, *d* the load-disturbance, and *y* the controlled variable. In this case, C(s) and Cf (s) are given by [12]:

$$C(s) = K_p\left\{1 + \frac{1}{T_i s} + T_D D(s)\right\} \tag{34}$$

$$C_f(s) = -K_p\{\alpha + \beta T_D D(s)\} \tag{35}$$

where α and β are controller weighting parameters ranging from 0 to 1, D(s) is the approximate derivative filter term given by:

$$D(s) = \frac{s}{1 + \tau s} \tag{36}$$

Where $\tau = T_d/K_d$

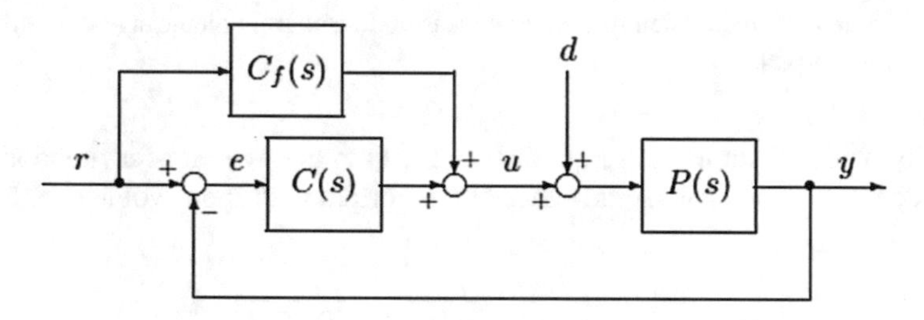

Fig. 6. Two-DOF PID feedforward type [12]

4 Simulation Results and Discussions

The performance of the controllers is analyzed based on the time domain parameters (M_p, t_s), error values like Integral of Time multiplied Absolute Error (ITAE), Integral of Squared Error (ISE), Integral of Absolute Error (IAE) and Integral of Time Multiply Squared Error (ITSE) [13, 14]. In the first phase, the conventional PID Controller is used for the self-balancing robot control as shown in Fig. 7.

In the second phase, the control of two wheeled self-balancing robot consists of two main loops as shown in Fig. 8. One of them for DC motors (inner loop control using 1-DOF PID controller) to maintain the required speed and torque for the pendulum (outer loop control using 2-DOF PID controller) which maintains the inclination angle of the robot. Each DC motor has its own PID controller and a feedback sensor where the two motors represents the inner control loop in which the speed and torque of the motors is controlled. The outer loop which control's the inclination of the two wheeled self-balancing robot consists of separate -DOF PID controller to control the inclination angle.

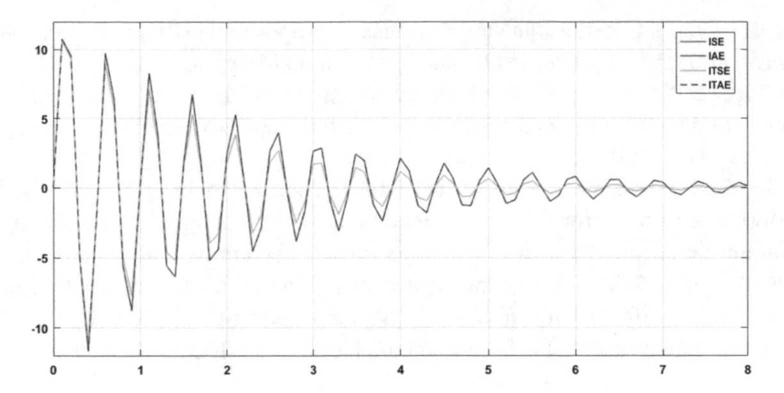

Fig. 7. 1 DOF PID controller for self-balancing robot control

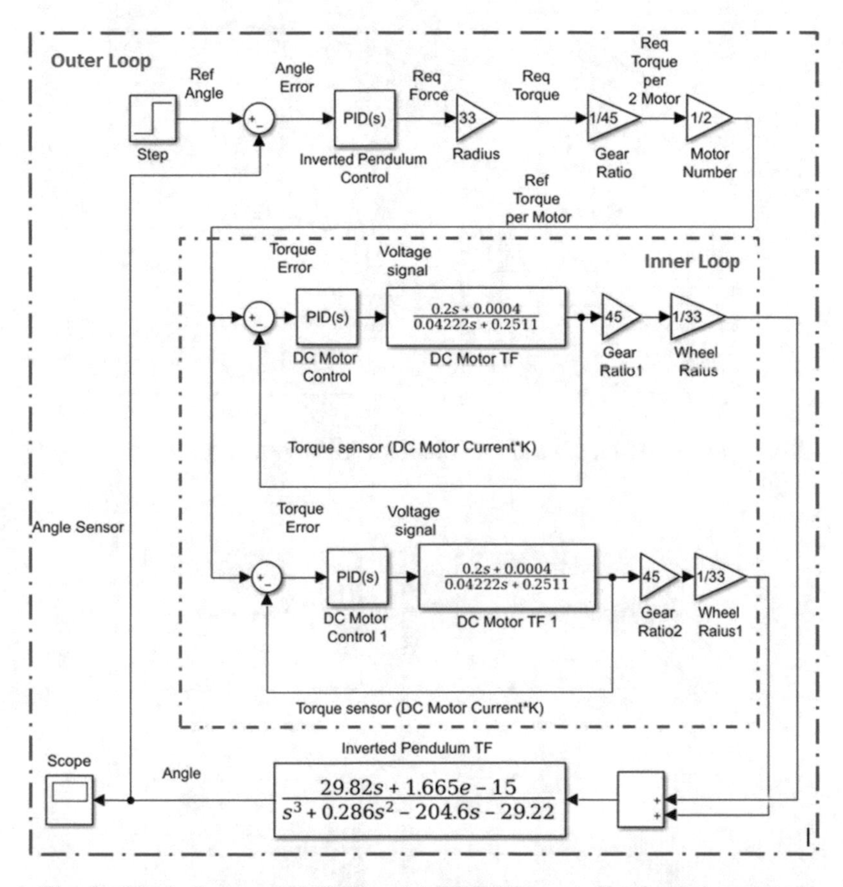

Fig. 8. Block diagram of 1-DOF and 2-DOF PID controller for Robot control

For the given DC motor, the step response of proportional (P), proportional-integral (PI), proportional-derivative (PD) and proportional-integral-derivative (PID) Controllers for speed and torque are compared as shown in Fig. 9. It's noted that the PID controller, in general, provides faster response and improved stability as compared to other types of controllers.

In the third phase, the control of two wheeled self-balancing robot consists of two main loops one of them for DC motor (inner loop control using 2-DOF PID controller) to maintain the required speed and torque for the pendulum. The outer loop control consists of another 2-DOF PID controller to maintain the inclination angle of the robot as shown in Fig. 10. For the given DC Motor response to 2-DOF P, PI, PD, PID controller as shown in Fig. 11. It can be concluded that 2-DOF PID controller is more

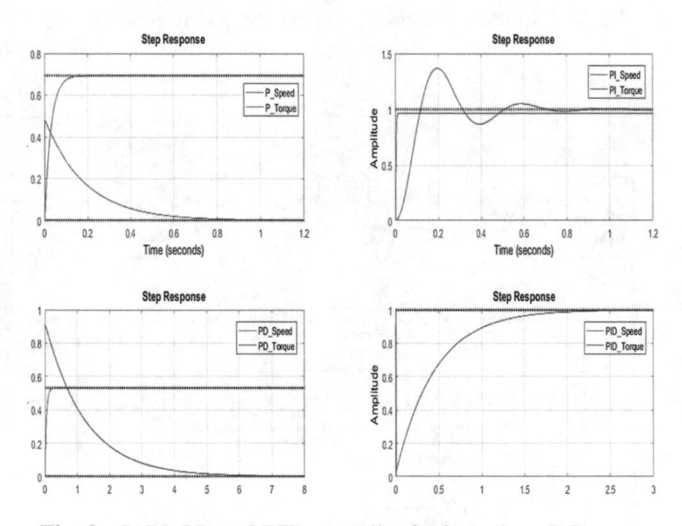

Fig. 9. P, PI, PD and PID controller for inner loop DC motor

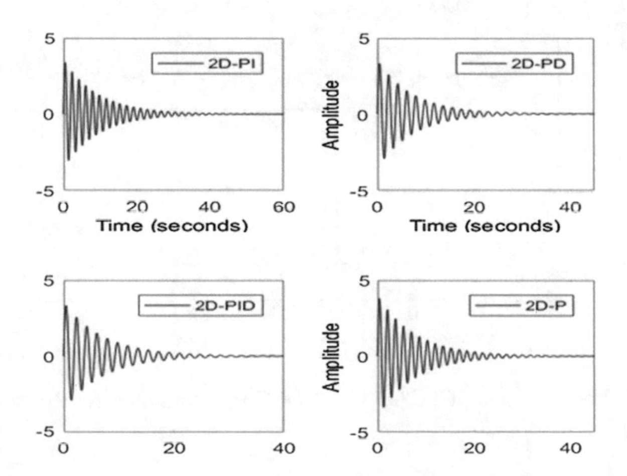

Fig. 10. Response of the pendulum to 2-DOF P, PI, PD, PID controllers

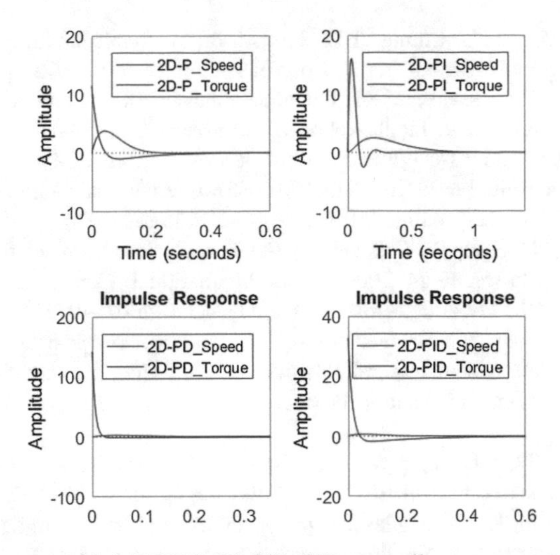

Fig. 11. DC motor with 2-DOF P, PI, PD, PID controller response to impulse input

effective than the traditional PID controller. The simulation results show that the setting value tracking characteristics of the system is obviously enhanced when the main controller is the two-degree-of-freedom with a feed-forward controller, compared to conventional PID controller.

5 Conclusion

In this paper, a model of a two-wheel self-balancing robot based on the theory of controlling the inverted pendulum is proposed. The model is obtained by two approaches, the mathematical approach and the system identification with parameter estimation approach based on experimental results. This study compares the results of two controllers conventional PID and 2-DOF PID controllers. The simulation results showed that the 2-DOF controller approach is much faster in response and disturbance rejection as well as smaller error than conventional PID controller while maintaining set-point performance.

References

1. Azar, A.T., Serrano, F.E.: Adaptive sliding mode control of the furuta pendulum. In: Azar, A.T., Zhu, Q. (eds.) Advances and Applications in Sliding Mode Control systems. Studies in Computational Intelligence, vol. 576, pp. 1–42. Springer, Heidelberg (2015)
2. Azar, A.T., Vaidyanathan, S.: Handbook of Research on Advanced Intelligent Control Engineering and Automation. Advances in Computational Intelligence and Robotics (ACIR). IGI Global, Pennsylvania (2015). ISBN 9781466672482
3. Deng, M., Inoue, A., Sekiguchi, K., Jiang, L.: Two-wheeled mobile robot motion control in dynamic environments. Robot. Comput. Integr. Manuf. **26**(3), 268–272 (2010)

4. Bui, T.H., Nguyen, T.T., Chung, T.L., Kim, S.B.: A simple nonlinear control of a two-wheeled welding mobile robot. Int. J. Control Autom. Syst. (IJCAS) **1**(I), 35–42 (2003)
5. Kim, Y., Kim, S.H., Kwak, Y.K.: Dynamic analysis of a nonholonomic two-wheeled inverted pendulum robot. J. Intell. Robot. Syst.: Theory Appl. **44**(1), 25–46 (2005)
6. Juang, H.S., Lurrr, K.Y.: Design and control of a two-wheel self-balancing robot using the arduino microcontroller board. In: 2013 10th IEEE International Conference on Control and Automation (ICCA), 12–14 June 2013, Hangzhou, China (2013)
7. Thibodeau, B.J., Deegan, P., Grupen, R.: Static analysis of contact forces with a mobile manipulator. In: Proceedings 2006 IEEE International Conference on Robotics and Automation, 15–19 May 2006, Orlando, FL, USA, pp. 4007–4012 (2006)
8. Sun, C., Lu, T., Yuan, K.: Balance control of two-wheeled self-balancing robot based on linear quadratic regulator and neural network. In: 2013 Fourth International Conference on Intelligent Control and Information Processing (ICICIP), 9–11 June 2013, Beijing, China (2013)
9. Lin, S.C., Tsai, C.C.: Develop of a self-balancing human transportation vehicle for the teaching of feedback control. IEEE Trans. Educ. **52**(1), 157–168 (2009)
10. Sugie, T., Fujimoto, K.: Controller design for an inverted pendulum based on approximate linearization. Int. J. Robust Nonlinear Control **8**(7), 585–597 (1998)
11. Takei, T., lmamura, R., Yuta, S.: Baggage transportation and navigation by a wheel inverted pendulum mobile robot. IEEE Trans. Ind. Electr. **56**(10), 3985–3994 (2009)
12. Araki, M., Taguchi, H.: Two-degree-of-freedom PID controllers. Int. J. Control Autom. Syst. **1**(4), 401–411 (2003)
13. Sánchez, H.S., Visioli, A., Vilanova, R.: Optimal nash tuning rules for robust PID controllers. J. Franklin Inst. **354**(10), 3945–3970 (2017)
14. Azar, A.T., Serrano, F.E.: Robust IMC-PID tuning for cascade control systems with gain and phase margin specifications. Neural Comput. Appl. **25**(5), 983–995 (2014)

Fractional Order Two Degree of Freedom PID Controller for a Robotic Manipulator with a Fuzzy Type-2 Compensator

Ahmad Taher Azar[1,2(✉)] and Fernando E. Serrano[3]

[1] Faculty of Computers and Information, Benha University, Benha, Egypt
ahmad.azar@fci.bu.edu.eg
[2] School of Engineering and Applied Sciences, Nile University, Sheikh Zayed District, 6th of October City, Giza, Egypt
ahmad_t_azar@ieee.org
[3] Central American Technical University UNITEC, Zona Jacaleapa, Tegucigalpa, Honduras
serranofer@eclipso.eu

Abstract. In this paper a novel strategy for the position control and trajectory tracking of robotic manipulators is proposed. This strategy consists of an independent two degree of freedom PID controller for a two links robotic arm. Due to the capability of two degree of freedom PID controllers to deal with disturbances, each link is controlled independently considering that the disturbance does not affect the system performance due to the robustness of the closed loop system. Then, a fuzzy type-2 centralized compensator is implemented to drive the orientation variables with the desired trajectory in order to improve the robustness and system performance. In this work, it is proved that the two degree of freedom fractional order PID controllers implemented with the fuzzy type-2 compensator improves the system performance in comparison with the results found in other studies, and one important issue is that the fuzzy type-2 system can be considered as a linear system emulating the capabilities of a linear compensator. To corroborate the theoretical results explained in this study, a numerical example is shown along with the respective discussion and conclusions.

Keywords: Intelligent control · PID control · Fuzzy type-2 systems
Two-DOF PID controllers · Robotic manipulators

1 Introduction

Robotic manipulators have been extensively used since decades ago because they fit in different applications especially in the industrial field (Spong et al. 2006). From the automatic control focus, one of the main concerns is the trajectory tracking control because there are some applications such as welding, screwing, moving cars or machinery parts and painting need a precise trajectory tracking to achieve these objectives. Even when there are a vast number of studies

A. E. Hassanien et al. (Eds.): AISI 2018, AISC 845, pp. 77–88, 2019.
https://doi.org/10.1007/978-3-319-99010-1_7

found in the literature, this problem can be solved in novel ways considering the advances of different automatic control sub-areas such as intelligent control and fractional order controllers. Nowadays there are some studies related to robotics and other applications such as Sharma et al. (2015). In Sharma et al. (2015), a two degree of freedom fractional order PID controller for a robotic manipulator with payload is designed and analyzed, where it is proved that the proposed approach solves some robustness and disturbance rejection issues to improve the closed loop system performance in an independent controller configuration. Then in Kumar and Kumar (2017), a type-2 fuzzy PID controller is implemented for a redundant 5-DOF robotic manipulator where interesting results were obtained in which the performance of the system is improved significantly measured by the ITSE (integral time squared error) but some disturbance rejection issues were not considered, problems that can be solved with a 2-DOF control approach. Another similar and important results are found in (Ghosh et al. 2014) where a 2 DOF PID compensator is implemented for a magnetic levitation system implementing a feedforward gain so the system performance can be designed in a more efficient way in comparison with a 1-DOF PID controller. Other examples of applications for the automatic generation control of power systems can be found in (Debbarma et al. 2014; Sahu et al. 2016; Panigrahi et al. 2017; Singh et al. 2018) in which 2-DOF fractional order PID controllers are implemented where the settling time and oscillations are reduced. Also, as an example for applications, in (Qiao et al. 2016) where an adaptive 2-DOF PI controller for a permanent magnet synchronous motor is proposed and in (Pachauri et al. 2017) where a 2-DOF PID inferential control of continuous bioreactor for ethanol production is presented. Other important studies about integer order and fractional order controllers can be found in (Li et al. 2016; Arrieta et al. 2015; Alfaro et al. 2009 Jin and Liu 2014) but some studies that are important for this work are found in (Taguchr and Arakr 2000; Inoue and Hirashima 2001).

In this work, an independent fractional order 2-DOF PID controller for a robotic manipulator with a fuzzy type-2 compensator is proposed (Azar and Serrano 2015). The fractional order controller is obtained by linearizing the robotic dynamic model (Spong et al. 2006) in the operating point and then by decoupling the system as shown in (Brogan 1991), the decoupled transfer functions for the link 1 and link 2 are obtained. The fractional order 2-DOF PID controllers are obtained according to a robustness index to obtain a robust and better disturbance rejection properties and then the fuzzy type-2 compensator improves the system performance, this means, settling time and reduced oscillations. The theoretical results obtained are tested later following a reference and applying a disturbance to the system measuring the integral squared error ISE to be compared with the results obtained in other studies. The paper is organized as follows: Problem Formulation is discussed in Sect. 2. Then in Sect. 3, Fractional Order 2-DOF PID Controller With Fuzzy Type-2 Compensation is discussed. Section 4 presents numerical simulation example. Finally, Sects. 5 and 6 includes the discussion and conclusion of the research work respectively.

2 Problem Formulation

In this section a brief description of the dynamic model of a two-links manipulator is shown along with the theoretical background of fuzzy type-2 systems and controllers and fractional order calculus.

2.1 Dynamic Model of the Robotic Manipulator

Consider the diagram of the two-links robotic manipulator shown in Fig. 1 (Spong et al. 2006) where the angles of each link are $q_1(t)$ and $q_2(t)$. The dynamic model can be represented as (Spong et al. 2006):

$$D(q(t))\ddot{q}(t) + C(q(t), \dot{q}(t))\dot{q}(t) + g(q(t)) = \tau(t) \tag{1}$$

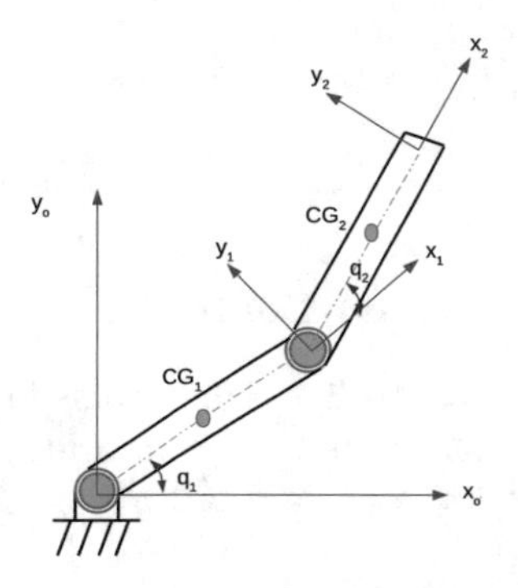

Fig. 1. Two-links robotic manipulator

where $D(q(t)) \in \mathbb{R}^{2\times2}$ is the inertia matrix, $C(q(t), \dot{q}(t)) \in \mathbb{R}^{2\times2}$ is the coriolis matrix, $g(q(t)) \in \mathbb{R}^2$ is the gravity vector and $\tau(t) \in \mathbb{R}^2$ is the torque vector. The values for these matrices and vectors are found in Spong et al. (2006) and then the system is linearized around the operating point $q_p = [0, 0]^T$ and making the following change of variable $q(t) = [x_1(t), x_2(t)]^T$, $\dot{q}(t) = [x_3(t), x_4(t)]^T$ and $X(t) = [x_1(t), x_2(t), x_3(t), x_4(t)]^T$ so the linearized system is (Brogan 1991):

$$X(t) = \begin{bmatrix} 0 & 0 & 0 & 0 \\ 0 & 0 & 0 & 0 \\ J_{31} & J_{32} & 0 & 0 \\ J_{41} & J_{42} & 0 & 0 \end{bmatrix} + \begin{bmatrix} 0 & 0 \\ 0 & 0 \\ J_{U31} & J_{U32} \\ J_{U41} & J_{U42} \end{bmatrix} \begin{bmatrix} \tau_1(t) \\ \tau_2(t) \end{bmatrix}$$

$$y(t) = \begin{bmatrix} 1 & 0 & 0 & 0 \\ 0 & 1 & 0 & 0 \\ 0 & 0 & 1 & 0 \\ 0 & 0 & 0 & 1 \end{bmatrix} X(t) \tag{2}$$

where the constants:

$$J_{31} = \frac{-d_{22}m_1gl_{c1} + d_{21}m_2gl_1}{\alpha}, \quad J_{32} = \frac{m_2gl_{c2}}{\alpha}$$
$$J_{41} = \frac{d_{12}m_1gl_{c1}}{\alpha} - \frac{m_2g(l_1 + l_{c2})}{\alpha}, \quad J_{42} = \frac{-d_{11}m_2g(l_1 + l_{c2})}{\alpha} \tag{3}$$

$$J_{U31} = \frac{d_{22}}{\alpha}, \quad J_{U32} = \frac{-d_{21}}{\alpha}, \quad J_{U41} = \frac{-d_{12}}{\alpha}, \quad J_{U42} = \frac{d_{11}}{\alpha} \tag{4}$$

where m_1 and m_2 are the link masses, l_{c1} and l_{c2} are the distance to the center of gravity, l_1 and l_2 are the link length and g is the gravity constant. The other constants are defined as:

$$\begin{aligned} d_{11} &= m_1l_{c1}^2 + m_2(l_1^2 + l_{c2}^2 + 2l_1l_{c2}) + I_1 + I_2 \\ d_{12} &= d_{21} = m_2(l_{c2}^2 + l_1l_{c2}) + I_2 \\ d_{22} &= m_2l_{c2}^2 + I_2 \\ \alpha &= d_{11}d_{12} - d_{12}^2 \end{aligned} \tag{5}$$

with I_1 and I_2 as the inertia of the link 1 and link 2.

2.2 Fuzzy Type-2 System and Controller

The fuzzy type-2 system and controller is defined as found in Serrano and Rossell (2017). The steps of the fuzzy type-2 system and controller are fuzzification, fuzzy inference, type reduction and defuzzification. Consider a fuzzy type-2 set defined as X

$$\tilde{A} = \int_{x \in X} \mu_{\tilde{A}}(x)/x = \int_{x \in X} \left(\int_{u \in J_x} f_x(u)/u \right) /x \tag{6}$$

where $J_x \subseteq [0,1]$, $\mu_{\tilde{A}}$ is the membership grade of $x \in X$ which is a type-1 fuzzy set in $[0,1]$. The footprint of uncertainty is given by:

$$FOU(\tilde{A}) = U_{x \in X} \left[\underline{\mu}_{\tilde{A}(x)}, \overline{\mu}_{\tilde{A}(x)} \right] \tag{7}$$

The firing strength for the k rule is:

$$F^k = \left[\underline{f}^k, \overline{f}^k \right] \tag{8}$$

where

$$\begin{aligned} \underline{f}^k &= min \left[\underline{\mu}_{E_1^k}, \ldots, \underline{\mu}_{E_p^k} \right] \\ \overline{f}^k &= min \left[\overline{\mu}_{E_1^k}, \ldots, \overline{\mu}_{E_p^k} \right] \end{aligned} \tag{9}$$

for the $p - th$ input. Then the type reduction is given by:

$$y_l = min_{\theta_i \in \left[\underline{f}^k(y_i), \overline{f}^k(y_i)\right]} \frac{\sum_{i=1}^{N} y_i \theta_i}{\sum \theta_i}$$
$$y_r = max_{\theta_i \in \left[\underline{f}^k(y_i), \overline{f}^k(y_i)\right]} \frac{\sum_{i=1}^{N} y_i \theta_i}{\sum \theta_i} \tag{10}$$

and finally the defuzzified output of the fuzzy type-2 system and controller is

$$y = \frac{y_l + y_r}{2} = \xi^T \Theta \tag{11}$$

2.3 Fractional Order Calculus

The Riemann-Liouville fractional derivative is given by Debbarma et al. (2014):

$${}_aD_t^{\alpha} f(t) = \frac{1}{\Gamma(n-\alpha)} \frac{d^n}{dt^n} \int_a^t (t-\tau)^{n-\alpha-1} f(\tau) d\tau \tag{12}$$

where $n - 1 \leq \alpha < n$, n is an integer and Γ is the Euler gamma function. The definition of the Euler integral is given in:

$${}_aD_t^{-\alpha} f(t) = \frac{1}{\Gamma(\alpha)} \int_a^t (t-\tau)^{\alpha-1} f(\tau) d\tau \tag{13}$$

Finally the Laplace transform is given by:

$$L\left\{{}_aD_t^{\alpha} f(t)\right\} = S^{\alpha} F(s) - \sum_{k=0}^{n-1} S_a^k D_t^{\alpha-k-1} f(t)|_{t=0} \tag{14}$$

3 Fractional Order 2-DOF PID Controller with Fuzzy Type-2 Compensation

Figure 2 shows the block diagram of the proposed control strategy and as can be noticed, there are two controllers for the 2-DOF controller strategy along with a centralized fuzzy type-2 compensator.

Before explaining the block diagrams, it is important to consider that the fuzzy type-2 compensator can be considered as a linear function, so consider (11) and the generalized Taylor series expansion centered in zero: $f(\Theta) = \xi^T \Theta$ to $f(\Theta) = \xi^T \Theta$

So $f(\Theta)$ can represent a linear function with a Laplace transform $F(s)$ given by a first order plus time delay transfer function.

$$F(s) = \frac{e^{-as}}{s+p} \tag{15}$$

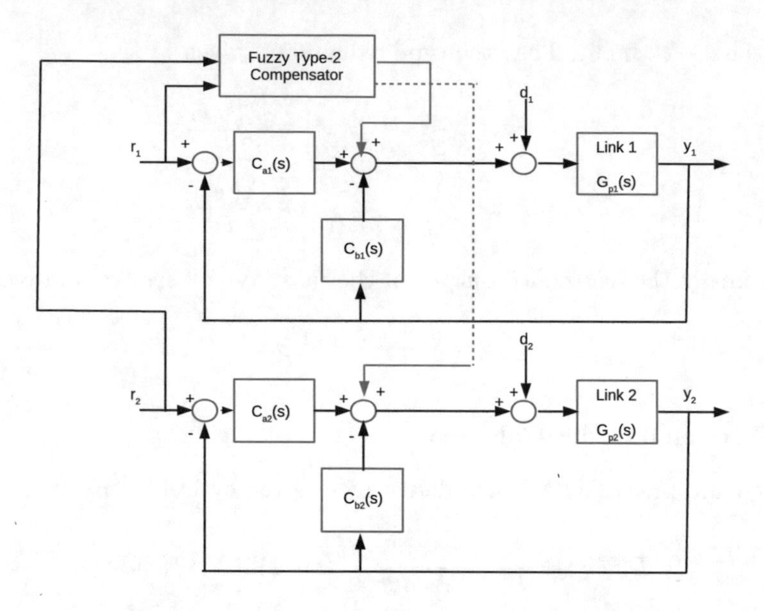

Fig. 2. Block diagram of the proposed approach

where a is the delay and p is an appropriate pole. So instead of implementing trial and error or other computational intelligence methods such as genetic algorithms, particle swarm optimization etc. the fuzzy type-2 compensator can be tuned following the dynamics of (16). The transfer functions of C_{ai} and C_{bi} for $i = 1, 2$ are given by:

$$\begin{aligned} C_{ai}(s) &= K_{pi} + \frac{K_{ii}}{s^{\lambda_i}} + K_{di}s^{\mu_i} \\ C_{bi}(s) &= K_{pbi} + K_{dbi}s^{\mu_{bi}} \end{aligned} \tag{16}$$

with the gains K_{pi}, K_{ii}, K_{di}, K_{pbi} and K_{dbi} along with their fractional orders λ_i, μ_i and μ_{bi} for $i = 1, 2$. The transfer functions from the references and disturbances are given:

$$\begin{aligned} \frac{y_i(s)}{r_i(s)} &= \frac{C_{ai}(s)P(s) + F(s)P(s)}{1 + C_{ai}(s)P(s) + C_{bi}(s)P(s)} \\ \frac{y_i(s)}{d_i(s)} &= \frac{P(s)}{1 + C_{ai}(s)P(s) + C_{bi}(s)P(s)} \end{aligned} \tag{17}$$

Considering that the fractional order Laplace transform can be given by an equivalent integer Laplace transform Li et al. (2016)

$$
\begin{aligned}
G(s) &= s^{\gamma} = K \prod_{j=-N}^{N} \frac{s+W_j}{s+W'_j} \\
K &= \left(\frac{\omega_b}{\omega_a}\right)^{-\gamma/2} \prod_{j=-N}^{N} \frac{\omega'_j}{\omega_j} \\
\omega_j &= \omega_a \left(\frac{\omega_b}{\omega_a}\right)^{\frac{K+N+0.5(1-\gamma)}{2N+1}} \\
\omega'_j &= \omega_a \left(\frac{\omega_b}{\omega_a}\right)^{\frac{K+N+0.5(1+\gamma)}{2N+1}}
\end{aligned}
\tag{18}
$$

in the frequency range $[\omega_a, \omega_b]$.

3.1 Robustness and Parameter Tuning

The gains K_{pi}, K_{ii}, K_{di}, K_{pbi} and K_{dbi} are found numerically in order to meet the robustness requirements as specified below:

$$
\begin{aligned}
&\left\| \frac{C_{ai}(s)P(s) + F(s)P(s)}{1 + C_{ai}(s)P(s) + C_{bi}(s)P(s)} \right\|_{\infty} < 1 \\
&\left\| \frac{P(s)}{1 + C_{ai}(s)P(s) + C_{bi}(s)P(s)} \right\|_{\infty} < 1
\end{aligned}
\tag{19}
$$

4 Numerical Simulation Example

The robot parameters used in this example is summarized in Table 1.

Table 1. Robot parameters

Parameter	Value
l_1	0.8 m
l_2	0.8 m
l_{c1}	0.4 m
l_{c2}	0.4 m
m_1	0.1 Kgs
m_2	0.1 Kgs
I_1	0.3 Kg.m^2
I_2	0.3 Kg.m^2
g	9.81 m/s^2

The robustness index used in this example is 0.05 and the fuzzy type-2 compensator is selected with two inputs and two outputs. In Fig. 3, the membership functions (negative, zero and positive are shown).

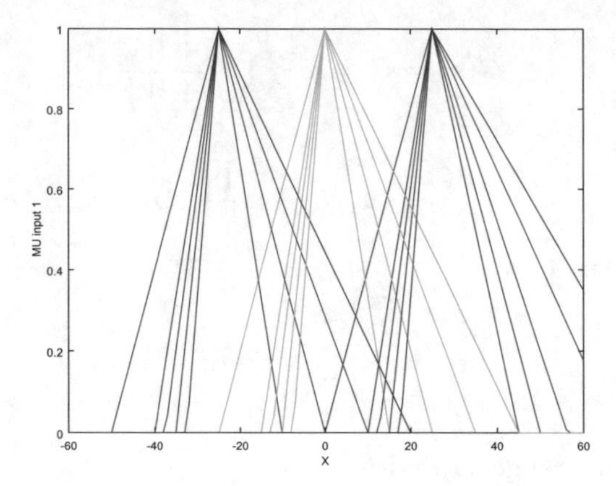

Fig. 3. Fuzzy type-2 membership function for the input 1

The rules for the fuzzy type-2 compensator are selected as:

$$\begin{array}{c}\textbf{IF x1=Z AND x2=N THEN Y1=N}\\ \textbf{IF x1=P AND x2=Z THEN Y2=P}\\ \textbf{IF x1=N AND x2=Z THEN Y2=N}\\ \textbf{IF x1=Z AND x2=Z THEN Y1=N}\end{array} \tag{20}$$

The results were obtained as shown in Figs. 4 and 5, the position follow the reference with a disturbance ($r(t) = 0.1sin(\omega t)$, $d(t) = 0.1sin(\omega_d t)$) with $\omega = 6.2832\ rad/s$ and $\omega_d = 6.2832 \times 10^9\ rad/s$ considering that $d(t)$ is the disturbance that can be vibrations added to the structure. In these figure, it's noticed that there is a small ISE (integral squared error) as shown in Table 2. There are better results obtained by the proposed strategy in comparison with the strategy shown in (Inoue and Hirashima 2001; Sharma et al. 2015).

Table 2. ISE for all the strategies

Strategy	ISE
Proposed	0.20346
Inoue and Hirashima (2001)	9.6138
Sharma et al. (2015)	0.53479

Then in Figs. 6 and 7, the velocities are illustrated showing an oscillatory behavior according to the position trajectory. Finally, in Figs. 8 and 9, the control torque for each actuator are shown in which it is noticed that there is a small control effort with the proposed control strategy.

5 Discussion

The theoretical results shown in this paper provides a novel and better solution for other results found in the literature. As explained before, even when there are other decentralized control strategies for robot manipulators, independent controller design still provides an effective, simple and non expensive solution for the trajectory tracking of robot manipulator. The two-link robotic manipulator used in this study provides a suitable model that can be extended later to other serial or parallel robot. This means that the control strategy shown in this study can be implemented later in other kind of mechanism. The 2-DOF fractional order PID controller strategy along with the fuzzy type-2 compensator provides the required robustness and disturbance rejection properties surpassing other results found in the literature. The ISE shown in the numerical simulation section

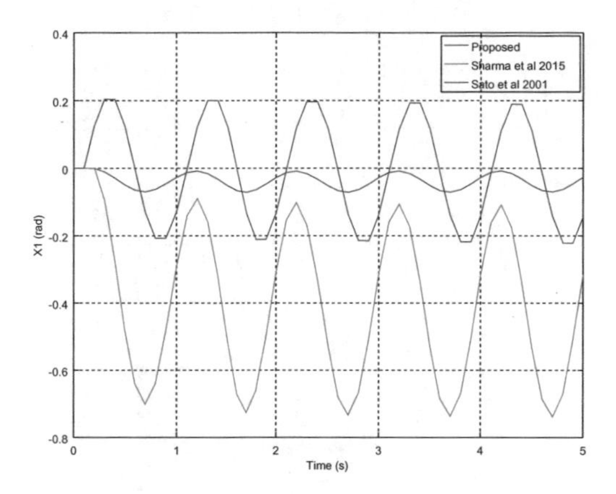

Fig. 4. Position variable $X(1)$

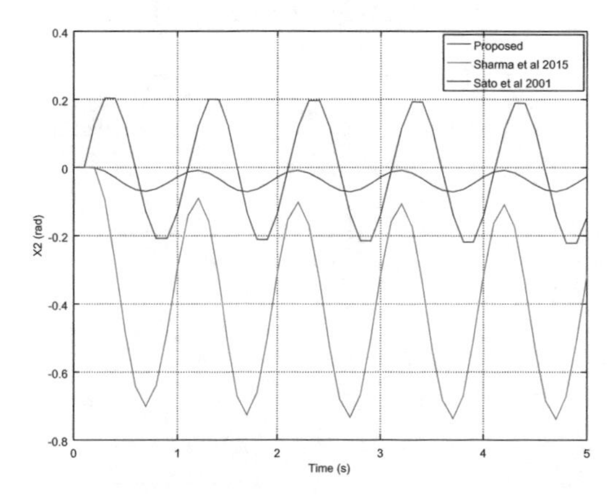

Fig. 5. Position variable $X(2)$

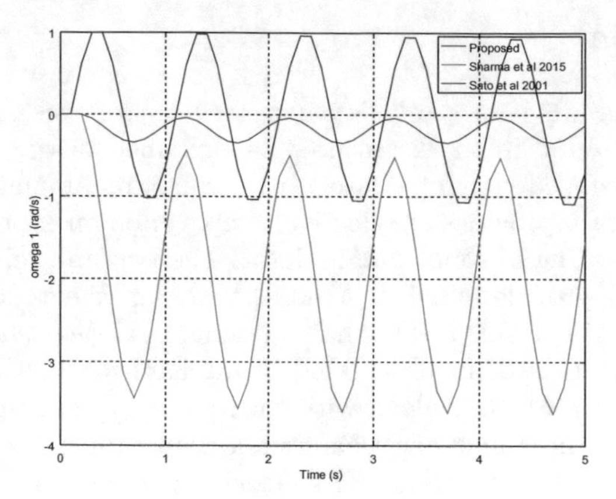

Fig. 6. Angular velocity $\dot{q}_1$

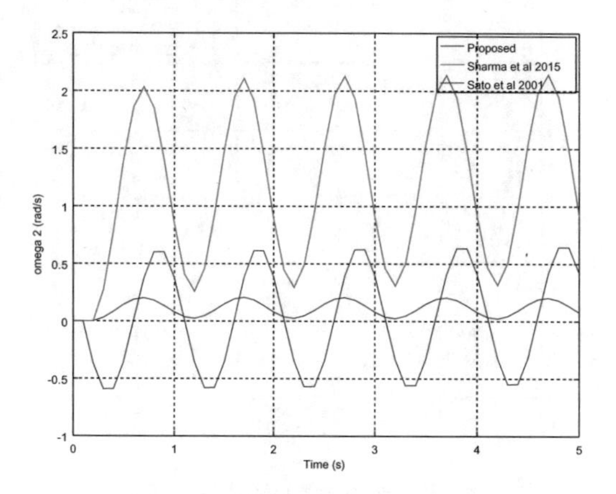

Fig. 7. Angular velocity $\dot{q}_2$

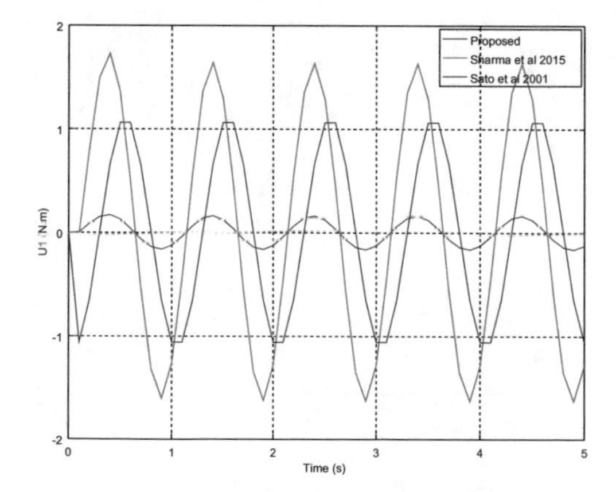

Fig. 8. Torque generated by the actuator 1

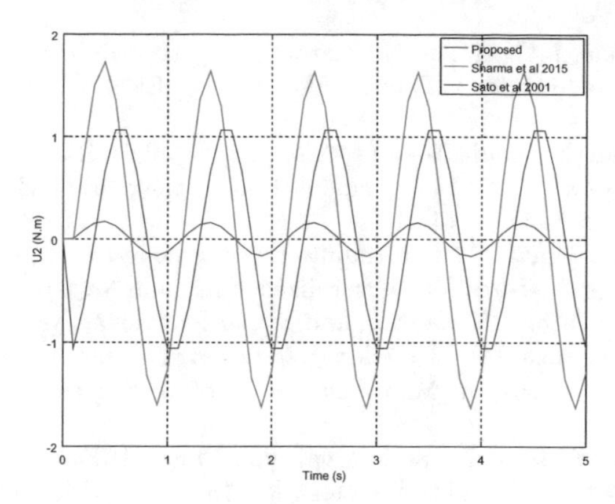

Fig. 9. Torque generated by the actuator 2

proves that the error between the reference and the position are smaller than other controller strategies, this means, integer order and fractional order.

6 Conclusion

In this study, the design of a 2-DOF PID controller with a fuzzy type-2 compensator for robotic manipulators is shown. The theoretical and numerical simulation results prove that the proposed controller/compensator improves the robustness and system performance even when disturbances are found in one or both links. It is important to remark that the disturbance rejection properties of the system are improved taking into account that in robotic systems the disturbances can be external loads or vibrations of the structure. One advantage of this approach is that the fuzzy type-2 compensator can be modeled taking the behavior of a first order plus time delay (FOPTD) system. Another advantage is that the 2-DOF fractional order PID parameters are tuned numerically by the robustness index in order to meet some requirements. In the future, the study of decentralized 2-DOF PID and other kinds of controllers can be implemented.

References

Alfaro, V., Vilanova, R., Arrieta, O.: Robust tuning of two-degree-of-freedom (2-DoF) PI/PID based cascade control systems. J. Process Control **19**(10), 1658–1670 (2009)

Arrieta, O., Wu, S., Vilanova, R., Rojas, J., Yu, L.: Development of a mobile application for robust tuning of one-and two-degree-of-freedom PI and PID controllers. IFAC-PapersOnLine **48**(29), 76–81 (2015)

Azar, A.T., Serrano, F.: Stabilization and control of mechanical systems with backlash. In: Handbook of Research on Advanced Intelligent Control Engineering and Automation, pp. 1 – 60 (2015)

Brogan, W.L.: Modern Control Theory. Prentice Hall, Upper Saddle River (1991)

Debbarma, S., Saikia, L.C., Sinha, N.: Automatic generation control using two degree of freedom fractional order PID controller. Inte. J. Electr. Power Energy Syst. **58**, 120–129 (2014)

Ghosh, A., Krishnan, R., Pailla Tejaswy, A.M., Pradhan, J., Ranasingh, S.: Design and implementation of a 2-DOF PID compensation for magnetic levitation systems. ISA Trans. **53**(4), 1216–1222 (2014)

Inoue, T.S.A., Hirashima, Y.: Self-tuning two-degree-of-freedom PID compensator based on two-degree-of-freedom generalized minimum variance control. In: IFAC Adaptation and Learning in Control and Signal Processing, pp. 207–212 (2001)

Jin, Q., Liu, Q.: Analytical IMC-PID design in terms of performance/robustness trade-off for integrating processes: From 2-Dof to 1-Dof. J. Process Control **24**(3), 22–32 (2014)

Kumar, A., Kumar, V.: Evolving an interval type-2 fuzzy PID controller for the redundant robotic manipulator. Expert Syst. Appl. **73**, 161–177 (2017)

Li, M., Zhou, P., Zhao, Z., Zhang, J.: Two-degree-of-freedom fractional order-PID controllers design for fractional order processes with dead-time. ISA Trans. **61**, 147–154 (2016)

Pachauri, N., Singh, V., Rani, A.: Two degree of freedom PID based inferential control of continuous bioreactor for ethanol production. ISA Trans. **68**, 235–250 (2017)

Panigrahi, T.K., Behera, A., Sahoo, A.K.: Novel approach to automatic generation control with various non-linearities using 2-degree-of-freedom PID controller. Energy Proc. **138**, 464–469 (2017)

Qiao, W., Tang, X., Zheng, S., Xie, Y., Song, B.: Adaptive two-degree-of-freedom PI for speed control of permanent magnet synchronous motor based on fractional order GPC. ISA Trans. **64**, 303–313 (2016)

Sahu, R.K., Panda, S., Rout, U.K., Sahoo, D.K.: Teaching learning based optimization algorithm for automatic generation control of power system using 2-DOF PID controller. Int. J. Electr. Power Energy Syst. **77**, 287–301 (2016)

Serrano, F., Rossell, J.M.: Hybrid passivity based and fuzzy type-2 controller for chaotic an hyper-chaotic systems. Acta Mec. Autom. **11**, 96–103 (2017)

Sharma, R., Gaur, P., Mittal, A.: Performance analysis of two-degree of freedom fractional order PID controllers for robotic manipulator with payload. ISA Trans. **58**, 279–291 (2015)

Singh, J., Chattterjee, K., Vishwakarma, C.: Two degree of freedom internal model control-PID design for LFC of power systems via logarithmic approximations. ISA Trans. **72**, 185–196 (2018)

Spong, M.W., Hutchinson, S., Vidyasagar, M.: Robot Modeling and Control. Wiley, Hoboken (2006)

Taguchr, H., Arakr, M.: Two-degree-of-freedom PID controllers their functions and optimal tuning. IFAC Digit. Control Past Present Future PlO Control Terrassa Spain **33**, 91–96 (2000)

Fuzzy Compensator of the Stator Resistance Variation of the DTC Driven Induction Motor Using Space Vector Modulation

Fouzia Benmessaoud(✉), Abdesselem Chikhi, and Sebti Belkacem

Department of Electrical Engineering,
University M. Benboulaid Batna 2, Batna, Algeria
fouzia.benmessaoud@yahoo.fr, chikhi_aslem@yahoo.fr,
belkacem_sebti@yahoo.fr

Abstract. This paper presents the contribution of a fuzzy controller to compensate the influence of stator resistance variation which can degrade the performance and stability of a direct torque control (DTC). Nevertheless, the original term DTC refers to a strategy that provides good performance, but it also has some negative aspects to the level of switching and inaccuracy in the engine model which recommends the use of a new technique the SVM which proposes an algorithm based on the modulation of the space vector in order to carry out a predictive regulation of the torque and flux of the induction motor and provides a fixed switching frequency, thus improving the dynamic response and the static behavior of the DTC.

Keywords: Direct torque control DTC · Induction motor · Fuzzy logic
Fuzzy estimator · Space vector modulation SVM

1 Introduction

DTC control was introduced for the first time in 1980, providing a very fast and accurate torque control response without the need for complex algorithms. However, the main advantage of DTC is its structure simplicity. Considerable research efforts have been made for about two decades to achieve the eradication of the inherent disadvantages of DTC. The slow transient response is primarily cited as the change of starting torque step. The latter can be considerably reduced by the use of Fuzzy Logic controllers [2]. To further overcome these problems, some researchers have proposed the DTC scheme using Space Vector Modulation (SVM) techniques. In [1] a control method has been discussed that allows constant switching frequency operation and uses two PI controller in order to generate the inverted reference voltage in the IM stator flux reference frame. In that control scheme, a PI speed controller is also used to obtain the torque reference signal. The control scheme in [1] is not robust to the stator resistance variations. In addition, To control the speed of an induction motor driven by the DTC-SVM. The PI controller is always the preferred choice [5]. This is because the implementation of the PI controller requires minimal information about the motor, where the controller gains are tunning until a satisfactory response is obtained. The ordinary voltage inverter has only seven independent voltage vector, which restricts the

A. E. Hassanien et al. (Eds.): AISI 2018, AISC 845, pp. 89–97, 2019.
https://doi.org/10.1007/978-3-319-99010-1_8

quantization precision of fuzzy control systems [2]. To solve this problem, SVM technology is proposed in this article. In the every sample period, reference voltage vector is found to compensate the current stator flux error and the torque error [4]. Moreover, in the direct torque control system, the estimate of stator flux is related to stator resistance. The errors of stator resistance cause all kinds of unreasonable error of stator flux estimate, which makes the control performance of controller bad. On the basis of the direct relation of stator current and stator resistance, the error and error change of stator current are used the online estimated method of input stator resistance of fuzzy estimator, which improves veracity of stator flux, so the response speed of the motor in the state of low speed will be satisfied. The objective of this article is primarily to reduce torque ripple and maintain constant switching frequency. In the second place, it is proposed to control the stator resistance of the asynchronous motor driven by DTC-SVM using a fuzzy logic controller in order to preserve the robustness of the system with respect to load variations and uncertainties eventual.

2 DTC-SVM

The block diagram of the DTC-SVM for induction motor is present in Fig. 1. In this structure two PI controllers are used for torque and flux regulation. The components of values $Vs\alpha$ et $Vs\beta$ are delivered to (SVM) which generates switching signals Sa, Sb, Sc of the inverter.

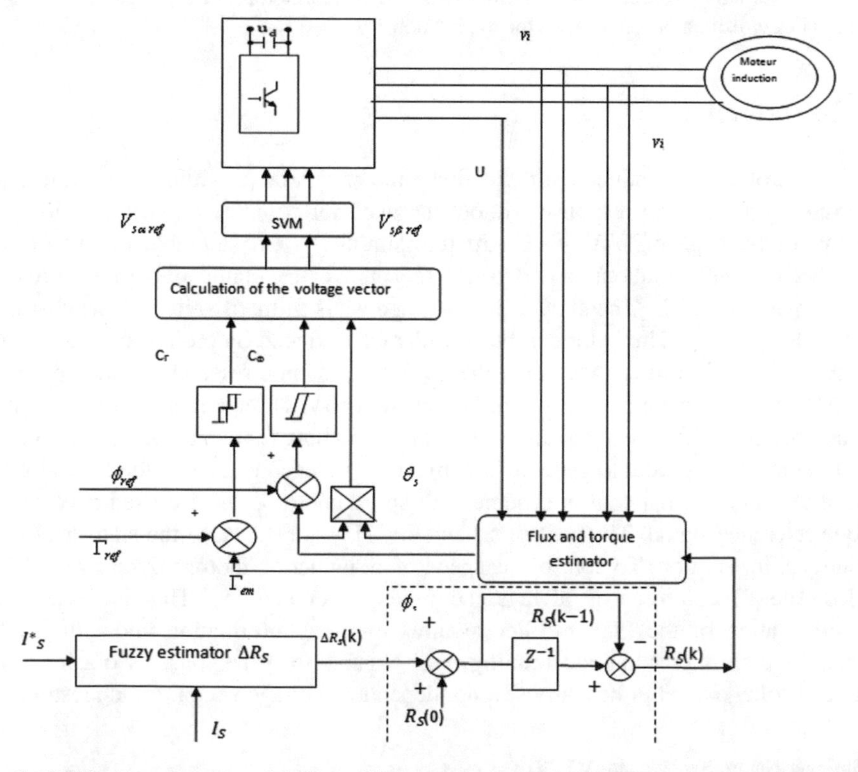

Fig. 1. DTC space vector modulation with fuzzy compensator of the stator resistance variation

3 Advantages of DTC-SVM

The structure DTC–SVM give very low torque and flux ripples and provide dynamic performance as better the DTC system. The DTC-SVM technic, though being a good performer, but introduce an extra complexity [2].

Torque fluctuation induction motor depend of the deviation from an ideal rotational stator flux vector. The difference between Φ_{sref} and Φ_s causes torque pulsation. The relation expressing the torque value $\Delta\Gamma_e$ as a function of the deviation of Φ_s from Φ_{sref} has been deduced as:

$$\frac{\Delta\Gamma_e}{\Gamma_{eref}} = k_s\frac{|\Delta\Phi_s|}{|\Phi_{sref}|} + k_\delta\Delta\delta \tag{1}$$

Where Γ_{eref} *is* the steady state torque $\Delta\Phi_s$ and $\Delta\delta$ are respectively the deviations from $\Delta\Phi_s$ and δ which are defined by:

$$\Delta\Phi_s = |\Phi_{sref}| - |\Phi_s|$$

$$\Delta\delta = \angle|\Phi_{sref}| - \angle|\Phi_s|$$

Where k_δ and k_s are the constants derived from the IM specifications.

The torque ripple is actually caused by $\Delta\Phi s$ and $\Delta\delta$ and the influence of the $\Delta\Phi s$ is considerably smaller than that of $\Delta\delta$. As a consequence the torque ripple can be almost removed if $\Delta\delta$ is kept close to zero [10].

In Fig. 2 one can notice that the torque error $\Delta\Gamma e$ and reference stator flux amplitude Φ_{sref} are delivered to voltage vector calculation which in its input gives the deviation of the reference stator flux angle from the α, β axes components of the stator reference voltage $V_{sref,}$ are calculated as [7].

$$V_{s\alpha ref} = \frac{\Phi_{sref}\cos(\delta+\Delta\delta) - \Phi_{sref}\cos(\delta)}{T_s} + R_s I_{s\alpha} \tag{2}$$

$$V_{s\beta ref} = \frac{\Phi_{sref}\sin(\delta+\Delta\delta) - \Phi_{sref}\sin(\delta)}{T_s} + R_s I_{s\beta} \tag{3}$$

Where the vector magnitude and angle are given as,

$$\delta = \arctan\left(\frac{V_{s\beta\, ref}}{V_{s\alpha\, ref}}\right) \tag{4}$$

where, T_s is the sample time of the system.

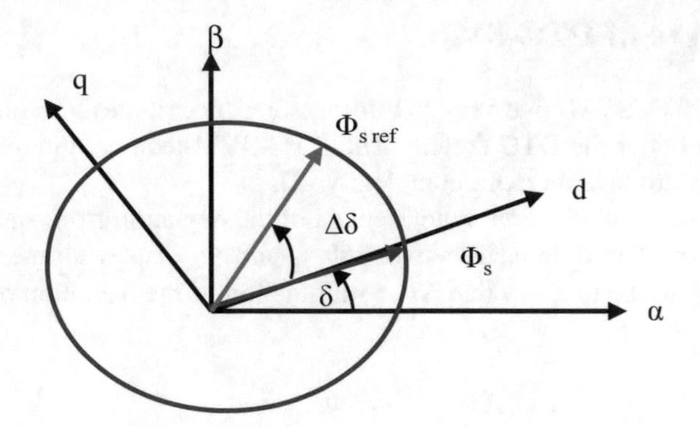

Fig. 2. Representation of stator flux vectors Φ_s and $\Phi_{s\ ref}$

4 Fuzzy Stator Resistance Observer

The error e(k) and error change Δe(k) of stator current is the input of the observer, and then

$$\Delta e(k) = e(k) - e(k-1) \tag{5}$$

$$e(k) = I_s^*(k) - I_s(k) \tag{6}$$

Where, k is switch state of inverter. $I_s^*(k)$ is given stator current value, $I_s(k)$ is actually measured stator current value [1] (Fig. 3).

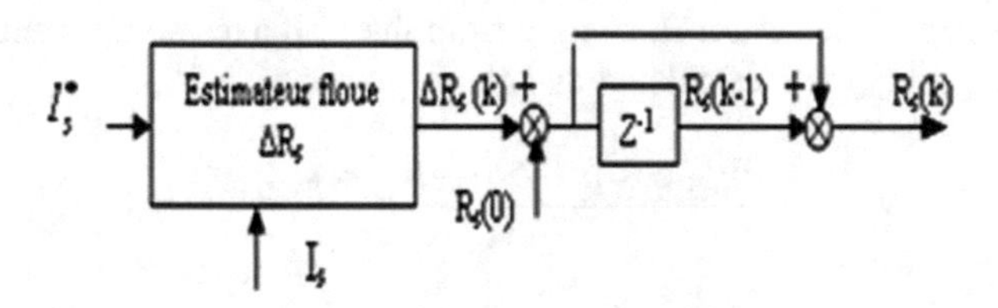

Fig. 3. Fuzzy estimators for tuning the stator resistance

The language variable of *e*(*k*) and Δe(k) are defined by five fuzzy language variables {NB, NS, ZO, PS, PB}.

The observer output is the stator resistance error ΔRs (k) [7].

$$\Delta R_s(k) = R_s(k) - R_s(k-1) \tag{7}$$

Where, R_s (*k*) is the value of stator resistance this time; R_s (*k* −1) is the value of stator resistance last time [1].

The language variable ΔR_s (*k*) is defined by five fuzzy language variables {NB, NS, ZO, PS, PB} [2].

4.1 Fuzzification

The regulator has two inputs, e(k), Δe(k) and for the fuzzyfied output R(k) as to Figs. 4, 5 and 6 [2].

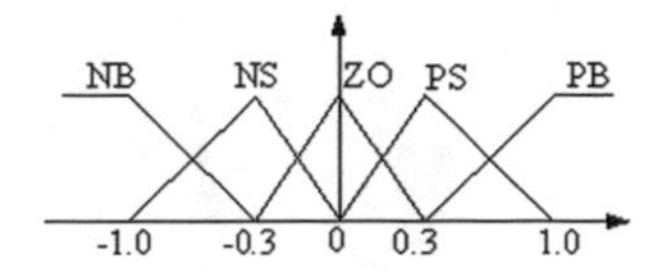

Fig. 4. The membership function of $e(k)$

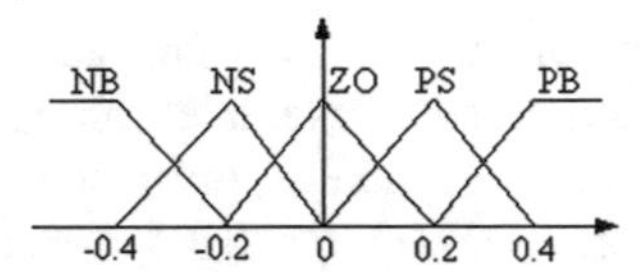

Fig. 5. The membership function of $\Delta e\ (k)$

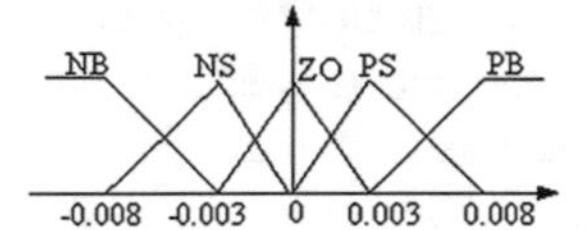

Fig. 6. The membership function of $R\ (k)$

The set of rules is described according to Mac Vicar with the format If- Thus, under the fuzzy rules table with two input variables according to [2] (Table 1).

Table 1. Determination of inference rules

$e(k)$ $\Delta e(k)$	NB	NS	ZO	PS	PB
NB	NB	NB	NB	NS	ZO
NS	NB	NB	NS	ZO	PS
ZO	NB	NS	ZO	PS	PB
PS	NS	ZO	PS	PB	PB
PB	ZO	PS	PB	PB	PB

4.2 Interfacing

The choice of the inference method depends upon the static and dynamic behavior of the system to regulate, the control unit and especially on the advantages of adjustment taken into account.

We have adopted the inference method Max–Min because it has the advantage of being easy to implement in one hand and gives better results on the other hand [8, 12].

4.3 Defuzzyfication

The most used defuzzification methods is that of the center of attraction of balanced heights, our choice is based on the latter owing to the fact that it is easy to implement and does not require much calculation [12].

5 Simulations

5.1 Positive Error on Rs

The value of the resistor Rs used by the control unit by the DTC is greater than the real asynchronous machine. In other words the actual resistance value is increased gradually from 0.2 (s) and 0.8 (s) of the order of 100% (the case of an increase of Rs 1.2 Ω to 2.4 Ω), by the influence of temperature example, this is what can generally happen during the increase of the load of the machine, or in the case of a low speed drive as in our case and which is of the order of 20 rad/s., However, the stator voltage is constant, the real current of the machine decreases as the value of the stator resistance of the machine increases.

This causes a decrease in the flux and therefore torque. And since the estimate of flux and torque using the nominal value of the stator resistance, the estimated flux will be higher than the actual flux, and speed is established with ripples and difficult. The variation of resistance stator deforms the trajectory of flux presented in the (α, β) reference (Fig. 8).

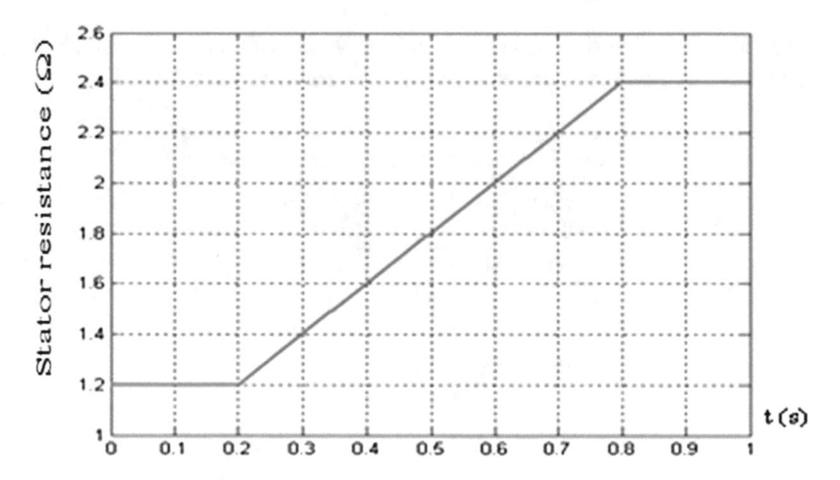

Fig. 7. Gradual variations of the stator resistance between (t = 0.2 s and t = 0.8 s)

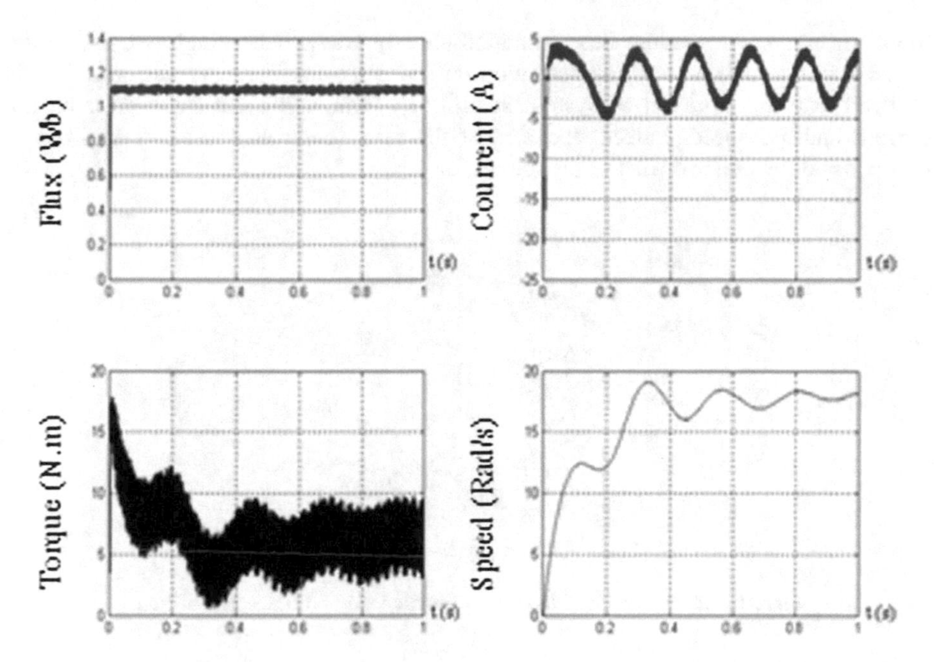

Fig. 8. Estimated flux, torque, current, speed and trajectory of flux for a gradual increase of 100% of Rs with t = 0.2 s to 0.8 s with a reduced speed about 20 rad/s.

5.2 Case of the Presence of the Fuzzy Estimator of the Resistor Rs for Low Speed Operation at 20 rad/s

In this simulation part, a fuzzy estimator stator resistance has been introduced to correct the stator flux estimate and torque, Fig. 9 illustrates the evolution of actual and estimated resistance delivered by the proposed fuzzy compensator. The two quantities are superimposed almost at the start of the regime. The variables estimated of flux, of torque, of current and speed are shown in Fig. 10. We note that they are almost

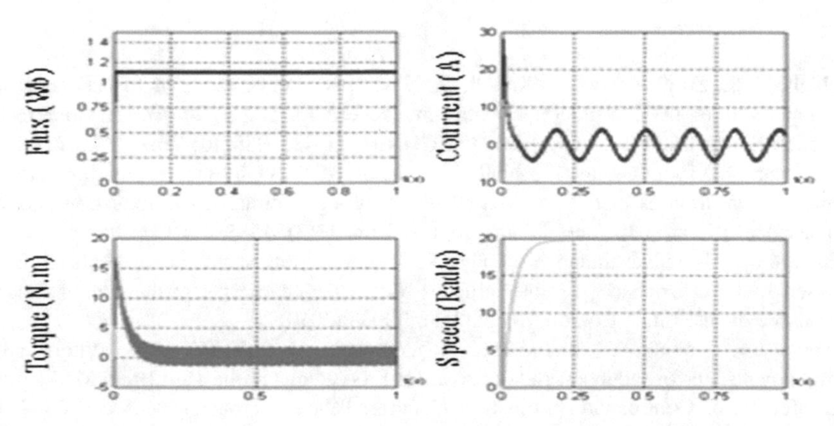

Fig. 9. Gradual variations of the stator resistance between 100% (t = 0.2 s and t = 0.8 s)

identical except the module flux estimated slightly wavy, accordingly we see in the Fig. 10 that a satisfactory compensation of the electromagnetic torque and flux is realized, and a restoration of system stability by eliminating the static error on the current and the speed. Indeed, the role of the fuzzy controller is easily realized to compensate for the error of the resistance Rs.

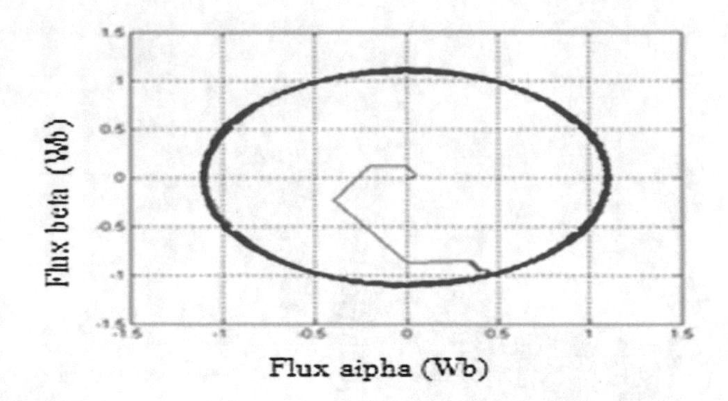

Fig. 10. Estimated flux, torque, current, speed and trajectory of flux.

6 Conclusion

It can be seen from the simulation results that the fuzzy stator resistance estimator is used for fuzzy direct torque control with space vector ùodulation, which clearly shows the interest and the role played by the fuzzy controller for an adequate operation of the induction machine for small speeds. Indeed, control system gets swifter response velocity. The ripple of stator resistance is obviously weakened, and velocity control accuracy of the system is improved. Nevertheless, there are certain reserves on the characteristics of this new control technique about its high performances when the operating conditions change in large band.

References

1. Haghbin, S., Zolghadri, M.R., Kaboli, S., Emadi, A.: Performance of PI stator resistance compensator on DTC of induction motor. In: Proceedings of the 29th Annual Conference of the IEEE Industrial Electronics Society IECON03, pp. 425–430, Roanoke, VA, USA (2009)
2. Mokhtari, B., Essounbouli, N., Nollet, F.: Modified direct torque control for permanent magnet synchronous motor drive based on fuzzy logic torque ripple reduction and stator resistance estimator. J. Control Eng. Appl. Inform. **15**(3), 45–52 (2013)
3. Bharatiraja, C., Jeevananthan, S., Latha, R.: A novel space vector pulse width modulation based high performance variable structure direct torque control evaluation of induction machine drives. Int. J. Comput. Appl. **3**(1), 33–38 (2010)
4. Abbou, A., Sayouti, Y., Mahmoudi, H., Akherraz, M.: dSPACE direct torque control implementation for induction motor drive. In: Proceedings in the 18th IEEE Mediterranean Conference on Control and Automation, Congress Palace, Marrakech, Morocco, 23–25 June pp. 1121–1126 (2010)

5. Wei, X., Chunyu, C.: Minimization of torque ripple of direct-torque controlled induction machines by improved discrete space vector modulation. Electr. Power Syst. Res. **72**, 103–112 (2004)
6. Rafa, S., Larabi, A., Manceur, M., Essounbouli, N., Hamzaoui, A., Barazane, L.: Fuzzy vector control of induction motor. In: 2013 IEEE International Conference on Networking Sensing and Control (ICNSC 13), IEEE, Paris, France (2013)
7. Liu, S.X., Wang, M.Y., Chen, Y.G., Li, S.: A novel fuzzy direct torque control system for three-level inverter-fed induction machine. Int. J. Autom. Comput. **7**(1), 78–85 (2010)
8. Aissaoui, A., Tahour, A., Essounbouli, N., Nollet, F., Abid, M., Chergui, N.: A fuzzy-PI control to extract an optimal power from wind turbine. Energy Convers. Manag. **65**, 688–696 (2013)
9. Markadeh, G.R.A., Soltani, J.: Robust direct torque and flux control of adjustable speed sensorless induction machine drive based on space vector modulation using a PI predictive controller. Electr. Eng. **88**(6), 485–496 (2006)
10. Lai, Y.S., Lin, J.C.: Fuzzy logic control for direct torque control induction motor drives with inverter controlled using space vector modulation technique. In: Proceedings of the Ninth European Conference on Power Electronics and Applications, EPE 2001, CD-ROM, EPE Assoc, Belgium (2001)
11. Miloudi, A., Al-Radadi, E.A., Draou, A.D.: A variable gain PI controller used for speed control of a direct torque neuro fuzzy controlled induction machine drive. Turk. J. Electr. Eng. **15**(1), 37–49 (2007)
12. Jiang, Z., Hu, S., Cao, W.: A new fuzzy logic torque control scheme based on vector control and direct torque control for induction machine. In: ICICIC 2008 (2008)

Classification Techniques for Wall-Following Robot Navigation: A Comparative Study

Sarah Madi(✉) and Riadh Baba-Ali

LRPE, USTHB, BP 32 El Alia, 16111 Bab Ezzouar, Algiers, Algeria
eng.s.madi@gmail.com, riadhbabaali@yahoo.fr

Abstract. Autonomous navigation is an important feature that allows the robot to move independently from a point to another without a teleoperator. In this paper, an investigation related to mobile robot navigation is presented. A group of supervised classification algorithms are tested and validated using the same dataset. Then focus will shift especially towards the k-Nearest Neighbors (KNN) algorithm. In order to improve the performance of KNN, an existing work related to genetic algorithms, local search, and Condensed Nearest Neighbors termed Memetic Controlled Local Search algorithm (MCLS) is applied to overcome the high running time of KNN. The results indicate that KNN is a competing algorithm especially after decreasing the running time significantly and combining that with existing algorithm features.

Keywords: Machine learning · Wall following · Robot navigation
Memetic Algorithm · KNN · Local search

1 Introduction

When complexity and non-linearity increase in a system in general or in a mobile robot system in particular; it becomes important to start thinking away from mathematical equations for system representation. Moreover, autonomous robots deal with a real-time environment that tends to be complex, non-linear and partially observed. According to [1], traditional techniques have failed to address the above-mentioned problems due to the non-linearity and the dynamic characteristics of the task. One can say that to control the actions of the mobile robot, means of supervised learning algorithms are needed. Recently, methods to learn models directly from data captured have become interesting tools as they allow straightforward and accurate model approximation in addition to avoiding pre-programming of all possible scenarios. This is more convenient especially when a mobile robot is present in unstructured and uncertain environments [2]. One of the common tasks a mobile robot should learn to be autonomous is wall following where the robot should follow a wall with all its curvatures, corners and turns while keeping a fixed distance to the wall whether it is to the left or the

A. E. Hassanien et al. (Eds.): AISI 2018, AISC 845, pp. 98–107, 2019.
https://doi.org/10.1007/978-3-319-99010-1_9

right of the robot. It lays the groundwork for more complex problem domains, such as maze solving, mapping, and full coverage navigation (i.e., vacuuming and lawn mowing applications) [3].

Intelligent control is the branch of control where the designer inputs the system targeted behavior and then the intelligent control system provides the model based on certain rules related to the choice of the method [4]. Usually, branches of Artificial Intelligence are the main inspiration for intelligent control as will be discussed later. Intelligent control systems shine in areas that are highly non-linear, or when classical control systems fail, or a model of the system is difficult or impossible to obtain [4]. This discussion about converting data or experience into knowledge or a computer program that perform some task is known as Machine learning. One can say that in Machine learning technologies, control strategies are emergent rather than predetermined [3].

Learning systems can be categorized based on their feedback into three main categories; supervised learning, unsupervised learning and reinforcement learning. In supervised learning, algorithms use labeled examples that consist of inputs along with their desired output for training. Algorithms learn by comparing actual outputs with correct outputs to find errors and modify the model accordingly. The structure that is discovered by studying the relationship between inputs and outputs is referred to as a model. Two main supervised learning models are mentioned in the literature; classification models and regression models [5].

In this paper, we study and compare each of the common supervised learning algorithms under the same conditions and see which one is more suitable for a mobile robot system to be controlled. A data set related to robot navigation is chosen as a benchmark to test some classification algorithms. The k-nearest neighbor technique will be the focus as we will study its competence with other algorithms. Its great usability will be proved after overcoming the main disadvantages using a new techniques termed: Memetic Controlled Local Search algorithm (MCLS).

2 Background and Preliminaries

This section will begin with introducing the main concept implemented in this paper, which is instance selection. After that, some algorithms that are deployed in this research are briefly explained. The K-nearest neighbor classification rule (KNN) is a powerful classification method that allows the classification of an unknown instance using a set of classified training instances. However, the process of computing distances is computationally high especially in large data sets. More on this algorithm is found later in this section. Set reduction is a preprocessing technique that allow lessening the problem of handling large amounts of data. Its main objective is to reduce the original dataset by selecting the most representative data. This way, it is possible to avoid excessive storage of data and excessive running time for supervised classification [6]. Instance selection is the process of finding representative patterns from data, which can help reduce the size of the data. This problem is classified as an NP-hard problem, since there is no polynomial algorithm able to find the optimal solution. Existing heuristics can however, give acceptable solutions in reasonable time. One of the techniques that has been used recently in the selection of instances is evolutionary algorithms [6]. Evolutionary algorithms (EA) are a family of algorithms belonging to meta-heuristics inspired by the theory of evolution dedicated to solve various problems. These algorithms evolve a set of solutions of a given problem in order to find the best results. Those are stochastic algorithms since they use random processes iteratively. The combination of EAs with local search (LS) was named Memetic Algorithm? (MA). Formally, a MA is defined as an EA such as genetic algorithms that includes one or more LS phases within its evolutionary cycle [6].

In this paper, the work of [6] and its extension in [7] will be used. The aim of their work was to present a new supervised memetic instance selection model based on the KNN algorithm for intrusion detection. In the proposed system, they have paired a controlled local search algorithm to an improved genetic algorithm. Their Memetic Controlled Local Search algorithm (MCLS) was successfully applied in intrusion detection systems, but in this paper, their algorithm will be tested and applied in the robot navigation field to help improve the performance. The algorithms presented next are related to classification methods and include decision trees, Neural Networks, Naive Bayes, JRipper, Support Vector Machines and k-Nearest Neighbors.

2.1 Decision Trees Induction C4.5

This is one of the most important supervised learning techniques that is presented in the shape of branches, nodes and leafs and decisions are taken from the root of the tree to the leaf. At the end of each branch is a leaf that presents the result obtained. Intermediate nodes in the tree contains a test on a particular attribute that distribute data in the different sub-trees [8]. Decision trees are widely acceptable because of their flexibility and applicability to wide range of problems. The resulting set of rules or tree paths are mutually exclusive and exhaustive which means that every instance is only covered by a single rule [9].

2.2 Neural Networks

A Neural Network is a computation model inspired by the structure of neural network in the brain, where it consists of a large number of computing devices/nodes/neurons connected to each other through links to consist a network. Each neuron will receive a weighted sum of the neurons outputs connected to its incoming links [10]. The main features of Neural Networks are their ability to utilize a large amount of sensory information, their capability of collective processing and finally their ability to learn and adapt to changes and new information.

2.3 Support Vector Machines

Support Vector Machines are a set of supervised learning methods that are applied in linear and non-linear classification problems, where they build a model that assigns new examples to one category or the other. They work by mapping the input to vectors into a high dimensional feature space and constructing the optimal separating hyper-plane through structural risk minimization [11]. After that, new examples are mapped to the same space where it is predicted that they belong to the gap they fall on [12]. According to [9] three main benefits are associated with SVMs. First, they are effective in high dimensional space. Second, they are memory efficient as they use a subset of training point in the support vectors or the decision functions. Finally, they are considered versatile since they can hold different kernel functions to be specified as decision functions. On the other hand, SVMs inherently do binary classification that means they only support two-class-problems where real life problems usually have more than two classes. Other procedures can be used to extend them to multiclass problems such as one versus the rest and one versus one approaches [13]. Please note that the Sequential Minimal Optimization (SMO) used in the comparison in this work is one way to solve Support Vector Machines training problems.

2.4 JRipper

JRipper is a supervised learning rule based learning algorithm that stands for Repeated Incremental Pruning to Produce Error Reduction. It builds a set of rules that identify the classes keeping in mind minimizing the error. It consists of two stages: the building stage and the optimization stage. Some references such as [14] add a third stage which is the delete stage. Another simple introduction to JRipper is given in [15] where they mention that all examples at start are treated as a class, and rules that cover all members of that class are found. After that, the algorithm moves to the class and does the same until all classes are covered. According to the creator of this algorithm, it was designed to be fast and effective when dealing with large and noisy datasets compared to decision trees [16].

2.5 Naive Bayes

The Naive Bayes classifier is a classical demonstration of how generative assumptions and parameter estimations simplify the learning process [10]. It assumes that the explanatory variables are independent conditionally to the target variable. This assumption contribute in reducing the training time complexity and helps the algorithm compete on numerous application areas. The performance of this classifier depends on the quality of the estimate of the uni-variate conditional distributions and on an efficient selection of the informative explicative variables. This approach as other approaches has some advantages such as its low training and prediction time complexity in addition to its low variance. It shines in classification problems where only few training examples are available.

This classifier if not used alone, is associated with other learning algorithms such as decision trees [8].

2.6 K-Nearest Neighbor

K-Nearest neighbor (KNN) is one of the most widely and oldest used methods for object classification especially when there is little or no prior knowledge about the distribution of the data as mentioned in [9]. Sometimes it is used as a replacement to support vector machines as it has the ability to handle more than two classes [17]. The nearest neighbor is calculated using a type of distance functions based on the value of k which specifies how many nearest neighbors are to be considered to define the class of a sample data point or a query. As [18] mentions in a comparison table between all nearest neighbors algorithms; KNN has some main advantages such as: (a) fast training time, (b) easy to learn and high algorithm simplicity, (c) robust to noisy training data and finally, (d) effective if training data is large. On the other hand, the same comparison table mentions some disadvantages of this algorithm such as slow running and memory limitations. To overcome some of the disadvantages related to the KNN algorithm, set reduction was proposed. It is a preprocessing technique that allow reducing the size of the dataset. Its main objective is to reduce the original dataset by selecting the most representative data points. This way, it is possible to avoid excessive storage of data points and excessive running time for supervised classification [18]. In a later section, the method of set reduction applied in this work is presented.

3 Datasets and System Properties

The authors in [19] prepared their own data by guiding the robot through a certain algorithm. A C++ routine was implemented to guide the behavior of following walls in a known situation. The algorithm basically is responsible for generating the decision the robot should take along its navigation. The information was collected as their robot SCITOS G5 navigate through the room following the walls during four rounds. The collection of data points was performed at a rate of 9 samples/s. Their generated database contain 5456 examples and it has three versions divided in three files. The first file contains sensor information from two sensors as in 2 attributes and one class, the second file contains information from four sensors as in 4 attributes and 1 class and the last file contains sensor information from 24 sensors as in 24 attributes and 1 class. The class contains four robot decisions; move forward, slight right turn, sharp right turn and slight left turn. This dataset is widely used in research that is related to robot navigation, machine learning, and classification algorithms such as [1,20]. The same dataset is used in our research for comparison between algorithms and set reduction. It is available in the UCI machine learning repository. The experiments were held on an Intel core i7, 64-bit operating system 2.20 GHz processor and 4 GB RAM.

4 Algorithm Description

This section presents a brief description on the technique applied in the paper to reduce the training files. As mentioned earlier, this work is found in [6,7]. Their aim was to set up a hybrid system between the genetic algorithm, local search, and they introduced the CNN algorithm to this hybridization, specifically to the genetic algorithm loop. This hybrid system is meant to reduce the size of the training files used in KNN to reduce the running time while maintaining the accuracy and the classification performance. To address this issue, they proposed an approach for instance selection using an algorithm that reduces the search space and determines a good subset of search Genes for classification. In fact, the instance ranking information is provided by the KNN classifier. The evaluation function takes into account both the Reduction and the accuracy of the classification calculated using KNN. Figure 1 presents the block diagram that summarizes the algorithm steps [7]. The genetic algorithm in loop is detailed with all its steps such as evaluation of individuals of initial population, selection, mutation, and crossover.

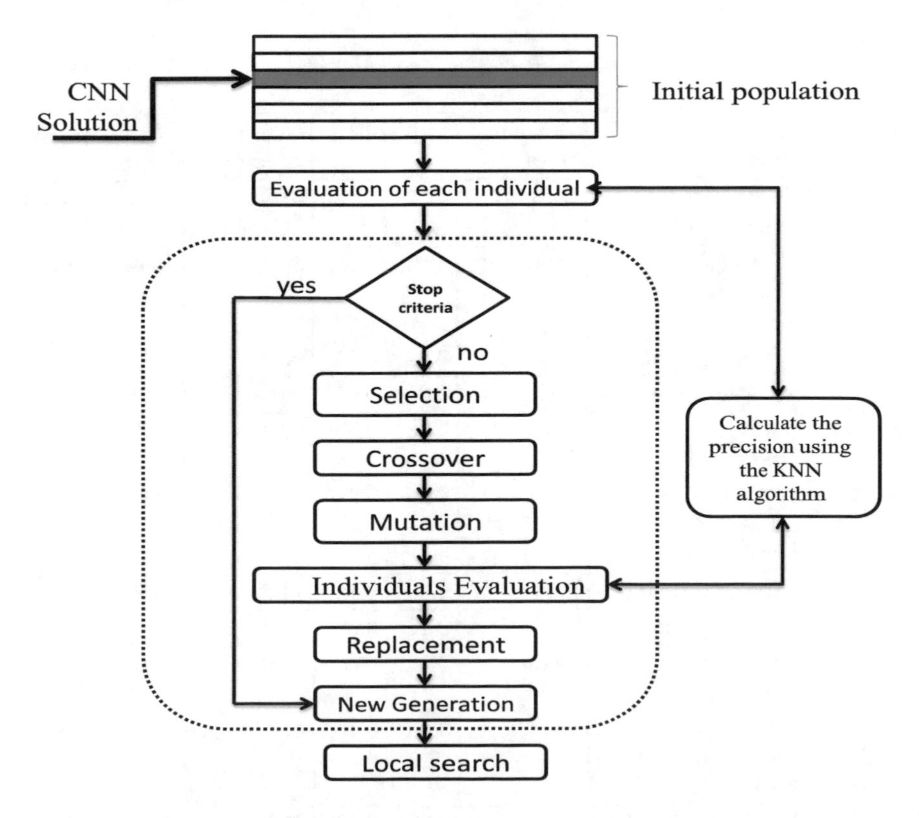

Fig. 1. The MCLS block diagram

To improve the results, they implemented a controlled local search algorithm, which is the algorithm of the descent, applied it in the genetic loop or at the end of it based on the users choice. Another step was integrated in the loop in order to further improve the results through integrating the solution of CNN (Condensed nearest neighbor) in the initial population pool after applying it to the original data set. Hybridizing the three proposed approaches; namely the genetic algorithm, local search and CNN, enabled them to reach their goal of reducing the classification time of large datasets (more than four hundred thousand instances). This system is here applied in the robot navigation field to test its usability and benefit from its features.

5 Results

Table 1 presents the comparison held between the different algorithms introduced earlier using the Robot navigation data set. The criterion included in the comparison are the accuracy, the number of correctly classified instances, training time, and testing time. This table presents the results of running the robot

Table 1. Comparison analysis

Data file used	Evaluation criteria	Decision trees	SMO	JRip	Neural network	Naive Bayes	KNN
Sensor 2 file	Accuracy (percent)	100	77.20	99.90	91.83	90.50	98.80
	Correctly classified (instances)	5456	4212	5453	5010	4942	5391
	Training time (s)	0.15	0.32	0.25	9.1	0.06	0.01
	Testing time (s)	0.17	0.25	0.01	0.02	0.12	3.91
Sensor 4 file	Accuracy (percent)	100	77.30	99.90	97.47	89.10	97.20
	Correctly classified (instances)	5456	4216	5453	5318	4862	5304
	Training time (s)	0.18	0.59	0.36	14.8	0.02	0.01
	Testing time (s)	0.14	0.15	0.02	0.07	0.13	4.05
Sensor 24 file	Accuracy (percent)	99.60	71.40	98.80	87.92	52.40	88.17
	Correctly classified (instances)	5437	3897	5392	4797	2862	4811
	Training time (s)	0.39	9.39	2.46	81.88	0.07	0.01
	Testing time (S)	0.03	0.09	0.02	0.12	0.42	6.88

navigation datasets in the WEKA software [21] for comparison purposes. The algorithms mentioned earlier were all applied to get the required results. As far as accuracy concerns, decision trees, JRip and KNN were the best algorithms to perform in almost all cases. On the other hand, KNN outperformed them all in terms of training time which is a point counted for this algorithm. However, the same algorithm recorded high testing time as expected after [18] discussed some disadvantages for the KNN algorithm such as slow running and memory limitation. From this result, the idea of applying the set reduction method using MCLS came. The reduced file was again tested in WEKA to obtain the new results.

Table 2. Comparison between original KNN and KNN with reduced files

Data file used	Evaluation criteria	K-nearest neighbors	KNN with reduced data
Sensor 2 file	File size (instances)	5456	417
	Accuracy (percent)	98.80	97.19
	Correctly classified (instances)	5391	5303
	Training time (s)	0.01	0
	Testing time (s)	3.91	0.1
Sensor 4 file	File size (instances)	5456	476
	Accuracy (percent)	97.20	94.11
	Correctly classified (instances)	5304	5135
	Training time (s)	0.01	0
	Testing time (s)	4.05	0.31
Sensor 24 file	File size (instances)	5456	948
	Accuracy (percent)	88.17	90.24
	Correctly classified (instances)	4811	4924
	Training time (s)	0.01	0
	Testing time (s)	6.88	1.63

Table 2 presents the evolution of the KNN algorithm after applying the MCLS technique to reduce training files. Note that the number of instances after reduction in the first file was 417/5456 with reduction rate of 92%, in the second file was 476/5456 with a reduction rate of 91% and in the third file was 948/5456 with a reduction rate of 84%. Some points should be noted regarding the performing of these tests. For the original KNN tests in Table 1, all results were obtained using the cross-validation method except for the testing time where it was obtained using testing with the original training file since cross validation yields no testing time. As for the KNN test using the reduced files, the testing was done using the original files as they will contain more information than the current reduced training file and their results will present an indication for the robot performance in real life after facing new situations. Additionally, if we test with the original file, the comparison will be valid as the file size in both situations will be the same. Moving to the results, the reduction in test time can be noticed clearly as the reduction exceeded 97% in the first file, 92% in the second file and exceeded 76% in the third file. The accuracy experienced some drop as well but it was very low between 1% and 3% in the case of the first two files. On the other hand, it increased in the case of the third file. As for the training time, it was less. The main explanation is related to the reduction in the training files. This reduction led to a simpler file construction, faster training, thus faster testing, learning and performing. This way, the major drawbacks of KNN will be solved while maintaining the great advantages of fast training time and learning.

6 Conclusion

The K-nearest neighbor algorithm is a supervised learning algorithm that is used widely in the classification problems. In this paper, the focus was on the robot navigation field where special robot navigation datasets were used to test multiple classification algorithms using WEKA. The results showed some limitations in terms of testing or running time in the KNN algorithm. Instance selection using a hybrid algorithm was applied to reduce the training files and the KNN algorithm was tested again to notice important improvements in the testing time and the accuracy in one case. It is concluded that KNN is an old yet still a promising algorithm as the major drawbacks of KNN will be solved while maintaining the great advantages of fast training time and learning.

References

1. Dash, T., Nayak, T., Swain, R.R.: Controlling wall following robot navigation based on gravitational search and feed forward neural network. In: Proceedings of the 2nd International Conference on Perception and Machine Intelligence, Kolkata (2015)
2. Nguyen-Tuong, D., Peters, J.: Model learning for robot control: a survey. Cogn. Process. **12**(4), 319–340 (2011)
3. Dain, R.A.: Developing mobile robot wall-following algorithms using genetic programming. Appl. Intell. **8**(1), 33–41 (1998)
4. Smith, B.: Classical versus Intelligent Control (2002). https://www.engr.mun.ca/~baxter/Publications/ClassicalvsIntelligentControl.pdf. Accessed 20 Dec 2016
5. Maimon, O., Rokach, L.: Introduction to supervised methods. In: Data Mining And Knowledge Discovery Handbook, 2nd edn, pp. 149–164. Springer, Boston (2010)
6. Miloud-Aouidate, A., Baba-Ali, A.R.: IDS false alarm reduction using an instance selection KNN-memetic algorithm. Int. J. Metaheuristics **2**(4), 333–352 (2013)
7. Benrabia, L., Sadouki, L.: Conception et realisation d'un algorithm de selection d'instance pour un IDS. Master thesis. USTHB, Algiers (2017)

8. Lemaire, V., Salperwyck, C., Bondu, A.: A survey on supervised classification on data streams. In: Lecture Notes in Business Information Processing, pp. 88–125, 16 April 2015
9. Muhammad, I., Yan, Z.: Supervised machine learning approaches: a survey. ICTACT J. Soft Comput. **5**, 946–952 (2015)
10. Shai, S.-S., Shai, B.-D.: Understanding Machine Learning: From Theory to Algorithms. Cambridge University Press, New York (2014)
11. Syed, N.A., Liu, H., Sung, K.K.: Handling concept drift in incremental learning with support vector machines. In: Proceedings of the ACM SIGKDD International Conference on Knowledge Discovery and Data Mining, California (1999)
12. Hormozi, H., Hormozi, E., Nohooji, H.R.: The classification of the applicable machine learning methods in robot manipulators. Int. J. Mach. Learn. Comput. **2**(5), 560–563 (2012)
13. Huang, P.X., Fisher, R.B.: Individual feature selection in each One-versus-One classifier improves multi-class SVM performance. In: Proceeding of the International Conference on Pattern Recognition, Stockholm (2014)
14. Shahzad, W., Asad, S., Khan, M.A.: Feature subset selection using association rule mining and JRip classifier. Int. J. Phys. Sci. **8**(18), 885–896 (2013)
15. Rajput, A., Aharwal, R.P., Dubey, M., Saxena, S., Raghuvanshi, M.: J48 and JRIP rules for e-governance data. Int. J. Comput. Sci. Secur. (IJCSS) **5**(2), 201–207 (2011)
16. Vijayarani, S., Divya, M.: An efficient algorithm for generating classification rules. Int. J. Comput. Sci. Technol. **2**(4), 512–515 (2011)
17. Gadepally, V.N.: Estimation of driver behavior for autonomous vehicle applications. Ph.D. dissertation. The Ohio State University, Ohio (2013)
18. Bhatia, N., Vandana: Survey of nearest neighbor techniques. Int. J. Comput. Sci. Inf. Secur. **8**(2), 302–305 (2010)
19. Freire, A.L., Barreto, G.A., Veloso, M.V., Varela, A.T.: Short-term memory mechanisms in neural network learning of robot navigation tasks: a case study. In: 6th Latin American Conference: Robotics Symposium, LARS (2009)
20. Karakus, M.O., Orhan, E.R.: Learning of robot navigation tasks by probabilistic neural network. In: Proceeding of the Second International Conference on Advanced Information Technologies and Applications (2013)
21. M. L. G. a. t. U. o. Waikato: Weka 3: Data Mining Software in Java. University of Waikato. http://www.cs.waikato.ac.nz/ml/weka/

A New Control Scheme for Hybrid Chaos Synchronization

Adel Ouannas[1], Giuseppe Grassi[2], Ahmad Taher Azar[3,4](✉), and Ahlem Gasri[5]

[1] Laboratory of Mathematics, Informatics and Systems (LAMIS), University of Larbi Tebessi, 12002 Tebessa, Algeria
ouannas.a@yahoo.com

[2] Dipartimento Ingegneria Innovazione, Università del Salento, 73100 Lecce, Italy
Giuseppe.grassi@unisalento.it

[3] Faculty of Computers and Information, Benha University, Benha, Egypt
ahmad.azar@fci.bu.edu.eg

[4] School of Engineering and Applied Sciences, Nile University Campus, 6th of October City, Giza, Egypt
ahmad_t_azar@ieee.org

[5] Department of Mathematics, University of Larbi Tebessi, 12002 Tebessa, Algeria
gasri.ahlem@yahoo.fr

Abstract. This paper presents a new hybrid chaos synchronization scheme, which assures the co-existence of the full-state hybrid function projective synchronization (FSHFPS) and the inverse full-state hybrid function projective synchronization (IFSHFPS) between wide classes of three-dimensional master systems and four-dimensional slave systems. In order to show the capability of co-existence approach, numerical example is reported, which illustrates the co-existence of FSHFPS and IFSHFPS between 3D chaotic system and 4D hyperchaotic system in different dimensions.

Keywords: Chaos
Full-state hybrid function projective synchronization
Inverse full-state hybrid function projective synchronization
Lyapunov stability

1 Introduction

Synchronization refers to a process wherein two dynamical systems (master and slave systems, respectively) adjust their motion to achieve a common behavior, mainly due to a control input [1]. By considering the historical time-line of the topic, it can be observed that a large variety of synchronization types has been proposed [2,3,5–8,11–13,17,23,25,26]. Among the different types, *full state hybrid projective synchronization (FSHFPS)* has been introduced, wherein each slave system variable synchronizes with a linear combination of master system variables [4]. On the other hand, when the inverted scheme is implemented,

A. E. Hassanien et al. (Eds.): AISI 2018, AISC 845, pp. 108–116, 2019.
https://doi.org/10.1007/978-3-319-99010-1_10

i.e., each master system state synchronizes with a linear combination of slave system states, the *inverse full-state hybrid function projective synchronization (IFSHFPS)* is obtained [9,16,19]. Recently, the topic of coexistence of several synchronization types between chaotic systems has recently started to attract increasing attention [10,14,15,17,18,20–22].

Based on these considerations, this paper aims to give a further contribution to the topic by considering the co-existence of FSHFPS and IFSHFPS between non-identical and different dimensions chaotic and hyperchaotic systems. Specifically, the paper illustrates new scheme, which proves the co-existence of the full-state hybrid function projective synchronization (FSHFPS) and the inverse full-state hybrid function projective synchronization (IFSHFPS) between a three-dimensional master system and a four-dimensional slave system in 4D and 3D, respectively.

The paper is organized as follow: Sect. 2 gives a formulation of the co-existence of FSHFPS and IFSHFPS. In Sect. 3, the main result is presented. Section 4 gives numerical examples with the aim to show the effectiveness of the approach developed herein. Finally, Sect. 5 concludes the paper.

2 Problem Formulation

We assume that master and slave systems can be considered as

$$\dot{X}(t) = f(X(t)), \tag{1}$$

and

$$\dot{Y}(t) = BY(t) + g(Y(t)) + U, \tag{2}$$

where $X(t) = (x_i(t))_{1 \le i \le 3}$, $Y(t) = (y_i)_{1 \le i \le 4}$ are states of the master system (1) and the slave system (2), $f : \mathbb{R}^3 \to \mathbb{R}^3$, $B = (b_{ij}) \in \mathbb{R}^{4\times 4}$, $g : \mathbb{R}^4 \to \mathbb{R}^4$ is a nonlinear function and $U = (u_i)_{1 \le i \le 4}$ is a controller to be designed.

Definition 1. *Let* $(\alpha_j(t))_{1 \le j \le 4}$, $(\beta_j(t))_{1 \le j \le 3}$, $(\gamma_j(t))_{1 \le j \le 4}$ *and* $(\theta_j(t))_{1 \le j \le 3}$ *be continuously differentiable and boundary functions, it is said that inverse full-state hybrid function projective synchronization (IFSHFPS) and full-state hybrid function projective synchronization (FSHFPS) coexist in the synchronization of the master system (1) and the slave system (2), if there exist controllers* u_i, $= 1, 2, 3, 4$, *such that the synchronization errors*

$$\begin{aligned} e_1(t) &= x_1(t) - \sum_{j=1}^{4} \alpha_j(t) y_j(t), \\ e_2(t) &= y_2(t) - \sum_{j=1}^{3} \beta_j(t) x_j(t), \\ e_3(t) &= x_3(t) - \sum_{j=1}^{4} \gamma_j(t) y_j(t), \\ e_4(t) &= y_4(t) - \sum_{j=1}^{3} \theta_j(t) x_j(t), \end{aligned} \tag{3}$$

satisfy $\lim_{t \longrightarrow +\infty} e_i(t) = 0, \quad i = 1, 2, 3, 4.$

Remark 1. *Here* e_1 *and* e_3 *stand for IFSHFPS error and* e_2 *and* e_4 *stand for FSHFPS error.*

3 Sufficient Conditions for Co-existence of IFSHFPS and FSHFPS

The error system (3) can be differentiated as follows:

$$\begin{aligned}
\dot{e}_1(t) &= \dot{x}_1(t) - \sum_{j=1}^{4} \dot{\alpha}_j(t)\, y_j(t) - \sum_{j=1}^{4} \alpha_j(t)\, \dot{y}_j(t), \\
\dot{e}_2(t) &= \dot{y}_2(t) - \sum_{j=1}^{3} \dot{\beta}_j(t)\, x_j(t) - \sum_{j=1}^{3} \beta_j(t)\, \dot{x}_j(t), \\
\dot{e}_3(t) &= \dot{x}_3(t) - \sum_{j=1}^{4} \dot{\gamma}_j(t)\, y_j(t) - \sum_{j=1}^{4} \gamma_j(t)\, \dot{y}_j(t), \\
\dot{e}_4(t) &= \dot{y}_4(t) - \sum_{j=1}^{3} \dot{\theta}_j(t)\, x_j(t) - \sum_{j=1}^{3} \theta_j(t)\, \dot{x}_j(t).
\end{aligned} \tag{4}$$

Furthermore, the error system (4) can be written as

$$\begin{aligned}
\dot{e}_1(t) &= \sum_{j=1}^{4} \alpha_j(t)\, u_j + R_1, \\
\dot{e}_2(t) &= u_2 + R_2, \\
\dot{e}_3(t) &= \sum_{j=1}^{4} \gamma_j(t)\, u_j + R_3, \\
\dot{e}_4(t) &= u_4 + R_4,
\end{aligned} \tag{5}$$

where

$$\begin{aligned}
R_1 &= f_1(X(t)) - \sum_{j=1}^{4} \dot{\alpha}_j(t)\, y_j(t) - \sum_{i=1}^{4} \alpha_i(t) \left(\sum_{j=1}^{4} b_{ij} y_j(t) + g_i(Y(t)) \right), \\
R_2 &= \sum_{j=1}^{4} b_{2j} y_j(t) + g_2(Y(t)) - \sum_{j=1}^{3} \dot{\beta}_j(t)\, x_j(t) - \sum_{j=1}^{3} \beta_j(t)\, \dot{x}_j(t), \\
R_3 &= f_3(X(t)) - \sum_{j=1}^{4} \dot{\gamma}_j(t)\, y_j(t) - \sum_{i=1}^{4} \gamma_i(t) \left(\sum_{j=1}^{4} b_{ij} y_j(t) + g_i(Y(t)) \right), \\
R_4 &= \sum_{j=1}^{4} b_{4j} y_j(t) + g_4(Y(t)) - \sum_{j=1}^{3} \dot{\theta}_j(t)\, x_j(t) - \sum_{j=1}^{3} \theta_j(t)\, \dot{x}_j(t).
\end{aligned} \tag{6}$$

To achieve the coexistence of IFSHFPS and FSHFPS between systems (1) and (2), we assume that $\alpha_3(t)\gamma_1(t) - \alpha_1(t)\gamma_3(t) \neq 0$. Hence, we have the following result:

Theorem 1. *The coexistence of IFSHFPS and FSHFPS between the master system (1) and the slave system (2) will occur under the following control law*

$$u_1 = \sum_{i=1}^{4} P_i \left(\sum_{j=1}^{4} (b_{ij} - c_{ij}) e_j(t) - R_i \right), \tag{7}$$
$$u_2 = \sum_{j=1}^{4} (b_{2j} - c_{2j}) e_j(t) - R_2,$$
$$u_3 = \sum_{i=1}^{4} Q_i \left(\sum_{j=1}^{4} (b_{ij} - c_{ij}) e_j(t) - R_i \right),$$
$$u_4 = \sum_{j=1}^{4} (b_{4j} - c_{4j}) e_j(t) - R_4,$$

where $(c_{ij})_{4\times4}$ *are control constants to be selected and* $P_1 = \frac{\gamma_3(t)}{\alpha_3(t)\gamma_1(t)-\alpha_1(t)\gamma_3(t)}$, $P_2 = \frac{\gamma_3(t)\alpha_2(t)-\alpha_3(t)\gamma_2(t)}{\alpha_3(t)\gamma_1(t)-\alpha_1(t)\gamma_3(t)}$, $P_3 = \frac{-\alpha_3(t)}{\alpha_3(t)\gamma_1(t)-\alpha_1(t)\gamma_3(t)}$, $P_4 = \frac{\gamma_3(t)\alpha_4(t)-\alpha_3(t)\gamma_4(t)}{\alpha_3(t)\gamma_1(t)-\alpha_1(t)\gamma_3(t)}$, $Q_1 = \frac{-\gamma_1(t)}{\alpha_3(t)\gamma_1(t)-\alpha_1(t)\gamma_3(t)}$, $Q_2 = \frac{\alpha_1(t)\gamma_2(t)-\alpha_2(t)\gamma_1(t)}{\alpha_3(t)\gamma_1(t)-\alpha_1(t)\gamma_3(t)}$, $Q_3 = \frac{\alpha_1(t)}{\alpha_3(t)\gamma_1(t)-\alpha_1(t)\gamma_3(t)}$ *and* $Q_4 = \frac{\alpha_1(t)\gamma_4(t)-\alpha_4(t)\gamma_1(t)}{\alpha_3(t)\gamma_1(t)-\alpha_1(t)\gamma_3(t)}$.

Proof. By substituting the control law (7) into (5), the error system can be described as

$$\dot{e}_i(t) = \sum_{j=1}^{4} (b_{ij} - c_{ij}) e_j(t), \; i = 1, 2, 3, 4, \tag{8}$$

or in the compact form

$$\dot{e}(t) = (B - C) e(t), \tag{9}$$

Construct the candidate Lyapunov function in the form: $V(e(t)) = e^T(t)e(t)$, we obtain

$$\begin{aligned} \dot{V}(e(t)) &= \dot{e}^T(t)e(t) + e^T(t)\dot{e}(t) \\ &= e^T(t)(B - C)^T e(t) + e^T(t)(B - C)e(t) \\ &= e^T(t)\left[(B - C)^T + (B - C)\right] e(t). \end{aligned}$$

If the control matrix C is chosen such that $(B - C)^T + (B - C)$ is negative definite matrix, we get $\dot{V}(e(t)) < 0$. Thus, from the Lyapunov stability theory, it is immediate that all solutions of the error system (9) go to zero as $t \to \infty$. Therefore, the systems (1) and (2) are globally synchronized in 4D.

4 Numerical Example

In this example, the master system is defined by the following new 3D system [24]

$$\begin{aligned} \dot{x}_1 &= a_1 (x_2 - x_1), \\ \dot{x}_2 &= x_1 x_3, \\ \dot{x}_3 &= 50 - a_2 x_1^2 - a_3 x_3. \end{aligned} \tag{10}$$

When $a_1 = 2.9$, $a_2 = 0.7$, $a_3 = 0.6$ and the initial conditions are taken as $(x_1(0), x_2(0), x_3(0)) = (0.6, 0.5, 0.4)$, system (10) exhibits chaotic attractors as shown in Fig. 1.

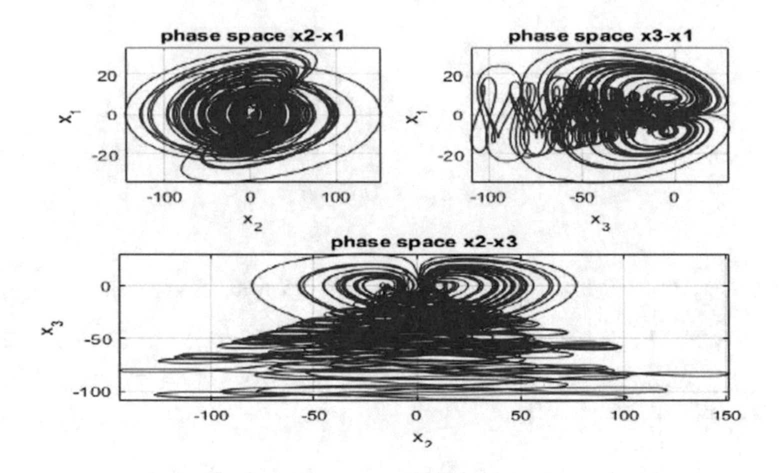

Fig. 1. Attractors of the master system (10) in 2-D.

The salve system is described by

$$\begin{aligned} \dot{y}_1 &= b_1 (y_2 - y_1) + y_2 y_3 + y_4 + u_1, \\ \dot{y}_2 &= b_2 y_1 + y_4 - b_3 y_1 y_3 + u_2, \\ \dot{y}_3 &= -b_4 y_3 + b_5 y_1 y_2 + u_3, \\ \dot{y}_4 &= -y_1 - y_2 + u_4. \end{aligned} \tag{11}$$

When the controllers $u_1 = u_2 = u_3 = u_4 = 0$, $(b_1, b_2, b_3, b_4, b_5) = (18, 40, 5, -3, 4)$ and the initial conditions are given as $(y_1(0), y_2(0), y_3(0), y_4(0)) = (0.5, 0.8, 0.2, 1.3)$, system (11) exhibits hyperchaotic attractors as shown in Fig. 2.

Based on the notations used in Sect. 3, the linear part B and the nonlinear part g of the slave system (11) are given as follows

$$B = \begin{pmatrix} -18 & 18 & 0 & 1 \\ 40 & 0 & 0 & 1 \\ 0 & 0 & -3 & 0 \\ -1 & -1 & 0 & 0 \end{pmatrix} \text{ and } g = \begin{pmatrix} y_2 y_3 \\ -5 y_1 y_3 \\ 4 y_1 y_2 \\ 0 \end{pmatrix}.$$

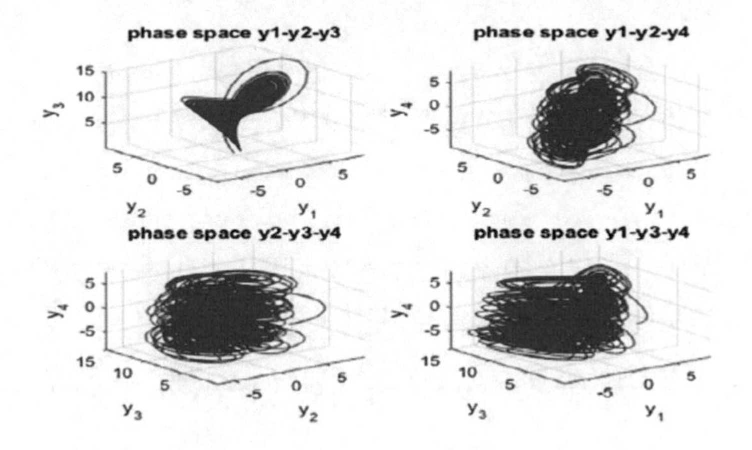

Fig. 2. Phase portraits of the slave system without control (11) in 3D.

According to the approach developed in Sect. 3, the synchronization errors between the master system (10) and the slave system (11) are defined as:

$$\begin{aligned} e_1 &= x_1 - \alpha_1(t)\, y_1 - \alpha_2(t)\, y_2 - \alpha_3(t)\, y_3 - \alpha_4(t)\, y_4, \\ e_2 &= y_2 - \beta_1(t)\, x_1 - \beta_2(t)\, x_2 - \beta_3(t)\, x_3, \\ e_3 &= x_3 - \gamma_1(t)\, y_1 - \gamma_2(t)\, y_2 - \gamma_3(t)\, y_3 - \gamma_4(t)\, y_4, \\ e_4 &= y_4 - \theta_1(t)\, x_1 - \theta_2(t)\, x_2 - \theta_3(t)\, x_3, \end{aligned} \tag{12}$$

where $\alpha_1(t) = \sin t$, $\alpha_2(t) = 1$, $\alpha_3(t) = \frac{1}{t+1}$, $\alpha_4(t) = 2$, $\beta_1(t) = 3$, $\beta_2(t) = \cos t$, $\beta_3(t) = 4$, $\gamma_1(t) = e - t$, $\gamma_2(t) = 2$, $\gamma_3(t) = 0$, $\gamma_4(t) = \frac{1}{t^2+1}$, $\theta_1(t) = \frac{t}{t+1}$, $\theta_2(t) = 0$, $\theta_3(t) = \sin 3t$. So, $\alpha_3(t)\gamma_1(t) - \alpha_1(t)\gamma_3(t) = \frac{1}{e^t(t+1)} \neq 0$.

The coexistence of IFSHFPS and FSHFPS, in this example, is achieved when the control matrix C is selected as

$$C = \begin{pmatrix} 0 & 18 & 0 & 1 \\ 40 & 1 & 0 & 1 \\ 0 & 0 & 0 & 0 \\ -1 & -1 & 0 & 1 \end{pmatrix}, \tag{13}$$

and the controllers u_i, $1 \leq i \leq 4$, are constructed according to (7) as follows:

$$\begin{aligned} u_1 &= -2e^t(-e_2 - R_2) + e^t(-3e_3 - R_3) - \frac{e^t}{t^2+1}(-e_4 - R_4), \\ u_2 &= -e_2 + 5y_1y_3 - 40y_1 - y_4 - R_2, \\ u_3 &= -(t+1)(-18e_1 - R_1) + \\ &\quad e^t(t+1)\left[-(2 + 2e_2 + 2R_2 + 3e_3 + R_3)\sin t + \left(\frac{\sin t}{t^2+1} - e^{-t}\right)(-e_4 - R_4)\right], \\ u_4 &= -e_4 + y_1 + y_2 - R_4, \end{aligned} \tag{14}$$

where

$$R_1 = 2.9(x_2 - x_1) - y_1 \cos t + \frac{1}{(t+1)^2} y_3 - \sin t (18(y_2 - y_1) + y_2 y_3) \tag{15}$$
$$+ \frac{1}{t+1}(4y_1 y_2 - 3y_3) - y_1 - y_2,$$
$$R_2 = -5y_1 y_3 + 40y_1 + y_4 + x_2 \sin t - 8.7(x_2 - x_1) - x_1 x_3 \cos t,$$
$$R_3 = 50 - 0.7x_1^2 - 0.6x_3 + e^{-t} y_1 + \frac{2t}{(t^2+1)^2} y_4 - e^{-t}(18(y_2 - y_1) + y_2 y_3) + 10y_1 y_3$$
$$+80y_1 - 2y_4 + \frac{1}{t^2+1}(y_1 + y_2),$$
$$R_4 = -y_1 - y_2 - \frac{t+1-t^2}{(t+1)^2} x_1 - 3x_3 \cos 3t - \frac{2.9t}{t+1}(x_2 - x_1) - (50 - 0.7x_1^2 - 0.6x_3)\sin 3t.$$

We can show that $(B-C)^T + (B-C)$ is negative definite matrix. Consequently, the error functions between systems (10) and (11) are described by

$$\dot{e}_1 = -18e_1, \tag{16}$$
$$\dot{e}_2 = -e_2,$$
$$\dot{e}_3 = -3e_3,$$
$$\dot{e}_4 = -e_4.$$

Numerical results plotted in Fig. 3 are obtained, indicating that the coexistence of IFSHFPS and FSHFPS is effectively achieved in 4D.

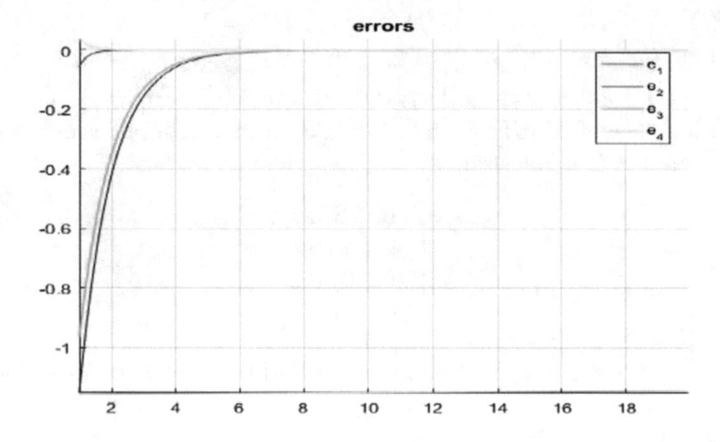

Fig. 3. Time evolution of the errors e_1, e_2, e_3 and e_4.

5 Conclusion

This paper has presented new result related to the co-existence of FSHFPS and IFSHFPS between a three-dimensional master system and a four-dimensional slave system. Specifically, the approach can be applied to: (i) wide classes of chaotic (hyperchaotic) master-slave systems; (ii) non-identical systems with different dimensions; (iii) scheme wherein the scaling factor of the linear combination can be any arbitrary differentiable function. Numerical example, describing the co-existence of FSHFPS and IFSHFPS between chaotic and hyperchaotic systems, has clearly highlighted the effectiveness of the approach proposed herein.

References

1. Azar, A.T., Vaidyanathan, S., Ouannas, A.: Fractional Order Control and Synchronization of Chaotic Systems. Studies in Computational Intelligence, vol. 688. Springer, Berlin (2017)
2. Azar, A.T., Volos, C., Gerodimos, N.A., Tombras, G.S., Pham, V.T., Radwan, A.G., Vaidyanathan, S., Ouannas, A., Munoz-Pacheco, J.M.: A novel chaotic system without equilibrium: dynamics, synchronization, and circuit realization. Complexity, Article ID 7871467, p. 11 (2017)
3. Azar, A.T., Ouannas, A., Singh, S.: Control of new type of fractional chaos synchronization, pp. 47–56. Springer, Cham (2018)
4. Ouannas, A.: On full-state hybrid projective synchronization of general discrete chaotic systems. J. Nonlinear Dyn. Article ID 983293, p. 6 (2014)
5. Ouannas, A.: A new generalized-type of synchronization for discrete-time chaotic dynamical systems. J. Comput. Nonlinear Dyn. **10**(6), 061,019–5 (2015)
6. Ouannas, A., Abu-Saris, R.: On matrix projective synchronization and inverse matrix projective synchronization for different and identical dimensional discrete-time chaotic systems. J. Chaos, Article ID 4912520, p. 7 (2016)
7. Ouannas, A., Al-sawalha, M.: On $\lambda - \phi$ generalized synchronization of chaotic dynamical systems in continuous-time. Eur. Phys. J. Spec. Top. **225**(1), 187–196 (2016)
8. Ouannas, A., Al-sawalha, M.M.: Synchronization between different dimensional chaotic systems using two scaling matrices. Optik - Int. J. Light Electron Opt. **127**(2), 959–963 (2016)
9. Ouannas, A., Grassi, G.: Inverse full state hybrid projective synchronization for chaotic maps with different dimensions. Chin. Phys. B **25**(9), 090, 503 (2016)
10. Ouannas, A., Grassi, G.: A new approach to study the coexistence of some synchronization types between chaotic maps with different dimensions. Nonlinear Dyn. **86**(2), 1319–1328 (2016)
11. Ouannas, A., Odibat, Z.: Generalized synchronization of different dimensional chaotic dynamical systems in discrete time. Nonlinear Dyn. **81**(1), 765–771 (2015)
12. Ouannas, A., Odibat, Z.: On inverse generalized synchronization of continuous chaotic dynamical systems. Int. J. Appl. Comput. Math. **2**(1), 1–11 (2016)
13. Ouannas, A., Azar, A.T., Abu-Saris, R.: A new type of hybrid synchronization between arbitrary hyperchaotic maps. Int. J. Mach. Learn. Cybern. **8**(6), 1887–1894 (2016)
14. Ouannas, A., Azar, A.T., Vaidyanathan, S.: A new fractional hybrid chaos synchronisation. Int. J. Model. Identif. Control **27**(4), 314–322 (2017). https://doi.org/10.1504/IJMIC.2017.084719

15. Ouannas, A., Azar, A.T., Vaidyanathan, S.: A robust method for new fractional hybrid chaos synchronization. Math. Methods Appl. Sci. **40**(5), 1804–1812 (2017). mma.4099
16. Ouannas, A., Azar, A.T., Ziar, T.: On inverse full state hybrid function projective synchronization for continuous-time chaotic dynamical systems with arbitrary dimensions. Differ. Eqns. Dyn. Syst. (2017)
17. Ouannas, A., Azar, A.T., Ziar, T., Radwan, A.G.: A study on coexistence of different types of synchronization between different dimensional fractional chaotic systems. In: Azar, A.T., Vaidyanathan, S., Ouannas, A. (eds.) Fractional Order Control and Synchronization of Chaotic Systems, pp. 637–669. Springer, Cham (2017)
18. Ouannas, A., Azar, A.T., Ziar, T., Vaidyanathan, S.: A new method to synchronize fractional chaotic systems with different dimensions. In: Azar, A.T., Vaidyanathan, S., Ouannas, A. (eds.) Fractional Order Control and Synchronization of Chaotic Systems, pp. 581–611. Springer, Cham (2017)
19. Ouannas, A., Grassi, G., Ziar, T., Odibat, Z.: On a function projective synchronization scheme for non-identical fractional-order chaotic (hyperchaotic) systems with different dimensions and orders. Optik - Int. J. Light Electron Opt. **136**, 513–523 (2017). https://doi.org/10.1016/j.ijleo.2017.02.068. http://www.sciencedirect.com/science/article/pii/S0030402617302115
20. Ouannas, A., Odibat, Z., Hayat, T.: Fractional analysis of co-existence of some types of chaos synchronization. Chaos Solitons Fractals **105**, 215–223 (2017)
21. Ouannas, A., Wang, X., Pham, V.T., Grassi, G., Ziar, T.: Coexistence of identical synchronization, antiphase synchronization and inverse full state hybrid projective synchronization in different dimensional fractional-order chaotic systems. Adv. Differ. Eqns. **2018**(1), 35 (2018)
22. Ouannas, A., Zehrour, O., Laadjal, Z.: Nonlinear methods to control synchronization between fractional-order and integer-order chaotic systems. Nonlinear Stud. **25**(1), 91–106 (2018)
23. Singh, S., Azar, A.T., Ouannas, A., Zhu, Q., Zhang, W., Na, J.: Sliding mode control technique for multi-switching synchronization of chaotic systems. In: 9th International Conference on Modelling, Identification and Control, ICMIC 2017, Kunming, China, 10–12 July 2017 (2017)
24. Vaidyanathan, S.: Global chaos synchronization of a novel 3-D chaotic system with two quadratic nonlinearities via active and adaptive control. In: Advances in Chaos Theory and Intelligent Control. Studies in Fuzziness and Soft Computing, vol. 337, pp. 481–506. Springer (2016)
25. Vaidyanathan, S., Azar, A.T.: A novel 4-D four-wing chaotic system with four quadratic nonlinearities and its synchronization via adaptive control method. In: Advances in Chaos Theory and Intelligent Control, pp. 203–224. Springer, Berlin (2016)
26. Vaidyanathan, S., Sampath, S., Azar, A.T.: Global chaos synchronisation of identical chaotic systems via novel sliding mode control method and its application to Zhu system. Int. J. Model. Identif. Control **23**(1), 92–100 (2015)

Integrated Multi-sensor Monitoring Robot for Inpatient Rooms in Hospital Environment

Lamia Nabil Mahdy[1,3], Kadry Ali Ezzat[1,3(✉)], and Aboul Ella Hassanien[2,3]

[1] Biomedical Engineering Department, Higher Technological Institute, 10th of Ramadan, Egypt
Kadry_ezat@hotmail.com
[2] Faculty of Computers and Information, Cairo University, Cairo, Egypt
[3] Scientific Research Group in Egypt (SRGE), Giza, Egypt
http://www.egyptscience.net

Abstract. The scope of this proposed system is to implement multi-sensor robot architecture. The reduction of the human activities in hospital environment is the main target of utilizing the self-governing portable robots in numerous applications. The executed robot is a self-governing four wheels system that is designed to determine the sound level, light intensity, humidity and high temperature then transmitting data to a remote location and visualized in mobile application. Bluetooth connection is established between the Arduino on the robot and the smart phone. The smart phone acts as the manual controller which is responsible for directing the robot and receive data from Arduino. The Arduino navigates the robot based on the feedbacks from ultrasonic sensor to detect the barriers.

Keywords: Robot · Sensor · Inpatient room · Humidity · Temperature

1 Introduction

A robot is a mechanically or practically virtual simulated operator, generally a mechanical electric machine that is routed through a computer or the electronic program and is subsequently prepared to do assignments all alone. The mechanical businesses affiliation [1] defines as follows: “The robot is a programmable multifunctional controller designed to transfer certain parts, devices or devices through programmed movements without variation to perform a variety of tasks.” Android ICS is an inexorably noticeable and vital part of the present-day business, particularly in clinics and hospitals. These robots, called “mechanical robots”, were only discarded in car paint factories 14–19 years ago [2]. But mechanical robots are currently being utilized as a part of labs, innovative work offices, storage houses, hospitality offices, vitality-oriented industries (oil, atomic power, and so forth), and other fields. This incorporates physically controlling a robot from one point to another during the task stages. The motivation of a task is characterized by the computer commands. This is alluded to as sub-level programming console. One of the key areas of research is to improve indirect programming that takes advantage of high-level languages, where robotic exercises are

A. E. Hassanien et al. (Eds.): AISI 2018, AISC 845, pp. 117–126, 2019.
https://doi.org/10.1007/978-3-319-99010-1_11

defined by errands or targets. The use of modern robots has become more widespread. The robot is utilized to check the temperature and humidity [3].

The person who has contracted the disease may be asked to remain in the healing center, to ensure proper medical attention. While the patient who is in the hospital may need certain room qualities to suit his statuses, for example, lighting, humidity, and temperature. These requirements must be of particular preference to allow for the ideal comfort conditions of the patient on his way to rehabilitation. In fact, these requirements may vary due to external variables, such as environment, and hospitals need the supplies to observe these rooms continuously [4]. The intended project is to construct a robot to observe these modifications in a built-in system.

One illustration where this incorporated control system might be beneficial in the state of neonatal hepatitis. Bilirubin, which occurs from the failure of red blood cells and excluded from the body with urinary excretion. Yet, the new infant has not completely gained the capability to manage bilirubin, so excess bilirubin may penetrate the body's tissues. When this occurs, yellowing of the eyes and skin occurs, directing to hepatitis [5]. If hepatitis progresses, the baby may require light healing. Throughout this method, blue fluorescent torches are utilized to suppress excess bilirubin in the skin [6]. In this method, the temperature in the around area must persist consistently, and this is where the proposed device can support this performance. At present, there are no interracial machines that observe and constrain the combination of temperature, light, sound, and humidity, composing this system novel.

After reviewing various papers, there is currently no article detecting a set of temperature, lighting, sound control and humidity in one interracial system and has the impulse to transform these frames. But, there are publications for automated disclosure and light administration. In one of those papers, the light sensor on the console identifies a variation in light strength and utilizes a radio frequency switch to adjust the light [7]. In addition, there another research article presented considering those three normal conditions. One of the advantages of this system sensor is that it is wireless, relies on Bluetooth technology, which consumes low power [8]. Another paper presented a method to efficiently control moisture in the incubator. Moisture control depended on the latent and powerful system. In the inactive system, the air was crossed over the water container, resulting in increased moisture in the air, before being scattered completely in the room.

The paper is sorted as follows: Sect. 2 discusses the materials (component description). Section 3 discusses the proposed system for the robot. Section 4 discusses experimental results. Section 5 discuss conclusion and future work.

2 Materials (Component Description)

The components of the proposed device discussed as follows:

2.1 Distance Ultrasonic Sensor

The ultrasonic signal is sent by ultrasonic sensor then received the echo signal. Depending on the time period taken by the signal to rebound the estimated distance of the barrier [9] can be computed using given formula

$$t = s/v \tag{1}$$

Where t is time, s is distance and v is the speed of sound.

This instrument was utilized to avoid the robot from hitting walls or other barriers in its way as shown in Fig. 1.

Fig. 1. Distance sensor

2.2 H Bridge

H Bridge is an electric circuit that empowers voltage to be implemented via opposite load as shown in Fig. 2. H Bridge is often utilized in robots to decrease foremost or backward motors [10].

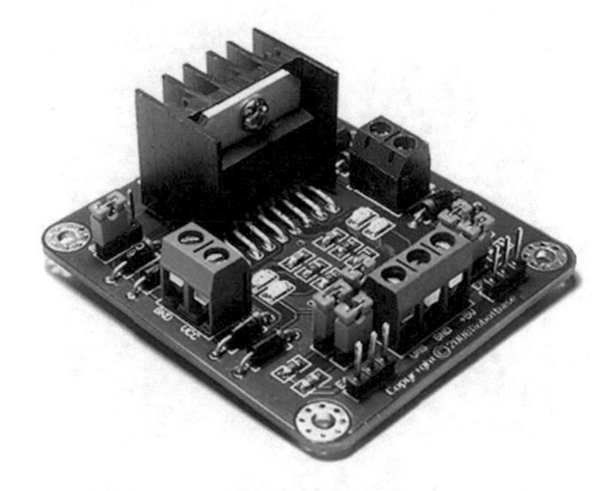

Fig. 2. H-bridge

2.3 Actuator

The "motor" can be described as a device that combines power in physical movement. Most motors provide a rotational or straight movement [11]. In our case, the motor is the DC motor as shown in Fig. 3. It is basically a DC motor with a gearbox that decreases motor velocity and increases torque.

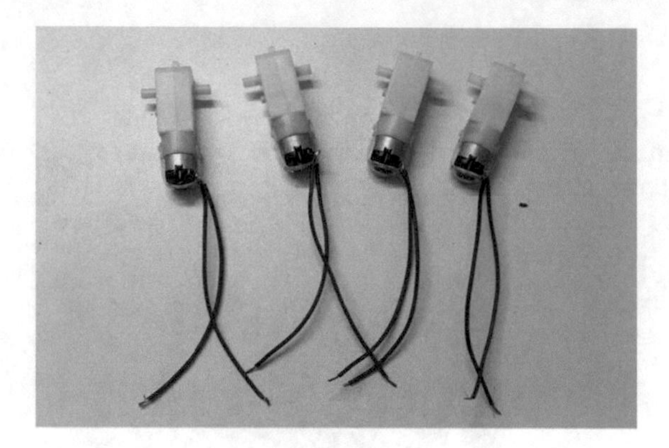

Fig. 3. Actuators

2.4 Humidity-and-Temperature-Sensor

The DHT11 temperature and humidity sensor calibrate the digital signal output with the temperature and humidity sensor system. It's accurate adequately for most impacts that require tracking moisture readings and temperature. The library is created especially for these sensors that present the code in a short and easy way to write [12] as shown in Fig. 4.

Fig. 4. Humidity and temperature sensor

2.5 Light Dependent Resistor (LDR) Sensor

LDR is a changeable resistor managed by light. Resist the form ridge resistance while improving the strength of the light dropping as shown in Fig. 5. Besides, it carries visual conductivity. Furthermore, the single LDR can cooperate very differently with photons in remarkable wavelength strips. LDR differs according to the quantity of light on which it is observed [13]. The association between RL resistance and light intensity of the LDR LUX is

$$Lux = 500/\mathrm{R_L} \tag{2}$$

Where Lux is light intensity and RL is resistance.

Fig. 5. LDR sensor

2.6 Sound Detection Sensor

The Sound Detection Sensor is a little board that joins an amplifier and some preparing hardware, it can recognize diverse sizes of sound as appeared in Fig. 6. Since the pressure numbers represented by Pascal are generally low and not easily managed, another, more commonly used scale, the decibel (dB) scale, was developed. This logarithmic scale more closely matches the response reactions of the human ear to the pressure fluctuations [14]. The amount of compression can calculate by using the following formula:

$$P = \mathit{voltage}\ (\mathrm{mV})/\mathit{sensitivity}\ (\mathrm{mV/Pa}) \tag{3}$$

Where P = pascals (Pa) and voltage is the preamplifier output peak voltage.

After determining the maximum pressure level that the microphone can sense at its peak voltage, you can convert this amount to decibels (dB) using the following logarithmic scale:

$$\mathrm{dB} = 20\log(p/po) \tag{4}$$

Where P = pressure in pascals
Po = reference pascals (constant = 0.00002 Pa).

Fig. 6. Sound detection sensor

3 System Description

The block diagram depicts the hardware modules under use: Arduino, H Bridge, 4 wheels, Wireless unit, temperature sensor, humidity sensor, Sound Sensor, Light Sensor, sets of batteries and Smart phone as control – monitor unit as shown in Fig. 7.

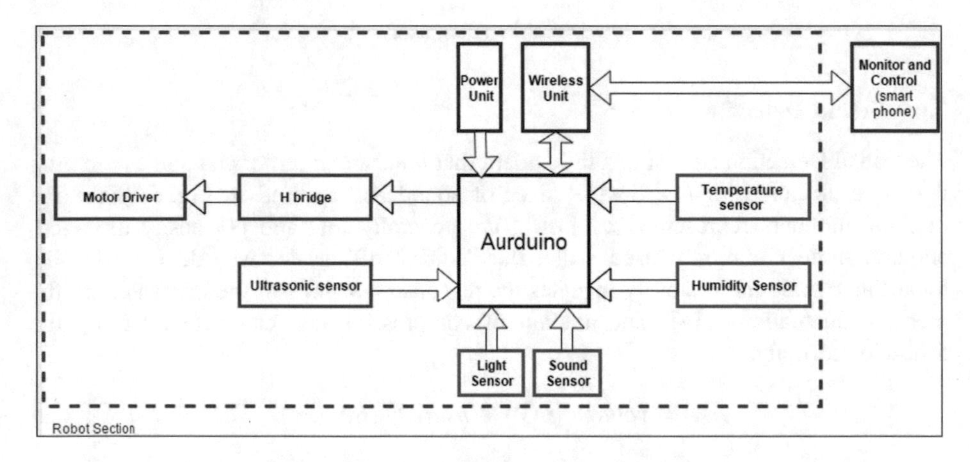

Fig. 7. The block diagram of the system

This system was formed by a Smart phone and a Bluetooth module for wireless communications among robot section and control & monitor section. The main working module of this robot consists of Arduino Uno processor. It is interfaced with ultrasonic sensor as shown in Fig. 8, temperature, light, sound and humidity sensors as shown in Fig. 9. The device framework of the robot works with H bridge and motor drivers to control the motion of the 4 wheels. Location of the robot is traced with ultrasonic sensor and the Bluetooth is used to send the data to the smart phone. The robot has two modes of motion, self-automatic motion and manual remote motion using the smart phone.

Fig. 8. Robot loaded with ultrasonic sensor

Fig. 9. Robot loaded with temperature, light, sound and humidity sensors

4 Results

The robot loaded with sensors measure five measurements for five different locations in each room of four inpatient rooms. The five measurements occurred for the four corners and the middle of the room to obtain accurate measurements. There is time delay for about 30 s between each measurement and the other one to give the robot the required time to move from one place to another. The results are shown in Tables 1, 2, 3, 4 and Fig. 10a, b, c and d.

Fig. 10. (a), (b), (c) and (d) The temperature, humidity, sound level and light intensity data gathered from four inpatient rooms

Table 1. The data gathered from first inpatient room

ROOM 1	Sound level (dB)	Light intensity (lx)	Temperature (°C)	Humidity (gm/m^3)
t0, Corner 1	317	315	21	31
t30, Corner 2	313	316	20	32
t60, Corner 3	313	319	20	32
t90, Corner 4	313	316	20	32
t120, Centre	315	315	20	32

Table 2. The data gathered from second inpatient room

ROOM 2	Sound level (dB)	Light intensity (lx)	Temperature (°C)	Humidity (gm/m^3)
t0, Corner 1	315	318	20	32
t30, Corner 2	317	319	20	32
t60, Corner 3	317	319	20	32
t90, Corner 4	317	317	20	32
t120, Centre	316	318	20	32

Table 3. The data gathered from third inpatient room

ROOM 3	Sound level (dB)	Light intensity (lx)	Temperature (°C)	Humidity (gm/m^3)
t0, Corner 1	315	317	25	42
t30, Corner 2	313	314	26	43
t60, Corner 3	315	321	25	45
t90, Corner 4	319	321	24	45
t120, Centre	320	323	22	45

Table 4. The data gathered from fourth inpatient room

ROOM 4	Sound level (dB)	Light intensity (lx)	Temperature (°C)	Humidity (gm/m^3)
t0, Corner 1	322	325	21	44
t30, Corner 2	323	325	21	43
t60, Corner 3	322	325	21	42
t90, Corncr 4	322	324	21	42
t120, Centre	322	322	21	42

5 Conclusion and Future Work

The proposed system was consists of smart mobile software and a Bluetooth for wireless communications between the robot section and control - monitoring section. The ultrasonic sensor is attached to the robot which continuously tracing the surrounding areas. The robot consists of Arduino interfaced with temperature, humidity, sound and light sensors. For robot movement the H Bridge is connected to the cortex-M3 port. The robot succeed in avoiding obstacles with automatic self-motion and transfer from one inpatient room to another by using the smart phone. Sensors were very accurate in measuring sound level, light intensity, temperature and humidity. The further extension of this project is to add more sensors like hazard gases sensors and adding two other wheels for fasting the robot.

References

1. Wu, C.-M., Lu, J.-T.: Implementation of remote control for a spraying robot. In: 2017 International Conference on Applied System Innovation, vol. 4, no. 2, pp. 1010–1013 (2017)
2. Block, V.A.J., et al.: Remote physical activity monitoring in neurological disease: a systematic review. PLoS One **11**(4), 1–41 (2016)
3. Cippitelli, E., Fioranelli, F., Gambi, E., Spinsante, S.: Radar and RGB-depth sensors for fall detection: a review. IEEE Sens. J. **17**(12), 3585–3604 (2017)
4. Erden, F., Velipasalar, S., et al.: Sensors in assisted living: a survey of signal and image processing methods. IEEE Signal Process. Mag. **33**(2), 36–44 (2016)
5. Lin, C.-S., et al.: The remote cruise method for the robot with multiple sensors. Measurement **118**(1), 194–201 (2018)
6. Kurkin, A.A., et al.: Autonomous mobile robotic system for environment monitoring in a coastal zone. Proc. Comput. Sci. **103**(4), 459–465 (2017)
7. Yuan, J., Zhang, J.: Cooperative localization for disconnected sensor networks and a mobile robot in friendly environments. Inf. Fusion **37**(5), 22–36 (2017)
8. Frommknech, A., et al.: Multi-sensor measurement system for robotic drilling. Robot. Comput. Integr. Manuf. **47**(3), 4–10 (2017)
9. Ilkka, K., Jero, A.: Estimating the specific heat capacity and heating of electronic sensors and devices. IEEE Inst. Meas. Mag. **21**(1), 54–62 (2018)
10. Liu, J., Shen, H.: Characterizing data deliverability of greedy routing in wireless sensor networks. IEEE Trans. Mob. Comput. **17**(3), 543–559 (2018)
11. Uthayakumar, S., Uma, G.: ANFIS-based sensor fault-tolerant control for hybrid grid. IET Gener. Transm. Distrib. **12**(1), 31–41 (2018)
12. Cong, W., Li, J., et al.: Combining solar energy harvesting with wireless charging for hybrid wireless sensor networks. IEEE Trans. Mob. Comput. **17**(3), 560–576 (2018)
13. Herrera, R.H., Tary, J.B., Van Der Baan, M., et al.: Body wave separation in the time-frequency domain. IEEE Trans. Biomed. Eng. **12**(4), 364–368 (2015)
14. Hadis, N.S.M.: Fabrication of fluidic-based mersister sensor for dengue virus detection. In: 2017 IEEE Asia Pacific Conference on Postgraduate Research in Microelectronics and Electronics (Prime Asia), vol. 1, no. 2, pp. 105–108. IEEE (2017)

Comparing Multi-class Approaches for Motor Imagery Using Renyi Entropy

Sahar Selim[1](✉), Manal Tantawi[2], Howida Shedeed[2], and Amr Badr[3]

[1] Computer Science Department, Modern Academy for Computer Science and Management Technology, Maadi, Cairo, Egypt
s.selim@grad.fci-cu.edu.eg

[2] Scientific Computing Department, Faculty of Computer and Information Science, Ain Shams University, Cairo, Egypt

[3] Computer Science Department, Faculty of Computers and Information, Cairo University, Cairo, Egypt
amr.badr@fci-cu.edu.eg

Abstract. One of the main problems that face Motor Imagery-based system is addressing multi-class problem. Various approaches have been used to tackle this problem. Most of these approaches tend to divide multi-class problem into binary sub problems. This study aims to address the multi-class problem by comparing five multi-class approaches; One-vs-One (OVO), One-vs-Rest (OVR), Divide & Conquer (DC), Binary Hierarchy (BH), and Multi-class approaches. Renyi entropy was examined for feature extraction. Three linear classifiers were used to implement these five-approaches: Support Vector Machine (SVM), Multinomial Logistic Regression (MLR) and Linear Discriminant Analysis (LDA). These approaches were compared according to their performance and time consumption. The comparative results show that, Renyi entropy demonstrated its robustness not only as a feature extraction technique but also as a powerful dimension reduction technique, for multi-class problem. In addition, LDA proved to be the best classifier for almost all approaches with minimum execution time.

Keywords: Brain Computer Interface · Motor imagery · Feature extraction Renyi entropy · Multi-class problem

1 Introduction

Brain Computer Interface (BCI) is an interface between human brain and Computer system. Motor Imagery (MI) based BCI system is one of the most popular systems in BCI area. It doesn't involve motor output from human, which is useful to help paralyzed people to achieve primitive tasks by themselves and be able to communicate with external environment [1].

Almost all MI-BCI systems suffer from high dimensionality of input data in the presence of small training dataset [2]. Most classifiers were designed to solve binary class problems. In multi-class MI based system, the binary classifiers must be extended to solve multi-class problem. Many approaches arose to help extending binary

A. E. Hassanien et al. (Eds.): AISI 2018, AISC 845, pp. 127–136, 2019.
https://doi.org/10.1007/978-3-319-99010-1_12

classifiers as OVO and OVR approaches, which divide the multi-class problem into smaller binary class problems [3].

Common Spatial Patterns (CSP) is one of the most commonly used feature extraction technique in MI based systems [2]. However, it offers reliable level of feature discrimination of input signal, yet it has some constraints. First constraint is that it works only on binary class problem. Even the extension of CSP to handle multi-classes didn't achieve good results [3]. Another constraint is that the two classes must have same number of records. Therefore, CSP is used with OVR and OVO approaches as in [4] with SVM classifier. Ang et al. [2] (the BCI Competition winners) used CSP by applying OVO, OVR and DC approaches to solve multi-class problem.

Renyi entropy has been used recently by few researchers in BCI-MI based systems [5]. Kee et al. [6] compared the use of Renyi entropy to CSP using three different datasets. Renyi outperformed CSP in binary class dataset, while there was no significant difference between CSP & Renyi for the multi-class datasets.

This study aims to compare the performance of five different multi-class approaches: One-vs-One (OVO), One-vs-Rest (OVR), Divide & Conquer (DC), Binary Hierarchy (BH), and Multi-class approaches. To our knowledge, no study compared all these five approaches simultaneously. The Renyi entropy was investigated as a feature extraction and dimension reduction technique for four-class problem using BCI Competition dataset IIa. Each approach was evaluated by three classifiers, which are Support Vector Machine (SVM), Multinomial Logistic Regression (MLR) and Linear Discriminant Analysis (LDA). Execution time of each approach was calculated to check the applicability of this approach for online systems.

The rest of the paper is organized as follows: Sect. 2 presents the proposed methodology and overviews multi-class approaches. In Sect. 3, the results of performed experiments are given. The discussion of obtained results is presented in Sect. 4. Finally, the conclusion is drawn in Sect. 5.

2 Materials and Methods

2.1 Proposed Methodology

Proposed approach is applied on dataset IIa of BCI Competition IV as illustrated in Fig. 1. Electrodes from 1 to 20 are selected. Band-pass filter is used to extract motor imagery frequencies of mu (μ) and beta (β) rhythms between 10 Hz and 30 Hz, using the 5th order of Butterworth filter. Epochs are extracted from the filtered signal for the time interval between 3.5 and 5.5 s. Discriminant features were obtained from epochs using Renyi entropy. These features were used as input to classifiers (SVM, MLR and LDA), according to the implemented approach. Addressing multi-class problem, these features were examined by five approaches. Four approaches are considered extension of binary problem (OVO, OVR, DC and BH), while one approach used multi-class classifiers.

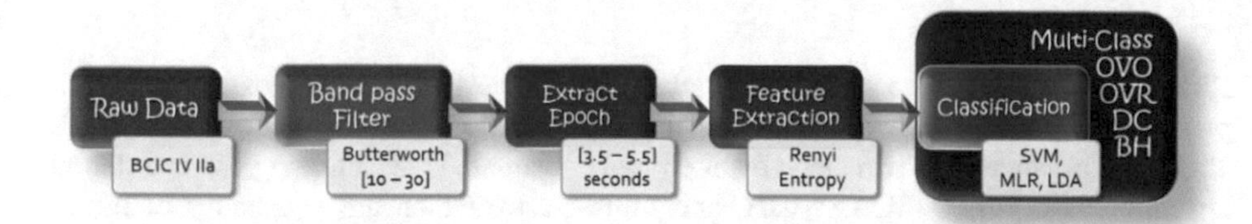

Fig. 1. Proposed BCI scheme

2.2 Data Acquisition

BCI Competition IV Data Set IIa. This dataset consists of data of nine subjects [7], performing four motor imagery actions (left hand, right hand, foot and tongue). The data was recorded from 22 electrodes based on the international 10–20 system. The signal was filtered by bandpass filter (0.5–100 Hz). The data was aggregated from two different sessions, where each session consists of 144 trials. Each class has 36 trials for either training or evaluation sessions.

2.3 Data Preprocessing

This study, aims to use Renyi entropy as an alternative to CSP for feature extraction in MI EEG signals. Renyi entropy has the advantage of reducing dimension of input signal of the 20 channels into only one feature. Another advantage is that it can be used in multiclass problem, on the contrary from CSP that is used for binary classes only.

Renyi Entropy. Renyi entropy uses brain electrical activity to characterize time series complexity [8]. It measures the chaoticity of EEG systems. Renyi entropy [9] is considered the generalization form of Shannon's entropy as defined in Eq. (1)

$$R_q = \frac{1}{1-q}\log_2\left(\sum_{i=1}^{n} p_i^q\right) \tag{1}$$

Renyi is of order q, where $q > 0$ and $q \neq 1$. p_i is the corresponding probability of distribution for $i = 1, \ldots n$.

Shannon's entropy is a special case of Renyi entropy when $q = 1$

$$\lim_{q \to 1} R_q = S \tag{2}$$

This results of the following Renyi's entropy equation

$$D_q = \lim_{\delta x \to 0} \frac{1}{1-q} \log_2 \frac{\log_2 \sum_{i=1}^{n} p_i^q}{\log_2 \delta x} \tag{3}$$

For all channels, the features extracted from Renyi entropy are concatenated into one feature vector to be input to the given classification approach.

2.4 Classification

Linear classifiers behave as nonlinear classifiers if the features were extracted appropriately [10]. Therefore, we examined three linear classifiers SVM, LDA and MLR. The performance of extracted features was evaluated by utilizing five different approaches (OVO, OVR, DC, BH, and multi-class), which are extensions of multi-class problem.

Classifiers. Generally, linear classifiers are considered more robust compared to non-linear ones. Linear classifiers have less probability to overfitting as they have few parameters to adjust. Yet, EEG signal which has strong noise may cause them to fail [11]. Three different linear classifiers with confidence score were used to examine the five approaches.

Support Vector Machine (SVM). It is one of the commonly used classifiers in MI applications. It tends to generalize the classification problem capability, in order to reduce learning model complexity. This is achieved by forming a hyperplane that tends to increase the gap between classes. The support vectors are formed from maximizing the distance between nearby points of unlike classes and the hyperplane [12].

Linear Discriminant Analysis (LDA). It is a generalized form of Fisher's linear discriminant. It has the ability to produce reliable output with few training data and low computational cost compared to other classifiers. It can be used to separate between two or more classes. Though this classifier achieved considerate achievement for MI systems, yet, it has disadvantage that it fails with non-Gaussian distribution [12].

Multinomial Logistic Regression (MLR). It is a generalized form of logistic regression to a multi-class problem. It models the relationship between explanatory variables and the effectiveness of their contribution. Positive regression coefficient indicates higher probability of the outcome, while negative coefficient decrease the probability of outcome [13].

Approaches to Resolve Multi-class Problem. Almost all real life problems are multi-class problems. Some classification approaches are naturally designed to classify more than two classes. While, most classification approaches address binary-problems. The simplest way is to divide multi-class problem into smaller binary-class classification problems. Figure 2 illustrates the use of these approaches for the 4-class problem of BCIC IV 2a.

One-vs-One. One of the best known methods is one-vs-one (pairwise) problem proposed by Hastie et al. [14]. In this approach every two classes are paired together to form binary-classification problem. Thus, an ***n-class*** problem will have $n(n-1)/2$ binary-problems. A given trial may be labelled for multiple classes. A Majority-voting scheme is applied to the output of these binary-problems, where each binary classifier votes for the given input trial. The class with the maximum number of votes for a given trial is classified as the winner class. It happens that majority voting fails for a trial when this trial is equally assigned to more than one class. In this case, confidence score of the classifier is required. It is used to assign these trials to only one class according the highest confidence score. Therefore, classifier used must output confidence score for its results; not only the class label. For dataset BCIC IV 2a, 4 * (4 − 1)/2 = *6 classifiers* have been examined.

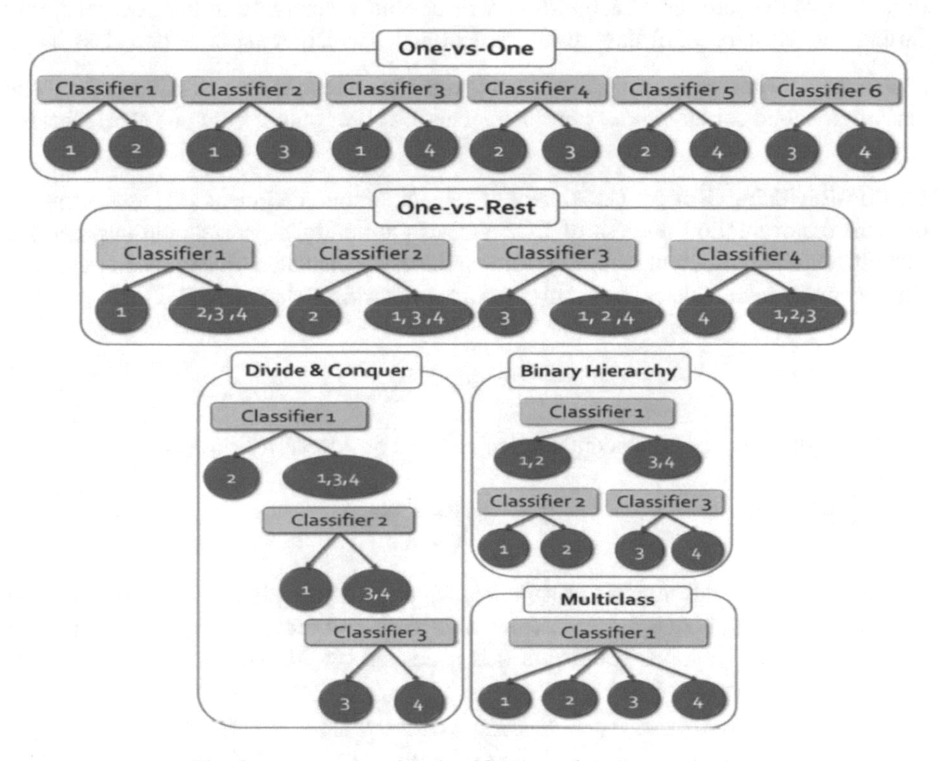

Fig. 2. Approaches for classification of 4-class problem

One-vs-Rest. This approach needs ***n*** classifiers, where each classifier differentiates between one class versus rest of classes. Like OVO, OVR suffers from having trials with the same number of votes [13]. Therefore, majority-voting scheme is applied to the output of these binary-classifiers to assign a trial to one label. Another problem with this approach is that the training sets are not equal in size, which might cause classifier to be biased towards larger size set, producing overfitting problem. For dataset BCIC IV 2a, *4 classifiers* will be employed.

Divide & Conquer. Zhang et al. [15] implemented a tree-like approach; consisting of ***n* − *1*** levels. It requires one binary classifier for each level. For dataset BCIC IV 2a, (4 − 1) = *3 classifiers* are required for the three levels as shown in Fig. 2. For example, 1st level will classify class 1 versus classes (2, 3, 4). Then, 2nd level will classify class 2 versus (3, 4). While last level classifier will classify given record to either class 3 or 4. This approach has disadvantage; that is the struggle to select which class to start with versus rest of classes for each level.

Binary Hierarchy. Binary Hierarchical Classifier was proposed by Kumar et al. [16]. This approach requires ***n* − *1*** binary classifiers arranged in a binary tree with ***n***-leaf nodes, where each leaf corresponds to a class. Starting from the root node, classes are divided into two clusters. Each cluster will be split recursively in the next level into further two clusters, until they reach the leaf node, which consists of one class only.

Multi-class. Some classifiers can naturally classify more than two classifiers. Only one classifier is used for any ***n-class*** problem. This classifier output only one class label for any given trial.

Evaluation Using Kappa Coefficient. Cohen's kappa coefficient [17] is commonly used in the evaluation process of BCI *N*-class problem. The correlation between the real classes and the output resulting from classifier is calculated by confusion matrix *H*. The accuracy is calculated from this confusion matrix as shown in Eq. (4).

$$ACC = p_0 = \frac{\sum_i H_{ii}}{N} \tag{4}$$

The probability of coincidental agreement is considered by Eq. (5)

$$p_e = \frac{\sum_i n_{oi} \times n_{io}}{N \times N} \tag{5}$$

where N is the overall number of samples, H_{ii} presents the elements of the confusion matrix H of the diagonal, n_{oi}, are the sums of each column while n_{io} are the sums of each row. Thus, the kappa coefficient κ is estimated by Eq. (6).

$$\kappa = \frac{p_o - p_e}{1 - p_e} = \frac{M \times p_o - 1}{M - 1} \tag{6}$$

3 Experimental Results

In this section, results have been presented to compare between the five multi-class approaches, using BCI Competition IV dataset IIa. The approaches were evaluated by calculating kappa coefficient and accuracy of the testing dataset. Execution time of classifiers was calculated and presented in milliseconds.

3.1 One-vs-One (Pairwise) Approach

LDA classifier gave the best accuracy and kappa values as shown in Table 1, while MLR gave worst execution time compared to other classifiers. Table 2 summarizes the results of One-vs-One approach for all nine subjects.

Table 1. Kappa coefficient of One-vs-One approach for the three classifiers

Classifier	1	2	3	4	5	6	7	8	9	Kappa
SVM	0.7203	0.3707	0.7677	0.3824	0.1934	0.2196	0.4306	0.5771	0.6296	0.4768
MLR	0.6970	0.3704	0.6748	0.3641	0.1706	0.1913	0.5602	0.5959	0.5417	0.4629
LDA	0.7483	0.3750	0.7352	0.4012	0.1661	0.2052	0.5972	0.5354	0.5972	0.4845

Table 2. Average kappa, accuracy and execution time of One-vs-One approach

Classifier	Kappa	ACC	Ex. time
SVM	0.4768	60.76%	0.4574
MLR	0.4629	59.71%	1.9218
LDA	0.4845	61.34%	0.6315

3.2 One-vs-Rest Approach

One advantage of Renyi is that it does not require number of trials of the two sets to be equal as in CSP, where the two groups of the training sets have to be equal in size. As shown in Tables 3 and 5, LDA gave best average kappa values with minimum time.

Table 3. Kappa coefficient of One-vs-Rest approach for the three classifiers

Classifier	1	2	3	4	5	6	7	8	9	Kappa
SVM	0.6595	0.3283	0.7584	0.3175	0.0660	0.1586	0.4954	0.6283	0.6204	0.4480
MLR	0.6924	0.3844	0.7073	0.3919	0.1518	0.2332	0.5926	0.6237	0.5880	0.4850
LDA	0.6597	0.3891	0.7630	0.3965	0.1654	0.2425	0.5648	0.6191	0.6065	0.4896

3.3 Divide & Conquer Approach

This approach has the disadvantage of struggling which classes to be classified against other classes. We have examined different combinations. In this paper, we present the combination, which gave best results presented in Table 4. This combination is shown in Fig. 2, beginning with Class 2 for level 1 versus (1, 3, 4). For level 2, we used Class 1 versus (3, 4). SVM classifier achieved minimum execution time as shown in Table 6.

Table 4. Kappa coefficient of Divide & Conquer approach for the three classifiers

Classifier	1	2	3	4	5	6	7	8	9	Kappa
SVM	0.7016	0.287	0.777	0.2945	0.1402	0.1767	0.4676	0.6468	0.537	0.4364
MLR	0.7343	0.2768	0.6051	0.3409	0.1513	0.2143	0.4769	0.5726	0.4537	0.4215
LDA	0.7343	0.2768	0.7166	0.3363	0.109	0.2097	0.5417	0.619	0.4722	0.4429

3.4 Binary Hierarchy Approach

This approach requires three classifiers divided into two levels. It has the same struggle as DC approach that is which classes to be classified against others. We have examined two combinations, where the average results of both combinations were almost the same. We present one of them as shown in Fig. 2. Tables 7 and 9 shows that LDA achieves best kappa values with minimum execution time, while MLR consumes highest time compared to SVM and LDA.

Table 5. Average kappa, accuracy and execution time of One-vs-Rest approach

Classifier	Kappa	ACC	Ex. time
SVM	0.4480	58.62%	0.4994
MLR	0.4850	61.38%	1.9238
LDA	0.4896	61.73%	0.3963

Table 6. Average kappa, accuracy and execution time of Divide & Conquer approach

Classifier	Kappa	ACC	Ex. time
SVM	0.4364	58.59%	0.0729
MLR	0.4215	56.88%	0.2178
LDA	0.4429	58.47%	0.0970

Table 7. Kappa coefficient of Binary Hierarchy approach for the three classifiers

Classifier	1	2	3	4	5	6	7	8	9	Kappa
SVM	0.6737	0.3566	0.7120	0.3965	0.1844	0.2382	0.4491	0.5446	0.5417	0.4443
MLR	0.6691	0.3751	0.6563	0.4058	0.1619	0.2239	0.4861	0.4287	0.5000	0.4259
LDA	0.6924	0.3563	0.7213	0.3966	0.1574	0.2285	0.5417	0.4565	0.5602	0.4438

3.5 Multi-class Approach

Table 8 presents the results of using multi-class approach. LDA classifier gave best result with minimum execution time compared to SVM and MLR. Table 10 summarizes the results of multi-class approach. Multi-class approach with only one classifier, unexpectedly, outperformed the four previous approaches for both kappa and time.

Table 8. Kappa coefficient of Multi-class approach for the three classifiers

Classifier	1	2	3	4	5	6	7	8	9	Kappa
SVM	0.6923	0.3660	0.7398	0.3825	0.2071	0.2195	0.4444	0.6050	0.6713	0.4809
MLR	0.6876	0.3939	0.6841	0.4058	0.1700	0.2378	0.5648	0.5681	0.6204	0.4814
LDA	0.6690	0.3703	0.7351	0.4290	0.2071	0.2518	0.537	0.6749	0.6343	**0.5009**

Table 9. Average kappa, accuracy and execution time of Binary Hierarchy approach

Classifier	Kappa	ACC	Ex. time
SVM	0.4443	59.13%	0.2634
MLR	0.4259	57.55%	2.5546
LDA	0.4438	59.25%	0.2926

Table 10. Average kappa, accuracy and execution time of multi-class approach

Classifier	Kappa	ACC	Ex. time
SVM	0.4809	61.07%	0.5388
MLR	0.4814	61.11%	3.1989
LDA	0.5009	62.58%	0.1559

4 Discussion

Figure 3 summarizes the comparison between approaches' kappa and Execution time. Among the five approaches, the OVR and Multiclass approaches gave the best kappa values. LDA proved to be the best classifier for almost all approaches with minimum execution time. MLR gave comparable results with SVM & LDA, yet, it gave the worst execution time for all approaches, which is considered a drawback for a real-time system. DC and BH approaches gave the lowest kappa values, especially with Multinomial LR classifier, compared to other approaches.

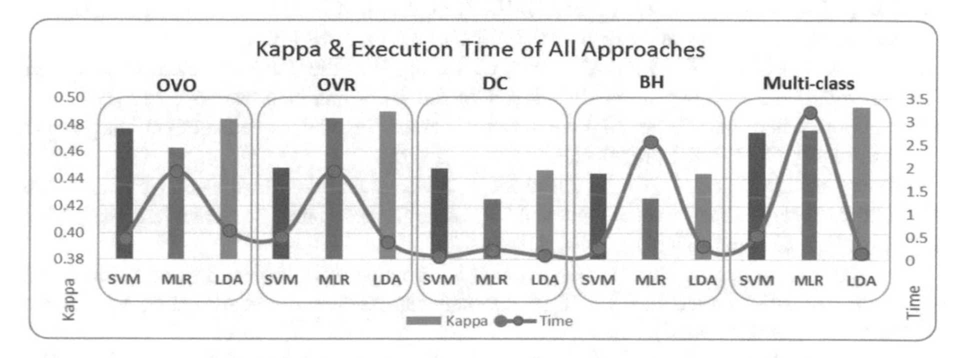

Fig. 3. Kappa values & execution time of the five approaches

Even though, results of Renyi were slightly less than results provided by winners of BCI Competition [2] using CSP, still Renyi is considered good alternative to CSP as it saves time and can deal with multi-class problem without having to divide it into binary sub problems which is time consuming.

5 Conclusion

This study aims to compare five multi-class approaches (OVO, OVR, DC, BH and multi-class) and evaluate the efficiency of using Renyi entropy as an alternative feature extraction technique and dimension reduction technique for MI signal, especially in multi-class problem. BCIC IV dataset IIa was used to emphasize the efficiency of examined approaches using three classifiers (SVM, MLR and LDA). CSP is a commonly used feature extraction technique in the field of BCI. Some of the barriers of using CSP are that it works on binary class problems only and the two classes must have same number of records. Probably, in OVR approach, one group is much greater in the number of records than the other group. Renyi can overcome these two drawbacks as it offers an advantage over CSP, which is that number of records of the two groups does not have to be the same, as it extracts features from each trial independently. In addition, Renyi can be used for multi-class problem without having to divide the problem into binary sub problem as in CSP.

As shown from results, multi-class approach gave comparable results with other approaches. Using multi-class approach with Renyi and LDA classifier achieved best kappa value with minimum execution time. This can fulfill time reduction, which is a crucial issue for online BCI systems. This study recommends employing Renyi entropy as an alternative to CSP with multi-class approach instead of having to divide multi-class problem into binary sub problems. In future, we need to apply our proposed approach to online system to check its robustness.

References

1. Kim, Y.K., Park, E., Lee, A., Im, C.-H., Kim, Y.-H.: Changes in network connectivity during motor imagery and execution. PLoS One **13**(1), 1–18 (2018)
2. Ang, K.K., Chin, Z.Y., Wang, C., Guan, C., Zhang, H.: Filter bank common spatial pattern algorithm on BCI competition IV datasets 2a and 2b. Front. Neurosci. **6**, 39 (2012)
3. Fang, Y., Chen, M., Zheng, X., Harrison, R.F.: Extending CSP to detect motor imagery in a four-class BCI. J. Inf. Comput. Sci. **9**, 143–151 (2012)
4. Dong, E., Li, C., Li, L., Du, S., Belkacem, A.N., Chen, C.: Classification of multi-class motor imagery with a novel hierarchical SVM algorithm for brain–computer interfaces. Med. Biol. Eng. Comput. **55**(10), 1809–1818 (2017)
5. Brockmeier, A.J., Santanna, E., Giraldo, L.G.S., Principe, J.C.: Projentropy: using entropy to optimize spatial projections. In: IEEE International Conference on Acoustics, Speech and Signal Processing (2014)
6. Kee, C.-Y., Ponnambalam, S.G., Loo, C.-K.: Binary and multi-class motor imagery using Renyi entropy for feature extraction. Neural Comput. Appl. (2016). https://doi.org/10.1007/s00521-016-2178-y
7. Brunner, C., Leeb, R., Müller-Putz, G., Schlögl, A., Pfurtscheller, G.: BCI Competition 2008—Graz Data Set A. Graz University of Technology, Graz (2008)
8. Andino, S.L.G., de Peralta Menendez, R.G., Thut, G., Spinelli, L., Blanke, O., Michel, C.M., Seeck, M., Landis, T.: Measuring the complexity of time series: an application to neurophysiological signals. Hum. Brain Mapp. **11**(1), 46–57 (2000)
9. Rényi, A.: On a new axiomatic theory of probability. Acta Math. Acad. Sci. Hung. **6**(3–4), 285–335 (1955)
10. Garrett, D., Peterson, D.A., Anderson, C.W., Thaut, M.H.: Comparison of linear, nonlinear, and feature selection methods for EEG signal classification. IEEE Trans. Neural Syst. Rehabil. Eng. **11**(2), 141–144 (2003)
11. Muller, K.R., Anderson, C., Birch, G.: Linear and nonlinear methods for brain-computer interfaces. IEEE Trans. Neural Syst. Rehabil. Eng. **11**, 165–169 (2003)
12. Bashashati, A., Fatourechi, M., Ward, R.K., Birch, G.E.: A survey of signal processing algorithms in brain–computer interfaces based on electrical brain signals. J. Neural Eng. **4** (2), R32–R57 (2007)
13. Bishop, C.M.: Pattern Recognition and Machine Learning. Springer, Berlin (2006)
14. Hastie, T., Tibshirani, R.: Classification by pairwise coupling. Ann. Stat. **26**(1), 451–471 (1998)
15. Zhang, D., Wang, Y., Gao, X., Hong, B., Gao, S.: An algorithm for idle-state detection in motor-imagery-based brain-computer interface. Comput. Intell. Neurosci. (2007). https://doi.org/10.1155/2007/39714
16. Kumar, S., Ghosh, J., Crawford, M.M.: Hierarchical fusion of multiple classifiers for hyperspectral data analysis. Pattern Anal. Appl. **5**, 210–220 (2002)
17. Cohen, J.: A coefficient of agreement for nominal scales. Educ. Psychol. Meas. **20**(1), 37–46 (1960)

Sensitivity Improvement of Micro-diaphragm Deflection for Pulse Pressure Detection

Amr A. Sharawi[1(✉)], Mohamed Aouf[2], Ghada Kareem[2], and Abdelhaleim H. Elhag Osman[1]

[1] Faculty of Engineering, Cairo University, Cairo, Egypt
amrarsha@engl.cu.edu.eg, ahh.elhag@gmail.com

[2] Biomedical Engineering Department, Higher Technological Institute, 10th of Ramdan, Cairo, Egypt
maoufmedical@yahoo.com, sayd_manar2000@yahoo.com

Abstract. Cardiovascular diseases are one of the leading causes of death. Globally, they underlie the death of one third of the world's population. The main cause of cardiovascular diseases is atherosclerosis which makes arteries less elastic (called "hardening of the arteries" or "arterial stiffness"). The optical Micro Electro Mechanical System (MEMS) pressure sensor has shown its potential in the diagnosis of arterial stiffness that can be conducted by detecting the pulse pressure in the radial artery. In this paper, we attempt to improve the sensitivity of micro-diaphragm deflection in optical Micro-electromechanical System (MEMS) sensors as applied to pulse pressure detection, thus aiming to determine the safety of a person's measured pulse of cardiovascular disease. The deflection sensitivity improvement was evidenced using Finite Element Analysis ANSYS software. Corrugation for periphery-clamped silicon nitride (Si_3N_4) micro-diaphragm based on the variation of the diaphragm thickness (t_d) and some corrugation factors such as the corrugation angle (β) and the corrugation depth (h_c) was implemented to reduce bending and tensile stresses which limit the micro-diaphragm deflection sensitivity. This was supported by calculating the von Mises stress. Analytic results show agreement with ANSYS software simulation with a static response of 1.27 μm maximum deflection under applied pressure of 300 mmHg in the case of the corrugated micro-diaphragm, compared to a 0.32 μm maximum deflection in the case of a flat micro-diaphragm, and for the same applied pressure, a maximum deflection sensitivity of 4.23×10^{-3} μm/mmHg for the corrugated micro-diaphragm compared to 1.07×10^{-3} μm/mmHg for the flat one, and the reduction of micro-diaphragm bending and initial tensile stresses exhibited by maximum equivalent stress (von Mises stress) of 159.99 MPa for the corrugated compared to 175.9 MPa for the flat one. Therefore, the implementation of corrugation presents the chance to control mechanical deflection sensitivity and compared to the film deposition process control it is often an easier way.

Keywords: Bending · Corrugated · Deflection sensitivity
Finite Element Analysis · Micro-diaphragm · von Mises stress

A. E. Hassanien et al. (Eds.): AISI 2018, AISC 845, pp. 137–151, 2019.
https://doi.org/10.1007/978-3-319-99010-1_13

1 Introduction

Cardiovascular diseases (CVDs) consider tobe number one of the major causes of death. Globally, one third of the world's population death is due to CVDs [1]. It has been shown that atherosclerotic changes in arterial wall include deposition of lipid, accumulation of collagen, smooth muscle cell proliferation, elastin, and proteoglycans [2, 3]. Less blood can flow through as an artery becomes more and more narrowed. It may also become less elastic (called "arterial stiffness" or "hardening of the arteries") [4].

The rigidity of the arterial wall is described by a generic term known as the arterial stiffness. Most of the research is based on the elastic theory in the hardening of the arteries; regardless of the fact that this requires the studied body to be elastic rather than viscoelastic, similar to the case of the arterial wall [5–9]. Measuring the shape of the arterial pressure waveform allows additional parameters of cardiovascular to be assessed. This include arterial stiffness that is a manifestation of atherosclerosis [10–13].

2 Background of Pressure Measurement Tools

In the recent years several techniques have been used. These include Computed Tomography (CT), Magnetic Resonance Imaging (MRI), angiography, Intravascular Ultrasound (IVUS), and biomedical pressure sensors [14, 15]. Using MRI for Diagnosing is considered to be time consuming even for a radiologists with very high skills because MRI images are noisy [16]. CT imaging, though very promising and noninvasive utilizes high radiation exposure, besides the effect of IV contrast material (CM), the requirement to suppress the heart rate to 65 bpm or less, and its low sensitivity for small vessels [17]. Although angiography is the best way for accurate assessment of the severity and the presence of vascular diseases, it is considered to be highly costly, compared to the other diagnosis techniques, and invasive [18, 19]. Regarding IVUS, only a technician who is trained in interventional cardiology techniques can perform with IVUS, and it is time consuming [20]. Biomedical sensors can range from noninvasive to invasive. Noninvasive sensors for example include radiant heat sensors or sensors of sound energy coming from an organism [21]. Nowadays, there is an urgent need for a miniature ultra-low semiconductor pressure transducer. The rapid expansion of Biological Micro-electromechanical Systems (BioMEMS) has allowed miniaturization of sensors for arterial pressure measurements, especially using sensing principles such as capacitive and piezoresistive [22].

2.1 MEMS Based Pressure Sensors

MEMS pressure sensors can be categorized primarily into capacitive, piezoresistive [23], and other sensing principles include fiber optic sensing. So far piezoresistive techniques are used in the most commercial MEMS pressure sensors due to the inherent linear resistance-pressure relation. Nevertheless, the main problem associated with the piezoresistive pressure sensor is its inherent cross sensitivity to temperature [24]. Other disadvantages are the very strict requirement of accurately placing the sensing resistors for maximum sensitivity and the large power consumption [24–26]. Capacitive sensors have a large nonlinearity. But they are less sensitive to the variation of the temperature due to the displacement based measurement mechanism [27, 28].

Fiber optic sensors used for pressure measurement (FOPS) hold a lot of inherent advantages including (i) wide range of measured potential, (ii) immunity to electromagnetic interference, (iii) remote sensing capability, (iv) high resolution, (v) avoidance of posing a spark source hazard for flammable environment applications, and (vi) high reliability [29]. We classify fiber optic sensors into four main groups including: wavelength-modulated, intensity-based, polarization-modulated, and phase-modulated (or interferometric). Each of these groups' applications are rapidly expanding [29–32]. Polarization-modulated fiber optic sensors mainstream a photoelastic and a Faraday effect. An external photoelastic pressure sensor may achieve 0.2% accuracy, with a very good self-compensation mechanism, which must be built at the same place as the external sensing element. This makes the head of the sensor very bulky [29]. The fiber grating based sensor is the most popular wavelength-modulated FOPS [33]. Its advantages are the immunity to optical power loss variation and the multiplexing capability of many sensors to share the same signal processing unit. However, fiber grating sensors long-term reliability has been a concern due to the degradation in optical properties and mechanical strength when the grating is exposed to a high pressure and temperature environment. Intensity-based (microbend) FOPS are simple and require only modest signal processing through a direct detection of the change in optical power whether in reflection or transmission. Although the reported resolution is typically better than 0.1%, the large hysteresis and the fluctuation of power associated with the optical source and fiber loss heavily reduce their [29]. There are three interferometric sensors that have been developed into pressure sensors with a very high resolution of 0.01% [29]. However, because of the very low stress-optic coefficients, a very long sensing fiber is necessary to obtain the desirable sensitivity, which unavoidably makes the sensor thermally unstable. A Fabry-Perot interferometer is considered to be a very simple device that is based on multiple beam reflections [34, 35].

2.2 Pressure Measuring Principle

A micro-scale pressure sensor can be most valuable for precise biomedical pressure measurements. It consists of an optical pulse sensor, a micro-diaphragm structure used as pressure transducer, and an optical fiber. Optical interferometry is used to transduce the deflection of the diaphragm, thus reducing long-term drift.

Diaphragm deformation is caused by the pulse pressure that is sensed from the radial artery surface, thus changed the reflected or transmitted spectrum. The reflected light strikes the diaphragm, which is in contact with the skin on the radial artery. The sensor cavity length, the corresponding pressure and deflection, all can be measured from the reflected spectrum. Information is provided from the reflected waveform based on the elasticity of the artery and the arterial stiffness [11, 13].

3 Aim of Study

The focus of this study is on optical corrugated micro-diaphragm modelling for the detection of human pulse pressure. The optical MEMS sensor has been chosen for its advantages regarding high sensitivity, high resolution, immunity to electromagnetic interference, and intrinsic electrical passivity [29–32]. The effects of diaphragm corrugation, thickness and radius on static and dynamic responses are analyzed.

4 Materials and Methods

Pressure is measured by revealing the reflected shift of transmitted spectrum of a light source (an LED), which emits light that will then be transmitted via optical fibers as shown in Fig. 1 [7].

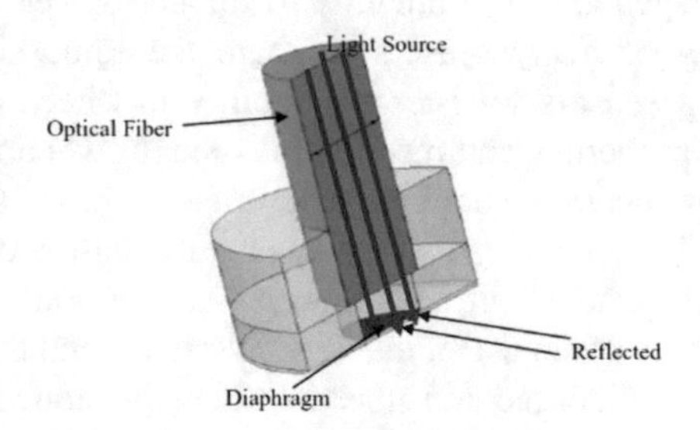

Fig. 1. Proposed design of the optical MEMS sensor (Reproduced from [7])

Sensor measurements of drift and hysteresis support the suitability of silicon nitride (Si_3N_4) for microdevices. Si_3N_4 is a fundamental material for the fabrication of MEMS, employed as structural layer for the device's micromechanical parts, as beams, or membranes, due to its good mechanical properties (i.e. large Young's modulus, internal stress, and high yield strength tunable by the process of deposition) [36]. Our sensor characteristics are shown in Fig. 2 [7, 37].

The diaphragm y out-of-plane deflection is a function of the radial distance r and the pressure P difference, as given by Eq. (1) [38, 39].

$$y = \frac{3(1 - \mathrm{v}^2)P}{16E \cdot t_d^3}\left(\mathrm{R}^2 - \mathrm{r}^2\right)^2 \tag{1}$$

Where v = Poisson's ratio, R = effective diaphragm radius, and t_d = diaphragm thickness.

At the center of the diaphragm the maximum deflection is ($r = 0$), given by y_c can be written as:

$$y_c - \frac{3(1 - \mathrm{v}^2)P}{16E \cdot t_d^3} \cdot R^4 \tag{2}$$

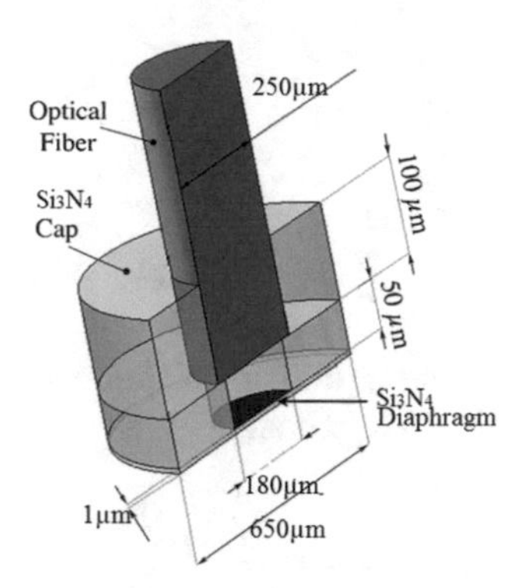

Fig. 2. The 3D cross-section schematic of sensor. The sensor consists of a Si_3N_4 cap and a fiber optic cable. The three layers of the cap form the 180 μm diameter, 1 μm thick diaphragm, a Fabry-Perot cavity approximately 50 μm in length (Reproduced from [7])

Clearly, from Eq. (2) the increment of the diaphragm deflection is closely dependent on radius increment and thickness decrement, but since a normal human artery diameter range from 2.5 to 3 mm, the sensor overall diameter must be smaller than the radial artery. Now that the overall proposed sensor diameter is about 650 μm as shown in Fig. 2, we fix the diaphragm radius to 90 μm in order to optimize the sensor corresponding to an appropriate radial artery diameter after sensor capsulation. Next, we proceed to vary the diaphragm thickness with other parameters, such as corrugation angle and depth to show their effects on the diaphragm deflection, which are the main objective of this paper.

The sensitivity of pressure (Y_c) can be defined as the ratio between the pressure difference and the deflection [21, 37]. Sensitivity is usually used in evaluating the diaphragm performance. The pressure sensitivity and the maximum deflection can be analyzed when the optical fiber is positioned facing the diaphragm center. The deflection at the diaphragm center, for a Si_3N_4 diaphragm can be calculated by Eq. (3) [7]:

$$y_c = 1.145 \times 10^{-8} \frac{R^4}{t_d^3} \cdot P \tag{3}$$

Where y_c, R, and t_d are in meters, P is measured in Pascal.

4.1 Stress Analysis

Diaphragm failure under pressure is best studied using stress analysis [29]. The bending stresses increase linearly with diaphragm thickness to the outer surfaces where their maximum values are reached. Because of the doubly curved surface the diaphragm is bent into, stress analysis must take into account the tangential and radial stresses. For any radial distance, the radial and tangential may be calculated from Eqs. (4) and (5), respectively:

$$\sigma_r = \pm\frac{3}{8}P\frac{R^2}{t_d^2}\left[(3+\nu)\frac{r^2}{R^2} - (1+\nu)\right] \tag{4}$$

$$\sigma_t = \pm\frac{3}{8}P\frac{R^2}{t_d^2}\left[(1+3\nu)\frac{r^2}{R^2} - (1+\nu)\right] \tag{5}$$

The radial and tangential stress maximum value occur at the edge and center respectively. They are given by Eqs. (6) and (7).

$$\sigma_{rmax} = \pm\frac{3}{4}P\frac{R^2}{t_d^2} \tag{6}$$

$$\sigma_{tmax} = \pm\frac{3}{8}P\frac{R^2}{t_d^2}(1+\nu) \tag{7}$$

Since the disadvantage of the periphery-clamped circular diaphragm is the possibility of a high stress in the film deposition which reduces the diaphragm sensitivity [40], the question to appear now is "how can these stresses be alleviated?"

5 The Modeling of a Corrugated Micro-diaphragm Excited by Pressure Pulse to Reduce the Diaphragm Deposition Stresses

The circular diaphragm is the critical part of an optical MEMS sensor, which is an out-of-plane movable microstructure driven by the force of pulse pressure. The disadvantage of a periphery-clamped circular diaphragm is the possibility of high stress in the film. Corrugation is considered an effective way to alleviate the diaphragm's high residual stress and to optimize its sensitivity of deflection [41, 42]. In our work an analytical model of the corrugation was mainly developed to validate the obtained results based on Finite Element Analysis (FEA). The diaphragm 3D model takes into consideration the corrugation factors to enhance the performance of the device. To model the solid interaction between the corrugated micro-diaphragm and optical device coupled-field analysis is carried out. Transient and static methods of analysis are implemented in the micro-diaphragm analysis. The relationship between different factors of corrugation and stress distribution is also addressed. Performance prediction allows to accurately estimate both yield and life time and to develop a better understanding of the device, while taking into account the device structural details.

Simulation of diaphragm based optical pressure sensors using FEA tools has become more popular. Alternatively, analytical models have also been used for the analysis of sensing mechanisms of the diaphragm. Both of these models facilitate the design of a model with optimal operating conditions prior to device fabrication [43].

The main characteristics of a diaphragm are ruggedness, excellent stability and reliability, low hysteresis and creep, and good dynamic response. Other properties such as the impact of environmental conditions, weight, size, diaphragm material, and fabrication technology vary greatly, depending on the principle of transduction in use, range, and actual application [43]. Due to diaphragm low stress and high displacement requirement the implementation of corrugation is considered to be an efficient way to mitigate the diaphragm residual stress [40]. Figure 3 shows the model of corrugated diaphragm.

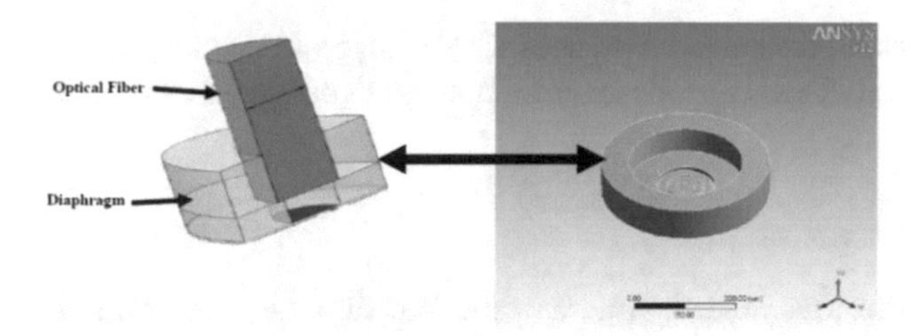

Fig. 3. The corrugated micro-diaphragm in optical MEMS sensor (Reproduced from [40])

Shallowly corrugated diaphragms have demonstrated improved sensitivity especially for high residual stress level compared to the conventional flat diaphragms [43]. Although shallow corrugation can adequately reduce diaphragm membrane and bending rigidity, these may increase largely because of the diaphragm increased "equivalent thickness" [43]. The improvement effect usually is limited by the trade-off, especially for the diaphragm of low or moderate stress level.

5.1 Corrugated Diaphragm Analysis

Corrugations in a diaphragm enable operation at larger displacements with improved linearity. They can have sinusoidal, triangular, rectangular, trapezoidal, and toroidal profiles. While this has a small influence on the behavior of the diaphragm, the depth of corrugation (h_c), material thickness (t_d), and wavelength (λ) are the main factors which contribute to the initial stress reduction of the Si_3N_4 microdiaphragm to improve its mechanical deflection sensitivity.

5.2 von Mises Stress

It is vital to gain knowledge about the yield strength σ_y (yield point), since it generally represents an upper limit to the load, which can be applied. When the material's von Mises stress reaches this critical value, then the material is said to start yielding. Therefore, analyzing this stress is preferred [44]. The von Mises stress is given by:

$$\sigma_{vm} = \sqrt{\left(\sigma_{xx} - \sigma_{yy}\right)^2 + \left(\sigma_{yy} - \sigma_{zz}\right)^2 + \left(\sigma_{zz} - \sigma_{xx}\right)^2 + 6\left(\sigma_{xy}^2 + \sigma_{yz}^2 + \sigma_{zx}^2\right)} \quad (8)$$

Equation (8) describes the calculation of the von Mises effective stress from the stress tensor components. The critical value for von Mises stress is [44]:

$$k = \frac{\sigma_y}{\sqrt{3}} \tag{9}$$

5.3 Corrugated Diaphragm Design

Figure 4 represent the corrugation used for the suggested optical MEMS sensor (graph plotted using ANSYS 12). The micro-diaphragm corrugation parameters are given in Table 1.

5.4 Static Response

The diaphragm was loaded with a static pressure ranging from 0 to 300 mmHg (40 kPa) [7, 45], in order to have deposited films with better sidewall coverage for the corrugated diaphragm [40]. The corrugation profile was positively tapered with an angle of 140° that can be realized by deep reactive ion etching (DRIE) in the process of fabrication [40] with optimized process parameters.

The residual stress reduction ratio (η) in the presence of corrugation is found by:

$$\eta = \frac{\sigma_o}{\sigma} = 1 + 6\sin(\beta)\left(\frac{h_c}{t_d}\right)^2 \cdot \frac{N_c w_c}{R - N_c(w_c + b_c)} \tag{10}$$

Where σ_o is the initial stress of a flat membrane without corrugation and σ is the equilibrium stress of the membrane with corrugations.

Finally, using the following formula the mechanical deflection sensitivity (Y_c) of a circular corrugated diaphragm can be calculated:

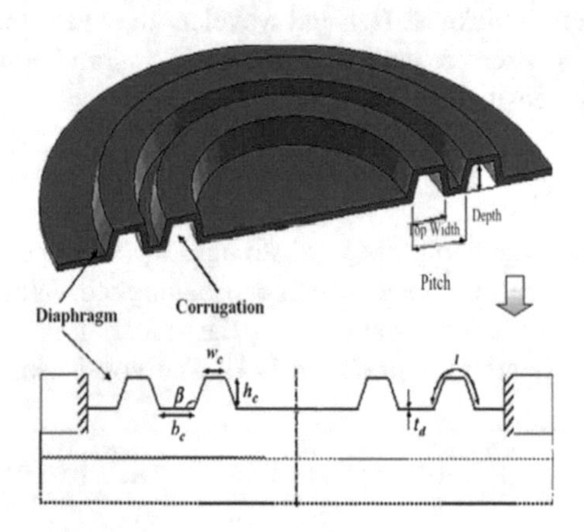

Fig. 4. Proposed profile of corrugated micro-diaphragm. b_c = top corrugation width, h_c = depth, β = corrugation angle (Reproduced from [40])

$$Y_c = \frac{dy}{dP} = \frac{R^2}{4t_d\left[\sigma_o \frac{b_p}{2.83} + \frac{a_p}{4}\left(\frac{t_d}{R}\right)^2\right]} \tag{11}$$

Where a_p and b_p are corrugation parameters defined in [40].

Table 1. Micro-diaphragm corrugation parameters

Symbol	Parameter	Typical value
R	Diaphragm radius	90 μm
t_d	Diaphragm thickness	1 μm
b_c	Width of bottom corrugation	7 μm
λ	Pitch of corrugation	13 μm
w_c	Width of top corrugation	3.14 μm
h_c	Depth of corrugation	12 μm
β	Profile angle of corrugation	140°
N_c	Number of corrugations	2

5.5 Dynamic Response

For the dynamic response, the input signal is given by a loaded pressure variation tabulated in Table 2 [46].

Table 2. Pulse pressure values for a normal and an atherosclerotic patient

Time (ms)	Normal case (mmHg)	Atherosclerotic patient (mmHg)
0	64	74
100	117	137
200	81	112
300	76	99
400	81	101
500	71	90
600	67	81
700	63	78
800	61	74

6 Results and Discussion

6.1 Static Loading

Figure 5 illustrates the increased deformation of the corrugated diaphragm compared to the flat diaphragm without corrugation, showing a maximum deformation of 0.0254 μm for the corrugated diaphragm compared to approximately 0.024 μm for the flat one at a thickness of 4 μm. But, since the diaphragm thickness is relatively large compared to its radius, this improved deformation is not strongly affected by the corrugation. Yet a relatively improved deflection can be obtained by decreasing the diaphragm thickness.

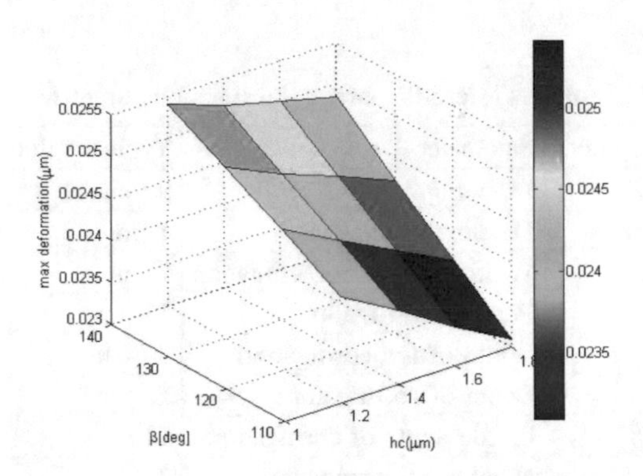

Fig. 5. The relationship between corrugation angle, depth and deformation at a thickness of 4 μm (produced using ANSYS 12)

At a thickness of 3 μm the maximum deformation for the corrugated diaphragm was 0.058 μm compared to 0.051 μm for the flat one. A maximum deformation of 0.184 μm was observed for the corrugated compared to 0.15 μm for the flat diaphragm at a thickness 2 μm. A thickness of 1 μm strongly exhibited a very good deformation for the corrugated diaphragm with maximum deformation of 1.27 μm compared to a maximum deformation of 0.32 μm for the flat diaphragm.

Internal stress reduction increases the sensitivity of the mechanical deflection of the small corrugation depths. Nevertheless, the mechanical sensitivity dramatically decreases from too large a corrugation depth because of the larger flexural rigidity in the tangential direction for a diaphragm with very deep corrugation. This increases the diaphragm stiffness [40, 47]. Thus, there is an optimum corrugation depth with maximum sensitivity. When these concepts were applied the resulting maximum deflection sensitivities turned out to be 3.73×10^{-4} μm/mmHg (2.80×10^{-3} μm/kPa), 4.48×10^{-4} μm/mmHg (3.36×10^{-3} μm/kPa), 0.000736 μm/mmHg (5.72×10^{-3} m/kPa), and 2.13×10^{-3} μm/mmHg (0.016 μm/kPa) at corrugation depths of 7, 5, 3, and 1 μm, respectively (Figs. 5, 6 and 7, produced using ANSYS 12). Clearly, the possibility to control mechanical deflection sensitivity can be reached by implementing corrugation, and compared to the approach of the film deposition process control; it is often consider an easier way.

The von Mises stress was used in FEM analysis to describe whether or not the material had reached a state of failure. It comes out that the reduction of the initial stresses was achieved by the von Mises stress distribution; the thinner the diaphragm the higher the stress and the deeper the corrugation the higher the reduction of the stress.

At a thickness of 4 μm von Mises stress has a maximum value 14.94 MPa at 1.2 μm depth and a proposed diaphragm angle of 130°. Reducing the thickness to 3 μm will increase the stress up to a value of 26.635 MPa at 1.2 μm depth and 110° angle. A diaphragm with thickness of 2 μm was shown to illustrate a relatively high initial stress, reduced by the corrugation and explored by von Mises stress that also reached a maximum of 45.5035 MPa at 1.2 μm depth and 120° angle of proposed diaphragm. The higher initial stress was reduced effectively at a thickness 1 μm; this was also implemented by corrugation. The maximum von Mises stress illustrating this reduction for the corrugated diaphragm was 159.99 MPa at 1.2 μm depth and 140° angle of proposed diaphragm, compared to 175.5 MPa for the flat diaphragm.

6.2 Dynamic Loading

Regarding dynamic response, as depicted in Fig. 8, micro-diaphragm deformations for the two cases according to applied pressures, showed maximum deflections of 0.54 and 0.63 μm for the normal and atherosclerotic cases, respectively.

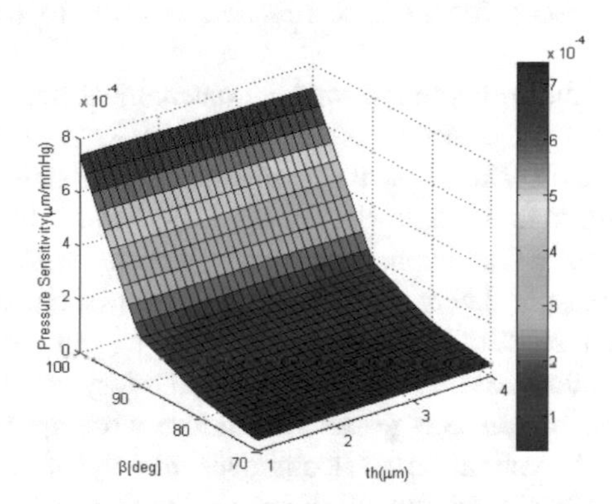

Fig. 6. The relationship between corrugation angle, thickness and deflection sensitivity at corrugation depth of 3 μm

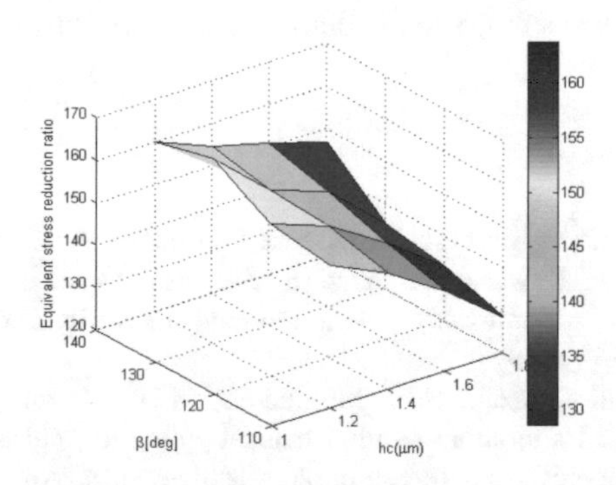

Fig. 7. The relationship between corrugation angle, depth and equivalent stress at a thickness of 1 μm

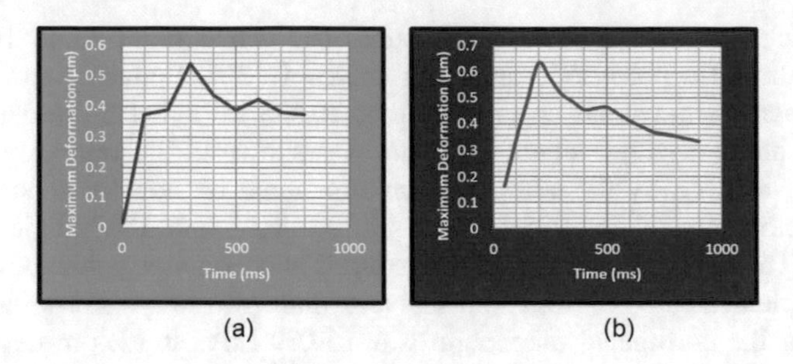

Fig. 8. Micro-diaphragm dynamic deformation for (a) normal and (b) atherosclerotic

7 Conclusions and Recommendations

The literature reports that most of the micro-sensors for pressure pulse detection still need to be incorporated into practical devices. This paper attempts to address these issues by designing a pressurized micro-diaphragm used in optical micro-sensors to detect human pulse pressure using a corrugation method to increase its deflection sensitivity.

The design of the micro-diaphragm and its mechanical analysis were carried out using simulation tools with a strong emphasis on FEA to analyze the mechanical behavior of the diaphragm. Based on the computational modeling results, a circular-shaped corrugated micro-diaphragm made of Si_3N_4 was able to generate larger deflections compared to a flat diaphragm. Hence, arterial stiffness diagnosis can be performed by pulse pressure detection. The development of corrugation techniques to improve the deflection sensitivity of the diaphragm of optical MEMS sensors for noninvasive pulse pressure measurement on the radial artery was successfully achieved to meet this study contribution. Suggestions for future work springing from this study may include: (1) the investigation of techniques for dynamic calibration, (2) the assessment of different types of optical signal processing to improve sensor system performance, (3) the investigation of the possibility of using MEMS sensors for different applications, such as acoustic imaging, (4) the fabrication and testing of device structures using optimal tools, and (5) the consideration of the possibility of using multiple sensors in one system, such as temperature and glucose test sensors.

References

1. WHO: Global Atlas on Cardiovascular Disease Prevention and Control. (2011)
2. Hirai, T., Sasayama, S., Kawasaki, T., Yagi, S.: Stiffness of systemic arteries in patients with myocardial infarction. A noninvasive method to predict severity of coronary atherosclerosis. Circulation **80**, 78–86 (1989)
3. Nguyen, P.H., Coquis-Knezek, S.F., Mohiuddin, M.W., Tuzun, E., Quick, C.M.: The complex distribution of arterial system mechanical properties, pulsatile hemodynamics, and vascular stresses emerges from three simple adaptive rules. Am. J. Physiol. Heart Circ. Physiol. **308**(5), 407–415 (2015)

4. Latifoglu, F., Sahan, S., Kara, S., Gunes, S.: Diagnosis of atherosclerosis from carotid artery Doppler signals as a real-world medical application of artificial immune systems. Expert Syst. Appl. **33**, 786–793 (2007)
5. Mac-Way, F., Leboeuf, A., Agharazii, M.: Arterial stiffness and dialysis calcium concentration. Int. J. Nephrol. **10**, 1–6 (2011)
6. Claridge, M.W., Bate, G.R., Hoskins, P.R., Adam, D.J., Bradbury, A.W., Wilmink, A.B.: Measurement of arterial stiffness in subjects with vascular disease: are vessel wall changes more sensitive than increase in intima-media thickness? Atherosclerosis **205**(2), 477–480 (2009)
7. Hasikin, K., Soin, N., Ibrahim, F.: Modeling of an optical diaphragm for human pulse pressure detection. WSEAS Trans. Electron. **5**, 447–456 (2008)
8. Claridge, M., Wilmink, T., Ferring, M., Dasgupta, I.: Measurement of arterial stiffness in subjects with and without renal disease: are changes in the vessel wall earlier and more sensitive markers of cardiovascular disease than intima media thickness and pulse pressure? Indian J. Nephrol. **25**, 21–26 (2015)
9. Ngajikin, N.H., Linga, L.Y., Ismail, N.I., Supaát, A.S.M., Ibrahim, M.H., Kassim, N.M.: CMOS-MEMS integration in micro perot pressure sensor fabrication. Jurnal Tknologi **64**(3), 83–87 (2013)
10. Nissilä, S., Sorvisto, M., Sorvoja, H., Vieri-Gashi, E., Myllylä, R.: Noninvasive blood pressure measurement based on the electronic palpation method. In: Proceedings of 20th Annual International Conference of the IEEE Engineering in Medicine and Biology Society, Chicago, USA, vol. 20, no. 4, pp. 1723–1726 (1998)
11. Wilkinson, I.B., Hall, I.R., MacCallum, H., Mackenzie, I.S., McEniery, C.M., Van der Arend, B.J., Yae-Eun, S., Mackay, L.S., Webb, D.J., Cockcroft, J.R.: Pulse wave analysis clinical evaluation of a noninvasive, widely applicable method for assessing endothelial function. Arterioscler. Thromb. Vasc. Biol. **22**, 147–152 (2002)
12. Sorvoja, H., Myllylä, R., Koskenka, P.K., Koskenkari, J., Lilja, M., Kesänimi, Y.A.: Accuracy comparison of oscillometric and electronic palpation blood pressure measuring methods using intra-arterial method as a reference. Mol. Quantum Acoust. **26**, 235–260 (2005)
13. Stoner, L., Young, J.M., Fryer, S.: Assessments of arterial stiffness and endothelial function using pulse wave analysis. Int. J. Vasc. Med. **2012**, 9 p. (2012)
14. Madjid, M., Zarrabi, A., Litovsky, S., Willerson, J.T., Casscells, W.: Finding vulnerable atherosclerotic plaques: is it worth the effort? Arterioscler. Thromb. Vasc. Biol. **24**, 1775–1782 (2004)
15. Larsen, P.J., Waxman, S.: Intracoronary thermography: utility to detect vulnerable and culprit plaques in patients with coronary artery disease. Curr. Cardiovasc. Imaging Rep. **2**, 300–306 (2009)
16. Hassan, M.A., El-Shabrawy, N., Mohamed, A.S., Youssef, A.M., Kadah, Y.M.: Signal processing in functional MRI: robust suppression of random and physiological noise components using threshold spectrum estimation and harmonics analysis. In: Proceedings of Cairo International Biomedical Engineering Conference (2006)
17. Morin, R.L., Gerber, T.C.: Radiation dose in computed tomography of the heart. Circulation **107**, 917–922 (2003)
18. Hoffmann, U., Ferencik, M., Cury, R.C., Pena, A.J.: Coronary CT angiography. J. Nucl. Med. **47**, 797–806 (2006)
19. Budoff, M.J., Achenbach, S., Duerinckx, A.: Clinical utility of computed tomography and magnetic resonance techniques for noninvasive coronary angiography. J. Am. Coll. Cardiol. **42**(11), 1867–1878 (2003)

20. Yamagishi, M., Terashima, M., Awano, K., Kijima, M., Nakatani, S., Daikoku, S., Ito, K., Yasumura, Y., Miyatake, K.: Morphology of vulnerable coronary plaque: insights from follow-up of patients examined by intravascular ultrasound before an acute coronary syndrome. J. Am. Coll. Cardiol. **35**, 106–111 (2000)
21. Dorf, R.C.: Sensors, Nanoscience, Biomedical Engineering, and Instruments. University of California Davis, Davis (2006)
22. Harsányi, G.: Sensors in Biomedical Applications: Fundamentals, Technology and Applications. CRC Press, Boca Raton (2000)
23. Eddy, D.S., Sparks, D.R.: Application of MEMS technology in automotive sensors and actuators. Proc. IEEE **86**, 1747–1755 (1998)
24. Fraga, M.A., Pessoa, R.S., Maciel, H.S., Massi, M.: Recent Developments on Silicon Carbide Thin Films for Piezoresistive Sensors Applications. University of Vale do Paraiba, São Paulo (2010)
25. Song, Y., Lee, H., Esashi, M.: A corrugated bridge of low residual stress for RF-MEMS switch. Sens. Actuators **135**, 818–826 (2007)
26. He, R., Yang, P.: Giant piezoresistance effect in silicon nanowires. Nat. Nanotechnol. **1**, 42–46 (2006)
27. Puers, R.: Capacitive sensors: when and how to use them? Sens. Actuators A **37–38**, 93–105 (1993)
28. Chena, L., Mehreganyb, M.: A silicon carbide capacitive pressure sensor for in-cylinder pressure measurement. Sens. Actuators A **145–146**, 2–8 (2008)
29. Xu, J., Wang, X., Cooper, K.L., Pickrell, G.R., Wang, A.: Miniature fiberoptic optic pressure and temperature sensors. In: Proceedings of SPIE, Optics East Boston, MA, vol. 6004 (2005)
30. Totsu, K., Haga, Y., Esashi, M.: Ultra-miniature fiber-optic pressure sensor using white light interferometry. J. Micromech. Microeng. **15**, 71–75 (2005)
31. Tohyama, O., Kohashi, M., Sogihara, M., Itoh, H.: A fiber-optic pressure microsensor for biomedical application. Sens. Actuators **66**, 150–154 (1998)
32. Pinet, É.: Medical applications: saving lives. Nat. Photonics **2**, 150–152 (2008)
33. Wang, A., He, S., Fang, X., Jin, X., Lin, J.: Optical fiber pressure sensor based on photoelasticity and its application. J. Lightwave Technol. **10**, 1466–1472 (1992)
34. Samoriski, B.P.: Fabry Perot Interferometers Theory (2000). http://www.burleigh.com/Pages/opticalTech.htm
35. Yu, B., Kim, D.W., Deng, J., Xiao, H., Wang, A.: Fiber Fabry Perot sensors for detection of partial discharges in power transformers. Appl. Opt. **42**(16), 3241–3250 (2003)
36. Cianci, E., Schina, A., Minotti, A., Quaresima, S., Foglietti, V.: Dual frequency PECVD silicon nitride for fabrication of CMUTs membranes. Sens. Actuators **127**, 80–87 (2006)
37. Hill, G.C., Melamudb, R., Declercq, F.E., Davenport, A.A., Chanc, I.H., Hartwell, P.G., Pruitt, B.L.: SU-8 MEMS Fabry-Perot pressure sensor. Sens. Actuators **138**, 52–62 (2007)
38. Scheeper, P.R., Olthuis, W., Bergveld, P.: The design, fabrication, and testing of corrugated silicon nitride diaphragms. J. Microelectromech. Syst. **3**, 36–42 (1994)
39. Wang, W.J., Lin, R.M., Li, X., Guo, D.G.: Study of single deeply corrugated diaphragms for high-sensitivity microphones. J. Micromech. Microeng. **13**, 184 (2003)
40. Ke, F., Miao, J., Wang, Z.: A wafer-scale encapsulated RF MEMS switch with a stress-reduced corrugated diaphragm. Sens. Actuators **151**, 237–243 (2009)
41. Kressmann, R., Klaiber, M., Hess, G.: Silicon condenser microphones with corrugated silicon oxide nitride electret membranes. Sens. Actuators **100**, 301–309 (2002)
42. Jeong, O.C., Yang, S.S.: Fabrication of a thermopneumatic microactuator with a corrugated p+ silicon diaphragm. Sens. Actuators **80**, 62–67 (2000)
43. Dissanayake, D.W., Al-Sarawi, S.F., Abbott, D.: Wireless interrogation of a micropump and analysis of corrugated micro-diaphragms. Recent Adv. Sens. Technol. **49**, 241–256 (2009)

44. Barkan, E.D.: Tool geometry effects in metal shearing using FEM, Thesis of Masters of Science in Mechanical Engineering, Montana State University Bozeman, Montana (2011)
45. Hasikin, K., Ibrahim, F., Soin, N.: Determination of design parameters of a biosensor for human artery pulse wave detection. Biomed. Proc. **21**, 707–710 (2008)
46. Mahmud, A., Feely, J.: Effects of passive smoking on blood pressure and aortic pressure waveform in healthy young adults—influence of gender. Br. J. Clin. Pharmacol. **57**, 37–43 (2003)
47. Beeby, S., Ensell, G., Kraft, M., White, N.: MEMS Mechanical Sensors. Artech House Inc., Boston (2004)

Active Suspension System Design Using Fuzzy Logic Control and Linear Quadratic Regulator

Ahmed A. Abdeen[1(✉)], Khalil Ibrahim[2], and Abo-Bakr M. Nasr[2]

[1] Arab Contractors Company, Assiut Branch, Cairo, Egypt
Success_is_my_target.eng@hotmail.com
[2] Department of Mechanical Engineering,
Faculty of Engineering, Assiut University, Assiut, Egypt
Khalil.ibrahim@aun.edu.eg, Abobakr.nasr@eng.au.edu.eg

Abstract. The motor vehicle industry has shown a mechatronics system with intelligent control systems. Mechatronics refers to a successful combination of mechanical and electronic systems. In mechatronics, traditional systems of mechanical engineering are combined together with components from computer science, mathematics and electrical engineering. This paper presents enhancing an active suspension for a quarter car model to improve its performance by applying a specific controller. Separating a vehicle's body from road abnormalities is the major purpose of a suspension system, in order to provide the maximum ride comfort for passengers and keep hold of continuous road wheel contact to provide road holding. First controller applied is fuzzy logic controller (FLC), and the second one is a Linear Quadratic Regulator, the car's behaviour such as car body displacement, suspension deflection, and wheel travel is considered to obtain maximum damping force in the actuator. A comparative study has been verified to get the best performance for comfort of passenger ride and road managing.

Keywords: Active suspension system · Fuzzy logic control (FLC)
Linear quadratic regulator (LQR) · Servo valve · Quarter car model
Sprung mass · Unsprung mass

1 Introduction

Luxury automobile companies like Volvo, Mercedes-Benz and BMW has been using active suspension system as a part of their designed manufacture. Because of the active suspension systems' high cost, power supply requirements and difficulty, that's why barely luxury cars manufacture it until nowadays. Allen constructed a quarter-car modeled test to apply different control methods on suspension systems [1]. Therefore, there were different designs applied since the early days of vehicles [2]. Methods of control for an active suspension system were developed by different scientists [3–8]. These research results can be classified depending on the used control methods. The definition of the word "Suspension" is referrers to the mechanical parts, that attaches a car chassis with axles and wheels. The performance of the suspension system depends on achieving passenger comfort while driving the vehicle, road holding, and handling [9].

A. E. Hassanien et al. (Eds.): AISI 2018, AISC 845, pp. 152–166, 2019.
https://doi.org/10.1007/978-3-319-99010-1_14

Passenger comfort and safety is the main task of a designed suspension system, therefore designs usually begin with this mission. In the last decades, active suspension systems have played a main task in the academic world and manufacture. Passive and active suspensions are the main types of an automobile suspension system. A Passive suspension system is reliable and easier and less expensive than active suspension, but its performance is limited, and it fails in the aspect of comfort. But it can properly be balanced through active suspension systems, because the active suspension system is able to regulate, control parameters in order to achieve road unevenness [10, 11].

Lately, a lot of theories were suggested to take control of the suspension problems In this paper, fuzzy logic control and Linear, quadratic regulator is planned as a tactic to control active suspension systems and simulation process was applied by MATLAB/SIMULINK. Numerous designers have chosen fuzzy logic control (FLC) to be an other control technique for the Active Suspension systems (ASS) [12–14]. In this paper, the considered model for the suspension system is the Quarter car model (QCM), the analysis for this system depends on two types of masses, sprung and unsprung mass [15]. Inactive suspension system Electro-hydraulic servo valves are the active parts in the system. Because of their high accuracy, they are used in applications which precise position control is occurring and they have the main effect on the whole electro-hydraulic servo systems' operation [16, 17].

This paper has studied the effect of road irregularities in order to achieve the optimal damping force as resultant with enhancement of passenger comfort, riding stability and road handling, and this is achieved by applying different control strategies which are FLC and LQR and a PID Controller.

2 Theoretical and Simulation Models

A QCM is tested in this paper as an alternative of a full car's model, two degrees of freedom is considered. The QCM shows the main characteristics of a full model and also simplifies the analysis, Fig. 1, shows a simplified block for the QCM and Table 1 shows the values for the system parameters. Assuming that the wheels have a continuous road contact and that Z_s and Z_u are calculated from static equilibrium point. We also assume that the driver is driving with constant speed, to design the road input. Equations of the systems are:

$$m_s\ddot{Z}_s = k_s(Z_u - Z_s) + b_s(\dot{Z}_u - \dot{Z}_s) + Fa \quad (1)$$

$$m_u\ddot{Z}_u = -k_s(Z_u - Z_s) - b_s(\dot{Z}_u - \dot{Z}_s) - k_t(Z_r - Z_u) - Fa \quad (2)$$

Where Fa stands for the input control variable from the Electro-Hydraulic Servo Valve and Z_r is the input disturbance from the road and it can be modelled as white noise. The state-space variables are identified as: $x_1 = Z_s - Z_u$ Sprung-Mass Displacement.

$x_2 = \dot{Z}_s$ Sprung-Mass Absolute Velocity.
$x_3 = Z_u - Z_r$ Unsprung-Mass Displacement (Tire Deflection).
$x_4 = \dot{Z}_u$ Unsprung-Mass Absolute Velocity.

Equation of motion is described as:

$$\dot{x} = Ax + Bu \tag{3}$$

$$y = Cx + Du \tag{4}$$

$$A = \begin{bmatrix} 0 & 1 & 0 & 0 \\ -ks/ms & -bs/ms & ks/ms & bs/ms \\ 0 & 0 & 0 & 1 \\ ks/mu & bs/mu & (-kt - kt)/mu & -bs/mu \end{bmatrix}, \quad B = \begin{bmatrix} 0 & 0 \\ 1/mb & 0 \\ 0 & 0 \\ -1/mu & kt/mu \end{bmatrix}$$

$$C = \begin{bmatrix} 1 & 0 & 0 & 0 \\ 1 & 0 & -1 & 0 \\ 0 & 0 & 1 & 0 \end{bmatrix}, \quad D = \begin{bmatrix} 0 & 0 \\ 0 & 0 \\ 0 & 0 \end{bmatrix}.$$

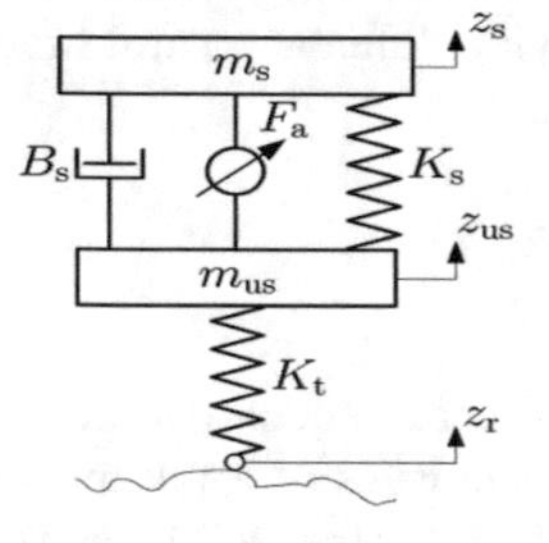

Fig. 1. Block diagram for the active suspension system of a QCM.

Table 1. Parameter values for suspension system.

Symbol	Parameter	Numerical value (unit)
M_s	Sprung-mass	250 (kg)
M_u	Unsprung-mass	50 (Kg)
K_s	Suspension stiffness	16812 (N/m)
K_t	Tire stiffness	190000 (N/m)
b_s	Damping coefficient	1000 (Ns/m)

2.1 Passive Suspension

It is mainly used in middle and low end cars. Springs and dampers are the main components of a passive system as shown in Fig. 2. No feedback or control algorithms are applied in this kind of suspensions. Passive systems are restricted systems, they have limited boundaries and do not enhance their behaviour over the environmental change.

As a result, they are generally successful with low limited input disturbances. Thus, they can't attain the optimal values for the passenger ride comfort nor road handling.

Therefore, many control theories have been applied on suspension systems to get the maximum ride comfort and road handling.

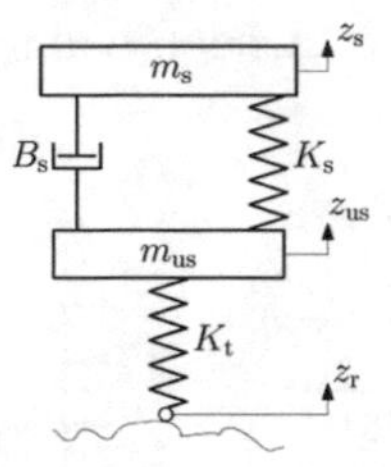

Fig. 2. A schematic diagram for passive suspension system.

2.2 Proportional–Integral–Derivative Controller (PID Controller)

PID controller depends generally on a feedback control loop theory, that is applied on many manufacturing industries. The main technique of this type of controller is that it tries to make the Error equals to a zero value, which the error is the difference between a set-point and the resultant value from the feedback of the plant.

Figure 3, shows a brief comparison between Passive suspension system and Active suspension system controlled by PID Controller, and the simulation results are shown in Figs. 4, 5, 6, 7, 8, 9 and 10, after applying different input variables.

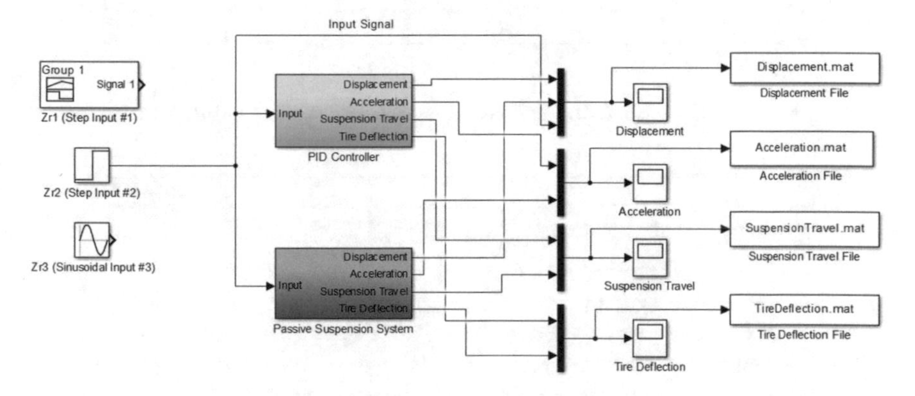

Fig. 3. Simulink model for passive suspension and active suspension system using PID.

After applying the step input, the system responce for passive suspension and PID will be shown in Figs. 4, 5, 6 and 7.

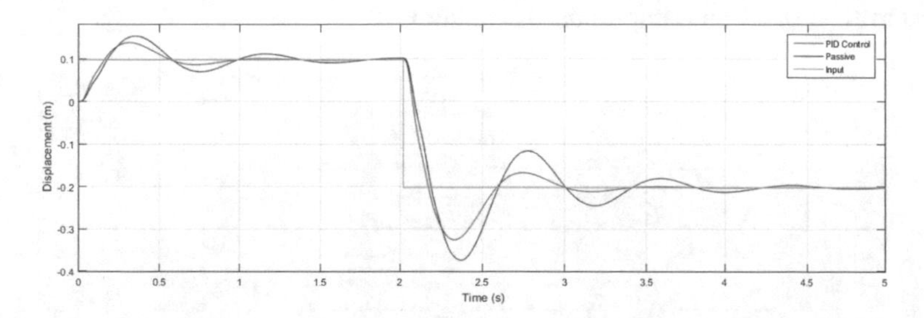

Fig. 4. Car body displacement for passive suspension system and PID controller.

Table 2 and Fig. 5, show the overshoot, settling time and rise time for two compared systems.

Table 2. Overshoot, settling time and rise time.

	Passive	PID
Overshoot	0.054	0.038
Settling time	4.6	3.7
Rise time	0.173	0.156

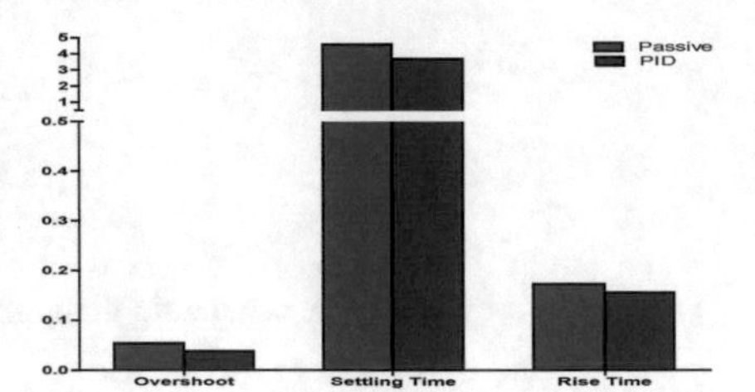

Fig. 5. Passive vs PID chart.

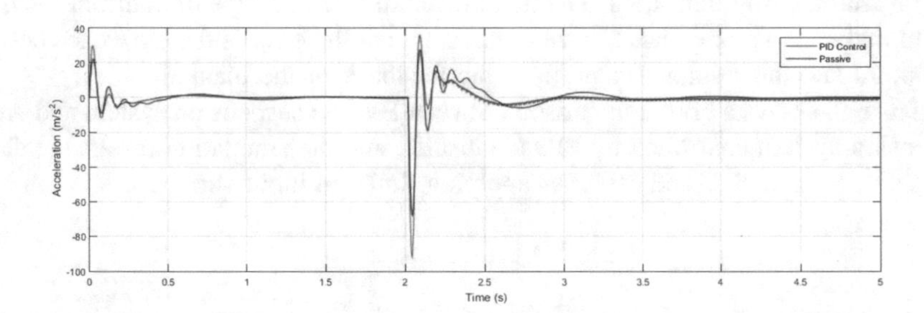

Fig. 6. Car body acceleration for passive suspension system and PID controller.

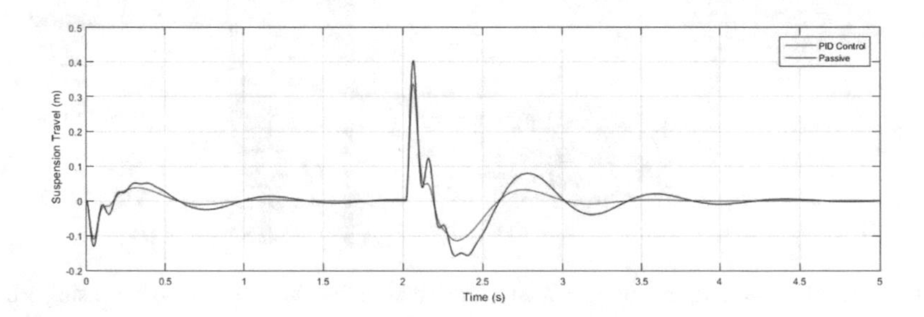

Fig. 7. Suspension travel for passive suspension system and PID controller.

Figure 8, show system responce for passive suspension and PID after applying step input step time 0.1 s and amplitude 0.5 m step.

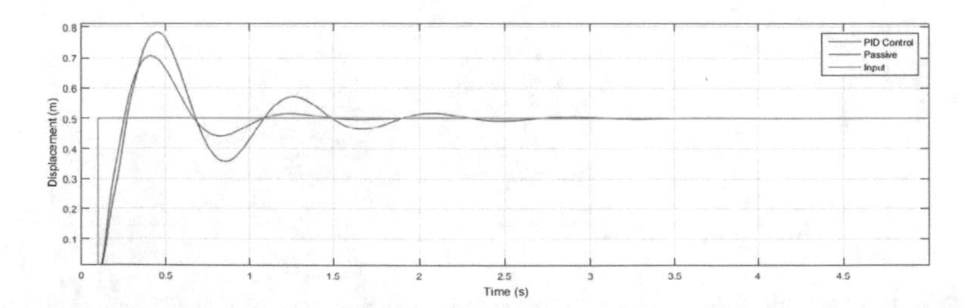

Fig. 8. Car body displacement for passive suspension system and PID controller.

Table 3 and Fig. 9, show the overshoot, settling time and rise time for two systems compared.

Table 3. Overshoot, settling time and rise time.

	Passive	PID
Overshoot	0.284	0.206
Settling time	3.9	2.3
Rise time	0.27	0.25

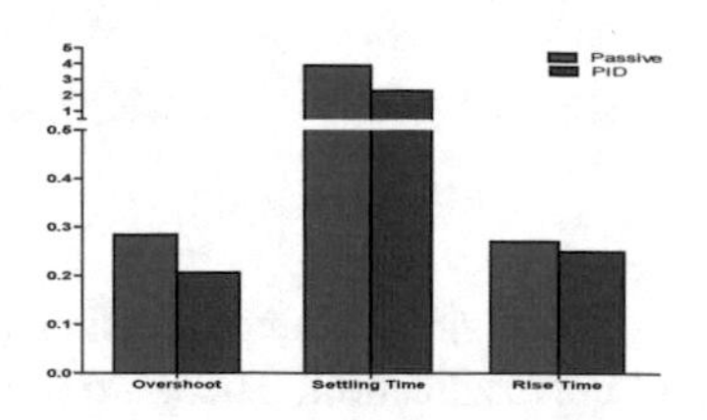

Fig. 9. Passive vs PID chart.

Figure 10, show system response for passive suspension and PID after applying sinusoidal wave frequency of 7.7 rad/s and amplitude 0.1 m.

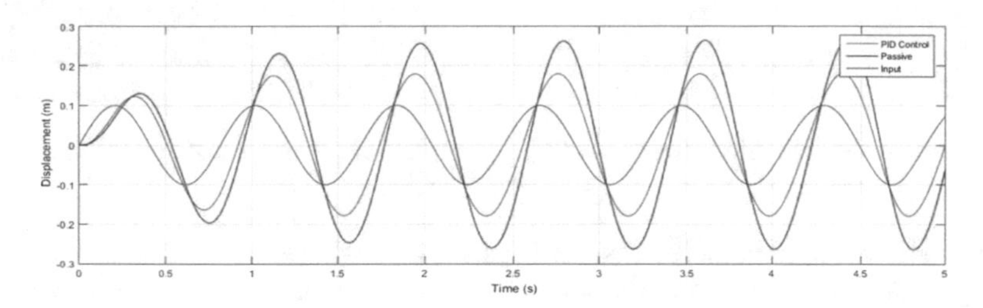

Fig. 10. Car body displacement for passive suspension system and PID controller.

2.3 Fuzzy Logic Control Design

In 1965, FLC was invented by Zadeh [19]. FLC consists of three main steps (1) **Fuzzification** (Use Membership-Functions to graphically illustrate a condition), (2) **Rule Evaluation** (Applying fuzzy rules on the system), and (3) **Defuzzification** (Achieving the crisp or actual results). The Simulink model and the designed model used in an active suspension system by applying FLC are shown in Figs. 11 and 12, respectively. In the designed Fuzzy model two input variables are obtained, which are Error and Error-Derivative, and one output variable as shown in Figs. 13, 14 and 15.

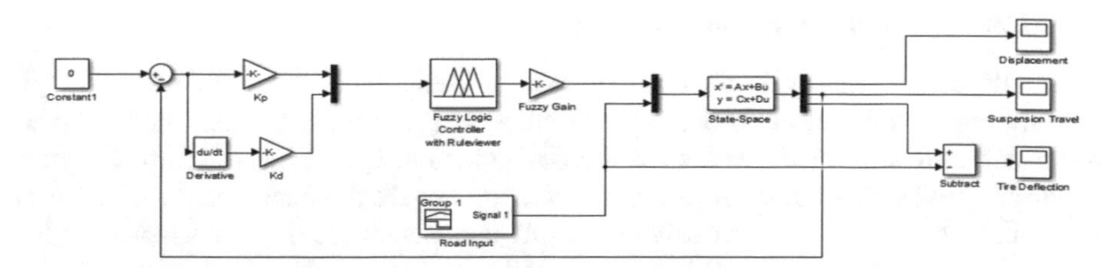

Fig. 11. Simulink model for active suspension system using FLC.

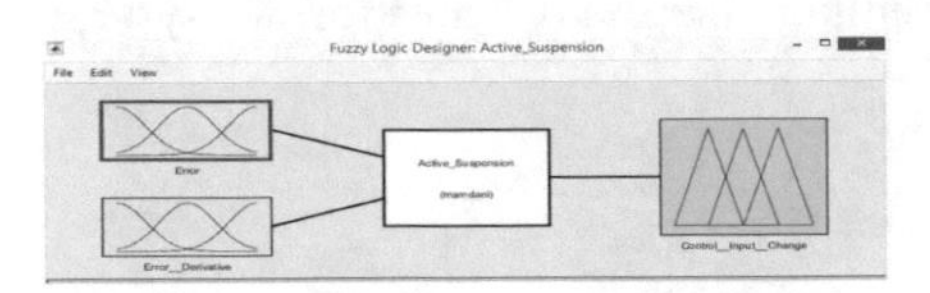

Fig. 12. FLC.

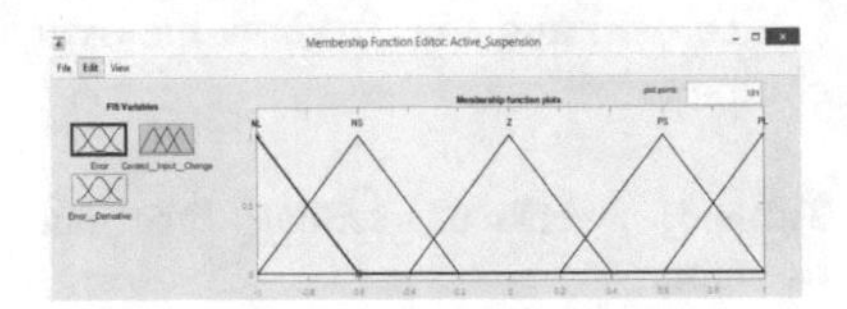

Fig. 13. Input variable "Error".

The control system consists of three stages: falsification, fuzzy inference process and defuzzification. There are five membership functions for both input and output, the membership functions are triangular-shaped with equaled width. Universe of discourse is from [−1 1].

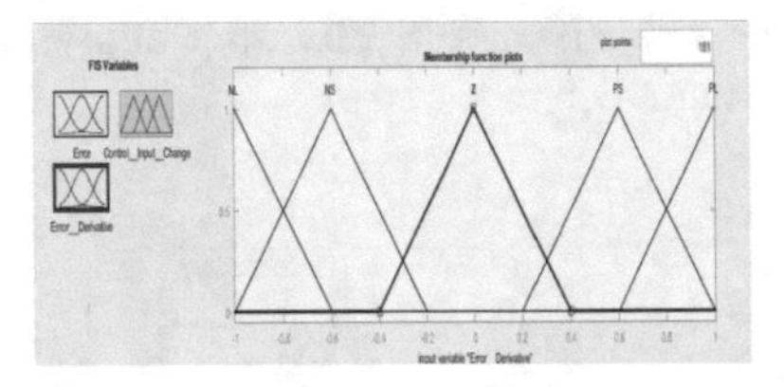

Fig. 14. Input variable "Error_Derivative".

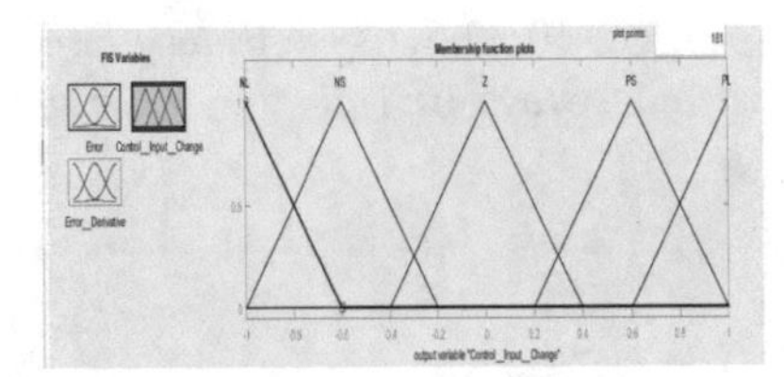

Fig. 15. Output variable "Control Output Change".

The rule base applied in the designed system is illustrated by the Table 4 with fuzzy terms gained from the designer's practice and knowledge. The output-surface of the FLC system is illustrated in Fig. 16.

Table 4. Rule base of FLC.

	Error-derivative					
Error		NL	NS	Z	PS	PL
	NL	NL	NL	NL	NS	Z
	NS	NL	NL	NS	Z	PS
	Z	NL	NS	Z	PS	PL
	PS	NS	Z	PS	PL	PL
	PL	Z	PS	PL	PL	PL

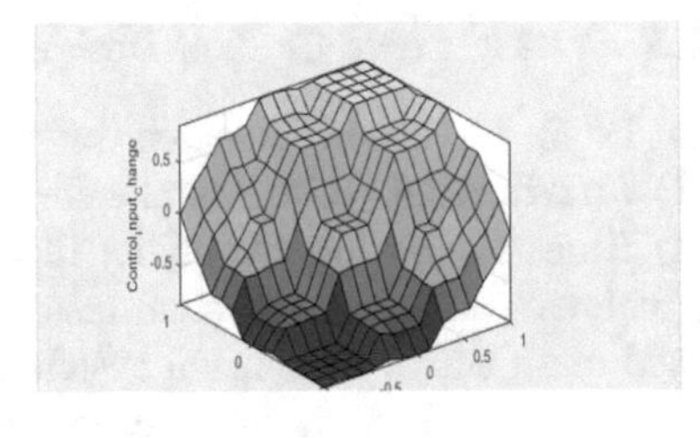

Fig. 16. The output surface of the fuzzy system.

2.4 Linear Quadratic Regulator Design

The LQR control theory is a great method for designing controllers with difficult systems that have strict performance necessities. It seeks for achieving the best controller parameters that reduces a specified cost-function. Q and R are weighs the state vector and the system input correspondingly, and they are the main vectors of the cost function. The methods used to solve the LQR parameters are the State-space model, figuring out the control law and calculating the feedback gain. The feedback gains applied in the LQR system will guide to the best results. In this research, the feedback gains and the state-feedback controller is designed using the LQR controller [16]

$$\dot{x}(t) = Ax(t) + Bu(t), \quad t \geq 0, x(0) = x_0 \tag{5}$$

Where: $x(t)$ is the state-vector and $u(t)$ is the input-vector. It is required to determine the matrix K also the static full state feedback control law,

$$u(t) = -Kx(t) \tag{6}$$

Satisfies the following criteria:

- the closed loop system is closely stable.
- the quadratic performance function is minimized.

$$J(K) = \frac{1}{2}\int_0^\infty \left[x^T(t)Qx(t) + u^T(t)Ru(t)\right] dt \tag{7}$$

Q is a nonnegative definite matrix that penalizes the departure of system states from the equilibrium and R is a positive definite matrix that penalizes the control input [15, 18]. LQR problem can be solved by achieving lagrange-multiplier method and is specified by:

$$\mathbf{K} = \mathbf{R}^{-1}\mathbf{B}^{\mathbf{T}}\mathbf{P} \tag{8}$$

where P is a nonnegative definite matrix satisfying the Riccati-matrix Equation,

$$A^T P + PA + Q - PBR^{-1}B^T P = 0 \tag{9}$$

The following LQR design algorithm is used to determine the optimal state feedback. To define the best possible state feedback, a designed LQR algorithm is applied as shown:

- Riccati matrix equation is first solved

$$-A^T P - PA - Q + PBR^{-1}B^T P = 0$$

- Optimal state $x^*(t)$ is determined from

$$\dot{x}^*(t) = \left[A - BR^{-1}B^T P\right]x^*(t) \tag{10}$$

Starting with $x(t_0) = x_0$ as initial conditions

- The optimal control $u^*(t)$ is obtained from

$$u^*(t) = -R^{-1}B^T Px^*(t) \tag{11}$$

- The optimal performance index is obtained from

$$J^* = \frac{1}{2}x^*(t)Px^*(t) \tag{12}$$

The weighting matrices Q and R are the main parameters developing any LQR system. The structure of Q and R values have a huge effect on systems reaction. The number of elements of Q depends on the number of state variable n, while the number of elements of R depends on the number of state variables of m [2, 14].

In the designed LQR system the state-space model is based on:

$$x_1 = \theta, x_2 = \dot{\theta}, x_3 = x, x_4 = \dot{x} \tag{13}$$

$$\dot{x}_1 = x_2$$

$$\dot{x}_2 = 62.43x_1 + 0.91x_4 - 9.09u$$

$$\dot{x}_3 = x_4$$

$$\dot{x}_4 = -2.67x_1 - 0.18x_4 + 1.8u \tag{14}$$

where θ is the pendulum angle measured from vertical reference, $\dot{\theta}$ is the rotational speed of pendulum, x is the cart position, $\dot{x}$ is the cart speed, and u is the input force. Matlab was used to solve Riccati's Equation to get the best gain vector. Simulink model for the designed LQR control system is shown in Fig. 17.

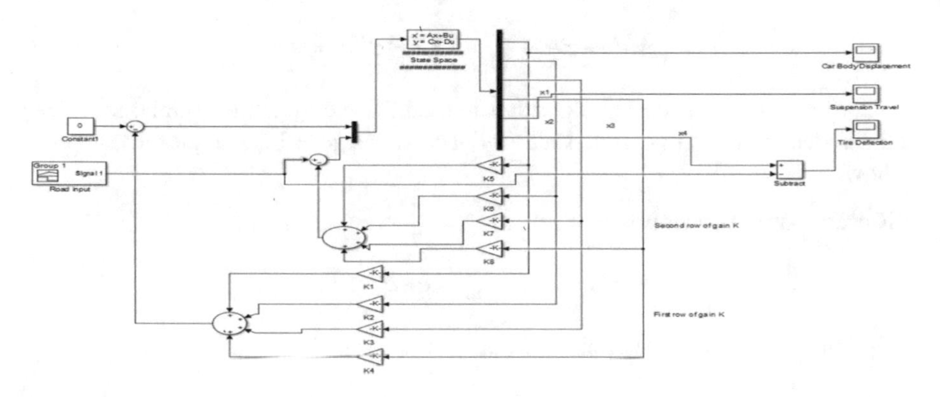

Fig. 17. Simulink model for active suspension system using LQR control system.

After applying the LQR equations on the designed system, these are the applied matrices for the designed system:

A = [0 1 0 0; [−ks −bs ks bs]/mb; 0 0 0 1; [ks bs −ks−kt −bs]/mw];
B = [0 0; 1/mb 0; 0 0; −1/mw kt/mw]; C = [1 0 0 0; 0 1 0 0; 0 0 1 0; 0 0 0 1];
D = [0 0; 0 0; 0 0; 0 0]; Q = [120 0 0 0; 0 1 0 0; 0 0 15 0; 0 0 0 0.06];
R = [100,0; 0,500];

3 Results and Discussion

The aim of this work is obtaining an optimal controller that achieves the best ride comfort and road handling qualities. In last few years, researchers have been trying to use different controllers instead of the passive system. The following results show an efficient active suspenstion system technique that uses different control theories for obtaining the best damping force and least settling time.

A brief comparison is described in this section between PID Controller, fuzzy logic controller and Linear Quadratic Regulator. Simulation results are introduced to evaluate the applied controllers. Different road profiles are applied such as step and sinusoidal disturbances, to check the performance of the controllers and make a comparison between them. Figure 18, shows a block diagram for the three control systems.

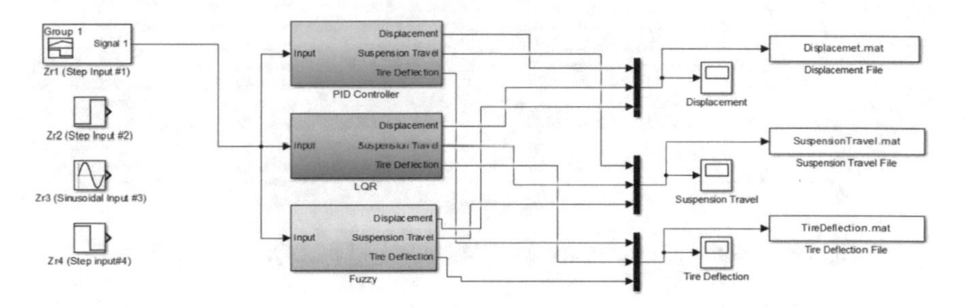

Fig. 18. Simulink model for active suspension system using PID, LQR and FLC.

a. **Step Disturbance**

Assuming the road disturbances are a step or Sinusoidal disturbances, The effect of car body displacement and suspension travel respectively in quarter car model after applying the FLC, LQR and PID controller with different inputs will be represented. After applying step input as 0.1 for 2 s and −0.2 after 2 s. The system response for LQR, FLC and PID will be shown in Figs. 19, 20 and 21.

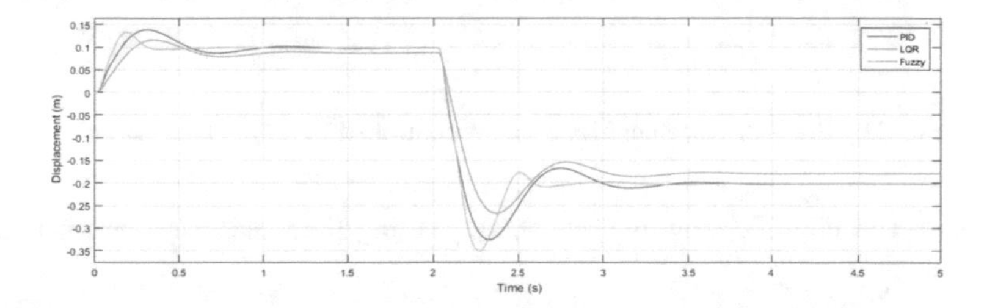

Fig. 19. Body displacement for a QCM controlled by LQR, FLC and PID.

Table 5 and Fig. 20, show the overshoot, settling time and rise time for two systems compared.

Table 5. Overshoot, settling time and rise time.

	PID	LQR	FLC
Overshoot	0.0385	0.015	0.033
Settling time	0.16	0.24	0.12
Rise time	4.4	4.5	3.8

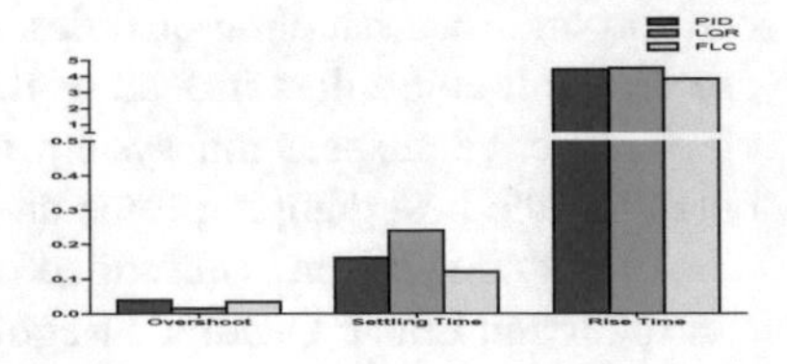

Fig. 20. PID vs LQR and FLC chart step input.

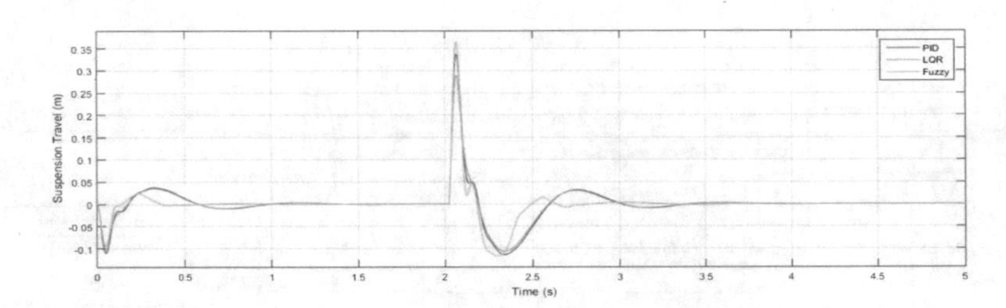

Fig. 21. Suspension travel for a QCM controlled by LQR, FLC and PID.

Figures 22, 23 and 24, show system response for LQR, FLC and PID after applying step input step time 0.1 s and amplitude 0.5 m step.

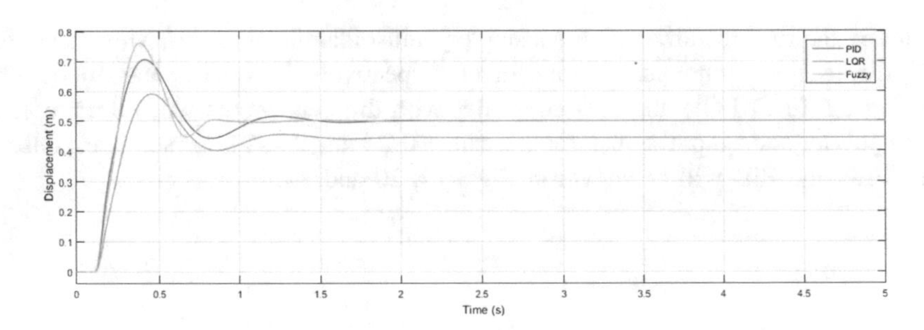

Fig. 22. Body displacement for a QCM controlled by LQR, FLC and PID.

Table 6 and Fig. 23, show the overshoot, settling time and rise time for two systems compared.

Table 6. Overshoot, settling time and rise time.

	PID	LQR	FLC
Overshoot	0.206	0.09	0.262
Settling time	1.8	2.4	1.6
Rise time	0.256	0.33	0.254

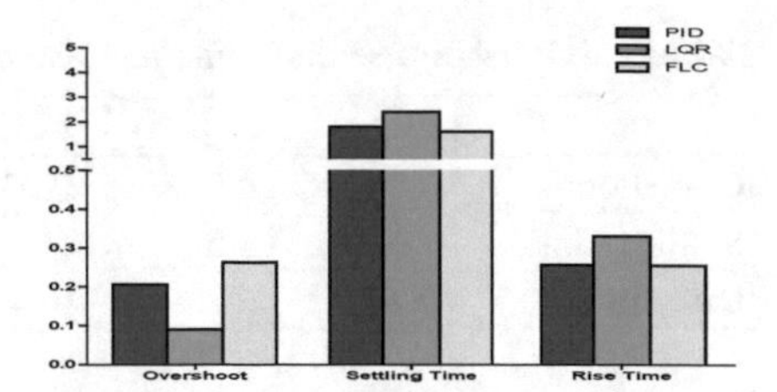

Fig. 23. PID vs LQR and FLC chart with step input 0.5.

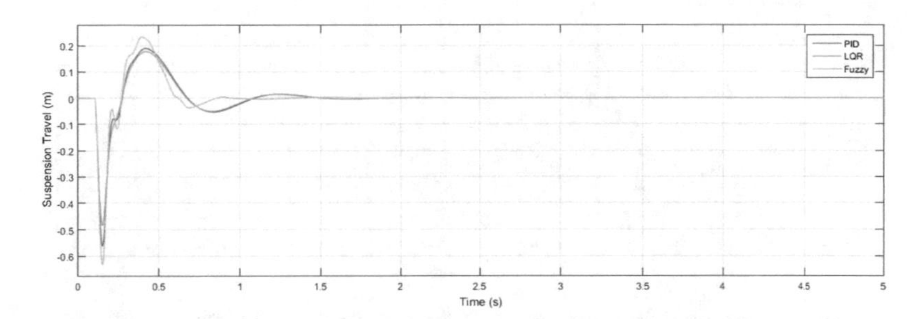

Fig. 24. Suspension travel for a QCM controlled by LQR, FLC and PID.

Figures 25, 26 and 27, show system response for LQR, FLC and PID after applying step input with a negative amplitude 0.1 m step.

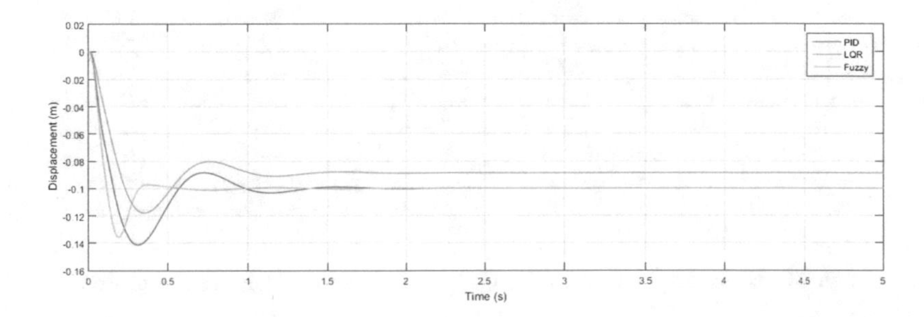

Fig. 25. Body displacement for a QCM controlled by LQR, FLC and PID.

Table 7 and Fig. 26, show the overshoot, settling time and rise time for two systems compared.

Table 7. Overshoot, settling time and rise time.

	PID	LQR	FLC
Overshoot	0.041	0.018	0.036
Settling time	1.6	1.5	1
Rise time	0.16	0.23	0.12

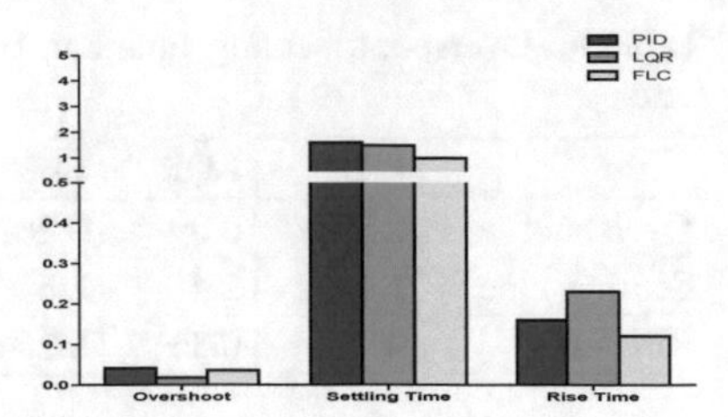

Fig. 26. PID vs LQR and FLC chart with step input 0.5.

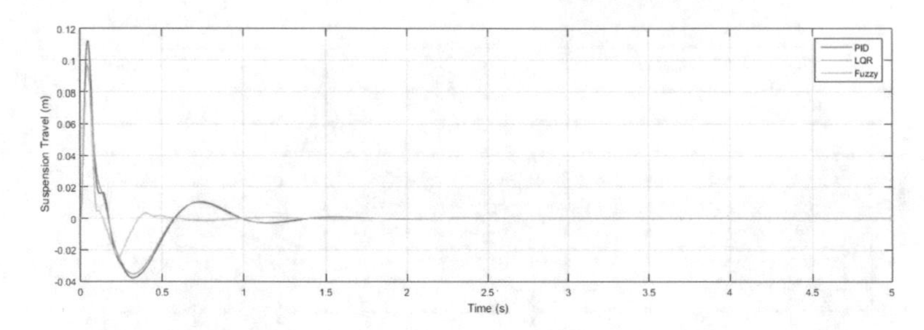

Fig. 27. Suspension travel for a QCM controlled by LQR, FLC and PID.

b. **Sinusoidal Disturbance**

Figures 28 and 29, show system response for LQR, FLC and PID after applying sinusoidal wave frequency of 7.7 rad/s and amplitude 0.1 m.

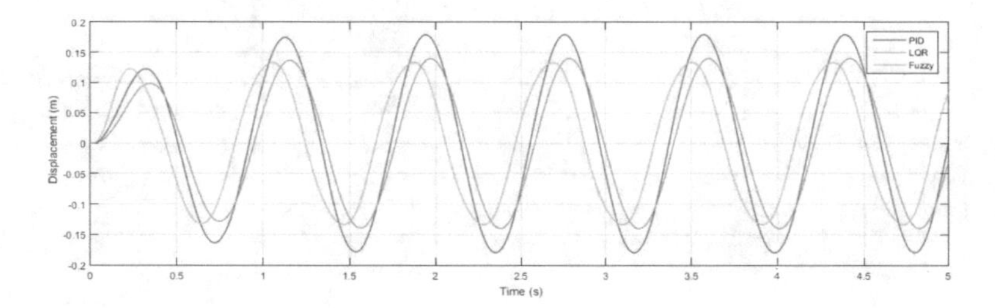

Fig. 28. Body displacement for a QCM controlled by LQR, FLC and PID.

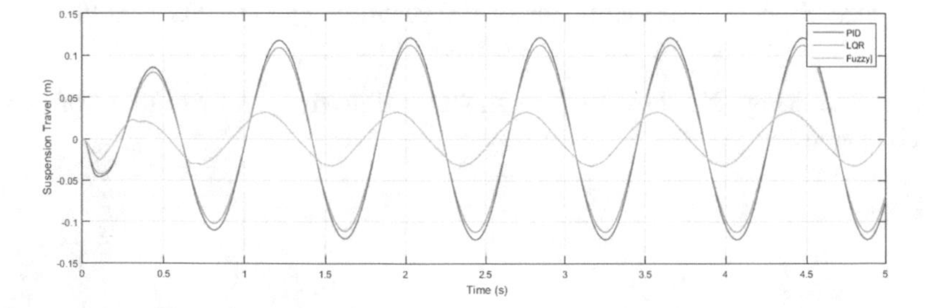

Fig. 29. Suspension travel for a QCM controlled by LQR, FLC and PID.

4 Conclusion

This paper has provided three design methods of controlling an active suspension system and a brief comparison between passive suspension system and PID Controller. The main three controllers applied in this paper are FLC, LQR and PID Controller. All methods were modelled and simulated using a Matlab/Simulink. After comparing a passive suspension system with an active suspension system controlled by PID controller, simulation results showed that the PID controller has shown better values in overshooting, settling time and rise time. Meanwhile, after comparing the FLC, LQR and PID together, LQR recorded the best overshoot values, while FLC recorded greatest results in settling time and rise time. So automobiles that will be designed in the future should apply active suspension controlled by FLC to achieve the best values for passenger ride comfort and road handling.

References

1. Allen, J.: Design of active suspension control based upon use of tubular linear motor and quarter-car model. Master's thesis. Texas A&M University, College Station (2008)
2. Yazar, G.: Design and analysis of helical coil spring forms for independent suspensions of automobiles. M.S. thesis. School of Natural and Applied Sciences of Middle East Technical University (2015)
3. Lauwerys, C., Swevers, J., Sas, P.: Design and experimental validation of a linear robust controller for an active suspension of a quarter car. Control Eng. Pract. **13**, 577–586 (2005)
4. Wang, J., Zolas, A.C., Wilson, D.A.: "Active suspension: a reduced-order" control design study. In: Mediterranean Conference on Control and Automation, vol. T31-051, pp. 27–29 (2007)
5. Savaresi, S., Spelta, C.: Mixed sky-hook and ADD: approaching the filtering limits of a semi-active suspension. J. Dyn. Syst. Meas. Control **129**(4), 382–392 (2007)
6. Savaresi, S., Spelta, C.: A single-sensor control strategy for semiactive suspensions. IEEE Trans. Control Syst. Technol. **17**, 143–152 (2009)
7. Jin, Y., Yu, D., Song, X.: An integrated-error-based adaptive neuron control and its application to vehicle suspension systems. In: Proceedings of IEEE International Conference Control Automatics, pp. 564–569 (2007)
8. Kou, F., Fang, Z.: An experimental investigation into the design of vehicle fuzzy active suspension. In: Proceedings of IEEE International Conference on Automation and Logistics, pp. 959–963 (2007)
9. Agharkakli, A., Sabet, G.S., Barouz, A.: Simulation and analysis of passive and active suspension system using quarter car model for different road profile. Int. J. Eng. Trends Technol. **3**(5), 2231–5381 (2012)
10. Hui, P., Wen-Qiang, F., Kai, L.: Stability analysis and fuzzy smith compensation control for semi-active suspension systems with time delay. J. Intell. Fuzzy Syst. **29**(6), 2513–2525 (2015)
11. Rao, T.R., Anusha, P.: Active suspension system of a 3 DOF quarter car using fuzzy logic control for ride comfort, 16–18 December 2013
12. Yoshimura, T.: Active suspension of vehicle systems using fuzzy logic. Int. J. Syst. Sci. **27**, 215–219 (1996)

13. Kuo, Y.P., Li, T.-S.S.: GA-based fuzzy PI/PD controller for automotive active suspension system. IEEE Trans. Industr. Electron. **46**, 1051–1056 (1999)
14. Kuo, Y.P., Li, T.H.S.: A composite EP-based fuzzy controller for active suspension system. Int. J. Fuzzy Syst. **2**, 183–191 (2000)
15. Gysen, B.L.J., van der Sande, T.P.J., Paulides, J.J.H., Lomonova, E.A.: Efficiency of a regenerative direct-drive electromagnetic active suspension. IEEE Trans. Veh. Technol. **60** (4), 1384–1393 (2011)
16. Li, H.R.: Hydraulic Control Systems. National Defense Industry Press, Beijing (1990)
17. Du, H., Zhang, N.: Fuzzy control for nonlinear uncertain electrohydraulic active suspensions with input constraint. IEEE Trans. Fuzzy Syst. **17**(2), 343–356 (2009)
18. Fischer, D., Isermann, R.: Mechatronic semi-active and active vehicle suspensions. Control Eng. Pract. **12**(11), 1353–1367 (2004)
19. Wilhelm, J.: Light weight suspension system for KTH research concept vehicle: design and construction of a composite suspension system with focus on application in KTH research concept vehicle with analysis of future solutions suitable for the automotive industry. Degree project, In light weight structures, second level, Stockholm, Sweden (2015)

Machine learning methodology and applications

Prediction of Football Matches' Results Using Neural Networks

Roger Achkar[1(✉)], Ibraheem Mansour[1], Michel Owayjan[2], and Karim Hitti[3]

[1] Department of Computer and Communications Engineering, AUST, Achrafieh, Lebanon
rachkar@aust.edu.lb
[2] Department of Mechatronics Engineering, AUST, Achrafieh, Lebanon
mowayjan@aust.edu.lb
[3] Department of Mathematics, Faculty of Sciences and Issam Fares Faculty of Technology, University of Balamand, El-koura, Lebanon

Abstract. In this paper, learning with a teacher artificial neural network to predict the results of football matches is presented. This type of networks requires training via examples, and when the training is complete, the network can be tested to check the results of new examples. In this application, the training examples are the results of previous matches which the network will use to predict the results of new ones.

Keywords: Neural network · Multilayer perceptron · Activation function Weights · Learning parameter

1 Introduction

The idea behind artificial neural networks is to produce a copy of human brain cells and their interconnections in order to make a network that behaves like the brain itself: learn new things, identify and classify patterns, and be able to take decisions just like human beings. Some types of neural networks need a teacher to learn, however, other types can learn by themselves without human intervention which makes neural networks applicable for various types of applications [1]. They are used to perform many tasks that include classification, clustering, recognition of patterns, etc. Neural networks are deep learning technologies; they resemble the human brain in two ways:

- A neural network gains knowledge through learning, whether it was supervised, unsupervised, or reinforced.
- This knowledge is stored in synaptic weights that are the inter-neuron connections [2–4].

Several types of neural networks exist today. The most popular one is the Multilayer Perceptron (MLP), which belongs to a general class of structures called feedforward neural networks. It is the type of network that will be implemented in the prediction application presented in this paper.

A. E. Hassanien et al. (Eds.): AISI 2018, AISC 845, pp. 169–178, 2019.
https://doi.org/10.1007/978-3-319-99010-1_15

2 The MLP's Structure

2.1 Neurons

The basic units of any neural network are the neurons. Similar to natural neurons, artificial neurons have a number of synapses (inputs), a nucleus (processor), and an axon (output). In an artificial neuron, each synapse carries a value called weight. When the neuron is activated, it sums up all the weights of the input neurons multiplied by their respective neuron value. This weighted sum is then passed through an activation function to produce the final output. Also, aside from the weights originating from other neurons, a neuron has an additional weight called the bias which allows the shifting of the activation function to the left or the right. Shifting could be essential sometimes in order to achieve successful learning. Figure 1 shows a block diagram of a neuron.

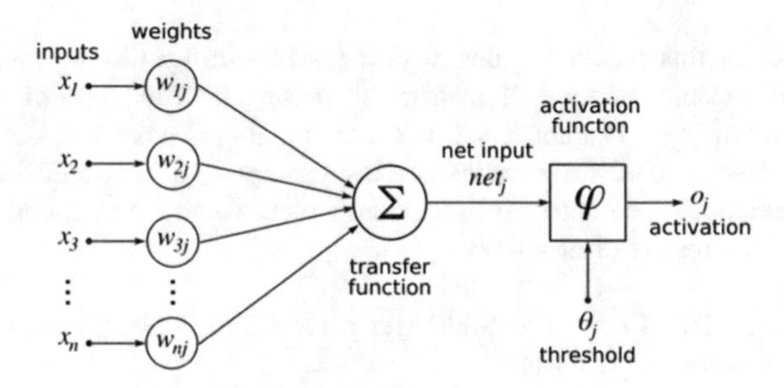

Fig. 1. Block diagram of a neuron

2.2 Activation Function

Several types of activation functions exist, but the MLP implements the sigmoid function, given by $\frac{1}{1+e^{-x}}$, which is bounded, continuous, monotonic, and continuously differentiable [5–7].

2.3 Layers

Neurons in a neural network are grouped into layers, where the first layer is the input layer, the last one is the output layer, and the layers in between are the hidden layers. In MLP, there's one input layer, one output layer, and at least one hidden layer.

3 Neural Network Learning

Neural network learning is the ability of a neural network to update itself to produce a certain output when it is provided with a certain input. In order for a neural network to learn, it should be trained. Training can be classified into three categories:

3.1 Supervised Learning

Supervised learning is the case where a neural network is presented with a set of training data pairs (an input and its desired output). Through these pairs, a network could learn to infer the relation which binds a neuron with another one. An algorithm that adjusts the neural network by updating the synaptic weights is the backpropagation method [8], which will be used in the presented application.

3.2 Unsupervised Learning

Unsupervised learning is based on training algorithms that alters the network's synaptic weights by relying only on input data. The results are networks that will learn to group received information with probabilistic methods.

3.3 Reinforced Learning

Reinforced learning doesn't depend on input data, but on agents which test input data produced by exploration algorithms. These agents will check the impact of the input data on the network, trying to determine the neural network performance on a given problem.

As stated above, the backpropagation algorithm is used in this work. It is a common method for training a neural network and consists of two phases:

Forward Computation: the input data is forwarded from the input layer to the output layer. In this phase, the synaptic weights remain the same, but the value of each neuron is calculated according to the summation and activation function explained above.

Backward Computation: it is the calculation of the deviation of the obtained values from the desired ones; the result is a delta factor which will propagate back to the input layer. At each layer, the neurons adjust their initial state according to the entity of error calculated. A constant, referred to as the learning parameter, is involved in order to speed or slow down the learning process. Upon repeating this process several times with different training sets, each time there will be a match between the obtained and the desired values. This will produce increasingly precise outputs, calibrating the weight of each network's components, and ultimately enhancing its ability to process the received data [9].

4 World Cup Application

The purpose of this application is to predict the winning team of a football match based on the results of previous matches. Seven parameters are used to represent each team. These parameters are actually the team's technical skills: shot power, shot accuracy, jumping, dribbling, tackling, passing, and speed. Each of these parameters can have a value from 1 to 8, thus each one will be represented by 3 bits. Combined together, their values will represent a team. Hence, each team will be represented by 3 bits $\times$ 7 parameters = 21 bits.

Moreover, the neural network will consist of three layers: input layer, hidden layer, and output layer. Since each team is represented by 21 bits and the input layer consists of the two teams involved in a football match, the number of input neurons will be 42. The output layer consists only of the game's winner, so it will have 21 neurons. The initial hidden neurons are also chosen to be 21. This value can be changed anytime later in case the learning was not successful.

Furthermore, the learning parameter is set to 0.1, and the initial values for the biases and weights are randomly generated between −1 and 1.

Also a windows application has been developed. It is designed to be very user-friendly. It is a Windows Forms application written in C#. The first form launched upon starting the program is the "Initialize Teams" form. In this form, the user will select the desired teams (see Fig. 2).

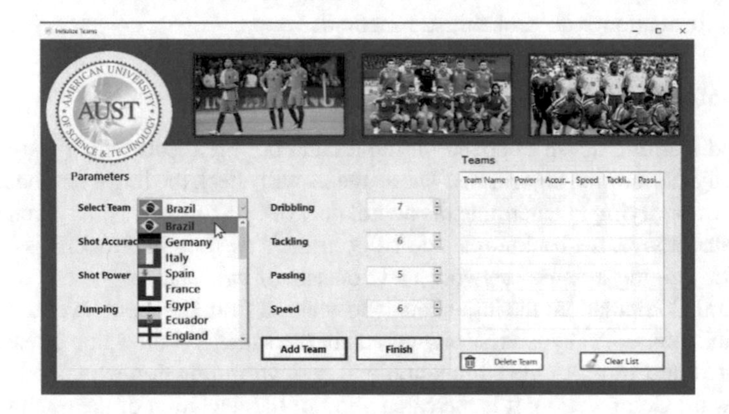

Fig. 2. "Initialize Teams" form

When a user selects a team and sets its skills, the "Add Team" button is clicked to add it to the list of teams, and it will be appended to the list view on the right of the form. The list view features a "Delete Team" button to delete a single selected team, and a "Clear List" button to delete all the teams added to the list. Figure 3 shows the addition of two teams.

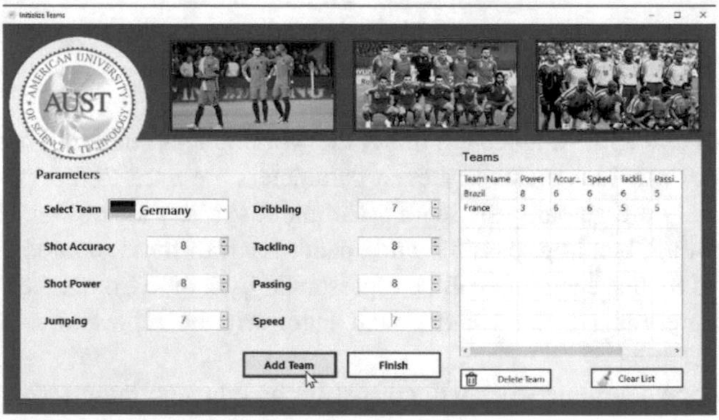

Fig. 3. "Initialize Teams" form showing two teams added

After adding a number of desired teams, the user can click the "Finish" button to proceed to the next form which is the "Initialize Network" form where the user sets the number of hidden neurons. Figure 4 shows this form where the number of neurons is set for each layer. The number of input neurons is fixed to 42 to represent 2 teams as explained above, the output layer consist of 21 neurons to represent the winning team, and the hidden layer is set to have 21 neurons initially.

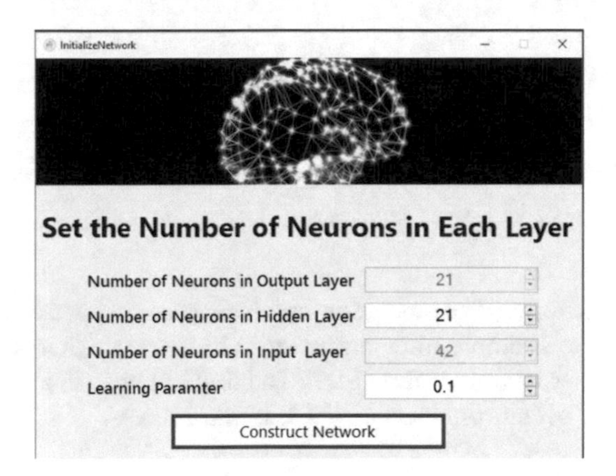

Fig. 4. "Initialize Network" form

After clicking the "Construct Network" button, the user will be transferred to the main form. Behind the scenes, the network is being constructed and the values for the weights and biases are being selected randomly. Usually, random numbers in C# are generated using the Random class. However, since the Random class will generate similar random values when generated in little timeframes (such as in our case), we will use a random number generator from cryptography services, to ensure real randomness.

In the main form, the training examples can be added and the network could be trained. The form has three dropdown lists: two to choose the game participants, and one to select the winner. The data for the first two dropdown lists will be populated from the teams selected by the user in the "Initialize Teams" form. Since the third dropdown is used to select the game's winner, it will only contain the two teams selected in the first and second dropdown lists. Note that a check function will be called after the user changes the value in either dropdown list one or two: in case the two dropdowns have the same value (a team playing against itself), the application will force the user to change one of them before adding this to the list of examples (see Fig. 5).

Furthermore, the form also includes a list view on the right to show the list of added examples. As mentioned above, a training example will consist of the two participating teams acting as an input, and the winning team acting as the desired output. Figure 6 shows some of the training sets added for this demonstration. Moreover, the number of iterations is set to 5000, using the numeric up-down located to the top of the "Examples" group box. This means that the network will undergo the two phases of

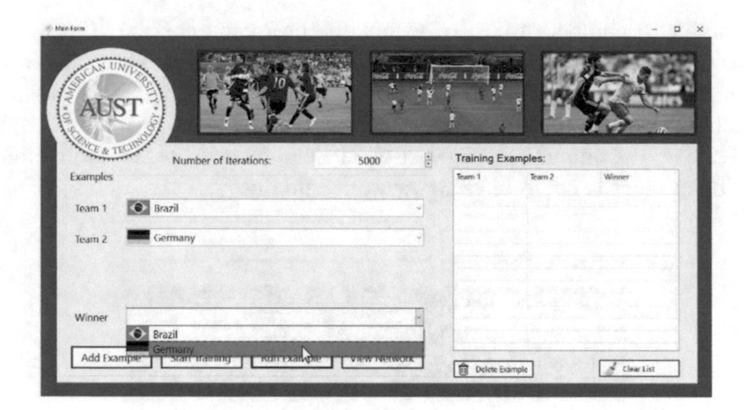

Fig. 5. “Main Form” where we add examples via three dropdown lists

supervised learning, forward computation and backward computation, for each of the added examples, one example after the other, 5000 times. After adding the needed training sets, the user can click the “Start Training” button and a progress bar will appear showing the overall progress of the learning process.

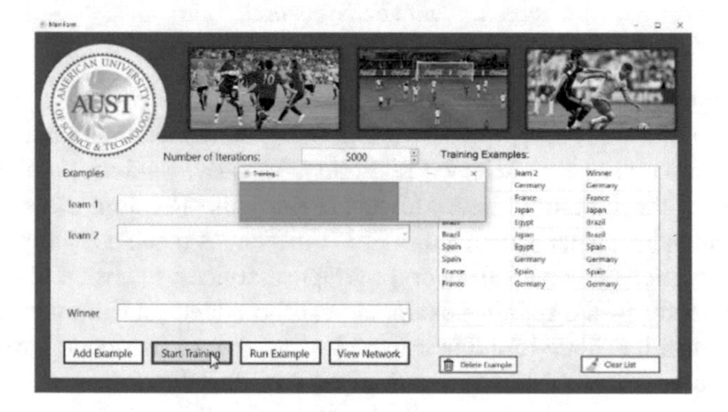

Fig. 6. “Main Form” showing the training progress bar

Once the training is complete, the progress bar is closed automatically and the user can now test new examples and check the obtained results. Also, the “Winner” radio button group will disappear since it will be not used anymore (see Fig. 7). To test a new example, the user selects the two participating teams and clicks the “Run Example” button, which will open a new form showing the result. The winner will be selected based on the Euclidean distance between each of the two teams and the results. The values of the bits of the obtained results should be closer to the winner, thus, the team that has a shorter Euclidean distance between its points and the points of the output, will be selected as the winning team.

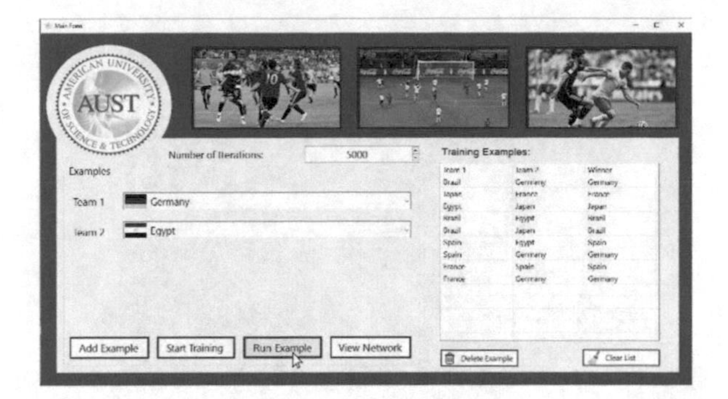

Fig. 7. To test a new example, click the "Run Example" button

As an example, a match involving Germany and Egypt is selected. The neural network predicts that Germany would win the game, as illustrated in Fig. 8. The form shows the flag and the name of the winning team, and the values of the 21 output neurons. Those neurons should be closer to those of Germany rather than to those of Egypt, based on the Euclidean distance.

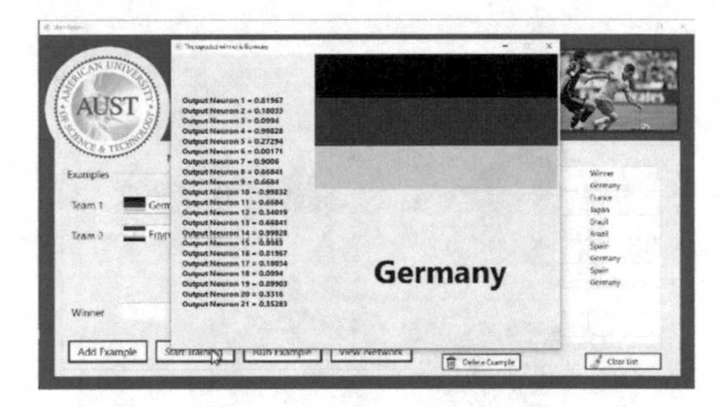

Fig. 8. Flag and name of the winning team

Moreover, another feature added to this application is the ability to draw the neural network constructed and display the values of its neurons and weights. In order to visually display the neural network, users can click the "View Network" button in the main form, and a new form will appear that draws the network. Figure 9 shows the drawing of the initialized neural network. To make the form load faster, the neurons and weights are drawn using the form's paint functions rather than adding images of circles and lines which slows down the painting process. Note that in order to fit all the neurons in one drawing, the size of neurons is chosen dynamically depending on the number of hidden neurons selected in the "Initialize Network" form.

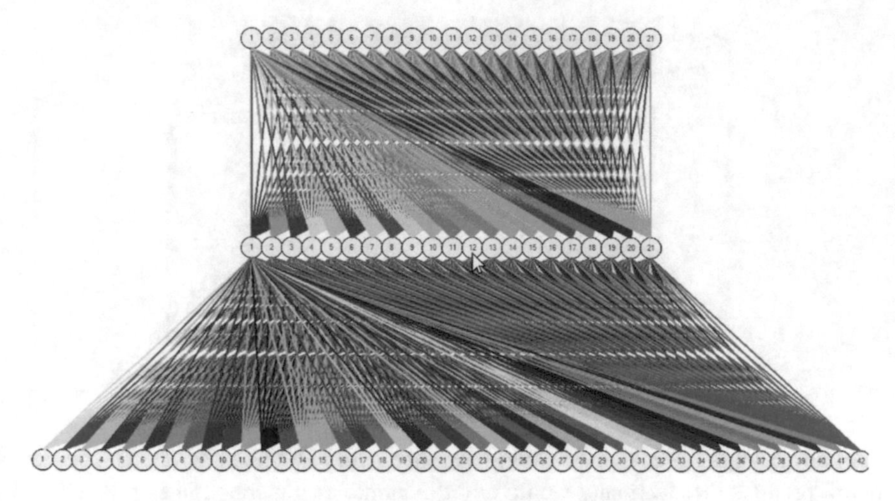

Fig. 9. Drawing the neural network: 42, 21, 21

Clicking on any neuron will show a new form displaying its value, bias, and weights (see Fig. 10).

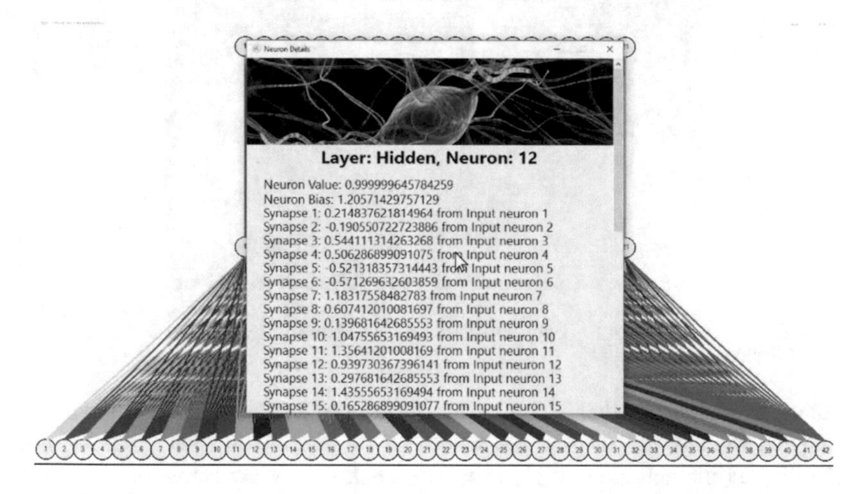

Fig. 10. Form showing the value, bias, and weights of hidden neuron 12

5 Validation

For a successful learning, the error should be decreased to an acceptable value. In the application, the error is being calculated in the backpropagation computation method according to formulae of the gradient error. Usually, this error decreases until reaching a minimum value. To illustrate this, the error value for output neuron 1 is shown. The formula is equal to the desired value – the calculated value of the neuron. Since the number of iterations is set to 5000, this method will be called 5000 times, thus, the error value will be printed 5000 times.

The first twenty values for this error are the following:

1. 0.451472891847536
2. 0.404541867523385
3. 0.361111405785288
4. 0.321522172486535
5. 0.285805738736867
6. 0.253839437729422
7. 0.225437248541106
8. 0.200383973114048
9. 0.178440499370849
10. 0.1593431125463231
11. 0.1428072829642354
12. 0.1285381673707032
13. 0.1162461515666864
14. 0.1056625910168237
15. 0.0965496854593152
16. 0.0887013947416231
17. 0.0819374823241772
18. 0.0760959152050235
19. 0.0710280209201444
20. 0.0665976575600615

While the last 20 values for this error are:

1. 0.00791178033714302
2. 0.00791119357187597
3. 0.00791060693681322
4. 0.00791002043190647
5. 0.00790943405710787
6. 0.00790884781236934
7. 0.00790826169764247
8. 0.00790767571287921
9. 0.00790708985803191
10. 0.00790650413305216
11. 0.00790591853789235
12. 0.00790533307250418
13. 0.00790474773683969
14. 0.00790416253085136
15. 0.00790357745449111
16. 0.00790299250771098
17. 0.00790240769046302
18. 0.00790182300269993
19. 0.00790123844437341
20. 0.00790065401543583

The error decreases from 0.451472891847536 to 0.00790065401543583 which is an acceptable value hence validating our methods.

6 Conclusion

An application predicting results of upcoming football matches based on the outcomes of previous games is designed. It is based on neural network featuring the MLP and the backpropagation algorithm. The application results showed that the training of the network was successful since the error has decreased to an acceptable value and that the usage of the mentioned algorithm was highly applicable in this sort of applications. Moreover, the time taken by the network to complete the training was satisfactory (less than one minute) given the fact that it is network with nine training examples used. This can encourage upgrading the whole procedure by adding more teams and training sets without worrying about a long training period. Also, the app can open the doors to a whole new world of prediction applications covering different sports fields.

References

1. Haykin, S.: Neural Networks—A Comprehensive Foundation, 2nd edn. Prentice Hall, Upper Saddle River (1998)
2. Nasr, G., Achkar, R., Fayad, F.: Stereo-Vision calibration by multi-layer perceptrons of an artificial neural network. In: IEEE 7th International Symposium on Image and Signal Processing and Analysis, ISPA 2011, Dubrovnik, Croatia, pp. 621–626 (2011)
3. Abou Kassm, G., Achkar, R.: LPR CNN cascade and adaptive deskewing. Procedia Comput. Sci. **114**, 296–303 (2017)
4. Achkar, R., El-Halabi, M., Bassil, E., Fakhro, R., Khalil, M.: Voice identity finder using the back propagation algorithm of an artificial neural network. Procedia Comput. Sci. **95**, 245–252 (2016)
5. Harkouss, Y., Mcheik, S., Achkar, R.: Accurate wavelet neural network for efficient controlling of an active magnetic bearing system. J. Comput. Sci. **6**(12), 1457–1464 (2010)
6. Saide, C., Lengelle, R., Honeine, P., Richard, C., Achkar, R.: Nonlinear adaptive filtering using kernel-based algorithms with dictionary adaptation. Int. J. Adapt. Control Signal Process. **29**(11), 1391–1410 (2015)
7. Liu, L., Han, B., Du, L., Gao, Z.: A Neural network structure and learning algorithms with the neuron output feedback. In: Third International Workshop on Advanced Computational Intelligence, Jiangsu, China (2010)
8. Hecht-Nielsen, R.: Theory of the backpropagation neural network. In: International Joint Conference on Neural Networks, Washington, D.C., USA (1989)
9. Musso, E.: Basis of Neural Networks in C#. https://social.technet.microsoft.com/wiki/contents/articles/36428.basis-of-neural-networks-in-c.aspx#Neurons_and_dendrite. Accessed 27 Apr 2018

Multi-class Support Vector Machine Training and Classification Based on MPI-GPU Hybrid Parallel Architecture

I. Elgarhy(✉), H. Khaled, Rania El Gohary, and H. M. Faheem

Faculty of Computer and Information Science, Ain Shams University, Cairo, Egypt
{islam_elgarhi, heba.khaled, dr.raniaelgohary, hmfaheem}@cis.asu.edu.eg

Abstract. Machine Learning (ML) is the process of extracting knowledge from current information to enable machine to predict new information based on the learned knowledge. Many ML algorithms aim at improving the learning process. Support vector machine (SVM) is one of the best classifiers for hyper-spectral images. As many of the ML algorithms, SVM training require a high computational cost that considered a very large quadratic programming optimization problem. The proposed sequential minimal optimization solve the highly computational problems using a hybrid parallel model that employs both graphical processing unit to implement binary-classifier and message passing interface to solve multi-class on "one-against-one" method. Our hybrid implementation achieves a speed up of 40X over the sequential (LIBSVM), a speed up of 7.5X over the CUDA-OPENMP for training dataset of 44442 records and 102 features size for 9 classes and a speed up of 13.7X over LIBSVM in classification process for 60300 records.

Keywords: Machine learning (ML) · Support vector machine (SVM)
Sequential minimal optimization (SMO) · Graphical processing unit (GPU)
Message passing interface (MPI) · Quadratic programming (QP)

1 Introduction

Machine Learning is the process that gives the ability for machines to learn. The goal is to have algorithm that take existing knowledge and make new inferences from the information automatically in two processes Training and Classification.

In many domains, training example is represented by a very large number of features and the number of training examples is very large which will increase the complexity of data analysis in both training and classification processes.

Due to limitations of the capabilities of semiconductor manufacturing, and the huge improvements done in parallel multi architecture in both high-scalable graphical processing unit (GPU) parallel implementations and Multi-Core CPU Architecture, there is a chance to enhance performance of problems that can be adapted to run in parallel architectures.

A. E. Hassanien et al. (Eds.): AISI 2018, AISC 845, pp. 179–188, 2019.
https://doi.org/10.1007/978-3-319-99010-1_16

In a Parallel Model, a group of nodes "processors" can work together to build a parallel solution. Parallel programming utilizes all available processing units which lead to decrease the program total elapsed time. Parallel architecture can be categorized to shared memory, distributed memory and hybrid memory architectures based on memory architectures between its nodes.

Solving SVM optimization problem is a time and memory consuming process as it is a quadratic programming (QP) solution. Our Hybrid Parallel implementation applied a two level parallelization to implement SVM optimization between multi-classes.

The rest of this paper is organized as follow. Section 2 introduces the compute unified device architecture (CUDA) and message passing interface (MPI). The introduction about the sequential SVM SMO algorithm and the parallel SMO SVM and our proposed technique design is represented and discusses in Sect. 3. Finally the experimental results and Dataset are described in Sect. 4, moreover the conclusion is conducted in Sect. 5.

2 Parallel Architecture

2.1 Compute United Device Architecture (CUDA)

NVIDIA released CUDA API to overcome the limitation of using GPU in general purpose computing which was needed to know OpenGL or DirectX to use before November 2006 the first time API supporting GPGPU was released [1]. CUDA Architecture can run multiple programs in various compute devices. Compute device GPU has its own memory is known as "Device Memory". A kernel is a program, which can run N times in parallel, by each individual streaming processing unit [2].

CPU is known as "host". It has its own memory "host memory", and it is responsible for initiating the work on the compute device GPU to run the kernel. The device Memory can be partitioned into three types (constant, shared and global) based on speed and role.

CUDA Stream achieves overlap between execution and data transfer operations. The streaming operations are executed by the sequence issued by the host. However, operations in different streams can be interleaved or run concurrently.

2.2 Message Passing Interface (MPI)

The hybrid memory architecture supports two level of parallelisms. It offers the benefits of the other two parallel memory architectures (shared and distributed). All threads per node share the same address space and run independently as shared memory architecture and, each node has the local address space and run independently as distributed memory architecture [3].

Collaboration and data exchange between nodes in distributed memory architectures is achieved through a communication network between the nodes. The standard programming interface for communication between different nodes in the cluster is MPI. MPI defines a set of functions that allow programmers to instruct their code moreover execute tasks in parallel, there are many implementations of this standard interface: two popular implementations are the MPICH and the LAMMPI [4]. Also MPI program is executed in the Single Program Multiple Data (SPMD) model where all the nodes run the same program, and each node may follow a different execution path.

3 Parallel SMO SVM and Proposed Implementation Design

3.1 Sequential Support Vector Machine (SVM)

SVM is a supervised learning algorithm, binary classifier technique. Assume the initial n samples X = {x1, x2... xn} where $x \in R^d$ and d size of input feature set, and Ω = {y1, y2} is the two binary class where any sample x belongs to one of y1 = 1 or y2 = −1 [5–8], assuming classes are linearly separable, a decision surface can always be found and the hyper-plane discriminating both classes as defined in [9, 10].

SVM is a Maximum-Margin Classifiers; that aims to find the maximum gap between training samples "support vectors" α and the margin. Kernel function gives advantage to SVM for non-linearly separable classes; the higher dimension new feature space was created using the kernel trick by means of a projection $\Phi(x)$.

```
SbinarySMO (Trainingdata   X , TrainingdataLabels Y , Trainingsize nsize, out Modal Alphas)
Input:   X is training dataset, Y its labels class, and nsize is training data size,
         where xi ∈ χ, yi ∈ Ω, i ∈ {1 ··· nsize}.
Output: Alphas are updated parameters was needed in classification process.
{
 1: Initialize:
      αi=0, fi=−yi , where i ∈ {1 ··· nsize}, αi ∈ Alphas
 2: Initialize:
      bhigh = −1, blow = 1,
      ihigh = min {i: yi = 1},
      ilow = min {i: yi = −1}
 3: Update: αilow, αihigh
 4: repeat
 5:  Update fi, where i ∈ {1 ··· nsize}
 6:  Compute: bhigh, blow, ihigh, ilow
 7:  Update αilow, αihigh
 8: until blow - bhigh ≤  2τ
}
```

$$\alpha_{i_{low}}^{new} = \alpha_{i_{low}} + y_{i_{low}} \frac{b_{high} - b_{low}}{\eta}$$

$$\alpha_{i_{high}}^{new} = \alpha_{i_{high}} + y_{i_{low}} y_{i_{high}} (\alpha_{i_{low}} - \alpha_{i_{low}}^{new})$$

$$\eta = K\left(x_{i_{high}}, x_{i_{high}}\right) + K\left(x_{i_{low}}, x_{i_{low}}\right) - 2 * K(x_{i_{low}}, x_{i_{low}})$$

$$f_i = f_i^{old} + (\alpha_{i_{high}}^{new} - \alpha_{i_{high}}) y_{i_{high}} K\left(x_{i_{high}}, x_i\right) + (\alpha_{i_{low}}^{new} - \alpha_{i_{low}}) y_{i_{low}} K\left(x_{i_{low}}, x_i\right)$$

$$I_{low} = \{i: 0 < \alpha_i < C\} \cup \{i: y_i > 0, \alpha_i = C\} \cup \{i: y_i < 0, \alpha_i = 0\}$$

$$I_{high} = \{i: 0 < \alpha_i < C\} \cup \{i: y_i > 0, \alpha_i = 0\} \cup \{i: y_i < 0, \alpha_i = C\}$$

$$i_{low} = \arg max \{f_i: i \in I_{low}\}\ ,\ i_{high} = \arg min \{f_i: i \in I_{high}\}$$

$$b_{low} = \arg max \{f_i: i \in I_{low}\}\ ,\ b_{high} = \arg min \{f_i: i \in I_{high}\}$$

Fig. 1. Sequential binary SMO

Sequential minimum optimization (SMO) solves SVM quadratic optimization problem [11]. SMO solves the smallest task at each step to find only two optimization variables and update two Lagrange multipliers under the Karush-Kuhn-Tucker (KKT) condition [12] as show in Fig. 1.

3.2 Parallel (SMO)

Binary SVM training process was solved sequential by SMO. It can be implemented in parallel on GPU to take advantage of its greater computing power and memory bandwidth. SMO run in GPU by launching thread per independent training data sample to achieve high parallelism from GPU.

There are three steps in SMO algorithm that can enhance its computational execution time when it is executed in parallel; the first of them is the KKT conditions update (iteration number 5 in Fig. 1), where each thread can compute fi independently. The second step is the calculation of bhigh, blow, ihigh, ilow (iteration number 6 in Fig. 1) that can be executed in parallel using reduction to perform a first order heuristic search. The third is calculated the offset b as in (1) also can be executed in parallel for each support vector.

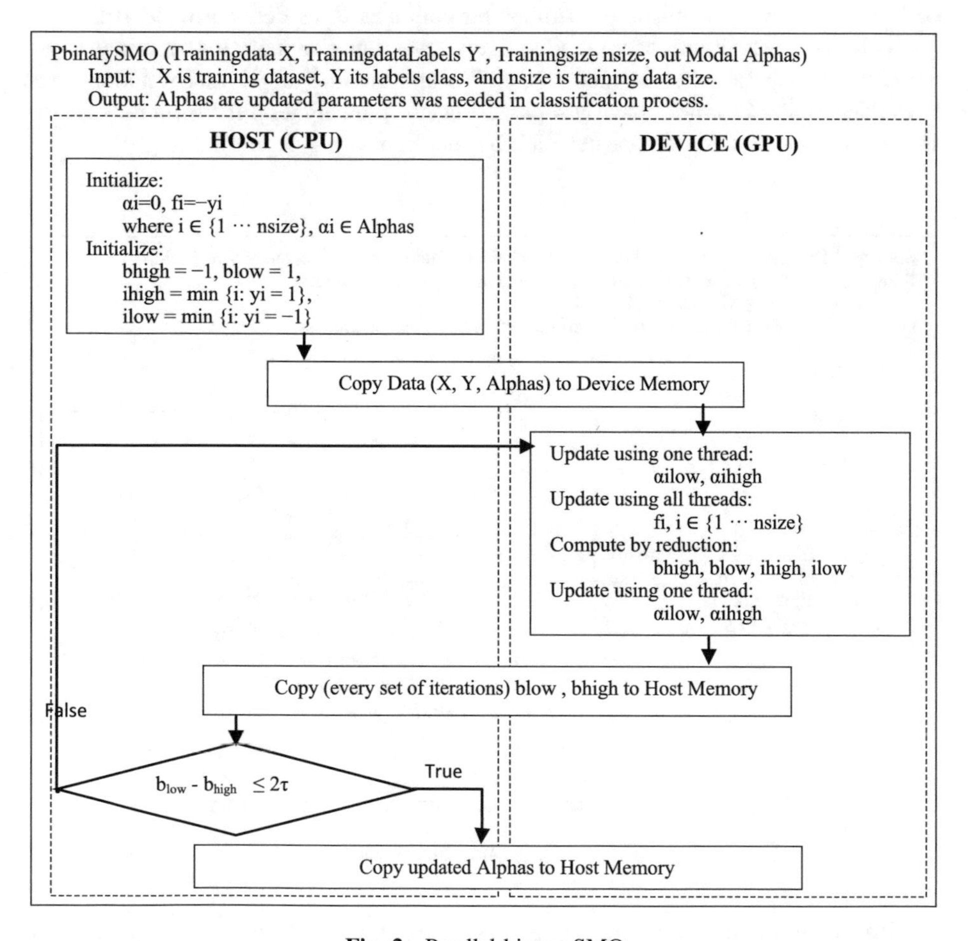

Fig. 2. Parallel binary SMO

Figure 1 shows that the other two steps are updating Lagrange multipliers and convergence checking. Both updating the lagrange multiplier (iterations number 3 and 7) and the convergence check (iteration number 8) are sequential steps. We execute the Lagrange multipliers update using one GPU thread to avoid latency occurred by copying the data from device to host. However, we execute the convergence check on the host for every set of training iteration on the device GPU.

Our proposed implementation of the binary SMO has a significant modification on the convergence check (iteration number 8). We reduced the communication between the host and device by applying the convergence check once at the host for a set of iterations rather than checking for convergence in each iteration and returning back to the device. So, we reduced the communication between the host and device. All the suggested modifications using the GPU parallel binary SMO version based on Fig. 1 that represents the sequential binary SMO represented in Fig. 2.

3.3 Parallel Multi-classes Parallel (SMO)

The SVM was mainly designed for binary classification; several methods are proposed to construct a multi-class classifier. "one-against-one" method was more suitable for practical use than the other methods. For m classes, need to run m(m − 1)/2 independent binary SMO. Cluster of GPUs can be used to execute the multiclass training [13]. Each GPU is responsible for running a group of binary SMO between different two classes.

As shown in Fig. 3, Parallel multi-class SMO can be achieved by running multiple of parallel binary SMO (from iteration4 to iteration7 in Fig. 3). Firstly, we divide the number of parallel SMO by the number of active processes (MPI working nodes); each node has its own GPU.

```
PmultiSMO()
{
    1:Initialize num_Of_Process = number of active process "MPI_Get_size ()"
    2: Initialize num_Of_BinarySMO= m(m − 1)/2
    3: Initialize num_Of_SMOPerProcess = num_Of_BinarySMO/num_Of_Process
    4: For (i=0; i< num_Of_SMOPerProcess;i++)
    5:        Initialize subX, subY, subnsize, subAlphas for current binarySMO
    6:        Call PbinarySMO (subX, subY, subnsize, subAlphas)
    7: End For
}
```

Fig. 3. Parallel multi-classes SMO

3.4 SVM Classification

Once SVM multi-classes model generated by SMO, it would be used to identify new data samples. Classifying new samples can be done by the majority "voting strategy" between training models generated from binary SMO on "one-against-one" approach. If the training model identified that the new sample belongs to the ith class, then, the vote for class label i is incremented by one. Finally, the new sample is labeled by the label with max votes.

After SVM training problem converged the variable b which represents the offset is calculated as shown in (1). The classification process can be calculated as shown in (2) where h ∈ Rn is the new sample to be classified [6].

$$b = \frac{1}{n_{sv}} \sum\nolimits_{j=1}^{n_{sv}} \left(\sum\nolimits_{i=1}^{n} \alpha_i y_i K(x_i, x_j) \right) - y_j \tag{1}$$

$$y(h) = sign\left(b + \sum\nolimits_{i=1}^{n} \alpha_i y_i K(x_i, h) \right) \tag{2}$$

Multiple GPUs used to execute "voting strategy" in parallel. Each GPU classify samples between two binary class based on training model. Each thread on GPU used to classifies a single sample where all thread executes the same instruction in their warp and minimizing thread divergence.

4 Dataset and Experimental Results

This section presents the dataset and the specs of hardware and software used to implement the hybrid MPI-CUDA implementation of SVM. It highlights the performance of our hybrid modal compared with sequential SVM library (LIBSVM) and OPENMP implementation with the same Gaussian kernel function, penalization parameter C and tolerance threshold value τ.

4.1 Dataset

The Hyperspectral image is a group of contiguous co-register spectral bands. These bands are very narrow, mostly on the range of 5 μm–20 μm depending on sensor that acquisitions data.

The Hyperspectral imaging systems can acquire data with greater than 200 spectral dimension channels [14, 15]. Due to the increase on this dimension, the complexity of data processing and analysis has also increased. There are many spectral dimension datasets that can be used to measure the performance of the hybrid multiclass SVM implementation. The dataset was over Pavia city center, Italy show in Fig. 4 it consists of 148155 data sample point in 1096 × 715 pixels with nine ground-truth classes labeled as: water, trees, grass, parking lot, bare soil, asphalt, bitumen, tiles, and shadow.

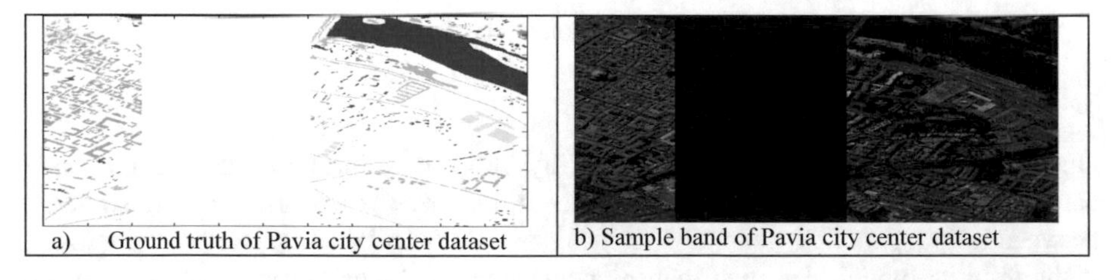
a) Ground truth of Pavia city center dataset b) Sample band of Pavia city center dataset

Fig. 4. Dataset

4.2 Experimental Results

The Experiments were carried out on a machine with an Intel (R) Core i5-4200 M CPU at 2.50 GHz and 4 GB of RAM. The GPU used was an NVIDIA GeForce 740M with 2 Stream Multiprocessor with 384 CUDA Cores. Multiclass SVM are implemented using CUDA Version 7.5 and MPICH2 on Windows 7 (64-bit).

Table 1 shows the training time for Pavia Centre dataset with 9 classes. Different training sample sizes on hybrid MPI-CUDA achieved a speed up of 5.5X at sample size 7404 and a max. Speedup of 40X at sample size 44442 over sequential (LIBSVM). As Fig. 5 shows, the speedup has a proportional increase to the training sample size.

Table 2 show the hardware specification comparison between our hybrid MPI-CUDA and OPENMP-CUDA. There is obvious difference in CPU and GPU specs.

Table 3 and Fig. 6 show the different training sample sizes using hybrid MPI-CUDA implementation. We achieved speed up from 1.2X to 7.5X over OPENMP-CUDA [16]. OPENMP-CUDA checks for learning process convergence per iteration.

Table 1. Training time (MPI-CUDA vs. LIBSVM).

Dataset (#Trainingsamples, FeaturesDimension, #classes)	Training time (s)		Speedup
	MPI-CUDA	LIbSVM	
(7404, 102, 9)	1.09	6	5.5X
(14811, 102, 9)	1.25	20	16X
(22118, 102, 9)	1.41	37	26.24X
(44442, 102, 9)	2.05	82	40X

Table 2. Hardware specs (MPI-CUDA vs. OPENMP-CUDA)

Hardware	MPI-CUDA	OPENMP-CUDA
Host (CPU)	CPU Core i5-4200M	CPU E5-2620
	2.50 GHz	2.50 GHz
	4 GB RAM	8 GB RAM
Device (GPU)	NVIDIA GeForce 740M	NVIDIA GeForce 780Ti
	2 SMP, 384 Cores	15 SMP, 2880 Cores
	CPU Core i5-4200M	CPU E5-2620

Kernel streams provide the ablity to overlap data transfer from/to device memory and computations operations. Using two streams one for SMO computation steps and another stream to retrieve data from the device memory periodically when checking the convergence.

Kernel Streams allow SMO steps and retrieve data to execute in parallel as execution done in different streams. Our proposed implementation used kernel stream to minimize the latency. It checks the convergence at the host every set of iterations done at the device rather than going back and forth between the host and device. This

Table 3. Training time (MPI-CUDA vs. OPENMP-CUDA)

Dataset (#Trainingsamples, FeaturesDimension, #classes)	Training time (s)		Speedup
	MPI-CUDA	OPENMP-CUDA	
(7404, 102, 9)	1.09	1.416	1.2X
(14811, 102, 9)	1.25	4.505	3.5X
(22118, 102, 9)	1.41	5.201	3.7X
(44442, 102, 9)	2.05	15.397	7.5X

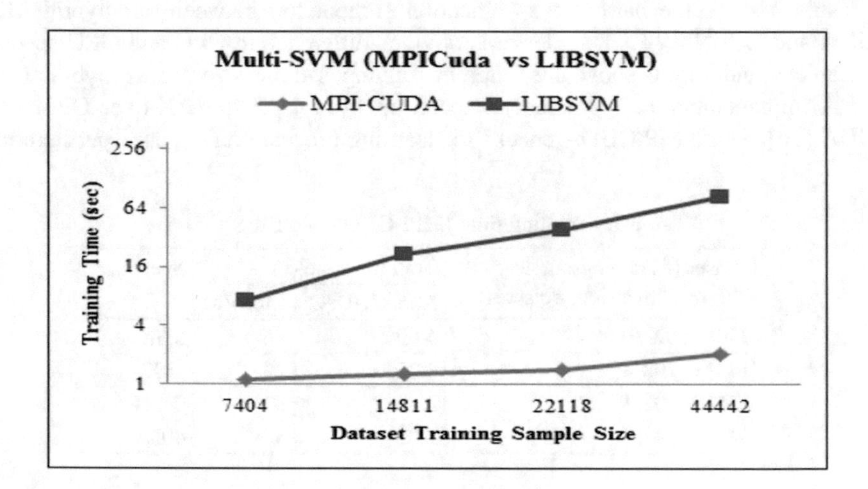

Fig. 5. Training time MPICUDA vs. LIBSVM

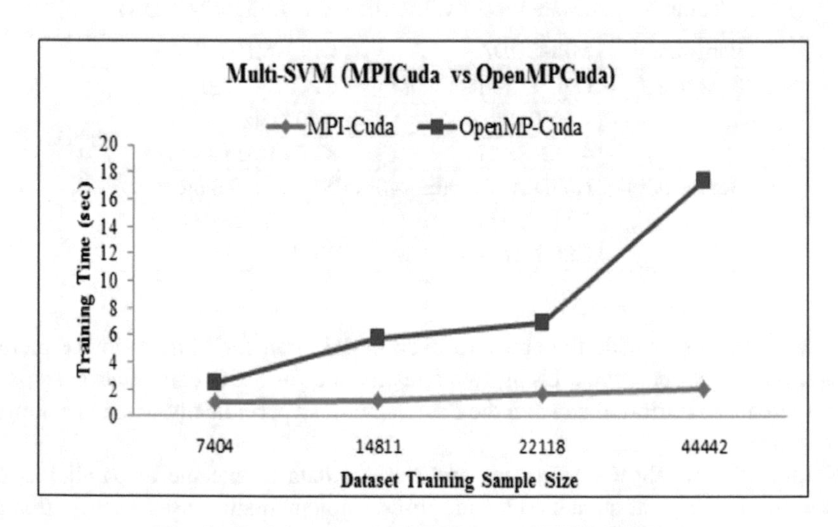

Fig. 6. Training time MPICUDA vs. OPENMPCUDA

implementation reduced the communication cost over the OPENMP-CUDA [16] implementation.

A small MPI communication overhead occurred between nodes but doesn't have impact on overall execution time because of it was needed for master node to initialize and transfer data to others only.

Table 4 shows the classification time for Pavia Centre dataset with 9 classes. Different testing sample sizes on hybrid MPI-CUDA. We achieved a speed up of 6.9X at sample size 24692 and a max. speedup of 13.7X at sample size 60300 over LIBSVM. The speed up has a proportional increase to the testing sample size.

Table 4. Classification time (MPI-CUDA vs. LIBSVM)

(#samples, FeaturesDimension, #classes)		SVM	Classification		Speedup
Training dataset	Testing dataset		Accuracy (%)	Time (s)	
(7404, 102, 9)	(24692, 102, 9)	LIBSVM	94.50	22.57	6.9X
		MPI-CUDA	95.01	3.27	
	(49384, 102, 9)	LIBSVM	92.04	56.01	11X
		MPI-CUDA	93.82	4.76	
	(60300, 102, 9)	LIBSVM	93.48	83.22	13.7X
		MPI-CUDA	94.40	6.07	

5 Conclusion

In the field of Earth observation, our hybrid parallel SVM implementation can be used to extract useful information form huge amount of hyperspectral image dataset. The proposed Hybrid Parallel SVM implementation presented in this paper enhanced the SVM high computational time for solving the optimization problem with SMO. The proposed SMO SVM implementation is designed as a binary classifier on GPU. The Multi classifier implemented the "one-against-one" approach with message passing interface (MPI), and a set of GPUs are used to complete SMO tasks. The training process reached a speedup of 42X over the sequential (LibSVM) and 7.5X over OPENMP-CUDA and a speedup of 13.7X over LIBSVM in Classification process for a dataset of 60300 samples with features size of 102 bands and 9 classes. We can also conclude that the speedup increases as the dataset increases in both training and testing process.

Experimental results derived from hyperspectral image dataset with a processing and analysis of heavy computation Classification Time and Accuracy burdens show that the hybrid parallel SVM can speed up the SVM training process.however it can increase the speed up with using kernel dot product caching feature and use another kernel function rather than gaussian kernel function to make svm implementation more general.

References

1. NVIDIA Corporation: NVIDIA CUDA C Programming Guide (2010)
2. Sanders, J., Kandrot, E.: CUDA by Example: An Introduction to General-Purpose GPU Programming, 1st edn. Addison-Wesley Professional, Reading (2010)
3. Khaled, H., Faheem, H.M., El-Gohary, R.: Design and implementation of a hybrid MPI-CUDA model for the Smith–Waterman algorithm. Int. J. Data Min. Bioinform. **12**(3), 313–327 (2015)
4. Aoyama, Y., Nakano, J., et al.: Rs/6000 sp: Practical MPI Programming. IBM Poughkeepsie, New York (1999)
5. Catanzaro, B., Sundaram, N., Keutzer, K.: Fast support vector machine training and classification on graphics processors. In: Proceedings of the 25th International Conference on Machine Learning, pp. 104–111 (2008)
6. Carlos, J., Ribeiro, B., Lopes, N.: Development of support vector machines (SVMs) in graphics processing units for pattern recognition (2012)
7. Lopes, N., Ribeiro, B.: GPU machine learning library (GPUMLib). In: 2015 Machine Learning for Adaptive Many-Core Machines - A Practical Approach. Springer, Cham, pp. 15–36 (2015)
8. Chang, C.-C., Lin, C.-J.: LIBSVM: a library for support vector machines. ACM Trans. Intell. Syst. Technol. **2**(3), 27:1–27:27 (2011)
9. Keerthi, S.S., Shevade, S.K., Bhattacharyya, C., Murthy, K.R.K.: Improvements to Platt's SMO algorithm for SVM classifier design. Neural Comput. **13**(3), 637–649 (2001)
10. Cortes, C., Vapnik, V.: Support-vector networks. Mach. Learn. **20**(3), 273–297 (1995)
11. Platt, J.: Sequential minimal optimization: a fast algorithm for training support vector machines (1998)
12. Fan, R.-E., Chen, P.-H., Lin, C.-J.: Working set selection using second order ınformation for training support vector machines. J. Mach. Learn. Res. **6**, 1889–1918 (2005)
13. Hsu, C.-W., Lin, C.-J.: A comparison of methods for multiclass support vector machines. IEEE Trans. Neural Netw. **13**(2), 415–425 (2002)
14. Goetz, A.F., Vane, G., Solomon, J.E., Rock, B.N.: Imaging spectrometry for earth remote sensing. Science **228**, 1147–1153 (1985)
15. Green, R.O., et al.: Imaging spectroscopy and the airborne visible/ınfrared ımaging spectrometer (AVIRIS). Remote Sens. Environ. **65**(3), 227–248 (1998)
16. Tan, K., Zhang, J., Du, Q., Wang, X.: GPU parallel implementation of support vector machines for hyperspectral image classification. IEEE J. Sel. Top. Appl. Earth Obs. Remote Sens. **8**(10), 4647–4656 (2015)

Supervised Classification Techniques for Identifying Alzheimer's Disease

Yasmeen Farouk(✉) and Sherine Rady

Faculty of Computer and Information Sciences, Ain Shams University, Cairo, Egypt
yfarouk.bakry@gmail.com

Abstract. Alzheimer's Disease is a serious form of dementia. With no current cure, treatments focus on slowing the progression of the disease and controlling its symptoms. Early diagnosis by studying the biomarkers found in structural MRI is the key. This paper proposes a method which combines texture features extracted from gray level co-occurrence matrix and voxel-based morphometry neuroimaging analysis to classify Alzheimer's disease patients. Different supervised classification techniques are studied, support vector machine, k-nearest neighbor, and decision tree, to obtain best identification accuracy. The paper explores as well the discriminative power for Alzheimer's disease of certain anatomical regions of interest. The proposed technique is applied on gray matter tissues, and managed successfully to differentiate between Alzheimer's disease patients and normal controls with accuracy 92%.

Keywords: Alzheimer's disease · Decision tree · K-nearest neighbor
Magnetic resonance imaging · Region of interest
Support vector machine

1 Introduction

Dementia is a slow decline in memory, problem solving skills and bodily functions. The most common form of dementia is Alzheimer's disease (AD), a chronic disorder that causes the loss of brain cells and eventually leads to death. The average life expectancy of AD patients is three to twelve years [24]. The cause of Alzheimer's disease is not known. However, researchers found that family history is one of the risk factors [9]. With no current cure, treatments focus on slowing the progression of the disease and controlling its symptoms. Available treatments help relieve AD symptoms and promise for better quality of life for the patient and his family. People at high risk for developing AD are encouraged to undergo periodic examinations for the early diagnosis of AD. Unfortunately, only less than half of all AD patients are diagnosed at an early stage.

Neuroimaging examination such as structural MRI (sMRI) can detect early changes in the brain showing evidence for early AD development. This is shown clearly as brains of people diagnosed with AD shrink significantly as the disease progresses specially the area of hippocampus [9]. Neuroimaging techniques

A. E. Hassanien et al. (Eds.): AISI 2018, AISC 845, pp. 189–197, 2019.
https://doi.org/10.1007/978-3-319-99010-1_17

involved in AD studies such as computed tomography (CT) and structural magnetic resonance imaging (sMRI), provides information about the shape, position, and volume of brain tissue. sMRI measures gray matter (GM), white matter (WM), and cerebrospinal fluid (CSF) atrophy related to the loss of nerve cells, synapses, and dendritic de-arborization.

Neurodegeneration in AD is proven to begin in the medial temporal lobe [20]. The loss of structure and function of neurons progresses to entorhinal cortex and then to the hippocampus. The former functions as a hub for memory and navigation and the latter is responsible for autobiographical events and spatial short term memory and its volume reduces abnormally in AD patients. Pre-specifying a set of anatomical regions of interest (ROIs) can lead to better classification results.

This paper proposes a method which combines Voxel-Based Morphometry (VBM) neuroimaging analysis and texture features extracted from gray level co-occurrence matrix. VBM is a technique that explores the focal differences in brain anatomy using statistical parametric mapping. It compares different MRI after spatially normalizing them. Additional gray level co-occurrence matrix (GLCM) texture feature analysis is employed on the gray level MRI to enhance the classification accuracy. Several classification techniques are investigated; Support vector machine (SVM), K-nearest neighbor (K-NN), and decision tree. SVM is a supervised statistical classification method. This binary classifier is used in several works to classify AD and mild cognitive impairment (MCI) [20, 27] or classify AD and normal controls (NC) [17]. KNN is another supervised instance-based learning technique [4] where hypotheses are constructed directly from the training set. Decision trees are successful predictive modeling approaches used in machine learning. The paper explores as well the discriminative power for Alzheimer's disease of certain anatomical ROIs. The work presented here compares to [12] where a whole brain MRI analysis is encountered.

The rest of this work is organized as follows; Sect. 2 gives an insight of the related work presented in AD classification. Section 3 gives a description of the scientific approach and methods used in this study. Section 4 explains data and tools used. It also presents the performance results obtained in the computational experiments performed on the data. Finally, Sect. 5 gives the conclusions of this work.

2 Related Work

Studies have shown that brain atrophy in AD and prodromal AD is spatially distributed over many brain regions [17]. Brain regions that AD can highly affects include the entorhinal cortex, the hippocampus, lateral and inferior temporal structures, and anterior and posterior cingulate. Depending on the type of features extracted from MRI; classification approaches can be categorized into regions of interest (ROI)-based, vertex-based or voxel-based approaches. The ROI-based approach applies nonlinear registration to register MRI to a predefined brain region template. One example of the studies based on volumetric measurements of ROI is [17]; where computing the ROIs from MRI is performed by combining image segmentation and a wrapped classification algorithm. Considerable efforts worked on studying the atrophy rate in volumetric measurements of regions of interest in AD [21] and in MCI [23]. Many researchers tried to investigate whether the atrophy in AD early stages is confined to the hippocampus and the entorhinal cortex or not [20]. In [19] a whole brain hierarchical network is employed to represent different subjects in the brain.

Several machine learning techniques are capable of dealing with high dimensional data encountered in MRI studies, such as support vector machine (SVM) [11], independent component analysis (ICA) [18], linear discriminant analysis (LDA) [5], wavelet transform [14], random forest [13], and Neural Network (NN) [6].

KNN proved robustness in many biomedical fields. Xie [26] presented an enhanced KNN approach for classifying AD patients using complex data that have heterogeneous views. In this study [7], pools of mass spectrometer saliva data are analyzed using an extended K-nearest neighbor algorithm for AD diagnosis and presents a modification of Euclidean distance formula as well. Demirhan [11] used SVM, KNN, and NN to discriminate AD and mild AD from healthy controls using the Open Access Series of Imaging Studies (OASIS) database. Sweety et al. [22] used decision tree classifier and Particle Swarm Optimization (PSO) for feature reduction. The research in [10] summarizes the approaches used for tree simplification in their size and complexity without compromising the accuracy.

Xiao et al. [25] combined three different sMRI features; GM volume, GLCM, and Gabor feature. The data set used was 112. The proposed method achieved 85.71% accuracy promoted to 92.86% by applying feature selection using SVM-RFE and covariance matrix. Aggarwal et al. [3] extracted statistical features from multiple trans-axial slices from hippocampus and amygdala regions. Their work extracted 14 features by applying first and second order statistics. Their method achieved 73.08% accuracy.

3 Proposed Supervised Learning Approach for AD Identification

MRI dataset is analyzed with two pipelined tracks. One track analyzes images using VBM analysis technique which generates one part of the feature descriptor. The other pipelined track uses GLCM to generate textural properties of the MRI deriving the other part of the feature descriptor.

The combined feature descriptor undergoes ROI masking process where the significance of different anatomical ROIs are explored.

Images are finally classified into two classes: AD patient or NC. The framework of the study is illustrated in Fig. 1.

VBM allows investigation of focal differences in brain structure. It can be applied to any measure of anatomy such as GM density. VBM of MRI data involves spatially normalizing all the images to the same stereotactic space, image segmentation where GM images are extracted, smoothing, and finally

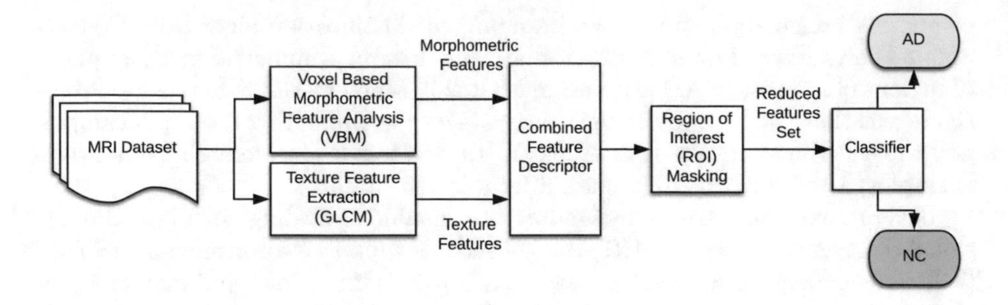

Fig. 1. Block diagram showing the processing for AD identification

performing a statistical analysis to localize differences between images [8]. The output from this analysis is statistical parametric map showing regions where GM concentration differs significantly between AD and NC.

Texture features reflect the regular changes of gray values in images. These changes in the values are correlated statistically and spatially. Textural feature descriptor is constructed from GLCM. GLCM estimates image properties related to second-order statistics. It accounts for the spatial inter-dependency of two pixels at specific neighboring positions. It is used to calculate five different texture features [16]; contrast, correlation, energy, homogeneity, and entropy.

The classifier block represents three different classification techniques; SVM, KNN, and decision tree [15]. SVM is a supervised learning model that can perform binary linear or non-linear classification. SVM supports both regression and classification tasks. It is based on the concept of decision planes that define decision boundaries. It maps the training set into a new maximal margin hyperplane where a clear gap between the two different classes can be created. SVM creates a model to classify new images. Linear SVM parameters define decision hyperplanes in the multi-dimensional feature space:

$$g(x) = w^T x + b \tag{1}$$

where x denotes the feature vector, w is the weight vector and b is the threshold. The hyperplane position is determined by vector w and b. The vector is orthogonal to the decision plane and b determines its distance to the origin. To construct an optimal hyperplane, SVM employs an iterative training algorithm, which is used to minimize an error function. According to the form of the error function, SVM classification models have two types C-SVM and nu-SVM. C and nu are regularization parameters which help implement a penalty on the mis-classifications that are performed while separating the classes. Thus helps in improving the accuracy of the output. The C-SVM model is chosen.

K-NN is a supervised instance-based learning model where hypotheses are constructed directly from the training set. The output of K-NN is a class membership determined by the dominant class label of its neighbors. The training phase of the algorithm stores the feature vectors and class labels of the training set. In the classification phase, the unlabeled test sample is assigned to the most frequent class label in the k training samples nearest to that query point. If k = 1, then the test sample is assigned to the class of that single nearest neighbor. The exhaustive comparison of K-NN can be time consuming. Moreover, storing the feature vector sets in the training phase consumes memory too. The Euclidean distance is used to compute distance between x and y samples:

$$d(x, y) = \sqrt{\sum_{i=1}^{k} (y_i - x_i)^2} \quad (2)$$

Decision tree represents the data in a tree-like structure, where each internal node denotes a test on a feature, each branch represents an outcome of the test, and each leaf node holds the class label. Given a test sample, X, with an unknown class label, its feature values are tested against the decision tree. The root node is tested first and continues down until reaching a leaf node holding the predicted class label. Decision tree can be converted to classification rules and can handle multidimensional data. Information gain is used as a feature selection measure. It is based on Shannon's information theory. The feature with the highest information gain is chosen as the splitting feature for the root node. This feature minimizes the number of tests needed to classify a given sample. The expected information needed to classify a sample in D is given by Eq. (3) where p_i is the probability that a sample in D belongs to class C_i.

$$Info(D) = -\sum_{i=1}^{n} p_i log_2(p_i) \quad (3)$$

4 Experiments and Results

4.1 Dataset

The dataset used in this work is obtained from Alzheimer's Disease Neuroimaging Initiative (ADNI) database [1]. Tissue can be characterized by two different relaxation times; T1 and T2. It is a measure of the time taken for spinning protons to lose phase coherence among the nuclei spinning perpendicular to the main field. The most common MRI sequences are T1-weighted and T2-weighted scans. In T1-weighted images GM appears darker than WM. In this study, a total of 275 T1-weighted MR images are considered. Table 1 shows the demographics of the dataset.

4.2 Experimental Work

AD patients and NC samples undergo a statistical t-test at the final step in VBM block in the proposed framework shown in Fig. 1. This statistical analysis shows the significant regions of GM loss between AD and NC in the whole brain. ROI binary masks are used on these regions to extract ROIs. Eight ROIs are defined; Hippocampus, Cerebelum Crus2 Left, Cerebelum Crus2 Right, Medulla,

Table 1. Demographics of the dataset

	Normal N = 113	AD N = 162
Male/female	72/41	71/91
Age: $\mu(\sigma)$	77.49(5.88)	73.82(7.63)
MMSE: $\mu(\sigma)$	25.74(7.74)	21.54(3.92)

Calcarine, Pons, Occipital Lobe, and Frontal Lobe. Performance tests of the three classifiers; SVM, KNN, decision tree, are evaluated on groups of ROIs. Each experiment uses different ROI feature vectors:

1. *roi1*: refers to Hippocampus region.
2. *roi2*: refers to Hippocampus, Cerebelum Crus2 Left, and Cerebelum Crus2 Right regions.
3. *roi3*: refers to Hippocampus, Cerebelum Crus2 Left, Cerebelum Crus2 Right, and Calcarine regions.
4. *roi4*: refers to Hippocampus, Cerebelum Crus2 Left, Cerebelum Crus2 Right, Calcarine, and Frontal Lobe regions.
5. *roi5*: refers to Hippocampus, Cerebelum Crus2 Left, Cerebelum Crus2 Right, Calcarine, Frontal Lobe, Pons, Occipital Lobe, and Medulla regions.

The k-fold cross validation strategy is adopted for evaluating the classification performance. Cross-validation is a technique used to evaluate predictive models by partitioning the original sample into a training set to train the model and a test set to evaluate it. Hence, all observations are used for both training and validation. In this work, k is set to 10.

Classification accuracy, sensitivity, and specificity are then calculated using the given formulas [15]:

$$Accuracy = (TP + TN)/(P + N) \tag{4}$$

$$Sensitivity, = TP/P \tag{5}$$

$$Specificity = TN/N \tag{6}$$

where TP, TN, P, N refer to the number of true positive, true negative, positive, and negative samples, respectively.

KNN classifier obtained the best accuracy 92%, SVM comes next with 88% and finally decision tree with 79%. KNN achieved the fastest running time with average 2.36 s for all ROIs. Decision tree obtained the slowest running time with average 21 s for all ROIs. SVM recorded 8.5 s average running time for all ROIs. Figure 2a, b, and c shows the details of the performance of each classifier.

ROI analysis proved to have more discriminative power than the whole brain methods for identifying AD. The third ROI recorded 4% increase in the accuracy compared to whole brain method [12] with only 7% of the whole brain feature vector and a 70X speedup in the running time. This ROI included Hippocampus,

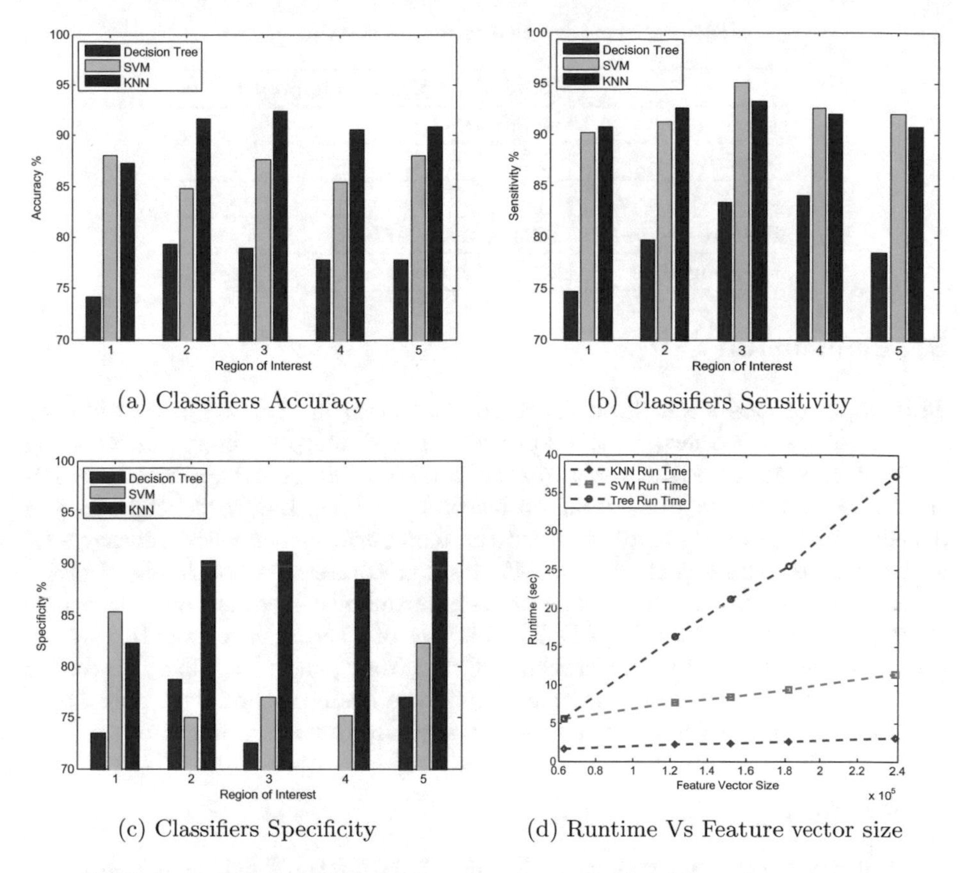

(a) Classifiers Accuracy

(b) Classifiers Sensitivity

(c) Classifiers Specificity

(d) Runtime Vs Feature vector size

Fig. 2. Comparing different classifiers; accuracy, sensitivity, specificity, and run-time vs feature vector size on different ROIs

Cerebelum Crus2 Left, Cerebelum Crus2 Right, and Calcarine regions. Performance results of the third ROI compared to the whole brain method is listed in Table 2. Figure 2d shows the relationship between the feature vector size of different RIOs and the running time. It is clearly shown that decision tree classifier running time is more sensitive to the feature vector size than other techniques. Its running time increases rapidly when the feature vector size increases. Other classifiers running time recorded slight increase.

All VBM data preprocessing was performed using DARTEL toolbox in SPM12 running on MATLAB R2014b [2]. The image is registered to the MNI space and smoothed by 8 mm FWHM Gauss kernel function. Selected spatial directions for GLCM computations are $0°$, $45°$, $90°$, and $135°$. SVM with a linear kernel implemented using MATLAB 'libsvm' toolbox is applied for classification. In SVM classifier, the coefficient C which affects the trade-off between complexity and proportion of non-separable samples is fixed with value 1 in all experiments. The presented work is done on a 64-bit Intel® Core^{TM} i5-4200M CPU @ 2.50 GHz x4 machine, with 11.5 GiB memory and ubuntu 14.04 operating system.

Table 2. Performance results of the third ROI

	ROI3	Whole brain approach [12]
Accuracy	92.36	88.00
Sensitivity	93.21	92.59
Specificity	91.15	81.42
Feature vector size	152371	2127297
Running time	2.3	140.9

5 Conclusion

This paper proposes a method which combines texture features extracted from gray level co-occurrence matrix and voxel-based morphometry neuroimaging analysis to classify Alzheimer's disease patients. Different supervised classification techniques are studied; support vector machine, k-nearest neighbor, and decision tree. KNN classifier obtained the best performance with accuracy 92% and the lowest running time 2 s. SVM classifier comes next and finally decision tree. ROI analysis proved to have more discriminative power than the whole brain methods for identifying AD. The best set of discriminator was; Hippocampus, Cerebelum Crus2 Left, Cerebelum Crus2 Right, and Calcarine. It recorded 4% increase in the accuracy compared to whole brain method with only 7% of the whole brain feature vector and a 70X speedup in the running time.

References

1. Alzheimer's disease neuroimaging initiative (adni). http://adni.loni.usc.edu/
2. Matlab software. http://www.mathworks.com/products/matlab/
3. Aggarwal, N., Rana, B., Agrawal, R.: Detection of Alzheimer's disease via statistical features from brain slices. In: FLAIRS 2013 - Proceedings of the 26th International Florida Artificial Intelligence Research Society Conference, pp. 172–175 (2013)
4. Aha, D.W., Kibler, D., Albert, M.K.: Instance-based learning algorithms. Mach. learn. **6**(1), 37–66 (1991)
5. Alam, S., Kwon, G.R.: Alzheimer disease classification using KPCA, LDA, and multi-kernel learning SVM. Int. J. Imaging Syst. Technol. **27**(2), 133–143 (2017)
6. Aljović, A., Badnjević, A., Gurbeta, L.: Artificial neural networks in the discrimination of Alzheimer's disease using biomarkers data. In: 2016 5th Mediterranean Conference on Embedded Computing (MECO), pp. 286–289. IEEE (2016)
7. Anyaiwe, D.E., Wilson, G.D., Geddes, T.J., Singh, G.B.: Harnessing mass spectra data using KNN principle: diagnosing Alzheimer's disease. ACM SIGBioinform. Rec. **7**(3), 2–9 (2018)
8. Ashburner, J., Friston, K.J.: Voxel-based morphometry—the methods. Neuroimage **11**(6), 805–821 (2000)
9. Association, A., et al.: Basics of Alzheimer's disease: what it is and what you can do. Alzheimer's Association (2012)
10. Breslow, L.A., Aha, D.W.: Simplifying decision trees: a survey. Knowl. Eng. Rev. **12**(1), 1–40 (1997)

11. Demirhan, A.: Classification of structural MRI for detecting Alzheimer's disease. Int. J. Intell. Syst. Appl. Eng. **4**(Special Issue–1), 195–198 (2016)
12. Farouk, Y., Rady, S., Faheem, H.: Statistical features and voxel-based morphometry for Alzheimer's disease classification. In: 2018 9th International Conference on Information and Communication Systems (ICICS), pp. 133–138, April 2018
13. Gray, K.R., Aljabar, P., Heckemann, R.A., Hammers, A., Rueckert, D., Initiative, A.D.N.: Random forest-based similarity measures for multi-modal classification of Alzheimer's disease. NeuroImage **65**, 167–175 (2013)
14. Hackmack, K., Paul, F., Weygandt, M., Allefeld, C., Haynes, J.D., Initiative, A.D.N.: Multi-scale classification of disease using structural MRI and wavelet transform. Neuroimage **62**(1), 48–58 (2012)
15. Han, J., Pei, J., Kamber, M.: Data Mining: Concepts and Techniques. Elsevier, New York City (2011)
16. Haralick, R.M., Shanmugam, K., Dinstein, I.H.: Textural features for image classification. IEEE Trans. Syst. Man Cybern. **6**, 610–621 (1973)
17. Hidalgo-Muñoz, A.R., Ramírez, J., Górriz, J.M., Padilla, P.: Regions of interest computed by SVM wrapped method for Alzheimer's disease examination from segmented MRI. Front. Aging Neurosci. **6**, 20–30 (2014)
18. Khedher, L., Ramírez, J., Górriz, J.M., Brahim, A., Illán, I.: Independent component analysis-based classification of Alzheimer's disease from segmented MRI data. In: International Work-Conference on the Interplay between Natural and Artificial Computation, pp. 78–87. Springer, Heidelberg (2015)
19. Liu, J., Li, M., Lan, W., Wu, F.X., Pan, Y., Wang, J.: Classification of Alzheimer disease using whole brain hierarchical network. IEEE/ACM Trans. Comput. Biol. Bioinform. **15**(2), 624–632 (2018)
20. Magnin, B., Mesrob, L., Kinkingnéhun, S., Pélégrini-Issac, M., Colliot, O., Sarazin, M., Dubois, B., Lehéricy, S., Benali, H.: Support vector machine-based classification of Alzheimer's disease from whole-brain anatomical MRI. Neuroradiology **51**(2), 73–83 (2009)
21. Rusinek, H., Endo, Y., De Santi, S., Frid, D., Tsui, W.H., Segal, S., Convit, A., de Leon, M.: Atrophy rate in medial temporal lobe during progression of Alzheimer disease. Neurology **63**(12), 2354–2359 (2004)
22. Sweety, M.E., Jiji, G.W.: Detection of Alzheimer disease in brain images using PSO and decision tree approach. In: International Conference on Advanced Communication Control and Computing Technologies (ICACCCT), pp. 1305–1309. IEEE (2014)
23. Tapiola, T., Pennanen, C., Tapiola, M., Tervo, S., Kivipelto, M., Hänninen, T., Pihlajamäki, M., Laakso, M.P., Hallikainen, M., Hämäläinen, A.: MRI of hippocampus and entorhinal cortex in mild cognitive impairment: a follow-up study. Neurobiol. Aging **29**(1), 31–38 (2008)
24. Todd, S., Barr, S., Roberts, M., Passmore, A.P.: Survival in dementia and predictors of mortality: a review. Int. J. Geriatr. Psychiatry **28**(11), 1109–1124 (2013)
25. Xiao, Z., Ding, Y., Lan, T., Zhang, C., Luo, C., Qin, Z.: Brain MR image classification for Alzheimer's disease diagnosis based on multifeature fusion. Comput. Math. Methods Med. **2017**, 13 (2017)
26. Xie, Y.: KNN++: An enhanced k-nearest neighbor approach for classifying data with heterogeneous views. In: International Conference on Hybrid Intelligent Systems, pp. 13–23. Springer, Heidelberg (2016)
27. Yang, S.T., Lee, J.D., Chang, T.C., Huang, C.H., Wang, J.J., Hsu, W.C., Chan, H.L., Wai, Y.Y., Li, K.Y.: Discrimination between Alzheimer's disease and mild cognitive impairment using SOM and PSO-SVM. Comput. Math. Methods Med. **2013**, 10 (2013)

Supervised Classification of Cancers Based on Copy Number Variation

Sanaa Fekry Abed Elsadek[1(✉)], Mohamed Abd Allah Makhlouf[2], and Mohamed Amal Aldeen[1]

[1] Computer and Systems Engineering Department, Faculty of Engineering, Zagazig University, Zagazig, Egypt
Sanaafekry2003@yahoo.com,
dr.mohamedamalzeidan@gmail.com

[2] Faculty of Computers and Informatics, Suez Canal University, Ismailia, Egypt
m.abdallah@ci.suez.edu.eg

Abstract. Genomic variation in DNA can cause many types of human cancer so the machine learning has important role in genomic medicine it can help to classify, predict and analysis of DNA sequence. Which is the most important biological characteristic? DNA copy number variations (CNVs) used to understand the difference between different human cancers and predict cancer causing from genetic sequence. But it's not easy due to the high dimensionality of the CNV features. This paper presents approach to computationally classify a set of human cancer types. We use machine learning to train and test various models on set of human cancer using the CNV level values of 23,082 genes (features) for 2916 instances to construct the classifier. Then the genes are selected according to their importance by the filter feature selection method. We compare the performance of seven classifiers Support vector Machine, Random Forest, j48, Neural Network, Logistic Regression, Bagging and Dagging with other benchmark using 10-fold cross validation. The best performance achieved accuracy value 0.859 and ROC value 0.965 which are promising results. The classification models developed in this research could provide a reasonable prediction of the cancer patients' stage based on their CNV level values. The proposed model confirmed that genes from chromosome 3 have in developing human cancers. It also predicted new genes not studied so far as important ones for the prediction of human cancers.

Keywords: DNA copy number variations · Support vector machine
Random Forest J48 · Neural networks · Logistic regression · Machine learning
Supervised classification

1 Introduction

CNV can be defined as deletions, amplification and insertion in one or both strands in DNA they will make DNA region be lager or fewer number of copies of genes this variation in human genes referred as DNA copy number variation (CNV). Many of studies done using CNV features by using many of algorithms Integrating multiple genomic data types for building predictive models or selecting features has been an

A. E. Hassanien et al. (Eds.): AISI 2018, AISC 845, pp. 198–207, 2019.
https://doi.org/10.1007/978-3-319-99010-1_18

important research focus in bioinformatics since several years ago [1, 2]. Li et al. [3] proposed a two layered Bayesian network approach to integrate relations from gene expressions. Cancer formation and progression are also associated with change in copy number [12]. Based on tree analysis of CNVs in breast cancer, Li et al. found that the genetic alteration of ErbB2 occurs early in breast cancer and the CNVs of PIK3CA, AKT2, RAS, PTEN and CCND1are late events [15]. DNA can be defined as normally two copies of each gene in double stranded of human genes. Also ample evidence proven that individuals with some CNVs might show cancer prone [4]. Currently, researches have shown that many human cancer diseases involve CNV that could alter the diploid status of particular locus of the genome. Cancer formation and progression are also associated with change in copy number [13]. Reasons for the relevance between CNVs and neurodevelopmental diseases could be the perturbation of gene pathways involving in neuron development [14]. Many studies integrated multiple types of genetic data for breast cancer clustering have been proposed. Curtis et al. [16] presented a novel molecular stratification of breast cancer by combining genome and transcriptome of 2000 breast cancer patients. Also, Ali et al. [17] classified breast cancers into ten subtypes based on the integration of genomic CNV and gene expression data. And in another study [18], proposed a computational method that combined gene expression and DNA methylation data to implement machine learning aided classification of breast cancer patients.

In this study, we developed a machine learning computational Method, which uses the information of gene CNV levels to classify different types of cancers, Breast. (BRCA), Bladder urothelial (BLCA), Colon (COAD), Glioblastoma. (GBM), Kidney (KIRC), and Head and neck squamous cell (HNSC) using 10-fold cross validation of 16382 genes are finally chosen to differentiate cancer types.

The remainder of the paper is organized as follows: Sect. 2 focuses on Materials and methods. Section 3 emphasizes on Experimental Results. Section 4 emphasizes on Comparison with Other Studies, finally Sect. 5 Conclusion and future work.

2 Materials and Methods

The general architecture of the proposed approach is depicted in Fig. 1. The details of the proposed approach are discussed in this section.

2.1 DNA Data Sets Characteristics and Preparation

The dataset in this research adopted from [10] to train-test the proposed model. It contains DNA CNVs level of 23082 genes for 2916 instances. The data used the CNV in variant types of cancers were downloaded from the cBioPortal for Cancer Genomics [5–7]. We used a six cancer types called Bladder urothelial carcinoma (BL), Breast invasive Carcinoma (BR), Colon and rectum (CO). Glioblastoma multiform (GB). Head and neck squamous cell (HN) and Kidney renal clear-cell (KI).

The feature descriptors were adopted to perform the classification task. The CNVs for 24,174 genes were used as features to construct the model. Also we adopted the discretization of the CNV values into 5 different values in the database. The database

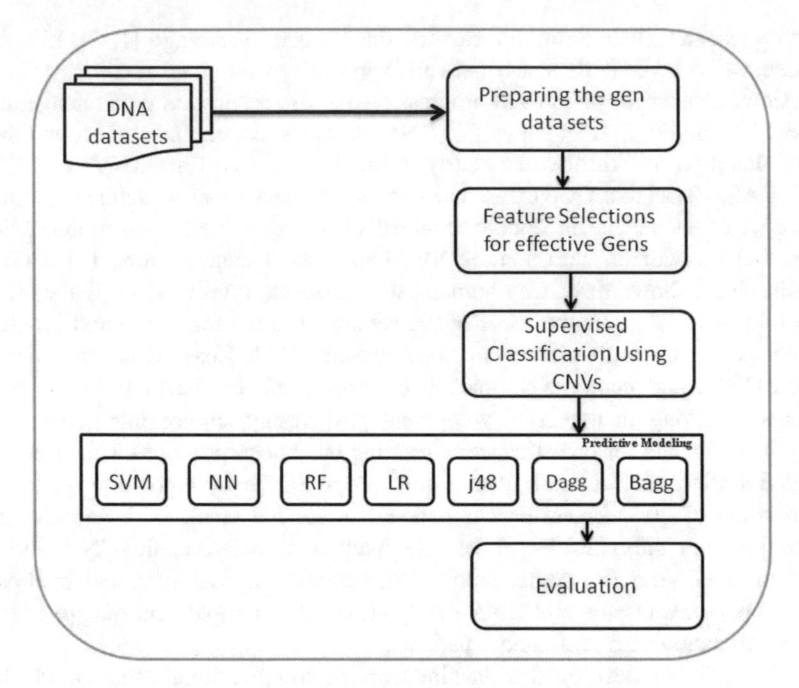

Fig. 1. Steps of the proposed approach

cBioPortal contains 11 cancer types were available. However, each cancer type has its samples number, in this study, we only used six cancer types of them. The cancer types and the sample numbers presented in Table 1. Totally there were 2914 samples in the six cancer types.

Table 1. The number of samples in the cancer types in our dataset

Index	Cancer type	Number of samples
1 (BR)	BRCA (breast invasive carcinoma)	847
2 (BL)	Bladder urothelial carcinoma (BL)	134
3 (CO)	COAD/READ (colon adenocarcinoma/rectum adenocarcinoma)	575
4 (GB)	GBM (glioblastoma multiforme)	563
5 (HN)	Head and neck squamous cell (HN)	305
6 (KI)	KIRC (kidney renal clear cell carcinoma)	490
Total		2914

2.2 Features Selection

Features Selection by filter methods were the earliest techniques in the machine learning which Rank the features in descending order according to their importance [19, 20]. Filters algorithms work to filter out features that have low chance to be useful in data analysis. Filter methods are less expensive in the computationally cost and also more generic than wrappers methods because they do not take classifiers into consideration. They break down characteristic properties of the data assess the significance of highlights, then the low scoring attributes are erased and the remaining subset will be the input to the classification algorithm. Information gain algorithm is one of the filter methods that evaluate the information about the class. Therefore a threshold for selecting process must be first determined, a feature is selected when the value of it is greater than the threshold value. In our experiment the threshold value was 0.0204 which filter and select 16381 genes from 23,082 genes.

2.2.1 Support Vector Machine (SVM)

SVM was first developed by Vapnik [8], SVM is a supervised machine learning technique for data analysis. SVM classifies data by nonlinearly mapping the data to high dimensional feature spaces first, and then finds the linear optimal hyperplane, a decision boundary, to separate the data set of one class from another. The hyperplane with the maximum margin between the classes is sought by SVM classifier. The hyperplane can be written as

$$\omega^T \cdot \phi(X) + b = 0 \tag{1}$$

where ω is the weight vector, $\phi(0)$ is the nonlinear mapping and b is the bias. The optimal hyperplane is defined by finding ω and b which minimize the following function [9].

$$\min \frac{1}{2} \omega^T \cdot \omega + C \sum_{i=1}^{n} \xi_i$$

2.2.2 Random Forest (RF)

RFs, are a popular ensemble algorithm that can be used to build predictive models for both classification and regression problems. Ensemble methods use multiple learning models to get better predictive in random forest, the model creates an entire forest of random uncorrelated decision trees to arrive at the best possible solution. RFs attempt to reduce the correlation issue by choosing only a subsample of the feature space at each split. Essentially, it aims to make the trees de-correlated and prune the trees by setting a stopping criteria for node splits.

2.2.3 Decision Trees (J48)

Decision Tree models used as a classifiers. It has a tree structure in which all the internal nodes represent an attribute and the leaf nodes represent the classification categories as per the goal class. J48 decision tree also called C4.5 first developed by Ross Quinlan. The basic concept by which classification is done in Decision tree is that it takes multiple linear decisions to perform a nonlinear classification. J48 tree uses information gain to decide on which attributes to branch at each node of the tree depending on information gain. A Decision Tree uses a top-down to divide and conquer as classification strategy and it partitions a set of given entities into smaller classes on the basis of automatically selected rules.

2.2.4 Neural Networks

Artificial Neural Network (ANN) is a computational system exploited from the analogy of biological neural networks. ANNs are physical systems, gain, and store and use experiential learning [19]. Each neuron has an interior state, which is called an activation function. The output signals delivered after combining the input signals and activation rule. Input layer represent the CNV for genes for every type of cancer then the hidden layer function to make learning for the system. So it can detect according to CNV feature the type of cancer.

2.2.5 Logistic Regression

Logistic regression is a regression model which is used to estimate or predict the probability of categorical variables. It has been widely used in binary classified variables where the output can take only two values, "0" and "1". LR was developed by statistician David Cox [11, 12] it is a discrete but linear continues. The logistic regression sigmoid activation is:

$$g(z) = \frac{1}{1 + e - z} \tag{3}$$

Given a set of features $X = (x_1, x_2, \ldots x_n)$ of the Genes, the probability $h_\theta\ (X)$ of the categorical dependent variable y (BR, BL, CO, GB, KI, HN) equals,

$$h_\theta(X) = g(\theta^T X) \tag{4}$$

where θ is the regression coefficient which is determined by minimizing the cost function of logistic regression.

3 Experimental Results

The features (genes) were ranked using Information Gain algorithm, as results shown in Table 4. The classification power of the different algorithms used in this study is evaluated by training/testing using 10-fold cross-validation method to examine the performances of the presented methods. The model performance was evaluated based on the actual and predicted classes [5–7]. Table 2 represents the different models performance measures like accuracy, Receiver operating characteristic (ROC) area, Matthew Correlation Coefficient (MCC) and precision. The equations from 6–10 represent these measures (Table 3).

$$Recall(I) = \frac{TP}{TP + FN} \tag{6}$$

$$\text{Precision } (I) = \frac{TP}{TP + FP} \tag{7}$$

$$ACC = \frac{TP + TN}{TP + TN + FP + FN} \tag{8}$$

$$\text{F - measure} = \frac{2 * \text{Precision} * \text{Recall}}{\text{Recall} + \text{Precision}} \tag{9}$$

$$MCC = \frac{TP \times TN - FP \times FN}{\sqrt{(TP + FN) \times (TP + FN) \times (FN + FP) \times (TN + FN)}} \tag{10}$$

Table 2. The performance of different algorithms.

Algorithms	Accuracy	MCC	ROC Area	Precision	Recall
Random forest	79.8	0.715	0.944	0.736	0.888
Support vector	80.38	0.721	0.903	0.804	0.799
J48	72.9	0.600	0.803	0.721	0.710
Logistic reg.	**85.9**	0.798	0.965	0.852	0.862
Neural network	**80.9**	0.757	0.874	0.825	0.829
Bagging	78.49	0.702	0.929	0.772	0.810
Dagging	77.50	0.695	0.936	0.722	0.867

Table 3. The performance of all algorithms for different types of cancer.

	BR	BL	CO	GB	KI	HN
Random forest						
TP rate	0.888	0.089	0.730	0.870	0.888	0.712
ROC area	0.944	0.815	0.948	0.978	0.948	0.942
Precision	0.736	0.667	0.800	0.894	0.867	0.724
F-measure	0.805	0.157	0.764	0.882	0.877	0.718
Recall	0.888	0.089	0.730	0.870	0.888	0.712
FP rate	0.130	0.002	0.045	0.025	0.028	0.031
Support vector						
TP rate	0.799	0.430	0.821	0.885	0.910	0.631
ROC area	0.903	0.841	0.935	0.974	0.964	0.880
Precision	0.804	0.397	0.795	0.931	0.869	0.675
F-measure	0.802	0.413	0.808	0.907	0.889	0.652
Recall	0.799	0.430	0.821	0.885	0.910	0.631
FP rate	0.080	0.032	0.052	0.016	0.028	0.036

(continued)

Table 3. *(continued)*

	BR	BL	CO	GB	KI	HN
J48						
TP rate	0.710	0.237	0.755	0.828	0.863	0.556
ROC area	0.803	0.593	0.850	0.894	0.922	0.769
Precision	0.721	0.246	0.706	0.860	0.882	0.599
F-measure	0.715	0.242	0.729	0.843	0.845	0.576
Recall	0.710	0.237	0.755	0.828	0.863	0.556
FP rate	0.113	0.035	0.077	0.032	0.036	0.044
Logistic reg.						
TP rate	0.862	0.481	0.889	0.924	0.918	0.748
ROC area	0.965	0.929	0.978	0.991	0.989	0.963
Precision	0.852	0.739	0.812	0.939	0.895	0.804
F-measure	0.857	0.583	0.849	0.931	0.906	0.775
Recall	0.862	0.481	0.889	0.924	0.918	0.748
FP rate	0.061	0.008	0.050	0.014	0.022	0.021
Neural network						
TP rate	0.757	0.310	0.740	0.869	0.858	0.664
ROC area	0.874	0.268	0.815	0.920	0.904	0.642
Precision	0.825	0.274	0.831	0.886	0.882	0.703
F-measure	0.070	0.019	0.065	0.023	0.024	0.036
Recall	0.829	0.411	0.758	0.902	0.882	0.696
FP rate	0.827	0.329	0.793	0.894	0.882	0.699
Bagging						
TP rate	0.810	0.133	0.819	0.837	0.892	0.673
ROC area	0.929	0.792	0.946	0.965	0.977	0.926
Precision	0.772	0.514	0.741	0.882	0.812	0.725
F-measure	0.790	0.212	0.778	0.859	0.850	0.698
Recall	0.810	0.133	0.819	0.837	0.892	0.673
FP rate	0.098	0.006	0.070	0.027	0.042	0.030
Dagging						
TP rate	0.867	00.00	0.777	0.844	0.871	0.578
ROC area	0.936	0.849	0.950	0.971	0.978	0.938
Precision	0.722	00.00	0.723	0.890	0.820	0.783
F-measure	0.790	00.00	0.749	0.866	0.845	0.665
Recall	0.867	00.00	0.777	0.844	0.871	0.578
FP rate	0.137	00.00	0.073	0.025	0.039	0.019

4 Comparison with Other Studies

Zhang [10] used the CNVs of 23,082 genes as features and construct a classifications using Dagging classifier after applying the Incremental Feature Selection methods, the model accuracy reached over 0.75 whereas our proposed model achieved the accuracy over 0.85 and ROC area 0.965 to evaluate the impact of our proposed method, we have analyzed the identified genes using published database, we also compared the significance of selected genes with the other studies. These analysis confirm that the identified genes are biologically relevant to contribute into molecular location on chromosome 3 genes. The study has found interesting results, one is that using filter feature selection has discovered the high ranked genes resides in the chromosome 3 as a number of previous studies. The difference is that in this study several genes were found significant that did not appear till now in cancer references.

Table 4. The top ranked genes according information gain value (IGV).

Gene symbol	Full name	IGV rank	Reference in cancer	Chromosome
IL17RC	Interleukin 17 receptor C	0.3871	[14, 22]	3
CRELD1	Cysteine rich with EGF like domains 1	0.3867	?	3
ARPC4-TTLL3	ARPC4-TTLL3 read through	0.3864	?	3
RPUSD3	RNA pseudouridylate synthase domain containing 3	0.3863	?	3
TTLL3	Tubulin tyrosine ligase like 3	0.3861	?	3
PRRT3	Proline rich transmembrane protein 3	0.3861	?	3
JAGN1	Jagunal homolog 1	0.386	?	3
CIDEC	Cell death inducing DFFA like effector c	0.386	?	3
CDC25A	Cell division cycle 25A	0.3075	[23]	3
ZMYND11	Zinc finger, MYND-type containing 11	0.2951	[25]	10
KBTBD6	Kelc repeat and BTB (POZ) domain containing 6	0.2813	[26]	13
EGFR	Epidermal growth factor receptor	0.2636		7
ZNF503-AS1	ZNF503 antisense RNA 1	0.2478	[26]	10
SEMA6A	Sema domain, transmembrane domain (TM), and cytoplasmic domain, (semaphorin) 6A	0.2155		5
CUL2	Cullin 2	0.2831	[21]	?
MIR4703	Microrna 4703	0.281	?	13
CDKN2A	Cyclin-dependent kinase inhibitor 2A	0.2831		9
CTBP2	C-terminal binding protein 2	0.2343	[22]	10
MMD2	Monocyte to macrophage differentiation-associated 2	0.1347	[24]	7
RPS15	Ribosomal protein S15	0.0746	?	19

IL17RC ranked the 1st in Table 4, is a transmembrane protein coding gene. In spite of its role in cancer is still dodgy, previous research presented that IL17RC is associated to prostate cancer. IL17RC isoforms are differently expressed in androgen-independent and androgen-dependent prostate cancers.

5 Conclusion and Future Work

In this research, we used a set of machine learning techniques to classify different cancer types using the genes CNV level values. Using the information gain filter method to filter the most important genes that achieve accuracy of over 0.85 by using CNV levels of 16381genes based on the information gain filter algorithm. Analysis of the top ranked genes may play important roles in differentiating human cancer types and help understand tumorigenesis of different cancers. The proposed model presented that genes from chromosome 3 have in developing human cancers. It also predicted new genes not studied so far as important ones for the prediction of human cancers. As future work, the model can be used with an optimization methods like genetic algorithm and PSO to apply feature selection techniques to improve the performance of the model and focus on the most effective genes.

References

1. Barutcuoglu, Z., Schapire, R.E., Troyanskaya, O.G.: Hierarchical multi-label prediction of gene function. Bioinformatics **22**(7), 830–836 (2006)
2. Tsuda, K., Shin, H.J., Scholkopf, B.: Fast protein classi_cation with multiple networks. Bioinformatics **21**(2), 59–65 (2005). Joint Meeting of the 4th European Conference on Computational Biology/6th Meeting of the Spanish-Bioinformatics-Network, Madrid, Spain, 28 Sept–01 Oct (2005)
3. Li, J., Li, X., Su, H., Chen, H., Galbraith, D.W.: Framework of integrating gene relations from heterogeneous data sources: an experiment on *Arabidopsis thaliana*. Bioinformatics **22** (16), 2037–2043 (2006)
4. Friedberg, E.C., Walker, G.C., Siede, W., Wood, R.D.: DNA Repair and Mutagenesis. American Society for Microbiology Press, Washington (2005)
5. Ciriello, G., Miller, M.L., Aksoy, B.A., Senbabaoglu, Y., Schultz, N., Sander, C.: Emerging landscape of oncogenic signatures across human cancers. Nat. Genet. **45**, 1127–1133 (2013)
6. Cerami, E., Gao, J., Dogrusoz, U., Gross, B.E., Sumer, S.O., Aksoy, B.A., Jacobsen, A., Byrne, C.J., Heuer, M.L., Larsson, E.: The cBio cancer genomics portal: an open platform for exploring multidimensional cancer genomics data. Cancer Discov. **2**, 401–404 (2012)
7. Gao, J., Aksoy, B.A., Dogrusoz, U., Dresdner, G., Gross, B., Sumer, S.O., Sun, Y., Jacobsen, A., Sinha, R., Larsson, E.: Integrative analysis of complex cancer genomics and clinical profiles using the cBioPortal. Sci. Signal. **6**, l1 (2013)
8. Cortes, C., Vapnik, V.: Support-vector networks. Mach. Learn. **20**(3), 273–297 (1995)
9. Schölkopf, B., Smola, A.J., Williamson, R.C., Bartlett, P.L.: New support vector algorithms. Neural Comput. **12**(5), 1207–1245 (2000)
10. Zhang, N., et al.: Classification of cancers based on copy number variation landscapes. Biochimica et Biophysica Acta (BBA) Gen. Subj. **1860**(11), 2750–2755 (2016)
11. Freedman, D.A.: Statistical Models: Theory and Practice, p. 128. Cambridge University Press, Cambridge (2009)
12. Walker, S.H., Duncan, D.B.: Estimation of the probability of an event as a function of several independent variables. Biometrika **54**(1–2), 167–179 (1967)
13. Frank, B., Bermejo, J.L., Hemminki, K., Sutter, C., Wappenschmidt, B., Meindl, A., Kiechle-Bahat, M., Bugert, P., Schmutzler, R.K., Bartram, C.R.: Copy number variant in the candidate tumor suppressor gene MTUS1 and familial breast cancer risk. Carcinogenesis **28**, 1442–1445 (2007)

14. Elia, J., Gai, X., Xie, H., Perin, J., Geiger, E., Glessner, J.: M. D'arcy, E. Frackelton, C. Kim, F. Lantieri, Rare structural variants found in attention-deficit hyperactivity disorder are preferentially associated with neurodevelopmental genes. Mol. Psychiatry **15**, 637–646 (2010)
15. Li, X.C., Liu, C., Huang, T., Zhong, Y.: The occurrence of genetic alterations during the progression of breast carcinoma. Biomed. Res. Int. **2016**, 5237827 (2016)
16. Curtis, C., Shah, S.P., Chin, S.-F., Turashvili, G., Rueda, O.M., Dunning, M.J., Speed, D., Lynch, A.G., Samarajiwa, S., Yuan, Y., Gräf, S., Ha, G., Haffari, G., Bashashati, A., Russell, R., McKinney, S., Langerød, A., Green, A., Provenzano, E., Wishart, G., Pinder, S., Watson, P., Markowetz, F., Murphy, L., Ellis, I., Purushotham, A., Børresen-Dale, A.-L., Brenton, J. D., Tavaré, S., Caldas, C., et al.: The genomic and transcriptomic architecture of 2,000 breast tumors reveals novel subgroups. Nature **486**, 346–352 (2012)
17. Ali, H.R., Rueda, O.M., Chin, S.-F., Curtis, C., Dunning, M.J., Aparicio, S.A., Caldas, C.: Genome-driven integrated classification of breast cancer validated in over 7,500 samples. Genome Biol. **15**, 431 (2014)
18. List, M., Hauschild, A.-C., Tan, Q., Kruse, T.A., Mollenhauer, J., Baumbach, J., Batra, R.: Classification of breast cancer subtypes by combining gene expression and DNA methylation data. J. Integr Bioinform. **11**, 236 (2014)
19. Hall, M.A.: Correlation-based feature selection for machine learning. Technical report, Department of Computer Science, University of Waikato (1998)
20. Chizi, B., Maimon, O.: Dimension reduction and feature selection. In: Data Mining and Knowledge Discovery Handbook, pp. 83–100. Springer, New York (2010)
21. Chinnadurai, G.: The transcriptional corepressor CtBP: a foe of multiple tumor suppressors. Cancer Res. **69**, 731–734 (2009)
22. Huang, M.-Y., Wang, J.-Y., Chang, H.-J., Kuo, C.-W., Tok, T.-S., Lin, S.-R.: CDC25A, VAV1, TP73, BRCA1 and ZAP70 gene overexpression correlates with radiation response in colorectal cancer. Oncol. Rep. **25**, 1297–1309 (2011)
23. Cristiana, L.N.: New insights into P53 signalling and cancer: implications for cancer therapy. J. Tumor 2 (2014)
24. Wen, H., Li, Y., Xi, Y., Jiang, S., Stratton, S., Peng, D., Tanaka, K., Ren, Y., Xia, Z., Wu, J.: ZMYND11 links histone H3. 3K36me3 to transcription elongation and tumour suppression. Nature **508**, 263–268 (2014)
25. Lorincz, A.T.: Cancer diagnostic classifiers based on quantitative DNA methylation. Expert. Rev. Mol. Diagn. **14**, 293–305 (2014)
26. Sengupta, N., Yau, C., Sakthianandeswaren, A., Mouradov, D., Gibbs, P., Suraweera, N., Cazier, J.-B., Polanco-Echeverry, G., Ghosh, A., Thaha, M.: Analysis of colorectal cancers in British Bangladeshi identifies early onset, frequent mucinous histotype and a high prevalence of RBFOX1 deletion. Mol. Cancer **12**, 1 (2013)

On Selection of Relevant Fuzzy Implications in Approximate Reasoning

Zbigniew Suraj(✉)

Faculty of Mathematics and Natural Sciences, University of Rzeszów, Rzeszów, Poland
zbigniew.suraj@ur.edu.pl

Abstract. The paper describes a methodology for selecting relevant fuzzy implications in forward and backward reasoning. The proposed methodology is based on the functional representation of fuzzy implications and dependencies between fuzzy implications. This can be useful for the design of an inference engine based on the rule knowledge for a given rule-based system.

Keywords: Fuzzy implication · Fuzzy production rule
Knowledge representation · Approximate reasoning
Intelligent system · Rule-based system

1 Introduction

The core of the use of soft computing methods in practice is based on the use of approximate reasoning methods. These are very general techniques that can be used in control, decision making, classification, pattern recognition and elsewhere, provided a description of the system (situation) is available. The latter is to be based on the use of a set of production rules, each of which is characterized by very detailed system behavior and can be expressed imprecisely. Impreciseness comes from various sources - too much complexity, insufficient amount of precise information, presence of a human factor, the need to save time or money, etc. Very often there is a combination of more similar factors [3,6].

One of the most popular approaches to knowledge representation are also the fuzzy production rules. In such situation, they are often presented in the form of IF-THEN and interpreted as Boolean implications. Boolean implications are employed in inference schemas like modus ponens, modus tollens, etc., where the reasoning is done with statements which truth-values are two-valued.

Fuzzy implications play a similar role in the generalizations of the above inference schemas, where reasoning is done with fuzzy statements whose truth-values belong to the unit interval $[0, 1]$ instead of the set $\{0, 1\}$. One of the best known application areas of fuzzy logic is approximate reasoning [4,10], wherein from imprecise inputs and fuzzy premises or rules we obtain, often, imprecise conclusions. Approximate reasoning with fuzzy sets [15] encompasses a wide variety

A. E. Hassanien et al. (Eds.): AISI 2018, AISC 845, pp. 208–218, 2019.
https://doi.org/10.1007/978-3-319-99010-1_19

of inference schemes and has been readily embraced in many fields, especially among others, decision making, expert systems and control [16]. Fuzzy implications can be represented in many ways. One of them is the functional representation [2]. There exist uncountably many implication functions in the field of fuzzy logic, and the nature of the fuzzy inference changes variously depending on the implication function to be used. The variety of implication functions existing in the fuzzy set framework has always been seen as a rich potential for modeling different shades of expert attitude in the inference process (e.g. [7]), although no precise, practical interpretation was provided for the different implication functions [9]. Moreover, it is very difficult to select a suitable implication function for actual applications.

In this paper we propose a methodology for selecting relevant fuzzy implications in forward and backward reasoning. This methodology is based on a functional representation of fuzzy implications and dependencies between fuzzy implications, treated as an external knowledge about fuzzy implications. Such knowledge can be very helpful when modeling a real system, which is described by fuzzy production rules interpreted as fuzzy implications. Taking into account the proposed methodology we can reduce the efforts related to a selection of a suitable implication function. In addition, it can clarify the nature of a given fuzzy production rule by better tuning a suitable fuzzy implication function. Therefore, the tuning of a suitable fuzzy implication function can be regarded as a kind of quantification of the meaning of the natural language sentences expressed in IF-THEN fuzzy production rules, representing fuzzy implications. In a sense, a similar approach to solving the problem considered here is presented in the papers [13, 14]. Nevertheless, the proposal presented in this paper is more general and concerns any set of fuzzy implications.

The organization of this paper is as follows. Section 2 serves as a brief introduction to ordered sets, fuzzy implications and fuzzy production rules. In Sect. 3, two important theorems concerning a selection of suitable implication function are formulated and proved. Section 4 contains descriptions of four algorithms formulated on the basis of two theorems presented in Sect. 3. The first (third) algorithm allows to select the suitable implication function based on information concerning a given set of fuzzy implications, their truth-values, and the truth value of premise (conclusion). The second (fourth) algorithm allows to select the greatest (least) fuzzy implication using the same input information as for the first (third) one. Examples illustrating the performance of these algorithms are also given in Sect. 4. Section 5 contains summary conclusions as well as comments on the directions of further research.

2 Basic Notions and Definitions

In this section we recall basic notions related to ordered sets, fuzzy implications and fuzzy production rules.

2.1 Ordered Sets

In mathematics, a partially ordered set formalizes and generalizes the intuitive concept of an ordering of the elements of a set.

A binary relation R on A is a *partial ordering on* A if, for any $x, y, z \in A$: (1) it is reflexive, i.e., $(x, x) \in R$, (2) it is transitive, i.e., if $(x, y) \in R$ and $(y, z) \in R$, then $(x, z) \in R$, (3) it is antisymmetric, i.e., if $(x, y) \in R$ and $(y, x) \in R$, then $x = y$. A partial ordering R on A is a *linear ordering on* A if, for any $x, y \in A$, at least one of the following conditions: $(x, y) \in R$, $(y, x) \in R$ or $x = y$ holds. If R is a partial ordering on A, then the pair $U = (A, R)$ is a *partially ordered set* (abbreviated poset). If R is a linear ordering on A, then the pair $U = (A, R)$ is a *linearly ordered set.* For a poset $U = (A, R)$ and $X \subseteq A$, the element $a_0 \in A$ is the *upper (lower) bound in* U of a subset $X \subseteq A$ if $(x, a_0) \in R$ $((a_0, x) \in R)$ for all $x \in X$. The upper (lower) bound in U of A is the *greatest* (*least*) element in U. An element $a \in A$ is *maximal* (*minimal*) in U if $(a, x) \in U$ (respectively $(x, a) \in R$) implies $x = a$.

A poset can be visualized by means of the Hasse diagram. This diagram is a graphic drawing on which the vertices are the elements of the poset and the ordering relation is indicated by both the edges and the relative positioning of the vertices. Orders are drawn from below: if the element x precedes y, then there is a path from x to y pointing up. For more information about ordered sets the reader is referred to [12].

2.2 Fuzzy Implications

Fuzzy implications are one of the main operations in fuzzy logic. They generalize the classical implication, which takes values in $\{0, 1\}$, to fuzzy logic, where the truth values belong to the unit interval $[0, 1]$.

A *fuzzy implication* is a function $I : [0, 1]^2 \to [0, 1]$ satisfying, for all $x, x', x'', y, y', y'' \in [0, 1]$, the following conditions: (1) if $x' \leq x''$, then $I(x', y) \geq I(x'', y)$); (2) if $y' \leq y''$, then $I(x, y') \leq I(x, y'')$); (3) $I(0, 0) = 1$, $I(1, 1) = 1$, $I(1, 0) = 0$.

Remark 1. Each fuzzy implication I is constant for $x = 0$ and for $y = 1$, i.e., I satisfies the following conditions, respectively: (1) $I(0, y) = 1$ for $y \in [0, 1]$, and (2) $I(x, 1) = 1$ for $x \in [0, 1]$.

If, for two fuzzy implications I' and I'', the inequality $I'(x, y) \leq I''(x, y)$ holds for all $(x, y) \in [0, 1]^2$, then we say that I' is *less than or equal to* I'' and we write $I' \leq I''$. We shall write $I' < I''$ whenever $I' \leq I''$ and $I' \neq I''$, i.e., if $I' \leq I''$ and for some $(x_0, y_0) \in [0, 1]^2$ we have $I'(x_0, y_0) < I''(x_0, y_0)$. In this case we also say that I' is comparable with I''.

Due to the fact that uncountably many fuzzy implications exist, in Table 1 we list only a few basic fuzzy implications, known from the literature, which are used in this paper.

Example 1. Let A be the set of basic fuzzy implications from Table 1 and let $<$ be the binary relation on A defined as above. It can be easily proved that $U = (A, <)$

Table 1. Basic fuzzy implications

Name	Year	Formula
Łukasiewicz	1923	$I_{LK}(x,y) = min(1, 1-x+y)$
Gődel	1932	$I_{GD}(x,y) = 1$, if $x \leq y$, $I_{GD}(x,y) = y$ otherwise
Reichenbach	1935	$I_{RC}(x,y) = 1-x+xy$
Kleene-Dienes	1938	$I_{KD}(x,y) = max(1-x, y)$
Goguen	1969	$I_{GG}(x,y) = 1$, if $x \leq y$, $I_{GG}(x,y) = \frac{y}{x}$ otherwise
Rescher	1969	$I_{RS}(x,y) = 1$, if $x \leq y$, $I_{RS}(x,y) = 0$ otherwise
Weber	1983	$I_{WB}(x,y) = 1$, if $x < 1$, $I_{WB}(x,y) = y$, if $x = 1$
Fodor	1993	$I_{FD}(x,y) = 1$, if $x \leq y$, $I_{FD}(x,y) = max(1-x, y)$ otherwise
Yager	1980	$I_{YG}(x,y) = 1$, if $x = y = 0$, $I_{YG}(x,y) = y^x$, if $x > 0$ or $y > 0$

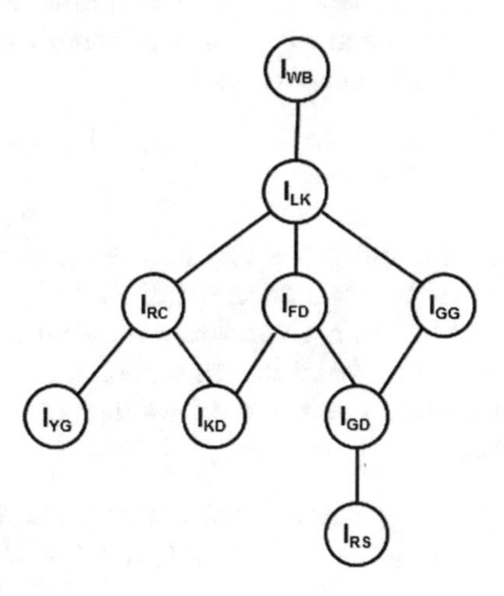

Fig. 1. The Hasse diagram of the poset from Example 1

is a poset. The Hasse diagram of this poset is shown in Fig. 1. In the poset U the Weber implication I_{WB} is the greatest element in U. This implication is also maximal element in U. However, the Yager implication I_{YG}, the Kleene-Dienes implication I_{KD} and the Rescher implication I_{RS} are minimal elements in U. There is no the least element in U. In addition, fuzzy implications from the set A form the following five chains: $I_{YG} < I_{RC} < I_{LK} < I_{WB}$; $I_{KD} < I_{RC} < I_{LK} < I_{WB}$; $I_{KD} < I_{FD} < I_{LK} < I_{WB}$; $I_{RS} < I_{GD} < I_{FD} < I_{LK} < I_{WB}$, $I_{RS} < I_{GD} < I_{GG} < I_{LK} < I_{WB}$.

For more information about fuzzy implications the reader is referred to [2].

2.3 Fuzzy Production Rules

One of the most popular approaches to uncertain knowledge representation is to use fuzzy production rules [5]. The general formulation of a fuzzy production rule r is as follows: IF s THEN s' (CF $= c$), where: (1) s and s' are statements; the truth degree of each statement is a real value between zero and one. (2) c is the value of the certainty factor (CF), $c \in [0, 1]$. The larger the value of c, the more the rule is believed in.

Fuzzy IF-THEN production rules serve also as a basis for approximate reasoning, which is a method for finding a conclusion in forward reasoning or a premise in backward reasoning on the basis of the imprecise initial information concentrated in the form of linguistic description and some new information. In this paper, we interpret any fuzzy production rule r in the form as above as a fuzzy implication $z = I(x, y)$, where values for z, x, y correspond to CF, the truth degree of a statement s (premise), and the truth degree of a statement s' (conclusion), respectively. The value of c is given by a domain expert [1], or generated from data base [11]. However, the value for x (or y) is given by the user of a rule-based system dependently on a selected reasoning method (x for forward reasoning, and y for backward one).

3 Theorems

In this section we formulate and prove two important in the paper theorems.

Theorem 1 suggests how to select from a finite set of fuzzy implications U the suitable implication function for a given fuzzy implication J in order to obtain the truth-value higher or equal to of its conclusion y in approximate reasoning taking into account information on the truth-value z of this implication and the truth-value of its premise x.

Theorem 1. *Let I and J be fuzzy implications such that $I < J$ on a set $D \subset [0, 1]^2$, and $x, y', y'' \in [0, 1]$ such that $I(x, y'') = J(x, y')$. Then $y'' \geq y'$ (Fig. 2(a)).*

Proof. Suppose $y'' < y'$. Then from the definition of fuzzy implication (condition 2) it follows that $I(x, y'') < I(x, y')$. From that and the equality $I(x, y'') = J(x, y')$ it follows that $I(x, y') > J(x, y')$. Since $I < J$, $I(x, y') \leq J(x, y')$. Thus, we have reached a contradiction. Therefore, we conclude that the theorem is true.

Remark 2. Theorem 1 geometrically adjudicates that every implication of A, which in the Hasse diagram of the poset U is placed below the implication J in accordance with the order relation $<$ meets the expected requirements. We can deduce from this theorem that the least element in U (if applicable) corresponds to the implication from A with the highest value of y relative to the implication I. If in U the least element does not exist, we should look for the implication I in among all the minimal elements of U.

A simple consequence of Theorem 1 is the following

Corollary 1. *Theorem 1 is false for $x = 0$.*

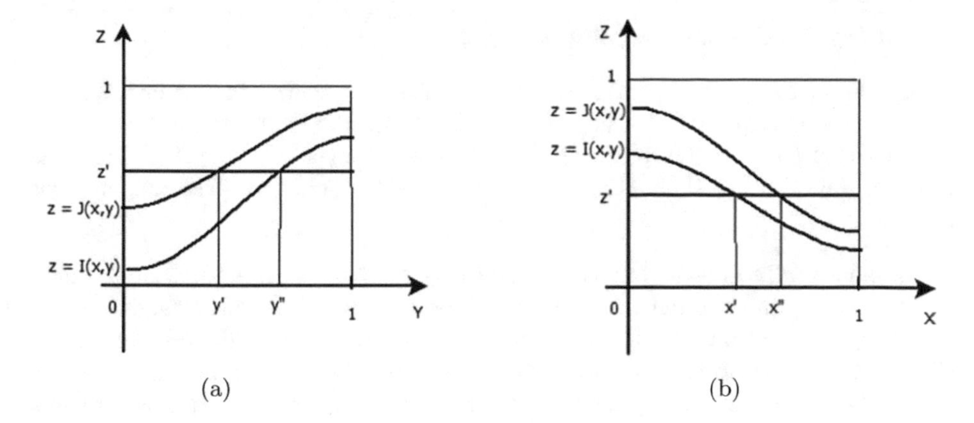

Fig. 2. A geometrical interpretation of: (a) Theorem 1, (b) Theorem 2

Proof. From the property of a fuzzy implication presented in Remark 1 we have $I(0, y) = 1$ for any $y \in [0, 1]$. It means that for any two fuzzy implications I and J the following double dependency $I(0, y'') = J(0, y') = 1$ is true for any $y', y'' \in [0, 1]$. Hence, we get that this equality is true not only for $y'' \geq y'$.

Theorem 2, similarly as Theorem 1, presents how to select from the set A the implication I corresponding to a given fuzzy implication J from A in order to obtain the truth-value less or equal to of its premise x in approximate reasoning taking into account information on the truth-value z of this implication and the truth-value of its conclusion y.

Theorem 2. *Let I and J be fuzzy implications such that $I < J$ on a set $D \subset [0, 1]^2$, and $x', x'', y \in [0, 1]$ such that $I(x'', y) = J(x', y)$. Then $x'' \leq x'$ (Fig. 2(b)).*

Proof. Suppose $x'' > x'$. Then from the definition of fuzzy implication (condition 1) it follows that $I(x'', y) < I(x', y)$. From that and from the equality $I(x'', y) = J(x', y)$ it follows that $I(x', y) > J(x', y)$. Since $I < J$, $I(x', y) \leq J(x', y)$. Thus, we have reached a contradiction. Therefore, we conclude that the theorem is true.

Remark 3. The geometric interpretation of Theorem 2 is similar to the interpretation of Theorem 1 (Remark 1). In this case, Theorem 2 answers the question of how to find the implication I in A in relation to the implication J with the value less or equal to x. We can also deduce from this theorem that the least element in U (if applicable) corresponds to the implication from A with the least value of x relative to the implication I. If in U the least element does not exist, we should look for the implication I in among all the minimal elements of U.

Similarly as in the case of Theorem 1, a simple consequence of Theorem 2 is the following

Corollary 2. *Theorem 2 is false for $y = 1$.*

Proof. From the property of a fuzzy implication presented in Remark 1 (item 1) we have $I(x, 1) = 1$ for any $x \in [0, 1]$. It means that for any two fuzzy implications I and J the following double dependency $I(x'', 1) = J(x', 1) = 1$ is true for any $x', x'' \in [0, 1]$. Hence, we get that this equality is true not only for $x'' \leq x'$.

It is generally known that there are many implication functions in the field of fuzzy logic, and the nature of fuzzy inference varies differently depending on the fuzzy implication function to be used. However, it is very difficult to choose the right implication function for real applications. But taking into account the theorems contained in this section, we formulate in the next section algorithms that reduce efforts related to the selection of an appropriate implication function.

4 Algorithms

In this section, we present four algorithms formulated on the basis of Theorems 1 and 2 (Sect. 3).

Let $U = (A, <)$ be a finite poset and $J \in A$. Algorithm 1 presented below makes simple use of Theorem 1.

Algorithm 1. Finding the implication $I \in A$ with greater value of y.

Input : A finite set A of fuzzy implications, $J \in A$.
Output: An implication $I \in A$.
Order the set A w.r.t. the order relation $<$;
Identify the implication J in a poset $U = (A, <)$;
if *there exists an implication $I \in A$ such that $I < J$* **then**
  Return I;

Example 2. Consider a set of fuzzy implications $A = \{I_{KD}, I_{RC}, I_{LK}, I_{WB}\}$ and the Łukasiewicz implication I_{LK}. After executing the first command of Algorithm 1 we get the chain $t : I_{KD} < I_{RC} < I_{LK} < I_{WB}$ (Fig. 1), to which the implication I_{LK} belongs. It is also easy to see that there are two implications less or equal to I_{LK} w.r.t. the order relation $<$ in this chain, i.e., the Reichenbach implication I_{RC} and the Kleene-Dienes implication I_{KD} (Fig. 1). Thus, the correct answer of Algorithm 1 is I_{RC} or I_{KD}.

Algorithm 2 using the same input information as Algorithm 1 for the argument x of the given fuzzy implication J and the truth value z of this implication finds the fuzzy implication of I in the set A with the greatest value y_{great} among all implications comparable with J in U.

Algorithm 2. Finding an implication in A with the greatest value y_{great}.

Input : A finite set A of fuzzy implications, $J \in A$, $x, z \in (0, 1]$, z is the truth-value of J.
Output: An implication $I \in U$.
Order the set A w.r.t. the order relation $<$;
Compute a set T of all maximal chains in A such that J belongs to each of them;
for *each chain* $t \in T$ **do**
  Find the least implication $I_{least}(t) < J$;
for *each implication* $I_{least}(t)$ **do**
  Compute a value $y(t)$ (if there exists) from the dependency $I_{least}(x, y(t)) = z$;
Find I corresponding to the great value y_{great}=max$\{y(t) : t \in T\}$;
Return I;

Example 3. Consider now a set of fuzzy implications $A = \{I_{YG}, I_{KD}, I_{RC}, I_{LK}, I_{WB}\}$, the Łukasiewicz implication I_{LK}, $x, z \in (0, 1]$. After executing the commands 1 and 2 of Algorithm 2 we get two maximal chains as follows: $t_1 = I_{YG} < I_{RC} < I_{LK} < I_{WB}$ and $t_2 = I_{KD} < I_{RC} < I_{LK} < I_{WB}$ (Fig. 1). We can identify the implication I_{LK} in these two chains. It is also easy to see that I_{YG} is the least element in the chain t_1, while I_{KD} is the least element in the chain t_2. Solving the equations $I_{YG}(a, y(t_1)) = z$ and $I_{KD}(x, y(t_2)) = z$, we get $y(t_1) = z^{\frac{1}{x}}$, $y(t_2) = z$, and $y(t_2) > y(t_1)$. Hence, we have $I = I_{KD}$.

Algorithms 3 and 4 are similar in operation to the Algorithms 1 and 2, respectively. Nevertheless, they are constructed using Theorem 2. The main difference in the input assumptions for these algorithms compared to the first is that in the present case at the input of both algorithms instead of the argument y we use the argument x. Therefore, these algorithms can be used in backward reasoning in order to select the appropriate interpretation of fuzzy implications. Algorithm 3 uses Theorem 2. In this case the implication found by Algorithm 3 gives a lower value of x than the implication J.

Algorithm 3. Finding the implication $I \in A$ with less value of x.

Input : A finite set A of fuzzy implications, $J \in A$.
Output: An implication $I \in A$.
Order the set A w.r.t. the order relation $<$;
Identify the implication J in a poset $U = (A, <)$;
if *there exists an implication* $I \in A$ *such that* $I < J$ **then**
  Return I;

Example 4. Consider again a set of fuzzy implications and the Łukasiewicz implication I_{LK} as in Example 2. It is obvious that the result of this algorithm is the same as Algorithm 1 in the context of seeking the implication function. However, in this case we are interested in calculating the value of x using the Reichenbach implication I_{RC} or the Kleene-Dienes implication I_{KD}.

Algorithm 4. Finding an implication in A with the least value y_{least}.

Input : A finite set A of fuzzy implications, $J \in A$, $y, z \in (0, 1]$, z is the truth-value of J.
Output: An implication $I \in U$.
Order the set A w.r.t. the order relation $<$;
Compute a set T of all maximal chains in A such that J belongs to each of them;
for *each chain* $t \in T$ **do**
  Find the least implication $I_{least}(t) < J$;
for *each implication* $I_{least}(t)$ **do**
  Compute a value $x(t)$ (if there exists) from the dependency $I_{least}(x(t), y) = z$;
Find I corresponding to the least value x_{least}=min$\{x(t) : t \in T\}$;
Return I;

Algorithm 4 using the same input information as Algorithm 3 for the argument y of the given fuzzy implication J and the truth value z of this implication finds the fuzzy implication of I in the set A with the least value x_{least} among all implications comparable with J in U.

Example 5. Consider again a set of fuzzy implications and the Łukasiewicz implication I_{LK} as in Example 3. It is obvious that after executing the commands 1 and 2 the result of this algorithm is the same as Algorithm 3. We get two maximal chains t_1 and t_2 in the form as in Example 3. Nevertheless, in this case we are interested in finding the implication I with the least value x in the set of all implications A. Solving the equations $I_{YG}(x(t_1), y) = z$ and $I_{KD}(x(t_2), y) = z$, we get $x(t_1) = log_y z$, $x(t_2) = 1 - z$, and $x(t_1) \leq x(t_2)$. Hence, we have $I_{least} = I_{YD}$.

5 Concluding Remarks

In the paper, we have presented an approach to selecting relevant fuzzy implications in forward and backward reasoning. To illustrate this approach, we used a set of basic fuzzy implications with the classical order relation "less than or equal to" defined in the set of real numbers. In a natural way, this approach can be extended to all other fuzzy implications. Their proper choice depends to a large extent on the experience and knowledge of the designer of the inference mechanism.

The tuning of fuzzy implications in the form IF-THEN can be regarded as an interpretation of the nature of the rules. We know that there are a lot of implication functions in the field of fuzzy logic, and the nature of the fuzzy inference changes variously depending on the implication function to be used. However, it is very difficult to select a suitable implication function for actual applications. But taking into account the methodology proposed in this paper we can reduce the efforts related to a selection of a suitable implication function. Moreover, the proposed approach looks promising for similar application problems that can be resolved in a similar way. It seems that this paper not only shows that the functional interpretation of fuzzy implications can help both practitioners and researchers to use the fuzzy production rules more effectively, but also it provides a stimulus for further deep analysis of the area and to develop a richer knowledge on inference mechanism models to help practitioners build more effective knowledge-based systems for intelligent decision making [1,3,6].

In this paper, we only considered the logic operator IF-THEN in terms of real numbers. It seems useful to study fuzzy implication functions in the context of the concept of fuzzy implications referring to more general mathematical structures, e.g. interval numbers [8]. In future work, we intend to deal with this problem, focusing in particular on the methodology presented here.

Acknowledgments. This work was partially supported by the Center for Innovation and Transfer of Natural Sciences and Engineering Knowledge at the University of Rzeszów. The author is also very grateful to the anonymous reviewer for giving him precious and helpful comments.

References

1. Avram, G.: Empirical study on knowledge based systems. Electron. J. Inf. Syst. Eval. **8**(1), 11–20 (2005)
2. Baczyński, M., Jayaram, B.: Fuzzy Implications, Studies in Fuzziness and Soft Computing, vol. 231. Springer, Berlin, Heidelberg (2008)
3. Dhar, V.: Data science and prediction. Commun. ACM **56**(12), 64–73 (2013)
4. Driankov, D., Hellendoorn, H., Reinfrank, M.: An Introduction to Fuzzy Control. Springer, Berlin, Heidelberg (1996)
5. Dubois, D., Prade, H.: What are fuzzy rules and how to use them. Fuzzy Sets Syst. **84**, 169–185 (1996)
6. Grosan, C., Abraham, A.: Intelligent Systems. A Modern Approach. Intelligent Systems Reference Library, vol. 17. Springer, Berlin, Heidelberg (2011)
7. Kaufmann, A.: Le Parametrage des Moteurs d'Inference. Hermes (1987)
8. Mayor, G., Torrens, J.: On a class of operators for expert systems. Int. J. Intell. Syst. **8**, 771–778 (1993)
9. Mendel, J.M.: Fuzzy logic systems for engineering. A tutorial. Proc. IEEE **83**(3), 345–377 (1995)
10. Muller, H., Serot, B.D.: Physical Review C52 2072 (1995)
11. Pawlak, Z., Skowron, A.: Rough sets and Boolean reasoning. Inf. Sci. **177**(1), 41–73 (2007)

12. Schröder, B.S.W.: Ordered Sets. An Introduction. Birkhäuser, Basel (2003)
13. Suraj, Z., Lasek, A.: Inverted fuzzy implications in backward reasoning. In: Proceedings of the International Conference on Pattern Recognition and Machine Intelligence (PReMI 2015). Lecture Notes in Computer Science, vol. 9124, pp. 354–364. Springer, Heidelberg (2015)
14. Suraj, Z., Lasek, A., Lasek, P.: Inverted fuzzy implications in approximate reasoning. Fundam. Inform. **141**, 69–89 (2015)
15. Zadeh, L.A.: Fuzzy sets. Inf. Control **8**, 338–353 (1965)
16. Zimmermann, H.-J.: Fuzzy Set Theory and Its Applications, 3rd edn. Kluwer Academic Publishers, Boston Dordrecht London (1996)

Sentiment Analysis and Arabic Text Mining

Expanding N-grams for Code-Switch Language Models

Injy Hamed(✉), Mohamed Elmahdy, and Slim Abdennadher

The German University in Cairo, El Tagamoa El Khames, New Cairo, Cairo, Egypt
{injy.hamed,mohamed.elmahdy,slim.abdennadher}@guc.edu.eg

Abstract. It has become common, especially among urban youth, for people to use more than one language in their everyday conversations - a phenomenon referred to by linguists as "code-switching". With the rise in globalization and the widespread of code-switching among multilingual societies, a great demand has been placed on Natural Language Processing (NLP) applications to be able to handle such mixed data. In this paper, we present our efforts in language modeling for code-switch Arabic-English. In order to train a language model (LM), huge amounts of text data is required in the respective language. However, the main challenge faced in language modeling for code-switch languages, is the lack of available data. In this paper, we propose an approach to artificially generate code-switch Arabic-English n-grams and thus improve the language model. This was done by expanding the relatively-small available corpus and its corresponding n-grams using translation-based approaches. The final LM achieved relative improvements in both perplexity and OOV rates of 1.97% and 16.36% respectively.

Keywords: Code-switching · Code-mixing · Arabic-English
Language modeling · Natural language generation

1 Introduction

Code-switching has become a worldwide common phenomenon, and thus a phenomenon of high interest to linguists as well as researchers in the NLP field. Code-switching involves the use of two or more distinct languages in speech or text. The primary language (also referred to as the matrix language) is spoken in majority, and the secondary language (also referred to as the embedded language) is used to introduce words or phrases into the main context. There are two types of code-switching:

– Inter-sentential code-switching: defined as switching languages from one sentence to another. For example, "Du musst immer optimistisch bleiben. Every cloud has a silver lining." (You must stay optimistic. Every cloud has a silver lining).

A. E. Hassanien et al. (Eds.): AISI 2018, AISC 845, pp. 221–229, 2019.
https://doi.org/10.1007/978-3-319-99010-1_20

- Intra-sentential code-switching (also known as "code-mixing"): defined as using multiple languages within the same sentence. For example, "Als Alice gesagt hat "that's great", waren sie alle zufrieden" (When Alice said "that's great", everyone was happy).

Code-switching evolved as a result of several factors, including globalization, immigration, colonization, the rise of education levels, and international business and communication. Code-switching can be seen in several bilingual/multilingual societies, such as: Mandarin-English in Singapore and Malaysia [17], Cantonese-English in Hong Kong [15], Mandarin-Taiwanese in Taiwan [9], Spanish-English in Hispanic communities in the United States [2], Turkish-German in Germany [3], Italian-French and German-Italian in Switzerland [4]. It has also become prevalent among many Arabic-speakers, especially those receiving high levels of education. For example, English is commonly used in Egypt, and French in Morocco, Tunisia, Algeria, Lebanon and Jordon.

In several NLP tasks, it is important to predict upcoming words (such as in spelling correction, handwriting recognition, and speech recognition), as well as to assign probabilities to sentences (such as in machine translation and augmentative communication). Language modeling is a widely used technique for both tasks. The goal of language modeling is to estimate the probabilities of: (1) a whole sequence of words, (2) a word following a sequence of words. The simplest and most commonly used technique for language modeling is the n-gram language modeling. The n-gram language modeling is a statistical modeling technique that estimates the probabilities of word sequences or the probability of the next word given history. This is done by calculating the relative frequency counts from a huge collection of text, referred to as a text corpus. In order to train an n-gram LM, large amounts of text are crucial to cover all the possible n-gram word sequences. Accordingly, the performance of n-gram LMs is tied with the availability of huge amounts of training data. N-grams have achieved great success for *resource-rich* languages, such as English, French, Spanish, Mandarin Chinese, Japanese and Modern Standard Arabic. However, according to [12], in 2008, less than 50 languages, out of a total of more than 6,900 languages in the world, had large-scale data collected. *Resource-poor* languages include less popular languages, in addition to dialectal and code-switch languages. In the case of code-switching, the available corpora are relatively small in size. This is due to the fact that it is a phenomenon more commonly used in spoken than written form. The published media is usually dominated by the standard languages, which makes available text resources for code-switch text scarce. Consequently, due to the absence of sufficient amount of text data for training n-gram models, word sequences can occur in the testing data without being present in the training data. This leads to Out-of-Vocabulary (OOV) words, which reduces the language model performance, thus affecting the overall system accuracy.

In this paper, we build on our work in [13], where a baseline LM for code-switch Arabic-English was presented. In this work we attempt to improve the language model by expanding the corpus and its corresponding n-grams with the use of glossaries and WordNet [10], and thus covering more possible n-grams.

The remainder of the paper is outlined as follows: Sect. 2 gives an overview on related literature. Section 3 briefly describes the Arabic-English code-switched corpus used in this study. In Sect. 4, a detailed description is given on the approach proposed for expanding the corpus and n-grams. Section 5 presents the evaluation of the language model resulting form the corpus and n-gram expansions. Section 6 provides a discussion on the model limitations as well as the factors that could have affected the performance of the proposed approaches, and suggests possible directions for future work. Finally, the work presented is concluded in Sect. 7.

2 Related Work

The approach used for code-switching language modeling depends on the extent of code-switching involved. For example, in the case of inter-sentential code-switching, where switching languages occurs at sentence boundaries, each sentence is language-homogeneous. Therefore, it is sufficient to use existing monolingual LMs of the languages involved [16,19,24], rather than building a truly multilingual LM. However, this approach is not suitable for handling code-switching within sentences. In the case of intra-sentential code-switching (code-mixing), a multilingual LM is needed. A basic approach to build a multilingual LM is to use a code-switch text corpus to train the LM [5,6]. However, this approach suffers from the limited amount of code-switch data available. In order to tackle this problem, researchers combined the separately trained monolingual LMs of the involved languages using linear interpolation [25,26]. An LM trained using the limited amount of code-switch text can also be interpolated with the monolingual LMs [20,22,23].

To compensate for the problem of limited resources, researchers have worked on expanding existing code-switch corpora. Semantic-based approaches [7] as well as translation-based approaches [7,8,14,22,23] have been recorded. In [23], Weiner et al. generated artificial Mandarin-English code-mixed texts using two monolingual corpora along with SEAME corpus (a Mandarin-English code-mixed corpus). Monolingual English sentences in the SEAME corpus were translated into Mandarin. The translated segments were looked up in the Mandarin monolingual corpus. Whenever the translated word was found, it was replaced by its corresponding original English segment. In [22], Vu et al. then further improved the work by placing a threshold on the minimum allowed count of the translated segment in the Mandarin corpus. Replacements were also restricted to those that follow a trigger word or a trigger part-of-speech tag. In [14], Li and Fung proposed a translation model to expand a monolingual Mandarin language model into a bilingual Mandarin-English language model. The constrained code-switch language model was built using parsing. The search paths were restricted to those permissible under the "Functional Head Constraint"; where "code-switching cannot occur between a functional head (a complementizer, a determiner, an inflection, etc.) and its complement (sentence, noun-phrase, verb-phrase)". In [7], Cao et al. addressed the problem of insufficient amount of Cantonese-English code-mixing text data by expanding a 3-gram LM obtained from training monolingual and code-mixed data. The generated n-grams were obtained from the original n-grams using translation-based and semantic-based approaches. In the translation-based approach, the Cantonese-English dictionary was used to map Cantonese n-grams to mixed-language n-grams. In the Semantic-based approach, the embedded English words were clustered into small semantic classes according to their meanings and syntactic functions. Using these semantic classes, unseen or low-frequency n-grams can then be mapped to present n-grams. The generated n-grams were added to the LM to reduce the unseen n-grams. In [8], Chan et al. also worked on language modeling for Cantonese-English code-mixing speech. The authors proposed a translation-based language model, where English segments were translated to their Cantonese equivalents. Since not all English terms have Cantonese equivalents, POS classes were used, where each English segment is mapped to one of 15 classes depending on their parts-of-speech (POS) or meanings. Other approaches were also proposed to generate code-switch text such as using the Universal Text Imitator [11] and Random Neural Networks LMs [1,21].

3 Corpus

In the scope of this work, the Arabic-English code-switched corpus gathered in [13] will be used. The text corpus was collected by harvesting documents from the web in the domain of computers. The documents were obtained by crawling on-line libraries and querying a search engine using certain keywords. The corpus contains a small portion of dialectal Arabic, however, it is mostly in Modern Standard Arabic. The corpus contains a total of 1,315,384 Arabic sentences ($Corpus_{Arabic}$), 804,744 English sentences ($Corpus_{English}$) and 240,874 code-mixed sentences ($Corpus_{CS}$). $Corpus_{CS}$ was divided into two data sets: $Corpus_{train}$ (228,831 sentences) and $Corpus_{test}$ (12,043 sentences).

4 Approach

The amount of gathered code-mixed sentences is considered to be relatively small. Therefore, many n-grams occurring in the testing data may not be seen in the training data, thus leading to Out-of-vocabulary (OOV) words. Our aim is to create artificial mixed-language n-grams based on the gathered corpus. In order to achieve this, two main resources were utilized: Glossaries and WordNet. In this section, the proposed methodology will be discussed. The approach is outlined in Fig. 1. For language modeling, SRI Language Modeling Toolkit [18] was used to build 3-gram LMs with Witten-Bell smoothing.

4.1 Glossary

Code-switching occurs often in technical contexts, where the technical terms are frequently embedded in the secondary language, most commonly the English language. Accordingly, our idea is to obtain an Arabic-English glossary for Computer terms. Whenever an Arabic term occurs in the text, it is replaced by

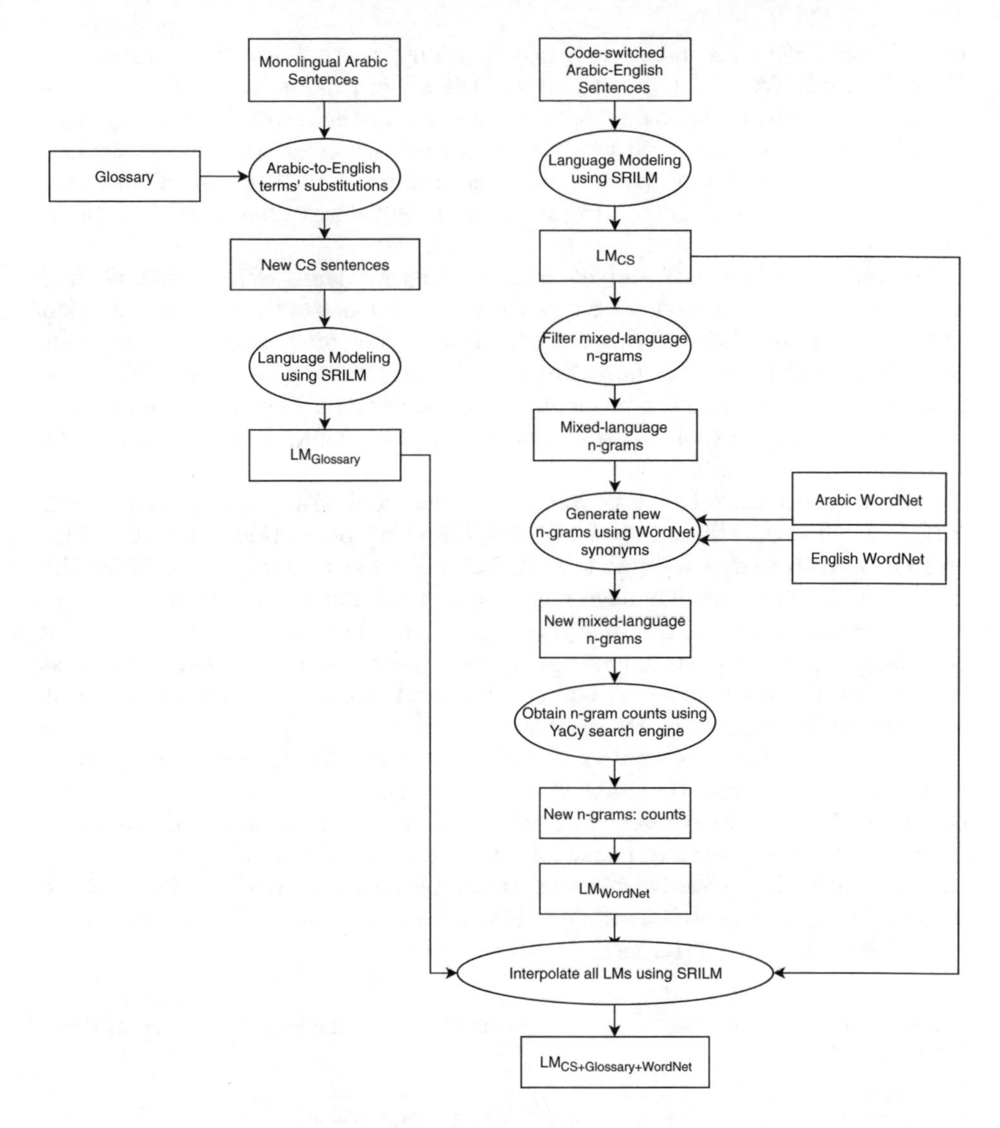

Fig. 1. An overview on the approach.

its corresponding English term. On-line glossaries were obtained and processed, providing a total of more than 4000 Arabic-English term translations. The translations were applied on $Corpus_{Arabic}$. The artificially-generated sentences were used to train an LM, which will be referred to as $LM_{Glossary}$.

4.2 WordNet

In this approach, WordNet is leveraged to paraphrase the previously collected mixed-language n-grams. The n-grams obtained from the $Corpus_{train}$ are filtered to consider solely the mixed-language n-grams; those containing Arabic and English words. WordNet is used to get the synonyms of each word in an n-gram. For each word, its synonyms are obtained for the respective language (i.e. Arabic synonyms for Arabic words and English synonyms for English words). After obtaining a list of synonyms for each word in an n-gram, all possible combinations are computed, and thus generating multiple new mixed-language n-grams.

A total of 2,424,079 new mixed-language trigrams and 1,827,606 monolingual and mixed-language bigrams were generated. It is expected that a huge portion of the generated mixed-language n-grams would not make sense, or would not be valid n-grams that are normally used in natural language (especially that inflection was not addressed when obtaining synonyms). Therefore, the web was used as a validation tool, where the number of hits found for an n-gram in the web indicates its validity.

For web search, YaCy[1] was used as it is free and with no limit restrictions. YaCy is a free distributed search engine based on peer-to-peer networks that can be used to search the public web. YaCy is used to query the web for the newly-generated mixed-language n-grams, and the number of hits would serve as the n-gram counts. When querying, the "exact phrase search" is used, where the query is placed between quotation marks. It is to be noted that the returned number of hits only serves as an estimate, and not the actual number, as it depends on the number of active peers.

In order to reduce the number of YaCy calls, the unigrams were first queried, and their counts were obtained. All trigrams containing unigrams with zero counts in YaCy were trivially filtered out, since the trigram counts returned from YaCy would certainly be zero as well. The remaining trigrams were used to generate their respective bigrams. Table 1 shows the number of artificially generated n-grams as well as the number of n-grams found in YaCy (where the count returned was greater than 0).

Table 1. The total number of artificially generated n-grams and the number of those found in YaCy.

	# artificially generated n-grams	# n-grams found in YaCy
Unigrams	88,145	35,590
Bigrams	1,827,606	19,421
Trigrams	2,424,079	1,664

The newly generated n-grams, along with their counts returned from YaCy, compose the new LM. The generated language model will be referred to as $LM_{WordNet}$.

[1] https://yacy.net/en/index.html.

5 Evaluation

In this section, the two LMs: $LM_{Glossary}$, $LM_{WordNet}$ will be evaluated. $LM_{WordNet}$ contains only the newly generated n-grams, and $LM_{Glossary}$ contains only the n-grams from the newly generated mixed-language sentences. Thus, each LM was interpolated with LM_{CS} (the LM built from $Corpus_{Train}$). Finally, all three LMs (LM_{CS}, $LM_{Glossary}$ and $LM_{WordNet}$) were interpolated. The interpolation weights were chosen to give the best perplexities. The final LMs were tested against $Corpus_{test}$. Table 2 presents the perplexities and OOV rates for the created LMs.

Table 2. The perplexities and OOV rates for the created LMs, as well as the interpolation weights.

LM (interpolation weights)	Perplexity	OOV
LM_{CS}	823.437	1.85%
$LM_{CS+Glossary}$ (0.78:0.22)	**788.309**	1.85%
$LM_{CS+WordNet}$ (0.98:0.02)	840.193	**1.55%**
$LM_{CS+Glossary+WordNet}$ (0.77:0.22:0.014)	807.212	1.55%

It can be seen that each approach has a different advantage. The Glossary achieved a relative improvement in perplexity of 4.27%, with no improvement in the OOV rate. On the other hand, the WordNet approach improved the OOV rate with a relative reduction of 16.36%, however, it led to an relative increase in perplexity of 2.03%. Both approaches were interpolated with the original LM_{CS}. The final LM achieved relative improvements in both perplexity and OOV rates of 1.97% and 16.36% respectively.

6 Discussion

In this section, the limitations of the proposed approaches as well as the factors that could have affected their performance will be discussed.

One of the limitations of the glossary approach described in Sect. 4.1 is that it would not easily applied on a corpus in any domain. The approach depends on the availability of a list of translations of words that are commonly used in the embedded language. In the scope of this work, the corpus is in the domain of computers. Therefore, an Arabic-English glossary of computer terms was used, under the hypothesis that code-switching occur frequently in technical terms, especially to the English language. However, in the case of general-domain corpora, the task becomes more challenging.

Several factors affected the performance of both approaches. Firstly, the number of hits returned by YaCy were usually significantly lower than other non-free search engines. This could have a direct effect on the respective language model, as the n-gram counts could be less accurate, and some valid n-grams might not be found and thus not considered in the language model. Therefore, it would be interesting to try the approach using other search engines. Secondly, the Arabic language is a rich morphological language, where words can have appended prefixes and suffixes. The synonyms of some of the words (with prefixes and/or suffixes) were not found on WordNet, while the stem of the word was found. It would be useful to look into the problem of obtaining the synonyms of words containing prefixes and/or suffixes. Finally, further evaluations need to be done, where the Arabic and English monolingual sentences ($Corpus_{Arabic}$ and $Corpus_{English}$) are involved in the language models.

7 Conclusion

Code-switching is a common phenomenon nowadays, found in bilingual and multilingual societies. Code-switching presents a challenge to NLP tasks, including language modeling. In order to train an n-gram language model, huge amounts of text corpora need to be present. However, since this behaviour is mostly found in speech, rather than text, available code-switched text is scarce. The aim of this work is to generate artificial mixed-language n-grams from a relatively-small code-switched corpus. We presented two translation-based techniques that use online glossaries and WordNet. The final LM achieved relative improvements in both perplexity and OOV rates of 1.97% and 16.36% respectively.

References

1. Adel, H., Kirchhoff, K., Vu, N.T., Telaar, D., Schultz, T.: Comparing approaches to convert recurrent neural networks into backoff language models for efficient decoding. In: Proceedings of the 15th Annual Conference of the International Speech Communication Association (Interspeech 2014), Singapore, pp. 651–655 (2014)
2. Ardila, A.: Spanglish: an anglicized Spanish dialect. Hispanic J. Behav. Sci. **27**(1), 60–81 (2005)
3. Auer, P.: A postscript: code-switching and social identity. J. Pragmat. **37**(3), 403–410 (2005)
4. Auer, P. (ed.): Code-Switching in Conversation: Language, Interaction and Identity. Routledge, London (1998)
5. Bhuvanagiri, K., Kopparapu, S.: An approach to mixed language automatic speech recognition. In: Proceedings of the Oriental COCOSDA, Kathmandu, Nepal (2010)
6. Bhuvanagirir, K., Kopparapu, S.K.: Mixed language speech recognition without explicit identification of language. Am. J. Sig. Process. **2**(5), 92–97 (2012)
7. Cao, H., Ching, P., Lee, T., Yeung, Y.T.: Semantics-based language modeling for Cantonese-English code-mixing speech recognition. In: Proceedings of the 7th International Symposium on Chinese Spoken Language Processing (ISCSLP 2010), pp. 246–250. IEEE, Tainan (2010)
8. Chan, J.Y., Cao, H., Ching, P., Lee, T.: Automatic recognition of Cantonese-English code-mixing speech. Comput. Linguist. Chin. Lang. Process. **14**(3), 281–304 (2009)
9. Chen, C.: Two types of code-switching in Taiwan. In: Proceeding of the 15th Sociolinguistics Symposium, Newcastle, UK (2004)

10. Fellbaum, C.: WordNet: An Electronic Lexical Database. MIT Press, Cambridge (1998)
11. Franco, J., Solorio, T.: Baby-steps towards building a Spanglish language model. In: Proceedings of the International Conference on Intelligent Text Processing and Computational Linguistics, pp. 75–84. Springer, Heidelberg (2007)
12. Fung, P., Schultz, T.: Multilingual spoken language processing. IEEE Sig. Process. Mag. **25**(3), 89–97 (2008)
13. Hamed, I., Elmahdy, M., Abdennadher, S.: Building a first language model for code-switch Arabic-English. In: Proceedings of The 3rd International Conference on Arabic Computational Linguistics (ACLing 2017), pp. 208–216. Elsevier, Dubai (2017)
14. Li, Y., Fung, P.: Code switch language modeling with functional head constraint. In: Proceedings of the International Conference on Acoustics, Speech and Signal Processing (ICASSP 2014), pp. 4913–4917. IEEE, Florence (2014)
15. Li, D.: Cantonese-English code-switching research in Hong Kong: a Y2K review. World Engl. **19**(3), 305–322 (2000)
16. Luján-Mares, M., Martínez-Hinarejos, C.D., Alabau, V.: A study on bilingual speech recognition involving a minority language. In: Proceedings of the Language and Technology Conference, pp. 36–49. Springer, Heidelberg (2007)
17. Lyu, D.-C., Tan, T.-P., Chng, E.-S., Li, H.: An analysis of a Mandarin-English code-switching speech corpus: SEAME. Age, vol. 21, p. 25-8 (2010)
18. Stolcke, A., et al.: SRILM-an extensible language modeling toolkit. In: Proceedings of the 7th Conference on Spoken Language Processing, Denver, Colorado, vol. 2, pp. 901–904 (2002)
19. Uebler, U.: Multilingual speech recognition in seven languages. Speech Commun. **35**(1), 53–69 (2001)
20. van der Westhuizen, E., Niesler, T.: Automatic speech recognition of English-isiZulu codeswitched speech from South African soap operas. Procedia Comput. Sci. **81**, 121–127 (2016)
21. Vu, N.T., Schultz, T.: Exploration of the impact of maximum entropy in recurrent neural network language models for code-switching speech. In: Proceedings of the 1st Workshop on Computational Approaches to Code Switching, Doha, Qatar, pp. 34–41 (2014)
22. Vu, N.T., Lyu, D.-C., Weiner, J., Telaar, D., Schlippe, T., Blaicher, F., Chng, E.-S., Schultz, T., Li, H.: A first speech recognition system for Mandarin-English code-switch conversational speech. In: International Conference on Acoustics, Speech and Signal Processing (ICASSP 2012), pp. 4889–4892. IEEE, Kyoto (2012)
23. Weiner, J., Vu, N.T., Telaar, D., Metze, F., Schultz, T., Lyu, D.-C., Chng, E.-S., Li, H.: Integration of language identification into a recognition system for spoken conversations containing code-switches. In: Proceedings of the 3rd Workshop on Spoken Language Technologies for Under-Resourced Languages, Cape Town, South Africa, pp. 61–64 (2012)
24. Weng, F., Bratt, H., Neumeyer, L., Stolcke, A.: A study of multilingual speech recognition. In: Proceedings of the 5th European Conference on Speech Communication and Technology (Eurospeech 1997), Rhodes, Greece, pp. 359–362 (1997)
25. Xu, R., Zhang, Q., Pan, J., Yan, Y.: Investigations to minimum phone error training in bilingual speech recognition. In: Proceedings of the 6th International Conference on Fuzzy Systems and Knowledge Discovery (FSKD 2009), vol. 4, pp. 486–490. IEEE, Tianjin (2009)
26. Yılmaz, E., van den Heuvel, H., van Leeuwen, D.: Investigating bilingual deep neural networks for automatic recognition of code-switching Frisian speech. Procedia Comput. Sci. **81**, 159–166 (2016)

A Sentiment Analysis Lexical Resource and Dataset for Government Smart Apps Domain

Omar Alqaryouti[1(✉)], Nur Siyam[1], and Khaled Shaalan[1,2]

[1] The British University in Dubai, Dubai, UAE
omar.alqaryouti@gmail.com, nur.siyam@gmail.com,
khaled.shaalan@buid.ac.ae

[2] School of Informatics, University of Edinburgh, Edinburgh, UK

Abstract. Sentiment resources are essential components for developing applications for Sentiment Analysis (SA). Common publicly available datasets such as products, restaurants and movies reviews usually fulfil the researchers needs in order to conduct their experiments. However, for specific domains, the needed dataset sources could be difficult to find. This signifies the need to construct domain specific datasets and lexicons which are vital to evaluate SA tasks. In this paper, we present the work that has been done in order to produce a unique dataset that consists of government smart apps domain aspects and opinion words. Additionally, we explain the approach that was carried out to measure the sentiment scores to opinion words and build the desired lexicons. A general-purpose data annotation and preparation tool was developed for facilitating the development of SA lexical resources and dataset for government smart apps domain.

Keywords: Dataset construction · Lexical resources · Sentiment Analysis
Aspect-based sentiment lexicons · Government smart apps · Manual annotation

1 Introduction

With the advance of technology and the dramatic increase of smart phones users, the provisioning of public sector services has evolved from the need of face-to-face interaction into offering powerful electronic and smart channels for providing services to the public. Government and business organizations invest massive amount of resources to capture their customers' feedbacks and opinions. Sentiment Analysis (SA) is very important for organizations as they tend to understand the level of their services from the customers' point of view through capturing and understanding opinions yet take it into consideration in their services development and improvements. Whereas, individuals consider the others opinions to take decision in buying a product or investing in a business. A study on Malaysian government by [1] pointed out the significance of SA in social media as a driver for understanding the citizens' needs and wants closely. Thus, government services will be customized and prioritized to meet the citizens' expectations rather than depending on the capability level of that government organization.

A. E. Hassanien et al. (Eds.): AISI 2018, AISC 845, pp. 230–240, 2019.
https://doi.org/10.1007/978-3-319-99010-1_21

It is vital in SA to discover the appropriate resources. This is essential to conduct the desired SA experiments. In this study, we aim to define the aspects in the government mobile applications that are important for aspect extraction. Moreover, we aim to produce a novel domain-specific annotated dataset that involves government apps in the United Arab Emirates (UAE) as well as its corresponding aspects terms and opinion lexicons.

This paper is divided into five sections. The next section discusses the previous work done on constructing resources for SA and presents the significant contributions in improving various tasks. The third section demonstrates the work that has been done in order to produce a unique dataset that consists of government mobile apps domain aspects and opinion words. Additionally, it explains the methodology that was carried out to assign the sentiment scores to opinion words and build the desired lexicons. The fourth section includes a discussion of this study. Finally, the conclusion gives a brief summary and critique of the findings and provides direction for future research areas.

2 Related Work on Approaches for Building Sentiment Analysis Lexical Resources

Researchers have described various approaches to construct resources in SA area. Tan and Wu [2] developed a domain-oriented approach based on random walks model to build sentiment lexicon that conveys four kinds of relationships between opinion words and documents. The authors approach utilized the PageRank and HITS graph-based concepts to represent the reviews as graphs and words as nodes. After that, a random walk over the nodes was performed in order to rank the nodes based on probabilities. The idea considered that the word polarity is discovered based on the adjacent words polarities. Likewise, the word polarity is related to the overall polarity of the document. Similarly, Baccianella et al. [3] produced version 3.0 of SentiWordNet. The researchers have utilized graph-based methods to automatically annotate the well-known WordNet Synsets considering its corresponding polarity.

Dang et al. [4] proposed an enhanced method for sentiment classification by combining both Machine Learning (ML) and Semantic Orientation (SO) approaches based on lexicon enhanced approach. The authors have collected product reviews from various online resources and merged them with the publicly available Blitzer's dataset.

Steinberger et al. [5] demonstrated a semi-automatic methodology in order to generate multi-language sentiment dictionaries. In their work, they started with gathering subjective terms and producing gold standard sentiment dictionaries in two languages, namely English and Spanish. Then, the resulted dictionaries were automatically translated using Google translator into a third language using the triangulation technique. Triangulation matches the intersections between the translated terms to produce the target language desired dictionaries. The outcome lists of triangulations are then passed to a manual process to filter and expand the lists in order to obtain a broader coverage.

Boldrini et al. [6] created a multilingual resource called EmotiBlog through developing a fine-grained annotation framework for classifying subjectivity at all levels. The researchers collected the blog posts that are covering three topics in

English, Italian and Spanish languages. An inter-annotator agreement was measured in order to verify and ensure the clarity for annotators.

Robaldo and Di Caro [7] developed an XML-based model to tag reviews that convey opinions towards certain objects. The authors formulated a structured standard formalism called OpinionMining-ML that would contribute in resolving issues related to emotion-based claims in the information retrieval arena as well as assisting and guiding annotators in tagging reviews in order to build comprehensive datasets. Furthermore, it provides an important option for researchers to evaluate SA existing approaches. On the other hand, Sarmento et al. [8] proposed a manual rule-based approach powered by sentiment lexicons to construct SA corpus based on posted online political comments.

The construction of SA datasets can be achieved through either automatic, semi-automatic or manual approaches. In this regard, the annotation process is essential to guarantee the quality of the dataset. Various automatic approaches were followed to build SA lexical resources such as graph-based approach [2, 3] as well as integrated ML and SO approach [4]. Conversely, other researchers followed semi-automatic approaches such as Triangulation technique [5] and fine-grained annotation approach [6]. Finally, constructing SA lexical resources also followed manual approaches such as XML-based model [7] and rule-based approach [8].

3 Constructing the Lexical Resources and Dataset

In this section, an illustration of the work has been done in order to build a unique dataset that consists of government mobile apps domain aspects and opinion words. Additionally, this section explains the approach that was carried out to measure the sentiment scores to opinion words and build the desired lexicons.

3.1 Methodology

A manual annotation approach has been chosen with the help of the developed Government Apps Review Sentiment Analyzer (GARSA) tool to facilitate the complete annotation process. Although this manual approach is time consuming and requires dedicated resources, it is considered to be more accurate than the automated methods [9]. Our proposed strategy starts with identifying the target government applications as source of customers' reviews and collecting the reviews for the selected target applications. Then, we define the aspects that need to be extracted and classified in aspect-based SA in the government mobile applications domain. After that, we analyze and filter the reviews that will be the scope for this study. Next, we developed GARSA that is used to simplify the annotation process of reviews aspects and opinion words. This is followed by an automatic assignment of sentiment rating to the extracted opinion words using SentiWordNet. Our methodology tool takes into consideration conflict resolutions. This was done at multiple levels. The first level, as explained in Sect. 3.5, is that the workflow has been configured to have two steps verification for any annotation before it is finally approved. Then, before the real annotation process started, an agreement analysis was conducted to evaluate the reliability of the annotation process and observe the agreement level between the annotators as illustrated in Sect. 3.6. Finally, the annotated reviews and extracted aspects and opinions as well as the assigned sentiment polarities have been also verified in order to produce the dataset and corresponding lexicons.

3.2 Data Analysis

This research scope covered the well-known government apps in UAE. We collected reviews from 60 smart apps in both Google Play and Apple Store. The reviews that have been collected included all reviews posted in the period from the beginning of 2013 to the end of 2016. The reason behind that is that, year of 2013 represents the year when His Highness Sheikh Mohammed bin Rashid announced the launch of the "Smart Government" initiative. The collected data consists of a set of reviews, each of which includes the app name, store name, app ID, review ID, language, star rating, date, author, subject, and body. The collected data consist of 11,912 reviews from both Google Play and Apple Store. The figures illustrated that 76% of the collected reviews were in English language, 20% in Arabic Language, and 4% in other languages. Moreover, the majority of the reviews were positively rated. For instance, 6438 of the reviews were rated by the users as five stars, 1068 as four stars, 691 as three stars, 693 as two stars, and 3022 as one star. According to the data analysis, it has been noticed that emoticons are not commonly used in apps reviews. Thus, the emoticons will not be considered in the experiments. Furthermore, it has been observed that slangs were used in a variety of the reviews. Also, there were a lot of spelling mistakes and incorrect grammar which will require special treatments. Interestingly, there were no spam reviews in the whole collected data. 1.62% of the reviews were blank and were excluded for not containing the subject or body fields.

The majority of the collected reviews for government mobile apps were written in English language. Therefore, for the sake of covering a larger segment of the government mobile app users and get conclusive results, the experiments in this research will be only limited to reviews written in English language.

3.3 Innovative Smart App Aspects Design Instruments

According to Gartner [10], the most widely used smart phones operating systems are iOS (developed by Apple) and Android (developed by Google Inc.) constituting of 99.6% of the total market share. For instance, Apple [11] considered the "Clarity", "Deference" and "Depth" themes. Moreover, Apple [12] defined a set of principles that should be taken into consideration to make the most of the impact and reach of the mobile app through its identity. These principals are "Aesthetic Integrity", "Consistency", "Direct Manipulation", "Feedback", "Metaphors" and "User Control". Further, Apple [12] defined a set of guidelines to be followed before releasing any app and deploying it to the app store. These guidelines are "Safety", "Performance", "Design" and "Legal". Similarly, Android [13] have defined a set of quality guidelines that should be taken into consideration when developing the app and should be tested against them before the app releases. These quality guidelines are "Visual Design and User Interaction", "Compatibility, Performance and Stability" and "Security".

Furthermore, Smart Dubai Government [14] have defined a set of components and guidelines in their Smart Website Excellence Model (SWEM) and Smart Services Delivery Excellence Model (SSDEM) in order to produce high quality smart applications and mobile websites that meet the customers' expectations and provide an exceptional customer experience. However, the customer experience aspect was not emphasized as a standalone component although it is one of the most important aspects if it is looked at from the customers' perspective rather than the entity's side. On a similar note, Mills [15] focused on following certain factors to design the desired customer experience according to who are the customers, what they want to do, and why. This can be concluded by answering the question "What will my customer remember about this experience?". Thus, a novel comprehensive model has been designed to serve this study. This model takes into consideration the state-of-the-art guidelines as well as the aspects according to the customers' perspectives and viewpoints. This model consists of five main type of aspects, as shown in Table 1.

Table 1. Government smart apps aspects

Aspect type	Description
User interface	This aspect conveys a unified design that follows a consistent pattern, clear navigation structure, and unambiguous user interface functionalities. The design elements should be simple, attractive, and easy to use
User experience	This aspect is generally concerned about the customers' experience and the level of satisfaction as well as their feelings about the service. This includes how convenient is the app for the users, how appropriate is the structure, and how the users are comfortably completing the service. Also, this aspect includes offering multilingual support as well as avoiding redundant features
Functionality and performance	This aspect states that the service should perform as indicated by the government entity and according to the customers' expectations. Likewise, the service should provide responses to the users in a timely manner and according to the expectations
Security	This aspect asserts that the service should provide secured user login and payment services as required. Additionally, it emphasizes on the user's privacy in a way that assures the user's collected and stored data is protected as per the terms and conditions
Support and updates	This aspect states that the service should offer effective channels to engage customers and gather their inputs to understand their pain and gain areas. The app should be reliable, stable and free of errors and should release updates when needed

3.4 GARSA for Reviews Annotation

GARSA is general-purpose data annotation and preparation responsive web tool that we have designed to facilitate the annotation process in an organized way. We have developed a tool using ASP.NET and HTML5 components and it currently covers the annotation part. GARSA is a flexible application that we designed in a way that will allow the annotators to perform the work efficiently with minimum errors. The tool was developed as a web application that can be accessed anytime and anywhere. This tool consists of a simplified user workspace, advanced search functionality, polarity assistant, simplified workflow, automatic rules definition, and instant dynamic graphical reports.

3.5 Manual Annotation Approach

A team of four that comprises three language experts in addition to the domain expert has been formed to perform the complete annotation process. The annotation process includes four main sub-processes, namely annotate, review, verify annotation, and approve annotation. Each of these phases comprises several activities. The team was divided into 3 groups. Two members play the "Annotator" role and the third member was assigned the "Reviewer" role. Whereas, the SA domain expert was assigned the "Administrator" role.

The annotators utilized the GARSA functionalities to perform the annotation by reading and understanding each review, selecting the opinion word, aspect term and assign it to the corresponding aspect. Moreover, the annotators assign a star rating for the review according to their understanding and judgement. Then, the annotated reviews are verified by the reviewer. The reviewer may update the assigned annotations as well as assign star rating for the review similar to the annotators. Additionally, the reviewer will mark the verified reviews as complete so it will be displayed in the domain expert page for the final review and approval. The domain expert may conduct the same functionalities as the reviewer.

3.6 Inter-annotator Agreement

A team of four members has been formed to perform the complete annotation process. Two of the team members are responsible for annotating the reviews. Before the real annotation process started, an agreement analysis was conducted by selecting a random 300 reviews from the database. The purpose of this analysis is to evaluate the reliability of the annotation process and observe the agreement level between the two annotators. Also, it ensures that the annotators have common and proper understanding of the requirements and assures high level of annotators' confidence of the annotation. The agreement measure provides an indication on the disagreement areas which may necessitate further explanation for the annotators. Similar to [16], Cohen's Kappa statistics are selected to measure the inter-annotator agreement. The Kappa agreement is only used when there are only two annotators. The Kappa (k) can be defined as per the following formula,

$$k = \frac{P_a - P_c}{1 - P_c} \tag{1}$$

where:
P_a is the qualified agreement that has been observed for the two annotators, and P_c is the theoretical probability of a random or chance agreement.

Table 2 illustrates the two annotators' agreement and disagreement matrix. The first annotator (A) agreed to annotate the same aspects with the second annotator (B) in 372 occurrences for the same reviews (TT). Also, A agreed with B not to annotate aspects in 748 occurrences for the same reviews (FF). On the other hand, A disagreed with B by annotating 12 aspects that have not been annotated by B (TF). And A disagreed with B by not annotating 13 aspects that were annotated by B (FT). Table 2 provides a breakdown details on the agreement and disagreement levels for all aspects along with the Kappa's measures. It is noticed that the "*User Experience*" aspects indicated the lowest k percentage due to the difficulties in indicating the opinion words and aspect terms that represent explicit or implicit aspects. However, the k percentage (82.63%) is still indicating high value with almost perfect results according to [16].

Table 2. Annotators agreement and disagreement matrix for all aspects

		Agreement and Disagreement				
	Aspect	**TT**	**TF**	**FT**	**FF**	***K***
Aspects	**User Interface**	45	2	4	174	**93.04%**
	User Experience	189	4	3	19	**82.63%**
	Functionality and Performance	39	3	2	181	**92.61%**
	Security	27	2	2	196	**92.09%**
	Support and Updates	63	1	2	178	**96.84%**
	Total	372	12	13	748	**95.11%**

According to the calculated results for Cohen's Kappa (k) measures, all results were greater than 81% for all aspects. As stated by Takala et al. [16], if the Kappa measure is greater than 81%, then the agreement level is almost perfect. Thus, the results confirmed that the annotators perceived a high level of confidence of the requirements understanding and annotation process.

3.7 Lexicon Generation

Prior to the data annotation, all annotated opinion words along with their aspects terms have been accumulated in "*SentimentAspectAnnotation*" database table to build the desired lexicon. First, the redundant opinion words and aspects terms are removed. Then, a process built on Python is utilized to go through all opinion words to obtain the polarity score from SentiWordNet. The polarity score ranges from −1.0 for negative words to 1.0 for positive words. SentiWordNet were chosen because it covers more words than any other available sentiment dictionaries. If the opinion word is not found in SentiWordNet, the process utilized WordNet Lemmatizer to remove grammatical markings. For example: "*flexibility*" is lemmatized to "*flexible*" and "*buggy*" is lemmatized to "*bug*". However, the sentiment in SentiWordNet is linked to the word meaning instead of the word itself unlike other lexicon resources such as SO-CAL [17]. This tolerates a word to have multiple sentiments towards each meaning. Therefore, the developed process has adopted similar approach to [4], where the following rules were applied to calculate the polarity score:

> ***If*** *the Average Positive Score (Opinion Word) > The Average Negative Score (Opinion Word)*
> ***Then*** *The process will set the average positive score as polarity*
> ***If*** *the Average Negative Score (Opinion Word) > the Average Positive Score (Opinion Word)*
> ***Then*** *The process will set the average negative score as polarity*

In the emerged dictionary, a variety of words were annotated as opinion words although they have zero polarity score in SentiWordNet. For instance, the word "*delay*" has zero average score in SentiWordNet. However, this word is considered as a domain-specific opinion and as agreed by annotators, it expresses negative opinion in the reviews. Thus, this kind of words was reviewed with the help of the GARSA in order to figure out the suitable polarity score.

Basically, the average of the user star ratings on the reviews that contain each of these words were calculated with approximate assumption that 5-star rating has 1.0 polarity score, 4-star rating has 0.5 polarity score, 3-star rating has 0 polarity score, 2-star rating has −0.5 polarity score and 1-star rating has −1.0 polarity score. For example, if the GARSA showed that the word "*delay*" occurred in 4 reviews where one review has 4-star rating, two reviews have 3-star rating, and one review has 1-star rating. Then, the resulted equation will be as follows:

$$Polarity\,Score\,(delay) = \frac{(1 \times 0.5) + (2 \times -0.5) + (1 \times -1)}{4} = -0.375$$

On the other hand, the dictionary contained some words with spelling mistakes, which were not found in the SentiWordNet as well. These words have been corrected and annotated during the dataset annotation phase and the polarity score of the correct word was assigned to them in the dictionary using the above approach. The misspelled words were linked to the parent corrected words and inherited its polarity. Finally, the remaining scored opinion words were also reviewed with the help of GARSA in order to verify and confirm the correct scoring with respect to Government Mobile Reviews domain. The resulted dictionary consists of four attributes: Opinion word, aspect term, aspect category and the polarity score.

3.8 Dataset Construction

As explained earlier, GARSA was used to annotate 7,345 reviews by manually identifying the opinion words, aspect terms and aspect categories. Then, each opinion word has been assigned with its polarity score by utilizing the generated lexicon. Next, the average of polarity scores associated with opinion words in each aspect category were calculated in order to identify the rating of each aspect in the review. Table 3 shows the criterion of the considered ratings in the dataset based on the resulted polarity score.

Finally, the dataset is constructed as XML format which contains a tag for each review. Each review tag consists of review subject, review body, user rating and all extracted aspect categories with their start rating, associated opinion words, aspect terms and the polarity score for each opinion word. The resulted dataset contains 7,345 reviews, each review contains an average of nine words. Table 4 demonstrates more statistics about the dataset.

Table 3. Star rating mapping to average polarity score criterion

Stars	Description	Average polarity score
1-Star	Very negative review	≤ -0.6
2-Star	Negative review	> -0.6 and ≤ -0.2
3-Star	Neutral review	> -0.2 and < 0.2
4-Star	Positive review	≥ 0.2 and < 0.6
5-Star	Very positive review	≥ 0.6

Table 4. Dataset content statistics

No. of reviews	7,345	Average words per review	9.23
No. of characters (no spaces)	307,978	No. of very positive reviews	2,001
No. of characters (with spaces)	366,797	No. of positive reviews	2,392
No. of words	67,777	No. of neutral reviews	1,435
No. of explicit aspects terms	279	No. of negative reviews	1,015
No. of implicit opinion-aspects terms	402	No. of very negative reviews	502
No. of opinion words	656		

4 Discussion

In the field of SA, it is crucial to find the suitable resources in order to conduct the desired experiments. In many cases, researchers use known publicly available datasets [18–22]. In other cases, it could be difficult to find a dataset in a specific domain or in a specific area. Tan and Wu [2] confirmed that domain dependent sentiment dictionaries are vital to evaluating SA tasks. Joshi et al. [17] confirmed that the annotated dataset comprises of two main components: the textual document and its labels. The textual document may consist of manually annotated reviews with its corresponding sentiment orientation called labels. But, there are various challenges in the area of constructing SA resources. For instance, Montoyo et al. [23] reported some of these challenges such as dealing with ambiguous words, building multi-language resources, dealing with words granularity in various ways the opinions can be expressed (as words, as sentences or as phrases), and dealing with variations in opinion expressions in the different data sources. Joshi et al. [17] declared that the sentiment resources such as lexicons and datasets are essential for any SA system and represent the core knowledge-base for its application. In this study, a manual annotation approach has been followed. GARSA tool was developed to assist in the annotation process. The manual approach is time consuming and requires dedicated resources. However, the manual approach is considered to be more accurate than the automated methods [9].

5 Conclusion and Future Prospects

This research aimed to produce a novel domain specific annotated dataset that involves government apps as well as its corresponding aspects terms and opinion lexicons. In this study, we proposed a government smart apps model based on written practices. Additionally, this study produced a unique government mobile apps domain specific dataset and lexicons. The final produced dataset consisted of government mobile apps domain aspects and opinion words. This was done with the help of the GARSA which is a responsive web tool that has been designed to facilitate the annotation process in a flexible, organized, efficient and tracked manner. The approach started with collecting the data, analyzing the data, designing the database, defining the domain specific aspects, building GARSA, annotating the reviews, and producing the anticipated dataset and lexicons. It is important to point out that GARSA can be also used for the annotation process in other domains. There is a potential to expand the produced dataset to cover more languages. Another area that can be applied on such dataset, is to identify the improvement areas and innovative ideas through analyzing the customer's reviews which may include suggestions, complaints, bugs, comparisons, among others. Finally, a comprehensive tool can be built to analyze any government mobile app reviews to evaluate the customer's satisfaction towards this mobile app as well as identifying the weak aspects in it. Such tool will provide the option to visualize the app reviews SA results and deliver accurate insights of what their customers feel towards the app.

References

1. Hasbullah, S.S., Maynard, D., Chik, R.Z.W., Mohd, F., Noor, M.: Automated content analysis. In: Proceedings of 10th International Conference on Ubiquitous Information Management and Communication, IMCOM 2016, pp. 1–6 (2016)
2. Tan, S., Wu, Q.: A random walk algorithm for automatic construction of domain-oriented sentiment lexicon. Expert Syst. Appl. **38**(10), 12094–12100 (2011)
3. Baccianella, S., Esuli, A., Sebastiani, F.: SentiWordNet 3.0: an enhanced lexical resource for sentiment analysis and opinion mining SentiWordNet. Analysis **0**, 1–12 (2010)
4. Dang, Y., Zhang, Y., Chen, H.: A lexicon-enhanced method for sentiment classification: an experiment on online product reviews. IEEE Intell. Syst. **25**(4), 46–53 (2010)
5. Steinberger, J., et al.: Creating sentiment dictionaries via triangulation. Decis. Support Syst. **53**(4), 689–694 (2012)
6. Boldrini, E., Balahur, A., Martínez-Barco, P., Montoyo, A.: Using EmotiBlog to annotate and analyse subjectivity in the new textual genres. Data Min. Knowl. Discov. **25**(3), 603–634 (2012)
7. Robaldo, L., Di Caro, L.: OpinionMining-ML. Comput. Stand. Interfaces **35**(5), 454–469 (2013)
8. Sarmento, L., Carvalho, P., Silva, M.J., de Oliveira, E.: Automatic creation of a reference corpus for political opinion mining in user-generated content. In: Proceeding 1st International CIKM Workshop on Topic Analysis for Mass Opinion, TSA 2009, p. 29 (2009)

9. Liu, B.: Sentiment analysis and opinion mining. Synth. Lect. Hum. Lang. Technol. **5**, 1–167 (2012)
10. Gartner: Gartner Says Worldwide Sales of Smartphones Grew 7 Percent in the Fourth Quarter of 2016. Gartner.com (2017). http://www.gartner.com/newsroom/id/3609817. Accessed 05 Mar 2017
11. Apple: App Store Review Guidelines - Apple Developer. Developer.apple.com (2017). https://developer.apple.com/app-store/review/guidelines/. Accessed 03 Jan 2017
12. Apple: Themes – Overview - iOS human interface guidelines. Developer.apple.com (2017). https://developer.apple.com/ios/human-interface-guidelines/overview/design-principles/. Accessed 05 Jan 2017
13. Andriod: Core app quality | android developers. Developer.android.com (2017). https://developer.android.com/develop/quality-guidelines/core-app-quality.html. Accessed 15 Jan 2017
14. Smart Dubai Government: Smart websites excellence model version 4.0. Smart Dubai Government (2016). http://dsg.gov.ae/en/OurPublications/Pages/StandardsGuidelines.aspx. Accessed 03 Jan 2017
15. Mills, A.: Certified Innovation Strategist, 1st edn. Global Innovation Institute, Boston (2016)
16. Takala, P., Malo, P., Sinha, A., Ahlgren, O.: Gold-standard for topic-specific sentiment analysis of economic texts. In: LREC, pp. 2152–2157 (2014)
17. Joshi, A., Bhattacharyya, P., Ahire, S.: Sentiment resources: lexicons and datasets. In: A Practical Guide to Sentiment Analysis, vol. 5, pp. 85–106. Springer, Berlin (2017)
18. Poria, S., Cambria, E., Ku, L.-W., Gui, C., Gelbukh, A.: A rule-based approach to aspect extraction from product reviews. In: Second Workshop on Natural Language Processing for Social Media, pp. 28–37 (2014)
19. Poria, S., Cambria, E., Gelbukh, A.: Aspect extraction for opinion mining with a deep convolutional neural network. Knowl. Based Syst. **108**, 42–49 (2016)
20. Liu, Q., Gao, Z., Liu, B., Zhang, Y.: A logic programming approach to aspect extraction in opinion mining, pp. 276–283 (2013)
21. Xianghua, F., Guo, L., Yanyan, G., Zhiqiang, W.: Multi-aspect sentiment analysis for Chinese online social reviews based on topic modeling and HowNet lexicon. Knowl. Based Syst. **37**, 186–195 (2013)
22. Mukherjee, A., Liu, B.: Aspect extraction through semi-supervised modeling. In: Proceedings of 50th Annual Meeting Association Computational Linguistics Long Paper, ACL 2012, vol. 1, no. July, pp. 339–348 (2012)
23. Montoyo, A., Martínez-Barco, P., Balahur, A.: Subjectivity and sentiment analysis: an overview of the current state of the area and envisaged developments. Decis. Support Syst. **53**(4), 675–679 (2012)

Pre-trained Word Embeddings for Arabic Aspect-Based Sentiment Analysis of Airline Tweets

Mohammed Matuq Ashi(✉), Muazzam Ahmed Siddiqui, and Farrukh Nadeem

Department of Information Systems, Faculty of Computing and Information Technology, King Abdulaziz University, Jeddah, Saudi Arabia
mashi0004@stu.kau.edu.sa, {maasiddiqui,fabdullatif}@kau.edu.sa

Abstract. Recently, the use of word embeddings has become one of the most significant advancements in natural language processing (NLP). In this paper, we compared two word embedding models for aspect-based sentiment analysis (ABSA) of Arabic tweets. The ABSA problem was formulated as a two step process of aspect detection followed by sentiment polarity classification of the detected aspects. The compared embeddings models include fastText Arabic Wikipedia and AraVec-Web, both available as pre-trained models. Our corpus consisted of 5K airline service related tweets in Arabic, manually labeled for ABSA with imbalanced aspect categories. For classification, we used a support vector machine classifier for both, aspect detection, and sentiment polarity classification. Our results indicated that fastText Arabic Wikipedia word embeddings performed slightly better than AraVec-Web.

Keywords: Data mining · NLP · Machine learning · Word embeddings Sentiment analysis · Aspect-based sentiment analysis

1 Introduction

In this paper, we have utilized vector-based features instead of hand-crafted features. We conduct separate experiments using two distributed word embeddings models which attempt to capture both semantic and sentiment information from the large corpus used in training. Namely, they are the fastText Wikipedia Arabic content model [1] and AraVec skip-gram model [2] trained on top of Arabic web pages. Word embeddings are very useful for morphologically rich languages [1], such as Arabic. That is, it does not ignore words morphology, through appointing a distinct vector to each word presented in a large dataset such as Wikipedia. In other words, it is very important especially for the Arabic language having a diverse vocabulary, and many rarely used words. We used both word embedding models to perform the two defined sub-tasks of ABSA being Aspect Detection, then, Aspect-Based Sentiment Classification.

Herein, we propose a data mining, more specifically, a text mining methodology that employs the microblogging site Twitter, identifying the most crucial components

A. E. Hassanien et al. (Eds.): AISI 2018, AISC 845, pp. 241–251, 2019.
https://doi.org/10.1007/978-3-319-99010-1_22

of airlines industry customer service. We use Twitter for it being the most frequently used data source in Arabic sentiment analysis articles [3]. We mine Saudi Airlines customers public Twitter commentaries unveiling the airline's customers' opinions regarding their products/services to determine which has caused customer satisfaction or dissatisfaction.

The rest of this research paper is arranged in the following order. The related work is described in Sect. 2. The data collection and labeling are discussed in Sect. 3. In Sect. 4, the experiment design and implementation are presented. Later, we discuss the obtained results for both sub-tasks of ABSA in Sect. 5. Eventually, we end this paper in Sect. 6 with a summary and a brief discussion of future work.

2 Related Work

According to [4], the opinion mining, sentiment analysis, or subjectivity analysis research field, is the body of work which deals with the computational treatment of opinion, sentiment, and subjectivity in text. Sentiment analysis has often been studied at three primary levels of classification either a document-level, a sentence-level, or at an entity- or aspect-level [5]. Aspect-level sentiment analysis (ABSA) is not only concerned with finding the overall sentiment related to an entity but the polarity associated with the specific feature/aspect of an entity [3].

The Arabic language is classified by its morphology, syntax, and lexical combinations into three different categories either classical Arabic, modern standard Arabic (MSA), and dialectal Arabic [3]. The Arabic language has a very complex morphology and structure. This has caused an inability to adapt existing research to Arabic contexts limiting progress in Arab sentiment analysis research, as opposed to the prevalent research conducted for the English language [6]. However, research on Arabic sentiment analysis has recently been getting more attention from the research community [7]. Most of the studied works in the literature on Arabic sentiment analysis have adopted hand-crafted features for their classification with prominent use of sentiment lexicons. Example of the reviewed lexicon-based sentiment analysis works are [8–12].

The process of manually extracting features is such a laborious task that heavily requires time and efforts. On the other hand, word embeddings are critical to lexical features being employed for sentiment analysis of Twitter data [13]. Most of these techniques represent each word of the vocabulary by a distinct vector. Specifically, in morphologically rich languages such as Arabic with a large vocabulary and many distinct forms and rarely used words. Word embeddings add useful information to the learning algorithms for achieving superior results in various NLP applications [14]. Arabic word embeddings have been used by many researchers as features for different Natural Language Processing tasks [2].

A number of works have been completed for Arabic sentiment analysis using word embeddings are as follows. In [7] trained their word embeddings model on top of a large corpus that was compiled of publicly available Arabic text collections such as newspaper articles and consumer reviews as well as others. Then, they trained and generated word vector representations (embeddings) from the corpus. They also trained several binary classifiers employing their word embeddings model using such vector-based features as feature representations detecting sentiment and subjectivity in three different datasets for dialectical Arabic. They used tweets and book reviews datasets and for subjectivity classification and MSA with news articles as a dataset. They achieved an accuracy of 77.87% that is slightly better than the major methods of manual feature extraction. Moreover, [15] compared the performance of different base classifiers and ensembles for polarity determination in highly imbalanced datasets of tweets in dialectical Arabic using features learned by word embeddings rather than hand-crafted features. Their results show that applying word embeddings with the ensemble and synthetic minority over-sampling technique (SMOTE) can achieve more than 15% improvement on average in the F1 score.

Also, [16] they used word embeddings to perform ABSA on English restaurant reviews dataset covering the tasks of aspect term extraction, aspect category detection, and aspect sentiment prediction achieving F1 scores of 79.9% for aspect term extraction, 86.7% for category detection, and 72.3% for aspect sentiment prediction. The features employed were the word2vec word embedding models trained on Yelp or GoogleNews. Their results indicate the impact of vector-based features that deliver the highest level of performance.

3 Dataset

We started this project by creating a Twitter API user to leverage Twitter's Search API obtaining the bulk of data 'tweets' required for this research. We used many keywords such as 'Saudia', 'Saudi Airlines', "الخطوط السعودية" or the studied airline customer service Twitter account 'Saudia_Care'. We used Python as the sole programming language; creating a script making the API calls, then, appending and storing the data into a csv file format. Abstracting data 'Tweets' from Twitter was iterated over a period of five months starting from May 2017. After removing the sum of 100k of duplicate tweets or retweets at every iteration, we ended up with a 5k tweets dataset containing the most relevant tweets. Our Arabic tweets dataset has the majority of tweets being in the Saudi dialect than MSA since the studied airline being the national carrier airline of Saudi Arabia. Moreover, as seen in Fig. 2 of the overall architecture of the project, we have performed tweets preprocessing removing all URLs, user mentions, and unrelated tweets, getting rid of the numerous irrelevant tweets such as advertisement and so on.

After observing the full dataset, the researcher, as well as annotators, were able to identify 13 aspect categories that encompass the various tweeted about topics of airline service. The dataset labeling was performed manually dividing the task with a group of native Arabic speakers who are working in related fields. Aspect categories with topics discussed in each aspect category are presented in Table 1. We ended up with a distribution plot of the final dataset of 5k tweets for the 13 specified aspects categories related to the airline service as seen in Fig. 1. The labeling technique conducted to label each tweet row with its aspect class and sentiment polarity being either positive or negative was compiled in a binary format where 1 indicates the tweet belongs to the aspect category and 0 means it does not. Then, the same technique was used for sentiment polarity labeling, 1 indicating positive and 0 negative and so on. We only classified predicted aspect tweets as being either positive or negative as we did not consider the neutral polarity class.

Table 1. Description of all aspect classes used to label the dataset.

Aspect	Tweets topics
Schedule	Flight schedule, rescheduling, timing, delays
Destinations	Airline destination and routes
Luggage & cargo	Luggage, air cargo, luggage allowance, luggage delays
Staff & crew	All staff such as pilots and flight attendants
Airplane	Airplane seating, cabin features, maintenance
Lounges	First-class and frequent flyer service and airport lounges
Entertainment	In-flight entertainment, other media, and wi-fi
Meals	In-flight meals and in-flight services
Booking services	Airline website, mobile app, and self-service machines
Customer service	Customer communications and complaint management
Refunds	Ticket refunds and compensations
Pricing	Ticket pricing and seasonal airline offers
Miscellaneous	Represents all tweets with gratitude or complaints about the airline as general with no relation to other 12 aspect categories

Also, for the other part of the study sentiment polarity detection, the collected tweets are distributed concerning sentiment polarity. For the 5k dataset, the majority of the tweets belonged to the negative class with the total of 3590 tweets and only 1410 positive tweets. Thus, some of the aspect classes are represented with very few positive examples as in Fig. 1. The reason is that most of Twitter users use it as a medium to

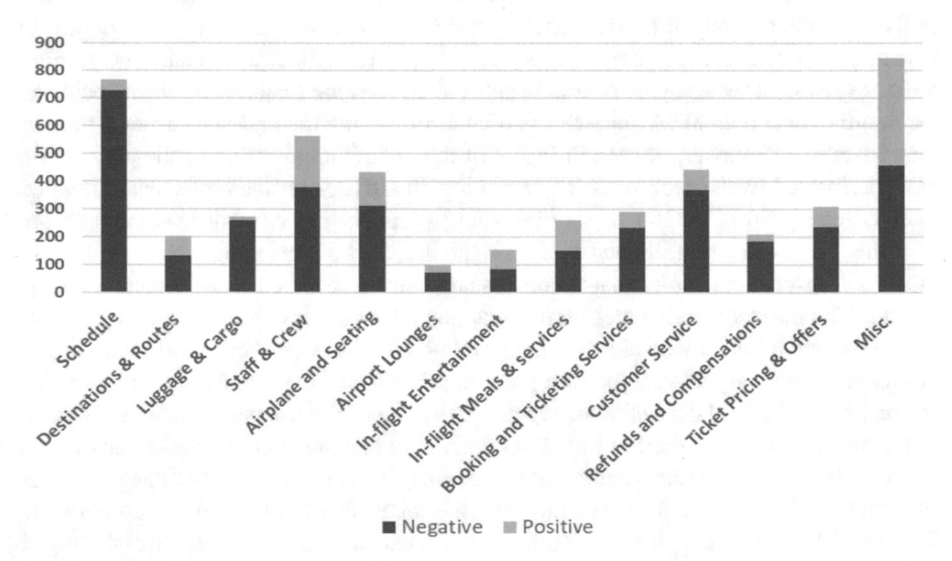

Fig. 1. A distribution plot of the dataset with every airline service aspect and its total tweet counts, in addition to the number of tweets for both sentiment polarity classes for each aspect category.

channel their complaints and negative energy obtained from negative experience and failed service. On the other hand, very few use it for positive commentaries, discussing their pleasant experiences or showing their constant gratitude to the airline. The Miscellaneous class was exceptional having an almost equal number of tweets for both polarity classes since it included all tweets that were generally referring to the airline.

4 Experiment Design and Implementation

Python is used as the sole programming language for this project because of its renowned machine learning library known as Scikit-learn. After loading the dataset, it was split into two sets, 80% of the data for training, and the other 20% for testing the performance of the classifier. In this research, we opted the Support Vector Machines (SVM) classifier due to its well-known superiority on other machine learning techniques as it has shown exceptional performance in text mining tasks in many works in the literature. The overall architecture of the proposed aspect-based sentiment analysis (ABSA) is described in Fig. 2.

4.1 Method

Applying the two pre-trained word embedding models, we used simple vector-based features to classify the 5k collected airline-related tweets to conduct ABSA two defined sub-tasks that are Aspect Detection and Aspect-Based Sentiment Detection. To clarify, we utilized the two distinct word embedding models as the lexical resource for the two ABSA sub-tasks.

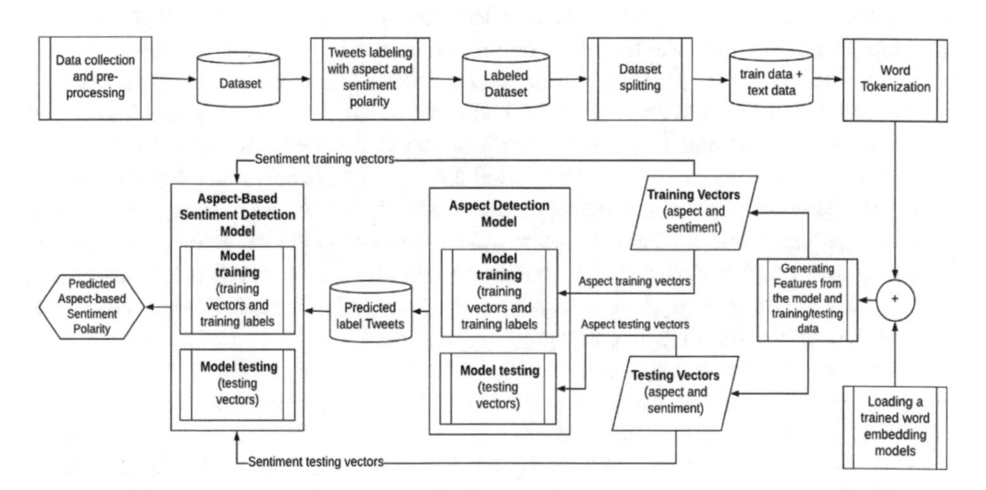

Fig. 2. The overall architecture of the proposed aspect-based sentiment analysis (ABSA).

With the creation of the Word2Vec toolkit, Mikolov et al. [14] has introduced the idea and brought considerable attention to the research community contributing to word embedding widespread use for it being easy to implement and tuned generating embeddings. Word2Vec models are shallow two-layer neural networks that have been presented for taking as input a large corpus of text to efficiently build word embeddings in a vector space where semantically similar words in the corpus have been assigned a corresponding vector in the space. Word2Vec use a probabilistic prediction approach which captures syntactic and semantic word relationships from huge datasets. The skip-gram model was introduced by [14] for representing words in a multidimensional vector space as it predicts the surrounding context words given the center word.

We were able to load and utilize the word embedding models using the 'gensim' [17] open-source vector space modeling toolkit implemented in Python; the applied two models are:

- **AraVec, Web skip-gram pre-trained word embedding model:** AraVec [2] is an Arabic language pre-trained distributed word embedding models where it was produced using the Word2Vec skip-gram technique trained on World Wide Web pages Arabic content with a dimension of 300 and a vocabulary size of 145,428.
- **fastText Arabic Wikipedia skip-gram pre-trained word embeddings model:** This model [1] was trained on top of Wikipedia Arabic articles using fastText with a vector embeddings of 300 and vocabulary size of 610,977. FastText is a library for learning of word embeddings created by Facebook's AI Research lab. [1] proposed a new approach based on the Word2Vec Skip-gram and CBOW architectures, where each word is represented as a bag of character n-grams to learn word representations considering sub-word information.

4.2 Aspect Detection

We trained a supervised SVM linear classifier using simple vector-based features on the labeled 5k tweets dataset to train it on unseen tweets to predict on which aspects of airline service they fall. Having a multiclass dataset, it required a multiclass classification task which refers to classifying data with more than two classes. Here, it assumes that each tweet is assigned to one and only one of the aspect classes, and not to two aspect classes at once. Therefore, we applied a one-vs-rest (one-vs-all) Support Vector Machine classifier for Aspect detection sub-task of ABSA. It is about fitting one classifier per class, and for each classifier, the class is fitted against all the other classes, which means training a classifier for each possible class. Hence, since only one classifier is representing each class solely, gaining knowledge regarding each class was feasible through inspecting its corresponding classifier.

Then, using word embedding models, the target word/label is the aspect category and that words in the tweet represented through the word vectors are the input to the classifier. In the simple setting, if one tweet contains the words, for example, "تأخرت رحلتي للرياض" with "الرياض" or Riyadh being the Arabic word for Saudi Arabia capital city. We use pre-trained word embeddings so that during training of our classifier, every word's feature vector is its Word2Vec or fastText word embeddings' vector. Therefore, during testing when a new word such as "الدمام" or Dammam another Saudi city is inferred which may or may not have occurred in training. We set its feature vector to its Word2Vec vector and predict its class since it is similar to "الرياض",

both are locations and cities. We can expect the SVM classifier to capture the features that are more specifically indicative of any Airline service aspect class or polarity class and predict words in a tweet being similar and more common in which class for both the Aspect Detection or the Aspect-Based Sentiment Classification sub-tasks of ABSA.

4.3 Aspect-Based Sentiment Detection

Later for the aspect-based sentiment detection sub-task, we used linear support vector classification for the aspect-based sentiment classification sub-task being a binary classification task. Every collected and labeled tweet consists of not only aspect specific words but also many positive and negative emotional words. Since we are tackling an ABSA problem, the aspects in the tweets were labeled with their sentiment polarity. That is, each of the aspect classes were manually labeled according to their sentiment polarity having either positive or negative but not neutral polarity.

Using word embedding models, we can classify similar polarity words for both polarity classes as being pre-trained in the word embedding models since it represents semantically similar words a corresponding vector in a vector space. As such simple vector-based features capture both semantic and sentiment information and this has improved the sentiment classification task, as seen in Table 3. In other words, the sentiment words are computed automatically through the selection of nearest word vectors neighboring the vectors of the word "ممتاز" meaning 'excellent' for positive or "سيء" meaning 'bad' for negative seeds as an example. For both parts of the study aspect detection and the aspect-based sentiment detection, we trained the SVM classifier on the produced training vectors along with the training labels. Then predicted and tested the classifier with the test vectors only for it to predict the labels for the test vectors to see how well it will generalize to measure its performance as a last step in the process as shown in Fig. 2.

For both ABSA sub-tasks, since we have the entire corpus of data labeled upfront, tweets were tokenized splitting into tokens being presented as a sequence of tokens or words for each of the aspect/sentiment polarity classes. With the mapping of the n-grams is being used to generate features from the training vectors along with the model word vector representation being associated to each n-gram. Feature vectors were combined with the input document of labelled airlines tweets dataset as a bag of words. With the mapping of the n-grams is being used to generate features allowing the model word vector embeddings being associated to each n-gram obtaining the simple vector-based features.

All tokens or words related to any of the 13 airline label classes and then to sentiment polarity classes were represented creating tweet/word-embeddings matrix for both classifier training and testing. This matrix includes all tweets text for a certain class in one column, then 300 dimension word embeddings representing word features in each tweet row. Then, such target word/label for each aspect/sentiment polarity class are words/vectors in tweets being represented through word embeddings as training and testing vectors are fed as input to the SVM classifier. Hence, the SVM classifier was able to separate each of the airline tweets classifying their related aspect of the 13 labeled classes, then, classify the predicted aspect classes sentiment polarity for the different experiments conducted. Hence, the SVM classifier was able capture new words occurring during testing since it was trained on such very broad set of similar words or word embeddings with proximity to word features representing any of the aspect or sentiment polarity classes that existed during training.

5 Results

We compare the two word embedding models' performances using various evaluation metrics. Namely, accuracy, precision, recall, and F1 with the macro-averaged precision, recall and F-measure at the bottom of Tables 2 and 3. The fastText Arabic Wikipedia word embeddings model uses character level information, thus, containing many word forms that are rarely occurring in the training corpus improving word vector representations for Arabic and on the two sub-tasks of ABSA results. Besides, it has four times the vocabulary size as the AraVec model. However, we can see the two models having almost closer results, and that can be due to the nature of Wikipedia having many words that carry a negative or a positive meaning in that in a spoken language having neutral meaning [2].

For the Luggage & Cargo aspect, the aspect-based sentiment polarity detection for the positive class results are zeros, as seen in below Table 3. That is due to the positive polarity class having very minimal examples as compared to the number of negative tweets for this same aspect class. Nevertheless, looking at the same Table 2 and Table 3 to compare the overall performances using the two word embedding models in terms of evaluation metrics. The adopted pre-trained word embedding models utilized for both ABSA sub-tasks has performed comparably well in comparison to existing techniques. The reason is that they bring extra semantic features that help in text classification as they rely not only on the word count but also save all required semantic and syntactic information [14]. However, one limitation that we faced was that our 5k tweets dataset was imbalanced with a significant difference in tweets counts between the 13 aspect classes as well as for the positive and negative polarity classes.

Word embedding is an essential resource for various Natural Language Processing applications such as sentiment analysis [14]. Hence, lack of high-quality word embeddings to capture the semantics of essential domain terms and their relations may prevent the state-of-the-art NLP techniques from being adopted in processing domain-specific texts [18]. To capture the distributional semantics of words, typical word embedding methods such as Word2Vec [14] rely on the cooccurrences of a target word and its context. Since robust inference can be achieved only with sufficient co-occurrences, this poses a challenge to applications where the domain texts are sparse [18]. That is, one limitation to using the traditional word embedding methods such as Word2Vec is that they do not perform very well in highly specialized domains such as the studied airline service where important domain concepts often do not occur many times. Also, for the aspect-based sentiment detection sub-task, specifically sentiment analysis of dialectical Arabic such as Saudi Arabic, where important Saudi cultural specific sentiment polarity terms often do not occur often as opposed to MSA.

Table 2. The experimental results of aspect detection sub-task using the two word embedding models.

Aspect	One-vs-the-rest (OvR) multiclass strategy (linear support vector classification)					
	AraVec, web skip-gram word embeddings model [2]			fastText skip-gram Arabic Wikipedia word embeddings model [1]		
	Precision	Recall	F1	Precision	Recall	F1
Schedule	0.67	0.73	0.70	0.74	0.84	0.79
Destinations & routes	0.56	0.59	0.58	0.74	0.62	0.68
Luggage & cargo	0.69	0.64	0.67	0.65	0.56	0.60
Staff & crew	0.65	0.63	0.64	0.63	0.71	0.67
Airplane & seating	0.60	0.60	0.60	0.68	0.74	0.71
Lounges	0.35	0.27	0.30	0.69	0.50	0.58
Entertainment	0.77	0.57	0.66	0.95	0.69	0.80
In-flight meals	0.63	0.56	0.59	0.70	0.58	0.63
Booking services	0.71	0.64	0.67	0.79	0.70	0.75
Customer service	0.52	0.60	0.56	0.64	0.61	0.63
Refunds	0.65	0.57	0.61	0.73	0.59	0.65
Pricing & offers	0.80	0.67	0.73	0.82	0.75	0.78
Misc.	0.61	0.67	0.64	0.64	0.70	0.67
Avg/total	0.64	0.64	0.63	0.74	0.84	0.79
Accuracy	0.64			0.70		

Table 3. The experimental results of aspect-based sentiment polarity detection sub-task using the two word embedding models.

Aspect	Polarity	Linear support vector machines classifier							
		AraVec, web skip-gram word embeddings model [2]				fastText skip-gram Arabic Wikipedia word embeddings model [1]			
		Precision	Recall	F1	Accuracy	Precision	Recall	F1	Accuracy
Schedule	Positive	0.67	0.33	0.44	0.96	0.67	0.22	0.33	0.94
	Negative	0.97	0.99	0.98		0.95	0.99	0.97	
Destinations & routes	Positive	0.67	0.57	0.62	0.76	0.73	0.67	0.70	0.83
	Negative	0.79	0.85	0.82		0.87	0.90	0.88	
Luggage & cargo	Positive	0.00	0.00	0.00	0.98	0.00	0.00	0.00	0.98
	Negative	0.98	1.00	0.99		0.98	1.00	0.99	
Staff & crew	Positive	0.85	0.87	0.86	0.89	0.83	0.93	0.88	0.93
	Negative	0.92	0.90	0.91		0.97	0.93	0.95	
Airplane & seating	Positive	0.80	0.65	0.71	0.81	0.84	0.73	0.78	0.89
	Negative	0.81	0.90	0.85		0.91	0.95	0.93	
Lounges	Positive	1.00	0.86	0.92	0.94	0.86	0.75	0.80	0.83
	Negative	0.92	1.00	0.96		0.82	0.90	0.86	

(*continued*)

Table 3. (*continued*)

Aspect	Polarity	Linear support vector machines classifier							
		AraVec, web skip-gram word embeddings model [2]				fastText skip-gram Arabic Wikipedia word embeddings model [1]			
		Precision	Recall	F1	Accuracy	Precision	Recall	F1	Accuracy
Entertainment	Positive	0.73	0.92	0.81	0.82	0.82	0.93	0.87	0.86
	Negative	0.92	0.75	0.83		0.91	0.77	0.83	
In-flight meals	Positive	0.56	0.56	0.56	0.69	0.94	0.65	0.77	0.80
	Negative	0.76	0.76	0.76		0.72	0.95	0.82	
Booking services	Positive	1.00	0.62	0.76	0.90	0.83	0.62	0.71	0.92
	Negative	0.88	1.00	0.94		0.93	0.98	0.95	
Customer service	Positive	0.93	0.78	0.85	0.94	0.86	0.67	0.75	0.90
	Negative	0.94	0.98	0.96		0.91	0.97	0.94	
Refunds	Positive	0.67	0.40	0.50	0.89	1.00	0.60	0.75	0.95
	Negative	0.91	0.97	0.94		0.94	1.00	0.97	
Pricing & offers	Positive	0.80	0.25	0.38	0.78	0.88	0.58	0.70	0.90
	Negative	0.78	0.98	0.87		0.90	0.98	0.94	
Misc.	Positive	0.76	0.80	0.78	0.80	0.87	0.81	0.84	0.84
	Negative	0.83	0.80	0.81		0.80	0.86	0.83	
Avg/total	Positive	0.73	0.59	0.63	0.86	0.78	0.63	0.68	0.89
	Negative	0.88	0.91	0.89		0.89	0.94	0.91	

6 Conclusion

We employed word embeddings to perform Aspect Based Sentiment Analysis on Arabic airline-related tweets. We used two pre-trained word vectors models producing vector-based features and showing their effectiveness for both aspect detection and aspect-based sentiment detection. Our vector space approach using these features and no hand-crafted features performs comparably well in comparison to techniques adopting manually engineered features. Nonetheless, our system results have presented that such word embedding features yet simple are powerful enough to achieve state of the art performance in performing ABSA for the Arabic language.

For future work, we would look at performing multi-label classification ABSA using Twitter where each tweet can fall into one or more classes/labels. Additionally, since the existing Arabic domain-specific word embedding models still insufficient or unavailable for us making it such a challenging task when dealing with domain-specific sentiment classifications tasks. Thus, constructing custom word embeddings resources is beneficial although can be time-consuming. So, the primary focus on future work of this research is to create a new large word embeddings resources for the Saudi dialect sentiment analysis.

References

1. Bojanowski, P., Grave, E., Joulin, A., Mikolov, T.: Enriching word vectors with subword information. Trans. Assoc. Comput. Linguist. **5**, 135–146 (2017)
2. Soliman, A.B., Eissa, K., El-Beltagy, S.R.: AraVec: a set of Arabic word embedding models for use in Arabic NLP. Procedia Comput. Sci. **117**, 256–265 (2017)
3. Biltawi, M., Etaiwi, W., Tedmori, S., Hudaib, A., Awajan, A.: Sentiment classification techniques for Arabic language: a survey (2016). http://ieeexplore.ieee.org/document/7476075. Accessed 14 Mar 2018
4. Pang, B., Lee, L.: Opinion Mining and Sentiment Analysis. Now Publishers Inc, Breda (2008)
5. Schouten, K., Frasincar, F.: Survey on aspect-level sentiment analysis. IEEE Trans. Knowl. Data Eng. **28**(3), 813–830 (2016)
6. El-Masri, M., Altrabsheh, N., Mansour, H.: Successes and challenges of Arabic sentiment analysis research: a literature review. Soc. Netw. Anal. Min. **7**(1), 54 (2017)
7. Altowayan, A.A.: Word embeddings for Arabic sentiment analysis. In: 2016 IEEE International Conference on Big Data (Big Data). IEEE (2016)
8. Abdul-Mageed, M., Diab, M.T., Korayem, M.: Subjectivity and sentiment analysis of modern standard Arabic (2011). http://dblp.uni-trier.de/db/conf/acl/acl2011s.html. Accessed 14 Mar 2018
9. Abdul-Mageed, M., Kuebler, S., Diab, M.T.: SAMAR: a system for subjectivity and sentiment analysis of Arabic social media (2012). http://dl.acm.org/citation.cfm?id=2392971. Accessed 14 Mar 2018
10. Mourad, A., Darwish, K.: Subjectivity and sentiment analysis of modern standard arabic and arabic microblogs. In: The 4th Workshop on Computational Approaches to Subjectivity, Sentiment and Social Media Analysis (2013)
11. Abdulla, N.A., Ahmed, N.A., Shehab, M.A., Al-Ayyoub, M.: Arabic sentiment analysis: Lexicon-based and corpus-based (2013). http://ieeexplore.ieee.org/xpls/abs_all.jsp?arnumber=6716448. Accessed 14 Mar 2018
12. Bouchlaghem, R., Elkhelifi, A., Faiz, R.: A machine learning approach for classifying sentiments in Arabic tweets. In: Proceedings of the 6th International Conference on Web Intelligence, Mining and Semantics. ACM (2016)
13. Ren, Y., Wang, R., Ji, D.: A topic-enhanced word embedding for twitter sentiment classification. Inf. Sci. **369**, 188–198 (2016)
14. Mikolov, T., Chen, K., Corrado, G., Dean, J.: Efficient estimation of word representations in vector space. arXiv: Computation and Language (2013)
15. Al-Azani, S., El-Alfy, E.-S.M.: Using word embedding and ensemble learning for highly imbalanced data sentiment analysis in short Arabic text. Procedia Comput. Sci. **109**, 359–366 (2017)
16. Alghunaim, A.: A vector space approach for aspect-based sentiment analysis. Dissertation, Massachusetts Institute of Technology (2015)
17. Rehurek, R., Sojka, P.: Software framework for topic modelling with large corpora. In: LREC 2010 Workshop on New Challenges for NLP Frameworks (2010)
18. Roy, A., Park, Y., Pan, S.: Learning domain-specific word embeddings from sparse cybersecurity texts. arXiv preprint (2017)

Segmentation Tool for Hadith Corpus to Generate TEI Encoding

Hajer Maraoui[1(✉)], Kais Haddar[2], and Laurent Romary[3]

[1] MIRACL Laboratory, Faculty of Sciences of Tunis, University of Tunis El Manar, Tunis, Tunisia
hajer.maraoui@fst.utm.tn

[2] MIRACL Laboratory, Faculty of Science of Sfax, University of Sfax, Sfax, Tunisia
kais.haddar@yahoo.fr

[3] Team ALMAnaCH, Inria, Berlin, Germany
laurent.romary@inria.fr

Abstract. A segmentation tool for a hadith corpus is necessary to prepare the TEI hadith encoding process. In this context, we aim to develop a tool allowing the segmentation of hadith text from Sahih al-Bukhari corpus. To achieve this objective, we start by identifying different hadith structures. Then, we elaborate an automatic processing tool for hadith segmentation. This tool will be integrated in a prototype allowing the TEI encoding process. The experimentation and the evaluation of this tool are based on Sahih al-Bukhari corpus. The obtained results were encouraging despite some flaws related to exceptional cases of hadith structure.

Keywords: Hadith corpus · Hadith structure · Segmentation tool TEI encoding

1 Introduction

A segmentation phase is necessary to analyse and standardize corpora especially for hadith. In fact, the automatization of the segmentation of large hadith corpora guarantees the treatment optimization in term of time and results precision. Moreover, it facilitates the processing and the manipulation for some tools that analyze and standardize each part of the text in hadith corpus.

However, this phase imposes some problems especially with the structure of large hadith corpora, which characterised with text particularity and variant features. Furthermore, these corpora are originally in Arabic language, which impose some difficulties that related with language specificities.

In this context, to segment a large corpus of hadith, such as Sahih al-Bukhari, we must first start with a deep study on hadith corpora specificities and identify the features of hadith text. Then, we must classify the separator terms that recognise the different units in hadith corpus and develop the necessary resources to realise the segmentation process. After that, we need to develop a tool for the segmentation of hadith corpus.

A. E. Hassanien et al. (Eds.): AISI 2018, AISC 845, pp. 252–260, 2019.
https://doi.org/10.1007/978-3-319-99010-1_23

In the present paper, we begin with a state of art on hadith in Sahih al-Bukhari corpus. Then, we illustrate the features of hadith text. After that, we present the segmentation tool for hadith text from sahih al-Bukhari. Then, we continue with by an evaluation step. We cloture our paper with a conclusion and some perspective.

2 Stat of Art on Hadith in Sahih al-Bukhari Corpus

Sahih al-Bukhari (or صحيح البخاري in Arabic) is one of the six major hadith collections of Sunni Islam. The Arabic word sahih (صحيح) translates as authentic or correct. The prophetic traditions (or hadiths), were collected by the Muslim scholar Muhammad al-Bukhari, after being transmitted orally for generations [4]. Al-Bukhari went to Mecca when he was 16 years old and learned hadith there. He traveled around other Islamic countries to collect hadiths. He had collected 600,000 hadiths of which he only considered 7,275 ones as authentic in his well-known work, Sahih al-Bukhari. He finished his work around 846/232 AH [5]. The book is considered as one of the two most authentic and trusted collections of hadith along with Sahih Muslim.

Sahih al-Bukhari covers almost all aspects of life in providing proper guidance of Islam such as the method of performing prayers and other actions of worship directly from the Prophet Muhammad. In Islamic terminology, the term hadith refers to reports of statements or actions of the Prophet Muhammad, or of his implicit approval or criticism of something said or done in his presence. Classical hadith specialist Ibn Hajar al-Asqalani says that the intended meaning of hadith in religious tradition is something attributed to Muhammad but that is not found in the Quran. The two major aspects of a hadith are the *Matn* which is the actual narration, and *Sanad* or *Isnad* which is the chronological list of narrators. Each *Isnad* reporter mentions the person from whom he heard the hadith all the way to the prime reporter of the *Matn* itself. The *Isnad* was an authentication of the hadith to verify that it is actually come from the Prophet Muhammad. The *Isnad* means literally ‘support’, so it is the support in determining the hadith authenticity or weakness (more details in [4, 6]).

Research in the *Isnad* is very important in the science of hadith. To define the hadith authentication, the traditional methods consist of following clear steps in the judgment on the *Isnad*. Currently, software tools allow hadith judging like electronic hadith encyclopedias and some websites. Moreover, information retrieval and search engines that related to semantic web can be used to serve in deciding the authenticity of hadith.

Many projects worked on hadith corpus. Indeed, these projects focused on several branches of researches such as hadith ontology, linguistic analyzing, Hadith segmentation, classification and the mining of information.

In [7], the researchers proposed a model named SALAH for the unsupervised segmentation and the linguistic analysis of the hadith texts. The model automatically segments each text unit in *Isnad* and *Matn*. After that, a personalized augmented version of the AraMorph morphological analyzer (RAM) examines and annotates lexically and morphologically the text content. The system generates as final output a graph with relations among transmitters and a lemmatized text corpus in XML format.

In [8], the author constructed an ontology-based Isnad Judgment System (IJS) that automatically generates a suggested judgment of hadith Isnad. This is based on the standard instructions followed by the hadith scholars to judge hadith Isnad. A prototype of the approach implemented to provide a proof of concept for the requirements and to verify its accuracy.

Authors of paper [9] built a domain specific ontology (Hadith Isnad Ontology) to support the process of authenticating Isnad. They evaluate the ontology through hadith example and DL-Queries.

In [10], the authors reported on a system that automatically generates the transmission chains of a hadith and graphically display it. They involve parsing and annotating the hadith text and identifying the narrators' names. They use shallow parsing along with a domain specific grammar to parse the hadith content.

3 Arabic Hadith Features: Sahih al-Bukhari Corpus

Sahih al-Bukhari is arranged like books in Islamic jurisprudence (or " فِقه ", fiqh). It contains also some additional sections focusing on different topics such as the origins of creation, the exegesis of the Quran and the Holy prophets. Al-Bukhari also expressed his own views of the issues in fiqh by classifying the sections of his book and assigned for each section a significant title which expresses the fiqh of al-Bukhari. These titles involve all the interpretations and explications of al-Bukhari to clarify the meaning of hadiths which are difficult to understand.

Sahih al-Bukhari corpus is structured into 97 chapters and 3450 sections. The chapters contain 7563 hadith in totality. This number includes repeated hadiths. It confirmed contains 7275 ones, and setting aside the repeated ones, as Nawawi says, it includes 4,000 hadiths, though Ibn Hajar says that on this count, it contains 2,761 hadiths. Moreover, the number of its hadiths is different in different versions of the book; for example, Firabri has cited 300 hadiths more than those cited by Ibrahim b. Ma'qil al-Nasafi, and the latter cited 100 hadiths fewer than those cited by Himad b. Shakir al-Nasawi.

3.1 Hadith Features

Each hadith in Sahih al-Bukhari is cited in a relative section, titled by the author to express his interpretation of the meaning of hadith, under a chapter combine the sections related with one topic. Moreover, each hadith is quoted with an order number and finish with the references. This structure in accurately respected in Sahih al-Bukhari. As an example, Fig. 1 present hadith number 3209 from Chap. 59 "beginning of creation", Sect. 6 "the reference to angels", as it occurs in sahih al-Bukhari.

In order to extract hadiths text from Sahih al-Bukhari, it is convenient to keep the relative coordination of each hadith to maintain the structure and authentication support.

59 - كتاب بدء الخلق (...)
6 - باب ذِكْرِ الْمَلاَئِكَةِ (...) 3209 - حَدَّثَنَا مُحَمَّدُ بْنُ سَلاَمٍ، أَخْبَرَنَا مَخْلَدٌ، أَخْبَرَنَا ابْنُ جُرَيْجٍ، قَالَ: أَخْبَرَنِي مُوسَى بْنُ عُقْبَةَ، عَنْ نَافِعٍ، قَالَ: قَالَ أَبُو هُرَيْرَةَ رَضِيَ اللَّهُ عَنْهُ، عَنِ النَّبِيِّ صَلَّى اللهُ عَلَيْهِ وَسَلَّمَ، وَتَابَعَهُ أَبُو عَاصِمٍ، عَنِ ابْنِ جُرَيْجٍ، قَالَ: أَخْبَرَنِي مُوسَى بْنُ عُقْبَةَ، عَنْ نَافِعٍ، عَنْ أَبِي هُرَيْرَةَ، عَنِ النَّبِيِّ صَلَّى اللهُ عَلَيْهِ وَسَلَّمَ، قَالَ: " إِذَا أَحَبَّ اللَّهُ العَبْدَ نَادَى جِبْرِيلَ: إِنَّ اللَّهَ يُحِبُّ فُلاَنًا فَأَحْبِبْهُ، فَيُحِبُّهُ جِبْرِيلُ، فَيُنَادِي جِبْرِيلُ فِي أَهْلِ السَّمَاءِ: إِنَّ اللَّهَ يُحِبُّ فُلاَنًا فَأَحِبُّوهُ، فَيُحِبُّهُ أَهْلُ السَّمَاءِ، ثُمَّ يُوضَعُ لَهُ القَبُولُ فِي الأَرْضِ ". [طرفاه في 6040, 7435].

59 – Chapter of beginning of creation (…)
6 – Section the reference to angels (…) 3209 - The Prophet (peace and blessing upon him) said, "If Allah loves a person, He calls Gabriel saying, 'Allah loves so and-so; O Gabriel! Love him.' Gabriel would love him and make an announcement amongst the inhabitants of the Heaven. 'Allah loves so-and-so, therefore you should love him also,' and so all the inhabitants of the Heaven would love him, and then he is granted the pleasure of the people on the earth". [Links in 6040, 7435].

Fig. 1. Hadith number 3209 from Chap. 59, Sect. 6 from Sahih al-Bukhari.

3.2 Hadith Structure

The typical hadith structure consists of two parts, the transmitter chain (*Isnad*) and the actual narration (*Matn*). Since just close sets of words separate both *Isnad* from *Matn* and the various transmitters inside *Isnad* one from another, this explicit organization allows detecting and retrieving information with a relatively small amount of ambiguity. Despite the regular structure of the majority of hadiths occurred in Sahih al-Bukhari, it still exist others with different structures. These cases are related with implicit parameters which are referenced or mentioned in implicit way. For example, it common to find a hadith text with only the *Matn*. But also cited with an expression of the reference where to find the appropriate isnad or came as a continuation of the Matn of a previous hadith. As an Example, Fig. 2 illustrates this case in hadith number 654 from the Sect. 32, Chap. 10 “Call to prayers”.

10 - كتاب الأذان (...)
32 – باب فضل التّهجير إلى الظّهر(...) 537 - « وَلَوْ يَعْلَمُونَ مَا في التَّهْجِيرِ لاَسْتَبَقُوا إِلَيْهِ ، وَلَوْ يَعْلَمُونَ مَا في الْعَتَمَةِ وَالصُّبْحِ لأَتَوْهُمَا وَلَوْ حَبْواً ». [أطرافه 615، 721، 2689 - تحفة 12570- 12577].

537 - “And if they knew the reward of offering the Zuhr prayer early (in its stated time), they would race for it and they knew the reward for 'Isha' and Fajr prayers in congregation, they would attend them even if they were to crawl”. [Links in 615, 721, 2689_ 12570 – 12577].

Fig. 2. Hadith number 654 from Chap. 10, Sect. 32 from Sahih al-Bukhari.

The hadith text in this case came to prove the previous one and occurs elsewhere with the reference indicated in the end of the *Matn*. In other context, other forms of hadith are more abstract and follow an irregular structure. The most of these hadith came to prove the previous hadith.

In Sahih al-Bukhari occurs also other form of hadiths. These traditions contain a complex structure in the part of mentioning the Isnad. Frequently, a hadith can have more than one chain of narrators as Isnad. These chains can occur in the front of the text cited after each other or divided in two sections before and after the Matn. As Example of this composition, Fig. 3 present hadith number 6 from Sect. 1 "How was the revelation on the Prophet", Chap. 1 "Revelation".

1- كتاب بدء الوحي (...)
1 - باب كَيْفَ كَانَ بَدْءُ الْوَحْيِ إِلَى رَسُولِ اللَّهِ (...) 6 - حَدَّثَنَا عَبْدَانُ قَالَ أَخْبَرَنَا عَبْدُ اللَّهِ قَالَ أَخْبَرَنَا يُونُسُ عَنِ الزُّهْرِىِّ ح وَحَدَّثَنَا بِشْرُ بْنُ مُحَمَّدٍ قَالَ أَخْبَرَنَا عَبْدُ اللَّهِ قَالَ أَخْبَرَنَا يُونُسُ وَمَعْمَرٌ عَنِ الزُّهْرِىِّ نَحْوَهُ قَالَ أَخْبَرَنِى عُبَيْدُ اللَّهِ بْنُ عَبْدِ اللَّهِ عَنِ ابْنِ عَبَّاسٍ قَالَ كَانَ رَسُولُ اللَّهِ صَلَّى اللهُ عَلَيْهِ وَسَلَّمَ، أَجْوَدَ النَّاسِ ، وَكَانَ أَجْوَدُ مَا يَكُونُ فِى رَمَضَانَ حِينَ يَلْقَاهُ جِبْرِيلُ ، وَكَانَ يَلْقَاهُ فِى كُلِّ لَيْلَةٍ مِنْ رَمَضَانَ فَيُدَارِسُهُ الْقُرْآنَ ، فَلَرَسُولُ اللَّهِ صَلَّى اللهُ عَلَيْهِ وَسَلَّمَ، أَجْوَدُ بِالْخَيْرِ مِنَ الرِّيحِ الْمُرْسَلَةِ. [أطرافه 1902، 3220، 3554، 4997 - تحفة 5840 - 1/5].

1 – Chapter of revelation (…)

1 – Section of how the revelation was for the prophet (…) 6 - Allah's Messenger (peace and blessing upon him) was the most generous of all the people, and he used to reach the peak in generosity in the month of Ramadan when Gabriel met him. Gabriel used to meet him every night of Ramadan to teach him the Qur'an. Allah's Messenger (peace and blessing upon him) was the most generous person, even more generous than the strong uncontrollable wind (in readiness and haste to do charitable deeds). [Links in 1902, 3220, 3554, 4997 _

Fig. 3. Hadith number 6 from Chap. 1, Sect. 1 from Sahih al-Bukhari.

Based on sahih al-Bukhari corpus study, we determinate that to realize a hadith normalization system, we must start with the prime unite of sahih al-Bukhari corpus which is the hadith text. In fact, to reach the corpus normalization, a hadith extraction and segmentation phase is essential to prepare to the succeeding phase for encoding hadith text with TEI [1]. This lead as to investigate the development of an extraction and segmentation tool integrated in the hadith encoding prototype to split out the hadith text for the next encoding step which is presented in the next section.

4 Hadith Segmentation Tool

To implement the segmentation method of hadith text from Sahih al-Bukhari corpus, we developed a tool for this process. This tool is made to prepare the hadith text for the hadith encoding prototype with TEI (more details in [2, 3]). The implementation is based on JAVA language and the API JDOM Library. Figure 4 presents the general architecture of this program. This tool is developed to extract automatically each hadith text from Sahih al-Bukhari then segment each one to *Isnad* and *Matn* which goes as input to the hadith encoding process.

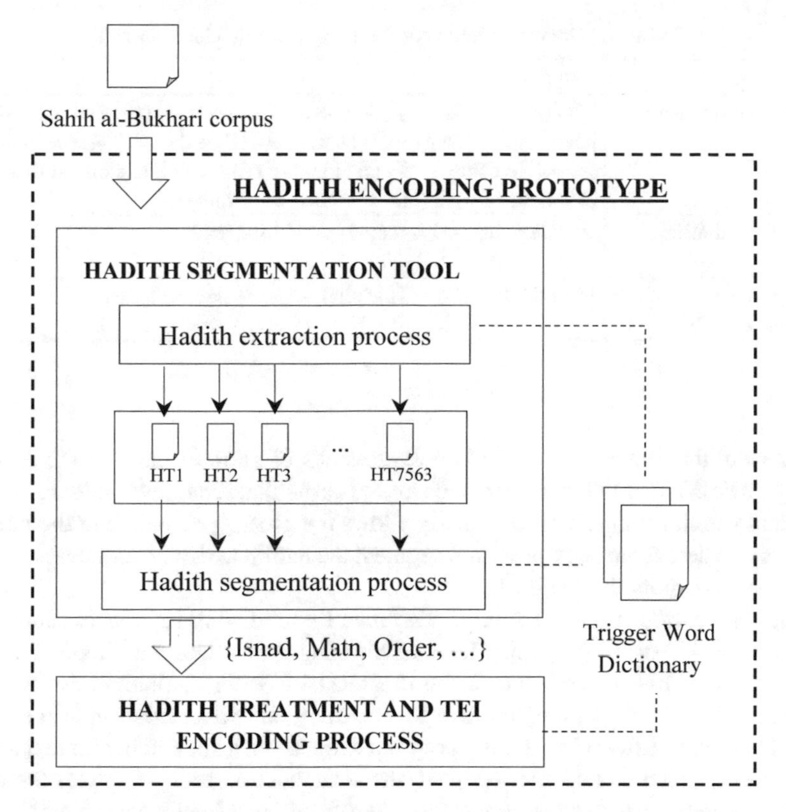

Fig. 4. General architecture of the segmentation tool of hadith corpus.

In the processing of the hadith encoding prototype, the system need as an input the *Isnad* and the *Matn* of each hadith in the used corpus. This input data is the output of the presented tool. To prepare hadith text for the encoding prototype, the system requests the user to choose an external file, with .txt extension which contains the hadith corpus in Arabic language, by selecting the file path. Then, the program reads the corpus and identify each hadith text. To detect hadith text in the corpus, we based on the fact that each hadith starts with a hadith number and a trigger word, as we name it, that express the beginning of the *Isnad*. We built for this reason a trigger word dictionary in XML format for the words that indicate the hadith beginning. Trigger word dictionary includes also the terms that separate the narrator names and indicates the commencement of the *Matn*. The trigger word identification allows the manipulation of hadith text. We classify these parts of speech to three classes: Trigger Word Before (TWB), Trigger Word After (TWA) and Trigger Word Between (TWT). Table 1 illustrates the different forms of these terms in the trigger words dictionary.

The first-class TWB comprise the terms that appears before each narrator name such as "حَدَّثَنَا مُحَمَّدُ بْنُ سَلاَمٍ" (Mohamed Ben Salam told us) or a person indication "عَنْ نَافِعٍ" (Transmitted from Nafaa). The second-class TWA is for the words that occur after a narrator name such as "عَبْدُ اللهِ قَالَ" (Abdu Allah said). The third-class TWT is for the

Table 1. Sammery table for the trigger words classification.

Trigger word	Example
Trigger word before (TWB)	حدّثنا (He told us) – أخبرني (He told me) – أخبرنا (He told us) – سمعت (I heard) – عن (from) – أنّ (That) – سأل (He asked) – زادنا (he enhanced) – جاء (He came) – ذكر (He mentioned) – زعم (He claimed that) – لقيت (I found) – وتابعه (He continued after him)...
Trigger word after (TWA)	قال (He said) – يقول (says) – قالت (She said) ...
Trigger word between (TWT)	أنّه (That he) – أنها (That she) – أخبرني (He told me)...

trigger word that appears between two expressions of narrator names such as "أَنّه قال" (That he said). The TWT class includes all the terms that relate two sections of trigger word and narrator name. This dictionary allows the prototype, including the integrated programs, to detect the right parts and segment the hadith text. The tool connects to this dictionary and spots the hadith text.

The termination of each hadith is identified by the hadith references numbers and the word "أطرافه" or "تحفه" (which have as meaning, references and links) or a flexion form of them. This method allows the treatment of Sahih al-Bukhari Arabic corpus with no need to modify the input file or to delete any other data exists in the corpus else then hadith texts. After this step, the program extracts each spotted hadith text and save it in the workspace as text file. As well, the segmentation tool completes the reading and the separation of the *Isnad* and the *Matn* from each hadith text. To achieve the hadith segmentation, we start by analyzing Sahih al-Bukhari corpus and identifying the terms that separate the two parts. These terms are included in the trigger word dictionary. Also, the segmentation process of *Isnad* and *Matn* is based on trigger word.

Hadith corpus segmentation tool is an integrated program in the hadith encoding prototype which allows the encoding of all the hadiths of Sahih al-Bukhari in one running action. This automatic preparation of hadith text supported the optimization of hadith corpus encoding using prototype. We have evaluated this tool based on Sahih al-Bukhari corpus. The results are presented in the following section.

5 Evaluation and Discussion

To evaluate the hadith corpus segmentation tool, we select as input a text file containing Sahih al-Bukhari corpus which include 7563 hadith. Consequently, the tool system creates respectively 7485 hadith text files and save them. Then the system segments each hadith and sends the Isnad and the Matn to the encoding process in the prototype core. Table 2 illustrates the obtained results.

Table 2 shows that sometimes for particular hadith texts, we can obtain erroneous segmentation. The total number of hadith text in Sahih al-Bukhari is 7563 hadith from 97 different chapters. The hadith segmentation tool succeeded to extract 7563 hadith

Table 2. Summary table for the hadith extraction and segmenation tool results.

Evaluated corpus	Extracted hadiths	Correct extraction	Segmented hadiths	Correct segmentation	Incorrect segmentation
Sahih al-Bukhari 7563 hadith 97 Chapter	7563	7563	7563	7259	304

correctly and save them as text files. Then, the segmentation process separate hadith fragments correctly for 7259 hadith texts. However, we found 304 hadith that were not segmented properly. These hadith have particular structure which require some rectification to cover these cases. Indeed, we estimate the capacity of this tool manually. Table 3 illustrates the obtained values of precision, recall and F-score.

Table 3. Summary table of the precision, recall and F-score.

Hadith corpus	Precision	Recall	F-score
7563 Hadith	0.96	0.96	0.96

According to the value of precision, we conclude that the value of precision is worth 0.96. Also, the recall value is 0.96. These values provide an F-measure equal to 0.96.

Consequently, we conclude that the obtained results are encouraging. Besides, the unsupervised tool system does not require a pretreatment or execute modification on the input corpus file. However, we handled some problems. Some of them are related with the particular hadith forms which need to integrate more specificity.

6 Conclusion and Perspectives

In order to optimize the processing of hadith text, we aimed to develop a segmentation tool for hadith corpus. To achieve that, first, we started with a deep study of hadith text structure from Sahih al-Bukhari. After that, we designed and created an integrated tool in the TEI encoding prototype for the segmentation of hadith corpus. Then, we tested this tool with a Sahih al-Bukhari corpus which englobe 7563 hadith texts from 94 chapters. As mentioned, the obtained values of measures show that the results obtained from the hadith segmentation tool are encouraging.

As perspective, the hadith segmentation tool can be used in others levels of analyses. Furthermore, we want to apply the proposed tool to other types of hadith corpora such as Sahih Muslim. Also, we want to extend the set of rules to cover the exceptional structures of hadith text.

References

1. Burnard, L., Sperberg-McQeen, C.M.: TEI P5: Guidelines for Electronic Text Encoding and Interchange, Text Encoding Initiative Consortium, Version 3.0.0. Revision 89ba24e (2016)
2. Maraoui, H., Haddar, K., Romary, L.: Modeling of Al-Hadith Al-Shareef with TEI. In: ICEMIS Conference, Monastir, Tunisia. IEEE (2017). 978-1-5090-6778-7/17/$31.00
3. Maraoui, H., Haddar, K., Romary, L.: Encoding prototype of Al-Hadith Al-Shareef in TEI. In: ICALP Conference, Fez, Morocco, CCIS 782, pp. 1–13. Springer (2017)
4. Abu Zaho, M.: الحديث والمحدثون, دار الفكر العربي, Riyadh, KSA, vol. 1 (1984)
5. Brown, J.A.C.: Hadith: Muhammad's Legacy in the Medieval and Modern World. Oneworld Publications, London (2009). ISBN 978-1851686636
6. Al-Ansari, S., المقنع في علوم الحديث, vol. 1. Fawaz Publishing House (1992)
7. Boella, M., et al.: The SALAH project: segmentation and linguistic analysis of hadīṯ Arabic texts. In: Proceeding of the Seventh Asia Information Retrieval Societies Conference. Springer, Heidelberg (2011)
8. Dalloul, Y.M.: An ontology-based approach to support the process of Judging Hadith Isnad. Thesis Submitted in Partial Fulfillment of the Requirements for the Degree of Master in Information Technology, March 2013
9. Baraka, R.: Building Hadith ontology to support the authenticity of Isnad. Int. J. Islam. Appl. Comput. Sci. Technol. **2**(1), 25–39 (2014)
10. Azmi, A., Badia, N.: iTree—Automating the Construction of the Narration Tree of Hadith. IEEE (2010)

ARARSS: A System for Constructing and Updating Arabic Textual Resources

Abdulmohsen Al-Thubaity and Muneera Alhoshan(✉)

The National Center for Artificial Intelligence and Big Data,
King Abdulaziz City for Science and Technology, Riyadh, Saudi Arabia
{aalthubaity,malhawshan}@kacst.edu.sa

Abstract. The growth of electronically readable Arabic content available on the web has become a rich source from which to build new corpora or update the existing ones. The availability of such corpora will be beneficial for Arabic corpus linguistics, computational linguistics, and natural language processing. In this paper, we present ARARSS, a tool capable of automatically constructing and updating textual corpora benefiting from the Rich Site Summary (RSS) feeds. ARARSS is capable of collecting the texts in a properly categorized manner according to user needs, in addition to their metadata (for example, location, time, and topic) as provided by RSS sources. We used ARARSS to construct a modern standard Arabic corpus comprising 117,819 texts and more than 28 million words. ARARSS is an open source tool and freely available to download (http://corpus.kacst.edu.sa/more_info.jsp) along with the constructed corpus.

Keywords: Language resources · Natural language processing
Computational linguistics · Corpus linguistics · Arabic corpora

1 Introduction

Texts in electronic or paper formats are repositories of human knowledge, which is usually codified through graphical representation and mathematical or chemical symbols and equations but mostly through the elaboration of natural language. The easy availability of electronically readable texts on the Internet and through private organizations has commanded the attention of researchers in areas such as natural language processing, computational linguistics, and corpus linguistics. Although these researchers may rarely study single texts, the majority of their efforts are expended towards studying large collections of texts: corpora.

The use and benefits of corpora vary among the above-mentioned research areas. For natural language processing and computational linguistics, corpora are mainly used to build language models and train and test machine learning algorithms used for different applications. It is well known that the accuracy of language models is heavily dependent on the corpus size and its various contents. The problem of most machine language algorithms is that they are data hungry and their performance markedly drops when the trained model is applied on different text types [1]. The obvious solution to overcome these problems is to expand the existent corpora or develop new ones.

A. E. Hassanien et al. (Eds.): AISI 2018, AISC 845, pp. 261–269, 2019.
https://doi.org/10.1007/978-3-319-99010-1_24

Retaining a large and regularly updated corpus is useful for many corpus linguistics studies, such as in lexicography, discourse analysis, special languages and terminology extraction, and diachronic and synchronic investigation of language. Sinclair [2], the well-known corpus linguistics scholar suggested that "a corpus should be as large as possible, and should keep on growing." Large and updated corpora are required to obtain useful empirical evidence regarding words usage, collocation behavior, and how they change over time.

In this paper, we describe ARARSS, an easy-to-use customizable tool for automatic corpus construction and updating based on Rich Site Summary (RSS) feeds. ARARSS is capable of collecting the texts in a properly categorized manner according to user needs in addition to their metadata (for example, location, time, and topic) as provided by RSS sources. The corpora constructed by ARARSS can be used as datasets for text classification and clustering, language models based on topics or language varieties, and comparable corpora for language usage among different languages or language varieties. Furthermore, ARARSS can be used to enlarge the size of the existing corpora.

2 Related Work

Given the importance of large and updated corpora for research enterprises, especially for poor-resourced languages, there is a need to automate the efforts of compiling corpora and ensuring their continuous growth and updates to maximize their benefits and broaden their usage. The obvious way to achieve this goal is through automatic text gathering from the World Wide Web. Generally, there are two methods that can be used to automatically construct large corpora from the World Wide Web: a method based on web crawling and another based on reading RSS feeds.

The first method crawls the Internet by using the Web addresses [3–5] or seed words [6, 7]. This method was used to construct several corpora, such as the TenTen corpus family in Sketch Engine [8] and SogouT-16, a Chinese web corpus [9].

Although this method has the advantage of being able to collect vast amounts of texts in a short time, the collected corpus (1) usually contains duplicate texts [4, 10] and (2) may have unusable texts containing graphs or advertising adds, very short texts (yields rate) [3, 10], or orthographic errors [11].

The second method relies on selecting specific Web content, namely reading RSS feeds [12–14]. For example, this method was used by Alzahrani [15] to construct an MSA corpus from newspapers and by Khoja [16] to construct a colloquial Arabic corpus from Arabic blogs.

The main advantage of this method is that it produces quality corpora as the probability of collecting duplicate texts, unusable texts, and very short texts is very low. Furthermore, as we know the sources of texts, we can control the balance and representativeness of the constructed corpora.

In this paper, we present ARARSS, a system that reads RSS feeds to construct corpora and overcome some of the limitations of the above-mentioned systems. ARARSS is distinguished from the conventional systems as follows.

1. ARARSS is not restricted to certain RSS feeds, such as newspapers or blogs, but was designed to read any user-specific RSS feeds.
2. The user has the ability to classify the location, medium, domain, and topic of the RSS feeds and save this information alongside the text date.
3. No text cleaning is required after extracting the text from the web page as unnecessary content, such as graphs or advertisings, are discarded automatically during text collection.
4. The user can ensure and control the representativeness and the balance of the corpus as all texts can be classified according to their metadata (location, time, medium, and topic).

3 System Description

1. ARARSS is free open-source software that can virtually run on any machine installed with JVM because it is a Java-based system.
2. It features an easy-to-use graphical user interface (GUI).
3. It uses a free database management system, namely SQLiteManager[1], to store the metadata of the RSS feeds.

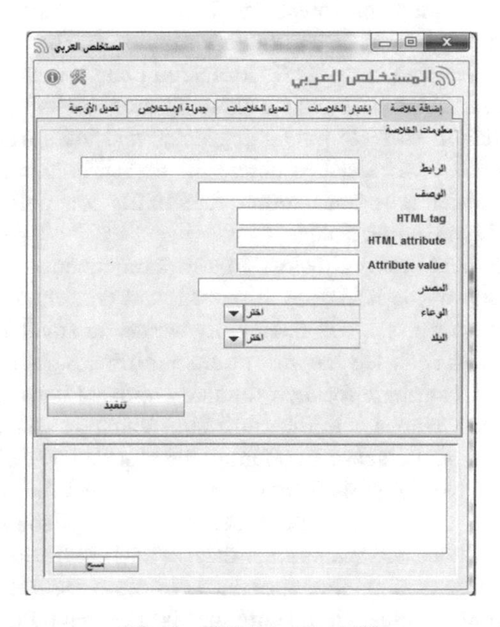

Fig. 1. GUI for ARARSS

[1] http://www.sqlabs.com/sqlitemanager.php.

4. It saves the texts in UTF-8 encoding. Each text in the corpus has a unique name comprising a sequence number, medium, source, and time.
5. It saves the corpus file in an easy-to-manage-folder hierarchy based on text medium, location, source, and topic.
6. It saves the URLs of all the corpus texts in a database so that they are not added to the corpus again.
7. The user can schedule the process of text gathering to work regularly without his/her interference. Therefore, if the user decides to set up the time of text gathering to every 12 h, the system will visit all RSS sources every 12 h and collect all the texts that not have been added to the corpus before.

Figure 1 illustrates the GUI for ARARSS.

4 System Architecture

ARARSS comprises five interrelated components: *RSS feed metadata management, RSS feed parser, web page parser and text extractor, text-collecting scheduler*, and *text storage.*

1. RSS feed metadata management: By using this component, the user can store all the necessary metadata for corpus resources, namely RSS feed URL, RSS feed description, location, medium, topic, and HTML tags. Moreover, this component enables the user to update or delete this information easily.
2. RSS feed Parser: RSS feed is in XML-formatted plain text that can be readable by human and machines. The RSS-feed-parsing process works by opening the link of the RSS feed. Then, it extracts publishing dates and links from all "item" nodes, after which each link is accessed to obtain the texts. For this component, we used the ROME tool[2], which is an open-source RSS utility containing a set of parsers for different types of syndication feeds.
3. Web Page Parser and text extractor: In this component, every web page link obtained from a file of the RSS feed is parsed, and then the main text is extracted from that page. Each web page is commonly written in standard HTML consisting of multiple tags with each tag having multiple attributes. Most websites normally use the same tag and attribute for their regularly updated content; thus, we searched for that specific tag with a specific attribute, which is stored along with other website metadata in our database, containing the required text. Note that each link is stored and used to avoid text duplication. Table 1 shows the algorithm we used to parse and fetch the texts. In this component we utilized Jsoup[3], which is an open source library to extract and manipulate data in an HTML document.
4. Text Storage: The goal of this component is to store the extracted text files in a hierarchical structure of folders indicating the text metadata. Furthermore, the stored

[2] http://rometools.github.io/rome/.

[3] https://jsoup.org.

text name shows the corresponding text's date and time obtained from the publish date node of the RSS feeds, as well as the topic and source description.

5. Texts collecting scheduler: This component is responsible for the automatic start-up of the text-collection process based on a user specification, where the user can schedule the process to work regularly every *n* hours after the first start-up of the system.

Table 1. Tool algorithm.

```
BEGIN
Fetch RSS link
Open link
Parse RSS feed file
Fetch N items
For N times
        Fetch link node
        IF link not already exists THEN
                Save it as web id
                Open link
                Parse the web page
                IF tag = stored tag AND attribute = stored attribute THEN
                        Extract text
                End if
        End if

End for
STOP
```

5 A Case Study

Although several Arabic corpora are available for exploring via the web or for direct download, to the best of our knowledge, no monitor corpus exists for Arabic. A monitor corpus can be simply defined as a continually growing collection of texts. While the growth of corpus size can be controlled (e.g., 1 million words per day), it is possible to not limit the growth rate of the corpus, in which case, the corpus will become an opportunistic corpus. ARARSS can be used to build both kinds of corpora. The availability of a monitor or opportunistic corpus would facilitate different kinds of studies in corpus linguistics, such as how language changes over time; computational linguistics, such as how the language models evolve over time; and text mining, such as developing classification models for each period. Furthermore, ARARSS can be used to build corpora for discourse analysis of certain topics, such as how different newspapers and agencies deal with presidential elections or political conflicts. Otherwise, we can simply use ARARSS to build a snapshot corpus, i.e., a corpus that covers a short period of time.

We utilized ARARSS to build a snapshot corpus of Modern Slandered Arabic (MSA) based on Arabic newspapers and RSS feeds of press Agencies. We undertook the following steps to build this corpus.

1. For each Arab country, we searched and identified newspapers providing RSS feeds. Our search revealed that not all Arabic newspapers utilize RSS feeds, and those that do vary across the Arab countries. RSS feeds are commonly used in Saudi and Egyptian newspapers but not in Jordanian newspapers. Moreover, for Algerian, Egyptian, Moroccan, Saudi, Sudanese, Tunisian, and Yemeni newspapers, we found Web sites that collect and classify nearly all newspapers articles for these countries and provide them through RSS feeds. For other countries, we selected the newspapers that provide RSS feeds and added them to the ARARSS database. However, we could not add any newspapers from Libya and Syria owing to the conflict situation in these countries. Additionally, during the phase of source identification, we could not find any Jordanian newspapers that provide RSS feeds and we could not reach any Qatari newspapers. Overall, we were able to identify the sources for newspapers of 13 Arab countries and 4 Arab press agencies. Table 2 illustrates the countries, major sources, and their website links.
2. Based on the search results and identification steps, we added the RSS feed links and their metadata to the ARARSS database, and then tested these links. Thus, the final composition of the database was 97 RSS feeds links.
3. We ran the ARARSS from 4 February 2018 until 8 May 2018. We scheduled it to collect new feeds every 12 h. However, we could not retrieve any text from the Bahraini newspaper, Al-watan, because the structure of its Web page was recently changed.

We were able to construct a corpus comprising 117,819 texts and 28,334,332 words, with the newspaper genre comprising 108,502 texts and 25,270,323 words (89%), and the press-agency genre comprising 9,317 texts and 3,064,009 words (11%). Table 3 illustrates the corpus content distribution for each country. The dataset shows eight classes for the collected texts: culture, economy, health, international, policy, religion, society, and sport. The content distribution across topics is illustrated in Table 4. As not all news articles were classified, we added the unclassified articles to the "unclassified" part of the dataset. Figure 2 shows, in logarithmic scale, the relation between words ranks and their frequency of our corpus. As Fig. 2 suggests, our corpus follows Zipf's distribution.

The availability of this snapshot corpus can promote different research studies. In the following direct examples of such studies:

1. Text mining: As 89% of the corpus is classified according to text topics, this corpus can be used in the classification, clustering, and topic modeling of Arabic text.
2. Corpus linguistics: As the corpus content is based on the publication origin, the corpus can be used to study language verities among Arab counties, for example, how the language of Saudi newspapers differs from that of Moroccan newspapers. As this corpus covered the presidential election in Egypt, it can be used for discourse analysis such as how the newspapers in Arab countries covered this case.
3. Computational linguistics: As the corpus was collected recently and covered various topics from different Arab countries, it can be used to build a language model of

Table 2. Countries, major sources, and their website link.

	Country	Major source	Link
1	Algeria	Djazairess	www.djazairess.com
2	Bahrain	Al-watan newspaper	www.alwatannews.net
3	Egypt	Maseress	www.masress.com
4	Iraq	Al-sabah newspaper	www.alsabaah.iq
5	Kuwait	Al-anba newspaper, Kuwait news agency	www.alanba.com.kw www.kuna.net.kw
6	Lebanon	National news agency	nna-leb.gov.lb
7	Morocco	Maghress	www.maghress.com
8	Oman	Al-watan newspaper	www.alwatan.com
9	Palestine	Al-ayyam newspaper	www.al-ayyam.ps
10	Saudi Arabia	Saudi press agency, SAURSS	www.spa.gov.sa www.sauress.com
11	Sudan	Sudaress	www.sudaress.com
12	Tunisia	Turess	www.turess.com
13	United Arab Emirates	Emarat Al-youm, Emirates news agency	www.emaratalyoum.com wam.ae
14	Yemen	Yemeress	www.yemeress.com

Table 3. Corpus content distribution across countries.

	Country	Text	Total number of words	Unique words
1	Algeria	13,677	3,683,399	269,466
2	Egypt	32,272	6,116,404	402,744
3	Iraq	1,007	403,789	72,351
4	Kuwait	3,435	965,423	104,094
5	Lebanon	1,479	603,736	92,668
6	Morocco	15,615	4,010,753	329,188
7	Oman	460	197,037	36,625
8	Palestine	353	214,030	54,470
9	Saudi Arabia	23,932	5,617,473	307,714
10	Sudan	5,738	1,670,035	217,859
11	Tunisia	5,978	1,041,086	122,079
12	United Arab Emirates	3,201	1,425,544	99,459
13	Yemen	10,672	2,385,623	210,728
	Total	117,819	28,334,332	1,052,693

MSA. This model can be compared with previously developed Arabic language models to reveal the similarities and differences. Furthermore, such a model can be used for Arabic Optical Character Recognition (OCR), information retrieval, or machine translation.

Table 4. Corpus content distribution across topics.

	Country	Text	Total number of words	Unique words
1	Culture	12,538	3,259,103	335,246
2	Economy	14,852	3,900,654	233,838
3	Health	5,914	1,326,937	128,588
4	International	16,089	3,835,354	260,631
5	Politics	19,925	4,854,134	331,484
6	Religion	3,033	1,001,034	157,416
7	Society	18,276	3,998,283	267,779
8	Sport	18,934	3,682,804	230,744
9	Unclassified	8,258	2,476,029	215,996
	Total	117,819	28,334,332	1,052,693

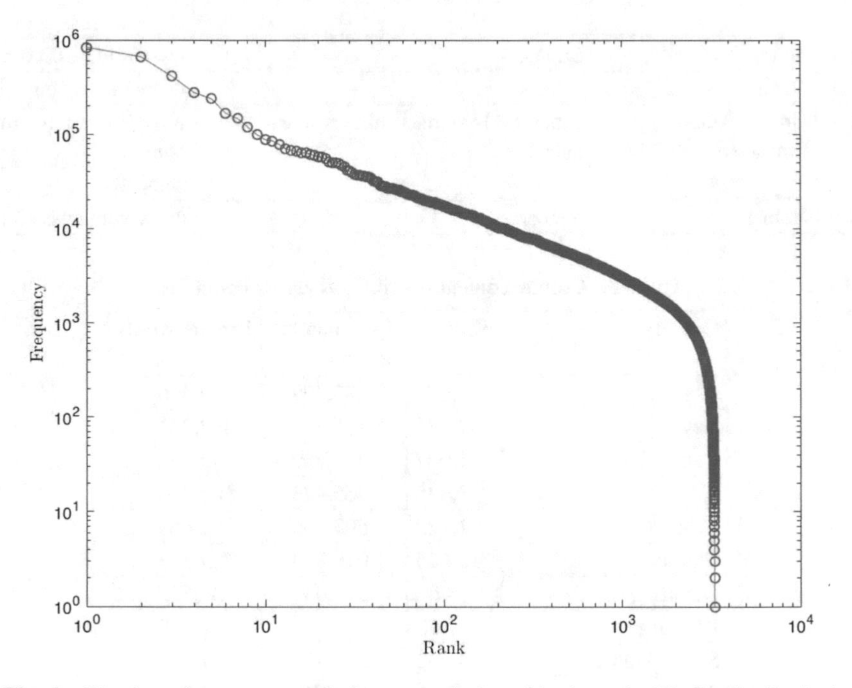

Fig. 2. Words ranks versus words frequency in logarithmic scale (Zipf's distribution)

6 Conclusion

In this paper, we present ARARSS, a tool capable of constructing and updating textual resources for the Arabic language from RSS feeds. ARARSS is open source and can be easily customized and used as a part of any other system. It can be utilized to build new corpora, as illustrated in Sect. 5, or autonomously update existing corpora. The main advantage of this tool is its ability to fetch clean texts from multiple resources to generate a very structured text corpus. The known limitation of ARARSS is that any update on RSS file link or HTML file style - tags and attributes - cannot be detected automatically which may lead to the failure of text extraction. The user needs to track each website and make sure that the tool is able to extract the texts for the designated RSS links. We utilized ARARSS to compile an MSA corpus comprising 117,819 texts and 28,334,332 words from 13 Arab countries. Both ARARSS and the corpus are freely available to download (http://corpus.kacst.edu.sa/more_info.jsp).

References

1. Manning, C.D.: Part-of-speech tagging from 97% to 100%: is it time for some linguistics? In: Computational Linguistics and Intelligent Text Processing, pp. 171–189. Springer, Heidelberg (2011)
2. Sinclair, J.: Corpus, Concordance, Collocation. Oxford University Press, Oxford (1991)
3. Suchomel, V., Pomikálek, J.: Efficient web crawling for large text corpora. In: Proceedings of the Seventh Web as Corpus Workshop (WAC7), pp. 39–43 (2012)
4. Schäfer, R., Bildhauer, F.: Building large corpora from the web using a new efficient tool chain. In: LREC, pp. 486–493 (2012)
5. Barbaresi, A.: Finding viable seed URLs for web corpora: a scouting approach and comparative study of available sources. In: Proceedings of the 9th Web as Corpus Workshop, WaC-9, Gothenburg, Sweden, pp. 1–8 (2014)
6. Baroni, M., Bernardini, S.: BootCaT: bootstrapping corpora and terms from the web. In: Proceedings of LREC, p. 1313. ELDA, Lisbon (2004)
7. Ueyama, M.: Evaluation of Japanese web-based reference corpora: effects of seed selection and time interval, Wacky, pp. 99–126 (2006)
8. Jakubíček, M., Kilgarriff, A., Kovář, V., Rychlý, P., Suchomel, V.: The TenTen corpus family. In: 7th International Corpus Linguistics Conference CL, pp. 125–127. UCREL, Lancaster (2013)
9. Luo, C., Zheng, Y., Liu, Y., Wang, X., Xu, J., Zhang, M., Ma, S.: SogouT-16: a new web corpus to embrace IR research. In: Proceedings of the 40th International ACM SIGIR Conference on Research and Development in Information Retrieval, pp. 1233–1236. ACM (2017). https://doi.org/10.1145/3077136.3080694
10. Schäfer, R.: Accurate and efficient general-purpose boilerplate detection for crawled web corpora. Lang. Resour. Eval. **51**(3), 873–889 (2017). https://doi.org/10.1007/s10579-016-9359-2
11. Ringlstetter, C., Schulz, K.U., Mihov, S.: Orthographic errors in web pages: toward cleaner web corpora. Comput. Linguist. **32**(3), 295–340 (2006)
12. Ojokoh, B.A.: Automated online news content extraction. Int. J. Comput. Sci. Res. Appl. **2**, 2–12 (2012)
13. George, A., Bouras, C., Poulopoulos, V.: Efficient extraction of news articles based on RSS crawling. In: International Conference on Machine and Web Intelligence, ICMWI, pp. 1–7. IEEE, Algiers (2010)
14. Qingcheng, L., Youmeng, L.: Extracting content from web pages based on RSS. In: 2008 International Conference on Computer Science and Software Engineering, vol. 5, pp. 218–221. IEEE2008))
15. Alzahrani, S.M.: Building, profiling, analysing and publishing an Arabic news corpus based on Google news RSS feeds. In: Information Retrieval Technology, pp. 488–499. Springer, Heidelberg (2013)
16. Khoja, S.: An RSS feed analysis application and corpus builder. In: The Second International Conference on Arabic Language Resources and Tools, pp. 01–04. The MEDAR Consortium, Cairo (2009)

Swarm Optimizations and Applications

Face Recognition Based on Grey Wolf Optimization for Feature Selection

Abd AL-BastRashed Saabia[1(✉)], TarekAbd El-Hafeez[2], and Alaa M. Zaki[2]

[1] College of Basic Education, Computer Science Department, University of Diyala, Diyala, Iraq
abdalbast.rsheed@gmail.com

[2] Computer Science Department, Faculty of Science, Minia University, EL-Minia, Egypt
{tarek,alaa.zaki}@mu.edu.eg

Abstract. Face recognition systems are progressively becoming popular as means of determining the people's identities. Moreover, face images are the only biometric trait that can be found in legacy databases and international terrorist watch-lists and can be taken without the need to the cooperation of subjects. In this paper, we present an effective face recognition system that consists of a set of steps: image preprocessing in which the person's face is detected and the median filter is applied for noise removal, feature extraction using Gabor filters, feature reduction using principle component analysis, feature selection using the grey wolf optimization (GWO) algorithm, and classification using k-NN classifier. The proposed system has been evaluated using Yale face database. The experimental results have revealed that the proposed system can achieve recognition accuracy up to 97%. Also, the performance of the proposed system is compared to the performance of other face recognition system and the obtained results have revealed that the proposed system is better in terms of both recognition accuracy and run time.

Keywords: Biometrics · Face recognition · Gabor filters · PCA
Grey wolf optimization · Feature selection

1 Introduction

In the last decades, automated personal identification systems have received an increasing attention due to the extra emphasis placed on security in our modern society. Biometric systems are automated recognition systems that depend on a scientific discipline which seek determining people's identities using their physiological and/or behavioral features. In other words, biometrics is recognizing people through their unique biological or behavioral characteristics [1]. Generally, there are two types of biometrics [2]: physiological and behavioral. Physiological biometric refers to the traits that are relevant to the body's shape, such as fingerprint, face recognition, DNA, hand and palm geometry, iris recognition, and retina [22]. On the other side, Behavioral biometrics refers to the traits that are relevant to the person's behaviour, such as gait, voice, signature, and keystroke. This type of biometric is also known as behaviometrics.

A. E. Hassanien et al. (Eds.): AISI 2018, AISC 845, pp. 273–283, 2019.
https://doi.org/10.1007/978-3-319-99010-1_25

The face is the leading part of the person head that expands from the brow to the chin. It includes the mouth, nose, cheeks, and eyes. The face is the most common and normal biometric trait through which people recognizes each other [1]. Because of the appearance of cheap digital cameras and social networking sites such as Facebook, millions of peoples are sharing their personal photos.

Face recognition is a non-intrusive recognition method in which the persons' identities are determined through their face characteristics. Face recognition is being employed in a various set of real world applications and ranges from controlled static "mug-shot" verification to uncontrolled dynamic identification based on group photos with highly cluttered background, such as in airports and subway stations [3]. Recently, many types of distinguishing features can be extracted from face images for recognition purposes. However, the extracted features may contain irrelevant/redundant features that can badly affect both of the recognition accuracy and the time needed for the recognition process. Therefore, the feature reduction and feature selection are paramount steps to overcome this problem.

In this paper, we introduce a face recognition system that consists of a set of steps: image preprocessing in which the person's face is detected and the median filter is applied for noise removal, feature extraction using Gabor filters, feature reduction using principle component analysis, feature selection using the grey wolf optimization (GWO) algorithm, and classification using k-NN classifier. The remaining sections of the paper are introduced as follows: Sect. 2 covers some of the efforts and relevant works which have been proposed in the face recognition field. Section 3 describes in detail the proposed face recognition system. Section 4 contains descriptions for the used dataset, the implementation environment, and the conducted experiments. The paper is concluded and the future work is recommended in Sect. 5.

2 Related Work

Generally, there are lot efforts that have been done in the field of Automatic Face Recognition (AFR). However, this field is still an active research area and more efforts are being done to improve the accuracy, efficiency, effectiveness of face recognition systems.

In [4], they proposed a face recognition system based on a single sample per person (SSPP), which can be considered a challenging problem, particularly, when the gallery image and the probe set are acquired under different conditions. They introduced an SSPP domain adaptation network (SSPP-DAN) in which the domain adaptation, extracting the features and the classification process are done by employing a deep architecture with domain-adversarial training. The conducted experiments have shown that their approach have effectively enhanced the recognition accuracy, giving a performance comparable to the state-of-the-art techniques on a benchmark dataset.

Also, another face recognition system based on a small number of training images is proposed in [5]. Their method consists of performing a novel alignment process followed by classification using a modified version of the Robust Sparse Coding (RSC) algorithm. They presented their recognition rates on a difficult dataset that represents real-world faces where they significantly outperformed state-of-the- art methods.

In [6], they were interested in determining the effectiveness of using adaptive directional wavelet transform for extracting the distinctive features of people's faces. They focused on extracting the discriminant directional multi-resolution facial features by employing the adaptive directional wavelet transform in addition to linear discriminant analysis (LDA). Their approach has been compared to other feature extraction methods, such as current subspace and local descriptor feature extraction methods, on number of well-known face databases such as ORL and Essex Grimace. The results have shown the effectiveness of their approach.

In [7], they proposed a face recognition system based on an improved local binary patterns (LBP) approach. In the proposed approach, the enhanced LBP considers more pixels and different neighborhoods during computing the features. The proposed approach has been validated and evaluated on UFI and FERET face datasets. The results have shown the effectiveness of their approach compared to other state-of-the-art approaches.

In [8], they proposed an approach to handle the challenges imposed by different illuminations, poses, and expressions found in unconstrained face recognition. The proposed approach seeks extracting "Multi-Directional Multi-Level Dual-Cross Patterns" (MDML-DCPs) from people's face images. Their approach employs the first derivative of Gaussian operator to decrease the effect of diverse illumination. After that, it calculates the DCP features at the holistic level and component level. Many face databases have been used for validating and evaluating the proposed approach, including FERET, CAS-PERL-R1, FRGC 2.0, and LFW. The results have shown that the proposed approach outperform the state-of-the-art local descriptors.

In [9], they suggested a novel supervision signal, known as center loss, for improving the distinguishing ability of features learned via a deep architecture to perform face recognition. Using the center loss, the center of deep features is learned for every class and the distance between different class centers is penalized. A number of face datasets have been used to evaluate the proposed approach including LFW, YTF, and MegaFace Challenge. The results have shown that the proposed approach can achieve better recognition accuracy compared to the state-of-the-are methods.

In [10], they proposed a Heterogeneous Face Recognition (HFR) approach which depends on a new graphical representation. They used Markov networks for separately representing heterogeneous image patches. Also, the similarity between the resulting graphical representations is measured using a coupled representation similarity metric (CRSM). A number of experiments have been conducted under many HFR scenarios. The results have revealed that the proposed approach outperforms state-of-the-art methods.

In [11], they proposed multiple pose-aware face recognition method using deep learning models. In their approach many pose specific deep CNNs are employed to process face images to produce multiple pose-specific features. The pose-specific CNN features highly decreased the impact of different poses in face recognition. IARPA's CS2 and NIST's IJB-A face databases have been employed for evaluation the performance of the approach, which achieved better accuracy recognition when compared to many of the state-of-the-art methods.

In [12], they have two contributions. First, they proposed a new adaptive cuckoo search (ACS) algorithm which can be used for optimization problems. Second, they suggested a face recognition approach based on ASC, PCA, and IDA. PCA + IDA are used to carry out the dimension reduction task, while ACS + IDA are employed to determine the best feature vectors for classifying the face images based on the IDA. YALE, ORL, and FERET face databases are used to analyze the performance of the proposed approach. The performance of the proposed approach has been compared to the state-of-the-art methods and achieved better results.

In [13], the proposed deep convolutional features based unconstrained face verification approach. Their DCNN has been trained using CASIAWebFace dataset and evaluated using IARPA Janus, IJB-A and LFW face datasets. Results of experimental evaluations on the IJB-A and the LFW datasets are provided.

In [14], they proposed an approach for adapting the variation in pose and expression in face images called High-fidelity Pose and Expression Normalization (HPEN). Using the proposed approach in addition to a3D Morphable Model (3DMM), a natural face image with normal frontal pose and natural expression can be generated regardless the nature of the original face images. Many evaluation experiments using Multi-PIE and LFW have demonstrated that the proposed method greatly improved face recognition performance and outperformed state-of-the-art methods in both constrained and unconstrained environments.

3 Proposed System

Face recognition is one of the most used biometric systems due to the many advantages it offers. In this section, we introduce a face recognition system which consists of a number of steps including: image preprocessing which applies RGB to gray scale conversion, face detection using viola-jones approach and noise removal using median filter, feature extraction using Gabor filters, feature reduction using PCA, feature selection using GWO, and k-NN based classification. The block diagram of the proposed system is shown in Fig. 1.

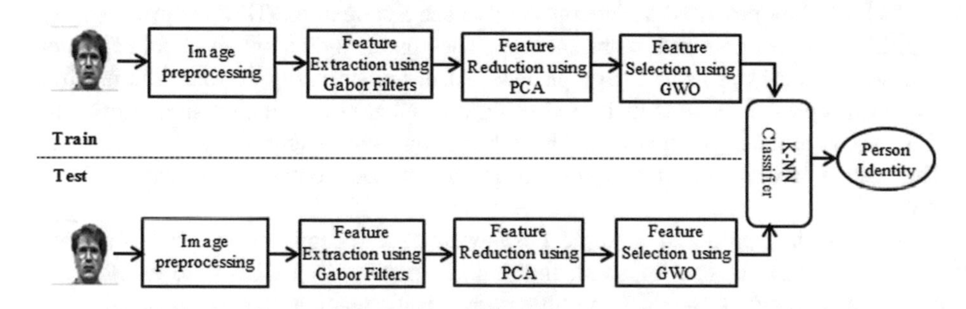

Fig. 1. The block diagram of the proposed face recognition system.

A detailed description for each step of the proposed system is presented in the following subsections.

3.1 Image Preprocessing

In this phase, a set of steps is performed to make face images ready for feature extraction with high accuracy. The image preprocessing phase starts with converting the RGB image into gray scale images. Then, the face is detected using viola-jones face detection method. Finally, the median filter is applied to remove the noise from the cropped image. A sample of images after applying the preprocessing step is shown in Fig. 2.

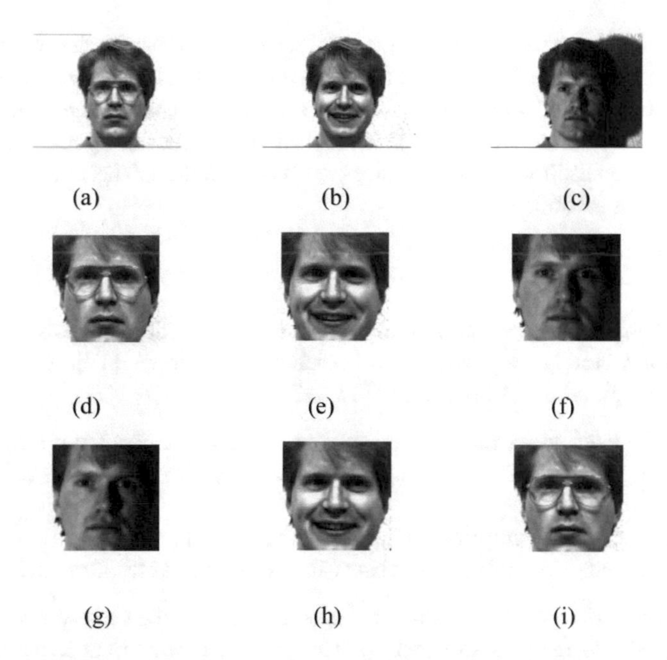

Fig. 2. A sample of image face images before and after the processing phase (Original images: a, b, and c, Corresponding images after applying viola-jones: d, e, and f, Corresponding images after applying the median filter: g, h, and i).

Converting the RGB face images into gray scale images is an important step in the proposed system, where the feature extraction step is based on Gabor filters, which work only on gray scale images.

3.2 Feature Extraction

Gabor filters is a well-known method to represent human faces. They have proved their effectiveness in many research fields such as face recognition [15]. The effectiveness Gabor filters comes from their robustness against rotation, scale, and translation variations. Also, they can deal with photometric disturbances, such as changes in illumination and noise. The Gabor filter-based features can be extracted directly from the grayscale images [16]. In the proposed method, a two-dimensional Gabor filter namely,

Gaussian kernel function modulated by a sinusoidal plane wave is used for feature extraction in the spatial domain and defined as follows:

$$G(x, y) = \frac{f^2}{\pi\gamma\eta} \exp\left(-\frac{x'^2 + \gamma^2 y'^2}{2\sigma^2}\right) \exp(j2\pi f x' + \phi) \tag{1}$$

$$x' = x cos\theta + y sin\theta \tag{2}$$

$$y' = -x sin\theta + y cos\theta \tag{3}$$

Where f, θ, ϕ, σ, and γ are the sinusoid frequency, the orientation of the normal to the parallel stripes of a Gabor function, the phase offset, the Gaussian envelope's standard deviation, and the spatial aspect ratio which specify the ellipticity of the support of the Gabor function, respectively. In our proposed system, we utilize 40 Gabor filters as a result of using 5 scales and 8 orientations. The number of extracted Gabor features is 11560.

3.3 Feature Reduction

PCA is probably the most commonly used and the oldest multivariate statistical technique which nearly employed by all scientific disciplines [17]. The objectives of PCA are summarized and presented below [17, 18]:

- PCA can be used to extract the most important information from a data-table.
- PCA can be used to reduce the data dimensionality by maintaining only significant information.
- PCA can be used to simplify the description of a data set.
- PCA can be used to analyze the observations and variables structure.

In the proposed work, PCA is mainly used to perform the feature reduction task to exclude the useless features in order to reduce the training time, avoiding the over-fitting problem, and reducing the time required to determine the age group of a certain person using his face image. After applying PCA, the number of Gabor Features is reduced from 11560 to 1560.

3.4 Feature Selection Using Grey Wolf Optimization (GWO)

Grey Wolf Optimizer (GWO) is a recent swarm intelligence algorithm inspired by the grey wolf community. It is developed by SeyedaliMirjalili et al. in 2014. Grey wolf is a very dangerous creature which belongs to Canidae family. It is categorized as apex predators, which means that it is one of the creatures that set at the top of food chain. Grey wolves usually live in groups of sizes range from 5 to 12 on average. Each group has social dominance hierarchy: alpha, beta, and omega, in order. The alphas are a male and female which represent the leaders of the group. They are responsible for decision making regarding the hunting process, where to sleep, when to wake up, etc. The betas are the second level of the pack management. They are considered the assistants of the alphas that help them in making the decisions relevant to the activities of the pack.

The omegas are the final level in the hierarchy which is dominated by the higher levels [19]. Group hunting is another interesting social behavior of grey wolves. The main phases of grey wolf hunting are as follows [19]:

- Tracking, chasing, and approaching the prey.
- Pursuing, encircling, and harassing the prey until it stops moving.
- Attack towards the prey.

In order to mathematically model the social hierarchy of wolves when designing GWO, we consider the fittest solution as the alpha (α). Consequently, the second and third best solutions are named Beta (β) and Delta (δ) respectively. The rest of the candidate solutions are assumed to be Omega (ω) [19]. The mathematical model of the encircling behavior is represented by the equations:

$$\vec{D} = \left|\vec{C}.\vec{X}p\ (t) - \vec{X}(t)\right| \tag{4}$$

$$\vec{X}(t+1) = \left|\vec{X}p\ (t) - \overrightarrow{A.}\vec{D}\right| \tag{5}$$

where t indicates the current iteration, $\vec{C} = 2.\vec{r}2$, $\vec{A} = 2\vec{a}.\vec{r}1 - \vec{a}$, $\vec{X}m$ is the position vector of the wolf, r1 and r2 are random vectors [0, 1] and a linearly varies from 2 to 1, C and A are coefficient vectors, Xp is the position vector of the prey. It should be noted here that, during optimization, the w wolves update their positions around α, β, and δ. Therefore, w wolves are able to reposition with respect to α, β, and δ. The mathematical model of ω wolves readjusting the positions is as follows [19].

$$\vec{D}\alpha = \left|\vec{C}1.\vec{X}\alpha - \vec{X}\right|, \vec{D}\beta = \left|\vec{C}2.\vec{X}\beta - \vec{X}\right|, \vec{D}\delta = \left|\vec{C}3.\vec{X}\delta - \vec{X}\right| \tag{6}$$

$$\vec{X}1 = \vec{X}\alpha - \vec{A}1.\left(\vec{D}\alpha\right), \vec{X}2 = \vec{X}\beta - \vec{A}2.\left(\vec{D}\beta\right), \vec{X}3 = \vec{X}\delta - \vec{A}3.\left(\vec{D}\delta\right) \tag{7}$$

$$\vec{X}(t+1) = \frac{\vec{X}1 + \vec{X}2 + \vec{X}3}{3} \tag{8}$$

With these equations, a search agent updates its position according to alpha, beta, and delta in an n-dimensional search space. In addition, the final position would be in a random place within a circle which is defined by the positions of alpha, beta, and delta in the search space. In other words alpha, beta, and delta estimate the position of the prey, and other wolves updates their positions randomly around the prey [19].

In the proposed system, GWO has been used to select the best minimal set of features that maximizes the classification accuracy. GWO has been adopted in the same way as in [20], where the fitness functions is computed using Eq. 9.

$$\text{Fitness} = \mu P + (1 - \mu)\frac{N - L}{N} \tag{9}$$

where P is the classification accuracy, L is the number of selected features, N is the total number of features in the dataset, μ is the classification accuracy weight, $(1 - \mu)$ is the feature selection quality weight, and $\mu \in [0, 1]$. The number of features after applying GWO becomes 110.

3.5 Classification

In the proposed system, a k-nearest neighbors algorithm (k-NN) classifier has been employed for performing the classification tasks with k = 1. The k-NN algorithm is a non-parametric method used for classification and regression [21]. In both cases, the input consists of the k closest training examples in the feature space. The output depends on whether k-NN is used for classification or regression. In classification, the output is a class membership. An object is classified by a majority vote of its neighbors, with the object being assigned to the class most common among its k nearest neighbors (k is a positive integer, typically small). If k = 1, then the object is simply assigned to the class of that single nearest neighbor [23]. The k-NN is a type of instance-based learning, or lazy learning, where the function is only approximated locally and all computation is deferred until classification. The k-NN algorithm is among the simplest of all machine learning algorithms [23].

4 Implementation and Experimental Results

In this section, a set of experiments have been performed for measuring the identification accuracy of the proposed system. All the experiments have been done using the same computer with Processor Intel(R) Zeon(R) CPU E5430@ 2.66 GHz (2 Processor), 32 GB installed memory (RAM), and 64 bit Windows 8.1 Enterprise operating system. The proposed method is implemented using Matlab 2016. Yale Face Database has been used for evaluating the performance of the proposed system. It is a challenging publicly available dataset that contains 11 gray scale face images for each subject among 15 subjects. Face images of each subject have been acquired under different configurations: center-light, with and without glasses, and different facial expressions. A sample of the used dataset is shown Fig. 3. Five images for each subject have been used for the training process and the rest for the testing process.

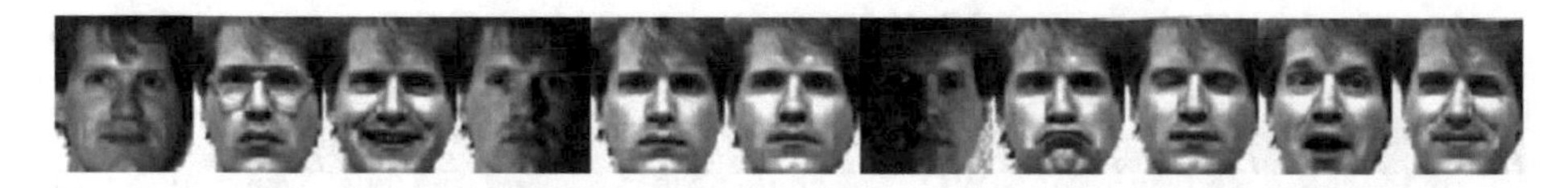

Fig. 3. A sample of the used dataset

GWO has a set of parameters that greatly affect the obtained results. The set of parameters includes the number of agents, upper bound (ub), lower bound (lb), and the maximum number of iterations (Imax). In the proposed system, the number of agents is 10, while a number of values have used for each parameters. The values of upper bound (ub) and lower bound (lb) ∈ ([−5, 5], [−10, 10], [−15, 15], [−20, 20]). The values of I_{max} ∈ (25, 50, 75, 100, 150, 200, 300, 400, 500). First, we have conducted an experiment to evaluate the effect of ub and lb on the proposed system with default values I_{max} = 25, and number of agents = 10. The accuracy of the proposed system, in addition to the run time in seconds, using different ub and lb values are shown in Table 1.

Table 1. The recognition accuracy and the run time of the proposed system using different lb and ub values.

[lb, ub]	Accuracy	Time (Sec.)
[−5, 5]	95%	5.35
[−10, 10]	96%	5.38
[−15, 15]	97%	6.49
[−20, 20]	95%	5.26

From Table 1, we can see that [−15, 15] gives the best recognition accuracy. Another experiment has been conducted to evaluate the effect of I_{max} on the performance of the proposed system. The accuracy of the proposed system, in addition to the run time in seconds, using different I_{max} values are shown in Table 2.

Table 2. The recognition accuracy and run time of the proposed system using different lb and ub values.

I_{max}	Accuracy	Time (Sec.)
25	97%	6.49
50	90%	10.49
75	87%	15.52
100	81%	21.94
200	73%	25.97
300	67%	40
400	51%	73
500	50%	91.72

From Table 2, we can see that increasing the number iterations leads to decreasing the recognition accuracy and increasing the run time. Also, the proposed system has been compared to the system proposed in [12] on the Yale face database in terms of recognition accuracy and run time as shown in Table 3.

Table 3. A comparison between the proposed system and the system proposed in [12].

System	Accuracy	Time (Sec.)
ACS-IDA [12]	88.9%	9.10
Proposed system	97%	6.49

From Table 3, we can that the proposed system is better that the system proposed in [12] in terms of both recognition accuracy and run time.

5 Conclusion

Face recognition systems are commonly used biometric system because of the many advantages it offers. In the paper, we have proposed an effective face recognition system that consists of a number of steps, including a number of image preprocessing, feature selection, feature reduction, feature selection, and classification. The proposed system has been evaluated through a number of conducted experiments on the challenging, publicly available Yale face database. The obtained results have shown that the proposed system has recognition accuracy of 97%. In the future, we intend to extend our work by adopting multiple biometric traits toward achieving higher recognition accuracy. Also, intent to deal with more challenging face images contained in lager face databases.

References

1. Dunstone, T., Yager, N.: Biometric System and Data Analysis: Design, Evaluation, and Data Mining. Springer, Heidelberg (2008)
2. Biometrics. http://searchsecurity.techtarget.com/definition/biometrics. Accessed 19 Feb 2018
3. Jain, A.K., Ross, A., Prabhakar, S.: An introduction to biometric recognition. IEEE Trans. Circuits Syst. Video Technol. **14**(1), 4–20 (2004)
4. Hong, S., Im, W., Ryu, J., Yang, H.S.: SSPP-DAN: deep domain adaptation network for face recognition with single sample per person. arXiv preprint arXiv:1702.04069 (2017)
5. Fontaine, X., Achanta, R., Süsstrunk, S.: Face recognition in real-world images. In: IEEE International Conference in Acoustics, Speech and Signal Processing (ICASSP), pp. 1482–1486 (2017)
6. Muqeet, M.A., Holambe, R.S.: Local appearance-based face recognition using adaptive directional wavelet transform. J. King Saud Univ.-Comput. Inf. Sci. (2017)
7. Král, P., Vrba, A.: Enhanced local binary patterns for automatic face recognition. arXiv preprint arXiv:1702.03349 (2017)
8. Ding, C., Choi, J., Tao, D., Davis, L.S.: Multi-directional multi-level dual-cross patterns for robust face recognition. IEEE Trans. Pattern Anal. Mach. Intell. **38**(3), 518–531 (2016)
9. Wen, Y., Zhang, K., Li, Z., Qiao, Y.: A discriminative feature learning approach for deep face recognition. In: European Conference on Computer Vision, pp. 499–515, Springer, Cham, (2016)
10. Peng, C., Gao, X., Wang, N., Li, J.: Graphical representation for heterogeneous face recognition. IEEE Trans. Pattern Anal. Mach. Intell. **39**(2), 301–312 (2017)
11. AbdAlmageed, W., et al.: Face recognition using deep multi-pose representations. In: IEEE Winter Conference in Applications of Computer Vision (WACV), pp. 1–9. IEEE (2016)
12. Naik, M.K., Panda, R.: A novel adaptive cuckoo search algorithm for intrinsic discriminant analysis based face recognition. Appl. Soft Comput. **38**, 661–675 (2016)
13. Chen, J.C., Patel, V.M., Chellappa, R.: Unconstrained face verification using deep CNN features. In: IEEE Winter Conference in Applications of Computer Vision (WACV), pp. 1–9. IEEE (2016)

14. Zhu, X., Lei, Z., Yan, J., Yi, D., Li, S.Z.: High-fidelity pose and expression normalization for face recognition in the wild. In: Proceedings of the IEEE Conference on Computer Vision and Pattern Recognition, pp. 787–796. IEEE (2015)
15. Gao, F., Ai, H.: Face age classification on consumer images with Gabor feature and fuzzy LDA method. In: International Conference on Biometrics, pp. 132–141. Springer, Heidelberg (2009)
16. Haghighat, M., Zonouz, S., Abdel-Mottaleb, M.: CloudID: trustworthy cloud-based and cross-enterprise biometric identification. Expert Syst. Appl. **42**(21), 7905–7916 (2015)
17. Abdi, H., Williams, L.J.: Principal component analysis. Wiley Interdisc. Rev. Comput. Stat. **2**(4), 433–459 (2010)
18. Kishor, N., Bhim, B., Rushikesh, B., Satish, K.: Facial expression identification by using principle component analysis. Int. J. Adv. Eng. Glob. Technol. **3**(5), 579–585 (2015)
19. Mirjalili, S., Mirjalili, S.M., Lewis, A.: Grey wolf optimizer. Adv. Eng. Softw. **69**, 46–61 (2014)
20. Li, Q., Chen, H., Huang, H., Zhao, X., Cai, Z., Tong, C., Liu, W., Tian, X.: An enhanced grey wolf optimization based feature selection wrapped kernel extreme learning machine for medical diagnosis. Comput. Math. Methods Med. (2017)
21. Altman, N.S.: An introduction to kernel and nearest-neighbor nonparametric regression. Am. Stat. **46**(3), 175–185 (1992)
22. Hassanien, A.E.: Hiding iris data for authentication of digital images using wavelet theory. Pattern Recognit. Image Anal. **16**(4), 637–643 (2006)
23. K-nearest_neighbors_algorithm. https://en.wikipedia.org/wiki/Knearest_neighbors_algorithm. Accessed 19 Feb 2018

Discrete Grey Wolf Optimization for Shredded Document Reconstruction

H. A. Badawy[1(✉)], E. Emary[2,3], Mohamed Yassien[1], and Mahmoud Fathi[1]

[1] Faculty of Engineering, Benha University, Benha, Egypt
hassan.badawy@outlook.com
[2] Faculty of Computers and Information, Cairo University, Giza, Egypt
[3] Faculty of Computer Studies, Arab Open University, Cairo, Egypt

Abstract. Shredded document reconstruction problem has gained more attention in the last few years. The reconstruction process is commonly involved in forensics, investigation sciences, and reconstruction of destroyed archaeological papyrus and manuscripts. Exhaustive search is common for handling such problem for small dimensions but with higher dimension, the problem becomes worse. Recently, a bio-inspired grey wolf optimizer (GWO) is a great algorithm for solving continuous problem but the nature of Shredded document reconstruction is a discrete problem, so a discrete version of GWO has been proposed and applied to it, This paper also proposed a general simple fitness function that can handle both text and image-based documents.

Keywords: Shredded paper · Exhaustive search
Bio-inspired optimizer · Discrete gray wolf optimizer
Document reconstruction

1 Introduction

The reconstruction of shredded Document problem gained the attention of the researchers during the last few years. The organizations like Defense Advanced Research Projects Agency (DARPA) held a competition in 2011 to reconstruct some shredded papers that have sensitive data, The challenge consisted of five shredded documents and the goal was to reconstruct them and extract their hidden messages [1].

There are two general methods to reconstruct shredded documents, the first one is semi-automated which requires human intervention with algorithm from time to time, the second is a fully-automated method which is fully controlled by algorithms and this one is our concern [2,3]. Schauer et al. [4] classify shredded documents into three categories: the manually torn documents, the cross-cut shredded documents, and the strip shredded documents.

The document reconstruction process has multiple stages that include preprocessing, feature extraction, matching, scoring, and merging. Many research literature discussed reconstruction of shredded Document problem differently,

A. E. Hassanien et al. (Eds.): AISI 2018, AISC 845, pp. 284–293, 2019.
https://doi.org/10.1007/978-3-319-99010-1_26

Lin and Fan-Chiang introduced an algorithm based on image feature matching [5]. Some literature used optical character recognition (OCR). Biesinger et al. [6] discussed the Cross-Cut reconstructing problem using the genetic algorithm (GA) without using any pattern recognition techniques.

In Sect. 2 we describe the problem and convert it to formulation form. In Sect. 3 we introduce the solution represented by discrete gray wolf optimizer (dGWO). In Sect. 4 we draw a comparison between discrete gray wolf optimizer (dGWO), genetic algorithm (GA) and random search algorithm (RSA) and discuss the results. In Sect. 5 we made a conclusion to the whole work from the point of view the problem and the results.

2 Problem Definition

The Reconstruction of strip-shredded documents (RSSD) problem came from cutting document along vertical side to identical rectangle pieces in height and width dimensions. We assume the orientation of each shred is known and is in the right orientation.

Assume we have shredded paper $S_r = \{s_1, s_2, \ldots, s_r\}$ where r is the number of strip-shreds cut by electrical shredder, the main challenge is to estimate the transformation function $f(S_r)$ that maps unordered shreds combination S_r to ordered one $\bar{S_r}$.

$$\begin{pmatrix} s_1 \\ s_2 \\ \vdots \\ s_r \end{pmatrix} \xrightarrow{f} \begin{pmatrix} \bar{s_1} \\ \bar{s_2} \\ \vdots \\ \bar{s_r} \end{pmatrix}$$

where s_r is the input shreds that shredded into r number of shreds, $\bar{s_r}$ is the output shreds that results from mapping unordered shreds combination s_r to ordered one using mapping function f.

The mapping function f is clearly a discrete mapping function of dimension r while as long as r increases the search space size increases dramatically -for $r = 15$ shreds the number of possible solution will be 1307674368000 possibility, this number is a results for the factorial r possible solution. For such huge search space an intelligent method should be proposed to adaptively search for the best mapping function. The mapping function has the role to map individual shreds from the input shreds into its appropriate position in the target space with the following two constraints:

- Individual shred must be be mapped to one and only one position in the target space.
- All shreds must have a mapping to the the target space.

$$x_r \in \boldsymbol{R} = \{1, 2, \cdots, r\} \tag{1}$$

3 Methodology

As mentioned in Sect. 2, to obtain the transform function f which is a discrete mapping function from input space to target space with huge possible solutions an exhaustive search for the different mappings seems to be impossible; Hence, an intelligent adaptive search for such optimal mapping is to be found, Fig. 1(b) shows a shredded document and a sample gradient fitness for a line. It can be noticed that at the edges of shreds we have a spike in gradient value, after reconstruction Fig. 1(a) most spikes at edges were disappeared. The fitness function considering the horizontal derivative of the proposed mapping can be formulated as per the below equation:

$$C(S_r(m,n)) = min \sum_{j=1}^{H} \sum_{i=1}^{W} |\frac{\mathrm{d}}{\mathrm{d}x} S_r(m,n)| \tag{2}$$

where H, W are the height and width of the constructed image $S_r(m,n)$ that consists of a number of shreds S_r mapped by a mapping function $f(s_r)$.

$$S_r(m,n) = \left[f(s_1)\ f(s_2)\ f(s_3)\ ...\ f(s_r)\right]$$

As can be noted from the formulation of the fitness function, the function is directly working of the dense content of the constructed image and hence it is not related to whether the content is a text, scenes, ...etc. The only condition for the

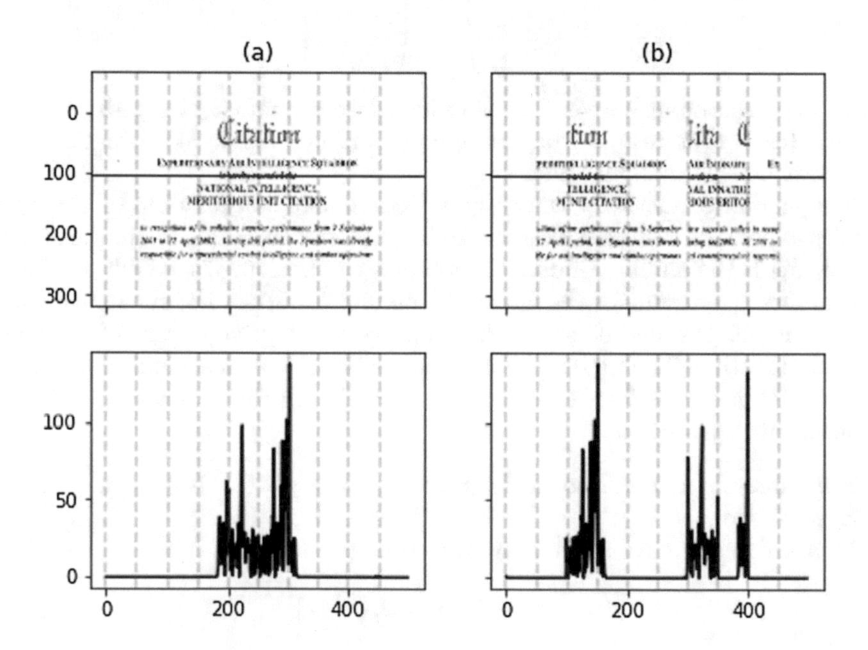

Fig. 1. (a), (b) Ordered and unordered shredded Document with gradient fitness respectively

success of such function is the existence of enough horizontally continuous objects to capture the horizontal continuity. The conditions for f are as mentioned in the past section to be valid for all s_i and $f(s_i) \neq f(s_j)\ \forall i, j$ and $i \neq j$ Although the above-mentioned function has the advantage of in dependency on the content of the image and seems to be still time costly.

Continuous Gray Wolf Optimizer(CGWO) is one of the most famous bio-inspired optimization techniques that proposed by [7] which mimic the leadership hierarchy and hunting process of gray wolves. In Gray Wolf's folks they have three leaders called alpha, beta, delta and any one of the folk called omega this hierarchy ordered from the senior leaders to juniors [8].

The mathematical model of GWO is based on the best solution α, the second and third fittest solution called β and δ, any candidate from the pack called ω and it can promote to substitute any one of best three solutions. the hunting method is based on three process searching, encircling and attacking prey [9]. the next equations represent encircling stage in hunting trip (3).

$$\boldsymbol{X}(t+1) = \boldsymbol{X}_p(t) + \boldsymbol{A}.\boldsymbol{D},\ \boldsymbol{D} = |\boldsymbol{C}.\boldsymbol{X}_p(t) - \boldsymbol{X}(t)| \tag{3}$$

where t is number of iterations, $\boldsymbol{X}_p(t)$ is the prey position, $\boldsymbol{A}$ and $\boldsymbol{C}$ are coefficient vectors.

$$\boldsymbol{A} = 2a.\boldsymbol{r}_1 - a,\ \boldsymbol{C} = 2r_2 \tag{4}$$

where a is linearly decreased from 2 to 0 over the t of iterations, $\boldsymbol{A}$ and $\boldsymbol{C}$ are coefficient vectors, $\boldsymbol{X}_p$ is the prey position, and $\boldsymbol{X}$ is the gray wolf position.the updating for the wolves positions is as in Eq. (5).

$$\boldsymbol{X}(t+1) = \frac{\boldsymbol{X}_1 + \boldsymbol{X}_2 + \boldsymbol{X}_3}{3} \tag{5}$$

where $\boldsymbol{X}_\alpha, \boldsymbol{X}_\beta, \boldsymbol{X}_\delta$ are the first three best solutions in the swarm at a given iteration t, $\boldsymbol{A}_1, \boldsymbol{A}_2, \boldsymbol{A}_3$ are defined using Eq. (6).

$$\boldsymbol{X}_1 = |\boldsymbol{X}_\alpha - \boldsymbol{A}_1.\boldsymbol{D}_\alpha|,\ \boldsymbol{X}_2 = |\boldsymbol{X}_\beta - \boldsymbol{A}_2.\boldsymbol{D}_\beta|,\ \boldsymbol{X}_3 = |\boldsymbol{X}_\delta - \boldsymbol{A}_3.\boldsymbol{D}_\delta| \tag{6}$$

And $\boldsymbol{D}_\alpha, \boldsymbol{D}_\beta, \boldsymbol{D}_\delta$ are defined using Eq. (7).

$$\boldsymbol{D}_\alpha = |\boldsymbol{C}_1.\boldsymbol{X}_\alpha - \boldsymbol{X}|,\ \boldsymbol{D}_\beta = |\boldsymbol{C}_2.\boldsymbol{X}_\beta - \boldsymbol{X}|,\ \boldsymbol{D}_\delta = |\boldsymbol{C}_3.\boldsymbol{X}_\delta - \boldsymbol{X}| \tag{7}$$

Updating of parameters a that controls the trade off between exploration and exploitation. The parameter a is linearly updated in each iteration to range from 2 to 0 according to the Eq. (8).

$$a = 2 - t\frac{2}{Maxitr} \tag{8}$$

where t is the iteration number and $Maxiter$ is the total number of iteration allowed for the optimization.

Discrete Gray Wolf Optimizer (dGWO) Is a novel discrete algorithm that cloned from the continuous gray wolf optimizer (CGWO) that has wolves continuously change their positions to any position in the space, but in our case we have two restrictions mentioned in Sect. 2 so wolves have to jump (not continuously walk) between a combination of solutions that $\in \boldsymbol{R} = \{1, 2, \cdots, r\}$ (Fig. 2).

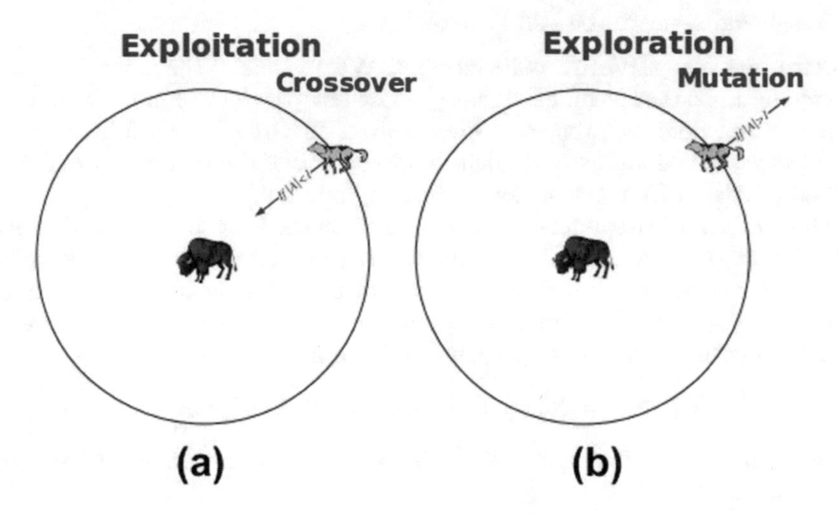

Fig. 2. (a) Exploitation by crossover process (b) Exploration using mutation process

3.1 Wolves Pack

The pool of solutions $\boldsymbol{X_t}$ that represents the pack of wolves x_t that has number of t agents, We will use the the same social hierarchy in which we have the best solutions (α), (β), (δ) and the rest of pack (ω), in first stage of our module we have to initialize population with a certain discrete shape that $\in \boldsymbol{R}$ and keep this shape during pool updating.

3.2 Encircling Prey

As mentioned before wolves jumps in discrete pattern, in the last step we started a discrete population and to sustain the discrete shape of population we have to update wolves positions in a discrete shape and taking into account the restrictions that mentioned before, Encircling prey stage oscillating between two phases, Exploration (searching for prey) and Exploitation (attacking the prey), we can mimic exploration phase as a mutation process with mutation rate $\boldsymbol{D}$ and mimic exploitation phase by crossover process that make a hybrid solution between (ω) and one of three best solutions x_p with crossover rate $\boldsymbol{D}$, the switching between exploration (mutation) and exploitation (crossover) is depending on $\boldsymbol{A}$. Equation (9) shows the prey encircling process.

$$X_{t+1} = \begin{cases} Crossover[\boldsymbol{X}_p(t), \boldsymbol{X}(t)]_D & 0 \leq \boldsymbol{A} \leq 2 \\ \\ Mutation[\boldsymbol{X}(t)]_D & -2 \leq \boldsymbol{A} < 0 \end{cases} \tag{9}$$

In CGWO the value of a decreases linearly from 2 to 0 using the update equation in CGWO in Exploration and Exploitation this equation provide a 50% in exploitation process and the same percentage for exploitation. Mittal, Singh and Sohi [10] introduced a good equation for searching curve the replacing linear curve by quadratic one, to let the wolves taking more time in searching and exploring than exploiting as in Eq. (10).

$$a = 2(1 - \frac{t^2}{T^2}) \tag{10}$$

3.3 Hunting

Is the process of convergence to solution (prey) in dGWO its a crossover combination from three solutions we got from Eq. (9), the cross over rate was taken equally between three solution as shown in Eq. (11) [9].

$$\boldsymbol{X}_{t+1} = Crossover[\boldsymbol{X_1}, \boldsymbol{X_2}, \boldsymbol{X_3}] \tag{11}$$

Validation: After Discretization we will find duplication in each agent set, as mentioned before in constrains part Eq. (1), to overcome out of range data issue we have to make boundary limiter that accept numbers in our range $[1, 2, \cdots, r]$ and replace out of range numbers by any number in our data set range r as in Eq. (12).

$$x_t^r = \begin{cases} x_t & x_t \in \boldsymbol{R} \\ \\ r & \text{otherwise} \end{cases} \tag{12}$$

begin dGWO Algorithm Main
  Randomly initialize population $\boldsymbol{X}_t \in \boldsymbol{R}$;
  Initialize the max numbers of iteration T_{max};
  Initialize a as equation (10);
  Initialize A and C as equations (4);
  while $T < T_{max}$ **do** for each iteration
    foreach *search agent* x_t *in the population* **do**
      Evaluate $fitt(x_t)$ using fitness function (2);
      find α, β and δ ;
      Update current wolf's position x_t according to equations (9), (11);
      validate solution using condition in equation (12);
    end
    Update a;
    Update A and C;
    Update α, β and δ;
  end
end

Algorithm 1. dGWO algorithm

4 Experimental Results

This section is composed of three main subsections covering data description, Evaluation Criteria indicators, numerical results, and discussion of results.

4.1 Data Description

Our experiment based on Eight images Six of them have English text with some images and Two of them are in the Arabic language all images have dimension $w \times h$ 500×700 pixels. We shredded images into n shreds equally. We assume that we know the right position and orientation of each shred then we calculate the fitness of original images and call it. Optimal fitness value. we take into account the rotation of the paper so we have an array of the optimal solution with n optimal fitness values.

4.2 Evaluation Criteria Indicators

The following set of assessment indicators have been used in the evaluation of individual optimizers:

- *Statistical mean* is the average of solutions acquired from running an optimization algorithm for different M running.
- *Statistical standard deviation (std)* is a representation of the variation of the obtained best solutions found for running a stochastic optimizer for M different runs.
- *Wilcoxon rank sum test* proposed by Frank Wilcoxon [11] as a nonparametric test. The test assigns a rank to all the scores considered as one group and then sums the ranks of each group.

4.3 Results and Discussion

In this experiment, we hold a comparison between three meta heuristic intelligent algorithms, Random Search Algorithm (RSA), Genetic Algorithm (GA) and discrete Gray Wolf Optimizer (dGWO). Table 1 shows general initialization settings for three algorithms dGWO, GA and RSA.

Table 1. Parameter setting for all algorithms

No of shreds	15
No of agents	225
No of repetition (M)	25
Max iteration	25000

Table 2 shows a big difference between the three algorithms at 15 shreds, dGWO shows less miss rate mean than GA and RSA. The difference in the performance can be interpreted by the fact that the problem becomes more challenging for the large size of the search space and the multiple local minima that exist in the fitness function. The results in this table show the capability of the proposed discrete algorithm to adaptively search the space of solution for an optimal solution with a good balance between exploration and exploitation. We can also remark that the standard deviation of the obtained fitness value is less for dGWO than GA which proves the repeatability of results and the convergence capability regardless of the used random factors.

Table 2. Mean and standard deviation for R = 15 miss rate

	dGWO	GA	RSA
F15-0	**0.760 ± 0.436**	0.920 ± 0.277	1.000 ± 0.000
F15-1	**0.520 ± 0.510**	0.720 ± 0.458	1.000 ± 0.000
F15-2	**0.760 ± 0.436**	0.920 ± 0.277	1.000 ± 0.000
F15-3	**0.120 ± 0.332**	0.560 ± 0.507	1.000 ± 0.000
F15-4	**0.000 ± 0.000**	0.800 ± 0.408	1.000 ± 0.000
F15-5	**0.080 ± 0.277**	0.720 ± 0.458	1.000 ± 0.000
F15-6	**0.120 ± 0.332**	0.720 ± 0.458	1.000 ± 0.000
F15-7	**0.240 ± 0.436**	0.600 ± 0.500	1.000 ± 0.000

Table 3 shows a big difference between the three algorithms at 15 shreds, dGWO shows less iteration rate mean than GA, and it can easily be noticed that at 15 shreds RSA is out of comparison.

Table 3. Mean and standard deviation for R=15 Iteration rate

	dGWO	GA	RSA
F15-0	**19037.560 ± 10829.335**	23713.800 ± 4843.203	25000.000 ± 0.000
F15-1	22514.080 ± 48909.452	**20247.040 ± 8925.401**	25000.000 ± 0.000
F15-2	**19189.920 ± 10563.888**	23156.720 ± 6382.540	25000.000 ± 0.000
F15-3	**5342.880 ± 8655.830**	17044.120 ± 10602.226	25000.000 ± 0.000
F15-4	**2454.680 ± 2358.713**	21759.760 ± 6994.819	25000.000 ± 0.000
F15-5	**5927.120 ± 7262.704**	20616.280 ± 8887.389	25000.000 ± 0.000
F15-6	**5442.520 ± 8340.937**	19307.480 ± 9356.430	25000.000 ± 0.000
F15-7	**9730.680 ± 11131.686**	18278.320 ± 9262.631	25000.000 ± 0.000

Table 4 shows a statistical test between dGWO and GA as well as dGWO and RSA based on t-test and Wilcoxon on 10 and 15 dimensions. It could be observed from the table the enhance of performance is significant considering

Table 4. dGWO algorithm comparison

Shreds no	10		15	
Algorithms	GA	RS	GA	RS
t-test	0.00	0.00	0.00	0.00
wil	0.00	0.00	0.00	0.00

both tests and on all used problem dimensions considering a significance level of 0.05.

Figure 3 shows the convergence performance of the different optimizers used on the average. It could be from the figure that the proposed dGWO converges -in the average- to the optimal/near optimal solution is much faster than the GA and RSA which proves its enhanced performance.

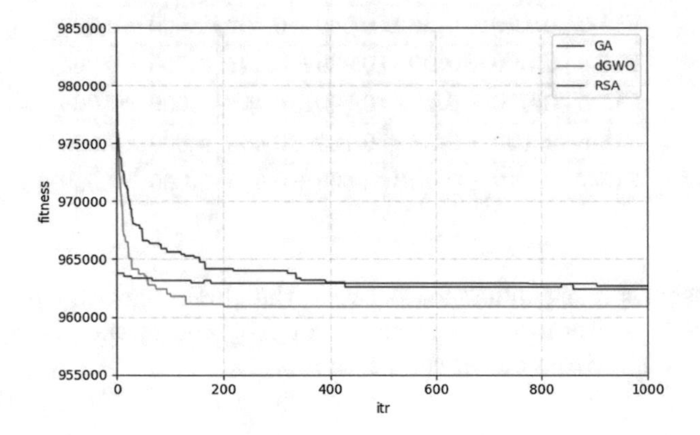

Fig. 3. dGWO, GA and RSA Convergence curve during course of iterations

5 Conclusion

The proposed system formulates the problem into an optimization problem where a suitable fitness function has been set with suitable constrains and variable. Based on that formulation, a discrete version of GWO has been proposed to tackle such optimization problem with suitable balance between exploitation and exploration. The proposed optimizer used has the capability to adaptively search the space of solution for an optimal/near optimal shred ordering minimizing the set fitness function. The proposed system has been benchmarked against the well known genetic algorithm as well as the random search over a set of horizontally shredded paper set at different dimensions and proves its significant enhancement in performance using a set of assessment indicators.

References

1. Atallah, A.S., Emary, E., El-Mahallawy, M.S.: A step toward speeding up cross-cut shredded document reconstruction. In: Proceedings - 2015 5th International Conference on Communication Systems and Network Technologies, CSNT 2015, pp. 345–349 (2015)
2. Perl, J., Diem, M., Kleber, F., Sablatnig, R.: Strip shredded document reconstruction using optical character recognition. In: 4th International Conference on Imaging for Crime Detection and Prevention 2011 (ICDP-2011), p. 6 (2011)
3. Xu, H., Zheng, J., Zhuang, Z., Fan, S.: A solution to reconstruct cross-cut shredded text documents based on character recognition and genetic algorithm. Abstr. Appl. Anal. **2014**, 1–11 (2014)
4. Schauer, C., Prandtstetter, M., Raidl, G.R.: A memetic algorithm for reconstructing cross-cut shredded text documents, pp. 103–117. Springer, Heidelberg (2010)
5. Lin, H.Y., Fan-Chiang, W.C.: Reconstruction of shredded document based on image feature matching. Expert Syst. Appl. **39**(3), 3324–3332 (2012)
6. Biesinger, B.: Enhancing an evolutionary algorithm with a solution archive to reconstruct cross cut shredded text documents, pp. 380–387 (2012)
7. Mirjalili, S., Mirjalili, S.M., Lewis, A.: Grey wolf optimizer. Adv. Eng. Softw. **69**, 46–61 (2014)
8. Hassanien, A.E., Emary, E.: Swarm Intelligence: Principles, Advances, and Applications. CRC Press, Boca Raton (2016)
9. Emary, E., Zawbaa, H.M., Hassanien, A.E.: Binary grey wolf optimization approaches for feature selection. Neurocomputing **172**, 371–381 (2016)
10. Mittal, N., Singh, U., Sohi, B.S.: Modified grey wolf optimizer for global engineering optimization. Appl. Comput. Intell. Soft Comput. **2016**, 1–16 (2016)
11. Wilcoxon, F.: Individual comparisons by ranking methods. Biometrics Bull. **1**, 80–83 (1945)

Chaotic Bird Swarm Optimization Algorithm

Fatma Helmy Ismail[1,4](✉), Essam H. Houssein[2,4], and Aboul Ella Hassanien[3,4]

[1] Faculty of Computer Science, Misr International University, Cairo, Egypt
fatmahelmy2000@yahoo.com
[2] Faculty of Computers and Information, Minia University, Minya, Egypt
[3] Faculty of Computers and Information, Cairo University, Giza, Egypt
[4] Scientific Research Group in Egypt (SRGE), Giza, Egypt
http://www.egyptscience.net

Abstract. Bird Swarm Algorithm (BSA) is a relatively new meta-heuristic optimization algorithm introduced to solve optimization problems. Birds have three types of conducts including searching for food (foraging), watchfulness (vigilance), and flying conduct. Foraging and vigilance behaviours are done by birds to improve their positions (exploitation) while flight behaviour is done to move from one location to another (exploration). This paper combines the chaotic-based methods with foraging and vigilance behaviours to improve the exploitation quality. The main privilege of chaotic-based methods is their capability for avoiding local minima. In order to assess the proposed chaotic BSA (CBSA) algorithm, a set of 7 unimodal benchmark functions are tested with ten different chaotic maps. The numerical results show that the performance of CBSA, with most of the chaotic maps, can clearly outperform the standard BSA.

Keywords: Bird Swarm Algorithm · Chaotic maps
Meta-heuristic algorithms · Global optimization

1 Introduction

Meta-heuristic techniques are the most commonly used algorithms to solve optimization problems and have been applied to diverse optimization problems in different areas that have been successfully applied in many real-world and complex optimization problems [1,2]. These techniques try to mimic social behavior to generate better solutions for optimization problem by using iterations and stochasticity [3]. Recently, a numerous meta-heuristic algorithms have been proposed to solve both of global search problem and larger problems on the other side to increase the computational efficiency and implement robust optimization codes [4–6], also a binary whale optimization algorithm for feature selection has been presented in [7].

A. E. Hassanien et al. (Eds.): AISI 2018, AISC 845, pp. 294–303, 2019.
https://doi.org/10.1007/978-3-319-99010-1_27

Chaos theory has been widely used in several applications. In the same context, one of the most famous applications is the introduction of chaos theory into the optimization methods [8]. Therefore, the chaos theory has been successfully combined with several meta-heuristic optimization such as Firefly algorithm with chaos [9], Chaos-enhanced Cuckoo search optimization algorithms for global optimization [10], Chaotic Krill Herd algorithm [11], Chaotic grey wolf optimization algorithm for constrained optimization problems [12], Chaotic bat algorithm [13], Chaotic antlion algorithm for parameter optimization [14], A chaos-based evolutionary algorithm for general nonlinear programming problems [15] and chaotic whale optimization algorithm for parameter estimation of photovoltaic cells [16].

All the above-aforementioned algorithms have revealed the ability of meta-heuristics to solve various optimization problems. In the other hand according to No-Free-Lunch (NFL) theorem [17], they are not able to solve all optimization problems. Therefore, the objective of this paper is to combine chaos theory with the original BSA to improve its exploitation behaviour. We use ten chaotic maps and a set of 7 unimodal benchmark functions. The results reveal that the developed CBSA algorithms can be applied to different complex optimization problems in various areas where it is expected to result in high quality solutions.

This paper proposes a Chaotic Bird Swarm Algorithm (CBSA) hybrid Chaos theory with the new meta-heuristic algorithm presented by Xian-Bing Meng et al. [18], called Bird Swarm Algorithm (BSA), that stimulates the behaviour of birds in bird swarms. This paper is structured as follows: the original BSA is presented in Sect. 2. Then, the chaotic maps and Chaotic Bird Swarm Algorithm (CBSA) are discussed in Sects. 3 and 4 respectively. The evaluation of the CBSA algorithm using 7 benchmarks is illustrated in Sect. 5. Finally, in Sect. 6, the conclusion and future work are introduced.

2 Bird Swarm Algorithm

Bird Swarm Algorithm (BSA) [18], was introduced to solve optimization problems. BSA originated from the swarm insights extricated from the social intuitive in bird groups. Birds do three types of behaviors: scrounging conduct (foraging behaviour), carefulness conduct (vigilance behaviour) and flight conduct. The social behaviors can be rearranged by a few rules as takes after:

1. Rule 1: Each bird can switch between the watchfulness conduct and scrounging conduct.
2. Rule 2: While searching for food, each bird records its past best position and the swarm's past best position around food patch. Social data is spread immediately among the entirety swarm.
3. Rule 3: When being watchfulness, each bird attempt to fly forward to the center of the swarm. The birds with the higher saves would be more likely to lie closer to the center of the swarm than those with the lower saves.

4. Rule 4: While birds are flying to another location, birds may makers or consumers. The bird with the most elevated saves would be a maker, while the one with the least saves would be a scrounger or a consumer.
5. Rule 5: Makers look for food while the scroungers haphazardly take after a maker to look for food.

Bird Swarm Algorithm can be described as follow; all N virtual birds, depicted by their position $x_i^t(i \in [1, ..., N]$ at time step t, forage for food and fly in a D-dimensional space.

2.1 Foraging Behaviour

The following set of equations translates Rule 2 that describes how birds search for food:

$$x_{i,j}^{t+1} = x_{i,j}^t + (p_{i,j} - x_{i,j}^t) \times C \times rand(0,1) + (g_j - x_{i,j}^t) \times S \times rand(0,1), \quad (1)$$

Where $j \in [1, ..., D]$, $rand(0,1)$ denotes independent uniformly distributed numbers in $(0,1)$. C and S are cognitive and social accelerated coefficients. $p_{i,j}$ is the best past position of the ith bird and g_j the best previous past position shared by the swarm of birds.

2.2 Vigilance Behaviour

Birds fly forward to the center of the swarm based on Rule 3, these motions can be formulated as follows:

$$x_{i,j}^{t+1} = x_{i,j}^t + A1(mean_j - x_{i,j}^t) \times rand(0,1) + A2(p_{k,j} - x_{i,j}^t) \times rand(-1,1), \quad (2)$$

$$A1 = a1 \times \exp\left(-\frac{pFit_i}{sumFit + \varepsilon} \times N\right), \quad (3)$$

$$A2 = a2 \times \exp\left(\left(\frac{pFit_i - pFit_k}{|pFit_k - pFit_i| + \varepsilon}\right) \frac{N \times pFit_k}{sumFit + \varepsilon}\right), \quad (4)$$

Where $k(k \neq i)$ is a positive integer, which is randomly chosen between 1 and N. $a1$ and $a2$ are two positive constants in $[0,2]$, $pFit_i$ denotes the ith bird's best fitness value and $sumFit$ represents the sum of the swarms' best fitness value. ε , which is a constant to evade divison by zero error. $mean_j$ presents the jth element of the average position of the whole swarm. The product of $A1$ and $rand(0,1)$ should not be more than 1. Here, $A2$ presents the direct effect of the interference of moving towards the center of the swarm.

2.3 Flight Behaviour

Birds fly to search for food and scrounge for food once more. A few birds acting as makers would look for food patches, while others attempt to feed from the food patch found by the makers. The makers and scroungers can be isolated from the swarm according to Rule 4. The behaviour of the makers and scroungers can be described mathematically as follows, respectively:

$$x_{i,j}^{t+1} = x_{i,j}^{t} + rand(0,1) \times x_{i,j}^{t}, \tag{5}$$

$$x_{i,j}^{t+1} = x_{i,j}^{t} + (x_{k,j}^{t} - x_{i,j}^{t}) \times FL \times rand(0,1), \tag{6}$$

Where $rand(0,1)$ presents Gaussian distributed random number with mean 0 and standard deviation 1, $k \in [1,2,3,...,N]$, $k \neq i. FL(FL \in [0,2])$ means that the scrounger would follow the maker.

3 Chaotic Maps

Chaos maps demonstarate excellent chaotic behaviours and can be utilized in truely random number generators. In this paper, ten distinguished maps are adopted to obtain chaotic sets. The mixing properties of chaos can perform the search process with higher speeds than the classical searching that is based on the standard probability distributions [19]. The adopted chaotic maps that are employed in this paper are defined in Table 1. Thus, chaotic maps have been used to propose a new version of BSA termed CBSA. In these chaotic maps, any number in the range [−1, 1] can be chosen as the initial value [20].

Table 1. Details of the ten adapted chaotic maps applied on CBSA.

NO.	Map Name	Equation	Range
CBSA1	Chebyshev	$x_{k+1} = \cos(k\cos^{-1}(x_k))$	(−1, 1)
CBSA2	Circle	$x_{k+1} = x_k + b - (\frac{a}{2\pi})\sin(2\pi x_k)\text{mod}$	(0, 1)
CBSA3	Gauss/Mouse	$x_{k+1} = \begin{cases} 0\; x_k = 0 \\ \frac{1}{x_k \bmod (1)}, otherwise \end{cases}$	(0, 1)
		$\frac{1}{x_k \bmod (1)} = \frac{1}{x_k} - \left[\frac{1}{x_k}\right]$	
CBSA4	Iterative	$x_{k+1} = \sin(\frac{a\pi}{x_k})$	(−1, 1)
CBSA5	Logistic	$x_{k+1} = ax_k(1 - x_k)$	(0, 1)
CBSA6	Piecewise	$x_{k+1} = \begin{cases} \frac{x_k}{P}, 0 \leq x_k \prec P \\ \frac{x_k - P}{0.5 - P}\; P \leq x_k \prec 0.5 \\ \frac{1-P-x_k}{0.5-P}, 0.5 \leq x_k \prec 1-P \\ \frac{1-x_k}{P}, 1-P \leq x_k \prec 1 \end{cases}$	(0, 1)
CBSA7	Sine	$x_{k+1} = \frac{a}{4}\; \sin(\pi x_k)$	(0, 1)
CBSA8	Singer	$x_{k+1} = \mu(7.86x_k - 23.31x_k^2 + 28.75x_k^3 - 13.3x_k^4)$	(0, 1)
CBSA9	Sinusoidal	$x_{k+1} = ax_k^2 \sin(\pi x_k)$	(0, 1)
CBSA10	Tent	$x_{k+1} = \begin{cases} \frac{x_k}{0.7}, x_k \prec 0.7 \\ \frac{10}{3}(1 - x_k), x_k \geq 0.7 \end{cases}$	(0, 1)

Figure 1 shows the visualization of these maps.

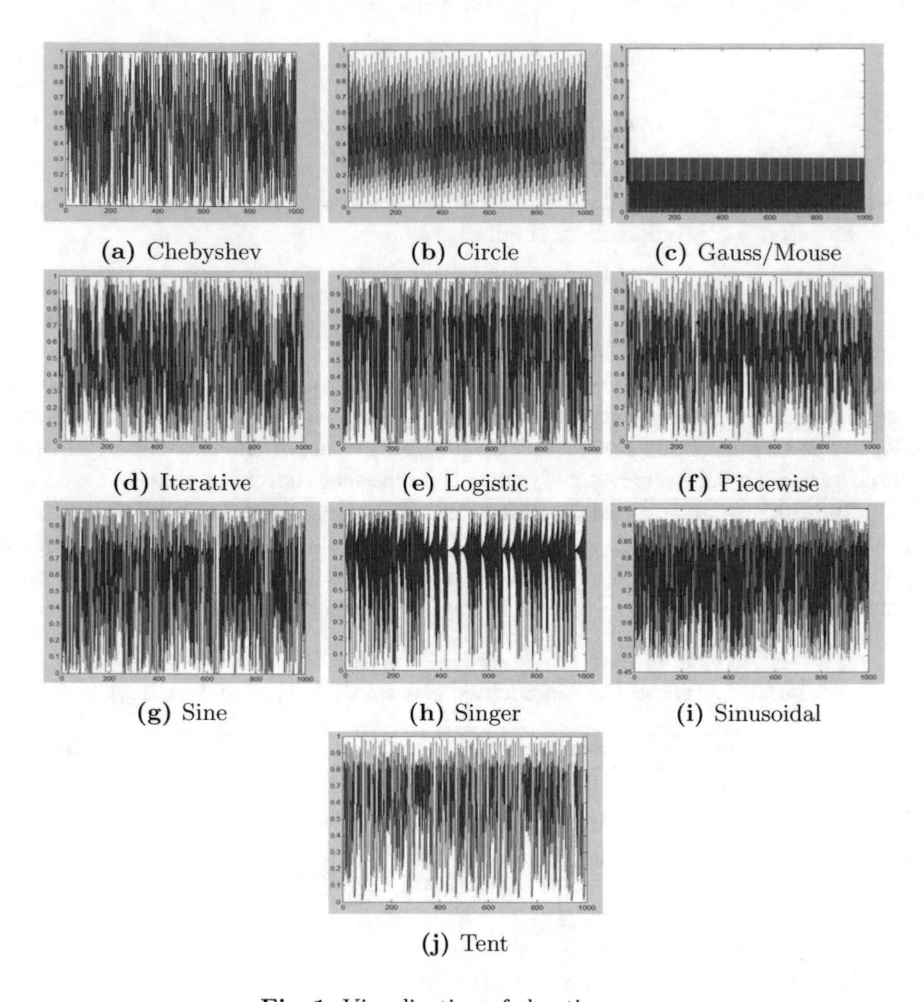

(a) Chebyshev (b) Circle (c) Gauss/Mouse

(d) Iterative (e) Logistic (f) Piecewise

(g) Sine (h) Singer (i) Sinusoidal

(j) Tent

Fig. 1. Visualization of chaotic maps.

4 Chaotic Bird Swarm Algorithm

The basic flowchart of the proposed CBSA for solving optimization problems is shown in Fig. 2. The random variables used for updating bird position are replaced with the values generated by a specific chaotic map within the range between 0 to 1. In this paper, ten different chaotic maps are used for the optimization process as shown in Table 1. In CBSA, a chaotic sequence is embedded in the exploitation phase only (Foraging and Vigilance behaviour). Mathematically, in order to propose the CBSA, Eqs. 1 and 2 in the original BSA have been reformulated to the following two Eqs. 7 and 8 as follows:

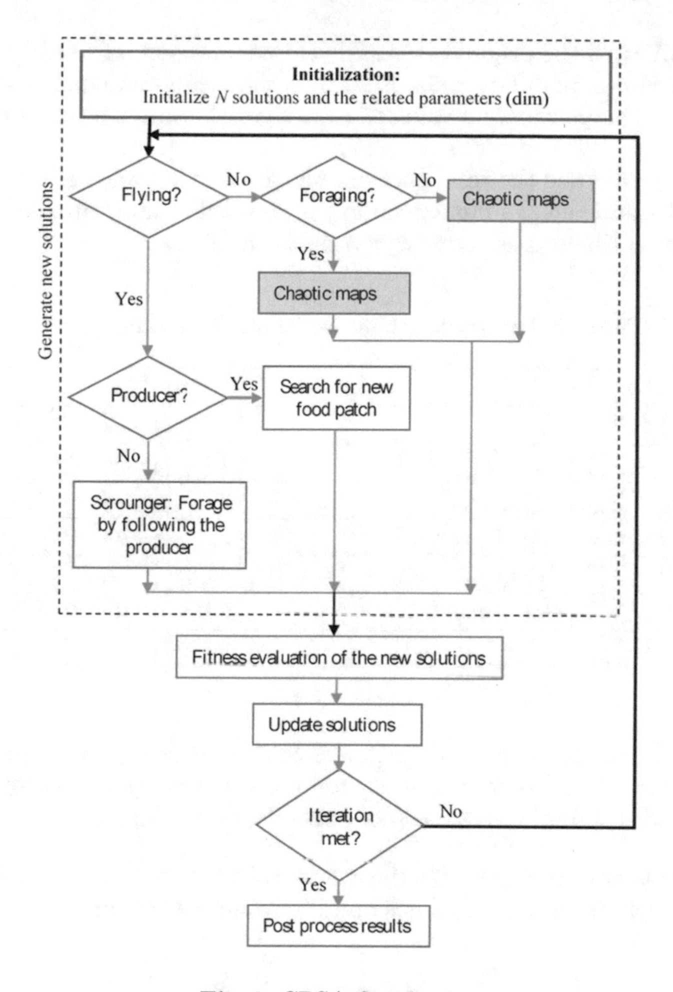

Fig. 2. CBSA flowchart.

$$x_{i,j}^{t+1} = x_{i,j}^{t} + (p_{i,j} - x_{i,j}^{t}) \times C \times chaotic(map, range) + (g_j - x_{i,j}^{t}) \times S \times chaotic(map, range), \quad (7)$$

$$x_{i,j}^{t+1} = x_{i,j}^{t} + A1(mean_j - x_{i,j}^{t}) \times chaotic(map, range) + A2(p_{k,j} - x_{i,j}^{t}) \times (2 \times chaotic(map, range) - 1). \quad (8)$$

5 Experimental Results and Discussion

The performance of the proposed algorithm is evaluated through comparing it with the original BSA using a set of 7 unimodal benchmark functions. In which, the aims of this section include, (1) determine the suitable chaotic map that improve the performance of the BSA. (2) Compare the performance of the

Chaotic BSA with the original BSA. For fair comparison, all the algorithms are running 100 times, population size is 200 and the maximum number of iterations is 1000, also, the parameters value of each algorithm are the same the original reference.

Table 2 summarize the test functions along with the range of their optimization variables and their minimum value f_{min} quoted in literature [21]. Note that V_no indicates the number of design variables in Table 2.

Table 2. Description of unimodal benchmark functions.

Function	V_no	Range	f_{min}
$F_1(x) = \sum_{i=1}^{n} x_i^2$	30	[−100, 100]	0
$F_2(x) = \sum_{i=1}^{n} \lvert x_i \rvert + \prod_{i=1}^{n} \lvert x_i \rvert$	30	[−10, 10]	0
$F_3(x) = \sum_{i=1}^{n} \left(\sum_{j-1}^{i} x_j\right)^2$	30	[−100, 100]	0
$F_4(x) = \max_i\{\lvert x_i \rvert, 1 \leq i \leq n\}$	30	[−100, 100]	0
$F_5(x) = \sum_{i=1}^{n-1} [100(x_{i+1} - x_i^2)^2 + (x_i - 1)^2]$	30	[−30, 30]	0
$F_6(x) = \sum_{i=1}^{n} ([x_i + 0.5])^2$	30	[−100, 100]	0
$F_7(x) = \sum_{i=1}^{n} ix_i^4 + rand[0, 1)$	30	[−1.28, 1.28]	0

The performance of each algorithm is evaluated using four different statistical measurements. These measurements are the best, the mean of the fitness function, and standard deviation (Std) and they are defined as:

1. **Statistical mean:** represents the average of the best fitness function F_* over running the algorithm N_R times and it is computed as in Eq. (9):

$$Mean = \frac{1}{N_R} \sum_{i=1}^{N_R} F_*^i \tag{9}$$

2. **Statistical best:** represents the minimum of the best fitness function value F_* over running the algorithm N_R times and it is computed as in Eq. (10):

$$Best = \min_{i=1}^{N_R} F_*^i \tag{10}$$

3. **Statistical standard deviation (Std):** represents the stability and robustness of the algorithm and it is defined as in Eq. (11):

$$Std = \sqrt{\frac{1}{N_R - 1} \sum (F_*^i - Mean)^2} \tag{11}$$

In order to determine the best chaotic map that must be combined with BSA, a set of 10 chaotic maps are used. The benchmark functions F1 to F7 are unimodal because they have one global minimum. The exploitation behaviour of any meta-heuristic algorithm can be tested by those unimodal functions. It is noted from Tables 3, 4 and 5 that Chaotic BSA is very competitive with other ordinary BSA. In particular, it is the most efficient optimizer for functions F5 and F6 in most chaotic maps used and compared to ordinary BSA in the other test problems. Adding chaos to the foraging and vigilance behaviors of BSA enhanced the exploitation capability of the algorithm. It must be pointed out that CBSA1, CBSA2, ..., CBSA10 in the Tables 3, 4 and 5 refers to the ten adopted chaotic maps as shown in Table 1. Tables 3, 4 and 5 compares CBSA with different chaotic maps with the original BSA in terms of mean, best and STD respectively.

Table 3. The mean statistical results.

Algorithm	F1	F2	F3	F4	F5	F6	F7
BSA	0	5.23e−198	0	6.035e−196	28.86	1.2842	6.19e−7
CBSA1	5.68 e -223	3.33e−99	1.11e−220	2.17e−110	19.85	0.0357	2.48e−6
CBSA2	0	3.13e−197	0	1.28e−197	28.8	0.3357	2.97e−7
CBSA3	2.06 e−313	2.12e−159	2.95e−314	1.806e−158	28.876	4.9735	3.704e−7
CBSA4	1.72e−281	1.24e−141	3.714e−272	2.07e−144	9.46	0.08	3.29e−7
CBSA5	1.66e−220	7.75e−106	4.92e−220	4.45e−111	24.9	0.084	5.75e−7
CBSA6	0	1.07e−161	0	2.56e−163	28.783	0.2368	5.77e−7
CBSA7	3.67e−241	2.45e−121	5.55e−243	5.86e−125	12.103	0.0923	9.16e−7
CBSA8	2.99e−195	2.52e−90	9.08e−186	8.92e−93	24.023	0.0207	6.04e−7
CBSA9	1.48e−143	1.45e−69	1.92e−144	1.52e−76	0.861	0.0251	1.79e−6
CBSA10	1.26e−309	1.53e−156	6.42e−316	2.84e−157	28.714	0.0147	6.213e−7

Table 4. The best statistical results.

Algorithm	F1	F2	F3	F4	F5	F6	F7
BSA	0	0	0	0	28.86	0.7347	1.149e−8
CBSA1	0	0	0	0	2.18e−9	2.515e−13	2.19e−9
CBSA2	0	0	0	0	28.77	1.38e−11	6.09e−8
CBSA3	0	0	0	0	28.87	4.923	1.15e−7
CBSA4	0	0	0	0	4.35e−15	5.02e−14	1.38e−7
CBSA5	0	0	0	0	22.98	2.67e−11	3.01e−7
CBSA6	0	0	0	0	28.586	7.706e−12	1.18e−7
CBSA7	0	0	0	0	6.98e−13	7.395e−12	1.22e−7
CBSA8	0	0	0	0	20.946	2.226e−21	1.68e−7
CBSA9	0	0	0	0	3.012e−13	2.42e−13	9.88e−7
CBSA10	0	0	0	0	28.707	3.27e−10	9.19e−8

Table 5. The standard deviation statistical results.

Algorithm	F1	F2	F3	F4	F5	F6	F7
BSA	0	0	0	0	5.002e−14	0.3585	1.527e−6
CBSA1	0	3.33e−98	0	2.17e−109	8.94	0.1764	5.14e−6
CBSA2	0	0	0	0	0.033	0.8262	1.49e−6
CBSA3	0	2.12e−158	0	1.806e−157	0.002	0.0814	3.108e−7
CBSA4	0	1.24e−140	0	2.07e−143	13.24	0.374	4.66e−7
CBSA5	0	7.75e−105	0	4.45e−110	2.272	0.3356	3.64e−7
CBSA6	0	1.07e−160	0	2.22e−162	0.1317	0.6303	2.13e−6
CBSA7	0	2.45e−120	0	5.86e−124	13.734	0.284	1.505e−6
CBSA8	0	2.52e−89	0	8.92e−92	2.838	0.1471	1.99e−6
CBSA9	1.48e−142	1.45e−68	1.92e−143	1.52e−75	4.922	0.1891	2.84e−6
CBSA10	0	1.53e−155	0	2.84e−156	0.0141	0.1268	1.714e−6

6 Conclusion and Future Work

Chaos theory was regarding as one of the most important techniques to enhance the performance of meta-heuristic algorithms, and this has become an active research topic in the recent optimization literature. A new Chaotic Bird Swarm Algorithm (CBSA) has been developed to improve the reliability of the global optimality for solving optimization problems. In fact, ten chaotic maps and seven different benchmark functions were applied to investigate its significant performance to improve the exploitation of the BSA. The simulation results revealed that the CBSA clearly improves the reliability of the global optimality in comparison with the original BSA. As future work, the proposed approach can be applied to a wide range of several application such as feature selection method, pattern recognition problems, image segmentation and renewable energy as a prediction method.

References

1. Yildiz, A.R.: Comparison of evolutionary-based optimization algorithms for structural design optimization. Eng. Appl. Artif. Intell. **26**(1), 327–333 (2013)
2. Tharwat, A., Houssein, E.H., Ahmed, M.M., Hassanien, A.E., Gabel, T.: MOGOA algorithm for constrained and unconstrained multi-objective optimization problems. Appl. Intell. **48**, 1–16 (2017)
3. Talbi, E.-G.: Metaheuristics: From Design to Implementation, vol. 74. Wiley, Hoboken (2009)
4. Hussien, A.G., Houssein, E.H., Hassanien, A.E.: A binary whale optimization algorithm with hyperbolic tangent fitness function for feature selection. In: Eighth International Conference on Intelligent Computing and Information Systems, ICICIS 2017, pp. 166–172. IEEE (2017)
5. Yang, X.-S.: Engineering Optimization: An Introduction with Metaheuristic Applications. Wiley, Hoboken (2010)

6. Hussien, A.G., Hassanien, A.E., Houssein, E.H.: Swarming behaviour of salps algorithm for predicting chemical compound activities. In: Eighth International Conference on Intelligent Computing and Information Systems (ICICIS 2017), pp. 315–320. IEEE (2017)
7. Hussien, A.G., Houssein, E.H., Hassanien, A.E., Bhattacharyya, S., Amin, M.: S-shaped binary whale optimization algorithm for feature selection. In: First International Symposium on Signal and Image Processing (ISSIP 2017), RCC Institute of Information Technology, Kolkata. Springer, Heidelberg (2017)
8. Yang, D., Li, G., Cheng, G.: On the efficiency of chaos optimization algorithms for global optimization. Chaos Solitons Fractals **34**(4), 1366–1375 (2007)
9. Gandomi, A.H., Yang, X.-S., Talatahari, S., Alavi, A.H.: Firefly algorithm with chaos. Commun. Nonlinear Sci. Numer. Simul. **18**(1), 89–98 (2013)
10. Huang, L., Ding, S., Yu, S., Wang, J., Lu, K.: Chaos-enhanced cuckoo search optimization algorithms for global optimization. Appl. Math. Model. **40**(5–6), 3860–3875 (2016)
11. Wang, G.-G., Guo, L., Gandomi, A.H., Hao, G.-S., Wang, H.: Chaotic krill herd algorithm. Inf. Sci. **274**, 17–34 (2014)
12. Kohli, M., Arora, S.: Chaotic grey wolf optimization algorithm for constrained optimization problems. J. Comput. Des. Eng. (2017)
13. Gandomi, A.H., Yang, X.-S.: Chaotic bat algorithm. J. Comput. Sci. **5**(2), 224–232 (2014)
14. Tharwat, A., Hassanien, A.E.: Chaotic antlion algorithm for parameter optimization of support vector machine. Appl. Intell. **48**(3), 670–686 (2018)
15. El-Shorbagy, M., Mousa, A., Nasr, S.: A chaos-based evolutionary algorithm for general nonlinear programming problems. Chaos Solitons Fractals **85**, 8–21 (2016)
16. Oliva, D., El Aziz, M.A., Hassanien, A.E.: Parameter estimation of photovoltaic cells using an improved chaotic whale optimization algorithm. Appl. Energy **200**, 141–154 (2017)
17. Wolpert, D.H., Macready, W.G.: No free lunch theorems for optimization. IEEE Trans. Evol. Comput. **1**(1), 67–82 (1997)
18. Meng, X.-B., Gao, X.Z., Lu, L., Liu, Y., Zhang, H.: A new bio-inspired optimisation algorithm: bird swarm algorithm. J. Exp. Theor. Artif. Intell. **28**(4), 673–687 (2016)
19. dos Santos Coelho, L., Mariani, V.C.: Use of chaotic sequences in a biologically inspired algorithm for engineering design optimization. Expert Syst. Appl. **34**(3), 1905–1913 (2008)
20. Saremi, S., Mirjalili, S., Lewis, A.: Biogeography-based optimisation with chaos. Neural Comput. Appl. **25**(5), 1077–1097 (2014)
21. Mirjalili, S., Lewis, A.: The whale optimization algorithm. Adv. Eng. Softw. **95**, 51–67 (2016)

Spherical Local Search for Global Optimization

M. A. El-Shorbagy[1,2,4](✉) and Aboul Ella Hassanien[3,4]

[1] Department of Basic Engineering Science, Faculty of Engineering, Menoufia University, Shebin El-Kom, Egypt
mohammed_shorbagy@yahoo.com
[2] Department of Mathematics, College of Science and Humanities Studies, Prince Sattam Bin Abdulaziz University, Al-Kharj, Kingdom of Saudi Arabia
[3] Faculty of Computers and Information, Cairo University, Cairo, Egypt
aboitcairo@gmail.com
[4] Scientific Research Group in Egypt (SRGE), Cairo, Egypt
http://www.egyptscience.net

Abstract. This paper proposed spherical local search (SLS) for solving unconstrained optimization problems in three dimensions. The algorithm begins with a randomly chosen point in the search domain. Then, spherical trust region around this point is defined by the radius of SLS; where any point in this region is feasible. Finally, SLS can move from current search point to obtain a new best point by using three strategies of search: radius, azimuth, and inclination. These strategies are modified during the search process. SLS is tested on the set of the CEC'2005 special session on real parameter optimization. Results show the robustness and effectiveness of the proposed method.

Keywords: Spherical local search · Local search · Unconstrained optimization Global optimization · Optimization

1 Introduction

Unconstrained optimization problem (UOP) considered one of an important problem in operations research. It appears in most real life applications such as structural optimization, engineering design, economics, allocation and many other applications [1, 2]. UOP consider the problem of minimizing or maximizing any objective function which depends on real variables.

The mathematical model of UOP in three dimensions is defined as: find $X = (x, y, z)^T \in F \subseteq S$ such that [1]:

$$\begin{array}{ll} \text{Min} & f(x, y, z), \\ \text{s.t.} & L_1 \le x \le U_1 \\ & L_2 \le y \le U_2, \\ & L_3 \le z \le U_3; \end{array} \tag{1}$$

where:

$X = (x, y, z)$: Decision variables vector,
S : The search domain,

A. E. Hassanien et al. (Eds.): AISI 2018, AISC 845, pp. 304–312, 2019.
https://doi.org/10.1007/978-3-319-99010-1_28

F : Feasible subspace,
f : The objective function,
L_1, L_2, L_3 : Lower boundaries of decision variables x, y and z, respectively,
U_1, U_2, U_3 : Upper boundaries of decision variables x, y and z, respectively.

One of the importance of solving UOPs is that most constrained optimization problem is solved as an unconstrained problem [1, 2]. Until now, Strenuous efforts are trying to present a way to determine the global optimal to the UOPs. Previously, UOPs are solved by two strategies which classified into direct methods and descent methods [3, 4]. These methods are called traditional optimization techniques, which have many limitations such as: depending on the existence of derivatives, local in scope, insufficiently robust in discontinuous and vast multimodal, and noisy search spaces [4].

These limitations prompted the researchers to rely on heuristics methods to solve UOPs such as: differential evolution (DE) [5], evolutionary programming (EP) [6], simulated annealing (SA) [7], tabu search [8], neural-network-based methods [9], genetic algorithm (GA) [10], particle swarm optimization (PSO) [11, 12], ant colony optimization (ACO) [13] and cuckoo search (CC) [14] etc.

One of the heuristic methods that used to solve solving hard optimization problems is Local search (LS). Algorithms of local search are moving from solution to other in the domain of search by applying local changes until the optimal solution is obtained or time bound is elapsed. LS algorithms have been applied successfully to many optimization problems plus it is applied on numerous hard computational problems [15].

Because of all of the above, spherical local search (SLS) is proposed in this paper to solve unconstrained optimization problems (UOPs) in three dimensions. Firstly, SLS start with randomly chosen of initial point in the domain of search. Secondly, spherical trust region around this point is determined; where any point in this region is feasible. Finally, three strategies of search (radius, azimuth, and inclination) are used to obtain a new best point. These strategies are modified during the search process. These procedures are repeated until the optimal solution is obtained. SLS is tested on the set of the CEC'2005 special session on real parameter optimization. Results show the robustness and effectiveness of the proposed method.

2 Spherical Local Search (SLS)

In this section, spherical local search for solving unconstrained optimization problems is proposed which is based on spherical coordinates. The search procedures depend on three strategies of search: radius r, azimuth θ and inclination φ. The main steps of SLS are described as follows:

Step 1. Start with an initial point $X_k = (x,\ y,\ z)_k$; which is randomly initialized and satisfying the search domain S (the lower and upper bounds for each variable), as:

$$X_k = \begin{pmatrix} x \\ y \\ z \end{pmatrix}_k = \begin{pmatrix} L_x \\ L_y \\ L_z \end{pmatrix} + \text{rand} \times \begin{pmatrix} U_x - L_x \\ U_y - L_y \\ U_z - L_z \end{pmatrix}; \tag{2}$$

where rand is random number distributed uniformly within the range [0, 1].

Step 2. Set $\theta = 0$ and $\varphi = 0$ and determine the maximum radius r of SLS which guarantee the feasibility of the spherical region; where it is the minimum distance between X_k and any of the bounds of the search domain.

$$r = \min \begin{cases} |x - L_x| & |x - U_x| \\ |y - L_y| \text{ or } |x - U_x| \\ |z - L_z| & |x - U_x| \end{cases} \tag{3}$$

Step 3. Determine the new solution X_{k+1} by using the following equation:

$$X_{k+1} = \begin{pmatrix} x \\ y \\ z \end{pmatrix}_{k+1} = \begin{pmatrix} x \\ y \\ z \end{pmatrix}_k + \begin{pmatrix} r\sin\varphi\cos\theta \\ r\sin\varphi\sin\theta \\ r\cos\varphi \end{pmatrix}; \tag{4}$$

Step 4. If $f(X_{k+1}) < f(X_k)$ then set $X_k = X_{k+1}$, and go to step 2.

Step 5. Azimuth strategy: if θ reach to the maximum value 2π go to step 6, Otherwise increasing the angle $\theta \rightarrow (\theta = \theta + \varepsilon_\theta | \varepsilon_\theta > 0)$ then go to step 3.

Step 6. Inclination strategy: if φ reach to the maximum value π go to step 7, Otherwise increasing the angle $\varphi \rightarrow (\varphi = \varphi + \varepsilon_\varphi | \varepsilon_\varphi > 0)$ then go to step 3.

Step 7. Radius strategy: if r reach to the minimum value (zero) go to step 8. Otherwise decreasing the angle $r \rightarrow (r = r - \varepsilon_r | \varepsilon_r > 0)$ then go to step 3.

Step 8. X_k is optimum and stop the procedures; where there is no better solution than X_k in the spherical region. The pseudo code of the proposed SLS is shown in Fig. 1.

3 Numerical Results

The performance of the proposed algorithm is evaluated for global optimization by testing 25 unconstrained test problems which appeared in special session on real parameter optimization (CEC'2005). This suite is composed of the following functions [16].

- Five unimodal functions
  - F1: Shifted Sphere Function.
  - F2: Shifted Schwefel's Problem 1.2.
  - F3: Shifted Rotated High Conditioned Elliptic Function.
  - F4: Shifted Schwefel's Problem 1.2 with Noise in Fitness.
  - F5: Schwefel's Problem 2.6 with Global Optimum on Bounds.
- 20 multimodal functions
  - Seven basic functions.
    - F6: Shifted Rosenbrock's Function.
    - F7: Shifted Rotated Griewank Function without Bounds.
    - F8: Shifted Rotated Ackley's Function with Global Optimum on Bounds.
    - F9: Shifted Rastrigin's Function.
    - F10: Shifted Rotated Rastrigin's Function.

- F11: Shifted Rotated Weierstrass Function.
- F12: Schwefel's problem 2.13.

– Two expanded functions.
  - F13: Expanded Extended Griewank's plus Rosenbrock's Function (F8 and F2)
  - F14: Shifted Rotated Expanded Scaffers F6.
– 11 hybrid functions (F15 to F25), each one has been defined through compositions of 10 out of the 14 previous functions (different in each case).

In all of these functions, their optima can never be found in the center of the search domain, the optima cannot be found within the initialization range, and the domain of search is not limited i.e. the optimum is out of the range of initialization.

SLS is coded in MATLAB 7.0, and the simulations have been executed on an Intel Core (TM)i7-4500cpu 1.8GHZ 2.4 GHz processor.

Start with $X_k = (x, y, z)_k \in F \subseteq S$

While X_k is not optimum?

  Set $\theta = 0$ and $\varphi = 0$

  Determine the radius r

  Determine the new solution X_{k+1}

  If $f(X_{k+1}) < f(X_k)$ set $X_k = X_{k+1}$, go to 1.

  Inclination strategy: If $\theta = 2\pi$ go to 6,

    Otherwise set $\theta \leftarrow (\theta = \theta + \varepsilon_\theta | \varepsilon_\theta > 0)$ go to 3.

  Azimuth strategy: If $\varphi = \pi$ go to 7,

    Otherwise set $\varphi \leftarrow (\varphi = \varphi + \varepsilon_\varphi | \varepsilon_\varphi > 0)$ go to 3.

  Radius strategy: If $r = 0$ **End while,**

    Otherwise set $r \leftarrow (r = r - \varepsilon_r | \varepsilon_r > 0)$ go to step 3.

End While

Fig. 1. The pseudo code of spherical local search.

3.1 Results and Discussion

To examine the capability of SLS for unconstrained optimization problems, Table 1 illustrates the comparisons between the solutions obtained by SLS and the global solutions of CEC'2005. As a result from the Table 1, for the problems (F1, F2, F4, F5, F6, F7, F9, F10, F12, F13, F14, and F17), SLS is succeeded in reaching the optimal solution. Also, the obtained solutions of SLS, for the problems (F3, F11, and F16), are near to the optimal solution. But it is failed to obtain the optimal solution to the remaining problems (F8, F15, F18, F19, F20, F21, F22, F23, F24, and P25). So, in most problems, SLS can access the optimal solution.

On the other hand, Figs. 2, 3, 4, 5, 6, 7, 8, 9, 10, 11, 12, 13, 14, 15, 16, 17, 18, 19, 20, 21, 22, 23, 24, 25 and 26 show the convergence curves of all test functions. These figures show that SLS has fastest convergence rate in most test functions specially in the first 20 iterations.

Table 1. Comparison between the global solution and SLS.

Problem	Optimal solution	SLS	Problem	Optimal solution	SLS
F1	−450	−450.0000	F14	−300	−300
F2	−450	−450	F15	120	220
F3	−450	−449.5489	F16	120	120.0002
F4	−450	−450	F17	120	120
F5	−310	−310	F18	10	410
F6	390	390	F19	10	510
F7	−180	−180	F20	10	210.0003
F8	−140	−119.9999	F21	360	560.0004
F9	−330	−330	F22	360	760.0071
F10	−330	−330	F23	360	560
F11	90	90.0008	F24	260	460
F12	−460	−460	F25	260	460
F13	−130	−130			

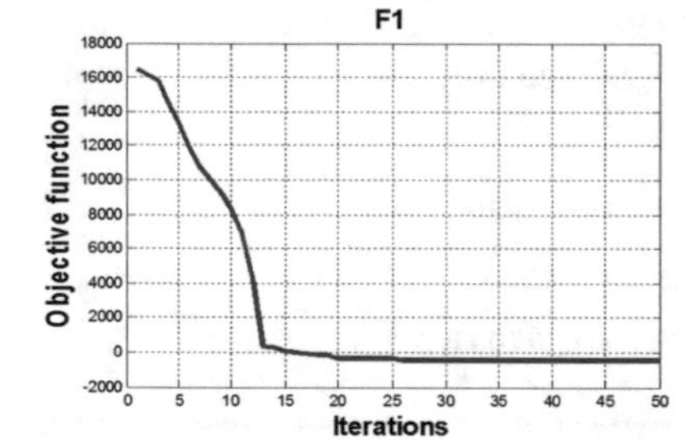

Fig. 2. Convergence curves of F1

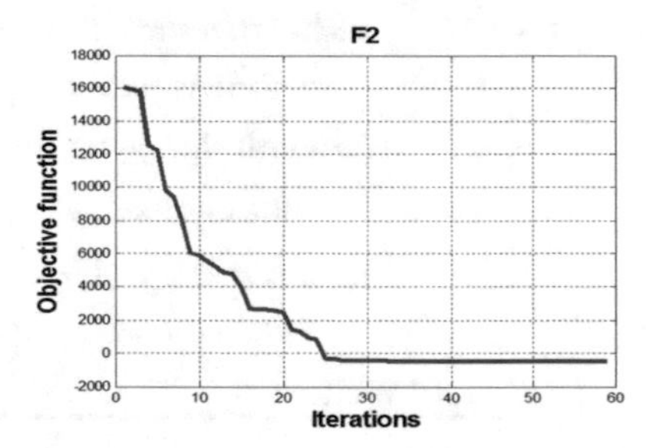

Fig. 3. Convergence curves of F2

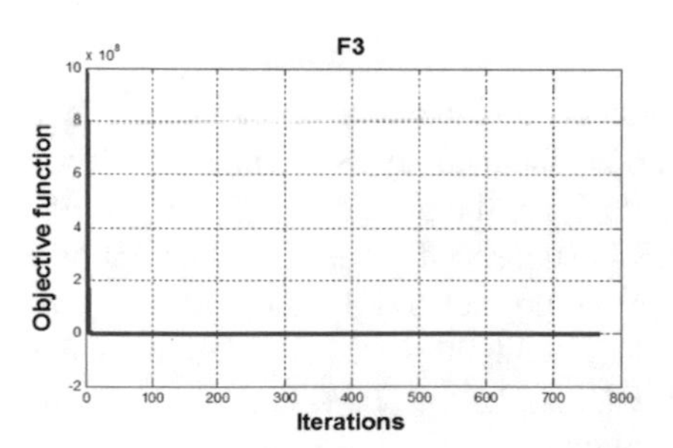

Fig. 4. Convergence curves of F3

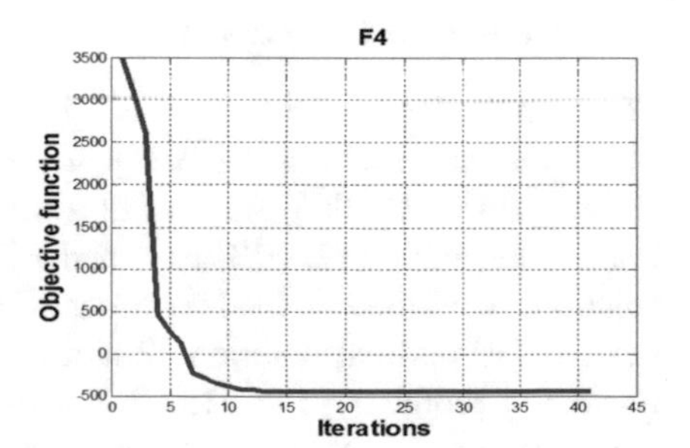

Fig. 5. Convergence curves of F4

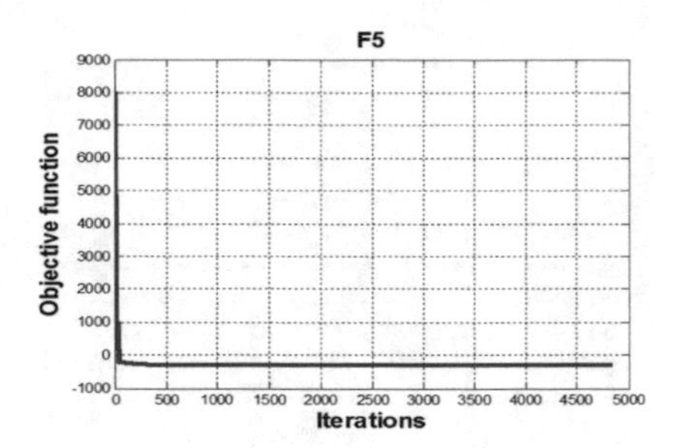

Fig. 6. Convergence curves of F5

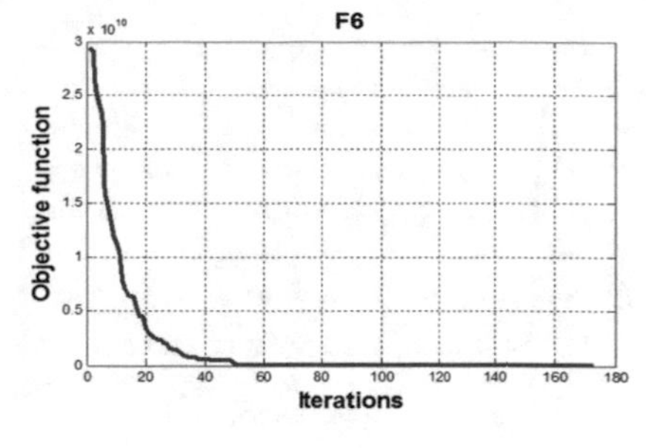

Fig. 7. Convergence curves of F6

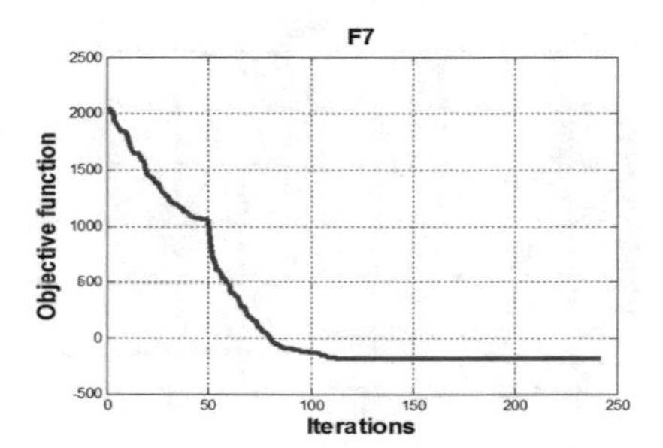

Fig. 8. Convergence curves of F7

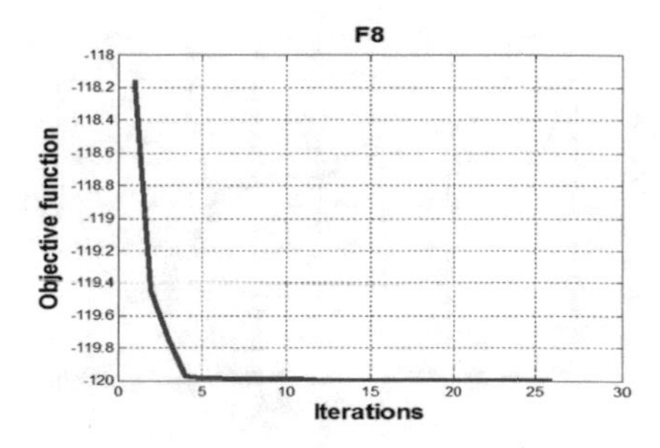

Fig. 9. Convergence curves of F8

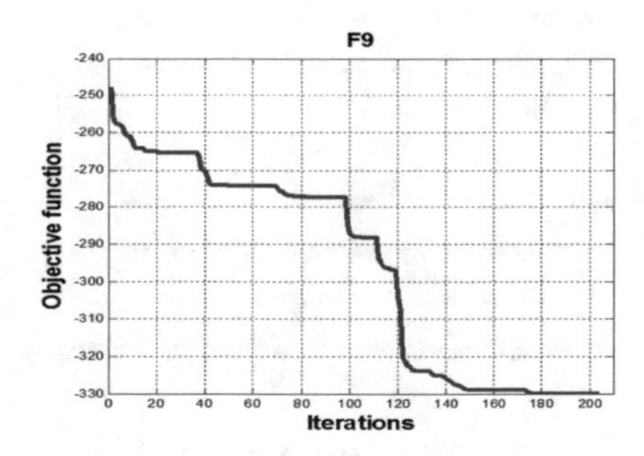

Fig. 10. Convergence curves of F9

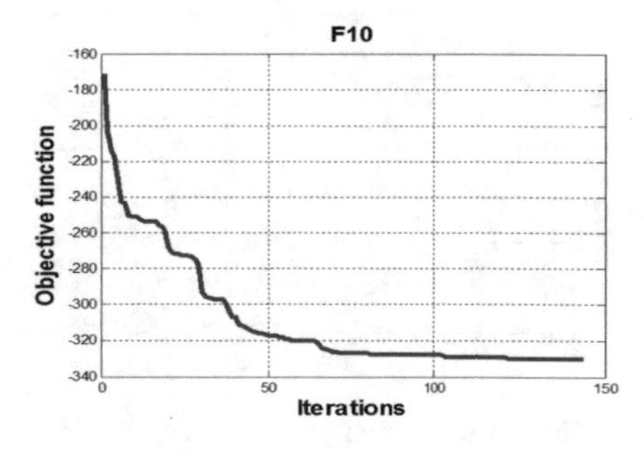

Fig. 11. Convergence curves of F10

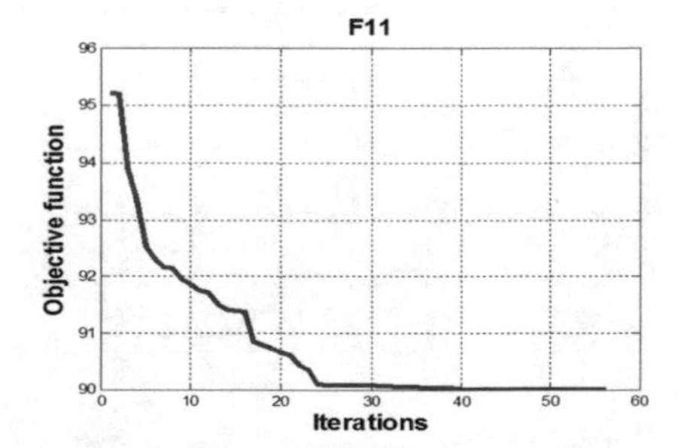

Fig. 12. Convergence curves of F11

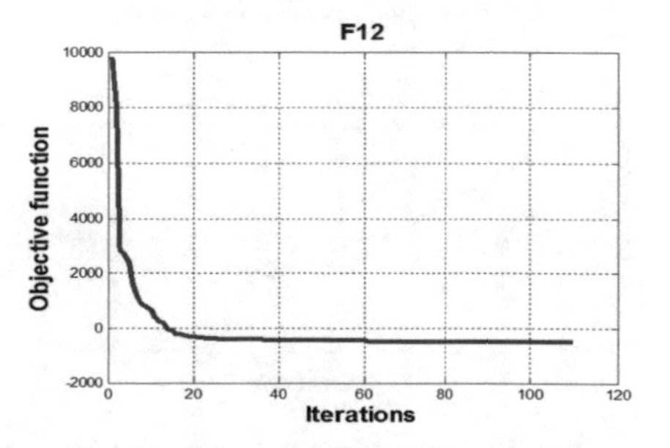

Fig. 13. Convergence curves of F12

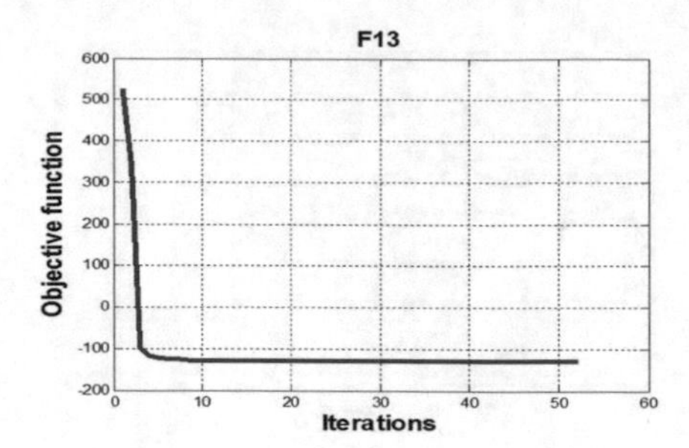

Fig. 14. Convergence curves of F13

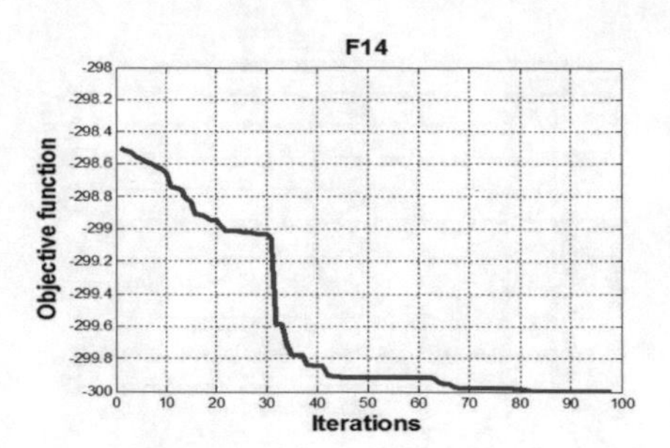

Fig. 15. Convergence curves of F14

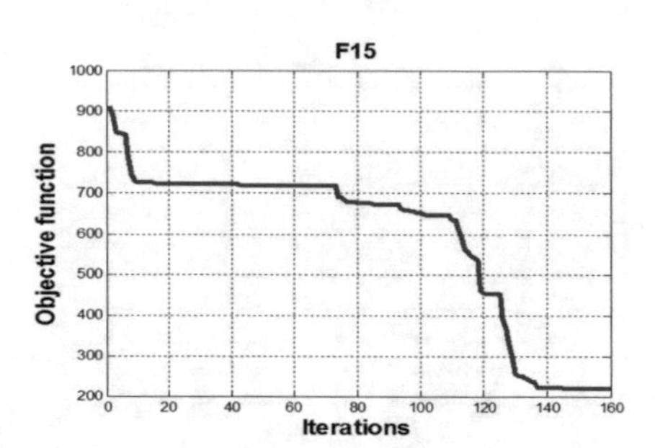

Fig. 16. Convergence curves of F15

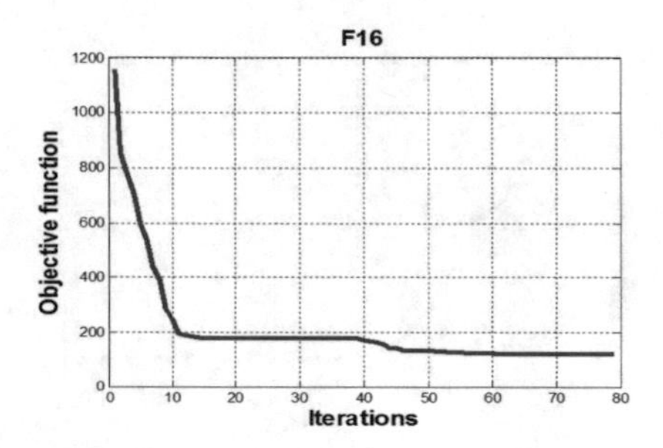

Fig. 17. Convergence curves of F16

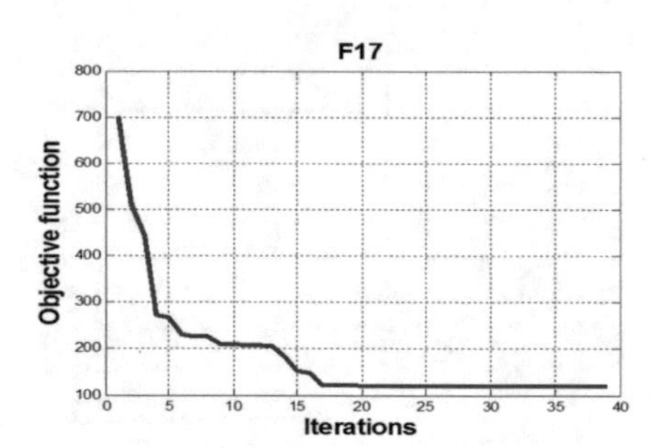

Fig. 18. Convergence curves of F17

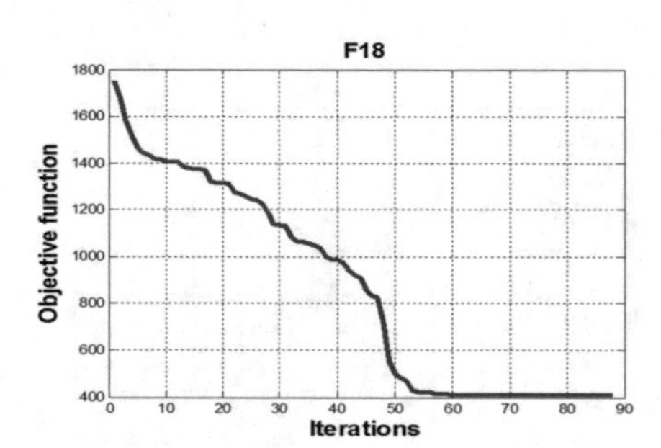

Fig. 19. Convergence curves of F18

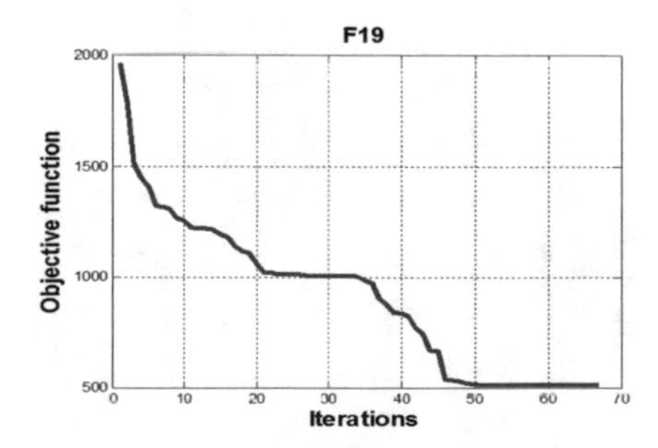

Fig. 20. Convergence curves of F19

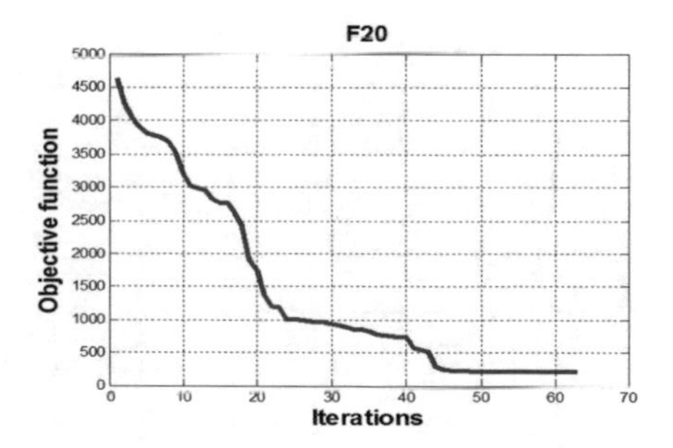

Fig. 21. Convergence curves of F20

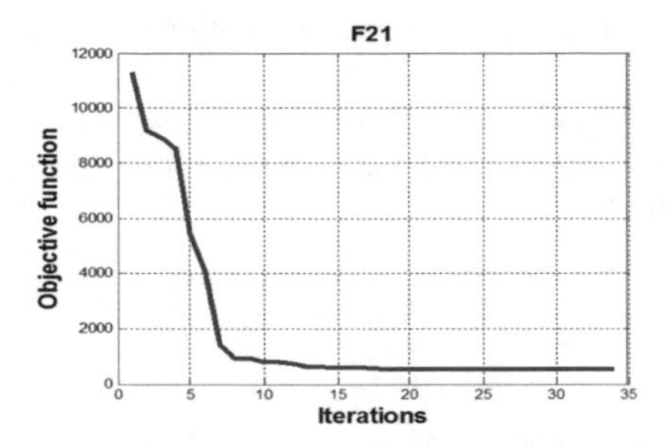

Fig. 22. Convergence curves of F21

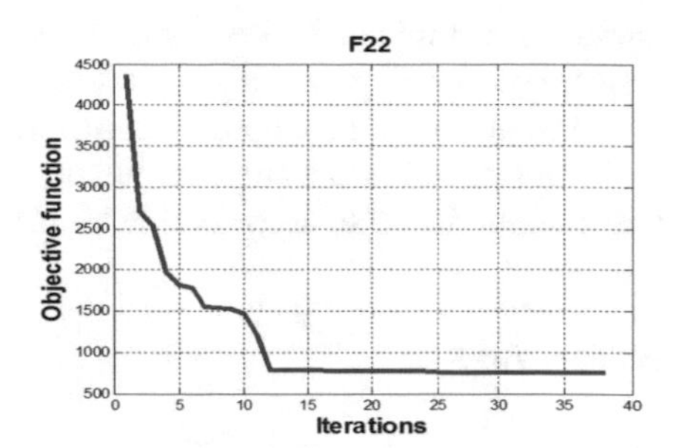

Fig. 23. Convergence curves of F22

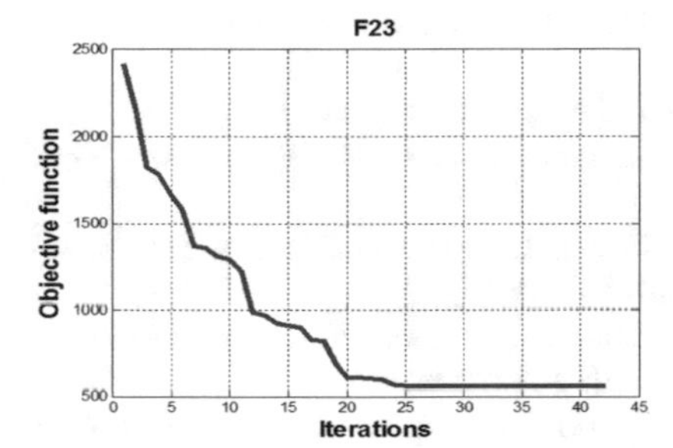

Fig. 24. Convergence curves of F23

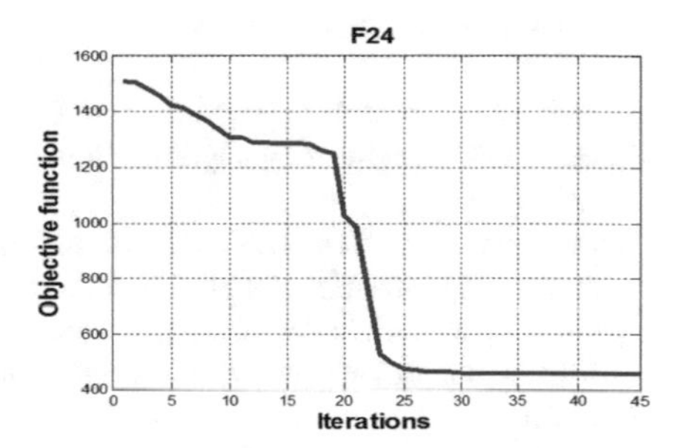

Fig. 25. Convergence curves of F24

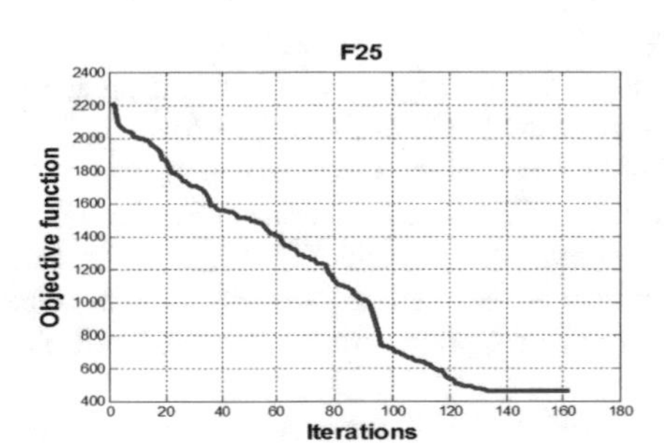

Fig. 26. Convergence curves of F25

4 Conclusion

In this paper spherical local search (SLS) for solving unconstrained optimization problems in three dimensions is proposed. It is based on the spherical coordinates. At first, a point in the search domain is randomly chosen. Secondly, spherical trust region around this point is determined; where any point in this region is feasible. Finally, by using three strategies of search including radius, inclination, and azimuth, SLS can move from current point to obtain a new best one. These strategies are modified during the search process. SLS is tested on the set of the CEC'2005 special session on real parameter optimization. Results show that SLS can access the optimal solution and has fastest convergence rate in most test functions.

In future work, he proposed algorithm will be applied to real world applications and multi-objective optimization problems. Also an extension of the proposed algorithm will be introduced to solve multivariable UOP.

References

1. Dennis, J.E., Schnabel, R.B.: Numerical Methods for Unconstrained Optimization and Nonlinear Equations. Classics in Applied Mathematics, vol. 16. SIAM, Philadelphia (1996)
2. Fiacco, A.V., Mccormick, G.P.: Nonlinear Programming: Sequential Unconstrained Minimization Techniques. Wiley, New York (1968). Reprinted, SIAM, Philadelphia, PA (1990)
3. Nocedal, J., Wright, J.S.: Numerical Optimization. Springer, New York (1999)
4. Rao, S.S.: Engineering Optimization: Theory and Practice, 3rd edn. Wiley, New York (2003)
5. Parouha, R.P., Das, K.N.: A memory based differential evolution algorithm for unconstrained optimization. Appl. Soft Comput. **38**, 501–517 (2016)
6. Dou, J., Wang, X.: An efficient evolutionary programming. In: International Symposium on Information Science and Engineering, vol. 2, pp. 401–404 (2008)
7. Lenin, K., Reddy, B.R., Suryakalavathi, M.: Hybrid Tabu search-simulated annealing method to solve optimal reactive power problem. Int. J. Electr. Power Energy Syst. **82**, 87–91 (2016)
8. Lai, X., Hao, J.-K.: A tabu search based memetic algorithm for the max-mean dispersion problem. Comput. Oper. Res. **72**, 118–127 (2016)
9. Zhou, A.: A genetic-algorithm-based neural network approach for short-term traffic flow forecasting. Adv. Neural Netw. **3498**, 965–969 (2005)
10. Abdelsalam, A.M., El-Shorbagy, M.A.: Optimization of wind turbines siting in a wind farm using genetic algorithm based local search. Renew. Energy **123**, 748–755 (2018)
11. El-Shorbagy, M.A., Mousa, A.A., Fathi, W.: Hybrid Particle Swarm Algorithm for Multiobjective Optimization: Integrating Particle Swarm Optimization with Genetic Algorithms for Multiobjective Optimization, Lambert Academic Publishing, Saarbrücken (2011)
12. El-Shorbagy, M.A.: Hybrid Particle Swarm Algorithm for Multi-Objective Optimization, Master of Engineering Thesis, Menoufia Univ., Egypt (2010)
13. Chena, Z., Zhou, S., Luo, J.: A robust ant colony optimization for continuous functions. Expert Syst. Appl. **81**, 309–320 (2017)
14. Yang, X.-S., Deb, S.: Cuckoo search via Lévy flights. In: 2009 World Congress on Nature & Biologically Inspired Computing (NaBIC), Coimbatore, pp. 210–214 (2009)
15. Cortinhal, M.J., Mourão, M.C., Nunes, A.C.: Local search heuristics for sectoring routing in a household waste collection context. Eur. J. Oper. Res. **255**(1), 68–79 (2016)
16. Suganthan, P.N., Hansen, N., Liang, J.J., Deb, K., Chen, Y.-P, Auger, A., et al.: Problem definitions and evaluation criteria for the CEC'2005 special session on real parameter optimization. Technical report, Nanyang Technological University (2005)

Automatic White Blood Cell Counting Approach Based on Flower Pollination Optimization Multilevel Thresholoding Algorithm

Shahd T. Mohamed[1,3](✉), Hala M. Ebeid[1](✉), Aboul Ella Hassanien[2,3](✉), and Mohamed F. Tolba[1](✉)

[1] Faculty of Computer and Information Sciences, Ain Shams University, Cairo, Egypt
shahd.tmohamed@gmail.com, {halam,fahmytolba}@cis.asu.edu.eg
[2] Faculty of Computers and Information, Cairo University, Giza, Egypt
srge1964@gmail.com
[3] Scientific Research Group in Egypt (SRGE), Giza, Egypt
http://www.egyptscience.net

Abstract. This paper presents an swarm optimization approach based on flower pollination optimization algorithm for multilevel thresholding by the criteria of Otsu minimizes the weighted within-class variance to make the optimal thresholding more effective. An application of microscopic white blood cell imaging has been chosen and the proposed approach has been applied to see their ability and accuracy to segment and count the blood cells. An adaptive watershed segmentation algorithm was applied that depends on a mask created from the required microscopic image to detect the minima points for segmenting the overlapped cells. The cell counting process depends on labeling the connected regions of the segmented binary image and count the labeled cells. The proposed approach archives promised results with respect to quality measures of accuracy, peak to signal-to-noise ratio (PSNR) and the root mean square error (RMSE) on microscopic images. Experimental results are recorded for the proposed approach over ten selected different images with accuracy of 98.4% that present better accuracy over the manual traditional techniques.

Keywords: Blood microscopic image · White blood cells
Multilevel thresholding · Watershed · Segmentation techniques

1 Introduction

Blood contains Plasma, Red Blood Cells, White Blood Cells and Platelets. Changing in blood cells' conditions or count shows the incompetence of the blood system, that helps in the diagnosis of blood cells disorder and indicates

A. E. Hassanien et al. (Eds.): AISI 2018, AISC 845, pp. 313–323, 2019.
https://doi.org/10.1007/978-3-319-99010-1_29

the development of diseases, as each type of the cells has its function in the body. This paper studies an algorithm for WBC types such as Neutrophils, Lymphocytes, Monocytes, Eosinophils and Basophils, as any change on shapes or number of WBC presents a disease, as the increase of the number of the cells indicate Cancer while the decrease reflects the viral infection. The shape of White Blood Cells is deformed because of cancers and their treatment with some medication [1–3]. Segmentation of WBC helps in counting and selecting the cells for further operations such as feature extraction and detection of infected and abnormal cells to determine the disease accurately but the segmentation and counting of complicated microscopic images represent a difficulty and have many research challenges such as the existing of particles on the background, cells' shape and texture, overlapping and close cells [4]. Researchers tackled segmentation and counting of WBC using traditional techniques such as thresholding, edge detection, region growing, morphological operations and Watershed with no efficient quality measures of accuracy and time [5]. WBC are considered an early signal of many blood diseases so it is important to segment these cells for analyzing the changes in texture, geometry and color and extracting useful information that leads to accurate and fast diagnoses [3]. The basic segmentation techniques are divided into main categories such as edge detection, region based segmentation, thresholding, Watershed and clustering [6,7].

Sobhy et al. [8] compared between the Otsu threshold and Watershed marker controlled segmentation methods and proved that the Watershed marker controlled out-perform the Otsu threshold with accuracy 99.3%. The comparison is done on 33 images and uses color correction technique as preprocessing and uses the shape features to mark single and grouped cells. It finally uses morphological operations to clean the image. However, the authors didn't mention if their experiment succeeded in segmenting the morphology of white blood cells such as Eosinophils and Basophils that have many nucleus and holes. Vaghela et al. in [9] proposed a technique for segmentation and counting Leukemia cells combining thresholding method and morphological operation and extracted some shape features such as area and perimeter to detect cells. They also compared between Watershed transform, k-mean clustering, edge detection and shape based features and concluded that shape based features give the best performance with accuracy 97.8%. Wei and Kangling in [10] presented a hybrid algorithm based on Multilevel Thresholding using particle swarm optimization algorithm, they compared their new algorithm with traditional Otsu method and the results proved that Multilevel Thresholding can be improved using PSO in the aspect of threshold selection and time. Allaoui and Barek in [11] proposed a marker-controlled Watershed technique that can solve the over segmentation problem of Watershed by generating markers by morphological reconstruction and image complementing that reduced the segmented region to achieve good performance. The algorithm was performed on many medical images and proved its efficiency.

Diego et al. in [12] introduced Multilevel Thresholding using harmony search algorithm to segment digital images. They used Otsu and Kapur functions and prove that Otsu function performs better than Kapur in their technique, and compared the new technique with Genetic Algorithms (GA), Particle Swarm Optimization (PSO) and Bacterial Foraging (BF) in the aspect of Peak to Signal-to-Noise Ratio (PSNR) and Standard Deviation (STD). The proposed technique proved its efficiency in performance and accuracy compared to the other techniques. Benson et al. in [13] reviewed and compared between the methods of detecting the markers for marker-controlled Watershed technique for different kind of medical image segmentation such as clustering based method, thresholding based method and morphology based method. The experiment shows that morphology based method outperforms the rest of the methods with 93.75% accuracy. The authors suggest using Neural Networks. Abdeldaim et al. [14] proposed an automated system to segment and classify White Blood cells to detect Leukemia. The authors used CMYK color conversion technique, Zack thresholding algorithm for segmentation and extract color, shape and texture features to classify the cells using different classifiers. The system proved its efficiency in segmentation and classification for normal and acute Leukemia cells.

This paper presents a new segmentation and counting algorithm that overcomes the drawback of the traditional techniques for WBC based on Multilevel Thresholding based on Flower Pollination Algorithm to extract White Blood Cells and modified Watershed algorithm to separate the connected cells accurately and count cells with higher accuracy results than traditional models The rest of the paper is organized as the follows. Section 2 gives a brief introduction to the Multilevel thresholding, flower pollination algorithm (FPA), and watershed algorithm. Section 3 presents the methodology and proposed counting algorithm. Section 4 discusses the experiment and results. Finally, Sect. 5 contains the conclusion.

2 Preliminaries

This section provides a brief explanation of the basic framework of segmentation and counting a white blood cells, Multilevel thresholding, flower pollination algorithm (FPA), and watershed algorithm, along with some of the key basic concepts.

2.1 Multilevel Thresholding: Otsu Algorithm

Otsu multilevel thresholding algorithm splits the image into m classes and calculate the thresholds by maximizing the variant between the classes to determine the number of optimum thresholds that minimizes the weighted within-class variance.

$$\sigma_B^2 = \sum_{j=0}^{m} w_j(\mu_j - \mu_t)^2 \tag{1}$$

where w is the classes probabilities and calculated by

$$w_j = \sum_{j=0}^{L} p_j, \quad Where \quad L = 255 \tag{2}$$

μ is the mean class levels and calculated by

$$\mu_j = \sum_{j=0}^{L} jp_j/w_j \tag{3}$$

μ_t is the total mean levels of the original image.

2.2 Flower Pollination Algorithm (FPA)

The Flower Pollination Algorithm (FPA) is developed by Yang [15]. FPA is a nature inspired algorithm based on pollination process which aims to optimally reproduce plants by keeping the best flowering species by pollination. The pollen is transferred by a Biotic process (cross pollination) and Abiotic Process (self-pollination). In the Biotic process, pollens transferred by insects, bats and birds between flowers in the different plants. It considered as a global pollination (Levy distribution) [16].

$$x_i^{t+1} = x_i^t + L(g^* - x_i^t) \tag{4}$$

where L is Levy distribution and $g*$ is the best solution.

While the Abiotic process pollinates the flower by the pollen from the same plant [17]. In this method, wind and the spread in water are the main pollinators and considered as local pollination.

$$x_i^{t+1} = x_i^t + \epsilon(x_j^t - x_k^t) \tag{5}$$

where $\epsilon \in [0, 1]$

2.3 Watershed Algorithm

Watershed algorithm is considered one of the region growing techniques that is based on image gradient and topographic surface. This algorithm is used in separating the connected components. The major problem of Watershed algorithm is the over segmentation because it detects many regions as there are many minima in the image rather than region of interest [11,18]. Marker controlled Watershed algorithm helps in solving the over segmentation problem by creating internal markers for interested objects and external markers for the background [11,19].

The main purpose of the markers is to detect the accurate minima in the images for the Watershed processing. Researchers present alternative solutions to modify the techniques of computing the markers, but morphological methods proved their efficiency compared to thresholding and clustering techniques [13].

Marker-controlled-Watershed algorithm outperforms the Watershed algorithms using many different medical image applications. However, it suffers from low quality performance with morphology for white cells' nucleus and this may cause false detection of the segmented regions [11,18,20].

3 The Proposed Automatic White Blood Cell Counting Approach

Thresholding algorithm is one of the most important approaches to image segmentation but Multilevel Thresholding is less accurate than single level thresholding because it is difficult to determine thresholds that separate objects of interest sufficiently. Hybrid Multilevel Thresholding algorithm based on inspired swarm techniques helps in selecting the thresholds accurately and improves the Multilevel Thresholding performance in the segmentation process. Watershed algorithm is considered a high accurate segmentation algorithm for connected components. However, it suffers from over segmentation. So marker-controlled Watershed algorithm tried to overcome this problem by detecting the appropriate marker that can reduce the marked region for better accuracy. The proposed segmentation and counting approach is comprised of the following two fundamental building phases:

- **Extract the white blood cells.** In the first phase, the multi-level thresholding based on flower pollination algorithm is used to extract the white blood cells
- **Counting.** In the second phase, an adaptive watershed algorithm is used to separate and count the connected and overlapped cells.

These two phases are described in detail in the following section along with the steps involved and the characteristics feature for each phase. Figure 1 also describe the overall architecture of the proposed approach.

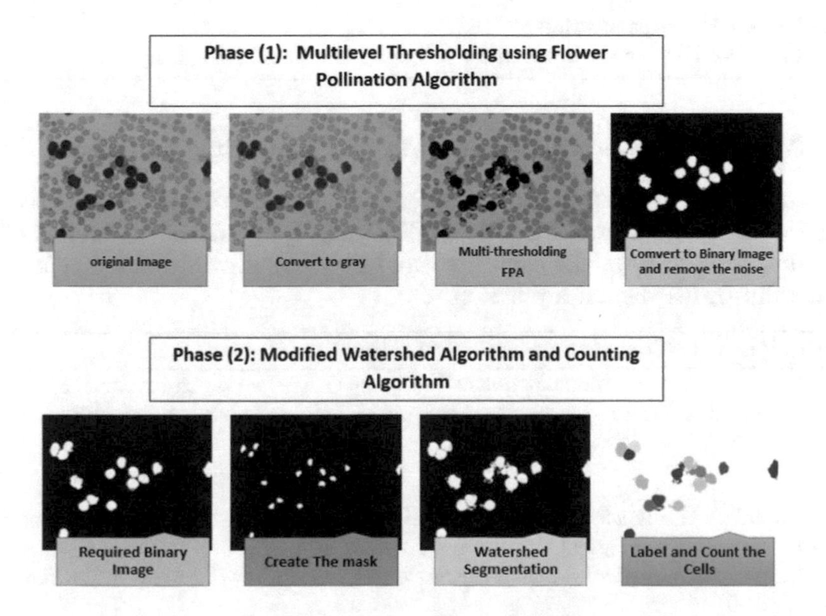

Fig. 1. The overall architecture of the proposed approach

3.1 Multilevel Thresholding Using Flower Pollination Algorithm Phase

Multilevel Thresholding algorithm has high computational complexity that affects its efficiency, so applying Flower Pollination Algorithm helps in solving the problem using Otsu method function as fitness function.

$$Fitness = \sum_{j=0}^{m} w_j(\mu_j - \mu_t)^2 \tag{6}$$

Algorithm 1 illustrates the Multilevel thresholding based on FPA Algorithm.

Algorithm 1. Multilevel thresholding based on FPA Algorithm

1- Convert original image to gray scale image and calculate the histogram;
2- Initialize the FPA parameters: iteration number (i), population size randomly (p) and acceptance rate (k) ;
3- Evaluate the solution by otsu fitness function ;
4- **while** $n < i$ **do**
    if $r > k$; `// r is random number in the range of [0,1]`
    then
        apply FPA global pollination in creating new solution;
    else
        apply FPA local pollination in creating new solution;
    end
    evaluate the new solution by otsu fitness function ;
    rank the solutions to find the best solution ;
end
4- Update the best solution ;
5- Generate the binary image after selecting the best histogram;

3.2 Modified Watershed Algorithm and Counting Algorithm Phase

The proposed Watershed algorithm depends on a mask created from the required image to detect the minima points for Watershed algorithm. The result of this algorithm is separating the connected and overlapped white blood cells. Algorithm 2 illustrates the mask creation.

Algorithm 2. Create the Mask Algorithm

1- Calculate the complement quasi-Euclidean distance transform for the complemented required image, where quasi-Euclidean distance calculated as ;

$$D = |x_i + x_{i+1}| + (\sqrt{2} - 1)|y_i + y_{i+1}| \tag{7}$$

2- Calculate the mask by extended minima for the quasi-Euclidean distance;
3- Perform morphological operation on the mask ;
4- Using morphological reconstruction to get the appropriate region for the watershed;
5- Perform watershed with the mask to segment the connected components;

The counting algorithm depends on labeling the connected regions of the segmented binary image and count the labeled cells. The result of this algorithm is counting the segmented white blood cells. Algorithm 3 illustrates the counting algorithm.

Algorithm 3. Count the segmented cells Algorithm

1- Label the connected component in the binary image ;
2- Convert the label matrix into rgb image to visualize the labeled cells;
3- Calculate the x and y coordinate the center of each labeled cell as ;

$$x = \sum_{j=0}^{m} x_j/m \quad , \quad y = \sum_{j=0}^{m} y_j/m \tag{8}$$

4- Count the coordinate of the labeled cells;

---

4 Experimental Results

The proposed FPA-Multilevel Thresholding algorithm compares with HM-Multilevel Thresholding. The experiment was implemented by "MATLAB 2013" on windows 8, 4 GB RAM and performed on samples of microscopic images that contain all morphological white cells; Neutrophils, Lymphocytes, Monocytes, Eosinophils and Basophils cells, connected, overlapped, infected and deformed cells. The algorithm detects and counts the cells at the boundary. Figure 2 show samples of blood microscopic images. The parameter values are population number (n) = 25 and number of iteration = 200. The experiment performance is measured using Peak to Signal-to-Noise Ratio (PSNR), the Root Mean Square Error (RMSE) and the count of white blood cells.

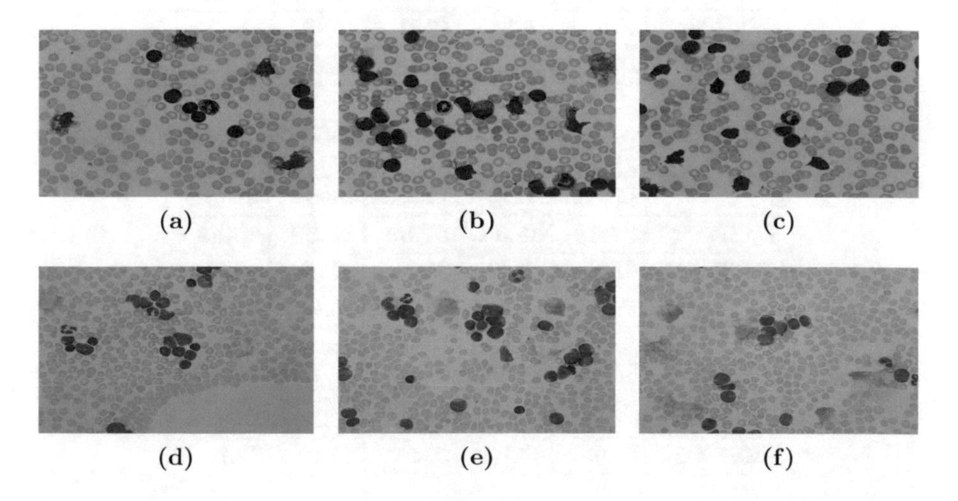

Fig. 2. Samples of blood microscopic images

Tables 1, 2 and 3 illustrate the results of the experiment in the aspect of count, (PSNR) and (RMSE). Figure 3 shows the segmentation and counting process on images 1, 4 and 10 that contain connected cells, leukemia cells and white morphological cells. The results show that the selected weights by FPA give better results in PSNR and less RMSR compared to HM-Multilevel Thresholding algorithm [12] and PSO-Multilevel Thresholding algorithm.

Table 1. Experiment results of proposed algorithm in the aspect of count

	Image1	Image2	Image3	Image4	Image5
Count-FPA	**12/12**	**35/36**	**24/24**	**20/20**	**5/5**
Count-HM	12/12	35/36	25/24	20/20	5/5
Count-PSO	12/12	39/36	24/24	20/20	5/5
	Image6	Image7	Image8	Image9	Image10
Count-FPA	**2/2**	**23/24**	**33/33**	**18/17**	**30/29**
Count-HM	2/2	23/24	33/33	18/17	30/29
Count-PSO	3/2	23/24	35/33	18/17	28/29

Table 2. Experiment results of proposed algorithm in the aspect of PSNR

	Image1	Image2	Image3	Image4	Image5
PSNR-FPA	**19.48**	**16.90**	**16.74**	**18.33**	**20.04**
PSNR-HM	17.90	17.11	17.71	18.13	18.86
PSNR-PSO	13	12.88	13.02	12.86	11.88
	Image6	Image7	Image8	Image9	Image10
PSNR-FPA	**16.05**	**20.03**	**18.26**	**19.20**	**18.29**
PSNR-HM	16.62	19.73	17.81	19.17	17.73
PSNR-PSO	13.33	14.55	13.13	14.30	14.02

Table 3. Experiment results of proposed algorithm in the aspect of RMSE

	Image1	Image2	Image3	Image4	Image5
RMSE-FPA	**27.18**	**36.44**	**37.32**	**30.98**	**25.58**
RMSE HM	31.59	35.62	33.23	31.69	29.56
RMSE-PSO	57.05	57.09	56.90	58.04	64.89
	Image6	Image7	Image8	Image9	Image10
RMSE-FPA	**40.13**	**25.46**	**31.40**	**28.05**	**31.09**
RMSE-HM	37.67	26.56	32.82	28.12	33.22
RMSE-PSO	54.95	47.76	56.25	49.15	50.77

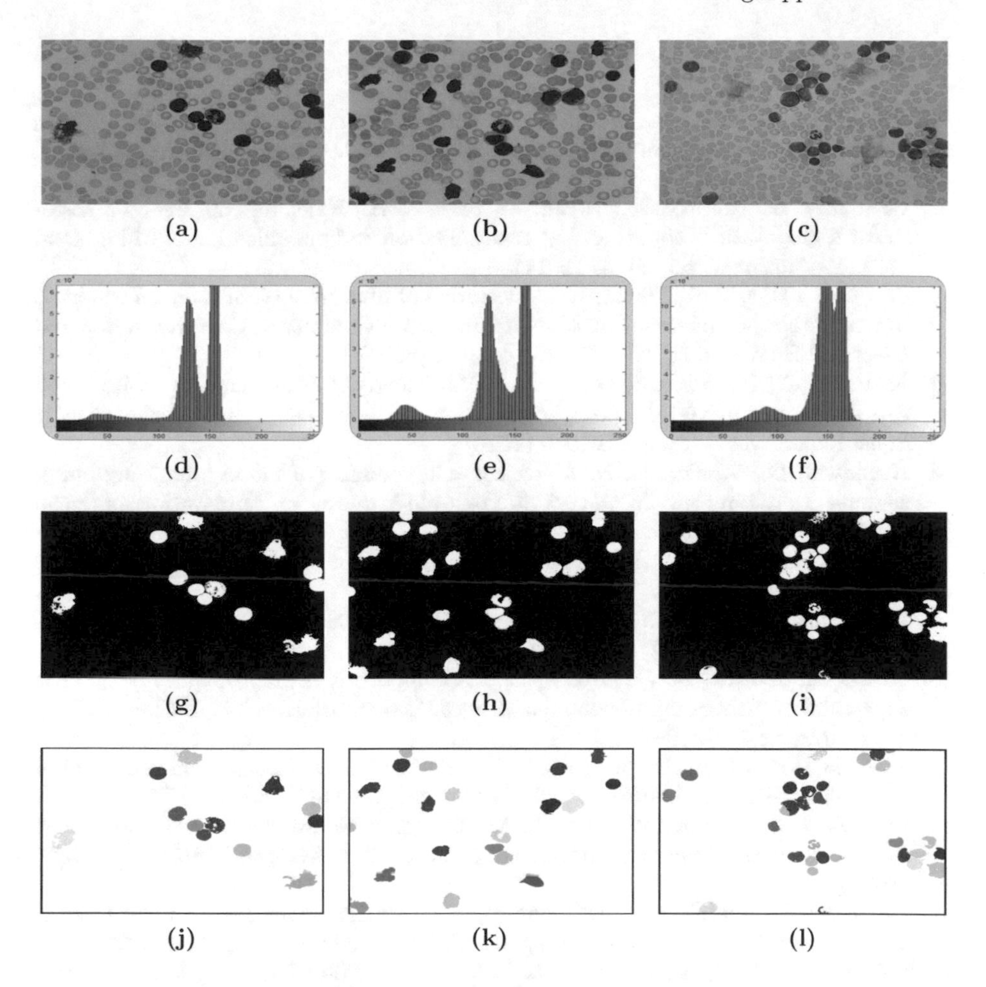

Fig. 3. Segmentation result sample of WBC microscopic images 1, 4 and 10.

5 Conclusion and Future Work

This paper proposes an efficient algorithm for segmentation and counting of White Blood Cells depending on Multilevel Thresholding based on Flower Pollination Algorithm and modified Watershed segmentation based on mask. The algorithm can detect and count the morphology of White Blood Cells that have holes and many nucleus, and can also detect the infected cells in addition to keeping the original size and shape of the cells with 98.4% accuracy. In the future, we will focus more on White Blood Cells that contain more than one nucleus and try to enhance the original images before segmentation process.

References

1. Malik, H., Randiwe, R., Patankar, J., Bhure, P.: Disease diagnosis using RBCs and WBCs cell structure by image processing. Int. J. Sci. Res. Sci. Technol. **3**(2), 2395–6011 (2017)
2. Nazlibilek, S., Karacor, D., Ercan, T., Sazli, M.H., Kalender, O., Ege, Y.: Automatic segmentation, counting, size determination and classification of white blood cells. Measurement **55**, 58–65 (2014)
3. Jha, K.K., Das, B.K., Dutta, H.S.: Detection of abnormal blood cells on the basis of nucleus shape and counting of WBC. In: Green Computing Communication and Electrical Engineering (ICGCCEE), pp. 1–5 (2014)
4. Fatichah, C., Purwitasari, D., Hariadi, V., Effendy, F.: Overlapping white blood cell segmentation and counting on microscopic blood cell images. Int. J. Smart Sens. Intell. Syst. **7**(3), 1271–1286 (2014)
5. Kolhatkar, D., Wankhade, N.: Detection and counting of blood cells using image segmentation: a review. In: Futuristic Trends in Research and Innovation for Social Welfare (Startup Conclave), pp. 1–5 (2016)
6. Rane, K.P., Zope, G.D., Rane, A.: Blood cell segmentation–a review. Int. J. Adv. Electron. Commun. Syst. (2014)
7. Patil, D.D., Deore, S.G.: Medical image segmentation: a review. Int. J. Comput. Sci. Mob. Comput. **2**(1), 22–27 (2013)
8. Salem, N., Sobhy, N.M., El Dosoky, M.: A comparative study of white blood cells segmentation using otsu threshold and watershed transformation. J. Biomed. Eng. Med. Imaging **3**(3), 15–24 (2016)
9. Vaghela, H.P., Modi, H., Pandya, M., Potdar, M.: Leukemia detection using digital image processing techniques. Int. J. Appl. Inf. Syst. **10**(1), 43–51 (2015)
10. Liu, Y., Mu, C., Kou, W., Liu, J.: Modified particle swarm optimization-based multilevel thresholding for image segmentation. Soft Comput. **19**(5), 1311–1327 (2015)
11. ElAllaoui, A., Nasr, M.B.: Medical image segmentation by marker-controlled watershed and mathematical morphology. Int. J. Multimedia Appl. **4**(3), 1–9 (2012)
12. Oliva, D., Cuevas, E., Pajares, G., Zaldivar, D., Perez-Cisneros, M.: Multilevel thresholding segmentation based on harmony search optimization. J. Appl. Math. (2013)
13. Cc, B., Rajamani, K., Lajish, V.: A review on automatic marker identification methods in watershed algorithms used for medical image segmentation. IJISET - Int. J. Innov. Sci. Eng. Technol. **2**(9), 829–839 (2015)
14. Abdeldaim, A.M., Sahlol, A.T., Elhoseny, M., Hassanien, A.E.: Computer-aided acute lymphoblastic leukemia diagnosis system based on image analysis. In: Advances in Soft Computing and Machine Learning in Image Processing, pp. 131–147. Springer (2018)
15. Yang, X.S.: Flower pollination algorithm for global optimization. In: International Conference on Unconventional Computing and Natural Computation, pp. 240–249. Springer (2012)
16. Chiroma, H., Shuib, N.L.M., Muaz, S.A., Abubakar, A.I., Ila, L.B., Maitama, J.Z.: A review of the applications of bio-inspired flower pollination algorithm. Proc. Comput. Sci. **62**, 435–441 (2015)
17. Yang, X.S., Karamanoglu, M., He, X.: Multi-objective flower algorithm for optimization. Proc. Comput. Sci. **18**, 861–868 (2013)

18. Benson, C., Lajish, V., Rajamani, K.: Brain tumor extraction from MRI brain images using marker based watershed algorithm. In: Advances in Computing, Communications and Informatics (ICACCI), pp. 318–323 (2015)
19. Gaikwad, V.J.: Marker-controlled watershed transform in digital mammogram segmentation. Int. J. Res. Appl. Sci. Eng. **3**, 18–21 (2015)
20. Hassan, M., Eko, N., Shafie, S.: Tuberculosis bacteria counting using watershed segmentation. Pertanika J. Sci. Technol. **25**, 275–282 (2017)

Deep Learning and Cloud Computing

Improving Citation Sentiment and Purpose Classification Using Hybrid Deep Neural Network Model

Abdallah Yousif[1(✉)], Zhendong Niu[1,2], Ally S. Nyamawe[1], and Yating Hu[1]

[1] School of Computer Science and Technology, Beijing Institute of Technology, Beijing, China
abdosif33@hotmail.com, zniu@bit.edu.cn, nyamawe@udom.ac.tz, huuting@hotmail.com
[2] School of Computing and Information, University of Pittsburgh, Pittsburgh, PA, USA

Abstract. Automated citation classification has received much attention in recent years from the research community. It has many benefits in the bibliometric field such as improving the methods of measuring publications' quality and productivity of the researchers. Most of the existing approaches are based on supervised learning techniques with discrete manual features to capture linguistic cues. Though these approaches have reported good results, extracting such features are time-consuming and may fail to encode the semantic meaning of the citation sentences, which consequently limits the classification performance. In this paper, a hybrid neural model is proposed, which combines convolutional and recurrent neural networks to capture local n-gram features and long-term dependencies of the text. The proposed model extracts the features automatically and classifies the sentiments and purposes of scientific citations. We conduct experiments using two publicly available datasets and the results show that our model outperforms previously reported results in terms of precision, recall, and F-score for citation classification.

Keywords: Recurrent neural network · Convolution · Citation sentiment Citation purpose · Citation classification

1 Introduction

With the rapid growth of academic databases which contain many publications, it is important to measure the quality of the published papers. Numerous citation impact indicators have been introduced for such purpose. For example, H-index has been used to measure the productivity of the researchers, and the impact factor has been used to assess the quality of the journals. However, these measurements are somehow unfair because they depend on the citations frequency only, which give equal weights to all citations and ignore the sentiments and purposes of the citations. To overcome this problem, Automated Citation Classification (ACC) provides an effective solution by considering the sentiments and purposes of citations rather than treating all citations

A. E. Hassanien et al. (Eds.): AISI 2018, AISC 845, pp. 327–336, 2019.
https://doi.org/10.1007/978-3-319-99010-1_30

equally. ACC is an emerging area of research in natural language processing that detects the citation purpose and sentiment in the citing papers as presented in the citation context [1]. Citation context is a sentence or number of sentences that convey authors' opinions and purposes about the cited works [2]. Citation sentiment and citation purpose classification are defined by Abu-Jbara et al. [3] as the main tasks of ACC for identifying the author's sentiment and purpose in citing works respectively.

The previous works on ACC are generally based on two techniques. The first one uses machine learning techniques, for example, Support Vector Machine (SVM), Naive Bayes (NB), and bayesNet classifiers with syntactic and semantic features to address citation sentiment and purpose classification [3–9]. Although they achieved good classification performance, they still suffer from the time-consuming feature engineering process and the complexity of choosing the optimal features for performing ACC. Also, it is difficult to compare their reported results due to the use of different citation schemes for the classification [10]. The second approach employed deep learning techniques for handling ACC tasks. For example, Jochim and Schutze [11] addressed the citation sentiment classification as domain adaptation based on the use of the Boltzmann machine algorithm. They used product review datasets to train the method and tested it in a citation dataset and improved the classification accuracy. Lauscher et al. [12] proposed a model based on convolution neural network (CNN) for citation sentiment and purpose classification. They used a public citation scheme which has six and three categories for citation purpose or function and sentiment respectively. Their model outperformed classical classifiers and achieved F-score of 0.79. Although their proposed CNN model manages to capture local n-gram features, however, the model fails to consider long-term relations in the text.

In this paper, we propose a hybrid neural network model based on CNN and LSTM (long short-term memory) network to classify sentiment and purpose of citations. The proposed model (CNN-BiLSTM) can handle the feature engineering process and capture local n-gram features and long sequence dependencies of the input citation context. To the best of our knowledge, the proposed approach is the first attempt to combine CNN and LSTM for performing citation sentiment and purpose classification. The main contribution of our work is to address the limitation of feature extraction process by using a hybrid deep learning approach. In addition, our study sheds light on a new way to improve the traditional citation impact indicators for evaluating the quality of publications. To evaluate our model, we conducted experiments by using two public datasets, and the experimental results show that our model outperforms the baseline models.

The rest of this paper is structured as follows: Sect. 2 presents the related work, Sect. 3 shows the proposed model, and Sect. 4 presents the experiments and evaluation. Finally, in Sect. 5, we conclude the paper and recommend future work.

2 Related Work

Several techniques for handling citation purpose and sentiment classification tasks have been proposed in the literature. For the Citation Purpose Classification (CPC), Nanba and Okumura [13] used a rule-based approach with the aim of reducing the functions into three functions to ensure easy annotation of the citations. Teufel and Siddharthan [1] randomly selected 116 articles and mapped the existing twelve citation functions into four categories to improve annotating the citations. Moreover, Nakagawa et al. [14] employed a supervised model based on the conditional random fields to detect citation functions in the Japanese language. Dong and Schäfer [4] proposed to use NB with features such as negation, cue words, POS-tag, location, and popularity. Their methods categorized the citations into four functions. Jochim and Schutze [15] used a maximum entropy classifier and a window size of three sentences, which captured the whole part of the cited work according to the citing work. Further, Meyers [16] proposed a hybrid method by using discourse as a tree model and analyzed Part of Speech (POS) tags to find out citation relations about contrast and corroboration functions. Tsai et al. [17] classified the citations into applications and techniques by using an unsupervised bootstrapping algorithm. Abdullatif et al. [18] proposed a rule-based approach to create citation classification scheme automatically. In their method, semantic role labeling was used to select the relevant verb that represents the citation sentence.

On the other hand, Citation Sentiment Classification (CSC) has been investigated using several techniques. Athar [5] categorized the citation sentences as either positive, negative or neutral by using various features, for example, scientific lexicon, dependency relations, and sentence splitting. SVM and NB were exploited as important classification methods and the author reported good classification results with SVM compared to NB. Likewise, to study citation sentiment detection, Athar and Teufel [19] recognized implicit citations as additional words with explicit citations and stated that extracting more opinionated words can help to improve the classification performance. Abu-Jbara et al. [3] applied SVM with negations, dependency relations and other features for citation classification and reported the accuracy of 81.4%.

Recently, Hernandez-Alvarez et al. [8] developed annotation scheme for coding the citation sentences to be sentiment or purpose with their functions or classes to handle the citation classification. Keywords and semantic patterns were used to discriminate the sentiment and purpose categories. They manually built a corpus and then used SVM for citation classification. Ma et al. [9] exploited author's reputation information and proposed to use different features such as uni-gram, polarity distribution, author information, and p-index. They reported best classification performance through combining author information and p-index features. Our work is different from the previous approaches as it focuses on the use of deep learning to detect the sentiment and purpose of citations by capturing the features automatically rather than using simple machine learning classifiers with manual extracted features.

3 The Proposed Hybrid Deep Learning Model

In natural language processing (NLP), a standard approach for text modeling is to transform a sequence of text as an embedding vector using neural network models such as CNN and LSTM. CNN has often been employed for syntactic and semantic representations of the text in various NLP tasks and has achieved good performance than traditional NLP methods. Usually, in CNN, the kernel or filter size is smaller than the input size. As a result, the output interacts with a narrow window of the input and highlights the local lexical connections of the n-gram. LSTM is well designed for sequence modeling and can effectively memorize long-sequence dependencies but may fail to capture the local n-gram context. Besides, it has been found that using a vector from either CNN or LSTM to encode an entire sequence is not enough to capture all the important information. Therefore, we propose a hybrid neural model, which is the combination of CNN and LSTM to exploit the benefits of both techniques.

Figure 1 shows the architecture of our model. First, the input sentence is represented as a matrix by using word embedding technique. Then, n-gram features are extracted by using the convolutional and max-pooling layers. Bi-LSTM is then used to compose the n-gram features to produce the results. The architecture can extract both local information within citations and long-distance dependency across citation context. The following is the detailed description of each part of the architecture.

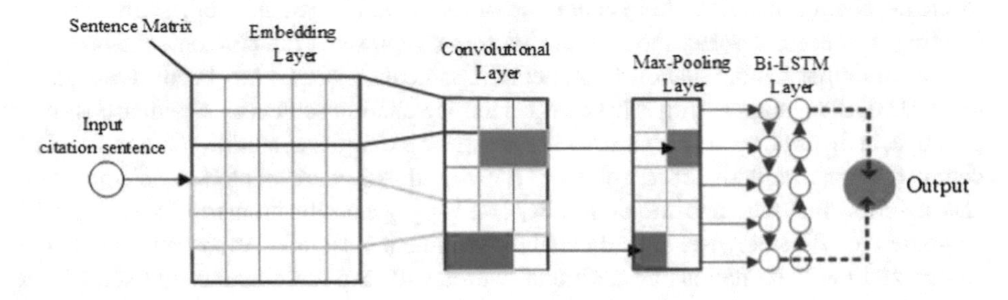

Fig. 1. The architecture of the proposed model

Embedding Layer. The input is a citation sentence S which is a sequence of words $w_1, w_2, \ldots, w_d$ where each word is drawn from a finite-size vocabulary V. Each word is transformed into a low-dimensional vector via a lookup table: $x_i = E_w(w_i)$, where $W \in \mathcal{R}^{d_w \times |V|}$ is the word embedding matrix and d_w is each word embedding dimension. As a result, the input sentence is represented as a matrix where each column corresponds to a word embedding.

Convolutional Layer. In this layer, we use a set of m filters to extract local n-gram features from the embedding matrix. The feature map output z_i^l can be produced by using a sliding window of size w, and a filter $F_l (1 \leq l \leq m)$ as follows:

$$z_i^l = f(X_{l;l+w-1} * W^l + b^l) \tag{1}$$

Where $*$ denotes a convolution operation, $W \in \mathbb{R}^{w \times d}$ is the weight matrix, b is a bias, and $X_{i;i+w-1}$ denotes the word vectors. The results of the filter F_l will be $z^l \in \mathbb{R}^d$, where z_i^l is the i-th element of z^l. Rectified Linear Units was used as activation function.

Max-Pooling Layer. We apply max-pooling operation for each filter to capture the most important features. The max value is extracted from each row of z^l, which will generate the next representation vector. To prevent the overfitting problem in CNN, a dropout strategy was applied.

Bi-LSTM Layer. Single direction LSTM is inadequate and suffers from not exploiting the contextual information from the subsequent words. Therefore, the Bi-LSTM was used to utilize both previous and future contexts by processing the sequence on both forward and backward directions. The following equations show how LSTM estimates the hidden states h_t and outputs o_t.

$$f_t = \sigma\left(W_f \cdot z_t + \mathrm{U}_f.h_{t-1} + b_f\right) \tag{2}$$

$$i_t = \sigma(W_i \cdot z_t + \mathrm{U}_i.h_{t-1} + b_i) \tag{3}$$

$$\widetilde{c}_t = tanh(W_c \cdot z_t + \mathrm{U}_c.h_{t-1} + b_c) \tag{4}$$

$$c_t = (i_t \otimes \tilde{c}_t + f_t^{\circ} c_{t-1}) \tag{5}$$

$$o_t = \sigma(W_o \cdot z_t + \mathrm{U}_o.h_{t-1} + b_o) \tag{6}$$

$$h_t = (o_t \otimes \tanh(c_t)) \tag{7}$$

Where z_t and h_t are the input to Bi-LSTM and output vectors at time t, respectively. W_f, W_i, W_o, U_i, U_f, and U_o are weight matrices and b_i, b_f, b_c, and b_o are bias vectors. The symbol $\otimes$ denotes element-wise multiplication and σ represents the sigmoid function.

Output Layer. We fed the outputs of the previous layer to the fully connected layer. The softmax function generates a distribution over the class labels for citation classification. This layer outputs the final classification results and is defined as:

$$p(y|x) = softmax(W_d.o + b_d) \tag{8}$$

Where o is the final representation word vector, y is the predicted label of the citation sentiment and purpose categories, and W_d and b_d respectively denote the weight and bias.

4 Experiments and Evaluation

4.1 Experimental Setup

To validate the effectiveness of our proposed model for citation classification, we performed experiments on two datasets. The first dataset (CSD) from [3, 20] has 3568 citation context examples and the second dataset (DFKI) as described in [4] is built based on randomly extracted citation sentences from the ACL corpus[1] and has 1768 instances. Table 1 shows the detailed statistics of the two datasets.

In all experiments, we followed the settings used in [12], which is applying 10-fold cross validation with the grid search method to select best hyper-parameters from the

[1] http://aclweb.org/anthology.

Table 1. Datasets statistics

Task	CSD		DFKI	
	Class	Ratio (%)	Class	Ratio (%)
CSC	Positive	32.6	Positive	10.75
	Negative	12.4	Negative	3.22
	Neutral	55.0	Neutral	86.03
CPC	Criticizing	16.3	Idea	7.18
	Comparison	8.1	Basis	23.81
	Use	18.0	GRelated	42.48
	Substantiating	8.0	SRelated	20.81
	Basis	5.3	MRelated	1.75
	Neutral	44.3	Compare	3.97

Table 2. Hyper-parameters settings

Layer	Parameter name	Parameter value
Lookup	Word embedding dimension	100, 200, 300
CNN	Window size	3, 4, 5
	Number of filters	50, 100, 150, 200
LSTM	Hidden units	50, 100, 150, 200
Dropout	Dropout rate	0.1, 0.2, 0.3, 0.4, 0.5, 0.6, 0.7
	Epochs	10, 20, 30, 50
	Batch size	16, 32, 64, 128

options presented in Table 2. We used Adam optimizer to train our model and the losses were calculated by using categorical cross entropy. For other baselines (SVM, and NB), each dataset was split by 80%, 10%, and 10% for training, development, and test sets respectively. For the SVM model, we performed a grid search using development set to select best hyperparameters for evaluation. Once the best settings were found, the final SVM model was learned on both the training and development sets and tested on the test set. We used the average-macro F-score, precision, and recall as the evaluation metrics and reported the performance on the test sets. Keras[2] and Sklearn[3] open source libraries were used to implement the proposed model and the baselines.

We compared our model with the following baseline models: NB with different features proposed in [4], SVM with rich features [3], and SVM with TF-IDF [12]. Moreover, SVM with word embedding [12] (in this model, the input is features generated using word2vec method), and CNN with word embedding [12]. We specifically chose these models because they were also evaluated on the same datasets. Furthermore, since our model is the combination of CNN and LSTM, we reported the performance of LSTM and Bi-LSTM.

[2] keras.io.

[3] http://scikit-learn.org/stable/index.html.

4.2 Experimental Results and Discussion

Table 3 shows the CSC results in terms of average macro-precision (P), recall (R), and F-score (F). Generally, the proposed model outperforms the baseline models on CSD and DFKI datasets. Moreover, our model achieves better results than traditional classifiers with an improvement of 7% on CSD dataset and 8% on DFKI dataset on F-score. Based on DFKI Dataset, it is noted that our model for the citation sentiment classification task generally gains the best results. Specifically, our model improves results by 3% in terms of F-score (85.27%) comparing with the CNN-embedding features F-score (82.86%). We further noted that compared with the traditional classifiers the proposed model is superior in its performance and reported an increase in F-score by more than 7%. Furthermore, for CSD dataset, the proposed model outweighs classical machine learning methods and CNN in all metrics. Our model achieves an F-score of 83.69% which is close to more than 4% increase over the CNN. Moreover, our model attains a higher than 6% increase of F-score over the rest of the baseline approaches.

Table 3. The experimental results of citation sentiment classification

Method	CSD			DFKI		
	P	R	F	P	R	F
NB with syntactic features [4]	69.0	62.5	64.4	–	–	72.00
SVM with features from [3]	67.1	70.6	68.8	70.08	56.66	54.48
SVM with TF-IDF [12]	77.9	76.3	77.1	75.13	61.71	59.53
SVM with embedding [12]	81.3	75.4	77.3	79.21	77.57	77.90
CNN with embedding [12]	82	75.9	78.8	85.21	81.02	82.86
LSTM	80.8	74.3	77.4	83.26	82.14	82.12
Bi-LSTM	80.4	77.56	79	84.76	83.41	83.97
CNN-BiLSTM model	85.17	82.45	83.69	86.14	84.47	85.27

Furthermore, the results for CPC task are presented in Table 4. Generally, our model outweighs the baseline methods on this task as well. On CSD Dataset, the proposed model achieves an F-score of 77.98%, which is equivalent to the performance increase of 3% over the F-score recorded (74.3%) for CNN. As well as for DFKI dataset the proposed model also achieves an increase of more than 3% for F-score. Moreover, comparing the baseline methods with each other, on both tasks, SVM with word embedding features leads to better results than NB with syntactic features. In addition, comparing the baselines with one another, on both tasks, SVM with word embedding features attained better results than NB with syntactic features. Note that the results of SVM on DFKI dataset for both tasks were worse than the proposed NB with syntactic features in [4] because we have only experimented by using n-grams features on this dataset. On the other hand, SVM with embedding features on DFKI dataset gained better results compared with NB.

Moreover, we conducted a series of experiments by using the dropout strategy to see its effect on the performance of our proposed model. Table 5 shows the impact of

Table 4. The experimental results of citation purpose classification

Method	CSD			DFKI		
	P	R	F	P	R	F
NB with syntactic features [4]	65.02	58.5	60.4	–	–	72.0
SVM with features from [3]	54.9	62.5	58.4	62.1	59.57	60.0
SVM with TF-IDF [12]	74.3	70.9	72.6	64.9	62.5	62.8
SVM with embedding [12]	86.8	64.7	74.1	79.6	67.6	73.1
CNN with embedding [12]	80.8	68.8	74.3	84.6	71.6	76.14
LSTM	79.87	67.8	73.21	82.3	69.2	74.1
Bi-LSTM	77.22	73.11	75.11	78.02	73.95	75.91
CNN-BiLSTM model	79.12	76.87	77.98	81.12	78.87	79.98

dropout on the experimental results of our model. We noted that using dropout strategy improves the classification performance of the model by 2 to 3% for both classifications. Figure 2 shows the performance of our proposed model at different dropout rate. As can be seen, the model achieves better F-score when the dropout value is equal to 0.5 for both classification tasks. Furthermore, we noted that the number of hidden nodes affects the classification accuracy. We observed that the big number of nodes leads to overfitting problem which consequently reduces the accuracy. As a result, we leveraged grid search technique to automatically set the optimal number of hidden nodes.

Table 5. The impact of dropout strategy on the performance of our model

Task	Option	P	R	F
Citation sentiment	Non-dropout	85.17	82.45	83.69
	Dropout	86.14	84.47	85.97
Citation purpose	Non-dropout	78.35	75.21	76.41
	Dropout	81.12	78.87	79.98

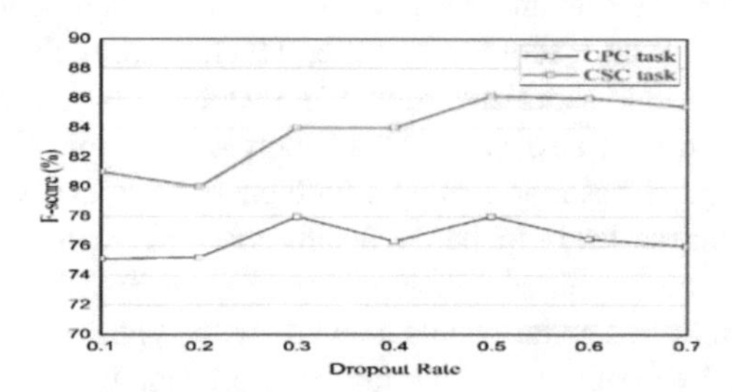

Fig. 2. The impact of dropout strategy on classification performance.

The findings of this study confirm that the proposed model produces more effective classification results than the handcraft feature-based, CNN, and LSTM methods.

Moreover, we believe that representing citation sentences using deep learning approaches is useful because they can capture the syntactic and the semantic information of the sentence unlike the count based methods, which are purely dependent on the statistics to represent the text. In addition to that, the used datasets are imbalance therefore, our model requires incorporating imbalance techniques to boost the classification performance.

5 Conclusion and Future Work

In this paper, we proposed a deep learning model based on CNN and LSTM algorithms for identifying author's sentiment in scientific citations and detecting purpose of citations by targeting the reasons/purposes behind citing a particular article. The presented network architecture is our main contribution which can represent the citations and improve the classification performance. Further, the proposed model was evaluated by using two public datasets and the results reveal that our model outperforms the baseline methods for citation sentiment and citation purpose classification. Further research will be investigating the benefits of multitask learning technique with our neural network architecture to jointly train the citation sentiment and purpose classification to improve the classification performance. In addition, assessing paper quality using citation classification would be an interesting issue.

Acknowledgements. This work is supported by the National Natural Science Foundation of China (No. 61370137), the National Basic Research Program of China (No. 2012CB7207002), the Ministry of Education - China Mobile Research Foundation Project No. 2016/2-7 and the 111 Project of Beijing Institute of Technology.

References

1. Teufel, S., Siddharthan, A., Tidhar, D.: Automatic classification of citation function. In: Proceedings of the 2006 Conference on Empirical Methods in Natural Language Processing, pp. 103–110. Association for Computational Linguistics (2006). 1610091
2. Small, H.: Interpreting maps of science using citation context sentiments: a preliminary investigation. Scientometrics **87**, 373–388 (2011)
3. Abu-Jbara, A., Ezra, J., Radev, D.R.: Purpose and polarity of citation: towards NLP-based bibliometrics. In: Proceedings of the North American Association for Computational Linguistics, NAACL-HLT 2013, pp. 596–606. Association for Computational Linguistics: Human Language Technologies (2013)
4. Dong, C., Schäfer, U.: Ensemble-style self-training on citation classification. In: Proceedings of the 5th International Joint Conference on Natural Language Processing, pp. 623–631. Association for Computational Linguistics (ACL) (2011)
5. Athar, A.: Sentiment analysis of citations using sentence structure-based features. In: Proceedings of the ACL 2011 Student Session, pp. 81–87. Association for Computational Linguistics (2011). 2000991
6. Li, X., He, Y., Meyers, A., Grishman, R.: Towards fine-grained citation function classification, pp. 402–407 (2013)

7. Sula, C.A., Miller, M.: Citations, contexts, and humanistic discourse: toward automatic extraction and classification. LLC **29**, 452–464 (2014)
8. Hernandez-Alvarez, M., Gomez S, J.M.: Citation impact categorization: for scientific literature. In: Ferreira, J.C. (ed.) IEEE 18th International Conference on Computational Science and Engineering, CSE, 21–23 October 2015, pp. 307–313. IEEE Computer Society, Los Alamitos (2015)
9. Ma, Z., Nam, J., Weihe, K.: Improve sentiment analysis of citations with author modelling. In: Proceedings of the Fifth Workshop on Computational Linguistics for Literature, NAACL-HLT 2016, pp. 122–127. Association for Computational Linguistics (ACL) (2016)
10. Alvarez, M.H., Gómez, J.M.: Survey in sentiment, polarity and function analysis of citation. In: Proceedings of the First Workshop on Argumentation Mining, pp. 102–103. Association for Computational Linguistics (ACL) (2014)
11. Jochim, C., Schutze, H.: Improving citation polarity classification with product reviews. In: Marcu, D. (ed.) 52nd Annual Meeting of the Association for Computational Linguistics, ACL, vol. 2, pp. 42–48. Association for Computational Linguistics (ACL), Baltimore (2014)
12. Lauscher, A., Glavaš, G., Ponzetto, S.P., Eckert, K.: Investigating Convolutional networks and domain-specific embeddings for semantic classification of citations. In: Proceedings of the 6th International Workshop on Mining Scientific Publications, pp. 24–28. ACM, Toronto (2017)
13. Nanba, H., Okumura, M.: Towards multi-paper summarization using reference information. In: IJCAI, pp. 926–931 (1999)
14. Nakagawa, T., Inui, K., Kurohashi, S.: Dependency tree-based sentiment classification using CRFs with hidden variables. In: Human Language Technologies: The 2010 Annual Conference of the North American Chapter of the Association for Computational Linguistics, pp. 786–794. Association for Computational Linguistics (2010)
15. Jochim, C., Schutze, H.: Towards a generic and flexible citation classifier based on a faceted classification scheme. In: 24th International Conference on Computational Linguistics, COLING, pp. 1343–1358. COLING 2012 Organizing Committee, Mumbai (2012)
16. Meyers, A.: Contrasting and corroborating citations in journal articles. In: RANLP, pp. 460–466 (2013)
17. Tsai, C.-T., Kundu, G., Roth, D.: Concept-based analysis of scientific literature. In: Proceedings of the 22nd ACM International Conference on Conference on Information & Knowledge Management, pp. 1733–1738. ACM (2013)
18. Abdullatif, M., Koh, Y.S., Dobbie, G., Alam, S.: Verb selection using semantic role labeling for citation classification. In: Proceedings of the 2013 workshop on Computational Scientometrics: Theory & Applications, pp. 25–30. ACM (2013)
19. Athar, A., Teufel, S.: Detection of implicit citations for sentiment detection. In: Proceedings of the Workshop on Detecting Structure in Scholarly Discourse, pp. 18–26. Association for Computational Linguistics (2012). 2391176
20. Jha, R., Jbara, A.-A., Qazvinian, V., Radev, D.R.: NLP-driven citation analysis for scientometrics. Nat. Lang. Eng. **23**, 93–130 (2016)

A Convolutional Neural Network Model for Emotion Detection from Tweets

Eman Hamdi(✉), Sherine Rady, and Mostafa Aref

Faculty of Computer and Information Sciences, Ain Shams University, Abbassia, Cairo, Egypt
{emanhamdi, srady, mostafa.aref}@cis.asu.edu.eg

Abstract. Sentiment analysis and emotion recognition are major indicators of society trends toward certain topics. Analyzing opinions and feelings helps improving the human-computer interaction in several fields ranging from opinion mining to psychological concerns. This paper proposes a deep learning model for emotion detection from short informal sentences. The model consists of three Convolutional Neural Networks (CNNs). Each CNN contains a convolutional layer and a max-pooling layer, followed by a fully-connected layer for classifying the sentences into positive or negative. The model employs the word vector representation as textual features, which works on random initialization for the word vectors, and are set to be trainable and updated through the model training phase. Eventually, task-specific vectors are generated as the model learns to distinguish the meaning of words in the dataset. The model has been tested on the Stanford Twitter Sentiment dataset for classifying sentiment into two classes positive and negative. The presented model achieved to record 80.6% accuracy as a prove that even with randomly initialized word vectors, it can work very well in text classification tasks when trained with CNNs.

Keywords: Deep learning · Sentiment analysis · Emotion detection
Social media networks

1 Introduction

Understanding emotions plays a main role in human-computer interaction development. It helps to further exploring the society trend towards different types of fields especially psychological concerns. Proven by psychology researches, language is a fundamental key to detect emotions as it is the main method used by humans for expression. A realistic source of textual data is needed to capture transformation in one's way for expression and transformation in feelings [1, 2]. Social media networks such as Facebook and Twitter act as a great source of this type of textual data. Artificial intelligence and machine learning methods can be applied in text classification tasks for investigating personal psychological behaviors and detecting emotions.

Machine learning methods, and most recently deep learning methods, have proven to outperform the traditional lexicon-based methods in many tasks, such as text classification tasks in [3]. While lexicon-based methods often employ explicit semantic relations for better performance [4], deep learning methods can capture the implicit

A. E. Hassanien et al. (Eds.): AISI 2018, AISC 845, pp. 337–346, 2019.
https://doi.org/10.1007/978-3-319-99010-1_31

semantic relations. Deep learning methods have the advantage of learning high-level features from low-level features. In text classification, this means they perform well even with the low-level features if large data sets are available for training [5].

Feature selection is one of the main concerns when working with machine learning methods on natural language processing tasks. On traditional methods, the words are treated as discrete values. Representing words as unique, discrete values causes the data to be sparse. This means more data may be needed to successfully train statistical models. For solving this issue, vector representations are suggested. Vector representation of words are dense representations instead of the sparse representation. Word vectors can capture the semantic relations between words as these words tend to appear in the same context. There are roughly two options for initializing word vectors: (1) using pre-trained word vectors, such as Word2Vec [6], or (2) using randomly initialized word vectors. In both cases, the word vectors can be either updated during the training phase or kept static. It is assumed logically that a model would employ randomly initialized word vectors which are updated during a training phase.

This paper introduces an emotion detection method from twitter's short informal text using deep learning modelling. More specifically, a Convolutional Neural Network (CNN) model that using word vectors as features. The proposed model works on randomly initialized word vectors which are used to represent each word in each sentence. The word vectors are fed to the model that consists of three CNNs followed by the output layer. The output layer is a fully-connected layer with the Sigmoid activation function. The proposed model is trained and validated on the Stanford Twitter Sentiment dataset [7]. The dataset originally contains 1.6 million tweets labelled as positive or negative. After training on 80k sentences and validated on 16k sentences, the model succeeded in achieving 80.6% accuracy to classify positive and negative sentiments.

The rest of the paper is organized as follows: In Sect. 2, the related work in text classification and sentiment analysis is given. Section 3 describes the main architecture of the deep learning proposed model, with descriptions and explanations for each layer and how it works. Experimental results are shown in Sect. 4, and finally, Sect. 5 summarizes the proposed work.

2 Related Work

Emotion detection and sentiment analysis studies have been explored using traditional lexicon-based and machine learning methods. In [4], the lexicon-based approach has been presented for determining sentiment from text using dictionaries of annotated words, where the words' polarity and strength are used to represent the semantic orientation, with a semantic orientation calculator to assign a negative or a positive label to opinions. In [8], a generic process to categorize sentiment polarity on the sentence-level is proposed using Naïve Bayesian, Random Forest, and Support Vector Machine (SVM) classifiers for the online product reviews from Amazon.com. Sentiment tokens and scores extracted from the dataset are used as features to classify sentiment into positive and negative. In [9] an unsupervised method is developed based on the expansion of concepts expressed in the tweets for polarity classification in

Twitter. The method recorded 64.90% accuracy when tested on the Stanford Twitter Sentiment Corpus.

Lexicon-based methods depend on creating dictionaries of words annotated with their semantic polarity and sentiment strength capturing the semantic orientation. A large dictionary will result in achieving outstanding results. However, the sentiment of a sentence is also related to the semantics of the context implicitly. This initiates the need for methods that can capture such implicit semantic relations from the source text.

Machine learning methods offer such solutions to capture such implicit semantic relationships. Recently, the focus is driven on the deep learning methods. Deep learning methods have proven to outperform the traditional methods in various tasks including text classification [3, 10–14], image processing [15, 16] and speech recognition [17, 18]. The main architecture of the CNN was originally developed to work on images. The convolutional layers are meant to perform the features' extraction from low-level to high level as it is going deep.

In text classification, CNNs have been also employed. The related work in such domain can be divided as working on character-level features [3, 10], word-level features [11–13], or both levels for the features [14].

In [3], a character-level CNNs for text classification are tested on several large datasets including several news datasets, DBPedia ontology dataset, Yelp reviews, Yahoo! Answers dataset, and Amazon reviews. The model achieved good performance in sentiment analysis and topic classification. In [10], the use of only character-level inputs is explored. The model consists of a CNN whose outputs are passed to a recurrent neural network. It is tested on many languages. For the English Penn Treebank dataset, the performance is comparative with the existing methods. For other languages, the model outperforms word-level/morpheme-level Long Short-Term Memory Recurrent Neural Networks. In [11], the performance of a single layer CNN is tested for sentence classification using word vectors, with four model variations: CNN-rand, CNN-static, CNN-non-static, and CNN-multichannel. The word vectors used are randomly initialized or pre-trained. The multichannel model can work on both word vectors via two channel selections. The model is tested on multiple datasets and the experiments showed improved accuracy on 4 different tasks out of 7. In [12], a semi-supervised CNN model for text categorization is proposed using unlabelled data to produce embeddings, which are next passed to a supervised CNN as 'one-hot' vectors. The error rates are improved for 3 classification tasks. Other works also recorded good performances for sentiment analysis. In [13], a deep CNN model is proposed for sentiment analysis of tweets. The model performance is explored using only initialized embeddings form an unsupervised neural language model. The word embeddings are tuned during the model training. The model is tested on two subtasks from Semeval-2015 Task 10 on Twitter Sentiment Analysis. The accuracy rate is 84.79 for the phrase-level subtask and 64.59% message-level subtask. In [14], a deep CNN followed by a Bidirectional Long Short-Term Memory network is introduced to perform sentence classification. The first neural network used character-level embeddings, while the second network produces sentence-level embeddings from word-level embeddings. The model is tested using three datasets and proved to outperform the previous works for the Stanford Twitter Sentiment Corpus with comparative performance for the other datasets.

3 CNN Model for Emotion Detection

This section explains the proposed model for detecting emotions from tweets. The architecture is shown in Fig. 1, and consists of the following processing modules: filtering and tokenizing, sentence indexing, word vectors, and the CNNs. The filtering and tokenization module performs some preprocessing to reduce the complexity of text by removing irrelevant words. The sentence indexing converts sentences into sequences of indices instead of sequences of words, where each index corresponds to a word. The setting of the embedding layer is done by feeding it with the indices together with the corresponding words in the vocabulary. The CNNs and max-pooling layers learn relations and extract the most important features from the sentences. The extracted features are concatenated to be fed to the fully-connected layer to perform classification.

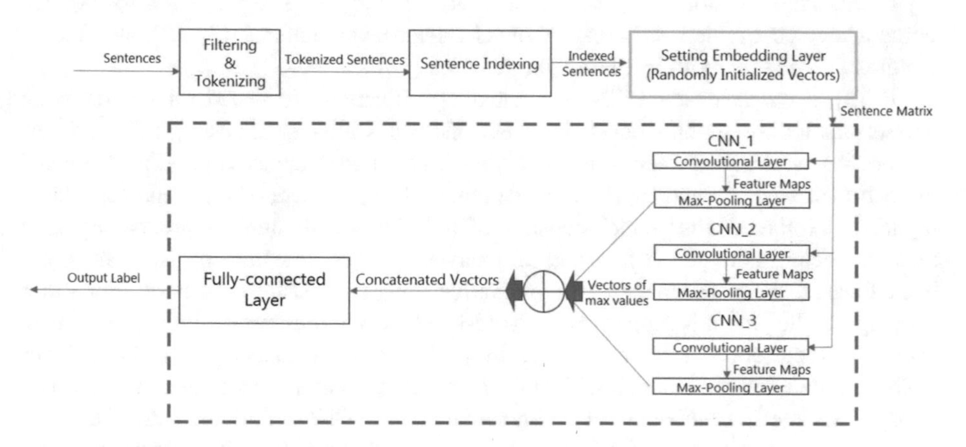

Fig. 1. A block diagram for the proposed CNN model for emotion detection

3.1 Text Pre-processing

In this processing, filtering and tokenizing are employed. Raw sentences are extracted from the dataset, then they are filtered by removing irrelevant words such as punctuation marks, stop words, links, mentions, hashtags, emoticons and any special symbols, only the textual data for each sentence is kept. The tokenizing splits each sentence into words. After tokenizing, each sentence is a sequence of words. A given sentence must be represented as a sequence of integer numbers-indices-corresponding to words to be fed to the embedding layer. Therefore, the indexing process is applied. A tokenizer is used to vectorize the sentences by turning each sentence into a sequence of integers as each integer represents the index of a word in the vocabulary. Then the sentences are fed to the embedding layer.

3.2 Embedding Layer

The embedding layer performs the data representation and acts as a look up table. It contains all the words in the vocabulary-as indices-and the corresponding word vectors for each word. The row number is an index and the row content are the word vector corresponding to that word. Initially, the embedding layer contains randomly initialized word vectors. Before feeding a sentence to the convolutional layer, it is pre-processed and passed to this layer to produce the sentences matrix. As a sentence is passed to the embedding layer, the sequence of indices-one by one-fetches the word vector in each index. This generates a sentence matrix containing the word vectors corresponding to the sentence. Given a sentence s that consists of n words, each word will have a corresponding word vector v, the sentence matrix is the concatenation of n word vectors as shown in below:

$$v_{1:n} = v_1 \oplus v_2 \oplus \ldots \oplus v_n \quad (1)$$

The embedding layer is set to be trainable such that during training phase, its values are updated to generate task specified word vectors that can be called semantically related words.

3.3 Convolutional and Max-Pooling Layers

This part consists of 3 CNNs, with each convolutional layer followed by a max-pooling layer. The CNNs use the same activation function on the convolutional layer and apply the same number of filters to the input sentence matrix; however, differ in the size of the convolutional filters and the size of the max-pooling filters. The convolutional layer role is to apply filtering to the sentence matrix. Filters slide over full rows of the matrix-words-to learn semantically meaningful representations. This layer produces a number feature maps equals to the filters' number as each filter produces a feature map, which are passed to the max-pooling layer.

The activation function in each convolutional layer is a ReLU, which gives output 0 if the input is less than 0, and raw output otherwise. Mathematically, it is applied by:

$$f(x) = \max(x, 0) \quad (2)$$

The output of the convolutional layer is passed to the max-pooling layer for sub-sampling, and reducing the dimensionality while keeping the most important features. The inputs to the Max-Pooling layer are the feature maps and the output is a vector containing the maximum values from each feature map.

Pool sizes differ from one CNN to another as the filter sizes differ to produce vectors of the same size from each CNN. The vectors are then concatenated to obtain a single output for this module, and which is fed to the fully-connected layer.

3.4 Fully-Connected Layer

The proposed model's output layer is a fully-connected layer with the Sigmoid activation function. It contains one neuron for classification as it classifies the sentences into two classes, positive or negative. The Sigmoid function is used to map the result of the output unit to be between 0 and 1. Mathematically, it is applied by the equation:

$$\sigma(\mathrm{z}) = \frac{1}{1 + \mathrm{e}^{-\mathrm{z}}} \quad (3)$$

where z represents the input to the output unit. For model generalization, a regularization technique for neural network models, which is Dropout, is applied at the output layer to prevent overfitting and co-adaptations for the training data [19]. During training, randomly selected neurons on a specific layer are dropped-out. The dropped-out neurons are temporally deactivated on the forward phase and no weight update is applied on the backward phase. In the proposed model, Dropout/this is applied on the Sigmoid fully connected layer, and the output is calculated by the equation:

$$y = w\,.\,(z \circ r) + b \quad (4)$$

Instead of:

$$y = w\,.\,z + b \quad (5)$$

In which $\circ$ is the element-wise multiplication operator, w is the weight matrix, z is the input to the fully-connected layer, b is the bias, and r is a masking vector that contains Bernoulli random variables with the probability of being 1.

4 Experimental Results

The proposed architecture of Fig. 1 is experimented on the labelled Stanford Twitter Sentiment dataset, which consists of 1.6 million tweets [7]. The data is automatically created by assuming tweets with positive emoticons were positive, and tweets with negative emotions were negative, and has been collected using Twitter Search API.

In experiments, 80K randomly selected sentences are collected for training, with another 16K sentences collected for validation. For testing, a variety of sentences is selected from the validation set which ranges from 500 to 16K to prove the generalization of the proposed model. The sentences are labelled with one value of two class labels: positive or negative emotion. The ratio of positive to negative labels is 50%. The maximum length of sentences has been adjusted to 8 and the vocabulary size is 50,485, fitting the experimented dataset. The picked top words are 50,485 representing the vocabulary size. In regard with the dimensions explained in Sect. 3, the randomly initialized word vectors have the dimensions [300 * 1], with the embedding layer and the sentence matrix dimensions as [50,485 * 300] and [8 * 300] respectively.

Figure 2 shows an illustrative example for the processing introduced in Sect. 3. As shown in the figure, each of the 3 CNNs applies 100 filters to the sentence matrix. The 3 CNNs run in parallel on the same data but process it differently. The first CNN applies filters each of size [3 * 300] generating 100 feature maps, each of size [6 * 1]. This is followed by the max-pooling with filter size 5, hence, generating 100 vectors that contain maximum features, each of size [2 * 1]. The second CNN applies filters each of size [5 * 300] generating 100 feature maps, each of size [4 * 1]. This is followed by the second max pooling of filter size 3, generating 100 vectors that contain the maximum features each of size [2 * 1]. The third CNN applies filters each of size [7 * 300] generating 100 feature maps, each of size [2 * 1]. This is followed by the final third max pooling of filter size 1 generating 100 vectors that contain the maximum features each of size [2 * 1]. The output of each CNN is of the same size [2 * 1], which are concatenated together. After Concatenating the outputs, the result becomes a matrix of the maximum features extracted at each CNN. This output is called the sentence matrix and has the size of [2 * 300]. The resulted matrix is then flattened and passed to the fully-connected layer with the Sigmoid activation function to perform the binary classification.

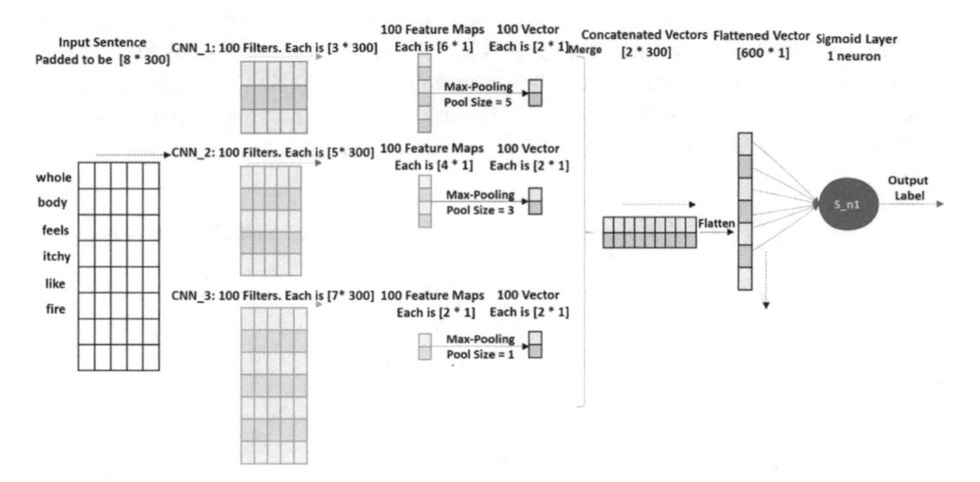

Fig. 2. Illustrative example on how the proposed model works

For measuring the model's performance, two evaluation measures are used: the accuracy measure and observed log loss as the loss function. As the testing is done on 50% positive samples and 50% negative sample, the accuracy measure is a real indicator of the model's performance. Accuracy is defined by the equation:

$$\text{Accuracy} = (\text{TP} + \text{TN})/(\text{TP} + \text{TN} + \text{FP} + \text{FN}) \tag{6}$$

Where TP, TN, FP and FN represent the true positives, true negatives, false positives, and false negatives respectively.

The log loss function describes the average log loss for all the samples. It is used to avoid the weights updating factor from getting smaller and training from stalling out as in mean square error loss. In experiments, the Sigmoid function is used and the classification problem is binary, therefore the used log loss function is the binary-binary-cross entropy that is given by the following equation:

$$E = y_i \log(\sigma(z)) + (1 - y_i)\log(1 - \sigma(z)) \tag{7}$$

Where y_i is the actual sentence label and $\sigma(z)$ is the predicted output from the model, and z is the input to the output unit.

The performances of the proposed deep learning model are shown in Fig. 3a and b. Figure 3a shows the accuracy performance versus different epochs, for both the training and testing phases. The figure shows that the accuracy starts low at the first 3 epochs as the model just starts recognizing the patterns. The values keep rising with higher rates till the 8th epoch, after which an almost stability is reached, that's when the training stops. The training stopping criteria is based on the number of epochs where stability is reached (i.e. after a specific number of epoch). This achieves an accuracy value for classification equal to 80.6%. In our experiments, the model is recorded to give the best accuracy at 15 epochs. Figure 3b shows the loss function versus the number of epochs. The training loss starts very high as it is the first time of exploring the data, on the other hand the testing loss starts lower as it follows the first training epoch. In the second epoch, the training loss drops dramatically, and the testing loss remains the same. After the first 3 epochs, training and testing losses keep getting lower until reaching the stable state, which is also after 15 epochs.

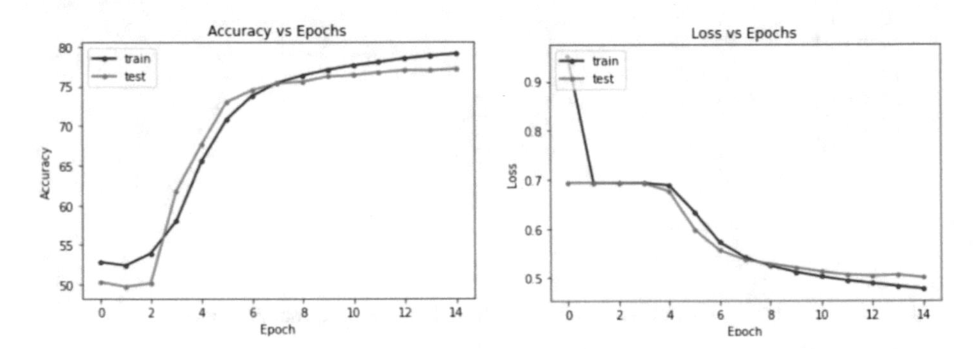

Fig. 3. Model's performance

The accuracy performance has also been tested versus different ranges of testing samples to test the generalization of the model. Figure 4 indicates a range for the accuracy for various numbers of testing samples from 76.8% to 80.6%. Despite using only randomly initialized word vectors, the obtained results are close to the previous text classification tasks that are accomplished using CNNs as mentioned in the related work section. The best accuracy is 80.6% obtained at a testing sample size of 500. This proves that the proposed model provides accuracy as well as generalization, as it is not overfitted over the training data. It remains within a relatively high range and doesn't dropout as the number of testing samples is getting larger.

The recorded experimented results are obtained using a PC with Intel(R) Core (TM) i7-5500U CPU @2.40 GHz with 16.0 GB RAM. The experiments have been developed using Keras deep learning framework, with TensorFlow backend and with CUDA enabled.

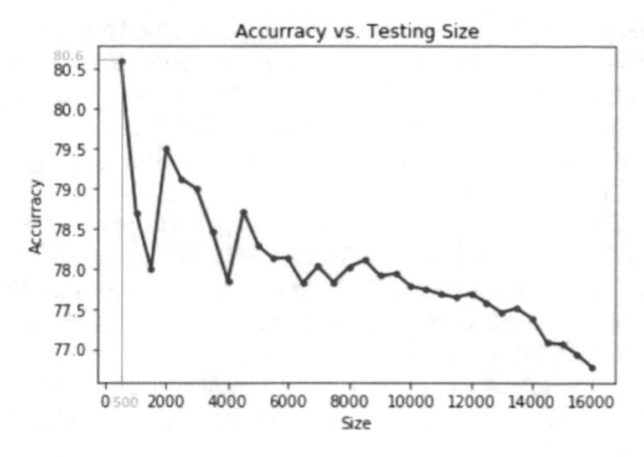

Fig. 4. Accuracy performance against various testing sample sizes

5 Conclusion

This paper presented a deep learning architecture for detecting emotions from social media. The proposed architecture contains modules for filtering and tokenizing sentences, sentence indexing, setting embedding layer of randomly initialized word vectors, and a CNN model. The CNN model has shown to work well on randomly initialized word vectors. The word vectors are updated during the training phase to generate task-specific word vectors that can map well towards emotion classification. It has achieved 80.6% accuracy in the sentiment analysis task, while using only randomly initialized vectors and experimented with Stanford Twitter Sentiment dataset. It has been shown that CNNs can work without a helper dataset or an extra model to provide word vectors and gives very good results. Future work will regard using other levels of embeddings as character-level and sentence-level embeddings and the performance of the model will be explored when using pretrained word vectors instead of randomly initialized ones.

References

1. Coppersmith, G., Hollingshead, K., Dredze, M., Harman, C.: From ADHD to SAD: analyzing the language of mental health on twitter through self-reported diagnoses. In: Proceedings of the 2nd Workshop on Computational Linguistics and Clinical Psychology: From Linguistic Signal to Clinical Reality, pp. 1–10 (2015)
2. Resnik, P., Armstrong, W., Claudino, L., Nguyen, T.: The University of Maryland CLPsych 2015 shared task system. In: The Conference of the North American Chapter of the Association for Computational Linguistic, pp. 54–60 (2015)
3. Zhang, X., Zhao, J., LeCun, Y.: Character-level convolutional networks for text classification. In: Advances in Neural Information Processing Systems, pp. 649–657 (2015). arXiv:1509.01626
4. Taboada, M., Brooke, J., Tofiloski, M., Voll, K., Stede, M.: Lexicon-based methods for sentiment analysis. Comput. Linguist. **37**(2), 267–307 (2011)

5. LeCun, Y., Bengio, Y., Hinton, G.: Deep learning. Nature **521**(7553), 436–444 (2015)
6. Mikolov, T., Chen, K., Corrado, G., Dean, J.: Efficient estimation of word representations in vector space. In: Proceedings of the International Conference on Learning Representations, pp. 1–12 (2013). https://doi.org/10.1162/153244303322533223
7. Go, A., Bhayani, R., Huang, L.: Twitter sentiment classification using distant supervision. Processing **150**(12), 1–6 (2009)
8. Paltoglou, G., Thelwall, M.: Twitter, MySpace, Digg: unsupervised sentiment analysis in social media. ACM Trans. Intell. Syst. Technol. **3**(4), 66 (2012)
9. Montejo-Ráez, A., Martínez-Cámara, E., Teresa Martín-Valdivia, M., Alfonso Ureña-López, L.: A knowledge-based approach for polarity classification in Twitter. J. Assoc. Inf. Sci. Technol. 414–425 (2014). https://doi.org/10.1002/asi.22984
10. Kim, Y., Jernite, Y., Sontag, D., Rush, A.M.: Character-aware neural language models. In: Association for the Advancement of Artificial Intelligence (AAAI), pp. 2741–2749 (2016). arXiv:1508.06615
11. Kim, Y.: Convolutional neural networks for sentence classification (2014). arXiv:1408.5882
12. Johnson, R., Zhang, T.: Semi-supervised convolutional neural networks for text categorization via region embedding. Neural Inf. Process. Syst. **28**, 919–927 (2015)
13. Severyn, A., Moschitti, A.: UNITN: training deep convolutional neural network for Twitter sentiment classification. In: Proceedings of the 9th International Workshop on Semantic Evaluation, pp. 464–469 (2015). https://doi.org/10.18653/v1/s15-2079
14. Nguyen, H., Nguyen, M.-L.: A deep neural architecture for sentence-level sentiment classification in Twitter social networking (2017). arXiv:1706.08032
15. Simonyan, K., Zisserman, A.: Very deep convolutional networks for large-scale image recognition. In: ICLR (2015). arXiv:1409.1556
16. Ren, S., He, K., Girshick, R., Zhang, X., Sun, J.: Object detection networks on convolutional feature maps. IEEE Trans. Pattern Anal. Mach. Intell. **39**(7), 1476–1481 (2017)
17. Zhang, I., Pezeshki, M., Brakel, P., Zhang, S., Laurent, C., Bengio, Y., Courville, A.: Towards end-to-end speech recognition with deep convolutional neural networks (2017). arXiv:1701.02720
18. Qian, Y., Woodland, P.C.: Very deep convolutional neural networks for robust speech recognition. In: IEEE Spoken Language Technology Workshop, SLT, San Diego, CA, USA, pp. 481–488 (2016)
19. Srivastava, N., Hinton, G., Krizhevsky, A., Sutskever, I., Salakhutdinov, R.: Dropout: a simple way to prevent neural networks from overfitting. J. Mach. Learn. Res. **15**(1), 1929–1958 (2014)

Aquarium Family Fish Species Identification System Using Deep Neural Networks

Nour Eldeen M. Khalifa[1,2](✉), Mohamed Hamed N. Taha[1,2], and Aboul Ella Hassanien[1,2]

[1] Information Technology Department, Faculty of Computers and Information, Cairo University, Giza, Egypt
{nourmahmoud, mnasrtaha, aboitcairo}@cu.edu.eg
[2] Scientific Research Group in Egypt (SRGE), Giza, Egypt
http://www.egyptscience.net

Abstract. In this paper, a system for aquarium family fish species identification is proposed. It identifies eight family fish species along with 191 sub-species. The proposed system is built using deep convolutional neural networks (CNN). It consists of four layers, two convolutional and two fully connected layers. A comparative result is presented against other CNN architectures such as AlexNet and VggNet according to four parameters (number of convolution and fully connected layers, the number of epochs in training phase to achieve 100% accuracy, validation accuracy, and testing accuracy). Through the paper, it is proven that the proposed system has competitive results against the other architectures. It achieved 85.59% testing accuracy while AlexNet achieves 85.41% over untrained benchmark dataset. Moreover, the proposed system has less trained images, less memory, less computational complexity in training, validation, and testing phases.

Keywords: Deep learning · Deep neural · Fish identification
Convolutional neural networks

1 Introduction

Fish species observation and identification in the aquarium are considered very informative for tourists. The aquarium is equipped with a camera and when a fish passes in front of it, an identification system is triggered to classify the fish and display information on the screen as illustrated in Fig. 1 and considered one of the main motivation of this research. Also, this research area is important for academic researchers like ocean scientists and biologists. Commercial applications like fish farming depend on fish species observation to achieve their benefits. This involves time-consuming and destructive measures to get physical samples and visual census. However, these approaches are still common.

Fish species recognition is a challenging issue for research. Great challenges for fish recognition appear in the special properties of underwater videos and images. Prior fish recognition, researchers were limited to constrained environments before fish recognition [1]. The focus of the most recognition research is on ground objects.

A. E. Hassanien et al. (Eds.): AISI 2018, AISC 845, pp. 347–356, 2019.
https://doi.org/10.1007/978-3-319-99010-1_32

However, there is a great demand for underwater object recognition. In the last two decades, many machine learning and image processing algorithms have been proposed for underwater species classification [2].

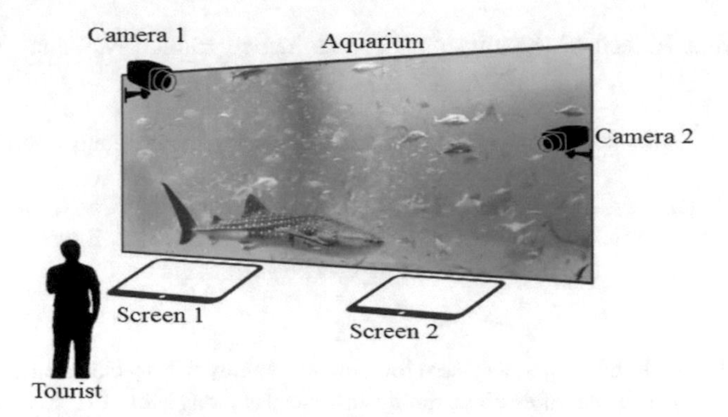

Fig. 1. The design of aquarium family fish species identification system

Convolution operation is famous in the computer vision and signals processing community. The convolutional operation is frequently used by conventional computer vision, especially for noise reduction and edge detection [3].

The idea of a Convolutional Neural Network (CNN) is not recent. In [4], CNN achieved great results for handwritten digit recognition. However, they slowly fell out of favor due to memory and hardware constraints, besides the lack of a large amount of training data. They were unable to scale to much larger images. With the huge increase in the processing power, memory size and the availability of powerful GPUs and large datasets, it was possible to train larger and more complex models. The machine learning Researchers had been working on learning models which included learning and extracting features from images. This leads to the start of the first deep learning model. AlexNet [5], Vgg-16 and Vgg-19 [6] are considered the famous deep convolutional neural networks.

Deep Learning has achieved significant results and a huge improvement in visual detection and recognition with a lot of categories [7]. Raw data images are used be deep learning as input without the need of expert knowledge for optimization of segmentation parameter or feature design.

Prior researchers do not achieve satisfying results. Firstly, most of the fish images were under constrained conditions. Secondly, the datasets were probably small. Thirdly, the accuracy is very unsatisfying under constrained and unconstrained conditions.

Early methods for fish species classification were performed in controlled environments only. In [8], dead fish samples in the laboratory were used for classifications based on color and shape. Storbeck and Daan in [9], proposed the use of laser light for 3D modeling of fish to measure fish features like height, length, and thickness of some species. Unconstrained classification of underwater fish is a very difficult and challenging task. The similarity in color, shape, and texture of different fish is considered another challenge in the classification of species. [9, 10] proposed two classical methods for fish species classification in unconstrained environments, based on texture and shape in nature.

2 Deep Convolutional Neural Networks

Deep learning is a data-driven method. Both the distinctive features and the classifier are trained simultaneously. Deep neural networks can learn the hierarchical representation of data. Besides that, data representation is improved with the increase in the number of layers [11].

A filter bank layer, a nonlinear transformation, and a feature pooling layer are the main stages of feature extraction. They are very common in several object recognition systems [12].

CNN typically consists of several layers that act as the layers mention before in object recognition systems. The convolutional filter bank can be used for local patterns extraction. Each convolutional layer in the CNN is followed by a nonlinearity processing layer [13]. A nonlinear processing layer works on forming a nonlinear complex model through capturing the nonlinearity dynamics of input data. The goal of feature pooling layer is to decrease the resolution of feature maps [14].

3 Related Works

Training on datasets with large variations of background and objects in the images gives the CNN ability to extract information for objects of interest based on their color, texture, and shape. So, any visual pattern could be captured and learned easily by the suitable network. As the number of example for specific object increases, the network generalization capability increases. This capability of generalization gives the trained network the ability to classify information that is not used for training [15]. AlexNet, Vgg-16, and Vgg-19 are examples of pre-trained deep CNN. The knowledge inside each of those deep CNN can be used by researchers for training and testing on more datasets. Knowledge transfer of deep CNN is one of the main advantages that improve the usability and accuracy of the deep neural [16].

3.1 Alexnet

Alexnet is considered one of the most famous high-performing deep convolution neural networks. It has been trained on 1.2 million images. It can classify 1000 different object. The network has nearly 60 million parameters with about 650,000 neurons [5]. Alexnet structure consists of five convolution layers in addition to two fully connected layers.

The first layer is a convolutional layer. It filters the $224 \times 224 \times 3$ input images with 96 kernels. The size of each kernel is $11 \times 11 \times 3$ with a stride of 4 pixels. After pooling and normalizing the output of the first layer, it becomes input to the second convolutional layer. It filters the input with 256 kernels of size $5 \times 5 \times 48$ then applies pooling and normalization on the output. The third layer has 384 kernels of size

$3 \times 3 \times 256$. It takes the response of the second layer after pooling and normalization on it. There are no pooling or normalization between the third, fourth and fifth layers. They are connected one to another. The fourth and fifth convolutional layers have 354 kernels of size $3 \times 3 \times 192$ and 256 kernels of size $3 \times 3 \times 192$ respectively. The output from the fifth layer is pooled and become input for the sixth layer. The sixth and seventh layers consist of 4096 fully connected neurons. The last layer has 1000 fully connected neurons.

3.2 VGG-16 and VGG-19

VGG-19 [6] is another famous example of a deep CNN. Stacks of smaller sized convolutional filters are considered interesting features in VGG design. The use of very small same size convolutional filters in all network layers is the unique thing in its architecture. VGG network depth was increased by adding more convolutional layers. The philosophy of deeper-is-better is applied in the VGG net design. One disadvantage of these very deep networks is that they become very difficult to train [17].

4 Fish Dataset

The dataset used in this research are taken from QUT Robotics fish dataset [18]. This dataset consists of 3,960 images and contains real-world images of fish captured in conditions defined as "controlled", "out-of-the-water" and "in situ." The "controlled" images consist of several types of fish specimens, with their fins spread, taken against a constant background with controlled illumination. The "in situ" images are taken underwater in the fish natural habitat with no control over background or illumination. The "out-of-the-water" images consist of fish specimens, which are taken out of the water with a varying background and limited control over the illumination conditions.

In this research, eight family species of fish were selected according to the availability of the captured images. The size of the images varies in width and height. Table 1 illustrates the fish dataset description with some sub-species. Also, some training, validation, and testing images were provided. Testing images are taken from a different benchmark dataset LifeClef2015 [19]. LifeClef'15 dataset contains more than 20,000 images of fish divided into 15 classes of species, details of which is given in [19]. For each species, this dataset has a different number of available images. In this research, the same eight training classes were selected.

5 The Proposed Deep CNN System

The architecture of the deep network proposed for the aquarium family fish species identification is introduced in detail in Figs. 2 and 3. Figure 2 illustrates an abstract graphical representation of the proposed system, while Fig. 3 provides a detailed architecture. The proposed system is a simple version of the AlexNet [5]. The selection of AlexNet as it contains a minimum number of layers and accepted accuracy for training and validation over 90%.

The proposed new version is adapted and reduced from AlexNet to limit the number of parameters, computational complexity (in training, validation and testing phases), and memory. It consists of 10 layers, made up of two convolutional layers for features extraction, followed by two fully connected layers for classification.

The first layer is the input layer. The second layer is considered the convolution layer. The third layer, a Rectified Linear Unit (ReLU) is which used as nonlinear activation function, followed by the fourth layer (convolution layer). A moderate pooling is performed, with subsampling applied after the previous convolution. The fully connected layer has 256 neurons, respectively, with ReLU activation function, while the last fully connected layer has 8 neurons and uses a soft-max layer to obtain class memberships as illustrated in Fig. 3.

Table 1. Fish family species distribution for training, validation, and testing with sample images.

Fish Family Species	Sample Image	Number of Sub-Species	Total Images	Training Set	Validation Set	Testing Set
			----------- *QUT dataset* -----------			*LifeClef '15*
Bodianus		9	111	64	18	29
Coris		8	96	67	19	10
Epinephelus		29	286	188	54	44
Halichoeres		16	215	132	38	45
Lethrinus		12	143	91	26	26
Lutjanus		20	325	204	58	63
Pseudanthias		16	201	133	38	30
Thalassoma		9	144	89	25	30
Total		**119**	**1521**	**968**	**276**	**277**

Visualizing the feature extraction phase in the proposed deep neural architecture will give a better understating, Fig. 4 shows the different images resulted from applying first convolution layer and RELU to the input image. The second Visualizing the feature extraction phase in the proposed deep neural architecture will give a better understating, Fig. 4 shows the different images resulted from applying first convolution

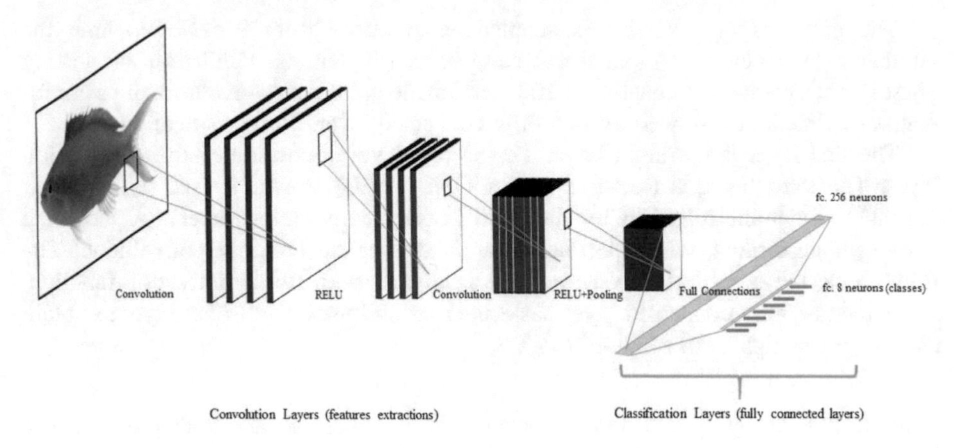

Fig. 2. Abstract view of the proposed deep CNN architecture

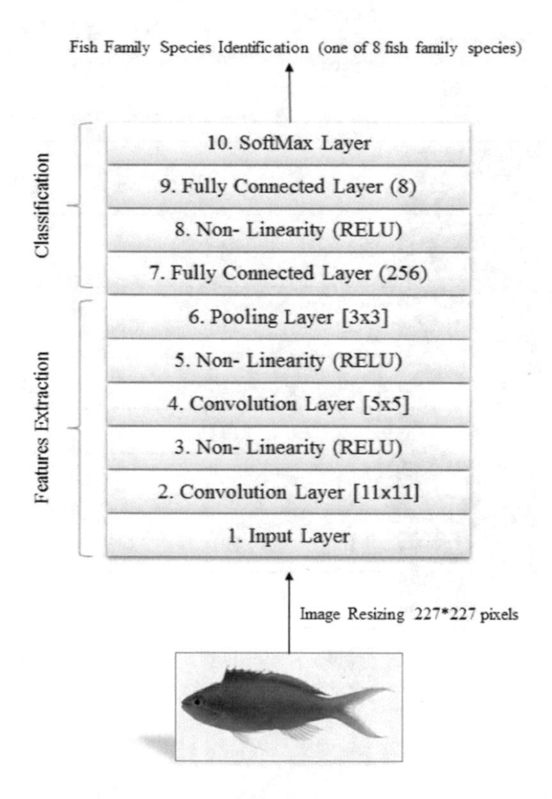

Fig. 3. Detailed component architecture for the proposed deep CNN system

layer and RELU to the input image. The second convolution layer and its RELU would produce more details (more features) from the output images after the first convolution and RELU layer as illustrated in Fig. 5.

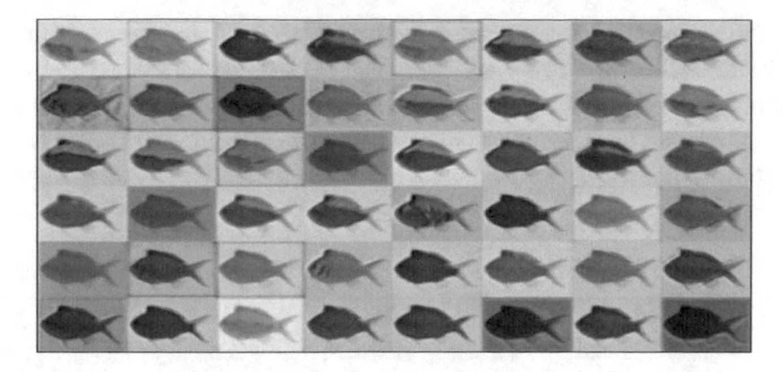

Fig. 4. Typical first convolutional and RELU layer features visualization

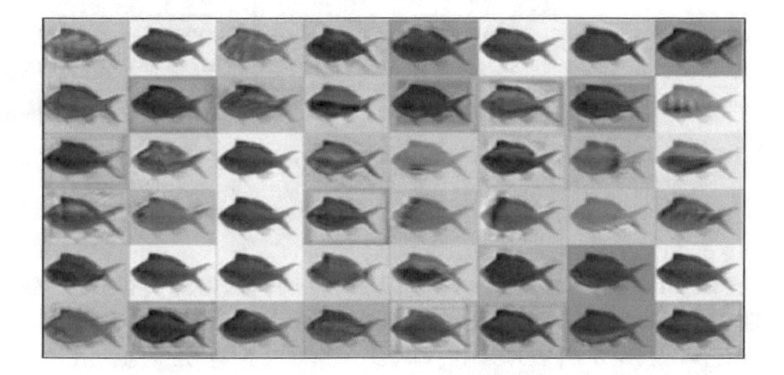

Fig. 5. Typical second convolutional and RELU layer features visualization.

6 Experiment Environment

The proposed system was implemented using a commercial software package (MATLAB), the implementation was GPU specific. All experiments were conducted on a server with Intel Xeon E5-2620 processor (2 GHz) and 96 GB Ram with Titan X GPU.

7 Experimental Results

To evaluate the proposed system, a different untrained fish benchmark dataset (LifeClef'15) was used for testing. It was compared against AlexNet, Vgg-16, and Vgg-19. The parameters used for comparison were (the number of convolution and fully connected layers, the number of epochs in training phase to achieve 100% accuracy, validation accuracy {QUT training dataset} and testing accuracy {LifeClef'15}). Table 2 illustrates the comparative results for family fish species identification.

The first comparative parameter is the number of convolution and fully connected layers; the proposed system has the least number of layers against other architectures that means less computational complexity (in training, validation and testing phases), and memory.

The second comparative parameter is the number of epochs in training phase to achieve 100% accuracy, the proposed system has the maximum number of epochs but the large number of epochs was expected as the proposed system network never trained before unlike the other architectures, so it takes more epochs to achieve 100% accuracy in training phase.

Table 2. Comparative results for fish family species identification

Model	Number of convolution and fully connected layers	# Epochs in training phase to achieve 100% accuracy	Validation accuracy (QUT training dataset)	Testing accuracy (LifeClef'15)
AlexNet	7	Epoch 13	98.63%	85.41%
Vgg16	16	Epoch 10	99.04%	87.86%
Vgg19	19	Epoch 8	99.64%	89.89%
Proposed system	4	Epoch 21	97.10%	85.59%

The third comparative parameter is the validation accuracy (QUT training dataset) which was used in training phase. The proposed system achieved 97.10%, this accuracy is very competitive with AlexNet which achieved 98.63% accuracy, but Alexnet was already trained before in a million images, same as vgg-16 and vgg-19.

The fourth comparative parameter is testing accuracy on (LifeClef'15 benchmark). The proposed system achieved 85.59% accuracy for untrained testing data. This accuracy outperforms AlexNet accuracy which achieved 85.41%. It is a small margin, but again all the other architectures were trained before and took a long time running their architectures for days in Matlab to build their network learning weights.

Although the proposed system doesn't outperform the other architectures validation accuracy, also, it doesn't outperform vgg-16 and vgg-19 in testing accuracy (as those architectures already loaded with learning), but it achieved better testing accuracy against AlexNet. The proposed system had less number of layers and tailored for fish family species identification and take less time in classification process and used in real time applications for fish family species identification in aquarium.

8 Conclusions

Real time aquarium fish identification system according to family species is an important topic. It will help tourists to know more information about fish pass in front of them. In this research, a proposed system for the deep neural network is introduced. The proposed system is a simple version for AlexNet. It consists of 4 layers, two convolution layers, and two fully connected layers. It can identify and classify eight family fish species with 119 sub-species. A comparative result is introduced, and it shows that the proposed system could not outperform vgg-16 and vgg-19 in validation and testing accuracy as they considered a large deep neural network with already trained million images while the proposed system trained on 1521 image. The proposed system outperforms AlexNet with a small margin in testing accuracy. It achieves 85.59% while AlexNet achieves 85.41% and have less trained images, less memory, less computational complexity in training, validation, and testing phases.

Acknowledgements. We gratefully acknowledge the support of NVIDIA Corporation with the donation of the Titan X GPU used for this research.

References

1. Fouad, M.M.M., Zawbaa, H.M., El-Bendary, N., Hassanien, A.E.: Automatic Nile Tilapia fish classification approach using machine learning techniques. In: 13th International Conference on Hybrid Intelligent Systems, HIS 2013, pp. 173–178. IEEE (2013)
2. Fouad, M.M., Zawbaa, H.M., Gaber, T., Snasel, V., Hassanien, A.E.: A fish detection approach based on BAT algorithm. In: The 1st International Conference on Advanced Intelligent System and Informatics, AISI 2015, pp. 273–283. Springer, Beni Suef (2016)
3. Dominguez, A.: A history of the convolution operation [Retrospectroscope]. IEEE Pulse **6**, 38–49 (2015). https://doi.org/10.1109/MPUL.2014.2366903
4. Lecun, Y., Bottou, L., Bengio, Y., Haffner, P.: Gradient-based learning applied to document recognition. Proc. IEEE **86**, 2278–2324 (1998). https://doi.org/10.1109/5.726791
5. Krizhevsky, A., Sutskever, I., Hinton, G.E.: ImageNet classification with deep convolutional neural networks. In: Proceedings of the 25th International Conference on Neural Information Processing Systems, pp. 1097–1105. Curran Associates Inc. (2012)
6. Russakovsky, O., Deng, J., Su, H., Krause, J., Satheesh, S., Ma, S., Huang, Z., Karpathy, A., Khosla, A., Bernstein, M., Berg, A.C., Fei-Fei, L.: ImageNet large scale visual recognition challenge. Int. J. Comput. Vis. **115**, 211–252 (2015). https://doi.org/10.1007/s11263-015-0816-y
7. Jiang, H., Learned-Miller, E.: Face detection with the faster R-CNN. In: 2017 12th IEEE International Conference on Automatic Face and Gesture Recognition, FG 2017, pp. 650–657. IEEE (2017)
8. Strachan, N., Kell, L.: A potential method for the differentiation between haddock fish stocks by computer vision using canonical discriminant analysis. ICES J. Mar. Sci. **52**, 145–149 (1995). https://doi.org/10.1016/1054-3139(95)80023-9
9. Rova, A., Rova, A., Mori, G., Dill, L.M.: One fish, two fish, butterfish, trumpeter: recognizing fish in underwater video. In: IAPR Conference on Machine Vision Applications, Tokyo, Japan, pp. 404–407 (2007)
10. Spampinato, C., Giordano, D., Di Salvo, R., Chen-Burger, Y.-H.J., Fisher, R.B., Nadarajan, G.: Automatic fish classification for underwater species behavior understanding. In: Proceedings of the First ACM International Workshop on Analysis and Retrieval of Tracked Events and Motion in Imagery Streams, ARTEMIS 2010, p. 45. ACM Press, New York (2010)
11. Khalifa, N.E.M., Taha, M.H.N., Hassanien, A.E., Selim, I.M.: Deep galaxy: classification of galaxies based on deep convolutional neural networks (2017). arXiv preprint arXiv:1709.02245
12. Sainath, T.N., Kingsbury, B., Mohamed, A., Ramabhadran, B.: Learning filter banks within a deep neural network framework. In: 2013 IEEE Workshop on Automatic Speech Recognition and Understanding, pp. 297–302. IEEE (2013)

13. Khalifa, N.E., Taha, M.H., Hassanien, A.E., Selim, I.: Deep Galaxy V2: robust deep convolutional neural networks for galaxy morphology classifications. In: 2018 IEEE International Conference on Computing Sciences and Engineering, ICCSE, pp. 122–127. IEEE (2018)
14. Bui, H.M., Lech, M., Cheng, E., Neville, K., Burnett, I.S.: Object recognition using deep convolutional features transformed by a recursive network structure. IEEE Access **4**, 10059–10066 (2017). https://doi.org/10.1109/ACCESS.2016.2639543
15. Scott, G.J., England, M.R., Starms, W.A., Marcum, R.A., Davis, C.H.: Training deep convolutional neural networks for land-cover classification of high-resolution imagery. IEEE Geosci. Remote Sens. Lett. **14**, 549–553 (2017). https://doi.org/10.1109/LGRS.2017.2657778
16. Lima, E., Sun, X., Dong, J., Wang, H., Yang, Y., Liu, L.: Learning and transferring convolutional neural network knowledge to ocean front recognition. IEEE Geosci. Remote Sens. Lett. **14**, 354–358 (2017). https://doi.org/10.1109/LGRS.2016.2643000
17. Srivastava, R.K., Greff, K., Schmidhuber, J.: Training very deep networks. In: Proceedings of the 28th International Conference on Neural Information Processing Systems, Montreal, Canada, pp. 2377–2385 (2015)
18. Anantharajah, K., Ge, Z., McCool, C., Denman, S., Fookes, C., Corke, P., Tjondronegoro, D., Sridharan, S.: Local inter-session variability modelling for object classification. In: IEEE Winter Conference on Applications of Computer Vision, pp. 309–316. IEEE (2014)
19. Joly, A., Goëau, H., Glotin, H., Spampinato, C., Bonnet, P., Vellinga, W.-P., Planqué, R., Rauber, A., Palazzo, S., Fisher, B., Müller, H.: LifeCLEF 2015: multimedia life species identification challenges. In: Mothe, J., Savoy, J., Kamps, J., Pinel-Sauvagnat, K., Jones, G., San Juan, E., Capellato, L., Ferro, N. (eds.) Experimental IR Meets Multilinguality, Multimodality, and Interaction. Lecture Notes in Computer Science. Springer, Cham (2015)

AMS: Adaptive Migration Scheme in Cloud Computing

Nesma Ashry, Heba Nashaat(✉), and Rawya Rizk

Electrical Engineering Department, Port Said University, Port Said 42523, Egypt
{nesma.sayed,hebanashaat,r.rizk}@eng.psu.edu.eg

Abstract. Due to the necessity of high availability for Cloud Systems, most organizations need cloud services which minimize system down-time. Virtual Machine (VM) Migration is a simple solution for what is called a Hot-Spot or highly utilized PM. Live Migrating VMs allows migration of VMs while they are running their applications with no shutdown or down-time. Therefore, it provides Hot-Spot relieve, helps in Business Continuity, and provides a high available system. In this paper, an adaptive migration scheme (AMS) is proposed in order to preserve the system's load balance state. It considers migrating VMs in groups, where each group includes a number of VMs increases from two to a maximum concurrent limit. AMS provides a way to find a maximum concurrent limit for the number of migrated VMs to be used to enhance the migration process. AMS is applied on a real case study. Two metrics are used to evaluate the improvements in migrations; concurrent time reduction and transfer rate improvement. The experimental results show that AMS achieves a time reduction percentage reaches 44%. This percentage is increased proportionally with the increasing of used memory. In addition, the migrated memory transfer rate is improved by a ratio ranges from 27% to 86%.

Keywords: Cloud computing · Live migration · Load balancing
Resource scheduling · VM migration

1 Introduction

Cloud computing has the facility to transform a large part of the Information Technology (IT) industry into services in which computer resources are virtualized and made available as a utility service, making the software smarter and defining how IT hardware is designed and purchased. Cloud computing also has the illusion of availability of as many resources like the processing time, memory and disk as demanded by the user. Moreover, users of cloud services can pay only for the resources used by them. From here comes the importance of resource allocation strategies in cloud systems [1, 2]. Cloud Services delivered to the users as Software as a Service (SaaS), Platform as a Service (PaaS), or Infrastructure as a Service (IaaS) [3].

Virtualization is considered the main enabling technology for cloud computing. It is a technique that allows running different Operating Systems (OSs) simultaneously on one physical machine (PM), isolated from each other and from the underlying physical infrastructure by the mean of a special middleware abstraction called virtual machine

A. E. Hassanien et al. (Eds.): AISI 2018, AISC 845, pp. 357–369, 2019.
https://doi.org/10.1007/978-3-319-99010-1_33

(VM). The piece of software that responsible for creating, running and managing these VMs is called Hypervisor or virtual machine kernel (VMkernel). It provides a mechanism for mapping VMs to physical resources from the cloud users.

VM placement in cloud systems whether for the first time or by changing its place depends on the resources needed by the VM in the available physical resources on each PM. In the first-time placement case, the VM is placed on a selected PM and gets its required resources from the PM's physical resources. In the second case, while there is a need to change the placement of VM from one PM to another, migration technique is needed.

Migration of VMs could be done offline or online. The offline migration is the default migration that needs the VM to be powered off. Then, it does not support high availability for the system [4]. The online migration is called live migration. It allows VMs to move from one PM to another within cloud transparently while VMs are still running. It is necessary for server virtualization as it provides the virtual and cloud systems with some benefits which include [5, 6]:

- **Load balancing**, accomplished by migrating VMs between PMs to balance physical resources utilization in the system and avoid hotspot or overloaded PMs.
- **Planned maintenance**, for high availability, when PMs need a hardware or software maintenance or update and there are important applications running on production VMs. These applications must be up and not be turned off while maintenance in order to preserve resource availability.
- **Consolidation**, where VMs on lightly loaded hosts can be packed onto fewer PMs with the consideration of meeting resource requirements and avoiding hotspot PMs. Then, the freed-up PMs can be turned off for power saving or supporting high resource availability for new VMs.

Resource Allocation in cloud computing has been studied from many different views. It could be at the application level which means finding the ways to assign the available resources to the needed cloud applications running on servers over the Internet [7–11]. Resource allocation could be used to balance and scale up and down virtualized computer environment by automatically migrating VMs among the system's PM [12–15].

One of the most effective resource allocation services is the Distributed Resource Scheduler (DRS) [16]. It is a VMware service which distributes the VMs among PMs depending on the system's load balance state. DRS depends on its operation on live migration technique and supports concurrent VM migrations. Concurrent migration is applied on VMs that are located on the same PM and shares some memory pages which are called co-located VMs. To enhance the system parameters such as memory, CPU, and network usage during migration, a technique called Live Gang Migration (LGM) is used for the concurrent migrations. LGM is used to handle the migration of co-located VMs. As shared memory pages between these VMs are needed to be transferred only once during the migration, LGM uses a de-duplication technique with a method called content-hashing to define the identical memory content on both page and sub-page levels to decrease the migration overhead. This LGM gives good results, but it lakes to the decision-making scheme that can be used to decrease the number of migrations by considering a holistic approach to the system with fast load balancing [17].

Migration-based Elastic Consolidation Scheduling (MECS) is presented in [18]. It is a scheduling mechanism that adaptively handles the variable workloads in cloud systems. The number of PMs is minimized while satisfying the VM Service Level Agreement (SLA). However, it does not support concurrent migrations. Therefore, applying this algorithm would lead to high memory and network usage.

In this paper, an Adaptive Migration Scheme (AMS) is proposed to support a maximum number of concurrent live migrations based on the available resources in each cloud system. The proposed algorithm is used to determine the best VMs' placement decision in order to obtain a fast load balancing of the system while minimizing the memory and network usage by applying the LGM technique. AMS is used to dynamically adapt the VM live migration. This done by considering the maximum number of migration per PM that does not violate the SLA by keeping the system load balance and avoiding server sprawl. The proposed scheme is experimentally tested on a real Vmware cloud environment for handling a maintenance case for the system.

The rest of this paper is organized as follow: Sect. 2 describes migration types, techniques, and the cases need live migration in cloud systems. The proposed AMS algorithm with LGM technique is presented in Sect. 3. Sections 4 and 5 introduce the experimental results of the maintenance case study and comparative analysis with the existing scheme; respectively. Finally, the main conclusions and future work are presented in Sect. 6.

2 VM Migration

VM Migration techniques have two major categories; offline migration or live migration. Offline migration is used in such cases where moving a powered off the machine, or using a suspend-and-copy technique which suspends a VM, copies all its memory pages, and then resumes running the VM on the target PM. In contrast, live migration allows migrating VMs among different PMs while VMs are still up and running. It uses one of two sub-techniques; pre-copy or post-copy. When using VM live migration, VM files must be located on a shared storage between PMs. This shared storage whether Network File System (NFS) share or Storage Area Network (SAN) Storage is not located locally on each PM's internal disks. The shared storage is used to eliminate the purpose of migrating the VMs' disk state since that, live migration only copies the in-memory state and the content of CPU registers between PMs.

Figure 1 shows live migration techniques. They are mainly two techniques [19]:

- **Pre-copy:** It transfers VM's memory pages iteratively from the source PM to the target PM while the VM is still up and running at the source PM. In the first iteration, the VM's entire memory is transferred, whereas in the subsequent iterations, only the pages modified during the preceding iteration are transferred. Upon converging on the VM's writable working set, the VM is suspended at the source, its writable working is set, and the execution state is transferred to the destination, where the VM resumes. If any disk migration is needed in case of storage migration which is the big part, it is performed prior to the memory migration. The time during which the VM remains suspended is called its downtime, whereas the time

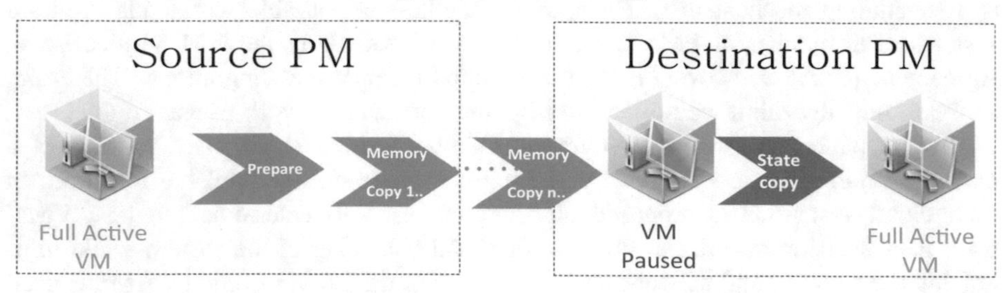

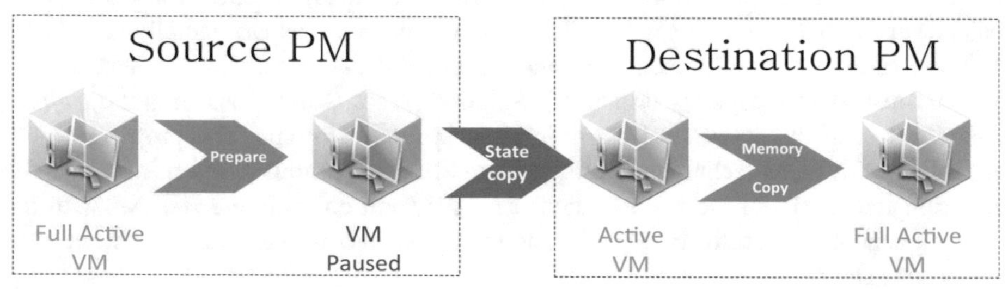

Fig. 1. Pre-copy versus post-copy migration

taken to complete the entire migration is called its total migration time. Since dirtied pages are retransmitted, write-intensive applications increase the VM's total migration time and the network bandwidth.

- **Post Copy:** It suspends the VM at the source PM, then copies its execution state to the destination PM and the VM is resumed there. Demand-paging is used over the network to fetch the remaining state. Post-copy has small total migration time and low network overhead since it transfers each page only once in contrast to pre-copy which retransmits dirtied pages. It is a fast technique since the VM resumes execution at the destination quickly after the migration starts. Therefore, the presented work is focused to use post-copy technique due to its advantages over pre-copy.

3 The Proposed AMS

The proposed AMS performs an adaptive post-copy live migration with minimum migration time. AMS migration is presented as a concurrent migration depends on the system resources "Memory". The number of concurrent migrations is determined by periodically load balancing check. The migratable VMs are selected to be migrated either individual or grouped in an adaptive way. The maximum number of VMs in each group is adapted by increasing it from two to a maximum concurrent limit. The maximum concurrent limit for migrations is decided depending on system utilization check. AMS is applied in real environments. It provides a mechanism for mapping

VMs to physical resources in order to adapt the number of concurrent live migrations based on the available resources "Memory". The applied case study in the real cloud environment is a physical PMs' maintenance case study.

3.1 System Architecture

The architecture of the system used to apply AMS is presented in Fig. 2. It consists of three PMs have the same processor vendor and model but differs from the other physical resources such as RAM, Disk, and Network Interface Cards (NICs). Each PM runs Esxi Hypervisor and there is a management server called VMware vCenter Server (VC) which gathers PMs into clusters. Each VM in the cluster shares physical resources of all PMs in the cluster such as CPU, Memory and NICs. All PMs share backend storage and all the VMs' files are located on it. In cloud computing, the Hypervisor is a piece of software that creates, runs, and manages VMs and provides a mechanism for mapping VMs to physical resources [20].

3.2 AMS Scheduling Operation

AMS is presented to manage the PMs' parameters of the physical resources automatically while considering the available memory and migration time. It is an adaptive scheduling algorithm that checks the system periodically and decides the best migration decesions. It includes two major phases:

- **The first phase:** AMS checks periodically the PMs' parameters in the cluster to reach a load balancing state. The load balancing state of the system will be determined by each PM volume. Then, the PM volumes are used to calculate the standard deviation for the entire system.

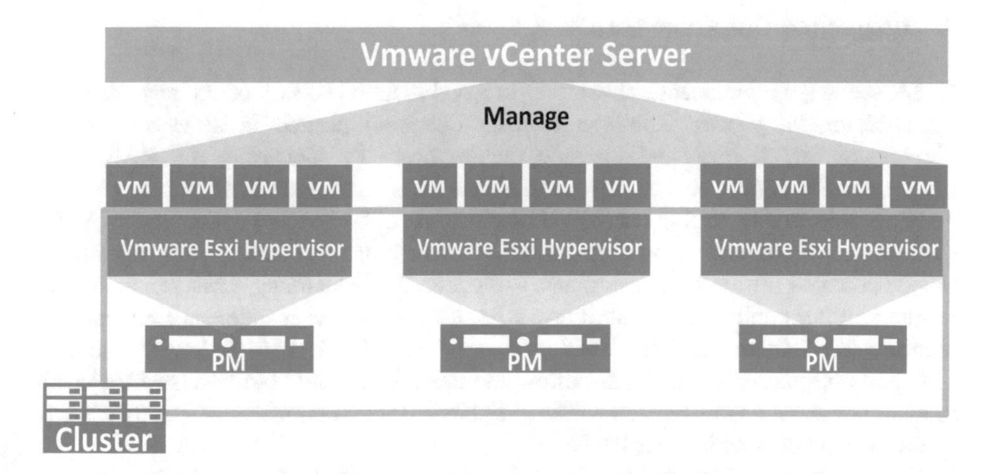

Fig. 2. Vmware virtualization architecture

The PM volume (PMV_i) is determined for each PM as follows [6]:

$$PMV_i = \frac{1}{1 - ram_i\%} \tag{1}$$

where i is the PM number. Then, based on each PMV_i, the system's current standard deviation (σ) is calculated as follows [20]:

$$mean = \frac{\sum_{i=1}^{no.\ of\ PMs} PMV_i}{no.\ of\ PMs} \tag{2}$$

$$\sigma = \sqrt{\frac{\sum_{i=1}^{no.\ of\ PMs} (PMV_i - mean)^2}{no.\ of\ PMs}} \tag{3}$$

σ is the main indicator for the system load balance state. It is calculated for the system in each migration decision as a main reference for the utilization check.

- **The second phase:** It determines the best migration decisions for VMs to PMs by system utilization check. This check is done by simulating the VMs placement from the source host; which will be maintained, to the rest of PMs which have enough resources for these VMs. The resulted σ after this simulation can be used to choose the best migration decision with the maximum concurrent limit for VMs. This maximum concurrent limit is simulated in a sequential order. First, the system allows migration one VM at a time. Second, it allows migration two or three VMs at a time. Finally, it allows migration the max concurrent limit for VMs by system utilization check based on σ values.

4 Experimental Case Study

In this section, a maintenance case study in a real system is developed to test our AMS by applying its phases. The maintenance operation needed is done to upgrade Hypervisor for all PMs and upgrade the hardware by increasing the RAM. The maintained PM should be shut down and all the VMs running on its Hypervisor should be either migrated or shut down also. Therefore, each PM has two collections of VMs; Collection 1 is containing the migratable VMs which are production VMs that should not be powered off for any reason, and Collection 2 is containing VMs which are not production ones and could be shut down for a while till the maintenance ended. The identification manner used for VMs' and PMs' in Table 1; PMs follow the naming scheme "mxegdam xx" and VMs follow the naming scheme "MXEGDAM xx" with the number at the end of the name. Table 1 shows PMs' volume as PMV and the VMs in the two collections in each PM.

In the scenario illustrated in Fig. 3, there are three PMs; "mxegdam09", "mxegdam10", and "mxegdam11", managed under one cluster called "DMZ cluster". For simplifying in figure; the PMs' and VMs' are identified only by its number. PMs in the system have the same amount of memory; 49142 MB, but not the same amount of used

Table 1. PMs' parameters and VMs collections

PMs in DMZ cluster	PMV	Migratable VMs collection 1		Shut down VMs collection 2	
		VM name	Used memory (MB)	VM name	Used memory (MB)
mxegdam09	2.4	MXEGDAM 33	3298	MXEGDAM 35	2441
		MXEGDAM 54	4138	MXEGDAM 38	749
		MXEGDAM 65	3217	MXEGDAM 43	977
		MXEGDAM 66	2549	MXEGDAM 44	635
		MXEGDAM 67	1201	MXEGDAM 55	3828
				MXEGDAM 56	2880
				MXEGDAM 59	651
				MXEGDAM 63	1047
				MXEGDAM 64	1048
mxegdam10	2	MXEGDAM 31	1909	MXEGDAM 42	0
		MXEGDAM 40	3101	MXEGDAM 49	0
		MXEGDAM 48	751	MXEGDAM 52	3624
		MXEGDAM 72	842	MXEGDAM 68	2076
		MXEGDAM 75	1898	MXEGDAM 83	0
		MXEGDAM 77	2953		
		MXEGDAM 81	1676		
mxegdam11	4.9	MXEGDAM 30	1044	MXEGDAM 36	2681
		MXEGDAM 37	2366	MXEGDAM 74	631
		MXEGDAM 39	1044	MXEGDAM 78	2393
		MXEGDAM 41	932		
		MXEGDAM 45	1570		
		MXEGDAM 46	4061		
		MXEGDAM 47	628		
		MXEGDAM 51	3286		
		MXEGDAM 53	1349		
		MXEGDAM 57	2484		
		MXEGDAM 58	1046		
		MXEGDAM 60	801		
		MXEGDAM 62	1404		
		MXEGDAM 69	2077		
		MXEGDAM 71	545		
		MXEGDAM 76	1739		
		MXEGDAM 79	3151		

memory as each PM has a different number of VMs with different parameters. As shown in Fig. 3(a) each PM has the run state for a collection of VMs, and the entire system in a load balance state. The Hypervisor installed on each PM will be upgraded sequentially by using the proposed AMS. The two phases of the proposed AMS are illustrated as follows:

Phase 1: In this phase, PMs' parameters in the cluster are calculated.

All the three PMs are up and running with PMV_i shown in Table 1. The load balance state of the system is determined by calculating σ using Eqs. (1)–(3). This value is evaluated before any migrations as in Fig. 3(a). After calculating σ, it's found to be 1.26.

Phase 2: In this phase, the best migration decisions are determined.

As mentioned before, the migratable VMs are selected to be migrated either individual or grouped in an adaptive way. The maximum number of VMs in each group is determined increasingly from 2 to maximum concurrent limit. In this scenario, the maximum concurrent limit is calculated based on utilization check. It is evaluated to be equal to 4.

In order to prove that the proposed AMS can keep the system in a load balance state during the maintenance procedure with VMs' migrations and shutdowns; σ is calculated for the system after migration for each PM shutdown.

- In Fig. 3(b), "mxegdam09" PM is powered off for maintenance. VMs in Collection 1 are migrated to "mxegdam10" PM and VMs in Collection 2 are shut down. As presented in Table 2, each VM in Collection 1 is migrated individually. The load balance state of the system is evaluated by using Eqs. (1)–(3) and it is determined by $\sigma = 0.07$. After upgrade process, "mxegdam09" PM is powered on again and all its VMs are back as before maintenance.
- In Fig. 3(c), "mxegdam10" PM is powered off for maintenance. VMs in Collection 1 are migrated to" mxegdam09" PM and VMs in Collection 2 are shut down. As presented in Table 2," MXEGDAM40" VM is migrated individually, however the others are categorized into two groups; Group_1 and Group_2. The load balance state of the system is determined by $\sigma = 0.98$. After upgrade process, "mxegdam10" PM is powered on again and all its VMs are back as before maintenance.
- In Fig. 3(d), "mxegdam11" PM is powered off for maintenance. VMs in Collection 1 are migrated to "mxegdam09" and "mxegdam10" PMs and VMs in Collection 2 are shut down. Collection 1 VMs are categorized into five groups; from Group_3 to Group_7. Group_3, Group_4 and Group_6 are migrated to "mxegdam09" PM with shutdown process for some VMs from Collection 1 in "mxegdam09"; {"MXEGDAM43", "MXEGDAM44", "MXEGDAM55", "MXEGDAM59", "MXEGDAM63"} to keep the load balance state. Then, Group_5 and Group_7 are migrated to "mxegdam10" PM. The load balance state of the system is determined by $\sigma = 0.1$. After upgrade process, "mxegdam10" PM is powered on again and all its VMs are back as before maintenance. After upgrade process, "mxegdam10" PM is powered on again and all its VMs are back as before maintenance.

The migration scenario is repeated and the average results for 100 iterations are calculated to get more real data. All migrations here occurred with respect to keep the system in a load balance state and avoid hotspot PMs. All migration steps' σ values as a load balance state indicator prove that AMS preserves the load balance state for the

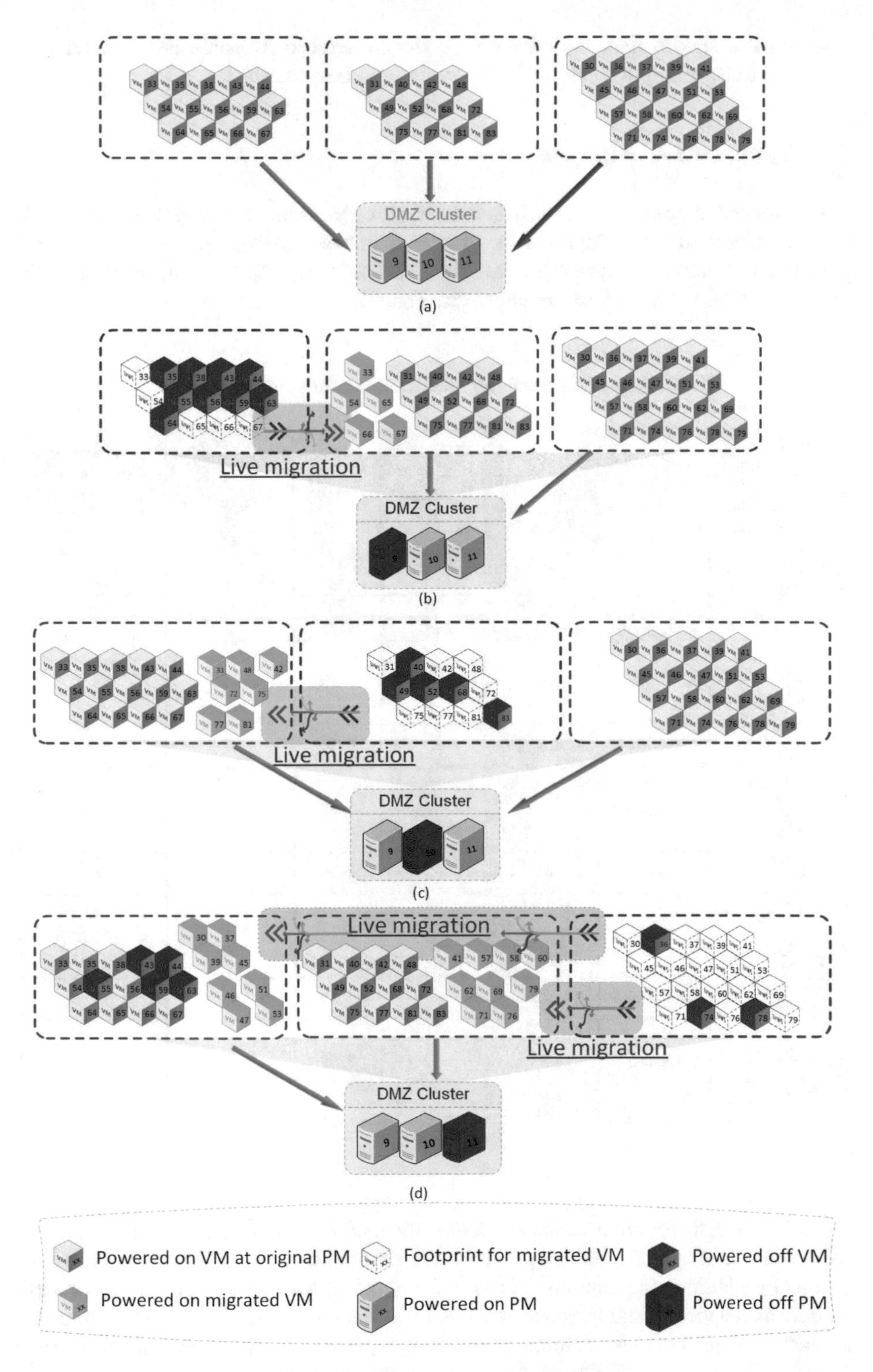

Fig. 3. Maintenance case study simulation

system. It is shown that, the values of σ do not exceed its value before migration. Whenever decreasing σ value indicates more load balance state.

5 Comparative Analysis

In this section, a comparative study is presented for VMs that are migrated individually and the others that are migrated in groups as in the existing scheme [12]. Table 2 presents in details all system's VMs in Collection 1. The used memory and the migration time for each VM are shown in Table 2.

Table 2. Maintenance migrations results for collection 1 VMs

VMs Name MXEGDAM XX		**Destination PM**	**Concurrent**	**Migration Time/VM (sec.)**	**Used Memory (MB)**	**Total Migration Time / VM (sec.)**	**Concurrent Time reduction Improvement %**
mxegdam09	33	mxegdam10	**Individual**	41	3298	41	-
	54		**Individual**	40	4138	40	-
	65		**Individual**	35	3217	35	-
	66		**Individual**	37	2549	37	-
	67		**Individual**	21	1201	21	-
Mxegdam10	40	mxegdam09	**Individual**	45	3101	45	-
	31	mxegdam09	Group_1	61	5400	62	35%
	48			40			
	72			49			
	75			62			
	77	mxegdam09	Group_2	47	4629	47	30%
	81			37			
mxegdam11	30	mxegdam09	Group_3	40	7958	67	44%
	37			56			
	39			67			
	45			50			
	46	mxegdam09	Group_4	55	4689	55	7%
	47			32			
	41	mxegdam10	Group_5	48	5263	65	29%
	57			65			
	58			55			
	60			45			
	51	mxegdam09	Group_6	50	4635	50	18%
	53			31			
	62	mxegdam10	Group_7	51	5765	56	41%
	69			56			
	71			22			
	76			56			
	79	mxegdam10	**Individual**	35	3151	35	-

In Table 2, the column which represents the total migrated memory is evaluated for individuals as it appears, but for groups is the summation of all VMs' memory in each group. In addition, the column called total migration time/VM is evaluated for individuals as it appears, and its values are calculated as the maximum migration time/VM in each group. The last column presents the percentage of improvement in time reduction when using the concurrent migration by AMS. As presented in the last column in Table 2, the percentages show that when using concurrent migration, there is a significant reduction in the migration time. The time reduction percentage varies from 7% to 44%. This percentage is increased proportionally with increasing the used memory.

The Migration Transfer Rate Improvement metric is another metric used to indicate the improvement in the proposed AMS. It can be calculated as follows:

$$\text{Migration Transfer rate} = \frac{\text{Used Memory}}{\text{Migration Time/VM}} \quad (4)$$

The Migration Transfer Rate Improvement is calculated with respect to the transfer rate for each group and the average transfer rate for the VMs in each group if the migration had done individually. Figure 4 shows how applying the AMS with concurrent migrations improve the Migration Transfer Rate especially when using the maximum concurrent limit for VMs in the group. The bars for Group 2, Group 4 and Group 6 represent the improvement percentage in groups with two VMs with a total used memory between 4629 and 46899 MBs. The improvement in this case ranges from 27% to 44%, and this considers a good improvement. The bars for Group 1, Group 3, Group 5 and Group 7 represent the improvement percentage in groups with four VMs; the maximum concurrent limit, with a total used memory between 5263 and 7958 MBs. The improvement in this case is between 49% and 86%. It is shown from these results that applying AMS with its maximum concurrent limit speeds up the migration process while preserving the load balancing state of the system.

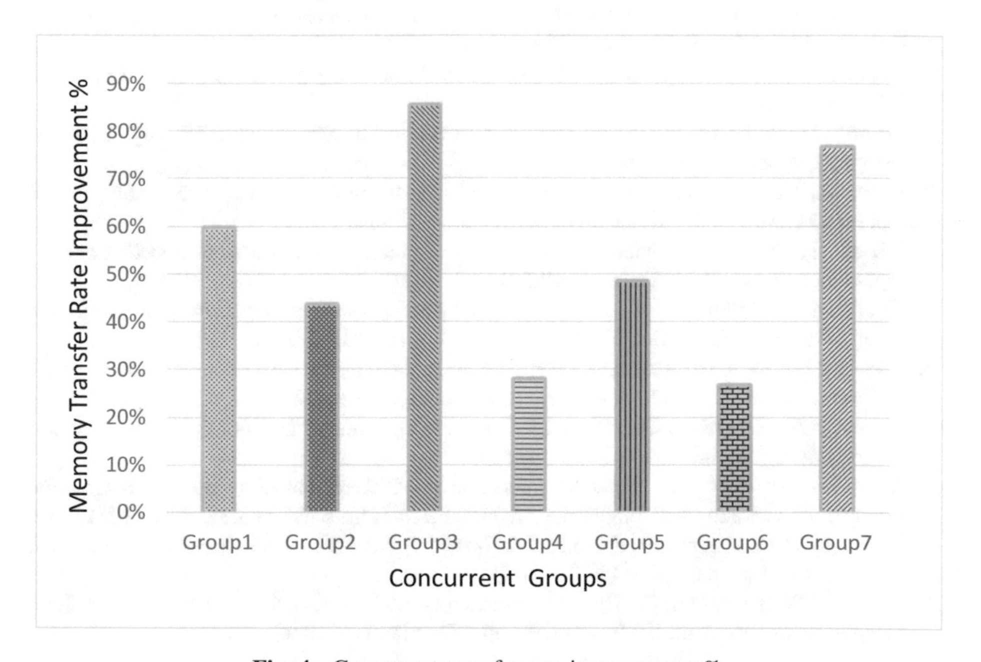

Fig. 4. Concurrent transfer rate improvement %

6 Conclusions

Live migration technique allows VMs to move from one PM to another within cloud transparently while VMs are still running. This is a critical issue in improving the cloud system performance. In this paper, the proposed AMS is used to dynamically adapt the VM live migration with minimal migration time and maximum migration transfer rate. AMS provides a way to find a maximum concurrent limit for the migrated VMs' number and this limit is used to enhance the migration process. All migrations occurred in the case study lead to an acceptable load balance state for the system. The concurrent time reduction percentages show that when using two VMs per group as concurrent migration, there is a significant reduction in the migration time up to 30% and when using the maximum concurrent limit that is equal to four VMs per group, the reduction reaches 44%. In addition, the migrated memory transfer rate is improved in a ratio between 27% and 86%. Therefore, it is proved that the proposed AMS causes an effective enhancement in migration performance in the system.

References

1. Hashem, W., Nashaat, H., Rizk, R.: Honey bee based load balancing in cloud computing. KSII Trans. Internet Inf. Syst. **11**(12), 5694–5711 (2017)
2. Gamal, M., Rizk, R., Mahdi, H., Elhady, B.: Bio-inspired load balancing algorithm in cloud computing. In: The International Conference on Advanced Intelligent systems and Informatics, AISI, Cairo, Egypt, pp. 579–589(2017)
3. Gorelik, E.: Cloud computing models, comparison of cloud computing service and deployment models. The MIT Sloan School of Management and the MIT Engineering Systems, Massachusetts Institute of Technology (2013)
4. López-Pires, F., Barán, B.: Many-objective virtual machine placement. J. Grid Comput. **15** (2), 161–176 (2017)
5. Strunk, A.: Costs of virtual machine live migration: a survey. In: IEEE Eighth World Congress conference on Services, Honolulu, HI, USA, pp. 323–329 (2012)
6. Ishra, M., Das, A., Kulkarni, P., Sahoo, A.: Dynamic resource management using virtual machine migrations. IEEE Commun. Mag. **50**(9), 34–40 (2012)
7. Sheng, D., Cho-Li, W.: Dynamic optimization of multiattribute resource allocation in self-organizing clouds. IEEE Trans. Parallel Distrib. Syst. **24**(3), 464–478 (2013)
8. Gouda, K.C., Radhika, T.V., Akshatha, M.: Priority based resource allocation model for cloud computing. Int. J. Sci. Eng. Technol. Res. **2**(1), 2015 (2013)
9. Abirami, S.P., Ramanathan, S.: Linear scheduling strategy for resource allocation in cloud environment. Int. J. Cloud Comput. Serv. Archit. **2**(1), 9 (2012)
10. Omara, F.A., Khattab, S.M., Sahal, R.: Optimum resource allocation of database in cloud computing. Egypt. Inf. J. **15**(1), 1–12 (2014)
11. Abar, S., Lemarinier, P., Theodoropoulos, G.K., O'Hare, G.M.P.: Automated dynamic resource provisioning and monitoring in virtualized large-scale datacenter. In: IEEE 28th International Conference on Advanced Information Networking and Applications, AINA, Victoria, BC, Canada, pp. 961–970 (2014)
12. Zhen, X., Weijia, S., Qi, C.: Dynamic resource allocation using virtual machines for cloud computing environment. IEEE Trans. Parallel Distrib. Syst. **24**(6), 1107–1117 (2013)

13. Yexi, J., Chang-Shing, P., Tao, L., Chang, R.N.: Cloud analytics for capacity planning and instant VM provisioning. IEEE Trans. Netw. Serv. Manag. **10**(3), 312–325 (2013)
14. Minarolli, D., Freisleben, B.: Distributed resource allocation to virtual machines via artificial neural networks. In: The 22nd Euromicro International Conference on Parallel, Distributed and Network-Based Processing, PDP, Torino, Italy, pp. 490–499 (2014)
15. Padala, P.: Resource management in VMware powered cloud: concepts and techniques. In: The 27th International IEEE Symposium on Parallel & Distributed Processing, IPDPS, Boston, p. 581 (2013)
16. Jie, Z., Ng, T.S.E., Sripanidkulchai, K., Zhaolei, L.: Pacer: a progress management system for live virtual machine migration in cloud computing. IEEE Trans. Netw. Serv. Manag. **10** (4), 369–382 (2013)
17. Deshp, U., Wang, X., Gopalan, K.: Live gang migration of virtual machines. In: The 20th International Symposium on High Performance Distributed Computing, San Joes, CA, USA, pp. 135–146 (2011)
18. Qingjia, H., Sen, S., Siyuan, X., Jian, L., Peng, X., Kai, S.: Migration-based elastic consolidation scheduling in cloud data center. In: IEEE 33rd International Conference on Distributed Computing Systems Workshops, ICDCSW, Philadelphia, PA, USA, pp. 93–97 (2013)
19. Deshp, U., Chan, D., Guh, T., Edouard, J., Gopalan, K., Bila, N.: Agile live migration of virtual machines. In: IEEE Symposium on International Parallel and Distributed Processing, IPDPS, Chicago, USA (2016)
20. Epping, D.: DRS Deepdive. http://www.yellow-bricks.com/drs-deepdive

Fully Homomorphic Encryption with AES in Cloud Computing Security

Yasmin Alkady, Fifi Farouk, and Rawya Rizk(✉)

Electrical Engineering Department, Port Said University, Port Said, Egypt
r.rizk@eng.psu.edu.eg

Abstract. With growing awareness and concerns in regards to cloud computing and information security with privacy protection, there is a need to increase a usage of security algorithms into data systems and processes. Homomorphic Encryption (HE) is the encryption scheme which accepts encrypted inputs and performs blind processing to achieve data confidentiality. HE is useful to transfer encrypted data in public area as it allows operations on the cipher text, which can provide the same results after calculations as working directly on raw data. In this paper, a Fully Homomorphic Encryption (FHE) system based on Advanced Encryption Standard (AES) is proposed. It can be applied to perform operations on encrypted data without decryption. The proposed scheme solves the problem of large cipher text usually associated with increased noise resulting from FHE usage.

Keywords: Advanced encryption standard cryptography · Cloud computing Confidentiality · Homomorphic encryption · Privacy protection

1 Introduction

Nowadays, the need of resources is highly increasing to process and maintain huge data in various fields. However, owning these resources and maintaining them are difficult tasks. Cloud computing [1, 2] is the recent efficient and effective technology which provides these resources on demand with minimal management effort. According to service model [3, 4], cloud computing can be categorized into three main categories: Infrastructure-as-a-Service (IaaS), Platform-as-a-Service (PaaS) and Software-as-a-Service (SaaS).

In 2009, Microsoft SIDEKICK service was interrupted for a week, a large number of users cannot access their emails and other personal data. More seriously, due to shortage of data backups, Microsoft could not recover lost data. Although the cloud storage service can realize multi copy of fault tolerance and backup automatically [5, 6], it also cannot guarantee 100% security. The sensitive data should be accessed only by the authorized users. Instead of being forced to fully trust the cloud provider, in cloud computing; security ensures encryption of the data stored. Traditionally, secure transmission from a local machine to cloud data store is performed with key exchanges. However, performing computations on that data stored in the cloud requires decrypting it first. This makes critical data available to the cloud provider. Data mining and other

A. E. Hassanien et al. (Eds.): AISI 2018, AISC 845, pp. 370–382, 2019.
https://doi.org/10.1007/978-3-319-99010-1_34

data analysis onto the encrypted database is a far distant thing to achieve by using encryption standards currently available.

The proposal in this paper is to encrypt data before sending to the cloud providers. Thereby, it enables a cloud computing vendor to perform all types of computations on clients' data at their request without exposing original data. To achieve this, it is necessary to use the cryptosystems based on Fully Homomorphic Encryption (FHE). The main feature of proposed scheme is to reduce the noise which is considered as a disadvantage in current FHE schemes, similar to bootstrapping and modulus switching. The dimension of ciphertexts of the proposed scheme does not change with homomorphic operations. So, the proposed scheme is compact. In addition, it can directly encrypt big integers, rather than only bit messages. The proposed scheme is based on matrix operations which are computationally "light". It uses AES symmetric keys of small size thereby making it suitable for many data centric applications. It derives its security from hardness of factorizing a large integer, which is the main disadvantage of many public key cryptosystem.

This paper is organized as follows: brief overview of homomorphic types and their related works are presented in Sect. 2. The proposed scheme is introduced in Sect. 3. Section 4 presents the performance and implementation results of the proposed scheme. Finally, the main conclusions are presented in Sect. 5.

2 Homomorphic Encryption for Data Security

Among cloud computing characteristics is sharing of conservation structures and data processing. One problem of this is the preservation of confidentiality between client and provider. Encryption could alleviate this issue, since the customer can decide to store only encrypted data. So if the client wants to perform calculations on his data in the cloud, the secret key to decrypt the data should be shared with the provider. Sharing the key would allow the cloud provider to access the data and the traditional way to solve Secure Multiparty Computation (SMC) problem is using Third Party Auditor (TPA), however, TPAs are particularly hard to achieve.

Homomorphic Encryption (HE) [7] gives solutions that accept encrypted inputs and performs blind processing. Suppose there is HE which can translate operations on integers to operations on polynomials of single variable. As shown in Fig. 1, two integers are encrypted into polynomials p1(x) and p2(x). Then, when these polynomials are added to give a third polynomial, it is required that the resultant polynomial when translated back should be equal to the sum of plaintext integers.

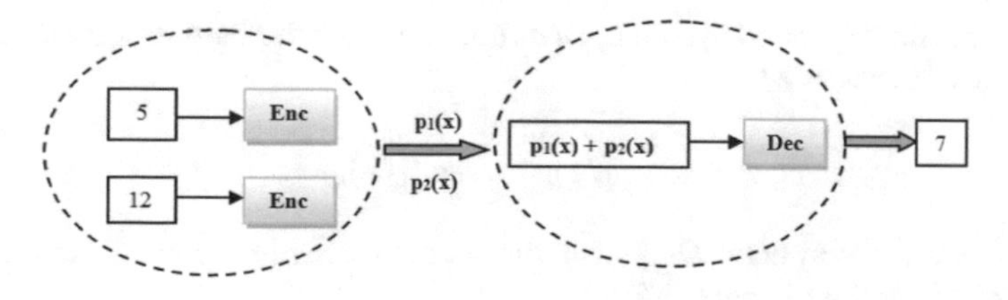

Fig. 1. Concept of HE.

The aim of HE is to ensure privacy of data in communication and storage processes, such as the ability to delegate computations to un-trusted parties. This makes cloud services more reliable and it offers security with lest overheads. The concept of computations on encrypted data without decryption was first introduced by Rivest, Adleman and Dertouzos in 1978 [8]. HE has the following four functions [Key generation (*KeyGen*), Encryption (*Enc*), Decryption (*Dec*), Evaluation (*Eval*)].

HE has been defined till now in public key system only. But in the proposed scheme, the definition is extended to allow the usage of HE in both symmetric as well as public key systems. HE scheme is a quadruple of probabilistic-polynomial time algorithms *HE = (KeyGen, Enc, Dec, Eval)*;

- *KeyGen:* In public key based systems the key generation function takes λ and generates keys *pk*, *sk* and *ek*.

$$(pk, sk, ek) \leftarrow keyGen(\lambda) \tag{1}$$

where λ is a security parameter, *pk* is a public key, *sk* is a private key and *ek* is evaluation key. In symmetric key system algorithm, the key generation function takes λ and generates *k* and *ek*.

$$(k, ek) \leftarrow keyGen(\lambda) \tag{2}$$

where *k* is a secret key.

- *Enc:* In public key based systems, *Enc* function takes *pk* and *M* to be encrypted and gives *C*.

$$(C) \leftarrow Enc_{pk}(M) \tag{3}$$

where *M* is a plaintext and *C* is a ciphertext. In symmetric key system, *Enc* takes *k* and *M* to be encrypted and gives *C*.

$$(C) \leftarrow Enc_k(M) \tag{4}$$

- *Eval:* It applies a function to ciphertext. Using evaluation key is optional. In public key based system, *pk* = *ek* and in symmetric system, *k* = *ek*.

$$(C') \leftarrow Eval_{ek}(f, C) \tag{5}$$

where function *f* is an arithmetic circuit or Boolean circuit and *C'* is a finally ciphertext.

- *Dec*: In public key based systems, *Dec* function takes the output of *Eval* function *C'* and *sk* and recovers *M*.

$$(M) \leftarrow Dec_{sk}(C') \tag{6}$$

In symmetric key system, *Dec* takes the output of *Eval* function *C'* and secret key *k* and recovers the plaintext M.

$$(M) \leftarrow Dec_k(C') \tag{7}$$

Therefore, the importance of homomorphism is to decouple the ability to perform computations from the necessity to view the data as clear text. This allows owners of sensitive data to manipulate encrypted secret data while it resides in an insecure location or to outsource computations on secret data to an un-trusted third party. HE can be broadly classified into Partial HE (PHE) and FHE.

2.1 Partially Homomorphic Encryption (PHE)

PHE properly perform only a limited number of operations. For example, either addition or multiplication on encrypted data due to an inability to properly decrypt after a certain threshold of noise is introduced by the operations:

Additive HE: HE is additive as the Pailler and Goldwasser-Micalli [9–11] cryptosystems, if:

$$Enc_k(M_1 \oplus M_2) = Enc_k(M_1) \oplus Enc_k(M_2) \tag{8}$$

Multiplicative HE: HE is multiplicative as RSA [12] and Elgamal [13] cryptosystems.

$$Enc_k(M_1) \otimes (M_2) = Enc_k(M_1 \times M_2) \tag{9}$$

2.2 Fully Homomorphic Encryption (FHE)

FHE [14–16] is a cryptosystem that performs additions and multiplications operations that can be performed on encrypted data without affecting the ring structure of plaintexts.

In 2009 Craig Gentry of IBM has proposed the first FHE that evaluates an arbitrary number of additions and multiplications and thus calculates any type of functions on encrypted data. The internal working of this adds another layer of encryption every few steps and uses an encrypted key to unlock the inner layer of scrambling. This decryption "refreshes" the data without exposing it. In [17], a state-of-art survey of FHE is presented and how it can be applied for delegation of computation. It raised an open question "Can HE be efficient enough to be practical?" Many researchers have done extensive work towards making FHE practical such as the following:

Smart FHE [18] presented an FHE scheme based on elementary theory of algebraic number fields (a finite extension of the field of rational numbers Q). To improve the efficiency, this scheme uses smaller key and ciphertext. One of the drawbacks of this scheme is that it takes longer time to generate the keys.

Marten FHE [19] presented simpler Somewhat (SWHE) by using the elementary modular arithmetic methods like the addition and multiplication over the integers and approximate Greatest Common Divisor (GCD). The efficiency of this scheme is found to be low due to the difficulty in preserving the hardness of the approximate GCD problem. However the public-key was in $O(\lambda^{10})$ which is too large for any practical system.

Zvika FHE [20, 21] proposed an SWHE scheme based on Learn with Errors (LWE). To improve the efficiency of the scheme a re-linearization technique and a dimension modulus technique are used to reduce ciphertext size and to simplify complex decryption circuit, respectively. However, it suffers from large size of keys and high time to be generated.

2.3 Drawbacks and Solutions

FHE performs huge number of operations as additions and multiplications operations and requires lattice based cryptosystem. Performing basic operations requires significantly complicated computations and huge ciphertext sizes. Therefore, existing FHE implementations remain impractical due to huge time and resource costs. Moreover, most FHE schemes have very large ciphertexts (millions of bits for a single ciphertext). This presents a considerable bottleneck in practical deployments. In addition, FHE allows arbitrarily large noises in its ciphertexts.

In this paper, an FHE scheme based on AES is proposed. It can perform operations on the ciphertext data instead of plaintext. It ensures query privacy, and also maintains performance time. The proposed scheme can perform additions and multiplications operations without changing the corresponding plaintext. Therefore, it can solve all mentioned problems in the existing schemes.

3 The Proposed FHE with AES Symmetric Algorithm

The proposed scheme is based on matrix operations which are computationally "light". It uses symmetric keys of small size there by making it suitable for many data centric applications. It derives its security from hardness of factorizing a large integer which is basis of many public key cryptosystems [22].

As shown in Fig. 1, the proposed scheme is used to design an efficient and practically feasible FHE that uses AES symmetric algorithm. It can handle arbitrary size of computations without the need of noise management and has scope of parallelization.

AES is a block cipher that works conjunctively with different key lengths (128, 192, and 256 bit). In the proposed scheme, AES presents 128 bit key length. AES is designed to be efficient in hardware and software through various platforms. AES is often used as a benchmark for implementations of protocols for secure Multi-Party Computation (MPC) [19, 20]. 128 bit AES refers to 2128 = (3.4 × 1038) possible keys. It means that, a PC that tries 255 keys per second will need 149.000 billion years to break AES (Fig. 2).

AES presents a good design space to investigate FHE techniques because it supports parallel nature of computations and algebraic nature of computations.

The basic concept is to translate operations on integers in a ring M_4 (Z_N); M_4 means that all operations are on square matrices of size 4 and Z_N means a set of integer numbers in algebra theory. Ring M_4 (Z_N) are sufficiently small to be used practically.

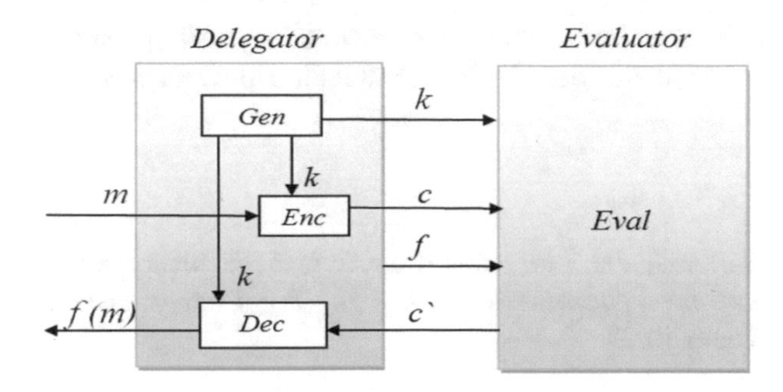

Fig. 2. FHE with AES symmetric key

FHE with AES is used to optimize communication with the cloud without bootstrapping [23]. In the context of realizing efficiency of the FHE scheme, the proposed scheme has four sets of operation, as follows:

3.1 Key Generation Process

In order for encryption to work effectively, it is important to manage the encryption keys securely. Even if a cloud service provider provides encryption, keys might be accessed. When encrypted data is stored in the cloud, the keys used for encryption should be kept separate and should only be accessed by the end user. Keys management involves the creation, use, distribution, and destruction of encryption keys.

The secret key of AES is 128-bit key. The key is arranged in the form of a matrix of 4×4 bytes in Z_N. Hence, it does not involve any computation theoretically. This approach is detailed in Appendix A.

3.2 Encryption Process

Formally, encryption process has been defined till now in context of public key systems only. In the proposed scheme, the encryption process extended the existing definition so as to incorporate both symmetric as well as public key systems. The encryption algorithm converts plaintext to ciphertext.

The plaintext $x \in Z_N$ is encrypted using symmetric scheme AES to produce a matrix $C \in M_4$ (Z_N). It's not needed to use a set of linear congruence by Chinese Remainder Theorem [24] because the construct of encrypted data by AES is also arranged in the form of a matrix of 4×4 bytes in Z_N. Encryption process algorithm is presented in Appendix B.

3.3 Evaluation Process

There is only one general evaluation function defined for computation *f*. It is expected that *f* is translated into basic operations on integers. To perform addition, subtraction, multiplication, and division of two numbers homomorphically, their ciphertexts are simply added/subtracted/multiplied/divided as two matrices. The proposed scheme evaluation function does not require any evaluation key. All operations on matrices are also performed within the ring $M_4(Z_N)$. Evaluation process algorithm is detailed in Appendix C.

3.4 Decryption Process

In decryption process, the new ciphertext $C' \in Z_N$ is decrypted using symmetric scheme AES to produce plaintext matrix $x \in M_4(Z_N)$. Decryption process algorithm is introduced in Appendix D.

4 Performance and Results

4.1 Testing Environment

This section presents the implementation environment. library Github is used for obtaining the source code for FHE with AES and other FHE schemes. All schemes were experimented in Oracle virtual machine Virtual Box 5.1.10. The Java implementation of this scheme is retrieved from Github. Ubuntu platform is selected for executing the Java implementation [25].

The memory of the OS (Ubuntu) in the VM is 1024 MB. It is implemented on a machine which consists of a processing unit of Intel Core i5-3320 M CPUs running at 2.60 GHz with 4 GB of RAM.

4.2 Implementation Results

This section presents the comparison of the proposed FHE with Smart, Marten and Zvika FHE. The execution time is shown in Figs. 3, 4 and 5 for key generation, encryption, and decryption; respectively at various file size 10 MB, 20 MB and 35 MB. It is shown that, the proposed FHE has the less execution time for key generation, encryption, and decryption process than the other algorithms since that the proposed FHE is based on matrix operations and factorizes the large integers.

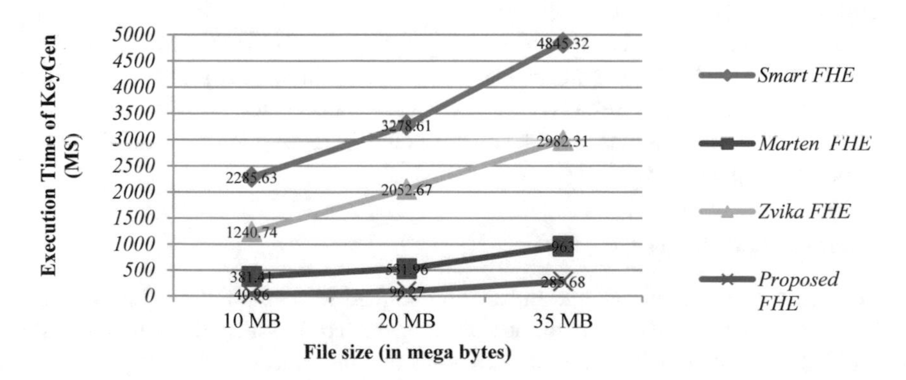

Fig. 3. Execution time of key generation

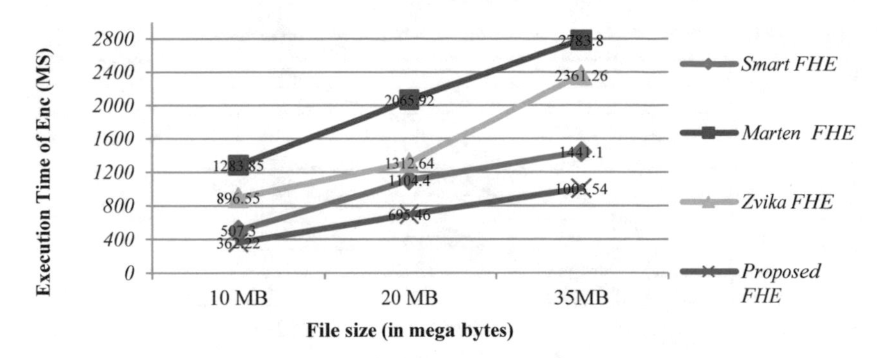

Fig. 4. Execution time of encryption

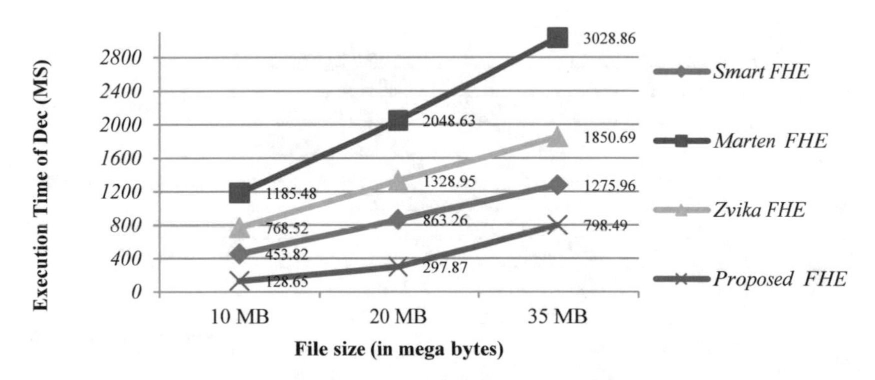

Fig. 5. Execution time of decryption

The data for homomorphic evaluations was gathered from running 1000 additions and 100 multiplications of randomly selected numbers of varying length. Figure 6 shows the execution time required for homomorphic of 1000 additions. It is shown that the proposed FHE is the second scheme after Marten FHE which has the least execution time for evaluation. Figure 7 shows the execution time required for homomorphic of 100 multiplications. It is shown that the proposed FHE has the least execution time for evaluation 1000 additions. This is due to that, the proposed scheme performs additions and multiplications operations without changing the corresponding plaintext.

4.3 Overhead, Complexity, and Plaintext Expansion

Homomorphic evaluation of a function is efficient if it has a low computation overhead. The overhead is defined as the ratio of the time taken for a computation homomorphically done on ciphertext to the time taken to compute on plaintext. If a computation consists only of addition, adding two integer's homomorphically in the proposed FHE implies adding two matrices. This gives a constant overhead of 16, since it is needed to add two matrices of size 4, containing 16 numbers. If a computation consists only of

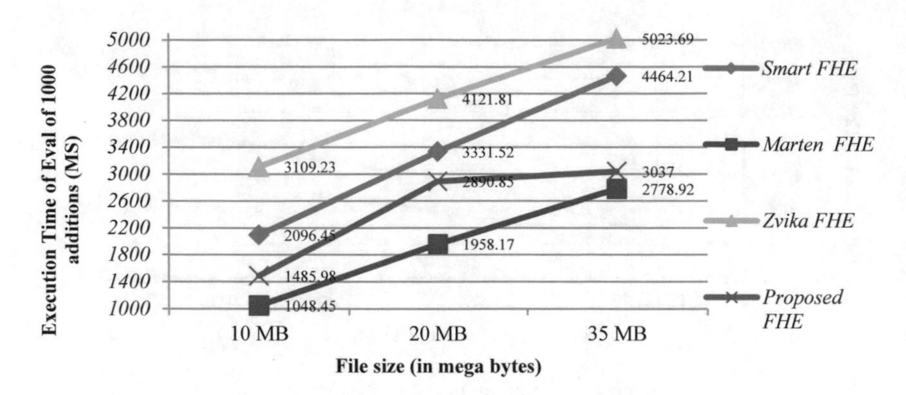

Fig. 6. Execution time of evaluation for running 1000 additions

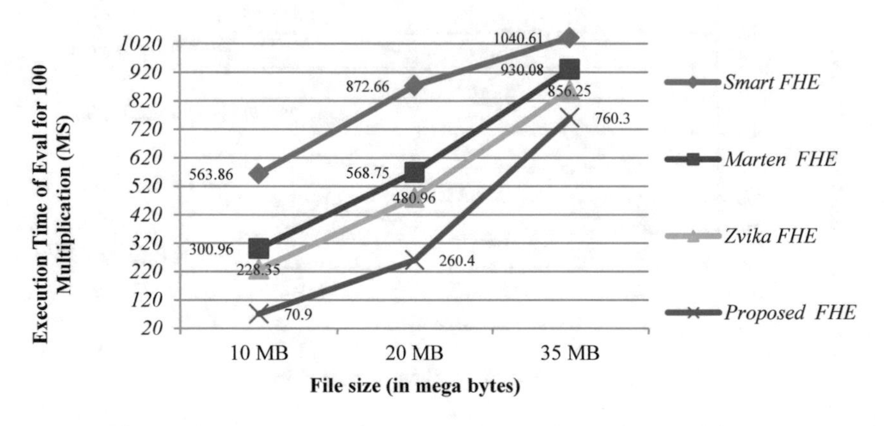

Fig. 7. Execution time of evaluation for running 100 multiplications

Table 1. Comparison of overhead, plaintext expansion and complexity

	Smart FHE	Marten FHE	Zvika FHE	Proposed FHE
Computation overhead	$O(\lambda^2)$	$O(\lambda^{3.5})$	$O(\lambda^2 + \log \lambda)$	$O(m\ \lambda)$
Plaintext expansion	$O(\lambda^3)$	$O(\log \lambda)$	$O(3 \log \lambda)$	$O(1)$ actually 16
Complexity	$O(\lambda\ (\log \lambda)^2)$	$O(\lambda^{10})$	$O(\log \lambda + \lambda^3)$	$O(m\ \lambda)$

multiplication, multiplying two integers homomorphically implies multiplying two matrices, which means 64 additions and 64 multiplications. Since N is b bit number, the cost of multiplying two numbers is given by $O(b^2)$. Thus giving computation overhead $O(b)$ or $O(m\ \lambda)$. Table 1 lists the comparison of overhead, complexity and plaintext expansion between the proposed scheme with Smart, Marten and Zvika FHE. It is clear from the results that the proposed FHE has the best performance since it has the least overhead, plaintext expansion and complexity.

5 Conclusions

Scope and promises of homomorphic cryptography in cloud computing environments cannot be ignored. Efforts have been made to propose ideas as to how symmetric keys and simple matrix-based operations could also lead to feasible schemes for cloud computing, specifically for delegation of computation and private data processing in clouds. To make up for this, the proposed scheme emphasizes on having low time complexity for cryptographic primitives, which are sufficiently small to be used practically. FHE has the advantage of performing all types of computations such as additions and multiplications. The proposed FHE scheme uses AES symmetric algorithm. It can handle arbitrary size of computations without the need of noise management.

Throughout the experiments, the results showed that the proposed FHE expressed less execution time than Smart, Marten and Zvika FHE. The only observed limitation was that the proposed FHE had more execution time than Marten FHE in evaluation of running 1000 additions. The proposed scheme achieves better performance since it has the least overhead, plaintext expansion and complexity.

6 Appendix A

6.1 Key Generation Algorithm

```
Keygen(){
1. int m,λ  ;
   /* m is number of sets in two selected prime numbers and λ is
   a security parameter*/
2. int  pi, qi ;
   /*  Choose  2m  odd  numbers  pi  and  qi,  1  ≤  i  ≤  m,
   which are mutually prime.*/
3. for (i=1;i≤ m; i++){
4. Let fi=piqi;
           m
5.  N= Σ (fi) ; }
           i
6. Pick an matrix k of size 4, k ∈M4(ZN)
7. intk_rows = maxSizeM(k[]);
 /*function to compute max no. of row of secret key k matrix */
8. int k_cols = maxSizeN(k[]);
   /*function to compute max no. of colmn of secret key k matrix
   */
9. for (int j=1;j≤ k_cols; j++){
   /*j is a column number of index of key k matrix*/
10. for (int r=1;r≤ k_rows; r++){
   /*r is a row number of index of key k matrix*/
11. k[r][j]=KeygenAES(m,λ )}}
12. output (f,N,k[r][j]);}
```

7 Appendix B

7.1 Encryption Algorithm

```
Enc (){
1. intx_rows = maxSizeM(x[]);
   /*compute max no. of row in plain matrix x */
2. intx_cols = maxSizeN(x[]);
   /*compute max no. of column in plain matrix x */
3. for (int j=1;j≤ x_cols; j++){
  /*j is a column number of index in plain matrix x */
4. C[r][j]=ENC_AES(x[r][w],k[w][j]);}}}
5. output (C[r][j]);} /*output is cipher matrix.*/
6. for (int r=1;r≤ x_rows; r++){
   /*r is a row number of index of plain matrix x */
7. Input(x[r][j]);}}
8. for (int j=1;j≤ 4; j++){
   /*j is a column number of index of cipher matrix C */
9. for (int r=1;r≤ 4; r++){
   /*r is a row number of index of cipher matrix C */
10. for (int w=1;w≤ 4; w++){
```

8 Appendix C

8.1 Evaluation Algorithm

```
Eval () {

1. intC_rows = maxSizeM(C[]);
   /*compute max no. of row of cipher matrix C */
2. intC_cols = maxSizeN(C[]);
   /*compute max no. of row of cipher matrix C */
3. for (intj=1;j≤ c_cols; j++){
   /* j is a column number of index of cipher matrix C */
4. for (intr=1;r≤ c_rows; r++){
   /*r is a row number of index of cipher matrix C */
5. C`[r][j]=Eval(C[r][j],fi);}}
6. output(C`[r][j]);} /*output is new cipher matrix.*/
```

9 Appendix D

9.1 Decryption Algorithm

```
Dec() {
1. for (int j=1;j≤ x_cols; j++){
   /* j is a column  number of index of plain matrix x */
2. for (int r=1;r≤ x_rows; r++){
   /* r is a to row number of index of plain matrix x */
3. for (int w=1;w≤ 4; w++){
4. x [r][j]=Dec (C`[r][w],k[w][j]);}}}
5. Output (x[r][j]);}/* output is plain matrix x */
```

References

1. Mell, P., Grance, T.: The NIST Definition of Cloud Computing, vol. 800. National Institute of Standards and Technology, Gaithersburg (2011). no. 145
2. Marston, S., Li, Z., Bandyopadhyay, S.: Cloud computing—the business perspective. Decis. Support Syst. **51**(1), 176–189 (2011)
3. Suciu, G., Halunga, S., Apostu, A.: Cloud computing as evolution of distributed computing —a case study for slapos distributed cloud computing. Inf. Econ. **17**(4), 109–122 (2013)
4. Ali, M., Khan, S.U., Vasilakos, A.V.: Security in cloud computing: opportunities and challenges. Inf. Sci. **305**, 357–383 (2015)
5. Vurukonda, N., Rao, B.: A study on data storage security issues in cloud computing. Procedia Comput. Sci. **92**, 128–135 (2016)
6. Wang, Q., Wang, C., Ren, K., Li, J.: Enabling public auditability and data dynamics for storage security in cloud computing. IEEE Trans. Parallel Distrib. Syst. **22**(5), 847–859 (2011)
7. Plantard, T., Susilo, W., Zhang, Z.: Fully homomorphic encryption using hidden ideal lattice. IEEE Trans. Inf. Forensics Secur. **8**(12), 2127–2137 (2013)
8. Rivest, R., Adleman, L., Dertouzos, M.: On data banks and privacy homomorphisms. In: Foundations of Secure Computation, pp. 169–179. Academia Press (1978)
9. Cheon, J.H., Kim, J.: A hybrid scheme of public-key en-cryption and somewhat homomorphic encryption. IEEE Trans. Inf. Forensics Secur. **10**(5), 1052–1063 (2015)
10. Coron, J.S., Naccache, D., Tibouchi, M.: Public key compression and modulus switching for fully homomorphic encryption over the integers. In: Pointcheval, D., Johansson, T. (eds.) Euro-Par 2012. LNCS, vol. 7237, pp. 446–464. Springer, Berlin (2012)
11. TFHE: Fast Fully Homomorphic Encryption Library over the Torus. https://github.com/tfhe/tfhe
12. Djatmiko, M., Cunche, M., Boreli, R., Seneviratne, A.: Heterogeneous secure multi-party computation. In: Bestak, R., Kencl, L., Li, L.E., Widmer, J., Yin, H. (eds.) NETWORKING 2012. LNCS, vol. 7290, pp. 198–210. Springer, Berlin (2012)
13. Orlandi, C., Nielsen, J.B., Nordholt, P.S., Burra, S.S.: A new approach to practical active-secure two-party computation. In: Safavi-Naini, R., Canetti, R. (eds.) Advances in Cryptology—CRYPTO 2012. LNCS, vol. 7417, pp. 681–700. Springer, Berlin (2012)

14. Sugumar, R., Imam, S.B.: Symmetric encryption algorithm to secure outsourced data in public cloud storage. Indian J. Sci. Technol. **8**(23) (2015)
15. Damgard, I., Keller, M.: Secure multiparty AES. In: Sion, R. (ed.) Financial Cryptography and Data Security. FC 2010. LNCS, vol. 6052, pp. 367–374. Springer, Berlin (2010)
16. Naehrig, M., Lauter, K., Vaikuntanathan, V.: Can homomorphic encryption be practical? In: 3rd ACM Workshop on Cloud Computing Security Workshop (CCSW 2011), pp. 113–124. ACM, New York (2011)
17. Vaikuntanathan, V.: Computing blindfolded: new developments in fully homomorphic encryption. In: 2011 IEEE 52nd Annual Symposium on Foundations of Computer Science, Palm Springs, CA, pp. 5–16 (2011)
18. Smart, N.P., Vercauteren, F.: Fully homomorphic encryption with relatively small key and ciphertext sizes. In: Nguyen, P.Q., Pointcheval, D. (eds.) PKC 2010. LNCS, vol. 6056, pp. 420–443. Springer, Berlin (2010)
19. Dijk, M.V., Gentry, C., Halevi, S., Vaikuntanathan, V.: Fully homomorphic encryption over the integers. In: Gilbert, H. (ed.) EUROCRYPT 2010. LNCS, vol. 6110, pp. 24–43. Springer, Berlin (2010)
20. Brakerski, Z., Vaikuntanathan, V.: Efficient fully homomorphic encryption from (standard) LWE. In: 2011 IEEE 52nd Annual Symposium on Foundations of Computer Science, pp. 97–106. ACM Press, Palm Springs (2011)
21. Brakerski, Z., Vaikuntanathan, V.: Fully homomorphic encryption from ring-LWE and security for key dependent messages. In: Rogaway, P. (ed.) CRYPTO 2011. LNCS, vol. 6841, pp. 505–524. Springer, Berlin (2011)
22. Rizk, R., Alkady, Y.: Two-phase hybrid cryptography algorithm for wireless sensor networks. J. Electr. Syst. Inf. Technol. **2**(3), 296–313 (2015)
23. Brakerski, Z., Gentry, C., Vaikuntanathan, V.: Fully homomorphic encryption without bootstrapping. In: 3rd Innovations in Theoretical Computer Science Conference, pp. 309–325. ACM Press, New York (2012)
24. Hilton, P., Holton, D., Pedersen, J.: A far nicer arithmetic. In: Mathematical Reflections, pp. 25–60. Springer, New York (1997)
25. Varia, M., Yakoubov, S., Yang, Y.: HEtest: a homomorphic encryption testing framework. In: A Homomorphic Encryption Testing Framework. Financial Cryptography and Data Security: FC 2015 International Workshops, pp. 213–230. Springer, Berlin (2015)

Deep Learning for Satellite Image Classification

Mayar A. Shafaey[1(✉)], Mohammed A.-M. Salem[1,2], H. M. Ebied[1], M. N. Al-Berry[1], and M. F. Tolba[1]

[1] Faculty of Computers and Information Sciences, Ain Shams University, Cairo, Egypt
mayar.al.mohamed@fcis.asu.edu.eg, {salem,maryam_nabil}@cis.asu.edu.eg, hala.m@outlook.com, fahmytolba@gmail.com
[2] Faculty of Media Engineering and Technology, German University, Cairo, Egypt

Abstract. Nowadays, large amounts of high resolution remote-sensing images are acquired daily. However, the satellite image classification is requested for many applications such as modern city planning, agriculture and environmental monitoring. Many researchers introduce and discuss this domain but still, the sufficient and optimum degree has not been reached yet. Hence, this article focuses on evaluating the available and public remote-sensing datasets and common different techniques used for satellite image classification. The existing remote-sensing classification methods are categorized into four main categories according to the features they use: manually feature-based methods, unsupervised feature learning methods, supervised feature learning methods, and object-based methods. In recent years, there has been an extensive popularity of supervised deep learning methods in various remote-sensing applications, such as geospatial object detection and land use scene classification. Thus, the experiments, in this article, carried out on one of the popular deep learning models, Convolution Neural Networks (CNNs), precisely *AlexNet* architecture on a standard sounded dataset, *UC-Merceed Land Use*. Finally, a comparison with other different techniques is introduced.

Keywords: Remote-sensing · Satellite image · Deep learning
Convolution Neural Networks (CNNs) · UC-Merceed Land Use
Parallel computing

1 Introduction

A Satellite Image is an image of the whole or part of the earth taken using artificial satellites. It can either be visible light images, water vapor images or infrared images [1]. The different types of satellites produce (high spatial, spectral, and temporal) resolution images that cover the whole Earth in less than a day. The large-scale nature of these data sets introduces new challenges in image analysis.

The analysis and classification of remote-sensing images is very important in many practical applications, such as natural hazards and geospatial object detection, precision

A. E. Hassanien et al. (Eds.): AISI 2018, AISC 845, pp. 383–391, 2019.
https://doi.org/10.1007/978-3-319-99010-1_35

agriculture, urban planning, vegetation mapping, and military monitoring [2]. Despite decades of research, the degree of automation for remote-sensing images analysis still remains low [3].

The main objective of this paper is to present a literature review on the recent deep-learning based techniques for satellite image classification and the available training and testing datasets. Moreover, testing results will present on one popular dataset using the *AlexNet* architecture of the Convolution Neural Networks (CNNs).

In the next section, a list of available datasets and their specifications are presented. A review on recent classification approaches applied on one or some of these datasets is presented in Sect. 3. The experimental work followed by results and evaluations are presented in Sect. 4. Finally, conclusions are highlighted in Sect. 5.

2 Review on Publicly Remote Sensing Images Datasets

In the past years, several high resolution remote-sensing image datasets have been introduced by different groups to enable machine-learning based research for scene classification and to evaluate different methods in this field. The authors will review some publicly available sets in this section, as given in Table 1. The table below shows the number of scene classes, images per class, total images, size of images, and spatial resolution.

Table 1. Comparison between the different remote-sensing datasets proposed

Data set	Scene classes	Images/class	Total images	Spatial resolution	Image sizes
AID [4]	30	200–400	10000	High	600 × 600
Patter Net [5]	38	800	30400	Up to 0.8	256 × 256
RSI-CB256 [6]	35	Various	34000	0.3–3	256 × 256
SAT_4 & SAT_6 [7]	Patches (500000 + 405000)			Low	28 × 28
UC-Merced Land Use [8]	21	100	2100	0.3	256 × 256
WHU-RS19 [9]	19	~50	1005	Up to 0.5	600 × 600
SIRI-WHU [10]	12	200	2400	2	200 × 200
RSSCN7 [11]	7	400	2800	–	400 × 400
RSC11 [12]	11	~100	1232	0.2	512 × 512
Brazilian Coffee [13]	2	1438	2876	Low	64 × 64
NWPU-RESISC45 [14]	45	700	31500	~30–0.2	256 × 256

The most images in these datasets are imported from Google Earth Engine and cover the areas of: agricultural, airplane, baseball diamond, beach, buildings, chaparral, dense residential, forest, freeway, golf course, harbor, intersection, medium density residential, mobile home park, overpass, parking lot, river, runway, sparse residential, storage tanks, and so on. Except the dataset in [13] “Brazilian Coffee Scene dataset”, cropped from SPOT satellite images and contains only two scene classes, which is appropriate for multi-class scene classification methods. In contradiction, the large number of classes and images in NWPU-RESISC45 [14] dataset, will impact positively the classification results.

However, the UC-Merced Land-Use [8] in Fig. 1 is the most popular and has been widely used for the task of remote-sensing image scene classification and retrieval so far. So, the authors will choose it to carry out the classification experiment.

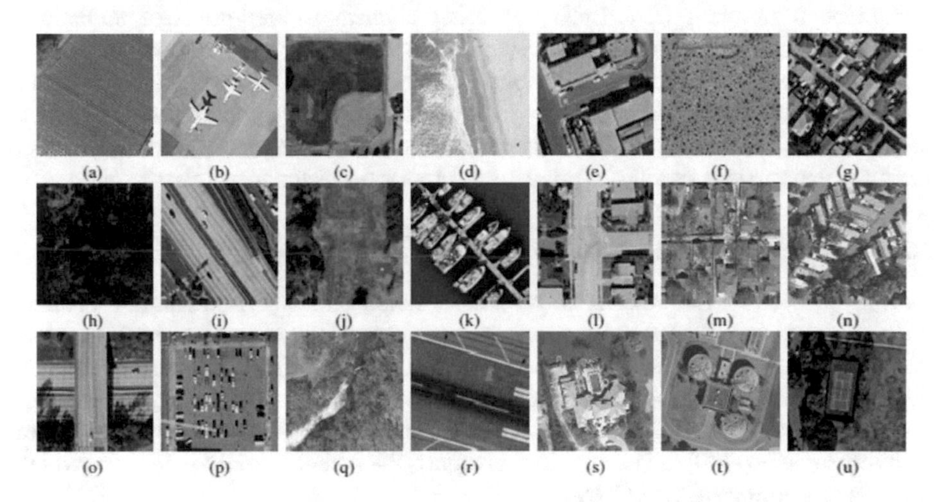

Fig. 1. 21 Classes representative [(a)–(u)] of the UC-Merced Land-Use dataset [34].

3 Remote Sensing Images Classification Methods

There are long and proud researches during the last and current decades that were carried out on the satellite images for the task of scene classification. From the vast publications of this topic, generally, the existing scene classification methods could summarized into four main categories according to the features they used: **manually feature based methods, unsupervised classification methods, supervised learning methods, and object-based methods**.

3.1 Manually Feature Based Methods

A fundamental step in image classification is based on handcrafted features. These methods measure the skills of researchers to design and extract important features, such as color, orientation, texture, shape, spatial and spectral information, or their combination. Some of the most common and essential features that are used for scene classification are: ***Color histograms*** - ***Texture descriptors*** – ***GIST***: *describe orientations of a scene* – ***SIFT***: *describe sub-regions of a scene* – ***HOG***: *describe gradient of objects* [15–17, 40].

3.2 Unsupervised Classification Methods

The limitations of manually feature based methods could be solved by self-learning features from images. This strategy is called unsupervised learning method. In recent years, unsupervised feature learning from unlabeled input data has become an attractive alternative to handcrafted features [18].

The idea behind that strategy is first grouping the image pixels into clusters based on their properties. By learning features from images instead of relying on manually designed features, we can obtain more discriminative feature that is better suited for the classification problem [19]. Such clustering algorithms are: principal component analysis (PCA) [20], *k*-means clustering [21], sparse coding [22], and so on.

In real applications, the aforementioned unsupervised feature learning methods have achieved good performance for land use classification, especially compared to handcrafted based methods. For example, authors in [23–25] applied unsupervised methods and made a significant progress for remote-sensing scene classification.

3.3 Supervised Learning Methods

Starting year 2006, the volcano of researches relied on supervised learning methods which need to use labeled data to extract more powerful features, especially, a deep learning method which made by *Hinton and Salakhutdinov* [26]. There exists different numbers of deep learning models, such as deep belief nets (DBN) [27], deep Boltzmann machines (DBM) [28], stacked auto-encoder (SAE) [29], Convolutional Neural Networks (CNNs) [30], and so on. In this article, the authors mainly review the widely used deep learning method CNNs.

The basic concept of CNN is to train huge multi-layer networks for giving impressive classification results of large scale input images. The CNN itself has different models like: *AlexNet, GoogleNet, ResNet, VGGNet, CaffeNet* … etc. [31]. Limited by the space, a short and highlight description of *AlexNet* architecture was given. The net consists of 25 layers: 5 convolution layers, max-pooling layers, dropout layers, and 3 fully connected layers, as shown in Fig. 2. It is trained on *ImageNet* data, which contained over 15 million annotated images from a total of over 22,000 categories and Used ReLU for the nonlinearity functions [32].

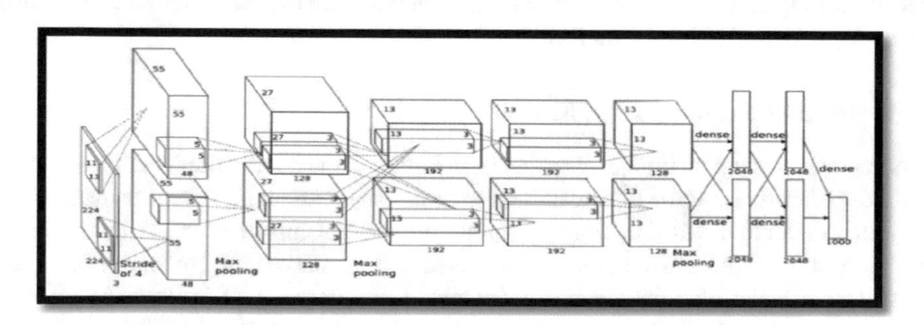

Fig. 2. *ImageNet* classification with *AexNet* CNN [32]

Table 2 represents some of authors who used the CNN models in their experiments for large scale image scene classification and gave the proud accuracy values which demonstrate the power of CNN learning model.

Table 2. Survey of recent publications applied CNNs in their experiments on large scale remote-sensing (RS) images, *UC-Mercced* dataset

References	Year	Application	Method	Accuracy
[33]	2015	Multi-spectral land use classification	Deep CNN	93.48%
[34]	2015	Land Use RS classification	GoogleNet, and CaffeNet	97%, and 95.48%
[35]	2016	RS scene classification	Large patch CNN	Effective results
[36]	2016	Large scale image classification	CNN	92.4%
[37]	2018	Remote sensing scene classification	CNN	92.43%

3.4 Object-Based Methods

Unlike pixel-based or image-based classification, object-based image classification groups pixels into representative shapes and sizes and assigns each group to a semantic object. This process relies on multi-resolution segmentation. Multi-resolution segmentation produces homogenous objects by grouping pixels. It generates objects with different scales in an image simultaneously. These objects are more meaningful because they represent features in the image [38, 41].

The question here is how to select the appropriate image classification techniques. It is based on common sense of the engineering. Let's say you want to classify water in a high spatial resolution image containing grasses. You decide to choose all pixels with low NDVI (Normalized Difference Vegetation Index) in that image. NDVI is used to analyze remote sensing measurements and assess whether the target being observed contains live green vegetation or not. But this could also misclassify other pixels in the image that aren't water i.e. pixels of the sky. For this reason, pixel-based classification as unsupervised and supervised classification gives a salt and pepper look.

As illustrated in this article, spatial resolution is an important factor when selecting image classification techniques. Hence, when you have low spatial resolution, both traditional pixel-based and object-based image classification techniques perform well. But when you have high spatial resolution, object-based image classification is superior to traditional pixel-based classification [39].

4 Experiments and Results

Taking advantages of the availability of *UC-Merceed* dataset [8], the *AlextNet* CNN approach was applied to represent the large scale image classification process. In this section, the experiment's steps will be described, i.e., software, hardware specification, results, comments, and comparisons.

4.1 Experimental Procedure

The experiment ran on two different computers. Machine 1 has a processor: Intel® Core™ i7-2670QM CPU @ 2.20 GHz–8 GB RAM. Machine 2 equipped with NVIDIA GTX 1050 4G cc: 6.1 GPU: Intel® Core™ i7-7700HQ @ 2.20 GHz–16 GB RAM. The time elapsed on machine 1 was 1800 s and on machine 2 was 14 s. Thanks to Graphical Processing Unit (GPU) for giving an impressive and significant execution time. The parallel computing optimizes the performance 100 times than the serial computations.

Hence, The experiment ran on Machine 2 and Matlab® software using *alexnet()* built-in function which is trained on a subset of the *ImageNet* database – ~1.2 million images - and can classify images into 1000 object categories. This function requires Neural Network Toolbox™ Model for *AlexNet* Network. The basic three steps are firstly resizing the image dimension from 256×256 to 227×227 as a required input for the CNN. The second step is to choose the training set percentage. And thirdly, train the multiclass SVM classifier, extract test features using the CNN, and pass them to the trained classifier to get the known labels. Finally, the classification results are given by computing the summation of main diagonal of the confusion matrix divided by the diagonal elements number.

A number of experiments were carried out to assess the performance of the CNN using the well-known *UC-Merceed* Land Use dataset. The *UC-Merceed* Land Use dataset contains 2100 images, 21 distinct classes and every class contains 100 different images. In the experiments, the size of training set ranged from 10 to 90% of the 100 different images per class and the remaining images where used for testing. Figure 3 shows the correct classification accuracy vs. the size of training set percentage.

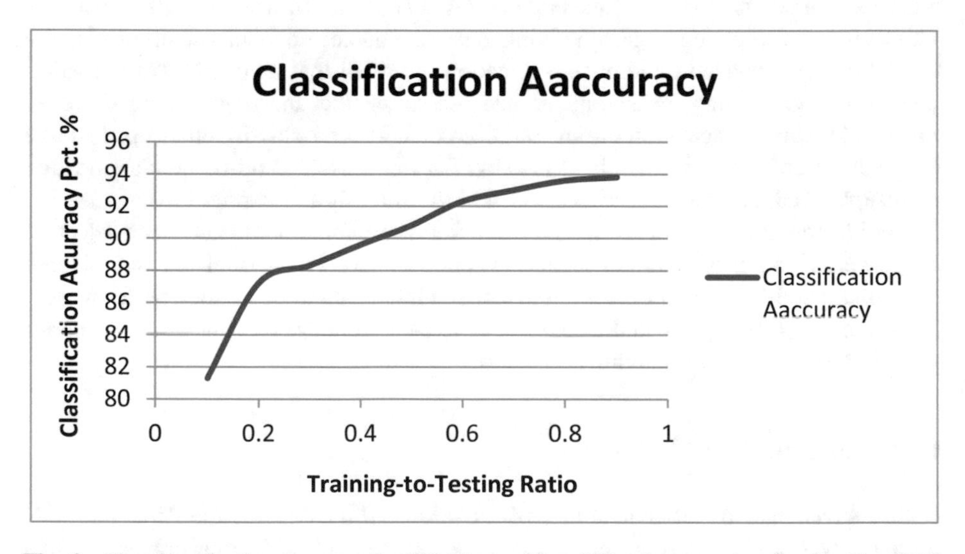

Fig. 3. The classification accuracy for *UC-Merceed* Land Use dataset using the *AlexNet* CNN. The *x-axis* represents the interval of training to testing set ratio [0.1–0.9]. The *y-axis* represents the classification accuracy.

The first trial started to split 10% of images into training set which gave 81.3% accuracy value. Then, repeated the experiment eight times up to 90% of images into training set which gave around 94% accuracy value. The figure below illustrates that the gradually increase of training images impacts positively the classification result.

4.2 Evaluation and Discussions

Compared with other CNN models, *GoogleNet* and *CaffeNet*, mentioned and discussed in [34], and applied also on UC-Merceed dataset, the authors observed that the classification accuracy gained by *GoogleNet* (~97%) was better than whose gained by *CaffeNet* (~94%) and *AlexNet* (~94%). However, The *AlexNet* is faster than *GoogleNet* model. The two models ran on the same GPU, as mentioned before, the *AlexNet* executed in only 14 s, but *GoogleNet* consumed 51 s, which is approximately 4 times slower.

On the one hand, in comparison with traditional handcrafted features that require a high mental thinking and skills, deep learning features are learned from data automatically via deep architecture neural networks. This is the key advantage of deep learning methods.

On the other hand, and compared with aforementioned unsupervised feature learning methods i.e. sparse coding, deep learning models can learn more powerful because it is composed of multiple processing layers which is more applicable for large scale and remote-sensing image scene classification. The deep feature learning methods act as a human brain in which every level uses the information from the previous level to learn deeply and accurately.

The following articles support our research. In [23], the high-resolution satellite scene classification using a sparse coding carried out on *UC-Merceed* dataset and reached about 91% accuracy. And in [24], the unsupervised feature learning via spectral clustering of multidimensional patches was carried out on the same dataset and achieved 90% right classification.

5 Conclusions

The automation target detection or recognition, and high resolution remotely sensed image classification are two hot topics nowadays. Hence, this paper firstly represented a comprehensive review of common and freely remote-sensing datasets to enable the community to develop the large scale image scene classification task. Then, it gave a summary of recent methods used for this task. Finally, the CNN deep learning method applied on *UC-Merceed* dataset evaluated and reported the results to compare against state-of-the-art and as a baseline for future research.

Deep learning methods can undoubtedly offer better feature representations for the related remote-sensing task, and there is a bright prospect of seeing more and more researchers dedicated to learning better features for the target detection and scene classification tasks by utilizing appropriate deep learning methods.

Thanks to parallel computing and GPUs for optimizing and enhancing the execution time 100× than the serial computations, our experiment ran in time not exceeding 14 s to classify one testing image out of 2100 images.

References

1. NASA: What Is a Satellite? NASA Knows! (Grades 5–8) series (2014)
2. Zhang, L., Xia, G., Wu, T., Lin, L., Tai, X.: Deep learning for remote sensing image understanding. J. Sens. **2016**, 1–2 (2016)
3. Marmanisad, D., Wegnera, J., Gallianib, S., Schindlerb, K., Datcuc, M., Stillad, U.: Semantic segmentation of aerial images with an ensemble of CNNs. ICWG **3**(4), 1–8 (2016)
4. AID Dataset. http://www.lmars.whu.edu.cn/xia/AID-project.html. Accessed 16 Feb 2018
5. PatternNet Dataset. https://sites.google.com/view/zhouwx/dataset?authuser=0. Accessed 16 Feb 2018
6. RSI Dataset. https://github.com/lehaifeng/RSI-CB. Accessed 16 Feb 2018
7. SAT_4 & SAT_6. http://csc.lsu.edu/~saikat/deepsat/. Accessed 16 Feb 2018
8. UC-Merceed Land Use Dataset. http://weegee.vision.ucmerced.edu/datasets/landuse.html. Accessed 16 Feb 2018
9. WHU-RS19 Dataset. https://www.google.com/url?q=http%3A%2F%2Fwww.xinhua-fluid.com%2Fpeople%2Fyangwen%2FWHU-RS19.html&sa=D&sntz=1&usg=AFQjCNFzrOnViW6TWOoFbN1IaIMfyLdJhQ. Accessed 16 Feb 2018
10. SIRI-WHU Dataset. http://www.lmars.whu.edu.cn/prof_web/zhongyanfei/e-code.html. Accessed 16 Feb 2018
11. RSSCN7 Dataset. https://www.dropbox.com/s/j80iv1a0mvhonsa/RSSCN7.zip?dl=0. Accessed 16 Feb 2018
12. RSC11 Dataset. https://www.yeastgenome.org/locus/ARP7. Accessed 16 Feb 2018
13. Brazilian Coffee Dataset. http://www.patreo.dcc.ufmg.br/downloads/brazilian-coffee-dataset/. Accessed 16 Feb 2018
14. NWPU-RESISC45 Dataset. https://www.google.com/url?q=http%3A%2F%2Fwww.escience.cn%2Fpeople%2FJunweiHan%2FNWPU-RESISC45.html&sa=D&sntz=1&usg=AFQjCNGs2uMeX7KT2QvEMzcD5uF4-aQChw. Accessed 16 Feb 2018
15. Cheng, G., Han, J., Lu, X.: Remote sensing image scene classification: benchmark and state of the art. Proc. IEEE **105**(10), 1–17 (2017)
16. Thomas, M., Farid, M., Yakoub, B., Naif, A.: A fast object detector based on high-order gradients and Gaussian process regression for UAV images. Int. J. Remote Sens. **36**(10), 2713–2733 (2015)
17. Aptoula, E.: Remote sensing image retrieval with global morphological texture descriptors. IEEE Trans. Geosci. Remote Sens. **52**(5), 3023–3034 (2014)
18. Mekhalfi, M., Melgani, F., Bazi, Y., Alajlan, N.: Land-use classification with compressive sensing multifeature fusion. IEEE Geosci. Remote Sens. **12**(10), 2155–2159 (2015)
19. Cheriyadat, A.: Unsupervised feature learning for aerial scene classification. IEEE Trans. Geosci. Remote Sens. **52**(1), 439–451 (2014)
20. Jolliffe, I.: Principal component analysis. Springer, New York (2002)
21. Zhao, B., Zhong, Y., Zhang, L.: A spectral–structural bag-of-features scene classifier for very high spatial resolution remote sensing imagery. Remote Sens. **116**, 73–85 (2016)
22. Olshausen, B., Field, D.: Sparse coding with an overcomplete basis set: a strategy employed by V1? Vision. Res. **37**(23), 3311–3325 (1997)

23. Sheng, G., Yang, W., Xu, T., Sun, H.: High-resolution satellite scene classification using a sparse coding based multiple feature combination. Int. J. Remote Sens. **33**(8), 2395–2412 (2012)
24. Hu, F., Xia, G., Wang, Z., Huang, X., Zhang, L., Sun, H.: Unsupervised feature learning via spectral clustering of multidimensional patches for remotely sensed scene classification. IEEE J. Select. Top. Appl. Earth Obs. Remote Sens. **8**(5), 2015–2030 (2015)
25. Daoyu, L., Kun, F., Yang, W., Guangluan, X., and Xian, S.: MARTA GANs: unsupervised representation learning for remote sensing image classification. National Natural Science Foundation of China (2017)
26. Hinton, G., Salakhutdinov, R.: Reducing the dimensionality of data with neural networks. Science **313**(5786), 504–507 (2006)
27. Hinton, G., Osindero, S., Teh, Y.-W.: A fast learning algorithm for deep belief nets. Neural Comput. **18**(7), 1527–1554 (2006)
28. Salakhutdinov, R., Hinton, G.: An efficient learning procedure for deep Boltzmann machines. Neural Comput. **24**(8), 1967–2006 (2012)
29. Vincent, P., Larochelle, H., Lajoie, I., Bengio, Y., Manzagol, P.-A.: Stacked denoising autoencoders: Learning useful representations in a deep network with a local denoising criterion. Mach. Learn. Res. **11**, 3371–3408 (2010)
30. Sermanet, P., Eigen, D., Zhang, X., Mathieu, M., Fergus, R., LeCun, Y.: OverFeat: integrated recognition, localization and detection using convolutional networks. In: Proceedings of the International Conference on Learning Representations, pp. 1–16 (2014)
31. Simonyan K., Zisserman, A.: Very deep convolutional networks for large-scale image recognition. In: Proceedings of the International Conference on Learning Representations, pp. 1–13 (2015)
32. Krizhevsky, A., Sutskever, I., Hinton, G.: ImageNet classification with deep convolutional neural networks. In: Proceedings of the Conference on Advances in Neural Information Processing Systems, pp. 1097–1105 (2012)
33. Luus, F., Salmon, B., Van Den Bergh, F., Maharaj, B.: Multiview deep learning for land-use classification. IEEE Geosci. Remote Sens. Lett. **12**(12), 2448–2452 (2015)
34. Castelluccio, M., Poggi, G., Sansone, C., Verdoliva, L.: Land Use Classification in Remote Sensing Images by Convolutional Neural Networks. Cornell University, Ithaca (2015)
35. Zhong, Y., Fei, F., Zhang, L.: Large patch convolutional neural networks for the scene classification of high spatial resolution imagery. Appl. Remote Sens. **10**(2), 025006–025006 (2016)
36. Marmanis, D., Datcu, M., Esch, T., Stilla, U.: Deep learning earth observation classification using ImageNet pretrained networks. IEEE Geosci. Remote Sens. Lett. **13**(1), 105–109 (2015)
37. Jingbo, C., Chengyi, W., Zhong, M., Jiansheng, C., Dongxu, H., Stephen, A.: Remote sensing scene classification based on convolutional neural networks pre-trained using attention-guided sparse filters. Remote Sens. **10**(290), 1–16 (2018)
38. Blaschke, T.: Object based image analysis for remote sensing. ISPRS J. Photogramm. Remote Sens. **65**(1), 2–16 (2010)
39. GIS Geography. http://gisgeography.com/image-classification-techniques-remote-sensing/. Accessed Feb 16 2018
40. Tahoun, M., Nagaty, K., El-Arief, T., A-Megeed, M.: A robust content-based image retrieval system using multiple features representations. In: Proceedings of IEEE Networking, Sensing and Control, pp. 116–122 (2005)
41. Mohammed, A-M.: Multiresolution Image Segmentation. Ph.D. Thesis, Department of Computer Science, Humboldt-Universitaet zu Berlin, Germany (2008)

Information Security, Hiding, and Biometric Recognition

Fake Reviews Detection Under Belief Function Framework

Malika Ben Khalifa[1(✉)], Zied Elouedi[1(✉)], and Eric Lefèvre[2(✉)]

[1] LARODEC, Institut Supérieur de Gestion de Tunis, Université de Tunis, Tunis, Tunisia
malikabenkhalifa2@gmail.com, zied.elouedi@gmx.fr

[2] Laboratoire de Génie Informatique et d'Automatique de l'Artois (LGI2A), EA 3926, Univ. Artois, 62400 Béthune, France
eric.lefevre@univ-artois.fr

Abstract. Online reviews have become one of the most important sources of customers' opinions. These reviews influence potential purchasers to make or reverse decisions. Unfortunately, the existence of profit and publicity has emerged fake reviews to promote or demote some target products. Furthermore, reviews are generally imprecise and uncertain. So, it is a difficult task to uncover fake reviews from the genuine ones. In this paper, we propose a fake reviews detection method using the belief function theory. This method deals with the uncertainty in the given rating reviews and takes into account the similarity with other provided votes to detect misleading. We propose numerical examples to intuitively evaluate our method. Then, to prove its performance, we conducted on a real database. Experimentation shows that the proposed method is a valuable solution for the fake reviews detection problem.

Keywords: Online opinions · Fake reviews · Uncertainty
Belief function theory

1 Introduction

During the last years, we notice the emergence of the opinions sharing websites such as Amazon.com, Tripadvisor.com, Yelp.com, PriceGrabber.com, Shopzilla.com and Resellerratings.com, which allow people to share their experiences, feelings, attitudes regarding products, services, business and even political issues. Such opinions, straightforwardly influence potential future customers and companies to make or reserve decisions.

Consequently, the increasing of positive reviews number will transform their readers to new customers which will provide significant financial gains. Similarly, negative reviews often cause financial losses.

Due to the reviews' dominance power, spammers create fake reviews to deteriorate the online reviews systems and confuse the consumers. This spam review does not reflect the real opinion reviewer's and it is intended for misleading

A. E. Hassanien et al. (Eds.): AISI 2018, AISC 845, pp. 395–404, 2019.
https://doi.org/10.1007/978-3-319-99010-1_36

reviewers' readers. It may be supportive in order to over qualify a product or a service or destructive to damage the e-reputation of competitors companies.

Therefore, it is crucial to detect fake reviews in order to protect e-commerce form fraudsters' activities, to ensure customers' confidence and to maintain companies' fair competition. As a consequence, spam reviews detection became one of the most challenging problems. Researchers have developed various spam detection techniques in which the major task is distinguishing between fake reviews and truthful ones. A largest number of methods and techniques is proposed to detect spam reviews. Most of these works tried to distinguish between fake and true opinions across the linguistic aspects and feeling [3], as well as the style of writing [2,8], and readability and subjectivity [13]. Some researchers try to catch the group spammers cause of their ability to manipulate the readers' desires, beliefs and consciousness for both product and service. In [12], the authors have studied this problem by defining a set of eight indicators that try to detect behavior of the group members such as time and rating deviation. Other studies [11] proposed the score candidate groups using the relationship between groups, individuals and products. Other techniques have focused on spammer detection. Most of them are graph based approaches [1,6,19] with tree types of nodes namely: review, reviewer and stores. The deviation from overall ratings was used as features in [14,16,20] and all these studies have bring a significant results. An algorithm to detect burst patterns in reviews was proposed in [7]. It generated five new spammer behavior features as indicators to used them in review spammer detection. In addition, an other method, proposed in [9], used three detection metrics (Context similarity authors' activeness, Authors' rating behavior) to score each review and detect spam ones.

Up to our knowledge, no one of the previous works is able to handle uncertainty in reviewers' votes. In fact, the fake reviews detection is an uncertain problem and involves imperfection concerning given reviews. In this context, we propose a novel method, the belief fake reviews detection (BFRD), based on the belief function theory. Indeed, this theory is able to handle uncertainty and allows to deal with partial and total ignorance. It can manipulate various pieces of information from different sources and also take into account their reliabilities through the discounting operation. Moreover, the similarity between the difference given reviews can be taken into account through the distances proposed under belief function framework and especially without forgetting its powerful on decision making under uncertainty. In addition, our proposed method takes into account the case where we have a lack of information and deals with only the overall rating reviews.

The remainder of this paper is structured as follows: In Sect. 2, we present the belief function theory concepts. Then in Sect. 3, our proposed method named BFRD, is detailed. Experimentation will be proven through the use of the numerical examples in Sect. 4. Finally, we conclude in Sect. 5.

2 Belief Function Theory

The belief function theory was introduced by Shafer [15] as a model to represent beliefs. It is considered as an efficient tool able to deal with uncertainty and to manage several types of imperfection. Various interpretations have been proposed from this theory, including the Transferable belief model (TBM) [17] that we adopt in our work. In this section, we elucidate the crucial basic belief concepts, the discounting operation, some combination rules and the decision process.

2.1 Basic Concepts

The universe of discourse Ω is a finite and exhaustive set of different events associated to a given problem. Its power set 2^{Ω} contains all possible hypotheses that formed union of events, and the empty set which represents the conflict, defined by: $2^{\Omega} = \{A : A \subseteq \Omega\}$. A basic belief assignment (*bba*) or a belief mass defined as a function from 2^{Ω} to $[0,1]$ such that: $\sum_{A \subseteq \Omega} m^{\Omega}(A) = 1$. Each subset A of Ω, having a strictly positive mass $m^{\Omega}(A) > 0$, is considered as the focal element of the *bba*. In order to express several types of imperfection, some special *bbas* were defined:

- A certain *bba* is defined as follows: $m^{\Omega}(\{\omega_i\}) = 1$ and $\omega_i \in \Omega$.
- A vacuous *bba* is defined as follows: $m^{\Omega}(\Omega) = 1$ $m^{\Omega}(A) = 0$ $\forall A \neq \Omega$. This function models the state of the total ignorance.
- A categorical *bba* is defined as follows: $m^{\Omega}(A) = 1$ $\forall A \subset \Omega$ and $m^{\Omega}(B) = 0$ $\forall B \subseteq \Omega$ $B \neq A$. This *bba* has a unique focal element A.
- A simple support function is defined as follows:

$$m^{\Omega}(X) = \begin{cases} \omega & \text{if} \quad X = \Omega \\ 1 - \omega & \text{if} \quad X = A \quad \forall A \subset \Omega \\ 0 & \text{otherwise} \end{cases} \tag{1}$$

  where A is the focal element and $\omega \in [0,1]$.

2.2 Discounting Operation

The discounting operation established by [15] allows us to weaken the masses by the *discount rate* $\alpha \in [0,1]$ such that $(1-\alpha)$ is the degree of reliability of a source.

Accordingly, the discounted *bba*, noted m^{α}, becomes:

$$\begin{cases} {}^{\alpha}m^{\Omega}(A) = (1-\alpha)m(A) & \forall A \subset \Omega, \\ {}^{\alpha}m^{\Omega}(\Omega) = \alpha + (1-\alpha)m(\Omega). \end{cases} \tag{2}$$

2.3 Combination Rules

Let m_1^{Ω} and m_2^{Ω} two different *bbas* defined on the same frame of discernment Ω providing by two distinct and cognitively independent reliable sources.

There are several combination rules proposed in the belief function framework, each rule has its specificities and its characteristics. Then, we will present some of the most used ones.

1. *Conjunctive rule (CRC)*: Introduced by Smets [18], it allows to combine two *bbas* induced by distinct and reliable sources of information. It is denoted by $\textcircled{\cap}$ and defined as:

$$m_1^{\Omega} \textcircled{\cap} m_2^{\Omega}(A) = \sum_{B \cup C = A} m_1^{\Omega}(B) m_2^{\Omega}(C) \quad (3)$$

The mass assigned to the empty set quantifies the degree of conflict between the two *bbas*.
2. *Dempster's rule of combination*: It is the normalized version of the conjunctive rule where it does not support the existence of a mass on the empty set [4]. It is denoted by $\oplus$ and defined as:

$$m_1^{\Omega} \oplus m_2^{\Omega}(A) = \begin{cases} \frac{m_1^{\Omega} \textcircled{\cap} m_2^{\Omega}(A)}{1 - m_1^{\Omega} \textcircled{\cap} m_2^{\Omega}(\emptyset)} & \text{if } A \neq \emptyset, \forall A \subseteq \Omega, \\ 0 & \textit{otherwise.} \end{cases} \quad (4)$$

3. *The combination with adapted conflict rule (CWAC)*: This combination [5] act as the conjunctive rule when the *bbas* are antonym (it keeps the conflict) and as the Dempster rule when the *bbas* are similar.
They use the notion of dissimilarity that is obtained through a distance measure, to ensure this adaptation between all sources. The CWAC is formulated as follows:

$$m_{\textcircled{\leftrightarrow}}^{\Omega}(A) = (\textcircled{\leftrightarrow} m_i^{\Omega})(A) = D_{max} m_{\textcircled{\cap}}^{\Omega}(A) + (1 - D_{max}) m_{\oplus}^{\Omega}(A) \quad (5)$$

where D_{max} represents the maximal value of all the distances, it can be used to find out if at least one of the sources is opposite to the others and thus it may be defined by:

$$D_{max} = max[d(m_i^{\Omega}, m_j^{\Omega})], \quad (6)$$

where $i \in [1, M]$, $j \in [1, M]$, M is the total number of mass functions and $d(m_i^{\Omega}, m_j^{\Omega})$ is the distance measure proposed by Jousselme [10]:

$$d(m_1^{\Omega}, m_2^{\Omega}) = \sqrt{\frac{1}{2}(m_1^{\Omega} - m_2^{\Omega})^t D (m_1^{\Omega} - m_2^{\Omega})}, \quad (7)$$

where D is the Jaccard index defined by:

$$D(E, F) = \begin{cases} 0 & \text{if } \quad E = F = \emptyset, \\ \frac{|E \cap F|}{|E \cup F|} & \forall E, F \in 2^{\Omega} \end{cases} \quad (8)$$

2.4 Decision Making

The decision making step chooses the most suitable hypothesis for a given problem. The Transferable Belief Model (TBM), proposed by [18], is composed by both the credal level where beliefs are defined by *bbas* then combined, and the pignistic level where *bbas* are transformed into pignistic probabilities denoted by $BetP$ and defined as follows:

$$BetP(B) = \sum_{A \subseteq \Omega} \frac{|A \cap B|}{|A|} \frac{m^{\Omega}(A)}{(1 - m^{\Omega}(\emptyset))} \quad \forall \, B \in \Omega \quad (9)$$

3 The Belief Fake Reviews Detection (BFRD)

The proposed method, in this paper, deals with fake reviews detection using the belief function tools. In order to distinguish between fake and genuine reviews, this method only requires the overall rating as an input. Given a dataset of N votes that have different values between 1 and 5 stars (respectively poor, below average, average, good and excellent), each vote V_i is associated to a reviewer denoted R_i where i is the *id* of the corresponding one. Our method follows four main steps detailed in-depth.

3.1 Modeling Reviewer's Vote by Mass Functions

As we adopt the belief function theory in order to model conveniently imperfections in votes, each vote V_i will be transformed into a mass function (i.e. bba) m_i^Ω with $\Omega = \{1, 2, 3, 4, 5\}$ where one of the elements in Ω represents the stars' number given by each reviewer R_i.

We think that the reviewer is either a spammer or a real customer gives an uncertain vote. We propose to model this uncertainty by considering the vote, the vote-1 and the vote+1 for each given overall rating review. In the upper extreme case (i.e. $V_i = 5$), we model the vote and the vote-1 and in the lower one (i.e. $V_i = 1$) we model the vote and the vote+1.

We transform each vote value V_i given by the reviewer R_i into *bbas* defined as follows:

$$m_{ik}^\Omega(\{k\}) = 1$$

where $k = \begin{cases} V_i \\ V_i + 1. \\ V_i - 1 \end{cases}$ In the upper extreme case, $k = \begin{cases} V_i \\ V_i - 1 \end{cases}$ and in the lower extreme case, $k = \begin{cases} V_i \\ V_i + 1. \end{cases}$

Example 1. Let us consider the case of five reviewers given an overall rating review for an hotel detailed in Table 1.

For R_1: $m_{14}^\Omega(\{4\}) = 1$; $m_{15}^\Omega(\{5\}) = 1$; $m_{13}^\Omega(\{3\}) = 1$. For R_3: $m_{35}^\Omega(\{5\}) = 1$; $m_{34}^\Omega(\{4\}) = 1$. For R_5: $m_{51}^\Omega(\{1\}) = 1$; $m_{52}^\Omega(\{2\}) = 1$.

We propose to model the reliability degree of the reviewer R_i by $(1-\alpha_i)$ where α_i is its discounting factor. Its value is between $[0, 1]$, if $\alpha_i = 0$ the reviewer is totally reliable and if $\alpha_i = 1$, it means that the reviewer is totally unreliable and it will not be taken into account. We calculate α_i as follows:

$$\alpha_i = \frac{\text{Number of votes different from the current vote of } R_i}{\text{Total votes' number}} \tag{10}$$

So, each vote transformed into mass functions is weakened by its relative reliability degree $(1 - \alpha_i)$ using the discounting operation and consequently changed into simple support functions. Thus, the reliability of the reviewer will be taken into consideration.

Table 1. Hotel reviews

Reviewer	Vote
R_1	4 stars
R_2	4 stars
R_3	5 stars
R_4	3 stars
R_5	1 star

Example 2. We continue with the previous Example 1, we calculate the reliability factor for R_1: $\alpha_1 = \frac{3}{5} = 0.6$.

After the discounting operation the *bbas* are transformed into a mass functions as follows: ${}^{\alpha_1}m_{14}^{\Omega}(\{4\}) = (1-0.6)*1 = 0.4$; ${}^{\alpha_1}m_{14}^{\Omega}(\{\Omega\}) = 0.6+(1-0.6)*0 = 0.6$. ${}^{\alpha_1}m_{15}^{\Omega}(\{5\}) = 0.4$; ${}^{\alpha_1}m_{15}^{\Omega}(\{\Omega\}) = 0.6$. ${}^{\alpha_1}m_{13}^{\Omega}(\{3\}) = 0.4$; ${}^{\alpha_1}m_{13}^{\Omega}(\{\Omega\}) = 0.6$.

In addition, we propose to model the distance between the ith given vote value denoted by V_i and its corresponding modeled values (vote, vote+1 and vote-1) denoted by k and represented by m_{ik}, in order to not consider them in the same way, by $(1-\beta_{ik})$ where β_{ik} is its discounting factor. Its value is between $[0, 1]$, if $\beta_{ik} = 0$ it means that the vote represents the current vote value and if $\beta_{ik} = 1$ it means that the vote is so far from the current vote value. The discounting factor β_{ik} is calculated as follows:

$$\beta_{ik} = \frac{|V_i - k|}{\text{The maximum vote value}} \tag{11}$$

Then, each simple support function associated to each vote given by a review is weakened by its relative reliability degree $(1 - \beta_{ik})$ using the discounting operation.

Example 3. Let us consider the same Example 1, we calculate the discount factor β for the R_1: $\beta_{14} = \frac{|4-4|}{5} = 0$; $\beta_{15} = \frac{|4-5|}{5} = 0.2$; $\beta_{13} = \frac{|4-3|}{5} = 0.2$.

After the second discounting operation, the *bbas* are transformed as follows:
${}^{\alpha_1\beta_{14}}m_{14}^{\Omega}(\{4\}) = 0.4$; ${}^{\alpha_1\beta_{14}}m_{14}^{\Omega}(\{\Omega\}) = 0.6$.
${}^{\alpha_1\beta_{15}}m_{15}^{\Omega}(\{5\}) = 0.32$; ${}^{\alpha_1\beta_{15}}m_{15}^{\Omega}(\{\Omega\}) = 0.68$.
${}^{\alpha_1\beta_{13}}m_{13}^{\Omega}(\{3\}) = 0.32$; ${}^{\alpha_1\beta_{13}}m_{13}^{\Omega}(\{\Omega\}) = 0.68$

Finally, we aggregate each three discounted *bbas* (two in the extreme cases) representing each given vote using the Dempster rule (Eq. 4) in order to represent each given vote by one *bba* containing four focal elements (three in the extreme cases).

Example 4. After aggregating the discounted votes corresponding to R_1 calculated in Example 3 using the Dempster rule, we found:
$m_1^{\Omega} = {}^{\alpha_1\beta_{14}}m_{14}^{\Omega} \oplus {}^{\alpha_1\beta_{13}}m_{13}^{\Omega} \oplus {}^{\alpha_1\beta_{15}}m_{15}^{\Omega}$
$m_1^{\Omega}(\{4\}) = 0.255$; $m_1^{\Omega}(\{5\}) = 0.180$; $m_1^{\Omega}(\{3\}) = 0.180$; $m_1^{\Omega}(\Omega) = 0.385$.
Then this *bba* will represent the vote (4 stars) given by R_1.

3.2 Distance Between the Current Reviewer's Vote and All the Other Votes' Aggregation

In order to evaluate the vote provided by each reviewer, we compared it to all other reviewers' vote as follows:

For each reviewer, we aggregate all the other reviewers' votes represented by *bbas* using the CWAC combination rule (Eq. 5) chosen, because it can cope with the conflict in different votes. The output of this combination is one *bba* m^{Ω}_{ci}, which represents the whole reviewers' votes except the current one as follows: $m^{\Omega}_{ci} = m^{\Omega}_{1} \circleddash m^{\Omega}_{2} \circleddash \dots \circleddash m^{\Omega}_{i-1} \circleddash m^{\Omega}_{i+1} \circleddash \dots \circleddash m^{\Omega}_{N}$.

Then, we calculate the distance $d(m^{\Omega}_{i}, m^{\Omega}_{ci})$ using the distance of Jousselme (Eq. 7), in order to measure the similarity between each review's vote and all others.

Example 5. We continue with the previous Example 4 where the current vote corresponds to the first reviewer R_1, so we aggregate all the reviewers' votes except the first one: $m^{\Omega}_{c1} = m^{\Omega}_{2} \circleddash m^{\Omega}_{3} \circleddash m^{\Omega}_{4} \circleddash m^{\Omega}_{5}$.

$m^{\Omega}_{c1}(\emptyset) = 0.16$; $m^{\Omega}_{c1}(\{1\}) = 0.04$; $m^{\Omega}_{c1}(\{2\}) = 0.25$; $m^{\Omega}_{c1}(\{3\}) = 0.15$; $m^{\Omega}_{c1}(\{4\}) = 0.25$; $m^{\Omega}_{c1}(\{5\}) = 0.15$.

Then, we apply the Jousselme distance between the first reviewer's vote and all the other ones, we found: $d(m^{\Omega}_{1}, m^{\Omega}_{c1}) = 0.155$.

3.3 Construction of a New *bba* Modeling the Vote into Fake or Not Fake

The distance measured in the previous step represents the degree of compatibility between the vote and all the other ones' which means that more the distance value decreases more the vote is reliable. So, we propose to transform each distance into a new *bba* with $\Theta = \{f, \bar{f}\}$ $f=$ fake and $\bar{f}=$ not fake as the following equation:

$$\begin{cases} m^{\Theta}(\{f\}) = \gamma * \frac{1}{1+e^{-a.ds+\frac{a}{2}}} \\ m^{\Theta}(\{\bar{f}\}) = \gamma * (1 - \frac{1}{1+e^{-a.ds+\frac{a}{2}}}) \\ m^{\Theta}(\Theta) = 1 - \gamma \end{cases} \tag{12}$$

with $ds = d(m^{\Omega}_{i}, m^{\Omega}_{ci})$, $a = 10$ and $\gamma = \frac{\text{The standard deviation of all votes}}{\text{The maximum value of the standard deviation}}$.

Example 6. We transform the Jousselme distance $d(m^{\Omega}_{1}, m^{\Omega}_{c1})$ calculated in the previous Example 5 into a new *bba* with $\Theta = \{f, \bar{f}\}$, we found: $m^{\Theta}(\{f\}) = 0.018$; $m^{\Theta}(\{\bar{f}\}) = 0.562$; $m^{\Theta}(\Omega) = 0.42$.

3.4 Decision Making

The decision process is made in this final step and assured by the pignistic probability $BetP$ (Eq. 10). The $BetP$ with the greater value is considered as the final decision.

Example 7. After applying the pignistic probability on the *bba* calculated in the previous Example 6, we found: $BetP(\{f\}) = 0.227$; $BetP(\{\bar{f}\}) = 0.773$.

Thus, we assume that the vote given by the first reviewer is a genuine one.

4 Experimentation and Results

In the fake reviews detection, the evaluation is one of the most challenging problems considering the unavailability of the labeled dataset because it is not obvious to distinguish between the fake and the real reviews, even with the knowledge of spammers and no spammers. In this paper, we propose to use the numerical examples to intuitively validate our method.

Table 2. The numerical examples' results

The numerical examples	Opinion		Fake rating stars	Not fake rating stars
	Rating stars	Number of reviewers		
Example 1	5*	5	-	5*, 4*, 3*, 2* and 1*
	4*	5		
	3*	5		
	2*	5		
	1*	5		
Example 2	5*	5	-	5* and 4*
	4*	5		
Example 3	5*	5	-	5* and 1*
	1*	5		
Example 4	4*	4	2* and 1*	4*, 5* and 3*
	5*	2		
	3*	2		
	2*	1		
	1*	1		

Figure 1 represents the obtained results in four different examples that are detailed in Table 2, where the reviewers judge a hotel by giving rating stars (denoted by * in Table 2). We deal with one standard case and three particular ones, where the given reviews represent five and two majority class rating stars. We can assume that our method provides logical results with different cases.

Then, in order to test our BFRD performance, we used a real database extracted from Tripadvisor. The dataset consists of a 1550 reviewers who express their opinion on "Melia Caribe Tropical hotel" by writing reviews, giving an overall, and vote other criteria. Our proposed method detects 967 fake reviews and 583 genuine ones.

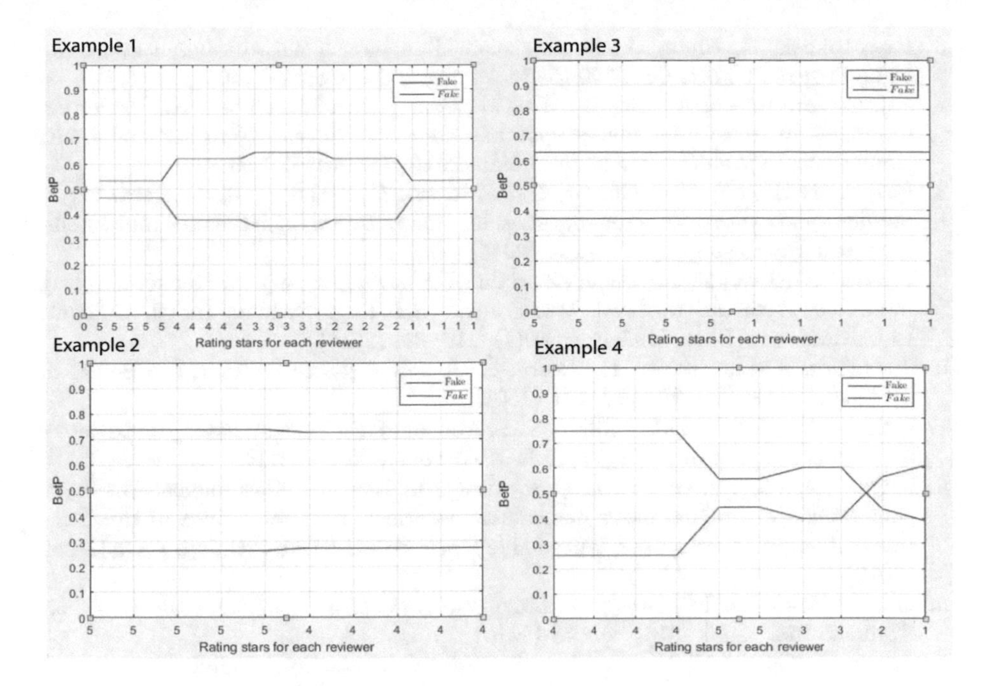

Fig. 1. The numerical examples' results

5 Conclusion

In this paper, we addressed the fake reviews detection problem in an uncertain context through the belief function tools. We proposed to handle uncertainty in the given rating reviews and to evaluate them through their compatibility with all others. In fact, our new method shows effectiveness in distinguishing between the fraudulent reviews and the honest ones. Moreover, this approach can be applied in several fields such as e-commerce and e-business. As future work, we will integrate some other notions like reviewers' trustiness.

References

1. Akoglu, L., Chandy, R., Faloutsos, C.: Opinion fraud detection in online reviews by network effects. Proceedings of the Seventh International Conference on Weblogs and Social Media, ICWSM **2013**, 2–11 (2013)
2. Banerjee, S., Chua, A.Y.K.: Applauses in hotel reviews: genuine or deceptive? In: Proceedings of Science and Information Conference (SAI), pp. 938–942 (2014)
3. Deng, X., Chen, R.: Sentiment analysis based online restaurants fake reviews hype detection. In: Web Technologies and Applications, pp. 1–10 (2014)
4. Dempster, A.P.: Upper and lower probabilities induced by a multivalued mapping. Ann. Math. Stat. **38**, 325–339 (1967)
5. Lefèvre, E., Elouedi, Z.: How to preserve the confict as an alarm in the combination of belief functions? Decis. Support Syst. **56**, 326–333 (2013)

6. Fayazbakhsh, S., Sinha, J.: Review spam detection: a network-based approach. Final Project Report: CSE 590 (Data Mining and Networks) (2012)
7. Fei, G., Mukherjee, A., Liu, B., Hsu, M., Castellanos, M., Ghosh, R.: Exploiting burstiness in reviews for review spammer detection. Seventh International AAAI Conference on Weblogs and Social Media **2013**, 175–184 (2013)
8. Fusilier, D.H., Montes-y-Gómez, M.M., Rosso, P., Cabrera, R.G.: Detection of opinion spam with character n-grams. In: Computational Linguistics and Intelligent Text Processing, pp. 285–294 (2015)
9. Heydari, A., Tavakoli, M., Ismail, Z., Salim, N.: Leveraging quality metrics in voting model based thread retrieval. World Acad. Sci. Eng. Technol. Int. J. Comput. Electr. Autom. Control Inf. Eng. **10**(1), 117–123 (2016)
10. Jousselme, A.-L., Grenier, D., Bossé, É.: A new distance between two bodies of evidence. Inf. Fusion **2**(2), 91–101 (2001)
11. Kolhe, N.M., Joshi, M.M., Jadhav, A.B., Abhang, P.D.: Fake reviewer groups detection system. J. Comput. Eng. (IOSR-JCE) **16**(1), 06–09 (2014)
12. Mukherjee, A., Kumar, A., Liu, B., Wang, J., Hsu, M., Castellanos, M.: Spotting opinion spammers using behavioral footprints. In: Proceedings of the ACM International Conference on Knowledge Discovery and Data Mining, pp. 632–640 (2013)
13. Ong, T., Mannino, M., Gregg, D.: Linguistic characteristics of shill reviews. Electr. Commer. Res. Appl. **13**(2), 69–78 (2014)
14. Savage, D., Zhang, X., Yu, X., Chou, P., Wang, Q.: Detection of opinion spam based on anomalous rating deviation. Expert Syst. Appl. **42**(22), 8650–8657 (2015)
15. Shafer, G.: A Mathematical Theory of Evidence, vol. 1. Princeton University Press, Princeton (1976)
16. Sharma, K., Lin, K.I.: Review spam detector with rating consistency check. In: Proceedings of the 51st ACM Southeast Conference, Article no. 34 (2013)
17. Smets, P.: The combination of evidence in the transferable belief model. IEEE Trans. Pattern Anal. Mach. Intell. **12**(5), 447–458 (1990)
18. Smets, P.: The transferable belief model for quantified belief representation. In: Smets, P. (ed.) Quantified Representation of Uncertainty and Imprecision, pp. 267–301. Springer, Dordrecht (1998)
19. Wang, G., Xie, S., Liu, B., Yu, P.S.: Review graph based online store review spammer detection. In: Proceedings of 11th International Conference on Data Mining (ICDM), pp. 1242–1247 (2011)
20. Xue, H., Li, F., Seo, H., Pluretti, R.: Trust-aware review spam detection. IEEE Trustcom/BigDataSE/ISPA **1**, 726–733 (2015)

On Mixing Iris-Codes

Randa F. Soliman[1(✉)], Mohamed Amin[1], and Fathi E. Abd El-Samie[2]

[1] Mathematics and Computer Science Department, Faculty of Science, Menoufia University, Shebin El-Koom 32511, Egypt
randafouad2010@yahoo.com, mohamed_amin110@yahoo.com
[2] Department of Electronics and Electrical Communications Engineering, Faculty of Electronic Engineering, Menoufia University, Menouf 32952, Egypt
fathi_sayed@yahoo.com

Abstract. In a cancelable iris recognition technique, all enrollment patterns are masked using a transformation function, and the invertibility for obtaining the original data should not be possible. This paper presents an approach for mixing multi-biometric features based on Double Random Phase Encryption (DRPE) to obtain a single protected IrisCode from different IrisCodes based on the Fractional Fourier Transform (FrFT). For the IrisCode generation, two encryption keys (RPM1 and RPM2) are utilized. The RPM2 is suggested to be the right iris feature vector of the same user. As a result, this feature level mixing of two different templates highly increases privacy and slightly enhances performance compared to its original counterpart. This proposed system achieves a high success rate in identification due to the fact that the iris authentication issue has been transformed to a key authentication process.

Keywords: Iris recognition · Cancelable biometrics
Double random phase encryption (DRPE) · Fractional fourier transform (FrFT)

1 Introduction

Recently, iris recognition systems peaked to a high level of statistical performance compared to other biometric systems [1]. The most-widely used iris recognition schemes were proposed by Daugman [2]. The best personal authentication form is the biometric recognition, but it suffers from privacy violation, because biometric characteristics can be re-constructed. Moreover, it is confirmed that the stolen iris patterns could be reissued from stored templates in the database [3]. Privacy schemes for biometric protection are classified as biometric cryptosystems and cancelable biometrics [4]. In general, the imposter distributions have a lower variability compared to the higher values of genuine ones, which seems to be a great challenge for protected IrisCodes. In order to face this challenge, protected multi-biometric techniques were proposed in [5, 6] as in the case of mixing various biometric features resulting in a higher recognition rate.

Nandakumar and Jain argued that security and privacy improvements rarely come without degradations to the authentication performance, and consequently, there is still a remarkable gap between theory and practice [7]. However, the proposed scheme satisfies an outstanding recognition performance with maintained cancelability.

A. E. Hassanien et al. (Eds.): AISI 2018, AISC 845, pp. 405–414, 2019.
https://doi.org/10.1007/978-3-319-99010-1_37

2 Related Work

Privacy concerns have evolved as a solution for the hacking attacks on databases in the last decades [8, 9]. If there is a database, which relates the user to his unique biometric template, it could be used illegally to monitor the activities of the user. These threats have to be treated, and one prospective solution is the cancelable biometrics. Ratha et al. [10] first introduced the notion of biometrics cancelability. In their works, they rearranged the fingerprint minutia in polar and Cartesian domains to obtain the cancelable template. Although their work renders a satisfactory accuracy performance, the non-invertibility was seen weak [11]. Zuo et al. [12] suggested the generation of cancelable iris templates by applying the row permutation on IrisCodes. A one more cancelable biometric method has been presented by Teoh et al. [13] for increasing the recognition rates of cancelable templates. They employed secret random numbers to pattern the original biometric features into transformed templates.

Rathgeb et al. [14, 15] utilized Bloom filter to construct cancelable templates from iris codes. They introduced a fast comparison between the transformed biometric templates using Bloom filters. Ouda et al. [16, 17] proposed a bio-encoding scheme. It was concluded that the accuracy performance is preserved compared to those of the unprotected counterparts. However, Lacharme [18] clarified that the non-invertible property of bio-encoding can be violated. Hämmerle-Uhl et al. [19] utilized block remapping methods to perform non-invertible transformations. Although this technique could not fulfill an acceptable recognition rate, Jenisch et al. [20] proved that most unprotected biometric data may be recovered. In [21, 22], a resistant cancelable biometric method for attacks was introduced. The associative memory is used for encoding cancelable transformation parameters with high recognition rates.

To increase the security of the encrypted iris verification systems, a new cancelable iris recognition system based on DRPE in the FrFT domain is presented in this paper. Two encryption keys (RPM1, RPM2) are used to enhance security. In the proposed algorithm, first, all iris images in the database are encrypted by the DRPE method in the FrFT domain. Then, an encrypted test template is used to evaluate the performance of the proposed method during verification. It should be noticed that train and test keys are identical for each person. The rest of this paper is organized as follows. In Sect. 3, preliminaries are explained. In Sect. 4, the proposed method is presented in more details. Section 5 shows experimental results and privacy threats. Section 6 gives the concluding remarks of the paper.

3 Preliminaries

3.1 Double Random Phase Encryption

The DRPE is the most-widely used optical encryption method. It was introduced in 1995 [23]. In this method, two random phase masks $RPM1(x, y)$ and $RPM2(u, v)$ assist in encoding the plain image, $f(x, y)$ with positive values into a white stationary noisy sequence template, $C(x, y)$. These encryption keys denote two independent white sequences uniformly distributed in $[0, 2\pi]$. The spatial and frequency domains are denoted by the coordinates (x, y) and (u, v), respectively. The primary image is multiplied by the first phase mask $RPM1(x,y)$ in the spatial domain to obtain the encrypted image. Then a multiplication process is applied on the output with the function $h(x,y)$. This phase mask represents the inverse Fourier transform of the $RPM2(u,v)$ as $(exp(2\pi i \times RPM2(u,v)))$. The encrypted function is complex with amplitude and phase, and it is given by:

$$C(x,y) = \{f(x,y) \times exp(2\pi i \times RPM1(x,y)))\} * h(x,y), \tag{1}$$

where the symbols * and × denote convolution and multiplication, respectively.

In the decryption process, the encrypted image $C(x,y)$ is Fourier transformed, multiplied by the complex conjugate of the second phase mask $RPM2(u,v)$ in the frequency domain, and then inverse Fourier transformed. Finally, the absolute value turns out the decrypted image $f(x,y)$. The second encryption key $RPM2(u,v)$ is the only required item for decrypting the iris patterns.

3.2 Fractional Fourier Transform (FrFT)

The Fractional Fourier transform (FrFT) was proposed in [24] as a general form of the conventional plane encoding. In this transform, a signal is rotated with any arbitrary angle. The FrFT is featured as a general rotation with angle $\alpha = a\pi/2$ in the conventional FT. The parameter a is denoted as the fractional order of the transform. Moreover, the parameter a is limited to the range of $0 \leq a \leq 1$ due to some restricted features of the FrFT. In the FrFT definition of the kernel-based integral transformation, $f_a(u)$ is considered as the FrFT of order a for the signal $f(x)$, while the hybrid time/frequency parameter is denoted by u. The parameter a is considered to be a time variable in the case of $a = 0$, but it is a frequency variable if $a = 1$. The gradual increase of a from 0 to 1 results in slight variation of u from "time" to "frequency". This yields temporary variations in the transformed signal frequency content. The kernel-based integral transformation is one of the commonly used definitions of the FrFT. Generally, the FrFT is defined as follows [25]:

$$k_\alpha(t,u) = \begin{cases} \sqrt{\frac{1-j\cot\alpha}{2\pi}}\,.\exp\left(j\frac{t^2+u^2}{2}\cot\alpha - j\frac{tu}{\sin\alpha}\right) & if\,\alpha \neq n\pi \\ \delta(u-t) & if\alpha = 2n\pi \\ \delta(u-t) & if\alpha = 2n\pi \end{cases} \tag{2}$$

The definition of the FrFT for a function $f(x)$ and an angle α is as follows:

$$X_\alpha(u) = \int_{-\infty}^{\infty} x(t)k_\alpha(t,u)dt \tag{3}$$

The kernel-based integral transformation form of the FrFT is given by:

$$f_\alpha(u) = F^a[f(x)] = C_\alpha \int f(x)\exp\left[i\pi\frac{u^2+x^2}{\tan\alpha} - 2i\pi\frac{ux}{\sin\alpha}\right]dx \tag{4}$$

where F^a represents the counter-clockwise rotation of the signal with angle α in the time frequency plane.

In this paper, the DRPE is implemented in the FrFT domain to increase the degree of security.

4 Proposed Method

This paper presents a cancelable iris verification system based on the DRPE algorithm in the FrFT Domain. A block diagram of the proposed algorithm is shown in Fig. 1. This algorithm has two steps; the enrolment step and the authentication step. In the enrollment step, the iris capturing device is used to acquire the iris images. We adopt the coarse-to-fine method in [26] for extracting the pupil and iris circles in the segmentation step. The DRPE is applied in the FrFT domain to encrypt iris codes for each person in both train and test phases with RPM1, and RPM2 keys, and the new template is saved in the database in the train or verified in the test. The keys must be uniformly distributed. For iris images with real-value pixels, only the second phase mask is needed for decryption. The second phase mask is proposed to be the right iris vector for the same user to increase privacy.

If the transformed pattern and the keys (i.e. RPM1 and RPM2) are stolen, it is difficult for the hacker to get a pattern close to the original iris pattern. It is an undetermined problem as the binary iris biometric feature vectors extracted from both eyes of a subject are mixed to a single protected template at feature level, which highly increases security while at the same time biometric performance is maintained. Also, even if a hacker steals the user's iris pattern, without knowing the keys, he/she cannot generate the transformed patterns required by the application.

During the verification stage, the application obtains the iris image and the second phase mask from the user, computes the transformed pattern, and compares it with the

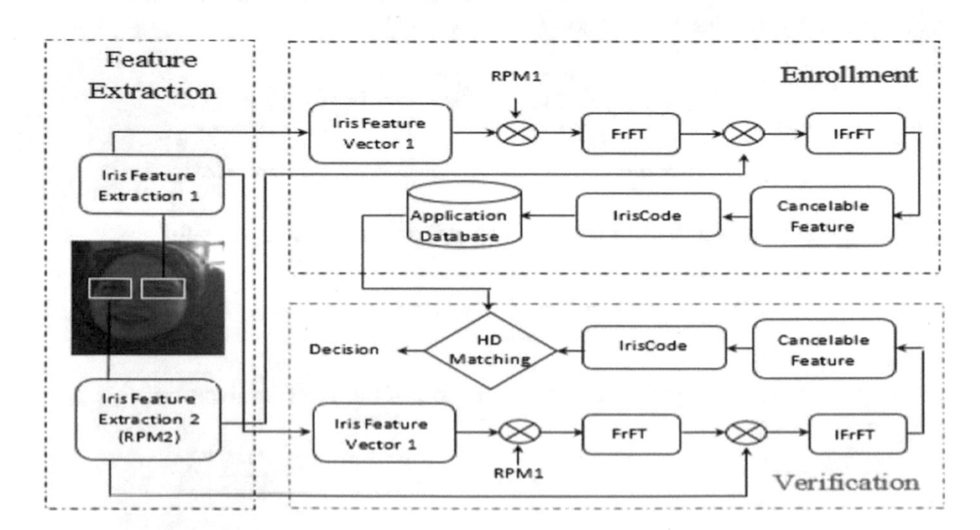

Fig. 1. The workflow layout of the proposed cancelable iris recognition system.

ones in its database using the Hamming distance to make a decision. In the case the second phase mask or the transformed patterns are compromised, one can create a new second phase mask and obtain a new transformed pattern which can be updated into the application database. Instead of the user providing the second phase mask during verification, the application can generate and store it along with the cancelable template in its database.

4.1 Generation of Encrypted IrisCodes

The proposed encryption algorithm used to generate the cancelable templates will be implemented with two keys, and the second key is related to the user's right iris image. The DRPE is based on the modification of the spectral distribution of the image. Without any prior information about this spectral modification or the target image at the receiver, the image decoding cannot be performed. This algorithm is summarized as follows:

Algorithm: Encrypted IrisCode Generation

Input: The upper unwrapped half of iris image.
Output: The cancelable IrisCode.
Step 1: Load the input image.
Step 2: Transform the input image into double format.
Step 3: Extract Gabor features from the upper unwrapped half of the normalized iris.
Step 4: The first key, i.e. the RPM1, is multiplied by only the high-frequency pixels in the left iris vector to be encrypted. This procedure introduces the first modification to the spectrum of the left iris vector.
Step 5: The second key, i.e. RPM2 (right iris vector of the same user), is directly inserted into the image spectrum in the Fractional Fourier plane. The multiplication of the RPM2 by the spectrum obtained in the fourth step introduces the second modification into the spectrum of the left iris vector.
Step 6: A second optical FrFT is carried out to obtain the encoded image in the original 2-D space of images.
Step 7: The concatenated output from step 5 and the low-frequency pixels of the upper unwrapped half of the normalized left iris gives the transformed features.
Step 8: The encoding process produces the IrisCode including a set of information bits equivalent to the angular resolutions multiplied by the radial resolutions.

---

5 Experimental Results and Privacy Threats

In the present section, the details of experimental results to illustrate the performance of the proposed scheme are presented. Moreover, a comparative study of the proposed scheme with the existing schemes is included.

5.1 Experimental Setup

Experiments have been carried out using the CASIA-IrisV3-Interval database [27]. At pre-processing, we adopt a coarse-to-fine method [26]. The coarse stage relies on image thresholding to separate dark parts. The iris of a given sample image is detected, and unwrapped to a rectangular texture as shown in Fig. 2a–f. In the feature extraction step, the unwrapped iris is convolved with a one-dimensional Gabor filter for extracting features. The feature extraction approach introduced by Masek in [28] is implemented, where the LogGabor filters are applied. Sample IrisCodes created by the feature extraction method are given in Fig. 2g.

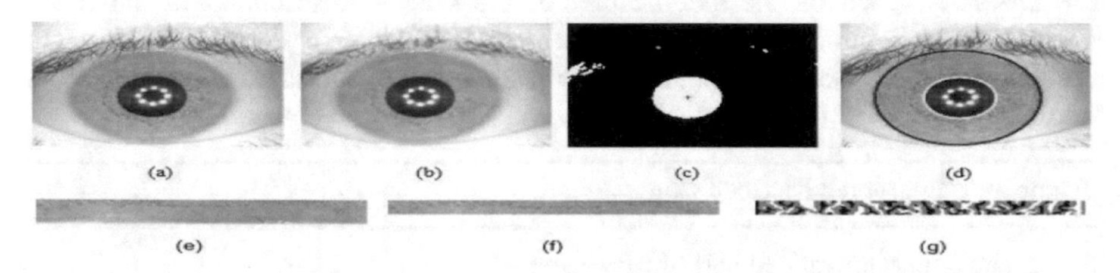

Fig. 2. Iris pre-processing (a) Eye image. (b) Eye image after grayscale closing. (c) Thresholded image. (d) Segmented iris. (e) Normalized image. (f) Upper half of the normalized image. (g) IrisCode with 1-D Log-Gabor filter.

5.2 Performance Evaluation

In this part, detailed results are given to describe the performance of the proposed scheme. Moreover, a comparison study between the proposed and traditional schemes is presented. The Hamming distance is used to determine the iris probe class. The Equal Error Rates (EER) and the Receiver Operating Characteristics (ROC) curve are used for evaluation [29]. The ROC curve is obtained by plotting True Positive Rates (TPR) versus versus the False Positive Rates (FPR). The EER is the point when the false rejection rate (i.e., 1 - TPR) and FPR hold equality at a particular threshold value. The TPR is also known as the sensitivity, while the TNR is known as the specificity. The FRR measures the probability of falsely rejecting an iris as an imposter (intra-class) iris pattern, and the FPR measures the probability of falsely accepting an imposter iris pattern as a genuine (inter-class) iris pattern. Negative and positive predictive values (NPV and PPV) are utilized to estimate the matching performance using Eqs. (5) and (6).

$$\mathrm{PPV} = \frac{\sum \text{True positive}}{\sum \text{Predicted condition positive}} \tag{5}$$

$$\mathrm{NPV} = \frac{\sum \text{True negative}}{\sum \text{Predicted condition negative}} \tag{6}$$

The values of these performance metrics are evaluated from the genuine and imposter scores. The EER is inversely proportional to the system performance. As seen in Table 1, the EER (%) for using the unprotected IrisCodes is 0.83%. On the other hand, it is 0.63% with the cancelable IrisCodes using DRPE in the FrFT domain. Lower EER values indicate better recognition performance. Also, the ability to distinguish the genuine and impostor distributions is measured by the decidability metric d' [15], which is computed by Eq. (7).

$$d' = \frac{\left|\mu_i - \mu_g\right|}{\sqrt{(\sigma_i^2 + \sigma_g^2)/2}} \quad (7)$$

where μ_i and μ_g are the means and σ_i^2 and σ_g^2 are the variances of the impostor and genuine distributions, respectively. As shown in Table 1, the obtained values of the decidability metric for the unprotected and proposed iris patterns are 4.31 and 5.18, respectively. The larger decidability values mean a larger deviation between genuine and impostor distributions, which indicates high recognition performance. The accuracy, NPV, PPV, specificity, and sensitivity of the proposed scheme are slightly increased with respect to its original counterpart. In general, encouraging results are collected by the proposed scheme.

Table 1. Summarized results of performance metrics for the original and proposed schemes.

Performance metric	Unprotected IrisCodes [26]	Proposed scheme
Sensitivity	99.5%	100%
Specificity	99%	99.5%
NPV	99.50%	100%
PPV	99.01%	99.50%
EER	0.83%	0.63%
Accuracy	99.25%	99.75%
Decidability	4.31	5.18

Figures 3a and b give the ROC curves for both the proposed and original patterns. It is noticed that the proposed scheme performs slightly better than the traditional unprotected one

A comparison between the proposed scheme and the other state-of-the-art cancelable biometric schemes (Teoh et al. [13]; Zuo et al. [12]; Uhl et al. [19]; Kumar et al. [30]; Ouda et al. [16]; Ouda et al. [17]; Rahulkar et al. [31]; Rathgeb et al. [14]; Rathgeb et al. [15]; Mayada et al. [21]; and Mayada et al. [22]) is presented in Table 2. The results in Table 2 illustrate a superior performance compared to the state-of-the-art schemes.

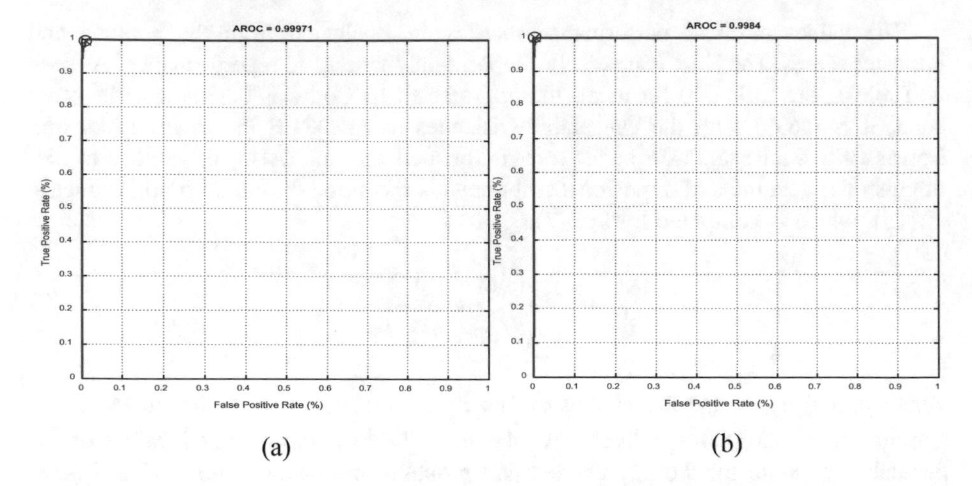

Fig. 3. ROC curves of (a) the original unprotected system, and (b) protected system with DRPE.

Table 2. Summarized results of EER for different competing techniques (CASIA-IrisV3-Interval database)

Scheme	Year	Performance (*EER %*)
Teoh et al. [13]	2004	4.81
Zuo et al. [12]	2008	4.41
Uhl et al. [19]	2009	1.30
Kumar et al. [30]	2010	1.48
Ouda et al. [16]	2010	5.54
Ouda et al. [17]	2010	6.27
Rahulkar et al. [31]	2012	1.91
Rathgeb et al. [14]	2013	1.14
Rathgeb et al. [15]	2014	8.98
Mayada et al. [21]	2016	3.56
Mayada et al. [22]	2017	2.001
Proposed scheme		0.63

5.3 Privacy Threats

The privacy in the proposed scheme is guaranteed through the utilization of a specific key. To guarantee privacy of the phase sequences utilized in the DRPE, the key is extracted from the right iris, while the IrisCodes are extracted from the left iris. Assuming that the algorithm for feature extraction from the iris is known, it is not possible for an adversary to capture the original iris feature vectors of users. This is attributed to the encryption process of iris features through the DRPE algorithm with a biometric-based key.

6 Conclusion

This paper presented a cancelable iris recognition scheme based on mixing multiple patterns of the biometric data. Feature vectors generated from both left and right iris of the same subject have been mixed in a single secure IrisCode. The DRPE method in FrFT domain is used to generate irreversible encrypted IrisCodes. The encryption keys; RPM1, and RPM2 are used in the proposed scheme, while the second phase mask is extracted from the right iris feature vector of the same subject. As a result, the proposed cancelable scheme avoids the necessity of transmitted keys and enhances the authentication privacy, because each user has unique keys for obtaining encrypted IrisCodes. Experimental results proved that the utilization of the proposed cancelable iris recognition scheme enhances the privacy, and at the same time maintains the authentication performance yielding an EER of 0.63% and an accuracy of 99.75%.

References

1. Daugman, J.: High confidence visual recognition of persons by a test of statistical independence. IEEE Trans. Pattern Anal. Mach. Intell. **15**(11), 1148–1161 (1993)
2. Daugman, J.: How iris recognition works. IEEE Trans. Circuits Syst. Video Technol. **14**(1), 21–30 (2004)
3. Venugopalan, S., Savvides, M.: How to generate spoofed irises from an iris code template. IEEE Trans. Inf. Forensics Secur. **6**(2), 385–395 (2011)
4. Rathgeb, C., Uhl, A.: A survey on biometric cryptosystems and cancelable biometrics. EURASIP. Inf. Secur. **2011**, 3 (2011)
5. Nagar, A., Nandakumar, K., Jain, A.: Multibiometric cryptosystems based on feature-level fusion. IEEE Trans. Inf. Forensics Secur. **7**(1), 255–268 (2012)
6. Rathgeb, C., Busch, C.: Multibiometric template protection: issues and challenges. In: New Trends and Developments in Biometrics, pp. 173–190. InTech (2012)
7. Nandakumar, K., Jain, A.K.: Biometric template protection: bridging the performance gap between theory and practice. IEEE Signal Process. Mag. **32**(5), 88–100 (2015)
8. Patel, V.M., Ratha, N.K., Chellappa, R.: Cancelable biometrics: a review. IEEE Signal Process. Mag. Spec. Issue Biom. Secur. Priv. **32**(5), 54–65 (2015)
9. Punithavathi, P., Subbiah, G.: Can cancellable biometrics preserve privacy? Biometr. Technol. Today **2017**(7), 8–11 (2017)
10. Ratha, N.K., Chikkerur, S., Connell, J.H., Bolle, R.M.: Generating cancelable fingerprint templates. IEEE Trans. Pattern Anal. Mach. **29**(4), 561–572 (2007)
11. Harjoko, A., Hartati, S., Dwiyasa, H.: A method for iris recognition based on 1D Coiflet wavelet. World Acad. Sci. Eng. Technol. **56**, 126–129 (2009)
12. Zuo, J., Ratha, N.K., Connel, J.H.: Cancelable iris biometric. In: Proceedings of the 19th International Conference on Pattern Recognition, Florida, USA, pp. 1–4 (2008)
13. Teoh, A.B.J., Ngo, D.C.L., Goh, A.: Biohashing: two factor authentication featuring fingerprint data and tokenised random number. Pattern Recogn. **37**(11), 2245–2255 (2004)
14. Rathgeb, C., Breitinger, F., Baier, H., Busch, C.: Towards bloom filter-based indexing of iris biometric data. In 15th IEEE International Conference on Biometrics, pp. 422–429 (2015)
15. Rathgeb, C., Breitinger, F., Busch, C., Baier, H.: On the application of bloom filters to iris biometrics. IET J. Biom. **3**(4), 207–218 (2014)

16. Ouda, O., Tsumura, N., Nakaguchi, T.: On the security of bioencoding based cancelable biometrics. IEICE Trans. Inf. Syst. **94-D**(9), 1768–1777 (2011)
17. Ouda, O., Tsumura, N., Nakaguchi, T.: A reliable tokenless cancelable biometrics scheme for protecting iriscodes. IEICE Trans. Inf. Syst. **E93-D**(7), 1878–1888 (2010)
18. Lacharme, P.: Analysis of the iriscodes bioencoding scheme. Int. J. Comput. Sci. Softw. Eng. **6**(5), 315–321 (2012)
19. Hammerle-Uhl, J., Pschernig, E., Uhl, A.: Cancelable iris biometrics using block re-mapping and image warping. In: International Conference on Information Security, Italy, pp. 135–142 (2009)
20. Jenisch, S., Uhl, A.: Security analysis of a cancelable iris recognition system based on block remapping. In: Proceedings of the IEEE International Conference on Image Processing, Belgium, pp. 3274–3277 (2011)
21. Tarek, M., Ouda, O., Hamza, T.: Robust cancelable biometrics scheme based on neural networks. IET J. Biom. **5**(3), 220–228 (2016)
22. Tarek, M., Ouda, O., Hamza, T.: Pre-image resistant cancelable biometrics scheme using bidirectional memory model. Int. J. Netw. Secur. **19**(4), 498–506 (2017)
23. Refregier, P., Javidi, B.: Optical image encryption based on input plane and fourier plane random encoding. Opt. Lett. **20**, 767–769 (1995)
24. Joshi, M., Shakher, C., Singh, K.: Color image encryption and decryption for twin images in fractional Fourier domain. Opt. Commun. **281**(23), 5713–5720 (2008)
25. Ozaktas, H.M.: Fractional Fourier domains. Signal Process. **46**, 119–124 (1995)
26. Soliman, N.F., Mohamed, E., Magdi, F., Abd El-Samie, F.E., AbdElnaby, M.: Efficient iris localization and recognition. Optik Int. J. Light Electron Opt. **140**, 469–475 (2017)
27. CASIA-IrisV3 Database. http://www.cbsr.ia.ac.cn/english/IrisDatabase.asp. Accessed Dec 2017
28. Masek, L.: Recognition of human iris patterns for biometric identification. M.Sc. Thesis, The University of Western Australia (2003)
29. Jain, A.K., Li, S.Z.: Handbook of Face Recognition. Springer, Berlin (2005)
30. Kumar, A., Passi, A.: Comparison and combination of iris matchers for reliable personal authentication. Pattern Recogn. **43**, 1016–1026 (2010)
31. Rahulkar, A.D., Holambe, R.S.: Half-iris feature extraction and recognition using a new class of biorthogonal triplet half-band filter bank and flexible k-out-of-n. A post classifier. IEEE Trans. IFS **7**, 230–240 (2012)

Interest Points Detection of 3D Mesh Model Using K Means and Shape Curvature

Mourad R. Mouhamed[1,3(✉)], Mona M. Soliman[2,3], Ashraf A. Darwish[1,3], and Aboul Ella Hassanien[2,3]

[1] Faculty of Science, Helwan University, Cairo, Egypt
mouradraafat@yahoo.com
[2] Faculty of Computers and Information, Cairo University, Giza, Egypt
[3] Scientific Research Group in Egypt (SRGE), Giza, Egypt
http://www.egyptscience.net

Abstract. Due to the great improvement of technology, the representation of the data in three dimensions widely used in many applications like scientific visualization, manufacturing, computer vision, engineering design, virtual reality, architectural walk through, and video gaming. The 3D objects consist of a huge number of components like thousands of vertices and faces, and that make researchers to detect the more interest components to deal with this objects instead of to deal with the whole object that for many applications like a watermark, and mesh simplification. This paper presents a model to detect the interest points of the object using k mean clustering algorithm depend on the curvature areas. The proposed model is applied to a set of benchmark data of 3D mesh models. The evaluation of the proposed model is verified using three evaluation measures, namely False Positive, False Negative Errors, and Weighted Miss Error. The evaluation is also verified by comparing the proposed model with the most popular interest point detecting techniques.

Keywords: 3D mesh model · Interest point · K-means
Shape curvature · Shape descriptors

1 Introduction

Graphics data are more and more widely applied to a variety of applications, including video gaming, engineering design, architectural walk through, virtual reality, e-commerce, and scientific visualization. Many three-dimensional (3D) objects are now represented in 3D meshes to truly reflect the topological structures of the object.

A large amount of research has been done in developing signatures to facilitate various tasks in computer vision, geometry processing, and data analysis. However, most of the signatures are presented without a rigorous explanation

A. E. Hassanien et al. (Eds.): AISI 2018, AISC 845, pp. 415–424, 2019.
https://doi.org/10.1007/978-3-319-99010-1_38

of their properties and their effectiveness. It is often illustrated only in specific applications through examples. Sun in [14] propose a point signature and argue from both theoretical and practical points of view that it has the following desirable properties: (1) It organizes information about the intrinsic geometry of a shape in an efficient, multi-scale way, (2) It is stable under perturbations of the shape, (3) It is concise and commensurable, but remains informative and (4) It can be estimated faithfully and efficiently.

On the other hand a point of signature defined as a point of interest, the detection of the point of interest on 3D mesh model is a challenging issue for a few reasons. Firstly, randomize of the topology in 3D mesh models. where this is hard to detect a local neighbour of the vertex due to the randomize of the number of neighbours vertex for each vertex. Secondly, the definition of the point of interest has no agreement between the researchers on it. Most of them used definition depends on the evaluation of interest with the level of bulge of extraordinary local structures. So, vertices on coarser level at most prominent area will have more interest. For example, in a human-shaped model, the detector of the point of interest should detect vertices on the hands, legs, and face. As inverse to vertices in smooth areas will have low interest.

One of the earliest work aiming to detect interest points presented in [1] that involves the use of integral volume descriptor which is related to the surface curvature and which is invariant to rotation and translation of the 3D object. Most of the 3D interest point detection algorithms developed in the last decade defined functions summarizing the geometrical content of localities on a 3D model in multiple scales, and selected the local extreme of those functions as interest points. helin [3] approach is in accordance with the fact that humans respond more to significant local changes on the surface. However, he aimed to investigate humans choices empirically on a number of 3D models, typically used in 3D shape research community.

Several methods have been proposed for different types of shape descriptors, such as view-based [5,7,8], transform-based [6]. Shape descriptors can also be classified as local and global shape descriptors. More details about the different methods can be found in the following review papers [9,10].

This paper presents a model to detect the points of interest of 3D mesh model. The proposed model composed of 4 steps. First, the normal of each vertex is calculated. Second step, is calculating the one ring neighbour of each vertex and calculate the angles between the normal of the every vertex and the normal of it's neighbour vertices. Thirdly, determine the summation of these angles of every neighbour individually. Finally using k-means to cluster of these summations to detect the local maximum of these summations, the local maximum express the high curvature area where the area of high curvature marked as interest points.

The rest of this article is organized as follows. Section 2 presents the background and related work, Sect. 3 describes the proposed model and algorithm then Sect. 4 introduces the evaluation of the proposed model and experimental results. Finally, Sect. 5 presents the conclusion and future work.

2 Backgrounds and Related Work

2.1 K_Means Clustering Algorithm

K_means algorithm introduced as an unsupervised clustering algorithm that parties the input data points into multiple classes based on their inherent distance from each other [13]. The algorithm assumes that the data features form a vector space and tries to find natural clustering in them. The k-means algorithm considers one of the popular clustering algorithms because of its simplicity in implement, and linearity of its time complexity. The points are clustered around centroid c_i, where i = 1, 2, 3, ..., k which are obtained by minimizing the objective.

$$V = \sum_{i=1}^{k} \sum_{x_j \in S_i} (x_j - c_i)^2 \tag{1}$$

where there are k clusters S_i, i = 1, 2, 3, ..., k and c_i is the centroid point of all the points $x_j \in S_i$.

2.2 Interest Points Detection Techniques

Many research directions try to introduce and implement different techniques that can help to detect the interest points of 3D models. For example, Castellani in [2] proposes a new methodology for the detection and matching of salient points over several views of an object. The technique is divided into three main steps: (1) detection is carried out by adopting a new perceptually-inspired 3D saliency measure. Such measure allows the detection of few sparse salient points that characterize distinctive portions of the surface, (2) a statistical learning approach is considered to describe salient points across different views. Each salient point is modeled by a Hidden Markov Model (HMM), which is trained in an unsupervised way by using contextual 3D neighborhood information, thus providing a robust and invariant point signature. (3) matching among points of different views is performed by evaluating a pairwise similarity measure among HMMs.

In [4] Ivan and Benjamin present an interest points detector for 3D mesh models based on Harris operator, which has been used with good results in computer vision applications. They propose an adaptive technique to determine the neighborhood of a vertex, over which the Harris response on that vertex is calculated. their method is robust to several transformations which can be seen in the high repeatability values obtained using the SHREC feature detection and description benchmark.

Nishino and Novatnack in [11] also built a scale-space representation of the model; however, they analyzed the scales independent of each other to detect scale-dependent corners. they focus on detecting scale-dependent geometric features on triangular mesh models of arbitrary topology. The key idea of their approach is to analyze the geometric scale variability of a given 3D model in the scale-space of a dense and regular 2D representation of its surface geometry encoded by the surface normals. they derive novel corner and edge detectors, as well as an automatic scale selection method, that acts upon this representation to detect salient geometric features and determine their intrinsic scales.

In [12] Afzal and Assem describe a new formulation for the 3D salient local features based on the voxel grid inspired by the Scale Invariant Feature Transform (SIFT). they use it to identify the salient keypoints (invariant points) on a 3D voxelized model and calculate invariant 3D local feature descriptors at these key points. they then use the bag of words approach on the 3D local features to represent the 3D models for shape retrieval. The advantages of the method are that it can be applied to rigid as well as to articulated and deformable 3D models.

Jian et al. [14] propose a novel point signature based on the properties of the heat diffusion process on a shape. this signature is obtained by restricting the well-known heat kernel to the temporal domain. Remarkably they show that under certain mild assumptions, HKS captures all of the information contained in the heat kernel, and characterizes the shape up to isometry. This means that the restriction to the temporal domain, on the one hand, makes HKS much more concise and easily commensurable, while on the other hand, it preserves all of the information about the intrinsic the geometry of the shape.

3 The Proposed Model for Interest Point Detection in 3D Object

The proposed model as shown in Fig. 1 aims to provide a set of interest points for each 3D object using k-means algorithm based on a normal calculation of one ring neighbor for each 3D vertex.

Fig. 1. The framework of the proposed model

The proposed model depends on the determination of the highest curvature area. Points in such area are considered as Points of Interest (POI). Such set of points can be considered an object signature that could be used later in the different set of applications. The algorithm starts by calculating the normal of every point and the normal of points on one ring neighborhood as shown in Fig. 2, where V_1 is the vertex with normal n_1, (V_2:V_7) are vertices of one ring neighbor of V_1 with normal values indicated by (n_2:n_7), the proposed model calculates the angle θ between n_1 (normal of V_1) and n_2 (normal of V_2). This calculation is repeated for all ring vertices. Equation 2 is used to calculate the summations of the angels between each normal n_i with the normal of a point of the one ring neighbor around. These summations values are clustered using k-means clustering algorithm, the cluster that has the highest values will be considered as the group of interest points.

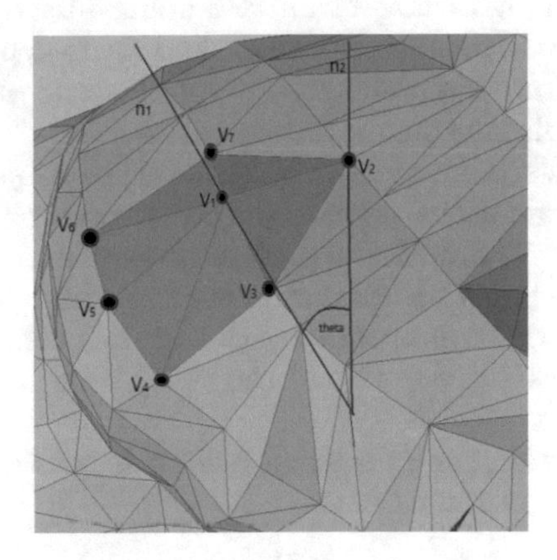

Fig. 2. The angle between normals of the ring vertices and the normal of center

Algorithm 1 provides a complete description of the proposed model.

Algorithm 1. The proposed detection POI in 3D object model

1: Input: The vertices v_i of the original model in cartesian form.
2: For each vertex v_i: compute the normal n_i.
3: For each vertex v_i: calculate the summation of angles between $v_i's$ normal and normal of each vertex belong to the one ring neighbor using the following equation.

$$sum = \sum_{i=2}^{n} \arccos \frac{n_i \cdot n_1}{\|n_i\|\|n_1\|} \tag{2}$$

where n_i is the normal of the vertices of the one ring a round n_1 as in Fig. 2.
4: Cluster the summations using K means clustering algorithm depends on the value of summations.
5: Output: The cluster with highest value marked as an interest points.

---

4 Experimental results

The evaluation of the proposed model is tested using a set of interest point detection benchmark [15]. Different 3D models with different structure and topology had been used in the evaluation Such as girl, chair, camel, cow, and Armadillo

Girl model which has facial features (especially, eyes, nose, and ears) are very important for human and animal models. Chair model with interest points should be on the corners. Models like the camel with convexity area on his back and concavity area on rest of his body.

The proposed model had been evaluated using a set of evaluation criteria proposed by Helin in [3]. He introduced three measures of evaluation, called False Positive Errors (FPE), False Negative Errors (FNE), and Weighted Miss Error (WME). These three measures are used to compare the proposed model and other well-known algorithms of interest point detection such 3D SIFT,3D Harris, Salient Points, HKS, Mesh Saliency, and Scale-dependent corners. Also, we compare our set of interest points result from the proposed model with ground truth points listed with benchmark data.

All the experiments done at $n = 11$ (number of subject [3]) and $\sigma = 0.05$. where helin in his evaluations method defined two criteria while constructing the ground truth: σ is The radius of an interest region, and n is the number of users n that marked a point within that interest region. False negative error and False positive error rate as proposed by Helin in [3] compare between ground truth POI defines it as a group G and the POI of the interest point detection algorithm D as S.

In false negative error, for an interest point g in set G the authors defines a geodesic neighborhood of radius r:

$$I_r(g) = \{p \in D \mid d(p, g) \leq r\}. \tag{3}$$

where $d(p, g)$ corresponds to the geodesic distance between points g and p. The parameter r controls the localization error tolerance. A point g is considered to be correctly detected if there exists a detected point s $\in S$in $I_r(g)$. and such that s is not closer to any other points in G. Denoting the number of correctly detected points in G as NC, we define the false negative error rate at localization error tolerance r as:

$$FNE = 1 - \frac{NC}{NG} \tag{4}$$

where NG is the number of points in G.

False positive error rate is proceed as follows: Each correctly detected point in $g \in G$ corresponds to a unique s, the closest point to g among the points in S. All points in S without a correspondence in G are declared as false positives.

$$NF = NS - NC \tag{5}$$

where NS is the number of detected interest points by the algorithm. The false positive error rate at localization error tolerance r is:

$$FPE = \frac{NF}{NS} \tag{6}$$

finally, To incorporate the prominence of an interest point into the evaluation, they introduce another miss error measure, which name as Weighted Miss Error (WME). Assume that within a geodesic neighborhood of radius r around the ground truth point $g_i \in G$, n_i subjects have marked an interest point. Then the Weighted Miss Error is defined as:

$$WME(r) = 1 - \frac{1}{\sum_{i=1}^{N_G}} \sum_{i=1}^{N_G} n_i \lambda_i \tag{7}$$

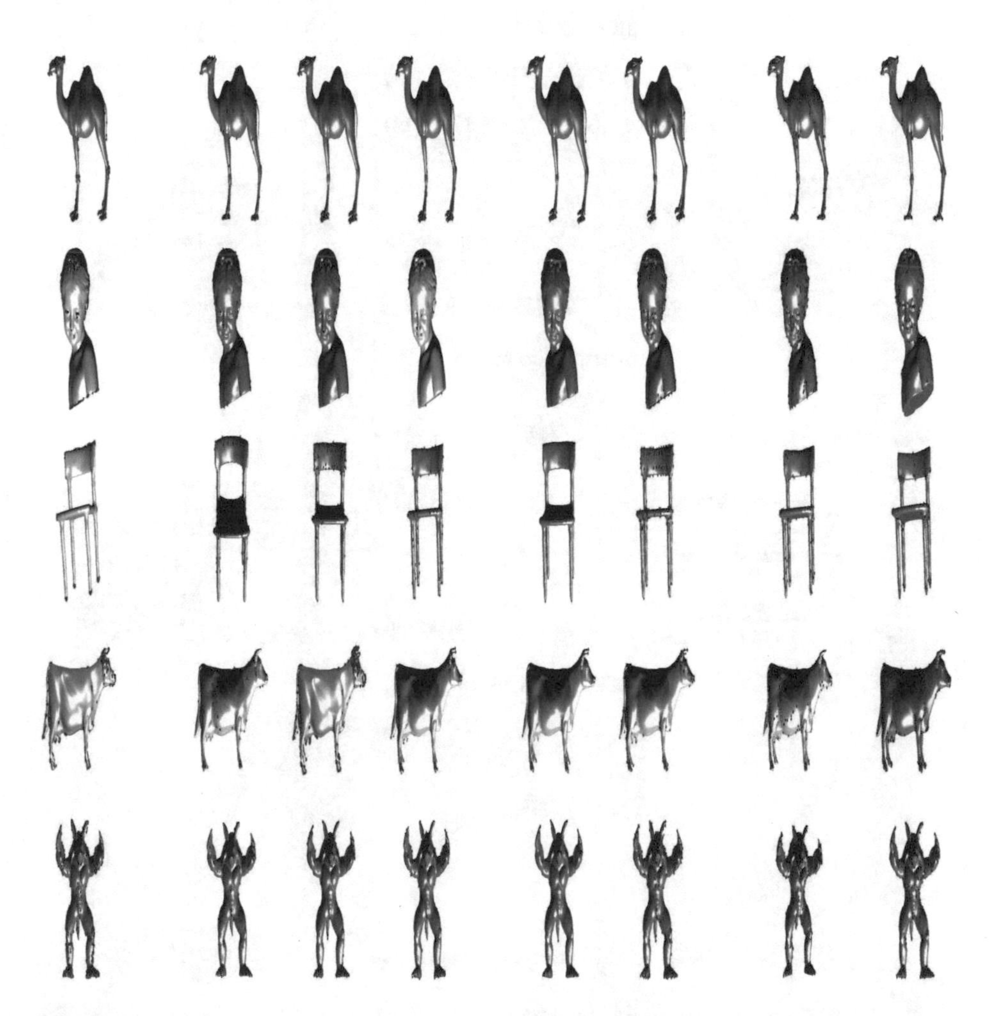

Fig. 3. From left to right first column is the interest points detected by the proposed model followed by the interest point by 3D SIFT, 3D Harris, Salient Points, HKS, Mesh Saliency, Scale-dependent corners and Ground truth.

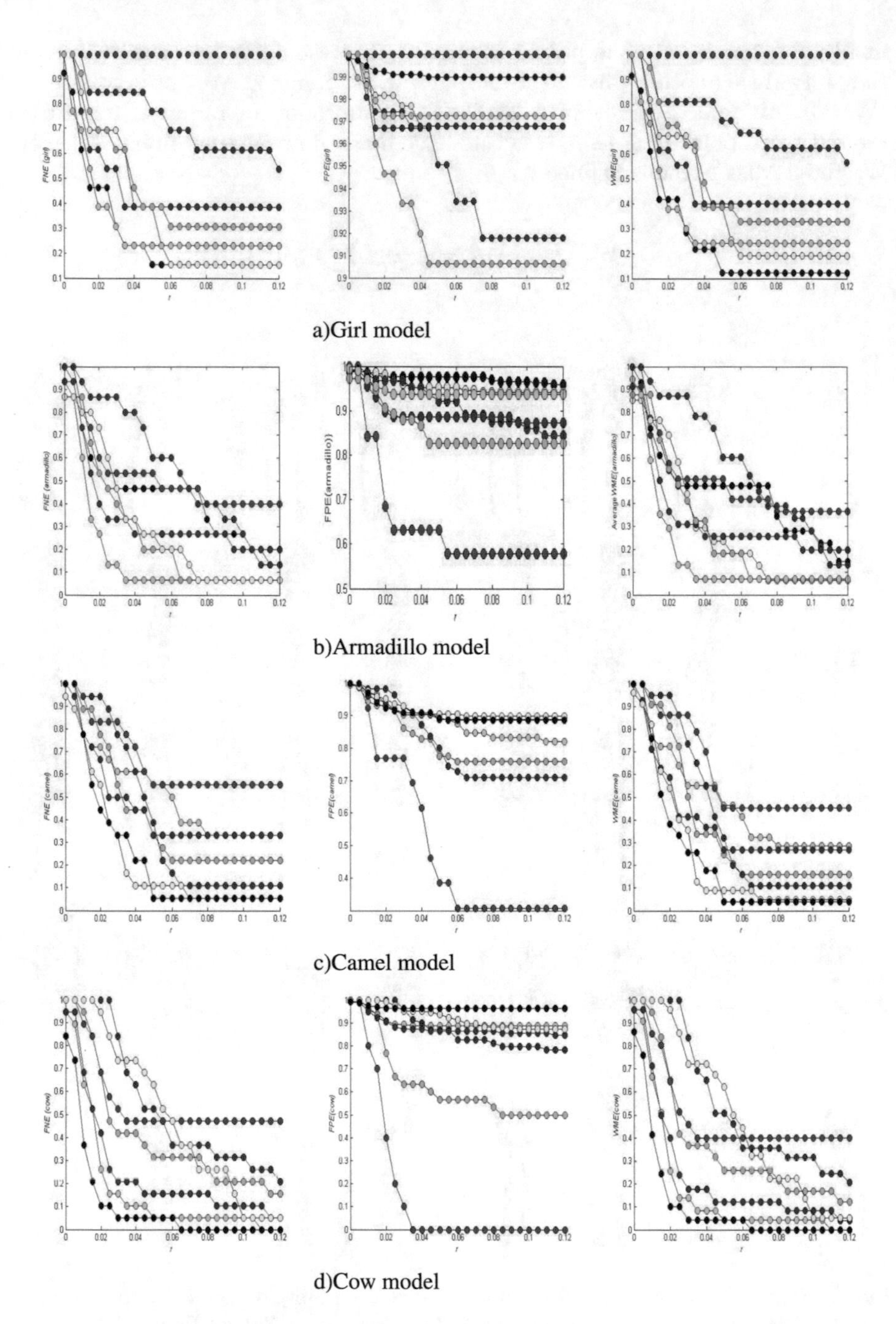

Fig. 4. From left to right first column is FNE, FPE and WME at n = 11, σ = 0.05. Where the proposed model (black curve), SD-corners (yellow curve), 3D SIFT (red curve), HKS (pink curve), salient point (green curve), mesh saliency (cyan curve) and Harris (blue curve) (Color figure online)

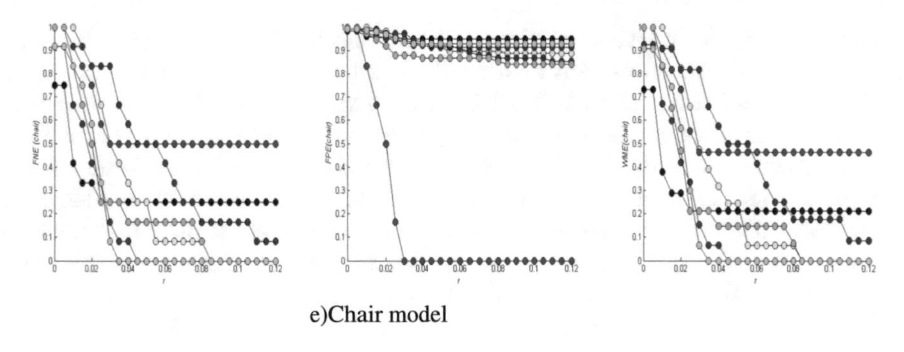

e)Chair model

Fig. 4. (*continued*)

where

$$\lambda_i = \begin{cases} 1, \text{ if } g_i \text{ detected by algorithm;} \\ 0, \text{ otherwise.} \end{cases} \tag{8}$$

The WME is a measure between 0 and 1, and an algorithm gets a good compensation if it manages to detect a point that is frequently voted by the human subjects.

When the FPE give a high value that means the algorithm detected large number of POI, but the WME and FNE when gave high value that mean the algorithm failed to detect a true POI compared with the ground truth.

Figure 3 presents a visual comparison between the interest points detected by the proposed model, set of other well known POI detection algorithms and the ground truth POI detected by the human. The marked points by red indicated the POI for each 3D model. The FNE, FPE, AND WME measures had been calculated for each model and compared with a set of well-known algorithms as shown in Fig. 4.

We can observe that both WME and FNE drop with high speed that means a good result to find the interest points as shown in cow model. The other techniques gave the best value of WME and FNE where most of the interesting point detected by the ground truth on its legs, muzzle, and ears where these areas are more prominent with high curvature, where that the proposed model is sensitive to more prominent areas it gave high FPE where the cluster of the highest value has large number of interest points, We can observe the proposed model was not the best in FNE nor WME for Armadillo model due to the nature of the interest points of armadillo which detected by the human where there are a lot of interest points on the smooth area like on its chest, where no any curve so the proposed model failed to detect all of them, however, it can detect some of them which lie on the hand, foot and head.

5 Conclusion and Future Work

This paper proposes a new method to detect the interest points of the 3D mesh models based on clustering technique. In this paper k-means cluster algorithm used to cluster the vertices of the 3D mesh model depend on the curvature area where the high prominent area marked as having most interest points, the model evaluated using the FNE, FPE, and WME. The model compared with most known interest points detection algorithms indicating a good performance and superiority for some models. Such interest points of a 3D model can be used later as model signature or shape signature that describes each model. Such shape signature can be used in different application such as 3D watermarking, compression and mesh simplification. In future we will improve the proposed model to be robust against the different types of attacks and try to use the optimization to improve the trade off between the FPE and WME.

References

1. Gelfand, N., Mitra, N.J., Guibas, L.J., Pottmann, H.: Robust global registration. In: Symposium on Geometry Processing, pp. 197–206 (2005)
2. Castellani, U., Cristani, M., Fantoni, S., Murino, V.: Sparse points matching by combining 3D mesh saliency with statistical descriptors. Comput. Graph. Forum **27**(2), 643–652 (2008)
3. Dutagaci, H., Pan, C., Godil, A.: Evaluation of 3D interest point detection techniques via human-generated ground truth. Vis. Comput. **28**, 901–917 (2012)
4. Sipiran, I., Bustos, B.: Harris 3D: a robust extension of the Harris operator for interest point detection on 3D meshes. Vis. Comput. **27**, 963–976 (2011)
5. Michael, K., Thomas, F., Szymon, R.: Progressive meshes. In: Proceedings of ACM SIGGRAPH, pp. 99–108 (1996)
6. Ding-Yun , C., Xiao-Pei, T., Yu-T, S., Ming, O.: Rotation invariant spherical harmonic representation of 3D shape descriptors. In: Proceeeding SGP, pp. 156–164 (2003)
7. Zhouhui , L., Afzal, G., Xianfang, S.: Visual similarity based 3D shape retrieval using bag-of-features. In: Proceedings of Shape Modeling, pp. 25–36 (2010)
8. Helin, D., Afzal, G., Blent, S., Ycel, Y.: View subspaces for indexing and retrieval of 3D models. Proc. SPIE **7526**, 75260–75271 (2010)
9. Tangelder, J.W.H., Veltkamp, R.C.: A survey of content based 3D shape retrieval methods. Multimedia Tools Appl. **39**, 441–471 (2008)
10. Benjamin, B., Daniel, A.K., Diemtar, S., Tobais, S., Dejan, V.V.: Feature-based similarity search in 3D object databases. ACM Comput. Surv. **37**, 345–387 (2005)
11. Novatnack, J., Nishino, K.: Scale-dependent 3D geometric features. In: ICCV, pp. 1–8 (2007)
12. Godil, A., Wagan, A.I.: Salient local 3D features for 3D shape retrieval. In: 3D Image Processing (3DIP) and Applications II. SPIE, Bellingham (2011)
13. Tatiraju, S., Mehta, A.: Image segmentation using k-means clustering, EM and normalized cuts. University of California Irvine (2008)
14. Sun, J., Ovsjanikov, M., Guibas, L.: A concise and provably informative multi-scale signature based on heat diffusion. In: Eurographics Symposium on Geometry Processing (SGP), pp. 1383–1392 (2009)
15. http://www.itl.nist.gov/iad/vug/sharp/benchmark/3DInterestPoint/

Two-Factor Authentication Scheme Using One Time Password in Cloud Computing

Samar H. El-sherif[1(✉)], Rabab F. Abdel-kader[2], and Rawya Y. Rizk[2]

[1] Faculty of Management Technology and Information Systems, Port Said University, Port Said, Egypt
samarhamdy@himc.psu.edu.eg

[2] Faculty of Engineering, Port Said University, Port Said, Egypt
{rababfakader, r.rizk}@eng.psu.edu.eg

Abstract. Cloud computing is widely used by many users, so any one must be authenticated before accessing any service into the cloud. In this paper, a two-factor authentication scheme is proposed to authenticate users in cloud computing. The first factor uses the traditional user name and password, while the second factor uses one time password, which is valid for only one login session or transaction. During the registration phase, user must choose a user name and a password in addition to specify 4 cells from a 3 × 3 grid to use its containing characters when logging into the system. At the login phase, the user inputs his user name and password then, a grid with random numbers is displayed to him, and he must input the sequence of characters chosen during the registration phase. These characters represent the one time password which is changed every login. The results show that the proposed scheme can resist practical attacks, easy for users, does not have strong constrains, and does not require specific extra hardware.

Keywords: Authentication · Cloud computing · One time password
Cells · Attacks

1 Introduction

Cloud computing provide three service model [1, 2]. These services can be explained as following. The first one is software as a service (SaaS) which enables user to use the application running on cloud infrastructure. This service offers safety standard authentication system (user name and password). The second service is platform as a service (PaaS) which enables user to deploy onto the cloud infrastructure his own application without installing any platform or tools on their local machines. The third one is infrastructure as a service (IaaS) which enables user to provision processing, storage, networks, and other fundamental. The IaaS performs all security function except availability.

Authentication in cloud is a very important step which proves that the accessing user is the real user. In general, there are three categories of authentication methods (something you know, something you have and something you are). Several schemes use two factors for authentication, in most schemes traditional user name and password

A. E. Hassanien et al. (Eds.): AISI 2018, AISC 845, pp. 425–434, 2019.
https://doi.org/10.1007/978-3-319-99010-1_39

are used as the first factor while the second factor differ from one technique to another. The second factor can be an image which is sent to the user via SMS, and then he can scan it manually on the web camera as produced in [3]. In [4], another scheme uses the idea of PASSMATRIX which one pass-square per pass image for a sequences of n images represent the password. Different two ideas for the second factor of authentication are presented in [5]. The first one represents each letter in the password by an icon. For an example, "I" is represented by an icon of ice cream. The other scheme represents each letter by braille icons which are displayed randomly on the screen. In [6], users represent his signature in the air using his mobile phone and use it as a password in each login. Users can also enter his password through special keyboard which depends on nomadic key algorithm [7]. In this algorithm, the numbers keys are showed freely on the screen but with some constraints. Another scheme which is depends on selecting sequential cells from a grid is proposed in [8]. Finger print can also be used for authentication as shown in [9]. Smart card can also be used to authenticate the user as represented in [10–12].

In this paper, a two-factor authentication scheme is produced to prove that the accessing user is the real user. This scheme uses one time password (OTP) as the second factor of authentication in addition to the traditional user name and password as the first factor.

The organization of this paper is as follows: brief overview of related work is presented in Sect. 2. The proposed two-factor authentication scheme is introduced in Sect. 3. Sections 4 and 5 present security analysis and performance analysis of the proposed scheme; respectively. Finally the main conclusion is presented in Sect. 6.

2 Related Work

Many authentication schemes were proposed to authenticate users in cloud, but a lot of them had some security troubles. In [13], user must enter his voice print and the image of his face after traditional user name and password to access the cloud. Another authentication scheme for distributed mobile cloud computing environment which used biometric information is produced in [14]. The two previous schemes used extra hardware to receive biometric information, so the authentication process requires extra time.

In [15], a private cloud system used OTP to authenticate the user. At the registration phase, some information provided to the cloud such as user name, 4-digit numbers which called pin code and a hexadecimal code which used to initialize the mobile phone and called INIT secret. A software application which ran on the user mobile phone was used to generate OTP based on the current time, the 4-digit pin code and the INIT secret value. At the login phase, user name and OTP send to the server which used the same parameters to generate OTP. If the received OTP was equal to the generated one, user could access the system. In this scheme, Mobile time and server time must be synchronized.

Another scheme uses a smart phone to generate OTP was proposed in [16]. Some operations were done by a mobile software application and finally generate the OTP. This application used some information like User name, password, date of birth, current date and time, and time stamp on its calculations. If the server and the mobile phone had the same factors to generate the same OTP, authentication would be succeeded. Mobile application was used in the two previous schemes to generate OTP which makes the authentication process more complicated and not easy for user.

A desktop application used to generate OTP was presented in [17]. At the registration phase, User creates an account onto the cloud. At the login phase, he is asked to input his traditional user name and password and the OTP. A software application was used at the user side to generate this password using some information which was represented by user name, password and account name. At the server side, based on user information, 1000 OTP is generated, and then saved in the temporary OTP database. If user's OTP matched one from the temporary OTP in the database, user would access the cloud. This scheme require more time to identity the user because it compares the OTP which user entered with 1000 OTP saved in the temporary OTP database. This was considered a weakness point in this scheme.

A special technique of authentication was presented to let user accesses the cloud service through his phone [18]. User registered himself by his personal information and voiceprint. At the login phase, there were two ways to authenticate the user. The first way was his voice print which was entered through the microphone of his phone and compared with the stored one. The other way was OTP which was generated by the system center as a CAPTCHA picture. User could access the system if the previous two steps succeeded. In this scheme, user took more time for authentication process. He used his voice print in addition to OTP to identify himself.

In [19], Image based authentication (IBA) was used, in addition to OTP which was sent to the user through the internet. In this scheme, user received the OTP through the internet and this was a weakness point if the message didn't reach to him for any reason.

All the previous schemes suffered from some problems. These problems can be summarized in using mobile phone application, using extra hardware, receiving message through the internet, and extra time to identity user. In this paper, a two- factor authentication scheme is presented to overcome the weaknesses found in reviewed schemes.

3 The Proposed Two-Factor Authentication Scheme (TFAS)

In this paper, a two-factor authentication technique is presented. The first factor depends on a traditional user name and password. The second factor depends on one time password technique. The process consists of four phases registration phase, login phase, encryption phase, and authentication phase.

3.1 TFAS Phases

In the registration page during the registration phase, the user enters his user name and password, and then he is asked to choose four cells from a 3 × 3 grid. For all four chosen cells the user enters the symbol which represents each cell and appears on it. For example, he can enter symbols like C1, C2, C3, C6 or C1, C2, C5, C4 or C3, C6,

C9, C8. The only restriction for the user is that he must choose four cells in any order and he must remember this order to use it at the login phase. At the end of registration phase, user name, password, and the order of the four cells which are chosen by the user are stored in the database for each registered user. This phase can be illustrated in Fig. 1.

At the login phase, a login page appears to the user, in which he must enter his user name and password which he had chosen at the registration phase. Then a grid of 3×3 cells will appear to the user. The numbers shown on the grid cells are randomly generated. Every cell contains a number ranging from 0 to 99. So the numbers can be formed from one digit or two digits. The length of the string formed from the four cells chosen at the registration phase may be 4, 5, 6, 7, or 8 digits. Each attempt by the user to login, the grid displays a random number in each cell. The user must remember the order of the four cells which he had chosen at the registration phase and enters the numbers listed on the cells corresponding to the chose pattern. These numbers are then encrypted using a special encryption application.

Next, during the encryption phase, a desktop application is used to encrypt the password. When the user runs this application, a window form is displayed. The form asks the user to enter his user name and password, and then the contents of the 4-cell which had been got at the login phase. If all information is entered correctly in the form, user can click on a button to run the code of the encryption process. The output of the encryption code may be 4, 5, 6, 7, or 8 digits. This output is used as OTP which is the second factor of authentication in this scheme.

Finally, the authentication phase, user must enter his username, password, and OTP into the login page. This OTP may be 4, 5, 6, 7, or 8 digits. Its length may be changed every login for the same user. All of this information is sent to the server. At the server side, the server has the same encryption code which exists at the user side. It is used to encrypt the contents of the cells which represent the user pattern. Comparison is done between the received OTP and the evaluated one. If the comparison process succeeds, user can access the system. Otherwise, error message is shown to the user. Authentication process is illustrated in Fig. 2.

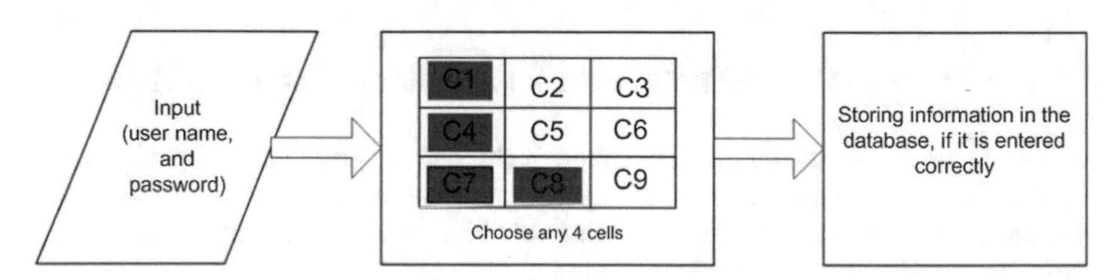

Fig. 1. New user registration process

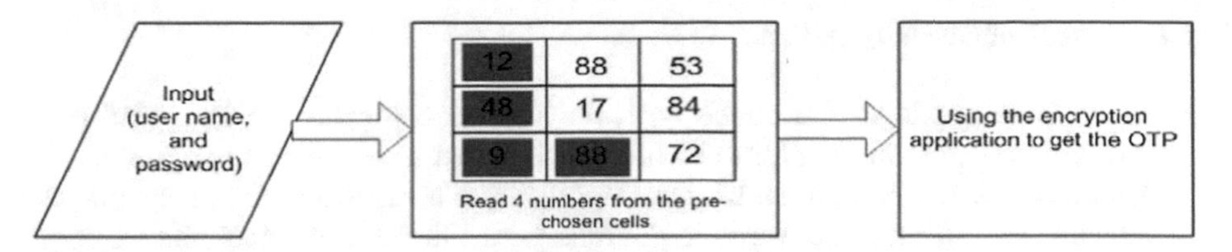

Fig. 2. User authentication process

3.2 TFAS Code

```
User name, password, and OTP (input)
access the cloud or not (output)
Declare uss as an object of the table of user
Declare rand_arr [L]: array of integers
Declare Enc_arr [4]: array of integers
Declare k
Initialize k to 0
Read numbers from the pre-chosen cells
For each cell
      For j=0 to L
            If cell content = rand_arr[j]
                  Enc_arr [k] = j
            Else
                  Enc_arr [k] = k+1
            End If
            Increment k
      End For
End For each
If user name = uss.user name && password = uss.password
then
      If 1st_cell of OTP = Enc_arr[0] &&
      2nd_cell of OTP = Enc_arr[1] &&
      3rd_cell of OTP = Enc_arr[2] &&
      4th_cell of OTP = Enc_arr[3] then
            Show message "welcome to the system "
      Else
            Show message "wrong OTP"
      End If
Else
      Show message "wrong user name or password"
End If
```

4 Security Analysis

The security of TFAS was analyzed. First, how it can resist several types of attacks [20], and how it can avoid several troubles which exist in other related works.

4.1 Types of Security Attacks

- Brute force attack: In this attack type, the attacker can crack the encrypted password, if he can steal the password file which stored in the operating system.
- Dictionary attack: In general, user uses words from his life to be his password. The attacker can collect these words in dictionary, and then he can crack the password by applying all these words.
- Shoulder surfing attack: Using different spying ways, attacker can get the user password. Attacker keeps his eye on keys which the user pressed to get the password. He can use binoculars or hidden camera to spy on the user.
- Phishing attacks: In this attack type, user enters his information in a fake page with the same interface of the original page. The attacker redirects the user to this web page to get the password.
- Key loggers attack: In this attack type, when the user enters his password, each key pressed is recorded by software program. The attackers setup this program into the user system. The collected information is stored into file which is sent to the attacker through his email. Now, the attacker can access the system.
- Video recording attack: In this attack type, the attacker uses miniature camera to spy on user and get the password.
- Replay attack: In this attack type, the attacker can spy the conversation between sender and receiver and gets the authenticated information.
- Impersonation attack: In this attack type, the attacker can successfully assume the identity of one of the legitimate parties in a system.
- Spoofing attack: In this attack type, one person or program successfully masquerades as another by falsifying data, thereby gaining an illegitimate advantage.
- Smudge attack: In this attack type, the attacker can detect the finger smudges, and the heaviest smudges can be used to infer the most frequent user input pattern (the password).
- Insider attack: This attack can be done by a person with authorized system access.
- Masquerade attack: In this attack type, the attacker can uses a fake identity, such as a network identity, to gain unauthorized access to personal computer information through legitimate access identification.

When the attacker can attack the password, he can use it again to access the system. In TFAS, he cannot use this password again because it will be changed for the next login. It also depends on encryption code which generate encrypted password to be used at the login phase. So, TFAS can resist all the previous types of attacks.

4.2 Security Troubles

Now, it is clear to us that TFAS can resist several attacks types and this is what the majority of other OTP techniques could do. But, TFAS has a lot of features which doesn't exist in other OTP techniques. These features will be explained in the next paragraph.

- Detecting wrong password quickly: In TFAS, Error message will be shown quickly to the user, if he enters wrong password.

- No strong restriction for password choice: In TFAS, there are no strong restrictions for choosing the cells which represent the OTP.
- No smartcard: In TFAS, User doesn't use any smart card.
- Two factors for authentication: In TFAS, traditional user name and password is used in addition to the OTP.
- Not more complicated: In TFAS, no more time will be wasted to authenticate the user.
- No mobile phone application: In TFAS, user doesn't use the mobile phone or any application run on it.
- No messages receiving through the internet: In TFAS, user doesn't receive any message through the internet. So, it avoids risks which appear as a result of receiving any message through the internet.

In Table 1, a comparison of security properties is displayed between TFAS and related work which uses OTP. This comparison show that how TFAS and other related schemes resist several types of security attacks and overcome security troubles.

Table 1. Comparison of security properties between TFAS and other related schemes

	Vishal et al. [15]	Sagar et al. [16]	Eman et al. [17]	Her-Tyan et al. [18]	Marim-uthu et al. [19]	TFAS
Resisting Brute force, Dictionary, Shoulder surfing, Phishing, Key loggers, Video recording, Replay, Impersonation, Spoofing, Smudge, and Insider attack	√	√	√	√	√	√
Resisting Masquerade attack	√	√	×	√	×	√
Detecting wrong password quickly	√	√	×	×	√	√
No strong restriction for password choice	×	×	×	√	√	√
No smartcard	√	√	√	√	√	√
Two factors for authentication	×	×	√	√	√	√
Not more complicated	×	×	×	×	×	√
No mobile phone application	×	×	√	×	√	√
No messages receiving through the internet	√	√	√	√	×	√

5 Performance Analysis

User chooses 4 cells which represent his personal identification pattern (PIP) from a 3×3 grid. Each cell displays a random number from 0 to 99. TFAS generates a password with length 4, 5, 6, 7, or 8 digits. Every login, the user reads the numbers which represent his PIP and enters these numbers into the encryption application to get the encrypted code which represents his OTP. Finally, user gets 4, 5, 6, 7, or 8 digits OTP. Each digit is a random number from 0 to 9. The maximum capacity for the 4, 5, 6, 7, and 8 digits one time password is 10^4, 10^5, 10^6, 10^7, and 10^8. In Table 2, a comparison between different OTP schemes is displayed. The maximum storage capacity is computed for every technique. From this comparison, it is obvious that the TFAS has different storage capacity from 10^4 to 10^8 and it has the minimum storage capacity compared to the other schemes.

From the previous analysis, it is obvious that TFAS has no strong limitation for being applied. Different systems can use TFAS to authenticate users in a secure manner.

Table 2. Comparison between different OTP techniques based on maximum data storage capacity

Different technique (for generating OTP)	Password length	Character set	Total number of different char	Maximum Capacity for different data values
TFAS	4, 5, 6, 7, or 8 -digit	Numeric (0:9)	10	10^4, 10^5, 10^6,10^7, or 10^8
Vishal et al. [15]	6-digit	Numeric (0:9)	10	10^6
Eman et al. [17]	30-digit	Alphanumeric (0:9, lower case letters, some special characters)	40	40^{30}

6 Conclusions and Future Work

In this paper, a two-factor authentication scheme is proposed in cloud computing. It is designed to resist several types of attacks and overcome some security troubles. This scheme uses one time password in addition to the traditional user name and password. A different technique is used to create OTP. User chooses 4 cells from a 3 × 3 grid when he registers to the system. This grid is displayed at the login page with random numbers. Numbers which represents user's choice of cells must be entered to a special encryption application. The output of this application is used as the OTP which is valid for only one login.

In the future, TFAS can be improved by using a password which is more than 8 digits or with alphabetic. This will increase the probabilities of the obtained password and improve the performance of the TFAS.

References

1. Hashizume, K., Rosado, D.G., Medina, E.F., Femandez, E.B.: An analysis of security issues for cloud computing. J. Internet Serv. Appl. **4**(1), 1 (2013)
2. Yan, X., Zhang, X., Chen, T., Zhao, H., Li, X.: The research and design of cloud computing security framework. In: Wu, Y. (eds.) Advances in Computer, Communication, Control and Automation. Lecture Notes in Electrical Engineering (LNEE), vol. 121. Springer, Heidelberg (2011)
3. Eminagaoglu, M., Cini, E., Sert, G., Zor, D.: A two-factor authentication system with QR codes for web and mobile application. In: 5th International Conference on Emerging Security Technologies, Turkey, 10–12 September, pp. 105–112. IEEE (2014)
4. Sun, H., Chen, S., Yeh, J., Cheng, C.: A shoulder surfing resistant graphical authentication system. J. Trans. Dependable Secure Comput. **99**, 1 (2016)
5. Devki, P., Rao, R.: A novel way of ICON based authentication methods. In: Proceedings of International Advance Computing Conference (IACC), India, 12–13 June, pp. 449–453. IEEE (2015)
6. Laghari, A., Rehman, W.U., Memon, Z.A.: Biometric authentication technique using smartphone sensor. In: 13th International Bhurban Conference on Applied Sciences and Technology (IBCAST), Islamabad, Pakistan, 12–16 January, pp. 381–384. IEEE (2016)
7. Cotta, L., Fernandes, A.L., Melo, L.T.C., Saggioro, L.F.Z., Martins, F., Neto, A.L.M., Loureiro, A.A.F., Cunha, I., Oliveira, L.B.: User authentication for smart devices based on nomadic keys. In: International Conference on Communication (ICC), Communication and Information Systems Security Symposium, Kuala Lumpur, Malaysia, 22–27 May, pp. 1–6. IEEE (2016)
8. Kim, T., Yi, J.H., Seo, C.: Spyware resistant smartphone user authentication scheme. Int. J. Distrib. Sensor Netw. **2014**, 1–8 (2014)
9. Kumari, S., Om, H.: Remote login authentication scheme based on bilinear and fingerprint. J. Trans. Internet Inf. Syst. KSII **9**(12), 4987–5014 (2015)
10. Kang, D., Jung, J., Mun, J., Lee, D., Choi, Y., Won, D.: Efficient and robust user authentication scheme that achieve user anonymity with a Markov chain. J. Secur. Commun. Netw. **9**(11), 1462–1476 (2016)
11. Reddy, A.G., Yoon, E., Das, A.K., Yoo, K.: Light weight authentication with key-agreement protocol for mobile network environment using smart cards. IET Inf. Secur. **10**(5), 272–282 (2016)
12. Leu, J., Hsieh, W.: Efficient and secure dynamic ID-based remote user authentication scheme for distributed systems using smart cards. IET Inf. Secur. **8**(2), 104–113 (2014)
13. Jeong, Y., Park, J.S., Park, J.H.: An efficient authentication system of smart device using multi factors in mobile cloud service architecture. Int. J. Commun. Syst. **28**(4), 659–674 (2015)
14. Amin, R., Hafizul, S., Biswas, G.P., Giri, D., Khan, M.K., Kumar, N.: A more secure and privacy-aware anonymous user authentication scheme for distributed mobile cloud computing environments. J. Secur. Commun. Netw. **9**(17), 4650–4666 (2016)
15. Paranjape, V., Pandey, V.: An approach towards security in private cloud using OTP. Int. J. Emerg. Technol. Adv. Eng. **3**(3), 683–687 (2013)
16. Acharya, S., Polawar, A., Pawar, P.Y.: Two factor authentication using smartphone generated one time password. IOSR J. Comput. Eng. (IOSR-JCE) **11**(2), 85–90 (2013)
17. Mohamed, E.M., Abdelkader, H.S., El-Etriby, S.: Data security model for cloud computing. In: 12th International Conference on Networks (ICN), pp. 66–74, Seville, Spain, 27 January–1 February 2013 (2013)

18. Yeh, H., Chen, B., Wu, Y.: Mobile user authentication system in cloud environment. J. Secur. Commun. Netw. **6**(9), 1161–1168 (2013)
19. Marimuthu, K., Gopal, D.G., Kanth, K.S., Setty, S., Tainwala, K.: Scalable and secure data sharing for dynamic groups in cloud. In: International Conference on Advanced Communication Control and Computing Technologies (ICACCCT), Ramanathapuram, India, 8–10 May, pp. 1697–1701. IEEE (2014)
20. Raza, M., Iqbal, M., Sharif, M., Haider, W.: A survey of password attacks and comparative analysis on methods for secure authentication. World Appl. Sci. J. **19**(4), 439–444 (2012)

Ear Biometric Recognition Using Gradient-Based Feature Descriptors

Hammam A. Alshazly[1(✉)], M. Hassaballah[2], Mourad Ahmed[1], and Abdelmgeid A. Ali[3]

[1] Department of Mathematics, Faculty of Science, South Valley University, Qena 83523, Egypt
hammam.alshazly@sci.svu.edu.eg
[2] Department of Computer Science, Faculty of Computers and Information, South Valley University, Luxor, Egypt
[3] Department of Computer Science, Faculty of Computers and Information, Minia University, Al Minia 61519, Egypt

Abstract. Recently, intensive research efforts are conducted on the human ear as a promising biometric modality for identity recognition. However, one of the main challenges facing ear recognition systems is to find robust representation for the image information that is invariant to different imaging variations. Recent studies indicate that using the distribution of local intensity gradients or edge directions can better discriminate the shape and appearance of objects. Moreover, gradient-based features are robust to global and local intensity variations as well as noise and geometric transformation of images. This paper presents an ear biometric recognition approach based on the gradient-based features. To this end, four local feature extractors are investigated, namely: Histogram of Oriented Gradients (HOG), Weber Local Descriptor (WLD), Local Directional Patterns (LDP), and Local Optimal Oriented Patterns (LOOP). Extensive experiments are conducted for both identification and verification using the publicly available IIT Delhi-I, IIT Delhi-II, and AMI ear databases. The obtained results are encouraging, where the LOOP features excel in all cases achieving recognition rates of approximately 97%.

Keywords: Identity verification · Ear recognition
Gradient features · Dense descriptors

1 Introduction

Automated personal identification from their physiological traits such as face, iris or fingerprint has been extensively studied in the inter-disciplinary field of biometrics. Among these biometric modalities is the human ear which attracted several studies due to the rich and reliable information it contains as well as its stable structure over a long period of age making the variability between two ears distinguishable enough even for identical twins [1,2]. Additionally, ear images can

A. E. Hassanien et al. (Eds.): AISI 2018, AISC 845, pp. 435–445, 2019.
https://doi.org/10.1007/978-3-319-99010-1_40

be easily acquired from a distance without individual's cooperation and are not subject to facial expressions. These characteristics have led to intensive research efforts on the applicability of using ear images to identify individuals and verify their identities. As a result, many ear recognition systems have been proposed in the literature exploiting the existence of several local feature descriptors and the success they achieved in other biometric applications. Examples of ear recognition systems employing local features are summarized in some recent surveys and comparative studies [3–6].

Furthermore, descriptor-based ear recognition techniques are divided into two main categories depending on the way of descriptor computation. First, techniques that extract local features by first detecting a set of interest points (a.k.a. keypoints) from images and then an independent descriptor is computed for each keypoint as described in [7]. Noticeable advantages of these approaches are their robustness to partial occlusion and various affine transformations. However, global information about the shape structure of the ears which is considered important for recognition is lost. Example of ear recognition systems from this group are given in [8,9]. Second, techniques that extract local feature densely for each pixel over the entire image. These techniques are found to give high recognition performance and are preferred due to their simplicity in computations.

Many dense feature descriptors have been proposed in the literature to capture the intrinsic and discriminative information from images [10]. Some of these descriptors have proven to be successful image representations and provide robustness against some imaging distortions. Among these features are the gradient-based features which exploit the gradient information to better discriminate image objects as the gradient magnitude indicates the level of intensity change and its orientation shows the direction of that change. Therefore, this paper concentrates on local features extracted from the gradient magnitudes and orientations computed densely over the entire image to build ear recognition systems. Though, some of these features like HOG [11] and WLD [12] have been used in ear recognition systems relying on their discriminating power, other features such as LDP [13] and LOOP [14] have not got the chance to be utilized in ear biometric recognition systems. To this end, this paper fills the gap by conducting several experiments to study the behavior of these features in ear recognition and highlight their strengths and weaknesses. In this respect, the identification and verification experiments are conducted separately to give recommendations about the best settings of these features in each case.

2 Feature Description

A common aspect of the evaluated feature descriptors used in this work is the computation of gradient orientations and building a histogram. However, each descriptor provides some unique properties in order to achieve the design goals and the required performance. This section explains the four types of features descriptors and highlights the main differences between them.

Histogram of Oriented Gradients (HOG): is a histogram-based feature descriptor that counts the occurrence of gradient orientations in localized image patches called cells [11]. Four main steps are involved in the computation of the HOG feature. First, computing the gradient images, horizontal and vertical gradients, through convolving the image with two filters, $[-1, 0, 1]$ and $[-1, 0, 1]^T$, and then calculating the gradient magnitude and orientation at every pixel. Second, computing a histogram of gradient orientations in 8×8 small regions called cells by considering the contribution of every pixel in the cell. Each histogram is a vector of 9 or 18 bins that correspond to angles 0–180 or 0–360° respectively. Third, the histograms are normalized in order to reduce the effect of illumination variations. The L2-norm is applied over larger spatial regions called blocks, which are formed using equal numbers of adjacent cells. Finally, concatenating all the normalized histograms to obtain the HOG feature vector representing the given ear image. This paper evaluates the recognition performance of HOG features derived from cells of size 8×8 pixels, 16×16 pixel blocks, blocks overlap of 8×8 pixels, and 9 bins histograms.

Weber Local Descriptor (WLD): is a dense local image descriptor that describes an image as a histogram of differential excitations and gradient orientations [12]. It is inspired by Weber's law which states that the ratio of the increment threshold to the background intensity is a constant. The three main steps for computing the WLD descriptor include calculating the differential excitations (ξ), finding gradient orientations (θ), and constructing a histogram. For each pixel in the given image, the differential excitation is computed first based on the intensity differences between a center pixel and its neighbors to extract the salient variations or patterns. The differential excitation $\xi(x_c)$ for a center pixel x_c using P neighborhood is computed as follows:

$$\xi(x_c) = arctan(\frac{v_s^{00}}{v_s^{01}}) = arctan\left[\sum_{i=0}^{p-1}(\frac{x_i - x_c}{x_c})\right], \tag{1}$$

where v_s^{00} represents the difference in intensity between the center pixel and its neighbors and v_s^{01} represents the intensity of the center pixel x_c.
The gradient orientation for the center pixel x_c is computed using

$$\theta(x_c) = arctan(\frac{v_s^{11}}{v_s^{10}}), \tag{2}$$

where v_s^{10} and v_s^{11} are the horizontal and vertical intensity differences respectively. Then, the gradient orientation is quantized into T dominant orientations in two steps. First, mapping θ to θ' according to:

$$\theta' = arctan2(v_s^{11}, v_s^{10}) + \pi, \tag{3}$$

and

$$arctan2(v_s^{11}, v_s^{10}) = \begin{cases} \theta, & if \ v_s^{11} > 0 and v_s^{10} > 0 \\ \pi + \theta, & if \ v_s^{11} > 0 and v_s^{10} < 0 \\ \theta - \pi, & if \ v_s^{11} < 0 and v_s^{10} < 0 \\ \theta, & if \ v_s^{11} < 0 and v_s^{10} > 0, \end{cases} \quad (4)$$

where $\theta \in [-\pi/2, \pi/2]$ and $\theta' \in [0, 2\pi]$.
Second, applying the quantization function defined as:

$$\phi_t = f_q(\theta') = \frac{2t}{T}\pi, \ where \ t = mod\left(\left\lfloor \frac{\theta'}{2\pi/T} + 0.5, T \right\rfloor\right) \quad (5)$$

Finally, the WLD histogram is constructed by incorporating the two components (ξ_i, ϕ_t) at each pixel. This study considers WLD features extracted from patches of 7×7 pixels with 24 neighbors around each pixel.

Local Directional Patterns (LDP): is a texture descriptor that encodes micro-level image information such as edges and other local features based on the abrupt intensity changes around pixels [13]. For each pixel in the given image, LDP operator uses a 3×3 image patch and computes the edge responses in eight possible directions and encode their relative changes as a binary code, where the corresponding decimal value of the code is the LDP value. The edge responses are computed using the Kirsch masks depicted in Fig. 1. Since the response values do not have the same importance in all directions, only the k most prominent directions are used to generate the 8-bit LDP codes. So, only the top k responses are set to 1 and the remaining (8-k) responses are set to 0. Formally, the LDP codes can be derived by:

$$LDP_k = \sum_{i=0}^{7} f(m_i - m_k) \times 2^i, \ f(x) = \begin{cases} 1, & if \ x \geq 0 \\ 0, & if \ x < 0, \end{cases} \quad (6)$$

where m_i is the edge response in the i-th direction and m_k is the k-th most significant directional response. The final feature descriptor is obtained by dividing the encoded ear image into blocks and the LDP histogram is computed from each block, then concatenating all blocks' histograms into a single feature histogram representing the ear image. This paper considers LDP features obtained using 3×3 image patches with k values of 3 and 4.

Local Optimal Oriented Patterns (LOOP): is a texture descriptor that encodes repeated local patterns in images as binary codes [14]. It is proposed to overcome the limitations in the LDP descriptor such as the orientation dependency resulting from the arbitrary sequencing of assigning binarization weights and limiting the number of possibly generated codes due to the restriction imposed on the number of bits allowed to have the value 1 (i.e., the parameter k). Thus, the LOOP integrates the strength of two texture descriptors LDP [13] and LBP [15] for assigning weights and finding the intensity differences, respectively. The LOOP feature is computed for each image pixel using a 3×3

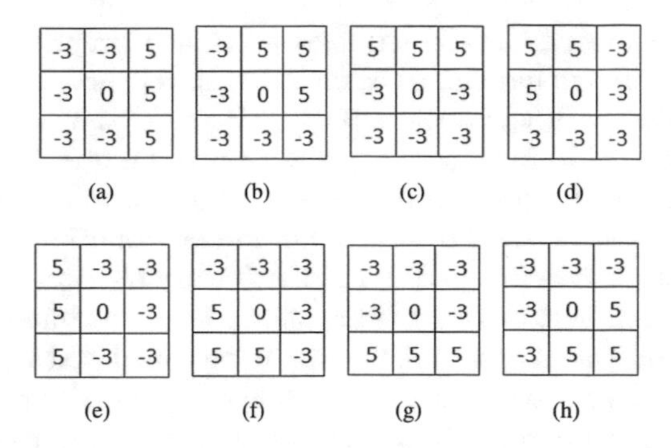

Fig. 1. Kirsch edge response masks for different directions: (a) East; (b) North East; (c) North; (d) North West; (e) West; (f) South West; (g) South and (h) South East.

neighborhood around each pixel according to the following steps. First, the eight responses of the Kirsch masks, $(m_i, i = 0, 1, ..., 7)$ corresponding to pixels with intensities $(g_i, i = 0, 1, ..., 7)$, are computed to obtain the intensity variation in the eight directions. Second, each pixel is assigned a weight w_i based on the rank of the magnitude value m_i. Third, computing the LOOP code for the center pixel (x_c, y_c) as follows:

$$LOOP(x_c, y_c) = \sum_{i=0}^{7} s(g_i - g_c) \times 2^{w_i}, \ s(x) = \begin{cases} 1, & if \ \ x \geq 0 \\ 0, & if \ \ x \ < 0, \end{cases} \tag{7}$$

where g_c represents the gray scale value of the center pixel. Finally, the frequency distribution of these codes over the entire image is used as an image descriptor. This paper considers LOOP features generated from 3×3 image patches.

3 Evaluation Framework

For any biometric recognition system the testing phase is so important to construct a reliable system. Many ear biometric systems have been developed and tested in the literature but each system has a subjective nature as experiments are conducted using different ear databases and custom classifiers inside individually developed frameworks. This brings variability into the evaluation process and masks the performance of feature extractors in ear recognition. Emeršič et al. [6,16] addressed the problem by building the AWE toolbox to measure the ear recognition performance and enable faster development in the field. Researchers can use the toolbox with their own implementation of feature descriptors, classifiers, and using their own benchmark ear databases to make their experimental results consistent and reproducible. So, we embed our implementation details for the evaluated feature descriptors into the toolbox and use different benchmark ear databases and report our results. Figure 2 illustrates the main phases in the ear recognition framework used in the evaluation process. First, the benchmark ear database is read and passed through an image preprocessing phase that ensures that every image has these characteristics: converted to gray scale if colored, spatial resolution is set to 100×100 pixels and obtain an enhanced version of the image by applying histogram equalization technique. Second, the feature extraction phase, each preprocessed image is subjected to the features extractor to extract the local structures from the image and counting their frequency of occurrence over the entire image to build the feature histogram representing the image. Third, in the features matching phase, the Chi-square similarity measure is used to find the distance between the obtained feature histograms and produces a similarity scores. For the identification experiments, the biometric feature vectors are compared in a one-to-many manner and the resulting scores are sorted and ranked to find the rank at which a true match occurs. While for the verification experiments, the biometric feature vectors are compared against each other (one-to-one) and genuine and impostor scores are generated. Finally, the generated scores are handed to the result representation phase to compute the performance metrics such as, rank-1 recognition rates, equal error rates and to draw the performance curves: CMC for identification and ROC for verification experiments.

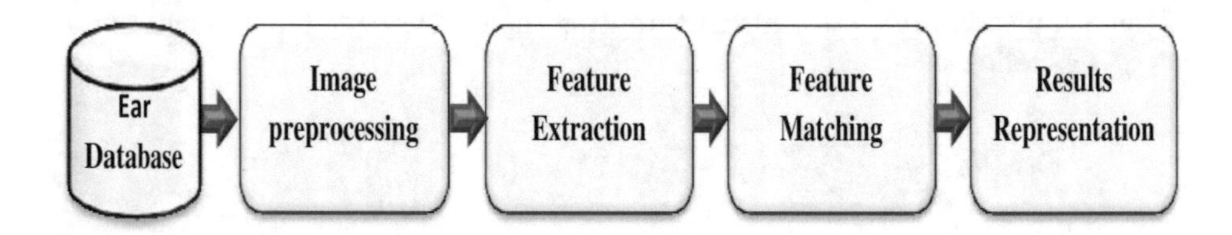

Fig. 2. A block diagram illustrating the different phases of the ear recognition framework.

4 Experimental Results

4.1 Ear Datasets

The experiments in this paper are conducted on three available ear databases that contain cropped ear images with wide variability reflecting different imaging conditions. The Indian Institute of Technology Delhi (IITD) ear database [17] which has two sub-datasets. The IITD I, has 493 ear images for 125 subjects and each subject has between three to six images. Images were taken from the same profile angle at different lighting conditions; while, the IITD II contains 793 normalized ear images for 221 subjects and each subject has between three to six images acquired under various lighting conditions. The AMI ear dataset [18] has 700 ear images for 100 subjects and each subject has seven images. Six images for the right ear and one for the left ear. The images have various poses with subjects looking forward, up, down, right and left. One image also is acquired with different focal length (zoomed). Sample images from the used datasets are shown in Fig. 3.

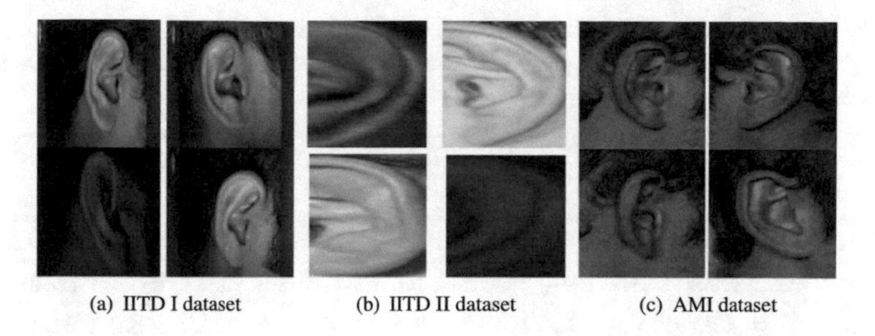

Fig. 3. Sample images from the used benchmark ear databases.

4.2 Identification Experiments

The identification experiments are conducted on the aforementioned ear datasets to highlight the performance of gradient-based features in ear identification problem. As a test protocol, we perform 5-fold cross validation experiments on all datasets and report the rank-1 (R1) recognition rate in the form of mean and standard deviation over the 5-folds. Besides, we generate the average cumulative match characteristic (CMC) curves over the 5-folds for each identification experiment to sample the fine differences in recognition performance. Table 1 summarizes the results of these experiments where the top R1 recognition rate for each feature descriptor on each dataset is highlighted in bold. Further, the corresponding CMC curves for each dataset are shown in Fig. 4.

For the IITD I dataset, the experimental results indicate that the texture descriptor LOOP performs the best with rank-1 recognition rate of approximately 94%. The WLD and LDP3 descriptors perform nearly in a similar manner with no significant performance differences as seen from Fig. 4(a). In addition, there is a slight improvement in the recognition rate when using LDP4 features (i.e., $k = 4$) as the number of generated patterns increases from 56 (with $k = 3$) to 70 patterns; thus, capturing more information and fine details. The HOG descriptor gives the lowest rank-1 recognition rate reaching approximately 83%. In the case of IITD II dataset which contains normalized images, a noticeable improvement in the recognition rates for all descriptors with HOG and LOOP features giving similar recognition rate of 97%. LDP3 and LDP4 features have almost similar recognition rate above 95%, while the WLD features give the lowest recognition performance above 93.5% as summarized in Fig. 4(b). Regarding the AMI dataset that contains cropped ear images with different pose variations and scaling, the descriptors work moderately well with an obvious decrease in the recognition performance ranging from 60% for WLD features and upto 70% achieved by LOOP descriptor as given in Table 1 and summarized in Fig. 4(c). Even though AMI dataset has ear images with good quality, pose variation has a major influence on the recognition performance for all the evaluated features.

Table 1. Results of the identification and verification experiments, where the performance metrics R1 and EER are given in the form of mean and standard deviation over the 5-folds.

Feature	IITD I		IITD II		AMI	
	R1	EER	R1	EER	R1	EER
HOG	82.5 ± 3.7	13.3 ± 4.4	$\mathbf{96.9 \pm 1.0}$	4.9 ± 1.7	66.0 ± 1.5	24.9 ± 4.0
WLD	91.2 ± 2.4	8.4 ± 2.0	93.5 ± 2.4	6.6 ± 2.0	59.7 ± 3.7	23.2 ± 2.5
LDP3	91.9 ± 3.4	7.7 ± 3.0	95.7 ± 1.1	5.7 ± 1.6	64.7 ± 3.0	24.5 ± 2.9
LDP4	92.3 ± 3.6	7.1 ± 2.7	95.2 ± 0.9	5.9 ± 1.7	66.7 ± 2.3	24.5 ± 2.9
LOOP	$\mathbf{93.9 \pm 2.3}$	$\mathbf{6.4 \pm 2.0}$	$\mathbf{96.9 \pm 0.9}$	$\mathbf{4.2 \pm 1.3}$	$\mathbf{70.2 \pm 1.9}$	$\mathbf{21.8 \pm 2.1}$

4.3 Verification Experiments

This section investigates the comparative performance for the verification experiments using the same test protocol discussed in Sect. 4.2. The results are reported

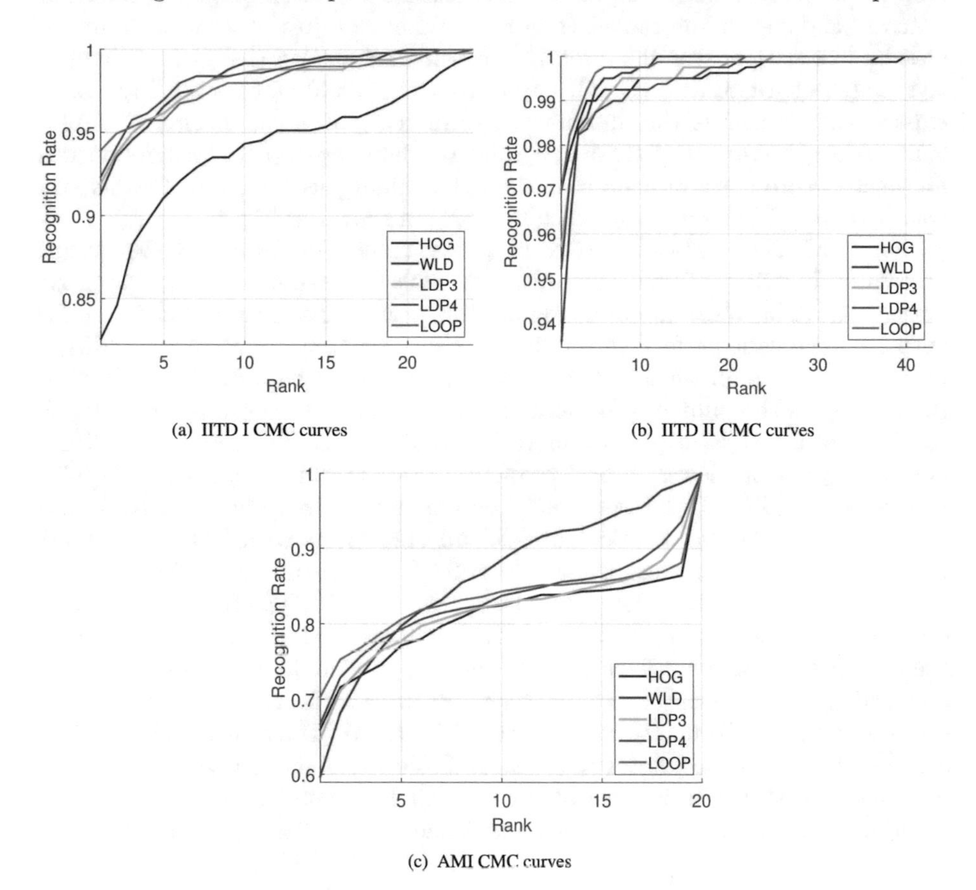

(a) IITD I CMC curves

(b) IITD II CMC curves

(c) AMI CMC curves

Fig. 4. The CMC curves generated in the identification experiments using 5-fold cross-validation for each ear database.

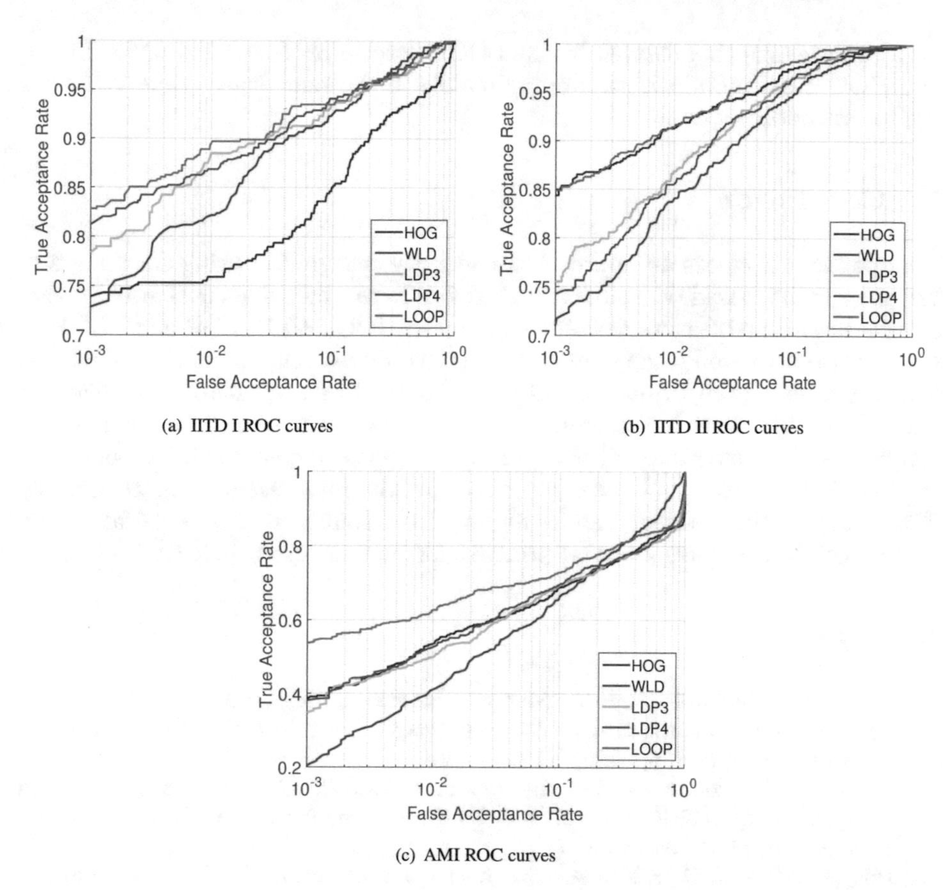

(a) IITD I ROC curves

(b) IITD II ROC curves

(c) AMI ROC curves

Fig. 5. The ROC curves generated in the verification experiments using 5-fold cross-validation for each ear database.

over the 5-folds using expected error rate (EER) and receiver operating characteristic (ROC) curves as given in Table 1 and shown in Fig. 5. The feature descriptors achieving the best performance on each dataset, i.e. the one that achieves the lowest EER, is highlighted in bold.

For the IITD I dataset, the best performance is achieved by the LOOP descriptor with EER of 6.4%. By considering the entire ROC curves in Fig. 5(a), we observe that, the WLD, LDP3 and LDP4 have nearly similar performance; while, the HOG descriptor gives lowest performance with EER of 13.3%. Also, LDP3, LDP4 and WLD descriptors perform similarly with EER differences less than 1%. With respect to the IITD II dataset, again the best performance is achieved by LOOP features with EER of 4.2%. The verification results depicted in Fig. 5(b) show that the difference in performance is somehow smaller than for that of the IITD I dataset, with the LOOP features performing the best and WLD descriptor gives the lowest performance. The conducted verification experiments on the AMI dataset summarized in Fig. 5(c) indicate that, the best performance is again achieved by the LOOP descriptor. Here, the LDP3, LDP4, and HOG descriptors perform approximately in the same level; while WLD has the lowest performance.

5 Conclusion

This paper introduces an approach for identity recognition using ear images of varying quality based on gradient features. Where, the behavior of state-of-the-art gradient-based feature descriptors namely: HOG, WLD, LDP and LOOP as ear feature extraction techniques is investigated for ear recognition. The analyzed features have proven to be successful candidates in discriminating ear features with slight differences in recognition performance in case of constrained imaging conditions. The experimental results indicate that the best results are obtained by LOOP features in all cases with recognition rate reaching approximately 97%. Besides, the obtained results highlight the influence of pose variations on the recognition performance and promote for further improvements.

References

1. Nejati, H., Zhang, L., Sim, T., Martinez-Marroquin, E., Dong, G.: Wonder ears: identification of identical twins from ear images. In: International Conference on Pattern Recognition, pp. 1201–1204 (2012)
2. Zhou, Y., Zaferiou, S.: Deformable models of ears in-the-wild for alignment and recognition. In: IEEE International Conference on Automatic Face & Gesture Recognition, pp. 626–633 (2017)
3. Pflug, A., Busch, C.: Ear biometrics: a survey of detection, feature extraction and recognition methods. IET Biometrics **1**(2), 114–129 (2012)
4. Pflug, A., Paul, P.N., Busch, C.: A comparative study on texture and surface descriptors for ear biometrics. In: International Carnahan Conference on Security Technology, pp. 1–6 (2014)
5. Benzaoui, A., Boukrouche, A.: Ear recognition using local color texture descriptors from one sample image per person. In: IEEE Conference on Control, Decision and Information Technologies, pp. 827–832 (2017)
6. Emeršič, Ž., Štruc, V., Peer, P.: Ear recognition: more than a survey. Neurocomputing **255**, 26–39 (2017)
7. Hassaballah, M., Aly, A.A., Alshazly, H.A.: Image features detection, description and matching. In: Image Feature Detectors and Descriptors: Foundations and Applications, vol. 630, pp. 11–45. Springer (2016)
8. Prakash, S., Gupta, P.: An efficient ear recognition technique invariant to illumination and pose. Telecommun. Syst. **52**(3), 1435–1448 (2013)
9. Anwar, A.S., Ghany, K.K.A., ElMahdy, H.: Human ear recognition using SIFT features. In: Third World Conference on Complex Systems, pp. 1–6 (2015)
10. Hassaballah, M., Awad, A.I.: Detection and description of image features: an introduction. In: Image Feature Detectors and Descriptors: Foundations and Applications, vol. 630, pp. 1–8. Springer (2016)
11. Dalal, N., Triggs, B.: Histograms of oriented gradients for human detection. In: Computer Vision and Pattern Recognition, pp. 886–893 (2005)

12. Chen, J., Shan, S., He, C., Zhao, G., Pietikäinen, M., Chen, X., Gao, W.: WLD: a robust local image descriptor. IEEE Trans. Pattern Anal. Mach. Intell. **32**(9), 1705–1720 (2010)
13. Jabid, T., Kabir, M.H., Chae, O.: Gender classification using local directional pattern (LDP). In: International Conference on Pattern Recognition, pp. 2162–2165 (2010)
14. Chakraborti, T., McCane, B., Mills, S., Pal, U.: LOOP descriptor: local optimal oriented pattern. IEEE Sig. Process. Lett. **25**(5), 635–639 (2018)
15. Ojala, T., Pietikäinen, M., Mäenpää, M.: Multiresolution gray-scale and rotation invariant texture classification with local binary patterns. IEEE Trans. Pattern Anal. Mach. Intell. **24**(7), 971–987 (2002)
16. Emeršič, Ž., Peer, P.: Toolbox for ear biometric recognition evaluation. In: International Conference on Computer as a Tool (EUROCON), pp. 1–6. IEEE (2015)
17. Kumar, A., Wu, C.: Automated human identification using ear imaging. Pattern Recogn. **45**(3), 956–968 (2012)
18. Gonzalez, E.: AMI Ear Dataset. http://www.ctim.es/research_works/ami_ear_database/. Accessed April 2018

A Secure Mutual Authentication Scheme with Perfect Forward-Secrecy for Wireless Sensor Networks

Mohamed M. Mansour(✉), Fatty M. Salem, and Elsayed M. Saad

Faculty of Engineering, Helwan University, 1 Sherif St., Helwan, Cairo, Egypt
Mohamed_mansor@hotmail.com,
Faty_ahmed@h-eng.helwan.edu.eg, elsayed012@gmail.com

Abstract. Recently, Internet of Things (IoT) is increasing pervasively, permits the unattended objects to be connected securely among themselves allowing highly classified processed data to be controlled. WSN has the most significant attention since transferring data is the most security challenge of the IoT environment. As well as, the previous messages which might be intercepted causing the established session key to be revealed. To mitigate such an attack, a lot of schemes were proposed, with an allegation of high efficient security features were achieved and powerful resilience against various attacks. After prudent studies have been made, we found that multiple proposed schemes do not support all the security requirements and are susceptible to some security attacks. In this paper, we propose a secure mutual authentication scheme with perfect forward-secrecy for WSN. Likewise, provides resilience against various types of known attacks in WSNs. Finally, the security of proposed scheme is proven.

Keywords: Internet of Things · Wireless sensor networks
Mutual authentication · Forward secrecy · Key agreement

1 Introduction

IoT is a new type of ubiquitous communication technology consists of worldwide interconnected objects and smart devices in different networks that are connected together, enabling them to exchange and transfer collected data among themselves. The IoT environment contains WSNs, RFID, smart devices, smart cities and homes, traffic observing, border patrol, etc. [1–4]. Many challenges engendered since sensed, collected and exchanged data might be altered, modified by unauthorized users. Consequently, reliable authentication schemes should be achieved to construct the security requirements CIA (Confidentiality, Integrity and authentication) [5]. WSN includes at least single sink node, called the gateway node GWN, and hundreds of unattended ubiquitous devices, sensor nodes, with unparalleled constraints, properties and characteristics which are limited in memory, processing capability, low power, low distance, and thus security, are deployed in different environments and interconnected together with the main goal of sensing, collecting, monitoring, detecting data from various areas and sharing this gathered data with authorized users only. Various

A. E. Hassanien et al. (Eds.): AISI 2018, AISC 845, pp. 446–456, 2019.
https://doi.org/10.1007/978-3-319-99010-1_41

security attacks might be launched against WSN to damage its routing scheme, broadcasting errors or harmful information to exhibit the node energy, consequently, decreasing the WSNs' existence [6].

WSN is defined as heterogeneous, since a lot of computationally-powerful devices can be built with different types of nodes, e.g. gateway nodes GWN. The heterogeneity of WSN is developed to become ad-hoc networks rather than infrastructure-based network, which enables interconnected nodes to communicate with each other to the global world directly instead of communicating directly with the base station. Consequently, a need of more security requirements should be achieved to withstand possible attacks. As a result, user authentication and key agreement is the most important issue that arose in WSN. Furthermore, all authorized users have to key their credentials, which may be revealed by any adversary A.

The idea of the past sessions protection against future compromises of session keys, known as forward-secrecy, is provided in many researches since the credentials of the registered user might be revealed causing the transmitted messages containing sensitive, collected, processed data might be disclosed as well as the user-based authentication scheme [7, 8]. Therefore, proposing a combination of forward security with a powerful mutual authentication and key agreement scheme yields a secure connection establishment.

The rest of this paper is organized as follows: Sect. 2 presents an overview of some existing of related works in this field. The proposed mutual authentication and key agreement scheme is presented in Sect. 3. We provide the analysis of the proposed scheme in Sect. 4. Finally we conclude the paper in Sect. 5.

2 Related Work

Since WSNs have limited-resource-constrained objects, the traditional security protocol cannot be applied such that they exhaust much more power. Hence, many researchers proposed a lot of lightweight security schemes for each possible security aspect of WSN [9, 10]. Most of these researches concentrated on the establishment of cryptographic key between user and a GWN using symmetric or asymmetric schemes, which requires much more computations and communicational power compel thereby unsuitable for resource constrained of the WSN.

Exploiting the power of the smart cards, two user authentication and key agreement schemes for WSN were proposed in [11, 12], giving the authorized remote user the ability to communicate with the nodes of the WSN effectively and securely, by providing security features and several attacks resilience. These features were achieved using only two simple computations, hash and XOR. Therefore, these schemes are considered lightweight and highly appropriate for WSN. Das presents an efficient and lightweight mechanism [13], since it was based on temporal credentials and gathered a lot of attentions. Nevertheless, it was proven insecure, vulnerable and had design flaws [14, 15]. Chen [16] clarified that the schemes proposed by Song [17] and Sood [18], claimed that their schemes can resist various attacks and provide perfect forward secrecy, have diverse security flaws being ignored, and proposed a robust remote user password authentication scheme. While, Li et al. [19] has mentioned that [16] failed to guarantee forward secrecy with password change phase of the scheme is inconvenient and inefficient. Li et al. likewise introduced an enhanced smart card based remote user password authentication scheme conquered the mentioned above problems. Furthermore, Kumari and Khan [20] proposed a password-based authentication protocol claiming that the proposed protocol can provide forward security and resist various types of attacks, as well as Jiang et al. [21]. Using a different authentication model based on [12], Turkanović et al. provided a new user authentication and key agreement scheme [22].

With the emergence of weaknesses in [22], Farash et al. [9] proposed another authentication scheme for heterogeneous WSN to eliminate such weaknesses. Unfortunately, both [9, 22] do not provide forward-secrecy since the authentication scheme is based on the password, which might be revealed by an adversary using off-line password guessing attack. Achieving many security requirements including forward-secrecy and user anonymity, meanwhile resisting to get sensitive real time sensor data, another authentication and key agreement schemes are presented by Lu et al. [23] and Jung et al. [24]. However although, a perfect forward security feature is not provided in [23, 24] was proven by Xiong et al. in [7], and they can suffer from SC loss attack. The resistance of Denial of Service (DoS) attack were discussed in many proposed schemes, while, Gope et al. focused on this type of attack and proposed a new scheme [25] for resilience of DoS attacks. Thereby, the registered users are required to store a large set of unlinkable pairs of shadow identities and emergency key in their memory. One of the most significant security requirements of authentication and key agreement schemes is the forward secrecy. Wei et al. [8] had proved that the proposed password-based authentication scheme provided by Kumari and Khan [20] did not achieve neither perfect forward security nor resist off-line password guessing attack. Meanwhile, with no user anonymity achievement, he proposed a mutual authentication and key agreement scheme achieving forward security. Hence, it is obvious that there is no scheme can provide all the essential security requirements for the WSNs.

3 The Proposed Scheme

This section provides a new user authentication and key agreement scheme distributed among user, sensor node and server/GWN which enables a connection to be established securely between a pre-registered user on demand and a registered sensor node in the WSN by applying the forward security feature such that even if the current session key is disclosed, the previous exchanged messages won't be revealed.

Our proposed scheme consists of four phases: a pre-deployment, registration, login and authentication and password change phase.

3.1 Pre-deployment Phase

The *WSN* is formed of hundreds of tiny resource constrained low-cost nodes that are limited in memory, energy, and processing capacity, and a proportional number of much powerful GWNs. This phase runs offline by the network administrator and uses the GWN to specify:

SID_j : The identification of each sensor node S_j.
p : Large prime number for DH-algorithm Algorithm.
B_j : The authentication parameter between the GWN and the sensor node S_j.

$$B_j = h(X_{GWN}||SID_j) \tag{1}$$

H : The authentication parameter of the GWN to the sensor node S_j

$$H = h(X_{GWN}||1) \tag{2}$$

The GWN is predefined with a randomly-generated highly secured password X_{GWN} known only to itself and is secretly stored in its memory. Consequently, SID_j, B_j and H are stored in the sensor node's memory. Furthermore, the GWN stores the corresponding secret credentials of each sensor node in its memory.

3.2 Registration Phase

The registration phase is done between the user U_i and the GWN which is initiated by the user in demand. In this phase the U_i registers his/her self to connect to a certain sensor node on demand and run the login and authentication phase later. The registration process is done using a smart card *SC*, which is personalized by the GWN at the end of the phase. The procedures of a detailed description of the User-GWN registration phase are as follows:

(1) The user U_i chooses its ID_i, PW_i and sends them to the GWN via a secure channel.
(2) The GWN computes parameters for user U_i as follows:
GWN: computes $B_i = \mathrm{h}(X_{GWN}||ID_i)$ to authenticate the ID_i with its secret master password X_{GWN}.
GWN: hides B_i in A_i using $U_i^{\prime}s$ information (ID_i, PW_i) as follows:

$$A_i = B_i \oplus \mathrm{h}(ID_i||PW_i) \tag{3}$$

GWN: saves all the previous parameters in a $SC = \{A_i, p, h(.)\}$, then sends the *SC* to the U_i via a secure channel.

3.3 Login and Authentication Phase

This phase is done through 4 transmitted messages between U_i, S_j, and the GWN as:

i. The *SC* verifies the legitimacy of the U_i before it computes its private data, selects a certain sensor S_j to connect with, and sends a connection request message (1st message) to it via an unsecure channel as:

U_i: keys its (ID_i, PW_i) into the device.
SC: computes B_i using (ID_i, PW_i) as follows:

$$B_i = A_i \oplus h(ID_i||PW_i) \quad (4)$$

SC: chooses $\alpha \in Z_p^*$ to establish its contribution part of SK_{ij}
SC: computes its contribution of the session key SK_{ij} using DH–Algorithm.

$$K_i = h(ID_i)^{\alpha} mod\ p \quad (5)$$

SC: authenticates its session key part K_i^* message to the GWN as follows:

$$K_i^* = K_i + B_i \quad (6)$$

SC: authenticates the user U_i to the GWN by computing M_{U_i} as follows:

$$M_{U_i} = \mathrm{h}(ID_i||K_i||T_1) \quad (7)$$

SC: edits an authentication message $AM_{U_i - S_j}$ and sends it to the GWN through the desired sensor node S_j via an unsecure channel as follows:

$$AM_{U_i - S_j} = \{ID_i, M_{U_i}, K_i^*, T_1\} \quad (8)$$

ii. The sensor S_j receives an authentication message $AM_{U_i - S_j}$ from U_i via an unsecure channel, then computes its private data before it sends an authentication message (2nd message) to the GWN as follows:

S_j: checks $|T_1 - T_c| < \Delta T$.
S_j: chooses $\beta \in Z_p^*$ to establish its contribution part of SK_{ij}
S_j: computes its contribution of the session key SK_{ij} using DH–Algorithm as:

$$K_j = h(ID_i)^{\beta}\ mod\ p \quad (9)$$

S_j: hides its session key part K_i^* message of the sensor node S_j by computing:

$$K_j^* = K_j + B_j + H \quad (10)$$

S_j: authenticates itself to the GWN by computing:

$$M_{S_j} = h(SID_j||H \oplus K_j||M_{U_i}||T_1||T_2) \quad (11)$$

S_j: edits an authentication message $AM_{S_j - GWN}$, sends it to the GWN via an unsecure channel as:

$$AM_{S_{j,}-GWN} = \{ID_i, SID_j, K_i^*, K_j^*, M_{U_i}, M_{S_j}, T_1, T_2\} \quad (12)$$

iii. The GWN receives an authentication message AM_{S_j-GWN} via an unsecure channel and it authenticates the user U_i and the sensor S_j to itself then authenticates them to each other, furthermore, it sends a reply message (3rd message) to the sensor S_j as follows:

GWN: checks $|T_2 - T_c| < \Delta T$.
GWN: extracts the session key part K_j' of the sensor S_j by computing:

$$K_j' = K_j^* - B_j - H \quad (13)$$

GWN: authenticates sensor S_j to itself by checking the following equality:

$$M_{S_j} = ?h(SID_j||H \oplus K_j'||M_{U_i}||T_1||T_2) \quad (14)$$

The GWN verifies the previous equality, if it holds, thereby identifying and authenticating the sensor node S_j to itself, otherwise, it terminates the session.
GWN: extracts the session key part K_i' of U_i by computing:

$$K_i' = K_i^* - B_i \quad (15)$$

GWN: authenticates user's U_i identification ID_i, to itself by checking the following equality as:

$$M_{U_i} = ?h(ID_i||K_i'||T_1) \quad (16)$$

The *GWN* verifies the previous equality, if it holds, thereby identifying and authenticating the user U_i to itself, otherwise, it terminates the session.
Finally, the sensor node S_j and the user U_i are authenticated to the GWN.
Now, the *GWN* needs to mutual authenticate itself, S_j, and U_i respectively to each other as follows:
GWN: authenticates $U_i's$ session key part to S_j as:

$$K_j^{**} = K_i' + B_j \quad (17)$$

GWN: authenticates $S_j's$ session key part to U_i as:

$$K_i^{**} = K_j' + B_i \quad (18)$$

GWN: mutual authenticates itself, S_j, and U_i at the same time as follows:

$$M_{GWN} = h(ID_i||SID_j||H \oplus K_i'||T_3) \quad (19)$$

GWN: edits a response authentication message RAM_{GWN-S_j} and sends it to the sensor S_j via an unsecure channel as:

$$RAM_{GWN-S_j} = \{SID_j, K_j^{**}, K_i^{**}, M_{GWN}, T_3\} \quad (20)$$

iv. The sensor node S_j receives a reply authentication message RAM_{GWN-S_j} from the GWN, to authenticate it and the user U_i simultaneously, before generates the complete session key SK_{ij}, furthermore, it sends a response authentication message (4th message) to the user U_i as follows:

S_j: checks $|T_3 - T_c| < \Delta T$.
S_j: extracts the U_i's session key part as:

$$K_i'' = K_j^{**} - B_j \quad (21)$$

S_j: authenticates the GWN to itself by computing:

$$M_{GWN} = ?h(ID_i||SID_j||H \oplus K_i''||T_3) \quad (22)$$

The sensor node S_j verifies the previous equality, if it holds, thereby identifying and authenticating the GWN and the user U_i's session key part K_i'' to itself, otherwise, it terminates the session.
S_j: computes the complete session key SK_{ij} between itself and U_i as:

$$SK_{ij} = (K_i'')^{\beta} \, mod \, p = (h(ID_i)^{\alpha})^{\beta} \, mod \, p \quad (23)$$

S_j: authenticates itself and the GWN to the U_i simultaneously as:

$$M_{ji} = h(ID_i||SID_j||K_i''||SK_{ij}||T_3||T_4) \quad (24)$$

S_j: edits a response authentication message $RAM_{S_j-U_i}$ and sends it to the user U_i via an unsecure channel as follows:

$$RAM_{S_j-U_i} = \{ID_i, M_{ji}, K_i^{**}, T_3, T_4\} \quad (25)$$

Finally, the SC of the user U_i receives a reply authentication message $RAM_{S_j-U_i}$ from the sensor node S_j, SC authenticates the sensor node S_j and the GWN to itself at the same time, then it establishes and verifies the complete session key SK_{ij}, furthermore, it sends a confirmation to the sensor S_j to start the session as follows:
SC: checks $|T_4 - T_c| < \Delta T$.
SC: extracts the S_j's session key part K_j'' by computing:

$$K_j'' = K_i^{**} - B_i \quad (26)$$

SC: computes the complete session key SK_{ij} between itself and the S_j as

$$SK_{ij} = (K_j'')^{\alpha} \, mod \, p = (h(ID_i)^{\beta})^{\alpha} \, mod \, p \tag{27}$$

SC: authenticates the GWN and the sensor S_j to itself at the same time as:

$$M_{ji} = ?h(ID_i||SID_j||K_i''||SK_{ij}||T_3||T_4) \tag{28}$$

The user U_i verifies the previous equality, if it holds, thereby identifying and authenticating the *GWN* and the sensor node $S_j's$ session key part K_j'' to itself, otherwise, it terminates the session.

SC: sends a confirmation message to the sensor S_j.

3.4 Password Change Phase

In this phase, we present a mechanism for when a user U_i needs to change his/her password PW_i with a new one PW_i^{new}. The specified procedures are performed as follows:

1) U_i mounts his/her smart card into a device reader, inputs its ID_i and PW_i, followed with the new password PW_i^{new}.
2) The PW_i^{new} is computed and is replaced with the old one by the smart card as:

$$A_i^{new} = A_i \oplus h(ID_i||PW_i) \oplus h(ID_i||PW_i^{new}) \tag{29}$$

Then U_i replaces A_i with A_i^{new}.

4 Security Analysis

In this section we elucidate the security analysis and evaluation of the proposed scheme. It is very hard to achieve secured authentication protocols; nevertheless, it is robustly substantial to combat adversaries. We prove that the proposed scheme maintains mutual authentication, a secure key agreement, forward security, password protection and change, and is resilient against impersonation attacks, replay attacks, off-line password guessing attacks, password change attacks, and denial-of-service attacks.

Mutual Authentication: It is a significant fundamental feature used in a user authentication scheme. An adversary $\mathcal{A}$ might execute a man-in-the-middle attack to access WSN thereby collects, inserts data into it, and takes an action. Thereby, any user tries to login into a WSN has to verify the GWN's legitimacy, since it might be masqueraded. On the other hand, the GWN has to verify the user's legitimacy, since it might be an adversary. This tow-way authentication technology is a process such that all parties authenticate each other and are sure of each other's identities.

Session Key Agreement: All parties in IoT environment should be involved independently and individually using a key agreement protocol achieving a randomly secure negotiable SK_{ij} (Eq. 23). Using hash function, XOR operation and exponentiation computations, a secure key agreement protocol between the user U_i and the sensor node S_j is provided ensuring better performance within the resource constrained environment of WSN. Consequently, a random SK_{ij} is computed using the asymmetric cryptography DH-algorithm. Thereby, it's very hard to recover it due to the discrete logarithm problem which is widely believed to be difficult.

Forward-Secrecy: Our scheme can provide perfect forward-secrecy, since the established SK_{ij} (Eq. 23) does not depend neither on the U_i's password nor the Master password. Furthermore, an $\mathcal{A}$ has to recover α from (Eq. 5) or β from (Eq. 9), which is a very hard computation since the discrete logarithm problem is widely believed to be difficult. Therefore, our proposed scheme prevents the $\mathcal{A}$ from recovering the established SK_{ij} by applying perfect forward-secrecy to protect past sessions against future compromises of secret keys.

Password Protection and Change: Since users could use the same password among different systems and exchanging user's password PW_i via insecure channel in authentication phase, these permit an $\mathcal{A}$ to intercept, extract and use it to impersonate a legitimate user U_i in a certain system. Furthermore, he could also use the same extracted password in order to masquerade the same user U_i in other systems, since that user might use it in different systems. Ensuring that $U_i's$ password will not be used to complete the authentication process, i.e., it will not be sent over the insecure channel. More and above, the authentication process cannot be executed outside the smart card. Furthermore, the $\mathcal{A}$ cannot reveal the stored $U_i's$ password in the smart card (Eq. 3) using smart card and/or offline password guessing attacks. Thereby, our proposed scheme achieves password protection and the user can change the password without affecting its secret credentials B_i.

Impersonation Attack: Such an attack takes place when the private and secure information of a legitimate user U_i could be obtained by an $\mathcal{A}$ which enables him/her to act as a legitimate user U_i. Although the $\mathcal{A}$ could capture a legitimate user's authentication message and could extract the session key portion, and retransmit it to the sensor node, he/she could not compute SK_{ij} since the discrete logarithm problem is believed to be hard problem to be solved. Furthermore, if a user U_i lost his/her smart card, the $\mathcal{A}$ without knowing the legitimate user's identification and password, he/she could not use it to impersonate the legitimate user.

Replay Attack: When an $\mathcal{A}$ intercepts, captures and modifies the WSN traffic then resends the modified traffic to its original destination, acting as the original sender, a replay attack takes place. To prevent such an attack to success, timestamps are added to each exchanged message between all parties, enabling the authentication system to accept the network packets that contain the appropriate timestamp. If the timestamp exceeds a certain time ΔT, the packet is discarded.

Offline Password Guessing Attacks: Assuming that the private data of a legitimate user U_i, $\{\mathrm{A_i, p, q, h(\cdot)}\}$, stored in his/her smart card, is extracted by an $\mathcal{A}$, who might select PW_i^* and compute $h(ID_i||PW_i^*)$, but could not reveal A_i to verify the correctness of PW_i^*. Since the master secret key X_{GWN} of the GWN is unknown, it makes the offline password guessing attack impossible for a passive $\mathcal{A}$. Consequently, our proposed scheme provides resilience against the offline password guessing attack.

Password Change Attack: This attack will succeed when an $\mathcal{A}$ has found a lost user's smart card SC or has stolen one, and will attempt to change the password PW_i of the legitimate user U_i. Therefore, to achieve such a success, the $\mathcal{A}$ needs to know the old password PW_i' first. Although, he/she could reveal the stored information A_i, it is still computationally infeasible for him/her to extract the PW_i' since the B_i could not be revealed or disclosed. Consequently, our proposed scheme provides resilience to password-change attack.

Denial-of-Service Attacks: DoS attacks typically flood servers or networks with traffic in order to overwhelm the victim resources and make it difficult or impossible for legitimate users to use them. Such an attack is mitigated in our proposed scheme, since it works on the concept of request-response communication; consequently, the legitimate user always gets a confirmation or rejection for his authentication request from the sensor node originated from the GWN. Timestamps have been introduced into the security scheme to mitigate any consequential request. Finally, our proposed scheme can resist against DoS attacks.

5 Conclusion

In this paper, we provided a secure mutual authentication scheme with perfect forward-secrecy for WSN using smart cards. This scheme focuses on forward-secrecy with a fully secured connection between a pre-registered user's desire to join a certain sensor node. The main goal is to protect the previous exchanged messages against future compromises of secret keys using a perfect session key created using DH-algorithm in the fact that the discrete logarithm problem is believed as a difficult problem to be solved. The proposed scheme uses a novel of distributed authentication scheme, mutual authentication among all parties, furthermore, a powerful key agreement protocol which achieving a shared secret session key between a user and a desired sensor node allowing he/she to aggregate sensitive data from the WSN securely. The GWN plays an important role in executing mutual authentication. This scheme uses a rare four-step authentication model which is, according to our research, the most suitable for such requirements and environment.

References

1. Alcaraz, C., Najera, P., Lopez, J., Roman, R.: WSNs and the internet of things do we need a complete integration? In: 1st International Workshop on the Security of the Internet of Things, SecIoT 2010. IEEE, Tokyo (2010)
2. Atzori, L., Iera, A., Morabito, G.: The internet of things: a survey. Comput. Netw. **54**, 2787–2805 (2010)
3. Romer, K., Mattern, F.: The design space of wireless sensor networks. IEEE Wirel. Commun. **11**, 54–61 (2004)
4. Akyildiz, E.A.: I.F.: Wireless sensor networks: a survey. Comput. Netw. **38**(4), 393–422 (2002)

5. Stallings, W., Brown, L.: Computer Security: Principles and Practice, 2nd edn. Pearson Education, London (2011)
6. Pathan, A.S.K., Lee, H.-W., Hong, C.S.: Security in wireless sensor networks: issues and challenges. Phoenix Park, 20 February 2006, pp. 1043–1048. Institute of Electrical & Electronics Engineers (IEEE) (2006)
7. Xiong, L., Peng, D., Peng, T., Liang, H., Liu, Z.: A lightweight anonymous authentication protocol with perfect forward secrecy for wireless sensor networks. Sensors **17**, 2681 (2017)
8. Wei, J., Liu, W., Hu, X.: Secure and efficient smart card based remote user password authentication scheme. Int. J. Netw. Secur. **18**(4), 782–792 (2016)
9. Farash, M.S., Turkanović, M., Kumari, S., Hölbl, M.: An efficient user authentication and key agreement scheme for heterogeneous wireless sensor network tailored for the internet of things environment. Ad Hoc Netw. Part I **36**, 152–176 (2016)
10. Yussoff, E.A.: Y.M.: A review of physical attacks and trusted platforms in wireless sensor networks. Procedia Eng. **41**, 580–587 (2012)
11. Das, A.K., Sharma, P., Chatterjee, S., Sing, J.K.: A dynamic password-based user authentication scheme for hierarchical WSNs. J. Netw. Comput. Appl. **35**, 1646–1656 (2012)
12. Xue, K., Ma, C., Hong, P., Ding, R.: A temporal-credential-based mutual authentication and key agreement scheme for wireless sensor networks. J. Netw. Comput. Appl. **36**, 316–323 (2012)
13. Das, M.L.: Two-factor user authentication in wireless sensor networks. IEEE Trans. Wirel. Comm. **8**(3), 1086–1090 (2009)
14. Wang, P.W.D.: On the anonymity of two-factor authentication schemes for wireless sensor networks: attacks, principle and solutions. Comput. Netw. **73**(14), 41–57 (2014)
15. Jiang, Q., Ma, J., Lu, X., Tian, Y.: An efficient two-factor user authentication scheme with unlinkability for wireless sensor networks. Peer-to-Peer Netw. Appl. **8**, 1070–1081 (2014)
16. Chen, B.L., Kuo, W.C., Wuu, L.C.: Robust smart-card-based remote user password authentication scheme. Int. J. Commun Syst **27**(2), 377–389 (2014)
17. Song, R.: Advanced smart card based password authentication protocol. Comput. Stand. Interfaces **32**(5), 321–325 (2010)
18. Sood, S.K., Sarje, A.K., Singh, K.: An improvement of Xu et al.'s authentication scheme using smart cards. In: Proceedings of the Third Annual ACM Bangalore Conference, p. 15 (2010)
19. Li, X., Niu, J., Khan, M.K., Liao, J.: An enhanced smart card based remote user password authentication scheme. J. Netw. Comput. Appl. **36**(5), 1365–1371 (2013)
20. Kumari, S., Khan, M.K.: Cryptanalysis and improvement of a robust smart-card-based remote user password authentication scheme. Int. J. Commun Syst **27**(12), 3939–3955 (2014)
21. Jiang, Q., Ma, J., Li, G., Li, X.: Improvement of robust smart-card-based password authentication scheme. Int. J. Commun Syst **28**(2), 383–393 (2015)
22. Turkanović, M., Hölbl, M.: A novel user authentication and key agreement scheme for heterogeneous ad hoc wireless sensor networks, based on the internet of things notion. Ad Hoc Netw. **20**, 96–112 (2014)
23. Lu, Y., Li, L., Peng, H., Yang, Y.: An energy efficient mutual authentication and key agreement scheme preserving anonymity for wireless sensor networks. Sensors **16**, 837 (2016)
24. Jung, J., Kim, J., Choi, Y., Won, D.: An anonymous user authentication and key agreement scheme based on a symmetric cryptosystem in wireless sensor networks. Sensors **16**, 1299 (2016)
25. Gope, P., Lee, J., Quek, T.Q.S.: Resilience of DoS attacks in designing anonymous user authentication protocol for wireless sensor networks. IEEE Sens. J. **17**(2), 498–503 (2017)

Personal Identification Based on Mobile-Based Keystroke Dynamics

Alaa Tharwat[1,2,6](✉), Abdelhameed Ibrahim[3,6], Tarek Gaber[4,6], and Aboul Ella Hassanien[5,6]

[1] Faculty of Engineering, Suez Canal University, Ismailia, Egypt
engalaatharwat@hotmail.com
[2] Faculty of Computer Science and Engineering, Frankfurt University of Applied Sciences, 60318 Frankfurt am Main, Germany
[3] Faculty of Engineering, Mansoura University, Mansoura, Egypt
[4] Faculty of Computers and Informatics, Suez Canal University, Ismailia, Egypt
[5] Faculty of Computers and Information, Cairo University, Giza, Egypt
[6] Scientific Research Group in Egypt (SRGE), Giza, Egypt
http://www.egyptscience.net

Abstract. This paper is addressing the personal identification problem by using mobile-based keystroke dynamics of touch mobile phone. The proposed approach consists of two main phases, namely feature selection and classification. The most important features are selected using Genetic Algorithm (GA). Moreover, Bagging classifier used the selected features to identify persons by matching the features of the unknown person with the labeled features. The outputs of all Bagging classifiers are fused to determine the final decision. In this experiment, a keystroke dynamics database for touch mobile phones is used. The database, which consists of four sets of features, is collected from 51 individuals and consists of 985 samples collected from males and females with different ages. The results of the proposed model conclude that the third subset of features achieved the best accuracy while the second subset achieved the worst accuracy. Moreover, the fusion of all classifiers of all ensembles will improve the accuracy and achieved results better than the individual classifiers and individual ensembles.

Keywords: Keystroke dynamics · RHU keystroke
Genetic algorithm · Bagging classifier · Personal identification

1 Introduction

The number of smartphone users in 2016 was 2.1 billion, and it is expected to reach around 2.5 billion in 2019 [1]. Along with sensing capabilities, the pervasive feature of smartphones has changed the prospective of people's everyday life. Smartphones are now full of personal information including bank account details, medical information (e.g., sugar level and heart rate), and other personal credentials required for different services and applications. In such context, smartphone's users become worried about the confidentiality and privacy

A. E. Hassanien et al. (Eds.): AISI 2018, AISC 845, pp. 457–466, 2019.
https://doi.org/10.1007/978-3-319-99010-1_42

of their data and information [2]. Thus, maintaining the privacy of such users' sensitive data and the information is a very vital issue.

Passwords, personal identification number (PIN), and pattern lock are considered the most common methods to protect the identity of smartphone's owners from illegal accesses. However, the PIN method is subject to the dictionary attack as well as shoulder surfing attack. Also, Pattern locks are subject to side channel attack. The most importantly, such commonly used methods for identifying a legitimate owner of a smartphone fail to recognize and detect an adversary when she/he has broken the PIN/Password [3] making these methods ineffective for continuous authentication.

Continuous authentication is a process in which user's behavioral biometrics are used for authenticating a smartphone user [3]. Mobile authentication based on behavioral biometrics depends on learning from user behaviors which do not change over a period. This includes gait patterns [4], signature [5], and touch-screen interactions [6,7]. Such characteristics are then implicitly used to identify a given smartphone user thus preventing an illegitimate user from accessing the device. The authentication decision in this context is taken based on distinctive features collected from the user's behaviors.

This paper describes an approach addressing the personal identification problem by using mobile-based keystroke dynamics. In the proposed approach, a keystroke dynamics dataset is used [8]. The most important features are selected using GA. Moreover, Bagging classifier used the selected features to identify persons by matching the features of the unknown person with the labeled features.

The rest of the paper is organized as follows; Sect. 2 presents the feature selection technique using a genetic algorithm and the Bagging classifier ensemble. Our proposed approach are presented in Sect. 3 while, the experimental results are given in Sect. 4. The discussion of the obtained results and the conclusion of the paper are presented in Sects. 5 and 6, respectively.

2 Preliminaries

2.1 Genetic Algorithm (GA)

Genetic Algorithm (GA) is one of the well-known optimization algorithms. The main idea of GA depends on evaluating and modifying the current population, i.e., a set of solutions, which are represented in binary numbers and selects the best solutions to generate new solutions from it. Hence, GA searches mainly in binary spaces to find the optimal solutions [9,10].

Mathematically, given the initial population set $P = \{p_1, p_2, \ldots, p_n\}$, where p_i is one candidate, agent or chromosome. Each chromosome consists of many genes and the value of each gene is zero or one. The population in GA is modified to create a new generation in each iteration. This modification includes three main processes, namely, (1) crossover, (2) mutation, and (3) selection. In the crossover process, the information of two chromosomes, i.e., parents, are exchanged to generate two new chromosomes, i.e., children. There are three methods of crossover between chromosomes: (1) single point, (2) multi-points, and (3) uniform. For example, in the single point crossover method, the information of two chromosomes 00100101 and 10111010 are exchanged and this can be achieved by selecting one cutting point and exchange one part from each parent to generate new chromosomes, 00101010 and 10110101. The cutting point is chosen randomly in many algorithms. In the mutation process, the original chromosome, i.e., parent, mutate randomly by changing the value of one or more bits to generate a new chromosome, i.e., child. Moreover, GA allows the chromosomes to mutate and generate a near identical chromosome from the original one. The current chromosomes and the new generated ones are evaluated using the fitness function and the best chromosomes are selected (survived) to the next generation, this is called selection step [11].

The performance of GA is controlled by many parameters such as the population size (N), crossover rate (P_c), mutation rate (P_m), and the number of iterations (T). The population size represents the total number of chromosomes while the crossover rate represents the probability of accepting an eligible pair of chromosomes for crossover. The mutation rate is the probability of switching bits in the chromosomes [12]. The steps of GA are summarized in Algorithm 1.

Algorithm 1. Genetic Algorithm

1: Initialize population $P = \{p_1, p_2, \dots, p_n\}$, where n represents the number of chromosomes in the populations, and the total number of generations is T.
2: **for all** (Generations $t < T$) **do**
3:    **for all** ($i = 1$ **to** $n/2$) **do**
4:       Select two parents p_i and $p_{n/2+i}$ from P.
5:       Offspring O_i=CrossOver($p_i, p_{n/2+i}$)
6:       Offspring O_i=Mutation (Offspring O_i)
7:    **end for**
8:    Combine the original population P and the new offsprings O_i into new $P_{New} = P + O_i$
9:    Evaluate all chromosomes in the new population P_{New}
10:    Select the best chromosomes to survive for the next generation
11: **end for**

2.2 Bagging Classifiers

Bagging is an acronym from *Bootstrap AGGregatING* and it is one of the well-known ensemble methods that creates its ensemble by training different classifiers on a random distribution of the training set. Many samples from the original training set (X) are drawn randomly with replacement to form a new training set X_{New}, which is then used for training one individual classifier C_i in the Bagging ensemble. The accuracy of the classifiers of each pattern is chosen with equal probability; i.e., not weighted as in AdaBoost [13,14], then combine the outputs of all classifiers using uniform averaging or voting over class labels. Since the Bagging resamples the training set with replacement, so some samples are chosen multiple times, while the other samples are left out [15,16]. The details of Bagging classifier are in Algorithm [2].

Algorithm 2. Bagging Algorithm

Input: Required ensemble size T, Training set $S = \{(x_1, y_1), (x_2, y_2), \ldots, (x_N, y_N)\}$
while $(t < T)$ **do**
  Build a dataset S_t, by randomly sampling N times from S (with replacement)
  Train a model h_t using S_t and add it to the ensemble
end while
for all (new test pattern $(\acute{x}, \acute{y})$) **do**
  for all (models h_t) **do**
    Classify the unknown or test pattern using the model h_t, the output of the classification is C_t
  end for
  Combine the outputs of the model $C_i, i = 1, 2, \ldots, T$
end for

---

3 Proposed Model

The proposed approach in this research consists of two main phases, namely feature selection and classification. As shown in Fig. 1, our dataset that is used in this research has four different sets (f_1, f_2, f_3, and f_4) of features. In the first phase, GA is applied for each feature subset separately to select the best features in each subset ($f_{new1}, f_{new2}, f_{new3}$, and f_{new4}). In the second phase, the data were partitioned into training set which is used to train the Bagging model and testing set which is used to evaluate the proposed model. More details about each phase in the next two sections.

3.1 Feature Selection Phase

In this phase, the GA was employed for selecting a small subset d from a large set of features D to remove independent and redundant features [17]. The large set of features is denoted by, $X = \{f_1, f_2, \ldots, f_D\}$. While the small subset of features denoted by, $Y = \{f_1, \ldots, f_d\}$, where $d \leq D$. GA first initializes a set of chromosomes coded in binary code randomly. Each chromosome represents a subset of features, and each bit of the chromosome denoting the presence, i.e., 1, or absence, i.e., 0, of that feature in this subset. Thus, the dimension of search space R^d for the features is determined by the total number of features d [18].

In GA, the fitness function should ensure that the fitter chromosome has a high probability to survive for the next generations. But, in feature selection application, the goal of the fitness function of the GA is to evaluate a subset of features requires an estimation of thc misclassification rate for that subset.

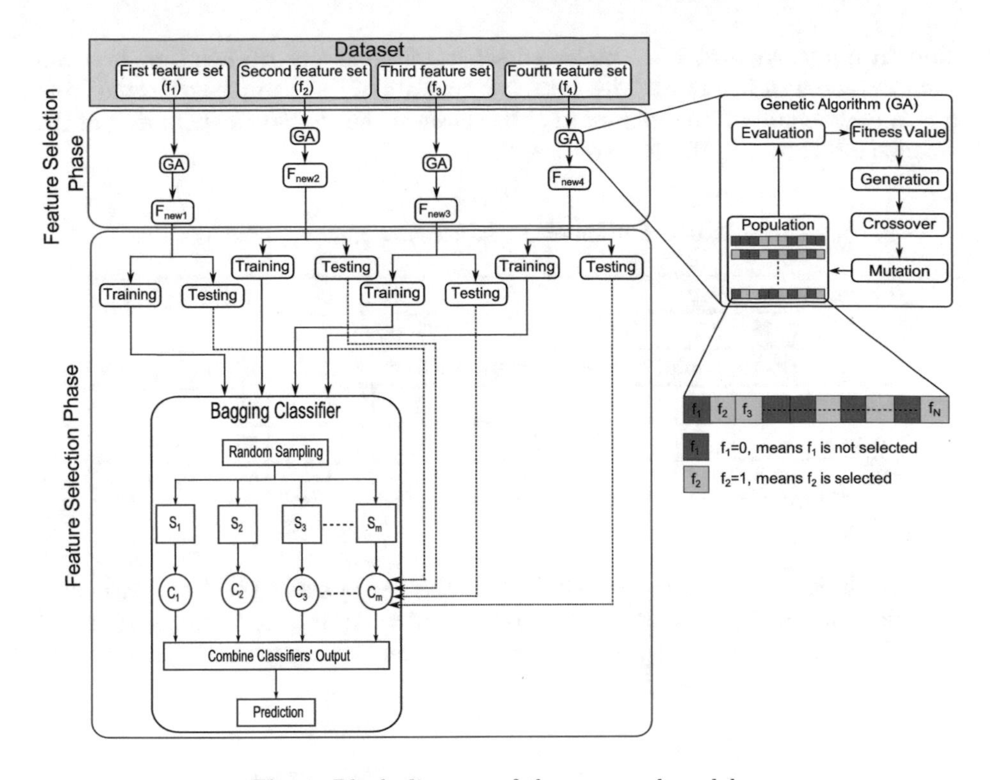

Fig. 1. Block diagram of the proposed model.

3.2 Classification Phase

In this phase, the new subset of features ($f_{newi}, i = 1, 2, 3, 4$) that were selected in the first phase are used. This phase consists of two main steps; training and testing. Each new subset of features is divided into training and testing sets. In the training step as shown in Fig. 1, each training set from the new subsets of features is used to train one Bagging model. Thus, four different models are built using the new four feature subsets that were selected. In the testing phase, each testing subset is matched or classified with their corresponding training model that is built using the corresponding feature subset. The outputs of the four classification processes are combined using majority voting to get the final decision.

4 Experimental Results

4.1 Dataset

In this experiment, a keystroke dynamics database for touch mobile phones was used. The database is collected from 51 individuals. Each individual is requested to type a password "rhu.university" 15 times during three different sessions [8].

The data in each session are collected separately and the average time between each session was five days. The database consists of 985 samples collected from males and females with different ages as shown in Fig. 2. The distribution of the database is shown in Table 1 and Fig. 2.

Table 1. Distribution of the keystroke dynamics database.

Gender	Age group						Total	
	7–18		19–29		30–65			
	Users	Samples	Users	Samples	Users	Samples	Users	Samples
Male	7	137	16	281	3	59	26	477
Female	4	75	14	263	7	140	25	478
Total	11	212	30	544	10	199	51	955

In the database there are four information were extracted from each user includes the timing between a key pressure and a key release, timing between a key release and a key pressure, the timing between two key pressures, and timing between two key releases [8].

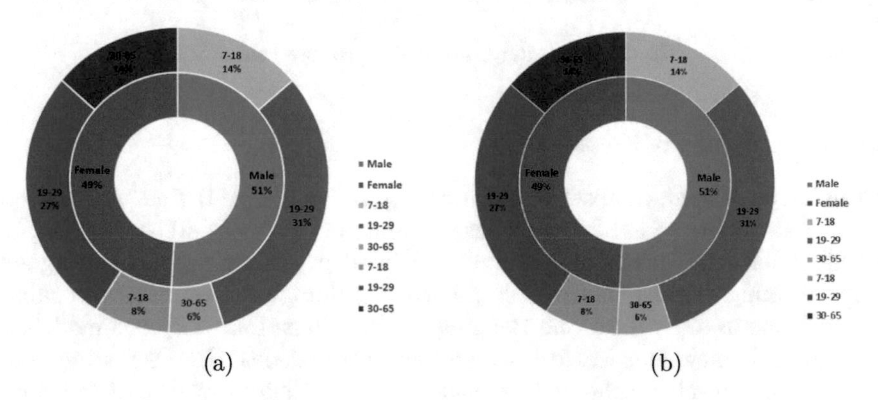

Fig. 2. A distribution of samples and users of dynamic keystrokes database, (a) distribution of the users over gender and age groups, (b) distribution of the samples over gender and age group.

In this research, two different experiments are used to test our proposed approach. The goal of the first experiment was to select the most important features using GA. While in the second experiment, the Bagging classifier was used to match the testing and training patterns.

4.2 Feature Selection Experiment

This experiment is conducted to select the most powerful features. As mentioned above, the dataset that was used in this research is consists of four subsets of features. In this experiment, GA is used to select the best subset of features from each subset. Each subset from the original features was used as an input to the GA algorithm, and the GA selects the best subset of features. The GA has many parameters that affect the performance of GA. In this experiment, the population size of GA is 50, crossover rate = 0.6, mutation rate = 0.01, and the number of iterations = 45. The results of this scenario are summarized in Table 2 and Fig. 3.

Table 2. The selected features (F_{newi}) using GA algorithm

Feature set	Selected features
$\mathbf{F}_{new1}$	{2, 3, 4, 5, 6, 7, 8, 10, 11, 13}
$\mathbf{F}_{new2}$	{2, 4, 6, 8, 10, 11, 13, 14}
$\mathbf{F}_{new3}$	{1, 3, 4, 5, 6, 7, 8, 9, 10, 13}
$\mathbf{F}_{new4}$	{1, 3, 6, 7, 8, 9, 10, 11, 12, 13}

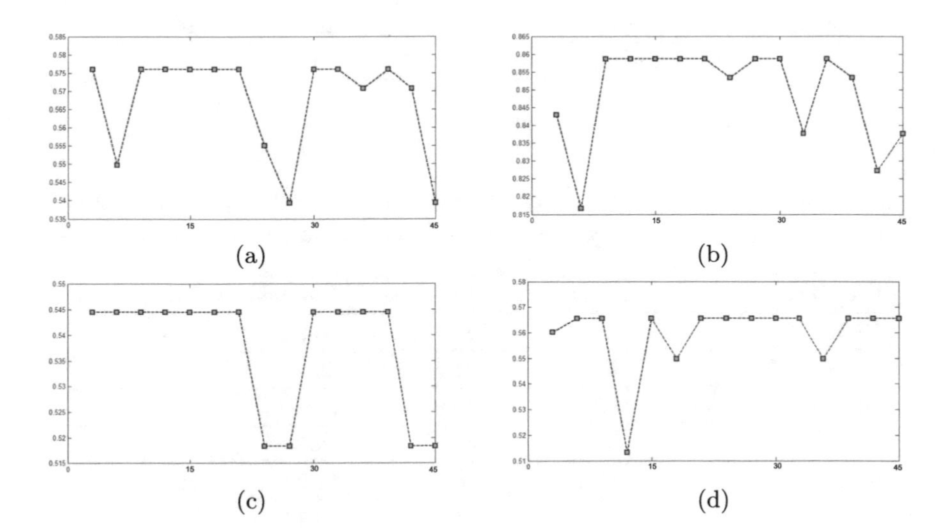

Fig. 3. Fitness values vs. generation number, (a) first feature subset, (b) second feature subset, (c) third feature subset, (d) fourth feature subset. chips.

4.3 Classification Experiment

In this experiment, the second phase of the proposed model shown in Fig. 1 is investigated. This experiment is consists of two steps. In the first step, four different Bagging classifiers are trained. Each Bagging classifier has five weak or individual decision tree classifiers. Each new features subset was divided into training and testing subsets as follows, ($f_{newi} = TR_{newi} + TS_{newi}$). Each subset of the training features ($TR_{newi}, i = 1, 2, 3, 4$) represents the input to one Bagging classifier. Each Bagging classifier generates a model, which will be used in the testing step to classify the unknown or test samples. The results of the training step are summarized in Table 3. In the second step, the testing samples were classified using the Bagging models that were calculated in the training step. Each testing subset was matched separately and the outputs of all Bagging classifiers were fused into one decision. The results of the testing step are summarized in Table 4.

Table 3. Accuracy (in %) of the individual classifiers of the Bagging ensemble, the accuracy of the ensemble, and the fusion of the four ensembles in the training step.

Classifier	Ensemble 1	Ensemble 2	Ensemble 3	Ensemble 4
Classifier 1	97.3	78.3	92.9	91.5
Classifier 2	92.4	76.3	92.1	90.8
Classifier 3	90.7	77	92.8	91.4
Classifier 4	92.1	78.4	91.4	94.7
Classifier 5	92.6	74.9	92.9	93.1
Bagging accuracy	100	99.2	100	100
Fusion of all ensembles	100			

Table 4. Accuracy (in %) of the individual classifiers of the Bagging ensemble, the accuracy of the ensemble, and the fusion of the four ensembles in the testing step.

Classifier	Ensemble 1	Ensemble 2	Ensemble 3	Ensemble 4
Classifier 1	58.7	37.3	55.9	56.4
Classifier 2	57.7	41.5	59.5	57.2
Classifier 3	53.5	39.9	60.6	59.2
Classifier 4	53	40.2	59.5	56.9
Classifier 5	52.2	40.4	57.2	55.6
Bagging accuracy	63.4	41	70.7	68.6
Fusion of all ensembles	83.8			

5 Discussions

From Table 2 many remarks can be seen. The number of selected features in all subsets was ten, except the second feature subset consists of eight features, which reflects that the second feature subset has redundant features more than the other feature subsets. From Fig. 3 we note that the fitness function fluctuated up and down and in the first and third subsets of features the fitness function value reached its minimum. In our experiment, in all runs (i.e. all feature subsets) the best solution is used to select the new subset of features.

Table 3 shows the experimental results of the training step in the classification phase and it has many findings. Firstly, the accuracy results of the individual classifiers in all subset of features are more than 90%, except in the second subset achieved accuracy ranged from 74.9% to 78.4%, which reflects that the second subset of features is not robust enough again identifying persons. Secondly, the first subset of features achieved the best accuracies. Thirdly, the accuracy of each Bagging, which represents the fusion of all individual classifiers in the Bagging classifier achieved an excellent accuracy ranged from 99.2% to 100%, which reflects that the individual classifiers are independent; and hence, the fusion rule (majority voting) achieved good results. The fifth remark is that the fusion of all Bagging achieved accuracy equal to 100%.

Table 4 presents the results of the testing step in the classification phase. As shown in the table, the third subset of the selected features achieved the best results while the second subset achieved the minimum accuracy. The accuracy of the Bagging classifier achieved accuracy better than all its individual classifiers, and it was ranged from 41% to 70.7%, while the fusion of all Bagging classifiers was 83.8%.

6 Conclusions and Future Work

In this study, a proposed approach is used to identify persons based on keystroke dynamics of touch mobile phone. A dataset of keystroke dynamics, which consists of four sets of features, is used. The proposed approach consists of two phases, namely, feature selection and classification. In the first phase, GA is used to select the best subset of features from each set of features. While, in the second phase, the features that were selected in the first phase are used to identify persons using Bagging classifier. The unknown person is identified based on each feature subset separately using an ensemble of Bagging classifiers. The outputs of all Bagging classifiers are fused to determine the final decision. The results of the proposed model conclude that the specific combination of the features, i.e., timing between a key pressure and a key release, timing between a key release and a key pressure, the timing between two key pressures, and timing between two key releases achieved the best accuracy while the second subset achieved the worst accuracy. Moreover, the fusion of all classifiers of all ensembles will improve the accuracy and achieved results better than the individual classifiers and individual ensembles. Future work includes, using different feature selection methods and also investigate other ensemble classifiers to increase the performance of the approach.

References

1. Smartphone users worldwide 2014–2020. www.statista.com. Accessed 11 Oct 2017
2. Baljit Singh Saini, N.K., Bhatia, K.S.: Keystroke dynamics for mobile phones: a survey. Indian J. Sci. Technol. **9**(6), 1–8 (2016)
3. Ali, M.L., Monaco, J.V., Tappert, C.C., Qiu, M.: Keystroke biometric systems for user authentication. J. Signal Process. Syst. **86**(2–3), 175–190 (2017)
4. Gafurov, D.: A survey of biometric gait recognition: approaches, security and challenges. In: Annual Norwegian Computer Science Conference, pp. 19–21 (2007)
5. Blanco-Gonzalo, R., Miguel-Hurtado, O., Mendaza-Ormaza, A., Sanchez-Reillo, R.: Handwritten signature recognition in mobile scenarios: performance evaluation. In: 2012 IEEE International Carnahan Conference on Security Technology (ICCST), pp. 174–179. IEEE (2012)
6. Zheng, N., Bai, K., Huang, H., Wang, H.: You are how you touch: user verification on smartphones via tapping behaviors. In: 2014 IEEE 22nd International Conference on Network Protocols (ICNP), pp. 221–232. IEEE (2014)
7. Buza, K.: Person identification based on keystroke dynamics: demo and open challenge. In: CAiSE Forum, 28th International Conference on Advanced Information Systems Engineering (2016)
8. El-Abed, M., Dafer, M., Khayat, R.E.: RHU keystroke: a mobile-based benchmark for keystroke dynamics systems. In: International Carnahan Conference on Security Technology (ICCST 2014), pp. 1–4. IEEE (2014)
9. Diaz, J.M., Pinon, R.C., Solano, G.: Lung cancer classification using genetic algorithm to optimize prediction models. In: The 5th International Conference on Information, Intelligence, Systems and Applications (IISA 2014), pp. 1–6. IEEE (2014)
10. Elhoseny, M., Tharwat, A., Farouk, A., Hassanien, A.E.: K-coverage model based on genetic algorithm to extend WSN lifetime. IEEE Sens. Lett. **1**(4), 1–4 (2017)
11. Li, S., Wu, H., Wan, D., Zhu, J.: An effective feature selection method for hyperspectral image classification based on genetic algorithm and support vector machine. Knowl. Based Syst. **24**(1), 40–48 (2011)
12. Elhoseny, M., Tharwat, A., Yuan, X., Hassanien, A.E.: Optimizing K-coverage of mobile WSNs. Expert Syst. Appl. **92**, 142–153 (2018)
13. Gaber, T., Tharwat, A., Hassanien, A.E., Snasel, V.: Biometric cattle identification approach based on webers local descriptor and adaboost classifier. Comput. Electron. Agric. **122**, 55–66 (2016)
14. Tharwat, A., Gaber, T., Ibrahim, A., Hassanien, A.E.: Linear discriminant analysis: a detailed tutorial. AI Commun. **30**(2), 169–190 (2017)
15. Breiman, L.: Bagging predictors. J. Mach. Learn. **24**(2), 123–140 (1996)
16. Tharwat, A.: Linear vs. quadratic discriminant analysis classifier: a tutorial. Int. J. Appl. Pattern Recognit. **3**(2), 145–180 (2016)
17. Tharwat, A.: Principal component analysis-a tutorial. Int. J. Appl. Pattern Recognit. **3**(3), 197–240 (2016)
18. Wojciech, S., Jack, S.: A note on genetic algorithms for large-scale feature selection. Pattern Recognit. Lett. **10**(5), 335–347 (1989)

Data Mining, Visualization and E-learning

Factors Affecting Students' Acceptance of E-Learning System in Higher Education Using UTAUT and Structural Equation Modeling Approaches

Said A. Salloum[1,2(✉)] and Khaled Shaalan[1]

[1] Faculty of Engineering and IT, The British University in Dubai, Dubai, UAE
ssalloum@uof.ac.ae, khaled.shaalan@buid.ac.ae
[2] Faculty of Engineering and IT, University of Fujairah, Fujairah, UAE

Abstract. One of the important revolutionary tools widely used and globally implemented by educational institutes and universities is none other than the electronic learning (E-learning system). The aim of this system is to deliver education. As a result, the users of an E-learning system can have enormous benefits. The developed countries are successfully implementing the E-learning system besides realization of its massive benefits. On the contrary, the developing countries have failed, either fully or partially, to implement the E-learning system. A main reason is that those countries do not have an absolute utilization and considered below the satisfactory level. For instance, in United Arab Emirate, one of the developing countries, a growing number of universities are investing for many years in E-learning systems in order to enhance the quality of student education. However, their utilization among students has not fulfilled the satisfactory level. Imagine the evidence that the behavior of user is mainly required for the successful use of these web-based tools, investigating the unified theory of acceptance and use of technology (UTAUT) of E-learning system used in practical education is the basic aim of this research study. A survey on E-learning usage among 280 students was conducted and by using the given responses, the assumptions of the research resulting from this model have been practically validated. The partial least square method was employed to examine these responses. In predicting a student's intention to use E-learning, the UTAUT model was strongly corroborated by the obtained results. In addition, the findings reveal that all important factors of behavioral intention to use E-learning system were reportedly found as the social influence, performance expectancy and facilitating conditions of learning. Remarkably, a significant impact on student intention towards E-learning system was not suggested by the effort expectancy. Consequently, The three key factors leading to successful E-Learning system are thought to be the good perception and encouraging university policy.

Keywords: E-learning
Unified theory of acceptance and use of technology (UTAUT)
System quality · Student's intention to use E-learning · UAE

A. E. Hassanien et al. (Eds.): AISI 2018, AISC 845, pp. 469–480, 2019.
https://doi.org/10.1007/978-3-319-99010-1_43

1 Introduction

Information and communication technologies bring a lot of opportunities to the higher educational settings [1–6]. One of such technologies is the E-learning. The students' acceptance of E-learning was emphasized by this study as an effective tool. Understanding of end-user acceptance process is basically required to effectively implement the E-learning. Hence, exploring the acceptance of E-learning approach among students within United Arab Emirates (UAE) universities, as a good example representing developing countries, was focused by this study. This research aims to deliver certain factors in line with the existing theories, which are to be considered, when the organization is recommending an E-learning activity to E-learners among the universities of Gulf region. In addition, the basic framework of this study is the unified theory of acceptance and use of technology (UTAUT) which was refined with the incorporation of some factors. The research goals can be successfully achieved by employing this model. In the higher educational institutions across the globe, the role of E-learning is effectively addressed by several research studies [7–18]. Prior to applying E-learning, the key factor that should be considered is the students' attitude. Regarding the literature and currently available researches, the Arab states of the Gulf universities have not considered those attitudes. In the higher educational environments within UAE context, students' attitudes towards the use of E-learning would be carefully examined. Therefore, exploring the factors in the acceptance of e-learning based on the unified theory of acceptance and use of technology (UTAUT) was the major goal of this study. This theory/model was proposed by [19]. In this study, the two universities covered, which have already implemented E-learning systems, were chosen to select the participants. Given below is the sequence of this paper. The UTAUT would be examined in the following section and we shall demonstrate the reason to accept it as the theoretical framework. Next the descriptions of the research model and methods would be described. Afterwards, the outcomes of the data analysis and hypotheses testing are produced. To conclude, the authors would take into account the implication of results and limitations.

2 Research Model and Hypotheses

As far as the E-learning perspective is concerned, learning activities are carried out by implementing the E-learning systems, and hence these systems have developed into an IT phenomenon, which impart itself to the UTAUT model. According to [19], practical suggestions including the UTAUT implementation have clearly described the IT behavior, and others are inspired to continue validating and testing their model. The implementation challenges of a new E-learning context can be addressed with the realization of the UTAUT model. Therefore, to assess the users' acceptance of E-learning, [19] UTAUT was adopted as a primary theoretical framework. Figure 1 is illustrative of the research model tested in this study. With reference to this model, performance expectancy (PE), effort expectancy (EE), social influence (SI), facilitating conditions (FC), and behavioral intention (BI) were hypothesized to be the factors of use behavioral (UB) to use E-learning. The earlier literature is in favor of the proposed constructs and hypotheses. The theory base is explained in the following section and the hypotheses are ultimately derived.

2.1 Performance Expectancy

As per [19], performance expectancy is defined as the degree to which an individual believes that s/he can have benefits realization in his/her job performance with the help of an information system. By adjusting the performance expectancy with an E-learning context, E-learning would be of great assistance for E-learners, because learners are thus enabled to instantly accomplish the learning activities, or this particular learning uplifts their education skills and performance. Consistent with the UTAUT and foregoing literature [7–16], a significant positive association was discovered between the two constructs. Therefore, the given below hypothesis is presented:

H1: Performance expectancy (**PE**) has a positive effect on behavioral intention to use E-learning system.

2.2 Effort Expectancy

The effort expectancy is defined as the degree of ease related to the information systems and their usage [19]. According to earlier research studies, constructs about effort expectancy will be the contributing factors of individuals' objectives [7–13, 15, 16]. Since E-learning is in its early stages, the effort expectancy is believed to be a vital element of behavioral intention to use E-learning. Therefore, along the lines of the UTAUT; it is anticipated that individual acceptance of E-learning will depend upon its ease and user friendliness, and the influence the effort expectancy has on behavioral intention. Accordingly, the following hypotheses were tested:

H2: Effort expectancy (**EE**) has a positive effect on behavioral intention to use E-learning system.

2.3 Social Influence

According to [19], social influence is defined as the degree to which a person realizes how the others believe that a new information system should be used by him or her. As per the previous studies, an individual's intention for using new technology is created through the social influence [7–9, 11–16, 20] Based on earlier studies and the UTAUT (e.g., [19, 21]), social influence is found to be a significant contributing factor of behavioral intention to use E-learning, and it is also learnt that how the behavioral intention is affected from the social influence. Consequently, we tested the following hypotheses:

H3: Social influence (**SI**) has a positive effect on behavioral intention to use E-learning system.

2.4 Facilitating Conditions

The environmental factors or behavior physical setting, which persuade a user's desire to carry out the tasks, are known as the "Facilitating conditions". The UTAUT has yielded in this construct. The inventor of UTAUT model discovered that FC is very important determinant, by which the use of information systems is influenced [19]. The current studies have corroborated the findings of [8, 9, 11–13, 15, 16, 22] In the existing research, FCs defines the degree to which individuals are of view that technical and organizational infrastructures endure to strengthen them. In this way an influence is exerted on student teachers' aspiration to use E-learning systems. The individual support, training, materials accessible to improve skills and knowledge and E-learning system accessibility are amongst the supporting facilities. From the above said discussion, the author has suggested the following hypotheses.

H4: Facilitating conditions (**FC**) will significantly and positively influence student to use E-learning system.

2.5 Behavioral Intention to Use

Determining the desire of a student in accepting E-learning is the main goal of BI items [23]. Moreover, [24] were of view that the intent of the learners in employing E-learning systems besides incorporating persistent use from the present to the future is referred to as the BI (Behavioral Intention). A number of authors (e.g. [9, 13–16, 25–31] have demonstrated that actual system use of E-Systems especially the E-learning ones is directly affected by the behavioral intention. A significant positive correlation was identified between the two constructs. Therefore, the given hypothesis is submitted:

H5: Intention to use the e-learning system (**BI**) has a positive effect on Actual use (**AU**) of the E-learning system.

The following research model is authenticated from the above hypotheses, which was developed in line with the UTAUT model for E-learning acceptance among students (Fig. 1).

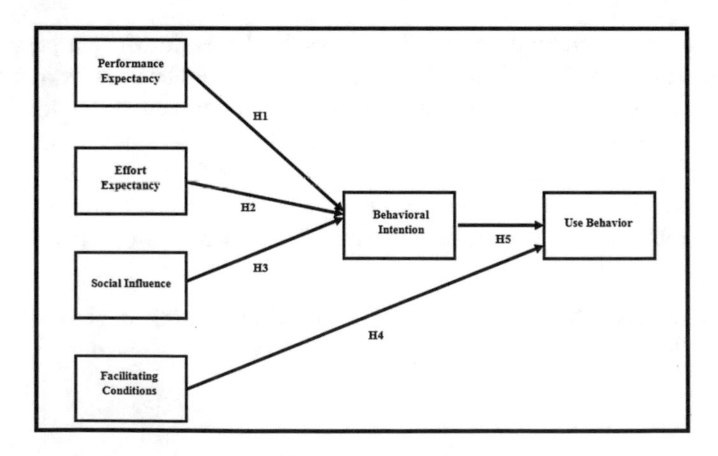

Fig. 1. Research model

3 Research Methodology

The data has been selected from two well-known educational institutions within UAE who had successfully implemented the E-learning systems. Collection of samples was major purpose in this regard. This online survey witnessed the contribution of a total of 300 respondents. The web-based E-learning systems were being used by these two universities. Moreover, the two different E-learning platform providers have developed these systems. Nearly four years ago, the academic institutes have implemented both of these E-learning platforms and this E-learning system was being used by the students of these institutes on a daily basis. The detailed sketch of the collected data is found in Table 1. Among all the responses gathered, 20 unfinished responses questionnaires were found, which were discarded. Rest, there were 280 complete questionnaires indicating a response rate of 93.3%. Generally speaking, the researchers only considered 280 responses with valid answers and the same were transformed into a sample size as recommended by [32]. Nearly 169 respondents make up the estimated sampling size for a population of 300. Afterwards, the conceptual model was used to analyze these responses. The acceptable sample size comes from structural equation modeling. Thus, a sample size of 280 was satisfying the situation to test the hypotheses in our study [33]. It is worth mentioning that the current theories were supporting the said hypotheses but the matter was observed in the E-learning background.

Table 1. Participants details

University	No. of students
The British University in Dubai (BUiD)	153
University of Fujairah	127
Total	280

3.1 Students' Personal Information/Demographic Data

Personal/demographic Information has been assessed and the results. The female student percentage was maintained at 54% and the male percentage was 46%. fifty-six percent of the respondents maintain a student age between the range of 18 and 29. There are 44% of the respondents who are more than 30 years age. Thirty-four percent of the students were from Business Administration major while students in Engineering and Information Technology, General Education, Humanities and Social Sciences and Mass Communication and Public Relations were 32%, 13%, 12%, and 9% respectively. Thirty-three percent of the respondents lived in Al Fujairah Emirate, and 28% of the respondents lived in Dubai. While Abu Dhabi, Sharjah, Ajman, Umm al-Quwain and Ras Al Khaimah have only 39% of the total respondents. Thirty-nine percent of the respondents had a bachelor degree, 40% had a master degree, and 21% had a doctoral degree.

4 Findings and Discussion

4.1 Measurement Model Analysis

A software has been developed by [34], known as the Smart PLS, which is commonly used for Partial Least Squares-Structural Equation Modeling (PLS-SEM). Within the current research, the measurement and structural models have been analyzed using the PLS-SEM [35]. The measurement model (outer model) includes the relationship which exists amongst the indicators. On the other hand, the structural model is the relationship present between latent constructs. The SEM-PLS has been applied with the highest probability method for the measurement of the current model [36]. There were various measurements carried out which include the Factor Loadings, Average Variance Extracted and Composite Reliability. These measurements help with the convergent validity and reliability. Factor loadings have been used to indicate each questionnaire variable's weight and correlation value. However, for the factors' dimensionality, the representation is made using the bigger load value. The Composite Reliability (CR) is applied to measure the reliability. There is a similar objective for CR since it brings forward a precise value by including factor loadings within the constructed formula. The average quantity of variance present within a specific variable, that states the latent construct, is known as the Average Variance Extracted (AVE). If the discriminate validity is higher than one factor, then the convergence of each factor is assessed through the AVE. After considering Table 2, it can be stated that the reliability and convergent validity condition is surpassed by the questionnaire reliability and convergent validity as part of the experiment consequence. Table 2 also indicates the questionnaire's validity and reliability after conducting an assessment upon each factor through the presentation of a variable attain from the questionnaire.

The indicators applied for the estimation of the convergent validity relative amount are factor loadings, variance extracted and reliability (consisting of Cronbach's Alpha and composite reliability). For each construct, the reliability coefficient and composite reliability (CR) [37] are higher than 0.7. This shows that the various construct measurements maintain internal consistency [38]. Table 2 indicates that 0.7 is the acceptable value that has been exceeded by the Cronbach's alpha scores [39, 40]. 0.704 to 0.850 is the construct range for the average variance extracted (AVE) values. Furthermore, the average variance extracted (AVE) values, which lie between the 0.613 to 0.769 range, are able to satisfy the explaining criteria of 50% of variance extracted from within an item set present within the latent construct [41]. Therefore, convergent validity can be attained using the construct evaluation scales.

Table 3 clearly indicates that the discriminant validity requirements have been satisfied as all AVE values are higher than the squared correlation present amongst the measurement model constructs [38, 42]. If the AVE value is above 0.5, the construct includes a measurement variance of minimum 50%. The AVE analysis is present in Table 3. The AVE scare square root is presented in the tables bold diagonal elements. The table also indicates that the AVE values square root is present within the 0.783 to 0.877 range and this is higher that than the 0.5 recommended value. As compared to other correlations in the construct, the AVE is higher within each construct and such an aspect indicates that the constructs would have a greater variance maintaining their own measures as compared to the model constructs that increase the discriminate validity.

Table 2. Convergent validity results which assures acceptable values (factor loading, Cronbach's Alpha, composite reliability ≥ 0.70 & AVE > 0.5).

Constructs	Items	Factor loading	Cronbach's alpha	CR	AVE
Behavioral intention	BEH_INT_1	0.789	0.704	0.826	0.613
	BEH_INT_2	0.763			
	BEH_INT_3	0.797			
Effort expectancy	EFF_EXP_1	0.820	0.788	0.875	0.700
	EFF_EXP_2	0.856			
	EFF_EXP_3	0.835			
Facilitating conditions	FAC_CON_1	0.869	0.757	0.861	0.674
	FAC_CON_2	0.825			
	FAC_CON_3	0.766			
Performance expectancy	PER_EXP_1	0.894	0.850	0.909	0.769
	PER_EXP_2	0.894			
	PER_EXP_3	0.841			
Social influence	SOC_INF_1	0.818	0.743	0.854	0.661
	SOC_INF_2	0.851			
	SOC_INF_3	0.768			
Use behavior	USE_BEH_1	0.819	0.795	0.880	0.709
	USE_BEH_2	0.865			
	USE_BEH_3	0.841			

Table 3. Fornell-Larcker scale

	Behavioral intention	Effort expectancy	Facilitating conditions	Performance expectancy	Social influence	Use behavior
Behavioral intention	**0.783**					
Effort expectancy	0.385	**0.837**				
Facilitating conditions	0.498	0.473	**0.821**			
Performance expectancy	0.397	0.685	0.514	**0.877**		
Social influence	0.440	0.570	0.606	0.565	**0.813**	
Use behavior	0.447	0.472	0.586	0.612	0.559	**0.842**

The coefficient of determination (R^2 value) measure is applied to analyze the structural model [43]. The model of predictive accuracy can be decided upon through the use of the coefficient. It is managed as the squared correlation present amongst the specific endogenous construct's actual and predicted values [43]. The coefficient is used to connote the exogenous latent variables that maintain a joined influence upon the endogenous latent variable. Amongst the actual and predicted values of the variables, the coefficient maintains a squared correlation. Therefore, within the endogenous constructs it maintains a variance degree which is defended using the exogenous construct that is associated with it. If the value is above 0.67, it is considered high even though direct 0.33 to 0.67 qualities within the scope are available. The scope qualities considered weak lie in the range of 0.19 to 0.33 [43]. It is inadmissible if the estimation is lower than 0.19. According to Table 4, the model includes a Moderate predictive power with a support of approximately 43 and 34% of the variance in the Behavioral Intention and Behavioral Intention respectively.

Table 4. R^2 of the endogenous latent variables.

Constructs	R^2	Results
Behavioral intention	0.432	Moderate
Use behavior	0.338	Moderate

4.2 Structural Model Analysis

The proposed hypotheses have been tested using a structural equation model which is embedded in the SEM-PLS software. There exists a maximum likelihood estimation that indicates the association between the theoretical constructs present within the structural model. Table 5 and Fig. 2 summarize the results. Out of 5 hypotheses, there are 4 hypotheses which are significant as observed in Table 5 and Fig. 2. Based on the data analysis hypotheses H1, H3, H4, and H5 were supported by the empirical data, while H2 was rejected. Performance Expectancy (PE), Social Influence (SI) has significant effects on Behavioral Intention (BI) ($\beta = 0.258$, $P < 0.05$), ($\beta = 0.286$, $P < 0.01$) respectively, but Effort Expectancy (EE) has insignificant effects on Behavioral Intention (BI) ($\beta = 0.114$, $P = 0.164$), hence, H4 and H5 are supported, but H2 is rejected. Behavioral Intention (BI), Facilitating Conditions (FC) has also significant effects on Use Behavior (AU) ($\beta = 0.206$, $P < 0.01$), ($\beta = 0.483$, $P < 0.001$) respectively, hence, H1 and H3 are supported.

Table 5. Results of structural model-research hypotheses significant at p** $\leq$ 0.01, p* = 0.05)

H	Relationship	Path	*t*-value	*p*-value	Direction	Decision
H1	Behavioral intention → use behavior	0.206	3.265	0.001	Positive	Supported**
H2	Effort expectancy → behavioral intention	0.114	1.395	0.164	Positive	Not supported
H3	Facilitating conditions → use behavior	0.483	7.356	0.000	Positive	Supported**
H4	Performance expectancy → behavioral intention	0.258	2.872	0.042	Positive	Supported*
H5	Social influence → behavioral intention	0.286	3.402	0.001	Positive	Supported**

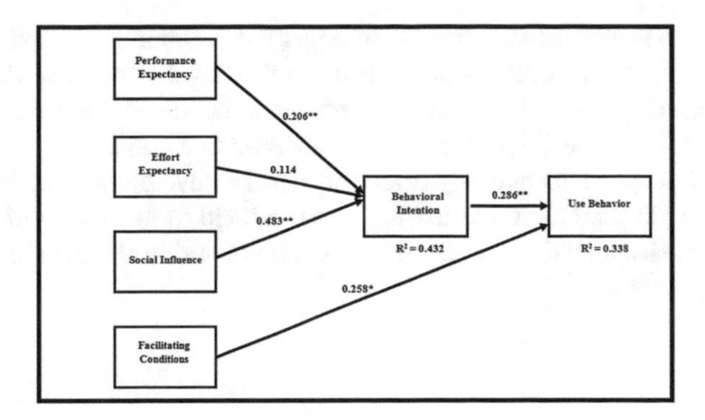

Fig. 2. Path coefficient results (significant at p** $\leq$ 0.01, p* < 0.05)

5 Conclusion and Future Works

5.1 Study Contributions and Discussion

The objective of the research is to extract factors which affect the perception regarding electronic learning (E-learning) keeping in mind the perspective of the UAE students. The data analysis results have been taken into account to state the proposed research model relevance and the hypothesis being used for the analysis of the behavioral intention to adapt E-learning system. The research hypotheses have been analyzed using the Structure equation modeling (PLS-SEM). In Fig. 2, the structural model has been presented which is assessed through the analysis of the structural paths, t-statistics and variance explained (R-squared value). The data analysis findings have been stated in Table 5. The five hypotheses mentioned earlier have been tested using the PLS technique. Evaluation was carried out upon the path significance of every hypothesized relationship that is part of the research model and the variance explained (R^2) by every path. Four hypotheses have been supported out of the proposed hypotheses. The hypotheses attained from UTAUT model (H1, H3, H4, and H5) have been supported. The research study indicates that Facilitating Conditions, Performance Expectancy, and Social Influence would help enhance the behavioral intention to adapt E-learning system, was also found in past studies [7–16]. It has also been observed that a positive influence was subjected upon use behavioral intention including the behavioral and facilitating conditions, thus supported H1 and H3. Positive influence was also subjected upon the Performance Expectancy and Social Influence on Behavioral Intention, supporting the H4 and H5. With the help of this outcome, it is possible to indicate the E-learning significance in terms of competency and high confidence levels when making use of online learning platforms.

5.2 Limitations and Future Directions

The present research suffers from limitation since other regions and their research results have not been considered. This information would have allowed for significant input and the points of the research could have been strengthened. Furthermore, it could help test the regional differences amongst the technology adoption perceptions and a robust model may be developed related to the developing nation's deterministic factors on technology adoption. The research includes developing models or frameworks which indicate the e-readiness. A comprehensive context is available for recognizing the concept of technology adoption within developing countries like UAE. The future strategies and their context is formed related to the ability of technology to bring forward positive business and social alterations within the developing countries.

References

1. Al-Emran, M., Mezhuyev, V., Kamaludin, A., AlSinani, M.: Development of M-learning application based on knowledge management processes. In: 2018 7th International Conference on Software and Computer Applications (ICSCA 2018) (2018)
2. Salloum, S.A., Al-Emran, M., Monem, A.A., Shaalan, K.: Using text mining techniques for extracting information from research articles. In: Studies in Computational Intelligence, vol. 740. Springer, Berlin (2018)
3. Al-Emran, M., Shaalan, K.: Academics' awareness towards mobile learning in Oman. Int. J. Comput. Digit. Syst. **6**(1), 45–50 (2017)
4. Al-Emran, M., Malik, S.I.: The impact of google apps at work: higher educational perspective. Int. J. Interact. Mob. Technol. **10**(4), 85–88 (2016)
5. Al-Emran, M., Mezhuyev, V., Kamaludin, A.: Technology acceptance model in M-learning context: a systematic review. Comput. Educ. **125**, 389–412 (2018)
6. Al-Emran, M., Shaalan, K.: Attitudes towards the use of mobile learning: a case study from the Gulf region. Int. J. Interact. Mob. Technol. **9**(3), 75–78 (2015)
7. Wang, Y.S., Wu, M.C., Wang, H.Y.: Investigating the determinants and age and gender differences in the acceptance of mobile learning. Br. J. Educ. Technol. **40**(1), 92–118 (2009)
8. Patil, S., Razdan, P.N., Chancellor, V.: Creating awareness acceptance and application of e-learning in higher education institutions: using technology acceptance model (TAM). Int. J. Multidiscip. Consort. **2**(4) (2015)
9. Tan, P.J.B.: Applying the UTAUT to understand factors affecting the use of English e-learning websites in Taiwan. Sage Open **3**(4), 2158244013503837 (2013)
10. Nassuora, A.B.: Students acceptance of mobile learning for higher education in Saudi Arabia. Am. Acad. Sch. Res. J. **4**(2), 1 (2012)
11. Yoo, S.J., Han, S., Huang, W.: The roles of intrinsic motivators and extrinsic motivators in promoting e-learning in the workplace: a case from South Korea. Comput. Hum. Behav. **28** (3), 942–950 (2012)
12. Alrawashdeh, T.A., Muhairat, M.I., Alqatawnah, S.M.: Factors affecting acceptance of web-based training system: using extended UTAUT and structural equation modeling. arXiv Prepr. arXiv:1205.1904 (2012)
13. Abu-Al-Aish, A., Love, S.: Factors influencing students' acceptance of m-learning: an investigation in higher education. Int. Rev. Res. Open Distrib. Learn. **14**(5), 82–107 (2013)
14. Pardamean, B., Susanto, M.: Assessing user acceptance toward blog technology using the UTAUT model. Int. J. Math. Comput. Simul. **1**(6), 203–212 (2012)
15. Raman, A., Don, Y.: Preservice teachers' acceptance of learning management software: an application of the UTAUT2 model. Int. Educ. Stud. **6**(7), 157–164 (2013)
16. Wong, W.-T., Huang, N.-T.N.: The effects of e-learning system service quality and users' acceptance on organizational learning. Int. J. Bus. Inf. **6**(2), 205–225 (2015)

17. Al-Maroof, R.A.S., Al-Emran, M.: Students acceptance of Google classroom: an exploratory study using PLS-SEM approach. Int. J. Emerg. Technol. Learn. **13**(06), 112–123 (2018)
18. Al-Emran, M., Salloum, S.A.: Students' attitudes towards the use of mobile technologies in e-evaluation. Int. J. Interact. Mob. Technol. **11**(5), 195–202 (2017)
19. Venkatesh, V., Morris, M., Davis, G., Davis, F.: User acceptance of information technology: toward a unified view. MIS Q. **27**(3), 425–478 (2003)
20. Alshurideh, M.: Customer Service Retention–A Behavioural Perspective of the UK Mobile Market. Durham University, Durham (2010)
21. Venkatesh, V.: Determinants of perceived ease of use: Integrating control, intrinsic motivation, and emotion into the technology acceptance model. Inf. Syst. Res. **11**(4), 342–365 (2000)
22. Alshurideh, M., Bataineh, A., Alkurdi, B., Alasmr, N.: Factors affect mobile phone brand choices-studying the case of Jordan universities students. Int. Bus. Res. **8**(3), 141 (2015)
23. Alia, A.: An Investigation of the Application of the Technology Acceptance Model (TAM) to Evaluate Instructors' Perspectives on E-Learning at Kuwait University. Dublin City University, Dublin (2016)
24. Liao, H.-L., Lu, H.-P.: The role of experience and innovation characteristics in the adoption and continued use of e-learning websites. Comput. Educ. **51**(4), 1405–1416 (2008)
25. Mou, J., Shin, D.-H., Cohen, J.: Understanding trust and perceived usefulness in the consumer acceptance of an e-service: a longitudinal investigation. Behav. Inf. Technol. **36** (2), 125–139 (2017)
26. Al-Gahtani, S.S.: Empirical investigation of e-learning acceptance and assimilation: a structural equation model. Appl. Comput. Inform. **12**(1), 27–50 (2016)
27. Mohammadi, H.: Investigating users' perspectives on e-learning: an integration of TAM and IS success model. Comput. Hum. Behav. **45**, 359–374 (2015)
28. Khor, E.T.: An analysis of ODL student perception and adoption behavior using the technology acceptance model. Int. Rev. Res. Open Distrib. Learn. **15**(6), 275–288 (2014)
29. Cheung, R., Vogel, D.: Predicting user acceptance of collaborative technologies: An extension of the technology acceptance model for e-learning. Comput. Educ. **63**, 160–175 (2013)
30. Cheng, Y.: Antecedents and consequences of e-learning acceptance. Inf. Syst. J. **21**(3), 269–299 (2011)
31. Al-dweeri, R.M., Obeidat, Z.M., Al-dwiry, M.A., Alshurideh, M.T., Alhorani, A.M.: The impact of e-service quality and e-loyalty on online shopping: moderating effect of e-satisfaction and e-trust. Int. J. Mark. Stud. **9**(2), 92 (2017)
32. Krejcie, R.V., Morgan, D.W.: Determining sample size for research activities. Educ. Psychol. Meas. **30**(3), 607–610 (1970)
33. Chuan, C.L., Penyelidikan, J.: Sample size estimation using Krejcie and Morgan and Cohen statistical power analysis: a comparison. J. Penyelid. IPBL **7**, 78–86 (2006)
34. Ringle, C.M., Wende, S., Will, A.: SmartPLS 2.0 (Beta). Hamburg (2005). http://www.smartpls.de
35. Chin, W.W.: The partial least squares approach to structural equation modeling. Mod. Methods Bus. Res. **295**(2), 295–336 (1998)
36. Anderson, J.C., Gerbing, D.W.: Structural equation modeling in practice: a review and recommended two-step approach. Psychol. Bull. **103**(3), 411 (1988)
37. Werts, C.E., Linn, R.L., Jöreskog, K.G.: Intraclass reliability estimates: testing structural assumptions. Educ. Psychol. Meas. **34**(1), 25–33 (1974)
38. Hair, J.F., Black, W.C., Babin, B.J., Anderson, R.E., Tatham, R.L.: Multivariate Data Analysis, vol. 5. Prentice hall, Upper Saddle River (1998). no. 3

39. Gefen, D., Straub, D., Boudreau, M.-C.: Structural equation modeling and regression: guidelines for research practice. Commun. Assoc. Inf. Syst. **4**(1), 7 (2000)
40. Nunnally, J.C., Bernstein, I.H.: Psychometric Theory. McGraw-Hill (1978)
41. Falk, R.F., Miller, N.B.: A Primer for Soft Modeling. University of Akron Press, Akron (1992)
42. Fornell, C., Larcker, D.F.: Evaluating structural equation models with unobservable variables and measurement error. J. Mark. Res. **18**(1), 39–50 (1981)
43. Lin, Y.-C., Chen, Y.-C., Yeh, R.C.: Understanding college students' continuing intentions to use multimedia e-learning systems. World Trans. Eng. Technol. Educ. **8**(4), 488–493 (2010)

Adoption of E-Book for University Students

Said A. Salloum[1,2(✉)] and Khaled Shaalan[1]

[1] Faculty of Engineering and IT, The British University in Dubai, Dubai, UAE
ssalloum@uof.ac.ae
[2] Faculty of Engineering and IT, University of Fujairah, Fujairah, UAE
khaled.shaalan@buid.ac.ae

Abstract. This paper employs the Technology Acceptance Model (TAM) to study the adoption of E-book amongst higher-education students in a well-known academic institute in the UAE, where E-book was being implemented. Computer self-efficacy, confirmation, innovativeness, satisfaction, and subjective norm are the five factors that this model embarks on to realize the influence on the university students as a result of the adoption of the E-book. This study was conducted among 350 university students through a survey which has used the quantitative evaluation to gain the optimum advantage from the subjective methods. The hypotheses were analyzed, and the model was assessed with the help of the statistical package for Structural Equation Modeling (SEM). The main findings that can be derived from the existing study are the factors that have positive impact on students' perceived ease of use and perceived usefulness of E-book. They are computer self-efficacy, confirmation, innovativeness, and subjective norm. As a result, it is imperative for legislators and managers of E-book applications to concentrate on the factors that are critical for encouraging learning and enhancing students' efficiency in developing and executing successful E-book applications.

Keywords: E-book · Technology Acceptance Model (TAM)
Computer self-efficacy · Confirmation · Innovativeness · Satisfaction
Subjective norm · Student's intention to use E-learning

1 Introduction

Educational technologies play a key role in facilitating the educational process [1–7]. The E-book (Electronic book) technology was introduced in the 1990s but was suppressed due to the anticipation that it would prove hazardous to the publishing industry [8]. It is an innovative technology which allows a learner to view electronic versions of books on devices like personal digital assistants (PDAs) and tablets PCs, among other portable gadgets [9, 10]. Researchers have weighed E-books in higher education perspectives as well as within the library limitations [9, 11–15]. The prospects that E-books hold in the market is debated vigorously and several robust user researches have been an active part of this debate. This field requires a vast amount of scrutiny with respect to user sentiments regarding E-books, and what they feel that the experience lacks. The E-book experience should be researched around what the user approach is regarding E-book accessibility, the senses that it evokes and the lacking aspects of

A. E. Hassanien et al. (Eds.): AISI 2018, AISC 845, pp. 481–494, 2019.
https://doi.org/10.1007/978-3-319-99010-1_44

E-books. This investigation utilizes a research model which helps in addressing these matters by discussing them under factors like computer self-efficacy, confirmation, innovativeness, satisfaction and subjective norm which will assume the driving forces for determining the user's likelihood of adopting E-books. By understanding the intentions and motivations of E-book users, this hiatus can be breached through this study. This will enlist a set of suggestions that will be of immense assistance to academic researchers in order to apprehend the cognitive aspects assumed by the user, the impact it has on the user intentions and behavioral intention that are influenced through these sentiments. The empirical intensity of the relationships is calculated to assume the amalgamation of uses. The TAM model, with the help of the structural equation modeling approach, is the tool we used to study the adoption of E-book amongst university students. The grounds for student intentions to adopt the modified form of digital books should be derived from the outcomes of these models and they should be useful to the academics. The future researches can benefit from this in applications and guidelines.

2 Literature Review

A thorough literature review is an important step before conducting any research study [16–25]. The student behavioral intention in utilizing E-books is the nucleus of this research. There have been umpteen researches with the aims of explaining and understanding the usage and IT adoption. The most accepted and frugal of these models is the Technology Acceptance Model (TAM) which elaborately discusses the differences between behavioral intentions of users when it comes to IT adoption and its possible applications across a wide spectrum of fields such as information systems, e-payment, mobile technology, web browsers and web retailing [26]. This research falls short on the influential factors of utilizing TAM for the purpose of E-book adoption [11]. The TAM reasons as to why the users have certain attitudes towards IT adoption and mainly discusses the contributing beliefs such as perceived usefulness and ease of use as the motivational factors of usage intentions. Computer self-efficacy and information technology were previously regarded as the essence of the constructs of TAM researches, which is acquired through computer self-efficacy of information technology [27]. Other integral variables in technology acceptance are explained through confirmation [28], innovativeness [29] and subjective norm [30]. The intention to use technology may differ from user to user and the extent of the impact of these factors may differ due to the various phases of adoption [11, 31]. The intentions of the university students in employing E-books for the purpose of analysis are the main objectives of that study. It reflects on the student behavior intention to adopt E-books and the scarcity of relevant material regarding the universal approach towards the subject. TAM will be the main groundwork of this research along with five other variables, including computer self-efficacy, confirmation, innovativeness, satisfaction, and subjective norm. The main objectives of this research are to:

(1) Research the students of a reputed university in the UAE regarding their intentions of adopting E-books.

(2) Recognize the factors that drive the E-books adoption that affect the intentions of the university students.
(3) Develop a model with the help of structural equation modeling (SEM) to understand the university student intentions of adopting E-books.

3 Research Model and Hypotheses

There are supporting E-book applications, and implementing these applications assists the learning process. The IT phenomenon evolves from these applications, divulging with the TAM model. The IT behavior is distinctly explained in many researches which favor the applied approach, including the TAM implementation [9, 11, 12, 15, 32–41]. This has motivated further researchers to validate and test their respective models. With the comprehension of the TAM model, we are able to understand and address the issues that a new E-book context can undergo. This is the reason for adopting TAM model as the chief theoretical foundation for understanding the users' adoption of the E-book [42]. The research model scrutinized in this study is demonstrated in Fig. 1. According to this model, computer self-efficacy (CSE), confirmation (CONF), innovativeness (INNO), satisfaction (SAT) and subjective norm (SUB), perceived ease of use (PEOU), perceived usefulness (PU) are hypothesized to be the factors of behavioral intentions (BI) to adopt E-book. These recommended hypotheses and constructs are supported by previous researches [9, 11, 12, 15, 32–34, 37–41, 43, 44].In the following subsections, hypotheses are derived as a result of discussing the theory's crux.

3.1 Computer Self-efficacy

The measurement of the judgment of one's ability to execute a behavior pattern is described as computer self-efficacy [45]. It has been used as a common parameter to assess behaviors like accepting ICT by mass media users [36]. Self-efficacy predicts technology adoption and use, this observation has been established in recent investigations [32, 37, 46, 47]. The hypotheses below an extract of the significant positive association that has been realized between CSE, PEOU, and PU.

H1a. Computer self-efficacy (CSE) significantly influences E-book students' perceived ease of use (PEOU).

H1b. Computer self-efficacy (CSE) significantly influences E-book students' perceived usefulness (PU).

3.2 Confirmation

The positive association amongst confirmation, perceived usefulness and perceived ease of use are verified in the study by [48, 49]. The positive influence of confirmation, on intention to adopt and use E-book is also evident for TAM with regards to the perceived usefulness and perceived ease of use, e.g. [46, 50, 51]. The following hypotheses have been derived as a result of the above-mentioned researches.

H2a. Confirmation (CONF) significantly influences E-book students' perceived ease of use (PEOU).

H2b. Confirmation (CONF) significantly influences E-book students' perceived usefulness (PU).

3.3 Innovativeness

In [52], the authors argued that innovativeness is "The degree to which an individual makes innovation decisions independently of communicated experience of others". In the literature, innovativeness is a vital contributing factor which impacts an individual's adoption behavior for new technologies [29]. The adoption of new technologies has been observed to take influence from "innovativeness" [53]. All the researches follow a similar consensus regarding the extent of individual innovativeness and the direct influence it holds on technology acceptance behavior [11, 54–57]. We derive the following hypothesis regarding the impact of external variables upon individual innovativeness, which is observed to be significantly positive on perceived usefulness and perceived ease of use.

H3a. Innovativeness (INNO) significantly influences E-book students' perceived ease of use (PEOU).

H3b. Innovativeness (INNO) significantly influences E-book students' perceived usefulness (PU).

3.4 Subjective Norm

In [42], subjective norm is defined as the degree to which a person realizes how others believe that a new information system should be used by him or her. Subjective norm helps in developing a person's intentions for adopting new technologies [26, 32, 37]. The hypotheses of [30] and similar other researches like the TAM model sums up the influence of perceived usefulness and perceived ease of use [42]. Therefore, we derive the following hypotheses:

H4a. Subjective norm (SUB) significantly influences E-book students' perceived ease of use (PEOU).

H4b. Subjective norm (SUB) significantly influences E-book students' perceived usefulness (PU).

3.5 Perceived Usefulness/Perceived Ease of Use

The adoption of new technology is assisted by perceived usefulness and perceived ease of use. These factors have a significant positive effect on behavioral intention to use. The extent of user's adoption to new technology and to realize its potency is also determined by the ease of understanding from the user's standpoint [36, 58]. The significant effect of E-book PEOU on perceived usefulness (PU) was evident in previous researches [12, 14, 37, 59, 60]. These researches also assumed the following hypotheses according to the available literature and demonstrated a significant positive relation amongst PEOU, PU and BI.

H5. Perceived ease of use has a positive effect on perceived usefulness.

H6. Perceived usefulness has a positive effect on behavioral intention to adopt E-books.

H7. Perceived ease of use has a positive effect on behavioral intention to adopt E-books.

3.6 Satisfaction

We can observe that satisfaction and behavioral intention to use e-book have a positive association [48]. The positive influence of satisfaction and behavioral intention to adopt or use E-books is also confirmed in identical researches performed recently, e.g. [15, 40, 46, 50]. Underneath are the acquired hypotheses of these researches:

H8. Satisfaction (SAT) has a positive effect on behavioral intention to adopt E-books.

With the help of these hypotheses, the research model depicted in Fig. 1 is derived, which was generated in compliance with the TAM model for E-book adoption by students.

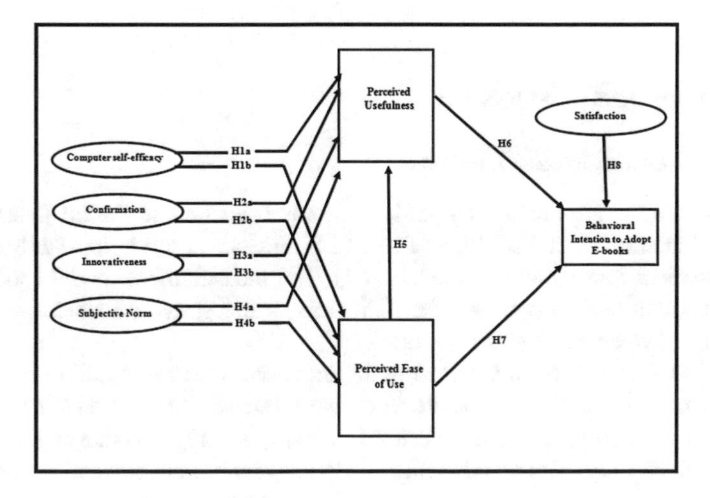

Fig. 1. Research model

4 Research Methodology

4.1 Procedure

The samples were obtained from the well-known academic institute in the UAE, where E-book was being implemented. The online survey was taken by 350 participants. A web-based E-book that was designed by one distinct E-learning platform/system providers were adopted by the university. This E-learning platform had been designed and employed around four years ago. The students of this institute were using these systems almost every day. In all, there were 40 uncompleted surveys, and these questionnaires were of no use. This meant that there were 310 complete questionnaires, indicating a response rate of 88.5%. According to [61], there wcrc a total of 310 responses comprising of valid responses that were contemplated and changed into a sample size. For a population of 300, the approximate sampling size is 169 respondents. The conceptual model was then used to assess these responses. When an analysis is performed with the help of structural equation modeling, it is approved as a sample size. Hence, in this study, a sample size of 310 was a lot higher than the insignificant requirements that were used for examining the hypotheses [62]. An important point to note is that the given hypotheses depended on the existing theories but were employed with reference to the E-book.

4.2 Participants

The survey was conducted on students from the University of Fujairah in the UAE. The size of the sample for this purpose was 310 undergraduates, 197 of which were females whereas 113 were males (63.5 and 36.4%, respectively). The age limit of 75% of these students ranged between 18 and 29. The rest of these participants (25%) were beyond 30 years of age. Of the 310 students, 300 (96.7%) believed to have a high aptitude in computer skills.

5 Findings and Discussion

5.1 Measurement Model Analysis

Smart PLS, a software created by [63], is widely used for the Partial Least Squares-Structural Equation Modeling (PLS-SEM). PLS-SEM was used in this study to analyze the measurement and structural models [64]. The measurement model (outer model) signifies the relationship between the indicators, while structural model refers to the relationship between the latent constructs.

PLS-SEM was employed to measure the proposed model in addition to the highest probability method. Various measurements were carried out to determine convergent validity and reliability, including Factor Loadings, Average Variance Extracted and Composite Reliability. The weight and correlation value of each questionnaire variable is explained as a perceived indicator using factor loadings, while factors' dimensionality can be determined using the larger load value. The Composite Reliability (CR) measure has been presented to measure reliability. CR has a similar objective as it provides an accurate value by using factor loadings in the formula developed. The average amount of variance in the provided variable is known as Average Variance Extracted (AVE) and this explains the latent construct.

AVE may be used to examine the convergence of each factor when the discriminate validity is higher than one factor. Table 1 shows that the experimental results for the questionnaire reliability and convergent validity are higher than the condition for reliability and convergent validity. Factor loadings, variance extracted and reliability (including Cronbach's Alpha and composite reliability) are employed as indicators to determine the corresponding degree of convergent validity. For all constructs, the reliability coefficient and composite reliability (CR) is more than 0.7, which suggests that the internal consistency is between multiple measurements of a construct [49, 65, 66]. Table 1 depicts that Cronbach's alpha scores are more than the acceptable value of 0.7 [67–70], while the range of values of composite reliabilities is between 0.728 to 0.954. Furthermore, the range of values of average variance extracted (AVE) is between 0.600 and 0.785, which fulfills the standard of explaining at least 50% of variance extracted between a group of items [71] based on the latent construct. Consequently, it is believed that the scales for assessing the constructs attain convergent validity.

Table 1. Assessment of the measurement model

Constructs	Items	Factor loading	Cronbach's alpha	CR	AVE
Behavioral intention	BI_1	0.799	0.807	0.728	0.718
	BI_2	0.785			
	BI_3	0.873			
Confirmation	CONF_1	0.725	0.888	0.778	0.600
	CONF_2	0.759			
	CONF_3	0.768			
Computer self-efficacy	SELF_EFF_1	0.765	0.977	0.762	0.773
	SELF_EFF_2	0.777			
	SELF_EFF_3	0.867			
Innovativeness	INNO_1	0.794	0.750	0.789	0.668
	INNO_2	0.893			
	INNO_3	0.899			
Perceived ease of use	PEOU_1	0.818	0.849	0.954	0.655
	PEOU_2	0.872			
	PEOU_3	0.898			
Perceive usefulness	PU_1	0.832	0.796	0.781	0.711
	PU_2	0.779			
	PU_3	0.844			
Satisfaction	SAT_1	0.742	0.797	0.857	0.751
	SAT_2	0.748			
	SAT_3	0.750			
Subjective norm	SUB_NOR_1	0.731	0.766	0.757	0.785
	SUB_NOR_2	0.844			
	SUB_NOR_3	0.849			

Table 1 shows that the discriminant validity requirements are met as all AVE values are more than the squared correlation among the constructs in the measurement model [65, 72]. It is suggested that the construct constitutes at least 50% of the measurement variance when the AVE value is greater than 0.5. The AVE analysis is shown in Table 1. The square root of the AVE scores is shown by the bold diagonal elements in Table 2. On the other hand, the correlations among the constructs are demonstrated by off-loading diagonal elements. The table shows that the square root of the AVE values is between 0.728 and 0.942, which is more than the recommended value of 0.5. It is suggested that the AVE is more than any of the correlations with the construct, for each construct, and this evidence shows a higher variance of all constructs with their own measures instead of other constructs in the model which improves the discriminate validity.

Table 2. Discriminant validity (inter-correlations) of variable constructs

	Behavioral intention	Confirmation	Computer self-efficacy	Innovativeness	Perceived ease of use	Perceive usefulness	Satisfaction	Subjective norm
Behavioral intention	**0.884**							
Confirmation	0.587	**0.933**						
Computer self-efficacy	0.467	0.480	**0.728**					
Innovativeness	0.595	0.686	0.613	**0.888**				
Perceived ease of use	0.343	0.675	0.504	0.555	**0.876**			
Perceive usefulness	0.542	0.372	0.587	0.432	0.557	**0.942**		
Satisfaction	0494	0.585	0.414	0.544	0.566	0.545	**0.876**	
Subjective norm	0.445	0.366	0.666	0.587	0.599	0.576	0.533	**0.831**

Usually, the structural model is examined using the coefficient of determination (R^2 value) measure [73]. The purpose of using the coefficient is to determine the model's predictive accuracy [73]. The coefficient is employed as the squared correlation between a certain endogenous construct's actual and predicted values. The coefficient indicates the joined impact of the exogenous latent variables on an endogenous latent variable. It signifies the squared correlation between the actual and predicted values of the variables; therefore, it shows the extent of variance in the endogenous constructs provided by every exogenous construct recognized with it. [64] suggested that when the value is higher than 0.67, it is considered as high. However, the qualities in the range of 0.33 to 0.67 are direct, while the qualities in the range of 0.19 and 0.33 are weak values [73]. In addition, the value is inadmissible when it is estimated to be less than 0. According to Table 3 and Fig. 2, the model has moderate predictive power, which supports almost 38, 43 and 54% of the variance in the Behavioral Intention and perceived ease of use, and perceived usefulness, respectively.

5.2 Structural Model Analysis

The proposed hypotheses are tested using a structural equation model, along with SEM-PLS and the maximum likelihood estimation, so as to examine how the theoretical constructs for the structural model are related to each other. Table 4 and Fig. 2

Table 3. R^2 of the endogenous latent variables.

Constructs	R^2	Results
Behavioral intention	0.381	Moderate
Perceived ease of use	0.432	Moderate
Perceive usefulness	0.537	Moderate

present a summary of the findings, and it can be seen that 4 out of 5 hypotheses are found to be significant. Out of 12 hypotheses, there are 11 hypotheses which are significant as observed in Table 4 and Fig. 2. Based on the data analysis hypotheses H1a, H1b, H2a, H2b, H3a, H3b, H4a, H4b, H5, H6, and H7 were supported by the empirical data, while H8 was rejected. The results showed that PEOU significantly influenced PU ($\beta = 0.101$, $P < 0.01$) and BI ($\beta = 0.244$, $P < 0.05$) supporting hypothesis H5 and H7, respectively. PU was determined to be significant in affecting BI ($\beta = 0.122$, $P < 0.05$), supporting hypotheses H6. Furthermore, PEOU was significantly influenced by four exogenous factors: CSE ($\beta = 0.127$, $P < P < 0.01$), CONF ($\beta = 0.214$, $P < P < 0.01$), INNO ($\beta = 0.626$, $P < 0.001$), and SUB ($\beta = 0.117$, $P < 0.05$), which support hypotheses H1b, H2b, H3b, and H4b. PU was found to be significantly influenced by four exogenous factors: CSE ($\beta = 0.227$, $P < 0.05$), CONF ($\beta = 0.245$, $P < 0.001$), INNO ($\beta = 0.142$, $P < 0.01$), and SUB

Table 4. Results of structural model-research hypotheses significant at p** = <0.01, p* z0.05)

H	Relationship	Path	*t*-value	*p*-value	Direction	Decision
H1a	Computer self-efficacy → perceived usefulness	0.227	3.234	0.015	Positive	Supported*
H1b	Computer self-efficacy → perceived ease of use	0.127	2.637	0.005	Positive	Supported**
H2a	Confirmation → perceived usefulness	0.245	6.144	0.000	Positive	Supported**
H2b	Confirmation → perceived ease of use	0.214	4.387	0.001	Positive	Supported*
H3a	Innovativeness → perceived usefulness	0.142	2.707	0.004	Positive	Supported**
H3b	Innovativeness → perceived ease of use	0.626	5.444	0.000	Positive	Supported**
H4b	Subjective norm → perceived usefulness	0.051	3.148	0.041	Positive	Supported*
H5b	Subjective norm → perceived ease of use	0.117	1.319	0.048	Positive	Supported*
H5	Perceived ease of use → perceived usefulness	0.101	2.675	0.006	Positive	Supported**
H6	Perceived usefulness → behavioral intention	0.122	2.110	0.035	Positive	Supported*
H7	Perceived ease of use → behavioral intention	0.244	2.244	0.023	Positive	Supported*
H8	Satisfaction → behavioral intention	0.117	0.161	0.571	Positive	Not supported

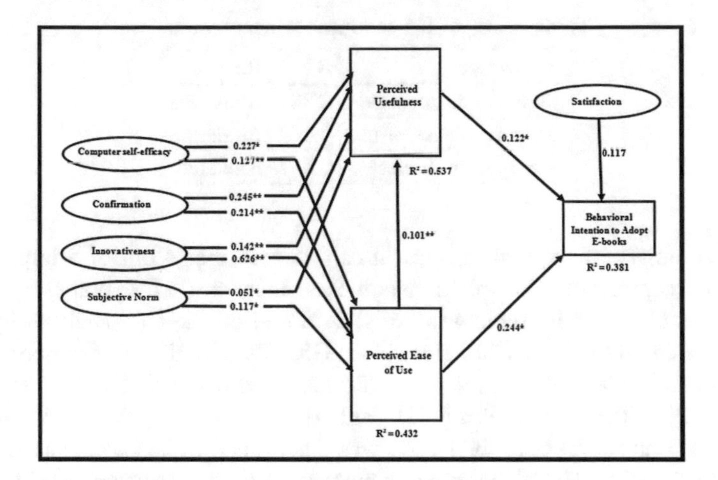

Fig. 2. Path coefficient results (significant at p** < = 0.01, p* < 0.05)

(β = 0.051, P < 0.05), supporting hypotheses H1a, H2a, H3a, and H4a, respectively. The relationship between SAT and BI (β = 0.117, P = 0.571) is statistically not significant, and hypotheses H8is generally not supported. A summary of the hypotheses testing results is shown in Table 4.

6 Conclusion and Future Works

The purpose of the study is to determine the factors that have an impact on the way electronic book (E-book) is viewed by the university students in the UAE. It can be deduced on the basis of the data analysis that the proposed research model and other hypotheses are appropriate for examining behavioral intention to adopt E-book technology in the UAE. The research hypotheses were assessed using structural equation modeling (PLS-SEM). Figure 2 presents the structural model which was examined by performing analysis of structural paths, t-statistics and variance explained (R-squared value). The results of data analysis are shown in Table 4. The twelve hypotheses presented above were examined with the help of the PLS method. The path significance of each hypothesized relationship involved in the research model and the variance explained (R^2) by each path was examined. Eleven hypotheses out of all the hypotheses put forward were supported. These included hypotheses obtained from TAM H5, H6, and H7. The study showed that computer self-efficacy, confirmation, innovativeness, and subjective norm caused an increase in the students' behavioral intention to adopt E-books. The computer self-efficacy, confirmation, innovativeness, and subjective norm were found to have a positive influence on perceived ease of use and perceived usefulness, and this supported the hypotheses H1a, H1b, H2a, H2b, H3a, H3b, H4b. This finding shows the significance of students' ability and confidence in using an E-book. A few previous studies also showed that there were a positive effect of computer self-efficacy, confirmation, innovativeness, and subjective norm on perceived ease of use, e.g., [9, 11, 12, 15, 32–34, 37–41]. As a result, it is imperative for legislators and managers of E-book applications in leaning to concentrate on the factors that are critical for encouraging learning and for enhancing students' efficiency in developing and executing successful E-book applications.

The limitation of the study was that it did not incorporate the findings of other regions which would have helped in either enhancing certain points included in this study or in evaluating the variations in regions with respect to perspectives on technology adoption. This would have been beneficial for establishing a robust model on the deterministic factors on technology adoption in developing countries. Valuable information was offered by the study with respect to establishing models or frameworks that facilitate e-readiness. A more extensive foundation is offered for comprehending technology adoption in developing countries like UAE. This serves as the basis for developing strategies in the future with respect to corresponding technology for giving rise to positive business and social adjustments in developing countries.

References

1. Al-Emran, M., Mezhuyev, V., Kamaludin, A., AlSinani, M.: Development of M-learning application based on knowledge management processes. In: 2018 7th International conference on Software and Computer Applications (ICSCA 2018) (2018)
2. Al-Emran, M., Mezhuyev, V., Kamaludin, A.: Students' perceptions towards the integration of knowledge management processes in M-learning systems: a preliminary study. Int. J. Eng. Educ. **34**(2), 371–380 (2018)
3. Al-Emran, M., Salloum, S.A.: Students' attitudes towards the use of mobile technologies in e-evaluation. Int. J. Interact. Mob. Technol. **11**(5), 195–202 (2017)
4. Salloum, S.A., Al-Emran, M., Monem, A.A., Shaalan, K.: Using text mining techniques for extracting information from research articles. In: Studies in Computational Intelligence, vol. 740. Springer, Berlin (2018)
5. Al-Emran, M., Shaalan, K.: Academics' awareness towards mobile learning in Oman. Int. J. Comput. Digit. Syst. **6**(1), 45–50 (2017)
6. Al-Emran, M., Malik, S.I.: The impact of Google apps at work: higher educational perspective. Int. J. Interact. Mob. Technol. **10**(4), 85–88 (2016)
7. Al-Emran, M., Shaalan, K.: Attitudes towards the use of mobile learning: a case study from the Gulf region. Int. J. Interact. Mob. Technol. **9**(3), 75–78 (2015)
8. Catenazzi, N.: The evaluation of electronic book guidelines from two practical experiences. J. Educ. Multimed. Hypermedia **6**(1), 91–114 (1997)
9. Shin, D.-H.: Understanding e-book users: uses and gratification expectancy model. New Media Soc. **13**(2), 260–278 (2011)
10. Lam, P., Lam, S.L., Lam, J., McNaught, C.: Usability and usefulness of eBooks on PPCs: how students' opinions vary over time. Austr. J. Educ. Technol. **25**(1), 30–44 (2009)
11. Poon, J.K.L.: Empirical analysis of factors affecting the e-book adoption—research agenda. Open J. Soc. Sci. **2**(05), 51 (2014)
12. Letchumanan, M., Tarmizi, R.: Assessing the intention to use e-book among engineering undergraduates in Universiti Putra Malaysia, Malaysia. Libr. Hi Tech **29**(3), 512–528 (2011)
13. Nelson, K., Webb, H.: Exploring student perceptions of an electronic textbook: a TAM perspective. In: AMCIS 2007 Proceedings, p. 107 (2007)

14. Ngafeeson, M.N., Sun, J.: The effects of technology innovativeness and system exposure on student acceptance of e-textbooks. J. Inf. Technol. Educ. Res. **14**, 55–71 (2015)
15. Tri-Agif, I., Noorhidawati, A., Ghalebandi, S.G.: Continuance intention of using ebook among higher education students. Malays. J. Libr. Inf. Sci. **21**(1), 19–33 (2016)
16. Al-Emran, M.: Hierarchical reinforcement learning: a survey. Int. J. Comput. Digit. Syst. **4** (2), 137–143 (2015)
17. Al-Emran, M., Mezhuyev, V., Kamaludin, A.: Technology acceptance model in M-learning context: a systematic review. Comput. Educ. **125**, 389–412 (2018)
18. Salloum, S.A., Al-Emran, M., Shaalan, K.: A survey of lexical functional grammar in the Arabic context. Int. J. Comput. Netw. Technol. **4**(3), 141–147 (2016)
19. Salloum, S.A., AlHamad, A.Q., Al-Emran, M., Shaalan, K.: A survey of Arabic text mining. In: Studies in Computational Intelligence, vol. 740. Springer, Berlin (2018)
20. Salloum, S.A., Al-Emran, M., Monem, A., Shaalan, K.: A survey of text mining in social media: Facebook and Twitter perspectives. Adv. Sci. Technol. Eng. Syst. J. **2**(1), 127–133 (2017)
21. Al Emran, M., Shaalan, K.: A survey of intelligent language tutoring systems. In: Proceedings of the 2014 International Conference on Advances in Computing, Communications and Informatics, ICACCI 2014, pp. 393–399 (2014)
22. Salloum, S.A., Al-Emran, M., Abdallah, S., Shaalan, K.: Analyzing the Arab Gulf newspapers using text mining techniques. In: International Conference on Advanced Intelligent Systems and Informatics, pp. 396–405 (2017)
23. Salloum, S.A., Mhamdi, C., Al-Emran, M., Shaalan, K.: Analysis and classification of Arabic Newspapers' Facebook pages using text mining techniques. Int. J. Inf. Technol. Lang. Stud. **1**(2), 8–17 (2017)
24. Salloum, S.A., Al-Emran, M., Shaalan, K.: Mining social media text: extracting knowledge from Facebook. Int. J. Comput. Digit. Syst. **6**(2), 73–81 (2017)
25. Al-Emran, M., Shaalan, K.: Learners and educators attitudes towards mobile learning in higher education: state of the art. In: 2015 International Conference on Advances in Computing, Communications and Informatics, ICACCI 2015, pp. 907–913 (2015)
26. Ngai, E.W.T., Poon, J.K.L., Chan, Y.H.C.: Empirical examination of the adoption of WebCT using TAM. Comput. Educ. **48**(2), 250–267 (2007)
27. Yang, K.C.C.: Exploring factors affecting the adoption of mobile commerce in Singapore. Telemat. Inform. **22**(3), 257–277 (2005)
28. Spreng, R.A., Chiou, J.: A cross-cultural assessment of the satisfaction formation process. Eur. J. Mark. **36**(7/8), 829–839 (2002)
29. Rogers, E.M.: Diffusion of Innovations, 4th edn. The Free Press, New York (1995)
30. Venkatesh, V., Davis, F.D.: A theoretical extension of the technology acceptance model: four longitudinal field studies. Manag. Sci. **46**(2), 186–204 (2000)
31. Alshurideh, M.: Customer Service Retention—A Behavioural Perspective of the UK Mobile Market. Durham University, Durham (2010)
32. Jin, C.-H.: Adoption of e-book among college students: the perspective of an integrated TAM. Comput. Hum. Behav. **41**, 471–477 (2014)
33. Ngafeeson, M.: E-book acceptance by undergraduate students: do gender differences really exist. Int. J. Web Based Learn. Teach. Technol. (2011)
34. Tsai, W.-C.: A study of consumer behavioral intention to use e-books: the technology acceptance model perspective. Innov. Mark. **8**(4), 55–66 (2012)
35. Al-Maroof, R.A.S., Al-Emran, M.: Students acceptance of Google classroom: an exploratory study using PLS-SEM approach. Int. J. Emerg. Technol. Learn. **13**(6), 112–123 (2018)
36. Davis, F.D.: Perceived usefulness, perceived ease of use, and user acceptance of information technology. MIS Q **13**(3), 319–340 (1989)

37. Smeda, A.M., Shiratuddin, M.F., Wong, K.W.: Factors affecting the e-book adoption amongst mathematics and statistics students at universities in Libya: a structural equation modelling approach. Int. J. e-Educ. e-Bus. e-Manag. e-Learn. **5**(4), 237 (2015)
38. Chiu, P.-S., Chao, I.-C., Kao, C.-C., Pu, Y.-H., Huang, Y.-M.: Implementation and evaluation of mobile e-books in a cloud bookcase using the information system success model. Libr. Hi Tech **34**(2), 207–223 (2016)
39. Huang, Y.-M., Pu, Y.-H., Chen, T.-S., Chiu, P.-S.: Development and evaluation of the mobile library service system success model: a case study of Taiwan. Electron. Libr. **33**(6), 1174–1192 (2015)
40. Park, E., Sung, J., Cho, K.: Reading experiences influencing the acceptance of e-book devices. Electron. Libr. **33**(1), 120–135 (2015)
41. Lai, J.-Y., Rushikesh Ulhas, K.: Understanding acceptance of dedicated e-textbook applications for learning: Involving Taiwanese university students. Electron. Libr. **30**(3), 321–338 (2012)
42. Venkatesh, V., Morris, M., Davis, G., Davis, F.: User acceptance of information technology: toward a unified view. MIS Q. **27**(3), 425–478 (2003)
43. Alshurideh, M., Masa'deh, R., Alkurdi, B.: The effect of customer satisfaction upon customer retention in the Jordanian mobile market: an empirical investigation. Eur. J. Econ. Financ. Adm. Sci. **47**, 69–78 (2012)
44. Alshurideh, M.T.: Do we care about what we buy or eat? A practical study of the healthy foods eaten by Jordanian youth. Int. J. Bus. Manag. **9**(4), 65 (2014)
45. Bandura, A.: Self-efficacy: toward a unifying theory of behavioral change. Psychol. Rev. **84** (2), 191 (1977)
46. Tri Agif, I., Noorhidawati, A., Siti Hajar, M.R.: Investigating continuance intention of using e-book among higher education students (2014)
47. Smeda, A.M., Shiratuddin, M.F., Wong, K.W.: Proposed framework of the adoption of e-book amongst mathematics and statistics students at universities in Libya. In: Proceedings of the 2nd International Virtual Conference on Advanced Scientific Results (SCIECONF) (2014)
48. Bhattacherjee, A.: Understanding information systems continuance: an expectation–confirmation model. MIS Q. **25**, 351–370 (2001)
49. Al-dweeri, R.M., Obeidat, Z.M., Al-dwiry, M.A., Alshurideh, M.T., Alhorani, A.M.: The impact of e-service quality and e-loyalty on online shopping: moderating effect of e-satisfaction and e-trust. Int. J. Mark. Stud. **9**(2), 92 (2017)
50. Chen, S.-C., Yen, D.C., Peng, S.-C.: Assessing the impact of determinants in e-magazines acceptance: an empirical study. Comput. Stand. Interfaces (2017)
51. Shin, D.-H., Shin, Y.-J., Choo, H., Beom, K.: Smartphones as smart pedagogical tools: implications for smartphones as u-learning devices. Comput. Hum. Behav. **27**(6), 2207–2214 (2011)
52. Midgley, D.F., Dowling, G.R.: Innovativeness: The concept and its measurement. J. Consum. Res. **4**(4), 229–242 (1978)
53. Lee, H., Kim, D., Ryu, J., Lee, S.: Acceptance and rejection of mobile TV among young adults: a case of college students in South Korea. Telemat. Inform. **28**(4), 239–250 (2011)
54. Kuo, Y.-F., Yen, S.-N.: Towards an understanding of the behavioral intention to use 3G mobile value-added services. Comput. Hum. Behav. **25**(1), 103–110 (2009)
55. Huang, T.K.: Investigating user acceptance of a screenshot-based interaction system in the context of advanced computer software learning. In: 2014 47th Hawaii International Conference on System Sciences (HICSS), pp. 4956–4965 (2014)
56. Lee, S.: An integrated adoption model for e-books in a mobile environment: Evidence from South Korea. Telemat. Inform. **30**(2), 165–176 (2013)

57. Aharony, N.: The effect of personal and situational factors on LIS students' and professionals' intentions to use e-books. Libr. Inf. Sci. Res. **36**(2), 106–113 (2014)
58. Venkatesh, V., Bala, H.: Technology acceptance model 3 and a research agenda on interventions. Decis. Sci. **39**(2), 273–315 (2008)
59. Torres, R., Johnson, V., Imhonde, B.: The impact of content type and availability on ebook reader adoption. J. Comput. Inf. Syst. **54**(4), 42–51 (2014)
60. Hyman, J.A., Moser, M.T., Segala, L.N.: Electronic reading and digital library technologies: understanding learner expectation and usage intent for mobile learning. Educ. Technol. Res. Dev. **62**(1), 35–52 (2014)
61. Krejcie, R.V., Morgan, D.W.: Determining sample size for research activities. Educ. Psychol. Meas. **38**(1), 607–610 (1970)
62. Chuan, C.L., Penyelidikan, J.: Sample size estimation using Krejcie and Morgan and Cohen statistical power analysis: A comparison. J. Penyelid. IPBL **7**, 78–86 (2006)
63. Ringle, C.M., Wende, S., Will, A.: SmartPLS 2.0 (Beta). Hamburg (2005). http://www.smartpls.de
64. Chin, W.W.: The partial least squares approach to structural equation modeling. Mod. Methods Bus. Res. **295**(2), 295–336 (1998)
65. Hair, J.F., Black, W.C., Babin, B.J., Anderson, R.E., Tatham, R.L.: Multivariate Data Analysis, vol. 5, no. 3. Prentice Hall, Upper Saddle River (1998)
66. Al Kurdi, B.: Healthy-Food Choice and Purchasing Behaviour Analysis: An Exploratory Study of Families in the UK. Durham University, Durham (2016)
67. Gefen, D., Straub, D., Boudreau, M.-C.: Structural equation modeling and regression: guidelines for research practice. Commun. Assoc. Inf. Syst. **4**(1), 7 (2000)
68. Alshurideh, M.: The factors predicting students' satisfaction with universities' healthcare clinics' services: a case-study from the Jordanian higher education sector. Dirasat Admin. Sci. **41**(2), 451–464 (2014)
69. Malik, S.I., Al-Emran, M.: Social factors influence on career choices for female computer science students. Int. J. Emerg. Technol. Learn. **13**(5), 56–70 (2018)
70. Al-Qaysi, N., Al-Emran, M.: Code-switching usage in social media: a case study from Oman. Int. J. Inf. Technol. Lang. Stud. **1**(1), 25–38 (2017)
71. Falk, R.F., Miller, N.B.: A Primer for Soft Modeling. University of Akron Press, Akron (1992)
72. Fornell, C., Larcker, D.F.: Evaluating structural equation models with unobservable variables and measurement error. J. Mark. Res. **18**(1), 39–50 (1981)
73. Lin, Y.-C., Chen, Y.-C., Yeh, R.C.: Understanding college students' continuing intentions to use multimedia e-learning systems. World Trans. Eng. Technol. Educ. **8**(4), 488–493 (2010)

A Predictive Model for Seminal Quality Using Neutrosophic Rule-Based Classification System

Sameh H. Basha[1,4], Alaa Tharwat[2,4](✉), Khaled Ahmed[3,4], and Aboul Ella Hassanien[3,4]

[1] Faculty of Science, Cairo University, Giza, Egypt
[2] Faculty of Computer Science and Engineering, Frankfurt University of Applied Sciences, 60318 Frankfurt am Main, Germany
engalaatharwat@hotmail.com
[3] Faculty of Computers and Information, Cairo University, Giza, Egypt
[4] Scientific Research Group in Egypt (SRGE), Giza, Egypt
http://www.egyptscience.net

Abstract. This paper presents a predictive model of sperm quality based on personal lifestyle and environmental factors. Predicting the fertility based on measuring the quality of sperm is an urgent research problem since the fertility rates have been decreased mostly in men. Infertility is the biggest problem which faces the married persons. A lot of tests are expensive and time-consuming. Hence, these tests are not suitable for evaluating the quality of sperm; as a consequence, there is a need for implementing computational models that can predict the quality of sperm. The proposed model consists of two stages. In the first stage, Neutrosophic Rule-based Classification System (NRCS) and genetic NRCS (GNRCS) are proposed to classify an unknown seminal quality into normal or abnormal. NRCS and GNRCS forecast fertility based on nine input parameters which cover person lifestyle. The NRCS and GNRCS models were compared with three well-known classifiers such as decision trees, Support Vector Machines and Multilayer Perceptron with quality measures of GM, sensitivity, specificity, and accuracy. The proposed model (GNRCS) achieved promising results of 98.03% accuracy while basic NRCS achieved 95.55% accuracy. In the second stage, three different sampling algorithms are used to obtain balanced data.

Keywords: Seminal quality · Neutrosophic classification system
Neutrosophic logic · Neutrosophic predictive model

1 Introduction

A man's fertility depends mainly on the quantity and quality of his sperm. If the quality of the sperm is low or if the number of sperm is low, it will be difficult for him to cause a pregnancy. The fertility problem occurs if the number of sperm is

A. E. Hassanien et al. (Eds.): AISI 2018, AISC 845, pp. 495–504, 2019.
https://doi.org/10.1007/978-3-319-99010-1_45

low; however, sometimes there is no sperm which is a severe problem. Generally, there are no clear signs of infertility. Intercourse, ejaculation, and erections will usually happen without difficulty, and the appearance and quantity of the semen appear normal to the naked eye.

In the past two decades, there are many studies reported that the quality of semen decreased due to many factors such as environmental factors, climate conditions, using alcohol and tobacco, lifestyle, or occupational factors. These factors affect the sperm production or sperm transportation; as a consequence, the male fertility is affected [1]. Another important reason for the small number of sperm is the obstructions in the tubes leading sperm away from the testes to the penis. Clinical tests are used for evaluating the quality of semen by comparing the results with the corresponding reference values established by World Health Organization [2]. However, the clinical tests are expensive and it is a high variability. Therefore, there is a need for an alternative for assessing the semen quality.

Machine learning technique can be used for predicting the semen quality [3,4]. Azam *et al.* used the optimized Neural Networks (NN) with Genetic Algorithm (GA) for predicting semen quality [5]. In another research, David *et al.* used Decision Trees (DT), Support Vector Machine (SVM), and Multilayer Perceptron (MLP) for predicting the quality of semen [1]. A Clustering-Based Decision Forests model was introduced for solving the imbalanced class learning problem in seminal quality prediction [6]. In this paper, due to the ability of Neutrosophic Logic (NL) to deal with deficient data and in most cases without a need to preprocessing steps, two neutrosophic-based models are proposed. The first model is the Neutrosophic Rule-based Classification System (NRCS) which generalize Fuzzy Rule-based classification system by using NL to represents a different form of the data. The second model is the Genetic NRCS (GNRCS) which is a hybrid model of NRCS and Genetic algorithm (GA) [7,8] where GA is used to refine neutrosophic rules generated by NRCS. Moreover, in our experiments, the dataset we used is imbalanced and hence in the proposed model, three sampling algorithms are proposed for obtaining balanced data.

The rest of the paper is organized as follows; Sect. 2 presents the theoretical background for our proposed model. Our proposed model is presented in Sect. 3 while, the experimental results and discussions are given in Sect. 4. The conclusions of the paper are presented in Sect. 5.

2 Preliminaries

2.1 Neutrosophic Set and Neutrosophic Logic

Neutrosophic Set: Neutrosophy is a philosophical theory concerned with the old roots and studying "the nature, origin, and scope of neutralities, as well as their relations with different conceptional visions" [9]. Neutrosophic set is a huge formal structure which generalizes the concept of all sets such as, fuzzy set, intuitionistic fuzzy set, and interval-valued fuzzy set [10]. In the neutrosophic theory, for every notion (<S>), there is exists a spectrum of "neutralities" (<Neut-S>)

which neither idea <S> nor its negation <Anti-S>. As a state of equilibrium, every idea tends to be neutralized by both its opposite <Anti-S> and <Neut-S>. According to the neutrosophic theory, to study any notion or idea <S> considers this idea with its opposite (<Anti-S>) as well as its "neutralities" (<Neut-S>). It is worth mentioning that the <Neut-S> and <Anti-S> ideas together are denoted by <Non-S> [9].

An element $e(T, I, F)$ belongs to the set A in the following way: it is true in A with a degree t, i degree indeterminate in A, and f degree false in A, where t, i, and f are real numbers taken from the sets T_A, I_A, and F_A with no restriction on T_A, I_A, and F_A, nor on their sum [11]. Given X is a space of elements, with a generic element x in X. A neutrosophic set A in X is represented by T_A, I_A, and F_A, where T_A, I_A, and F_A represent the truth-membership, indeterminacy-membership, and falsity membership functions. These membership functions are real standard/non-standard subsets of $]^-0, 1^+[$. That is: T_A, I_A, and $F_A : X \rightarrow]^-0, 1^+[$. There is no restriction on the sum of $T_A(x)$, $I_A(x)$ and $F_A(x)$, so $^-0 \leq supT_A(x) + supI_A(x) + supF_A(x) \leq 3^+$ [10].

Neutrosophic Logic: Neutrosophic logic was introduced for representing a mathematical model of redundancy, vagueness, ambiguity, contradiction, uncertainty, incompleteness, imprecision, and inconsistency [12]. All factors which are stated in neutrosophic logic are very integral to human thinking. The imprecision of human is due to the imperfection of knowledge/data which are observed from the external environment [10]. As neutrosophic set operators, there are many methods for constructing neutrosophic logic connectives according to each application or problem under consideration [12].

Neutrosophic logic is a logic in which each statement is estimated to have three percentages which are one of the truth in a subset T, one of indeterminacy in the subset I, and one of falsity in a subset F [11]. In neutrosophy and their corresponding mathematics theories, T, I, and F are called neutrosophic components, where T represents the truth, I is the indeterminacy, and F indicates falsehood values. In real applications, for T, I and F components, the standard real interval [0,1] can be used instead of using the non-standard interval $(]^-0, 1^+[)$ [10].

2.2 Imbalanced Datasets

The imbalanced data problem happens when one class (is called the majority class (S_{maj})) has samples more than the other class (is called the minority class (S_{min})) [13]. This problem decreases the classification performance because any learning algorithm cannot learn from the small size of the minority class and hence the trained model cannot accurately classify minority class samples. In this problem, there are different metrics which are not sensitive to the imbalanced data such as sensitivity and specificity metrics [13]. There are many algorithms that are utilized for obtaining balanced data such as sampling methods [13], Kernel-Based methods [14], and Cost-Sensitive methods [15]. In this paper, sampling methods are employed for obtaining balanced data.

In sampling methods, the goal is to obtain balanced data in all classes by modifying the prior distribution of the majority and minority classes. There are three well-known sampling methods: *Random Over-sampling* (ROS), *Random Under-sampling* (RUS), and *Synthetic Minority Over-sampling Technique* (SMOTE). In the RUS method, the goal is to select randomly a small set of majority class samples for training a learning algorithm while preserving all minority class samples. Thus, the training data become smaller which reduces the required computational time. However, the removed samples may have discriminative information and hence RUS may reduce the classification performance [13]. In the ROS method, the aim is to replicate minority class samples which improve the minority class recognition. However, making exact copies of minority class samples increases the learning time and may lead to the overfitting problem [13]. In the SMOTE method, minority samples are created/synthesized based on the similarities between existing minority samples. For each minority sample ($x_i \in S_{min}$), k nearest samples are selected and a synthetic sample can be generated as follow,

$$x_{new} = x_i + r_{ij} \times \rho = x_i + (\hat{x_{ij}} - x_i) \times \rho \tag{1}$$

where $x_i \in S_{min}$ is a minority class sample, $\hat{x_{ij}}$ is one of the k-nearest samples for $x_i : \hat{x_{ij}} \in S_{min}, j = 1, 2, \ldots, k$, k is the number of selected neighbors, $\rho \in [0, 1]$ is a random number, and x_{new} is a sample along the line joining x_i and $\hat{x_{ij}}$ [16].

3 The Proposed Predictive Seminal Quality Model

3.1 Neutrosophic Rule-Based Classification System (NRCS)

The proposed Neutrosophic Rule-based Classification System is a rule-based system which generalizes fuzzy rule-based classification system by using neutrosophic logic. Neutrosophic logic is utilized for representing different forms of knowledge as well as modeling the interactions and relationships that exist between its attributes [10]. As any rule-based classification system, NRCS uses "IF-THEN" rules which are generated using three neutrosophic membership functions which will be extracted from fuzzy-Trapezoidal membership function [10]. The proposed NRCS contains four main phases: information extraction phase, Neutrosophication phase, Generate set of rules phase, and the Classification/matching phase. These four phases are described in detail in the following section, and the overall architecture of the introduced model is described in Fig. 1.

Information Extraction Phase: In this phases, important information are extracted from the given dataset such as the number of classes, classes' names, number of features, and the maximum and the minimum values for each feature. This information will be the inputs for the next phase.

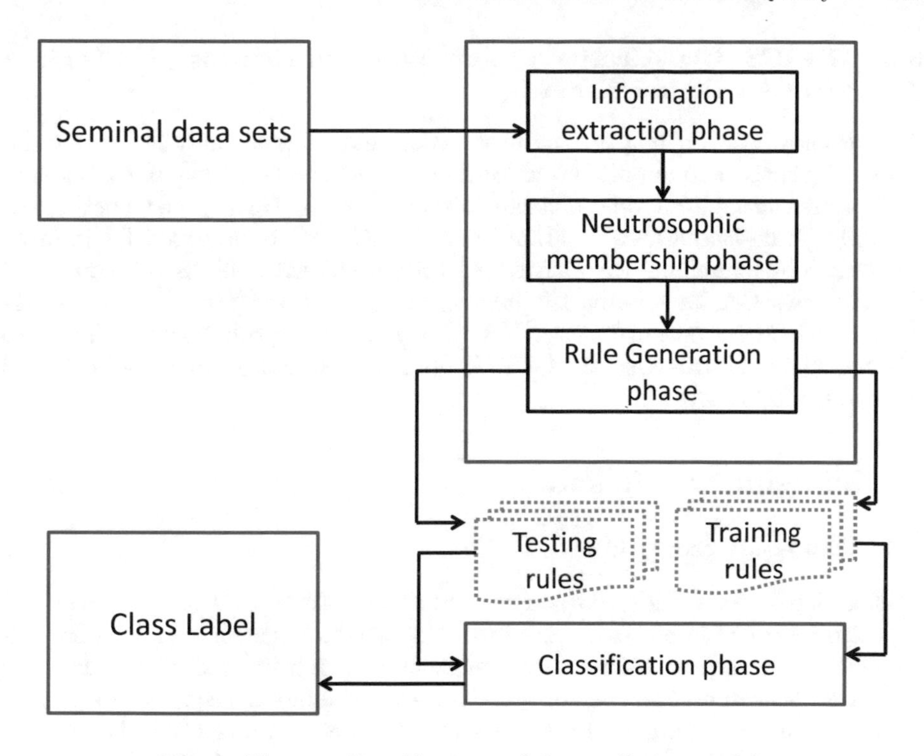

Fig. 1. The overall architecture of the predicting model.

Neutrosophication Phase: The aim of this phase is to convert each value for each feature to a neutrosophic form which is the form presented by three neutrosophic components $<T, I, F>$. In this phase, the three neutrosophic membership functions are applied on the dataset to represent each value in the neutrosophic form.

Generate Set of Rules Phase: The main task in the proposed NRCS is to generate a set of rules for training and testing data. In each attribute, the value in NRCS rules has three components which describe the degrees of truth, indeterminacy, and falsity, respectively.

Classification Phase: In this phase, the testing rules which are generated in the previous phase are used to construct the testing rule matrix without class labels. Then, to predict the class label, i.e., classify, any testing sample, we compare the rule of that sample with all training rules and assign the class label with the training rule which has the maximum number of attribute intersected with the current testing rule. If there the matching rate between the training rules and the current testing rule is less than 50%, the class label for this testing rule is determined from the original dataset and this rule is then added to the training rules. At the end of this phase, the predicted class labels are compared with true class labels to evaluate our model.

3.2 GNRCS: Classification System Based on Neutrosophic Logic and Genetic Algorithm

The problem of searching for the best set of neutrosophic rules can be considered as an optimization problem. Evolutionary techniques are capable to explore large search space for finding optimal solutions [17–19]. Hence, the Genetic Neutrosophic Rule-Based Classification System (GNRCS) is proposed for improving NRCS by combining the Genetic Algorithm GA with NRCS. The proposed GNRCS uses GA for refining the neutrosophic "IF-THEN" rules which generated in NRCS by optimizing the KB. This model is approximately similar to the NRCS model. However, the GNRCS optimizes the rules which are generated from the NRCS model.

4 Experimental Results

4.1 Dataset Description

In this paper, the fertility dataset which is available at UCI machine learning repository is used in our experiments [1]. In this data, semen samples were collected from more than 100 persons. Each sample is represented by nine features as follows: season number, personage, diseases in childhood, accident or seriously previous life event, surgery, fever rates in last years, the rate of alcohol, sitting hours, smoking; the data covers environmental, health, and lifestyle factor to predict the fertility. The data has 88 normal cases and only 12 abnormal cases and hence the data is imbalanced. Figure 2 shows the plot-matrix of the dataset. As shown, the data is imbalanced with a high imbalance ratio.

In this section, two experiments were conducted. The goal of the first experiment is to test the proposed NRCS and GNRCS models for predicting the

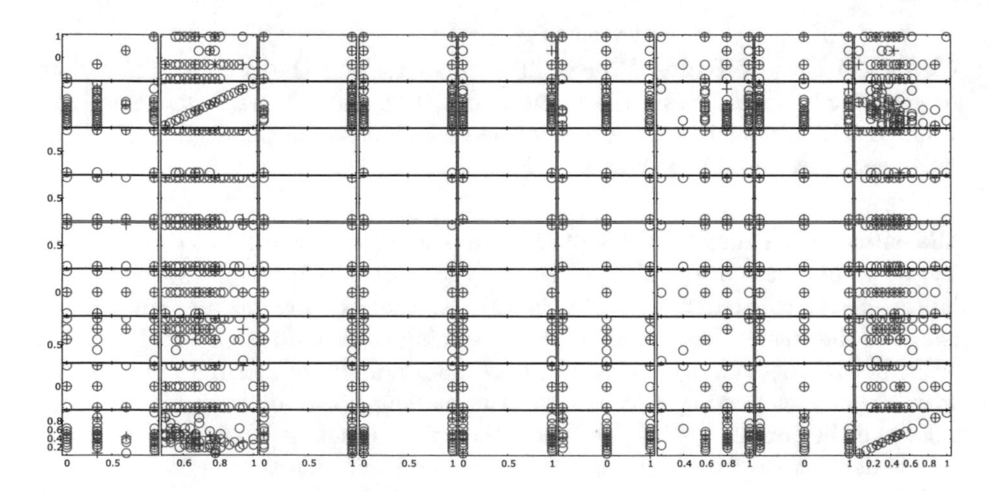

Fig. 2. Plot-matrix of our dataset.

seminal quality. In this experiment, the NRCS and GNRCS algorithms were compared with conventional classifiers such as decision trees [20], Support Vector Machines [21,22], and Multilayer Perceptron [23]. The aim of the second experiment is to obtain balanced data for improving the results of the proposed model. In this experiment, three sampling methods were employed for obtaining balanced data, namely, RUS, ROS, and SMOTE algorithms.

4.2 Seminal Quality Classification Using the NRCS and GNRCS Models

The aim of this experiment is to test the two proposed models NRCS and GNRCS for predicting the seminal quality. In this experiment, the results of the NRCS and GNRCS algorithms were compared with classical classifiers such as DT, MLP and SVM that were employed for predicting of the seminal quality. The results of this experiment are summarized in Table 1.

From Table 1 it is clear that the NRCS and GNRCS algorithms obtained the best accuracy. Moreover, the two proposed algorithms, i.e., NRCS and GNRCS, obtained the best specificity results. Due to the imbalanced data, the sensitivity and GM results of all algorithms were much lower than the specificity and the sensitivity results ranged from 12.5% to 70%. However, the NRCS and GNRCS algorithms achieved competitive sensitivity results better than the other algorithms. This is because the GNRCS as NRCS algorithm can determine the more significant, indeterminacy or neutral and non-significant features without using any feature selection method. Additionally, GNRCS obtained results better than NRCS because GA in the GNRCS algorithm is used to refine the generated neutrosophic rules to get high prediction accuracy with less number of training rules and hence GNRCS reduces also the required computational time. The results of the NRCS and GNRCS algorithms also were compared with the results that were presented in [1], where three different artificial intelligence techniques were used, and our proposed models obtained competitive results.

Table 1. A comparison between NRCS, GNRCS, MLP, DT, and SVM.

Metrics	MLP	SVM	DT	NRCS	GNRCS
Accuracy	69	69	67	95.6	98.0
Specificity	72.5	73.9	71.7	97.6	98.0
Sensitivity	25	12.5	12.5	66.7	70.0
GM	42.6	30.4	29.9	80.7	82.3

4.3 Experimental 2 (Obtaining Balanced Data)

This experiment aims to improve the results by dealing with the problem of imbalanced data for improving the sensitivity results. In this experiment, three sampling algorithms were employed: RUS, ROS, and SMOTE. In ROS, the samples of minority class were randomly oversampled until obtaining balanced data. In RUS, majority class samples were randomly removed until obtaining balanced data. While In the SMOTE algorithm, a set of minority samples were created to equalize the two classes. The results of this experiment are summarized in Table 2. From the table, it is clear that all sampling algorithms improved the sensitivity results dramatically. Moreover, the specificity results of the RUS and ROS algorithms reduced. This is because in the RUS algorithm, according to our data, 76 samples were removed and hence $\frac{76}{88} \approx 86.4\%$ from the data of the second class were removed. Hence, the specificity results reduced. According to the ROS algorithm, many minority samples were added/replicated; however, this addition will not extend the search space of the minority class because the samples are replicated in the same position. On the contrary, the SMOTE algorithm improved the sensitivity without reducing specificity because SMOTE generates minority samples close to the original minority samples without removing data.

To conclude, despite the small number of samples and the imbalanced data, our proposed model obtained competitive results.

Table 2. A comparison between NRCS GNRCS after applying ROS, RUS, and SMOTE algorithms.

Metrics	NRCS				GNRCS			
	Orig.	RUS	ROS	SMOTE	Orig.	RUS	ROS	SMOTE
Accuracy	95.6	96.7	97.2	98.5	98.0	98.2	98.6	99.3
Specificity	97.6	91.2	94.3	98.2	98.0	92.5	96.9	98.4
Sensitivity	66.7	93.2	90.3	95.6	70	91.4	94.3	96.7
GM	80.7	85.7	93.2	96.8	82.3	91.9	95.2	97.6

5 Conclusions and Future Work

This paper presents two models to predict the quality of semen based on personal lifestyle which is important research area as millions of person in a need for such highly accurate prediction model which is applicable in the medical domain. The first one is the Neutrosophic Rule-based Classification System (NRCS) and the second is a hybrid genetic algorithm (GA) with NRCS (GNRCS). The proposed models applied in two scenarios. The first scenario which applied the proposed models with the original data without any preprocessing steps. In this scenario, GNRCS and NRCS models achieved the best results in all measures against three learning algorithms, namely, DT, MLP and SVM, and GNRCS obtained the best results. In the second scenario, three data sampling algorithms, namely, Random Under-Sampling (RUS), Random Over-Sampling (ROS), and Synthetic Minority Over-sampling Technique (SMOTE) were used for obtaining balanced data and then applied the proposed models with balanced data. The balanced data improved the results of the proposed model especially the sensitivity. In future work, different optimization algorithm will be used for optimizing the NRCS model. Moreover, counting semen and tracking movement from microscope medical imaging are two attractive points in this research area.

References

1. Gil, D., Girela, J.L., De Juan, J., Gomez-Torres, M.J., Johnsson, M.: Predicting seminal quality with artificial intelligence methods. Expert Syst. Appl. **39**(16), 12564–12573 (2012)
2. World Health Organisation: WHO Laboratory Manual for the Examination of Human Semen and Sperm-Cervical Mucus Interaction. Cambridge University Press, Cambridge (1999)
3. Tharwat, A.: Principal component analysis-a tutorial. Int. J. Appl. Pattern Recognit. **3**(3), 197–240 (2016)
4. Tharwat, A., Gaber, T., Ibrahim, A., Hassanien, A.E.: Linear discriminant analysis: a detailed tutorial. AI Commun. **30**(2), 169–190 (2017)
5. Bidgoli, A.A., Komleh, H.E., Mousavirad, S.J.: Seminal quality prediction using optimized artificial neural network with genetic algorithm. In: 9th International Conference on Electrical and Electronics Engineering (ELECO), pp. 695–699. IEEE (2015)
6. Wang, H., Xu, Q., Zhou, L.: Seminal quality prediction using clustering-based decision forests. Algorithms **7**(3), 405–417 (2014)
7. Elhoseny, M., Tharwat, A., Yuan, X., Hassanien, A.E.: Optimizing K-coverage of mobile WSNS. Expert Syst. Appl. **92**, 142–153 (2018)
8. Elhoseny, M., Tharwat, A., Farouk, A., Hassanien, A.E.: K-coverage model based on genetic algorithm to extend WSN lifetime. IEEE Sens. Lett. **1**(4), 1–4 (2017)
9. Basha, S.H., Sahlol, A.T., El Baz, S.M., Hassanien, A.E.: Neutrosophic rule-based prediction system for assessment of pollution on Benthic Foraminifera in Burullus Lagoon in Egypt. In: 12th International Conference on Computer Engineering and Systems (ICCES), pp. 663–668. IEEE (2017)
10. Basha, S.H., Abdalla, A.S., Hassanien, A.E.: NRCS: Neutrosophic rule-based classification system. In: Proceedings of SAI Intelligent Systems Conference, pp. 627–639. Springer (2016)
11. Ansari, A.Q., Biswas, R., Aggarwal, S.: Neutrosophic classifier: an extension of fuzzy classifer. Appl. Soft Comput. **13**(1), 563–573 (2013)
12. Hassanien, A.E., Basha, S.H., Abdalla, A.S.: Generalization of fuzzy c-means based on neutrosophic logic. Stud. Inf. Control **27**(1), 43–54 (2018)
13. He, H., Garcia, E.A.: Learning from imbalanced data. IEEE Trans. Knowl. Data Eng. **21**(9), 1263–1284 (2009)
14. Tharwat, A., Moemen, Y.S., Hassanien, A.E.: A predictive model for toxicity effects assessment of biotransformed hepatic drugs using iterative sampling method. Sci. Rep. **6**, 38660 (2016)
15. Tharwat, A., Moemen, Y.S., Hassanien, A.E.: Classification of toxicity effects of biotransformed hepatic drugs using whale optimized support vector machines. J. Biomed. Inf. **68**, 132–149 (2017)
16. Chawla, N.V., Bowyer, K.W., Hall, L.O., Kegelmeyer, W.P.: SMOTE: synthetic minority over-sampling technique. J. Artif. Intell. Res. **16**, 321–357 (2002)

17. Tharwat, A., Houssein, E.H., Ahmed, M.M., Hassanien, A.E., Gabel, T.: MOGOA algorithm for constrained and unconstrained multi-objective optimization problems. Appl. Intell. **48**(8), 1–16 (2017)
18. Metawa, N., Hassan, M.K., Elhoseny, M.: Genetic algorithm based model for optimizing bank lending decisions. Expert Syst. Appl. **80**, 75–82 (2017)
19. Ibrahim, A., Tharwat, A., Gaber, T., Hassanien, A.E.: Optimized superpixel and Adaboost classifier for human thermal face recognition. Signal Image Video Process. **12**, 1–9 (2017)
20. Tharwat, A.: Linear vs. quadratic discriminant analysis classifier: a tutorial. Int. J. Appl. Pattern Recognit. **3**(2), 145–180 (2016)
21. Tharwat, A., Hassanien, A.E.: Chaotic antlion algorithm for parameter optimization of support vector machine. Appl. Intell. **48**(3), 670–686 (2018)
22. Tharwat, A., Hassanien, A.E., Elnaghi, B.E.: A BA-based algorithm for parameter optimization of support vector machine. Pattern Recognit. Lett. **93**, 13–22 (2017)
23. Yamany, W., Fawzy, M., Tharwat, A., Hassanien, A.E.: Moth-flame optimization for training multi-layer perceptrons. In: 11th International Computer Engineering Conference (ICENCO), pp. 267–272. IEEE (2015)

Robust Simulation and Visualization of Satellite Orbit Tracking System

Kadry Ali Ezzat[1,4(✉)], Lamia Nabil Mahdy[1,4], Aboul Ella Hassanien[2,4], and Ashraf Darwish[3,4]

[1] Biomedical Engineering Department, Higher Technological Institute, Cairo, Egypt
kadry_ezat@hotmail.com
[2] Faculty of Computers and Information, Cairo University, Cairo, Egypt
[3] Faculty of Science, Helwan University, Helwan, Egypt
[4] Scientific Research Group in Egypt (SRGE), Cairo, Egypt
http://www.egyptscience.net

Abstract. In order to obtain the optical tradition of the satellite path, traditional orbital elements must be obtained. In recent years, models and simulation of spacecraft have been developed. The aim of this research paper is to demonstrate the detailed reliability and adequacy of simulations, which can be obtained using cable or neutral factors rather than the Cartesian' coordinates accordingly to the state-consistent pattern of the spacecraft. The solid body pattern of the Multi-body's standard library is expanded by adding equations that determine the centre of the body's transformation of cluster coordinates from Keplerian variables to Cartesian coordinates, and by assigning identified cases. The remaining parts of the model, including the gravitational field pattern, are left intact, ensuring a severe reuse of the system. The results shown in the paper show the super-precision and the calculation of the six elements in the case of reference to the gravitational field of the cluster point and the strong visualization of satellite tracking.

Keywords: Simulation model · Satellite · Orbit parameter
Keplerian orbit elements

1 Introduction

Over the past convention, simulation, modeling, science and computer designing have become the most significant tools in development and research scheme. The design stage ought to be diminished in time and cost when the utilization of new concepts and toolkits becomes conceivable [1, 2]. This is likewise the direction in space applications where genuine tests are impractical or at least expensive. Novel applications have increased on aeronautical engineering, supervision and must be able to respond to requirements. Orbits or orbital simulations in general, are programming tools that can be utilized by scientists, engineers, understudies or all to resolve and evaluate system processes and to answer questions about the phenomenon. Simulations are fundamental toolkits in mission design and spacecraft surveillance. The status of a satellite mass

A. E. Hassanien et al. (Eds.): AISI 2018, AISC 845, pp. 505–514, 2019.
https://doi.org/10.1007/978-3-319-99010-1_46

center in space requires six parameters to be determined [3]. These parameters may take numerous equal structures. Regardless of the shape, the coincidence of these parameters either case vectors or orbital elements are important [4]. Group of quantities is referred to a specific scope of indication and totally distinguishes the orbit of the two bodies from a whole set of starting conditions for understanding a primary issue category of differential conditions. In the following sub-sections, the paper will deal with the spacecraft as a theme only for the appeal of the Earth's gravity, which is a bulleted mass (uncluttered) [5]. The paper also will refer primarily to earth centred inertia (ECI), which are described as follow. The root of these axes is located in the center of the earth. X axis is in parallel to the contract line [6]. The Z axis is parallel to the north-south geographical axis of the earth and points to the Nordic. The Y axis supplements a solid triple orthogonal. There are no pads ready to be used. Here, cannot prove the algorithms operation and control systems in the genuine process but can use simulation medium [7]. Presently, there are many simulation tools accessible for satellite plan. For example, the Satellite Tool Kit (STK), created by Analytical Graphics Incorporated, is a phenomenal tool that can be utilized for various portions of the satellite configuration process. Yet, one of the drawbacks of STK and numerous other business test systems created by MATLAB/Simulink or C++ is that it provides clients with only limited customization capability. Other configurable emulators, such as CubeSim, exist before digital integration errors, resulting in reduced simulation precision as the simulation period rise. Include a small model of current simulators [8]. In this paper, space education curricula and human resources are explained and experienced to back up the National Aeronautics Program. One of the basic aspects of space innovation is the information of orbital objects, particularly Earth-orbiting satellites. An artificial trial is done to understand the base of the Keplerian elements that depend on the space mechanics basis, and the basic constellation of satellites is presented utilizing the Walker constellation concept in theory utilizing Matlab to deliver visual development of satellite orbit.

2 Methods

At first, the initial velocity and position of satellite are given, and then the six Keplerian elements are computed. After that Euler angle transformations takes place producing orbit prediction.

2.1 Position and Velocity Coordinates

In the ECI indication framework, the spacecraft positions and velocities affected just by the gravitational fascination of the earth being taken in the form of dotted blocks shall be indicated as follows:

$$r = [x \quad y \quad z]^{\mathrm{T}} \tag{1}$$

Where r is the position of satellite, (x, y, z) are the three coordinates of the satellite position.

$$v = \begin{bmatrix} v_x & v_y & v_z \end{bmatrix}^{\mathrm{T}} = \frac{\mathrm{d}r}{\mathrm{d}t} \tag{2}$$

Where v is the velocity of satellite, (v_x, v_y, v_z) are the three components of the satellite velocity.

The speeding up of such a satellite fulfills the condition of two-body movement

$$\frac{\mathrm{d}^2 r}{\mathrm{d}t^2} = -\mu \frac{r}{\|r\|^3} \tag{3}$$

Where μ is the earth attraction factor. The special setting for this differential equation is called second-order orbit that can be elliptical, equivalent or superfluous, depend upon the underlying estimations of the satellite position vectors and the velocity vectors $r(t0)$ and $v(t0)$. Only the elliptical path is beheld in this investigation [9, 10]. Time t is ever related with the case vector, what's more, is frequently viewed as a seventh segment. Time is called as an indication to the state vector.

2.2 Keplerian Elements

The traditional orbital elements are the six Kepler elements as shown in Figs. 1 and 2. When viewed from an inertial frame, two orbital objects follow distinct paths. All of these paths has its concentrate in the core of the common mass. When sight from a clear frame centered on an object, the path of the corresponding object is only visible; the cable elements describe these inertial paths [11, 12]. The orbit contains two groups of Keplerian elements based on the body utilized as a reference point. The body is known as the essential reference, another is named the minor body. The base level does not

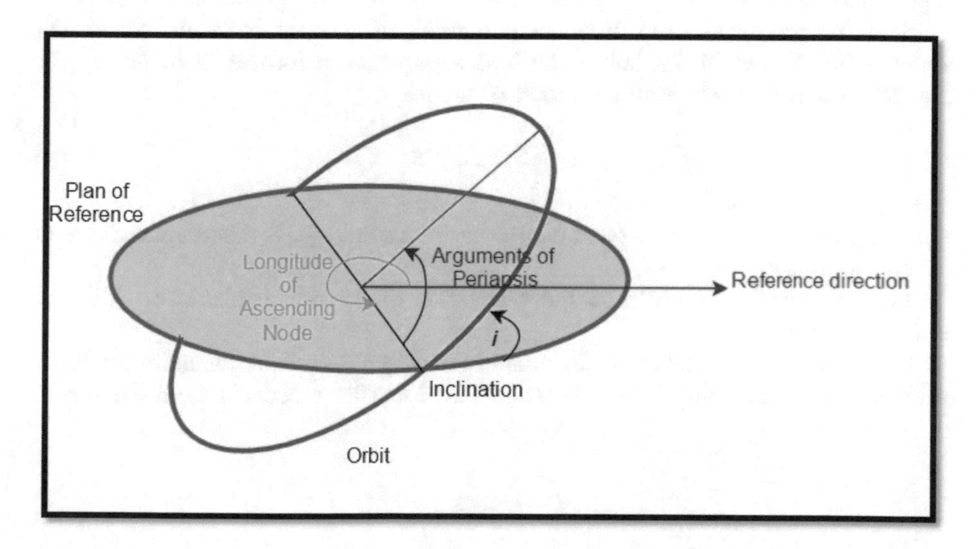

Fig. 1. The first three Keplerian elements: inclination angle, arguments of periapsis and longitude of ascending node

require to contain a mass greater than minor level, and even when the objects are similar in mass, the orbital components depend upon the selection of the base.

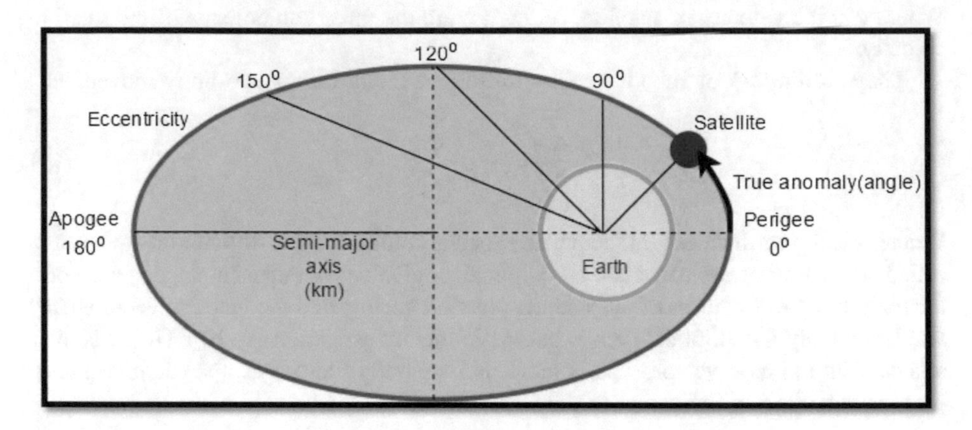

Fig. 2. The second three Keplerian elements: eccentricity, true anomaly and semi-major axis

The two main components that determine the form and dimension of the ellipse:

Eccentricity (e) - a form of ellipse, which describes how much is spread relative to a circle, and then a vector of decentralized direction

$$\vec{e} = \frac{(v^2 - \mu/r)\vec{r} - (\vec{r}.\vec{v})\vec{v}}{\mu} \quad (4)$$

Where, r is position, v is velocity vectors and μ is standard gravitational parameter.

Semi-main axis (a) Total distances periapsis and apoapsis isolated by two. For round orbits, the semimajor hub is the separation between focuses of items, not the distance between objects from the center of the mass.

$$a = \frac{r_{\max} - r_{\min}}{2} \quad (5)$$

Where $r_{\max}$ is maximum distance of the ellipse from a and $r_{\min}$ is minimum distance of the ellipse from a.

Two components characterize the orientation of the orbital plane in which the ellipse is installed:

Inclination (i) - The default inclination of the ellipse relative to the indication level, estimated at the climbing hub. The inclination i can be processed from the orbital energy vector h as

$$i = \arccos \frac{h_z}{|\mathrm{h}|} \quad (6)$$

Where h is orbital momentum vector and h_z is the z-component of h.

Interprets the ascending node of the ellipse horizontally relative to the angle point of the indication frame. In astronomical dynamics, the length of the ascending node can be computed as follows:

$$n = k \times h \tag{7}$$

Where n is a vector that point out to the rising hub, k is the unit vector (0, 0, 1), which is the typical vector to the x-y indication plane and h is orbital momentum vector.

Longitude of the ascending node (Ω) horizontally orients the ascending node of the ellipse with respect to the reference frame's vernal point

$$\Omega = \begin{cases} \arccos \frac{n_x}{|n|}, n_y \geq 0 \\ 2\pi - \arccos \frac{n_x}{|n|} < 0 \end{cases} \tag{8}$$

Here, the reference plane is thought to be the x-y plane, and the longitude origin is taken to be the positive x-axis.

And finally, Argument of periapsis (ω) the direction of the elliptical shape at the orbital level is defined as an edge estimated from the climbing hub to the deviation. In astrodynamics the contention of periapsis ω can be calculated as follows:

$$\omega = \arccos \frac{n.e}{|n||e|} \tag{9}$$

Where n is a vector pointing towards the ascending node and e is the eccentricity vector.

True anomaly (ν, θ) at epoch (M0) characterizes the position of the orbiting body along the ellipse at a particular time (the "epoch"). For elliptic orbits, the true anomaly ν can be computed from orbital state vectors as:

$$\nu = \arccos \frac{e.r}{|e||r|} \tag{10}$$

Where v is the orbital speed vector of the circling body, e is the eccentricity vector, r is the orbital position vector of the circling body. Normal conflict is a numerically advantageous "point" that differs directly with time, however does not compare to a true geometric edge. It can be changed over into the true anomaly v, which meets to a true geometric point at the orbital level, between the difference and the position of the orbital object at any known time. Tilt edges, line length of the increasing axis, and deviation contention can likewise be represented as Euler angles that determine trajectory of the relative orbit of the indication coordinate framework [13, 14].

2.3 Euler Angle Transformations

The angles Ω, i, ω are the Euler edges portraying the orientation of the coordinate framework $\hat{x}$, $\hat{y}$, $\hat{z}$ from the inertial arrange outline $\hat{I}, \hat{J}, \hat{K}$ as shown in Fig. 3.

Where, Î, Ĵ is in the tropical plane of the focal body. Î is toward the vernal equinox. Ĵ is opposite to Î and with Î characterizes the indication plane. K̂ is opposite to the indication plane. Orbital components of bodies in the solar system usually use the ecliptic as that plane [15, 16]. x̂, ŷ are in the orbital plane and with x̂ toward the path to the periapsis. ẑ is opposite to the plane of the circle. ŷ is commonly opposite to to x̂ and ẑ [17].

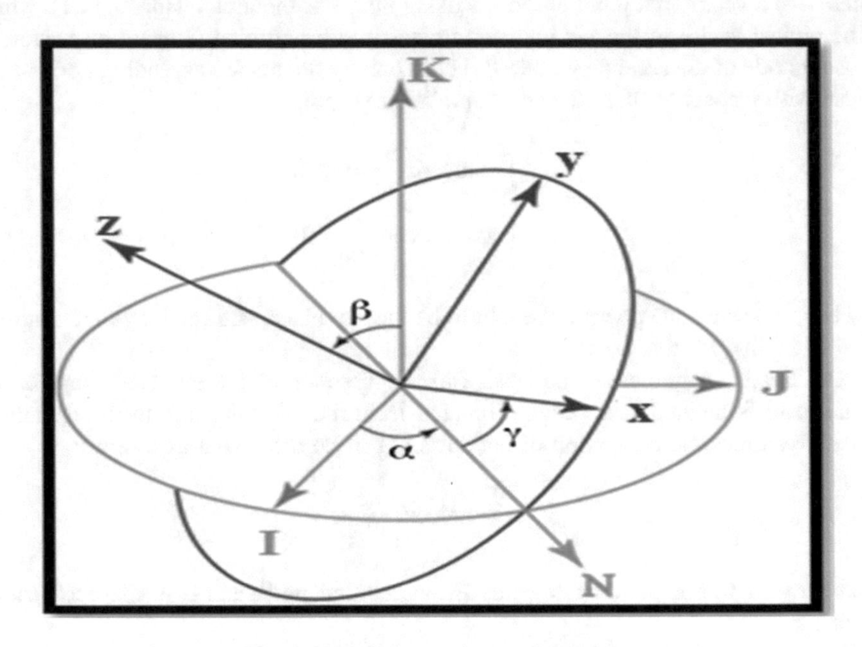

Fig. 3. The Euler angle transformations

At that point, the change from the Î, Ĵ, K̂ arrange casing to the x̂, ŷ, ẑ outline with the Euler angles Ω, i, ω is

$$x_1 = \cos\Omega.\cos\omega - \sin\Omega.\cos i.\sin\omega \tag{11}$$

$$x_2 = \sin\Omega.\cos\omega + \cos\Omega.\cos i.\sin\omega \tag{12}$$

$$x_3 = \sin i.\cos\omega \tag{13}$$

$$y_1 = -\cos\Omega.\sin\omega - \sin\Omega.\cos i.\cos\omega \tag{14}$$

$$y_2 = -\sin\Omega.\sin\omega + \cos\Omega.\cos i.\cos\omega \tag{15}$$

$$y_3 = \sin i.\cos\omega \tag{16}$$

$$z_1 = \sin i.\sin\Omega \tag{17}$$

$$z_2 = -\sin i.\cos\Omega \tag{18}$$

$$z_3 = \cos i \tag{19}$$

$$\begin{pmatrix} x_1 & x_2 & x_3 \\ y_1 & y_2 & y_3 \\ z_1 & z_2 & z_3 \end{pmatrix} = \begin{pmatrix} \cos\omega & \sin\omega & 0 \\ -\sin\omega & \cos\omega & 0 \\ 0 & 0 & 1 \end{pmatrix} \cdot \begin{pmatrix} 1 & 0 & 0 \\ 0 & \cos i & \sin i \\ 0 & -\sin i & \cos i \end{pmatrix} \cdot \begin{pmatrix} \cos\Omega & \sin\Omega & 0 \\ -\sin\Omega & \cos\Omega & 0 \\ 0 & 0 & 1 \end{pmatrix} \tag{20}$$

Where

$$\widehat{x} = x_1\widehat{\mathrm{I}} + x_2\widehat{\mathrm{J}} + x_3\widehat{\mathrm{K}} \tag{21}$$

$$\widehat{\mathrm{y}} = \mathrm{y}_1\widehat{\mathrm{I}} + \mathrm{y}_2\widehat{\mathrm{J}} + \mathrm{y}_3\widehat{\mathrm{K}} \tag{22}$$

$$\widehat{\mathrm{z}} = \mathrm{z}_1\widehat{\mathrm{I}} + \mathrm{z}_2\widehat{\mathrm{J}} + \mathrm{z}_3\widehat{\mathrm{K}} \tag{23}$$

The backwards transformation, which registers the 3 facilitates in the I-J-K framework given the 3 (or 2) coordinates in the x-y-z framework, is represented by the opposite grid. As per the standards of grid polynomial math, the opposite framework of the result of the 3 turn matrices is acquired by altering the request of the three lattices and exchanging the indications of the three Euler edges.

The transformation from $\hat{x}, \hat{y}, \hat{z}$ to Euler angles Ω, i, ω is:

$$\Omega = \arg(-\mathrm{z}_2, \mathrm{z}_1) \tag{24}$$

$$i = \arg\left(\mathrm{z}_3, \sqrt{z_1^2 + z_2^2}\right) \tag{25}$$

$$\omega = \arg(\mathrm{y}_3, \mathrm{x}_3) \tag{26}$$

Where arg(x, y) implies the polar contention that can be figured with the standard function atan2(y, x) accessible in numerous programming dialects.

2.4 Orbit Prediction

Under normal conditions of a remarkable round focal object and zero irritations, every single orbital component in the mean inconsistency is constants. The mean peculiarity changes directly with time, scaled by the mean movement using:

$$n = \sqrt{\frac{\mu}{a^3}} \tag{27}$$

Where n is the mean motion, a is the semi-major axis or mean distance, and μ is a constant for any particular gravitational system.

3 Simulation Results

According to the implementation of Keplerian elements equations, the three axes of the center of earth were constructed, the earth is then placed on the coordinates spinning around its axis. The orbit is constructed in three dimensions around the earth, then the satellite is placed on the orbit and move around the earth, as shown in Fig. 4.

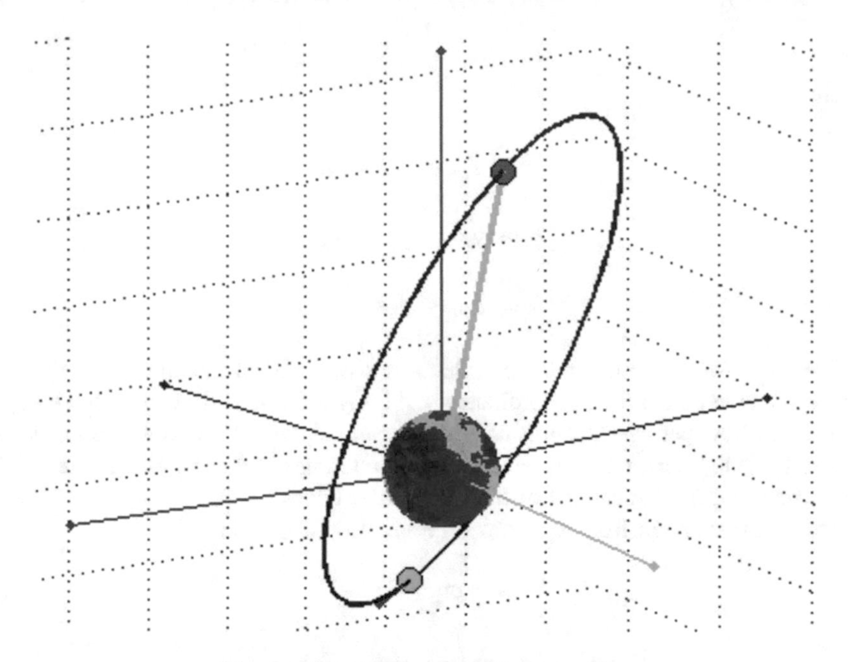

Fig. 4. The visualization of orbit satellite

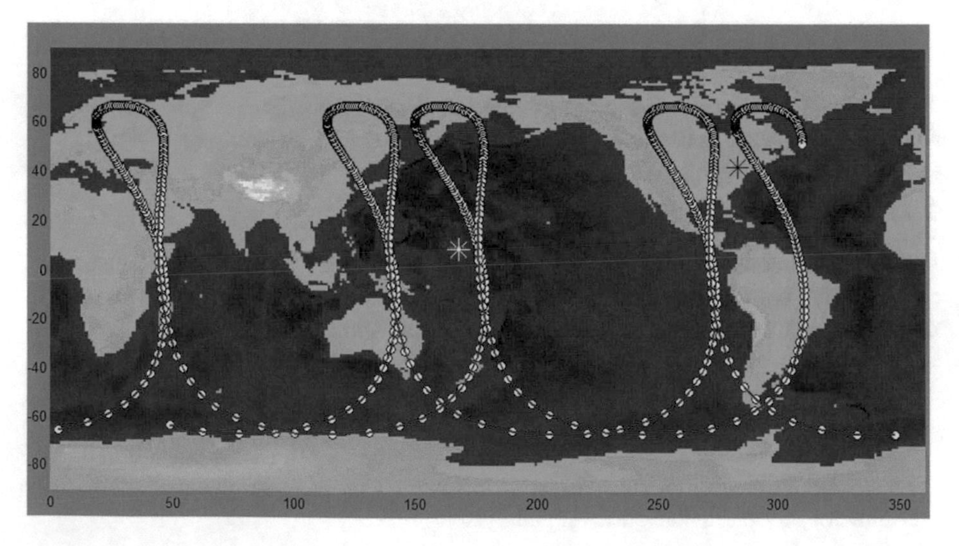

Fig. 5. Tracing of the satellite around earth

The map of the world was build and the satellite tracked through it as shown in Fig. 5.

The following outputs were computed from the system presented in the Table 1.

Table 1. The output elements from the orbital simulation system

Computed element	Value
Eccentricity e (%)	0.497%
Right ascension of the ascending node RAAN (degree)	32.005°
Argument of periapsis ω (degree)	256.289°
Inclination i (degree)	67.023°
Semi major a (km)	31139.026 km
Semi minor b (km)	27014.916 km
Orbital Period (h)	54685.039 h
Altitude_low (km)	9281.475 km
Altitude_high (km)	40254.577 km
GMST greenwich mean sidereal time (s)	0.0074 s

4 Conclusion

The paper has provided a simulated model for the satellite orbit for tracking based on Keplerian system. The solid body pattern of the Multibody's standard library is expanded by adding equations that determine the centre of the body's transformation of cluster coordinates from Keplerian variables to Cartesian coordinates, and by assigning identified cases. The results show the super-precision and the calculation of the six elements in the case of reference to the gravitational field of the cluster point and show the calculated elements from the orbital simulation system and the strong visualization of satellite tracking.

Acknowledgment. This research was supported by a TEDDSAT Project grant, Egypt.

References

1. Kim, W., Chinn, J.Z., Katz, I., Garrett, H.B., Wong, K.F.: 3-D NUMIT: a general 3-D internal charging code. IEEE Trans. Plasma Sci. **45**(8), 2298–2302 (2017)
2. Wang, J.Z., Chen, H.F., Yu, X.Q., Zou, H., Shi, W.: Study on internally dielectric charging of multilayer circuit board. Sci. Sinca Tech. **45**(3), 330–337 (2015). (in Chinese)
3. Beecken, B.P., Englund, J.T., Lake, J.J., Wallin, B.M.: Application of AF-NUMIT2 to the modeling of deep-dielectric spacecraft charging in the space environment. IEEE Trans. Plasma Sci. **43**(9), 2817–2827 (2015)
4. Li, T., Zhou, H., Luo, H., You, I., Qi, X.: SAT-FLOW: multi-strategy flow table management for software defined satellite networks. Access IEEE **5**(2), 14952–14965 (2017)
5. Mijumbi, R., Serrat, J., Gorricho, J., Bouten, N., De Turck, F., Boutaba, R.: Network function virtualization: state-of-the-art and research challenges. Commun. Surv. Tutor. IEEE **18**(1), 236–262 (2015)

6. Yang, B., Wu, Y., Chu, X., Song, G.: Seamless handover in software-defined satellite networking. IEEE Commun. Lett. **20**(9), 1768–1771 (2016)
7. Zhou, J., Ye, X., Pan, Y., Xiao, F., Sun, L.: Dynamic channel reservation scheme based on priorities in LEO satellite systems. J. Syst. Eng. Electron. **26**(1), 1–9 (2015)
8. Liao, M., Liu, Y., Hu, H., Yuan, D.: Analysis of maximum traffic intensity under pre-set quality of service requirements in low earth orbit mobile satellite system for fix channel reservation with queueing handover scheme. IET Commun. **9**(13), 1575–1582 (2015)
9. Musumpuka, R., Walingo, T., MacGregor Smith, J.: Performance analysis of correlated handover service in LEO mobile satellite systems. IEEE Commun. Lett. **8**(11), 515–523 (2016)
10. Moscholios, I., Logothetis, M., Boucouvalas, A.: Blocking probabilities of elastic and adaptive calls in the Erlang multirate loss model under the threshold policy. Telecommun. Syst. **62**(1), 245–262 (2016)
11. Stasiak, M.: Queuing systems for the Internet. IEICE Trans. Commun. **E99-B**(6), 1234–1242 (2016)
12. Moscholios, I., Vassilakis, V., Logothetis, M., Boucouvalas, A.: A probabilistic threshold-based bandwidth sharing policy for wireless multirate loss networks. IEEE Wirel. Commun. Lett. **5**(3), 304–307 (2016)
13. Glabowski, M., Hanczewski, S., Stasiak, M.: Modelling load balancing mechanisms in self-optimizing 4G mobile networks with elastic and adaptive traffic. IEICE Trans. Commun. **E99-B**(8), 1718–1726 (2016)
14. Nunes, B., Mendonca, M., Nguyen, X., Obraczka, K., Turletti, T.: A survey of software-defined networking: past, present, and future of programmable networks. Commun. Surv. Tutor. IEEE **16**(3), 1617–1634 (2014)
15. Ferrús, R., Koumaras, H., Sallent, O., Agapiou, G., Rasheed, T., Kourtis, M.-A., Boustie, C., Gélard, P., Ahmed, T.: SDN/NFV-enabled satellite communications networks: opportunities, scenarios and challenges. Phys. Commun. **18**(3), 95–112 (2016)
16. Chen, T., Matinmikko, M., Chen, X., Zhou, X., Ahokangas, P.: Software defined mobile networks: concept, survey, and research directions. IEEE Commun. Mag **53**(11), 126–133 (2015)
17. Jeannin, N., Castanet, L., Radzik, J., Bousquet, M., Evans, B., Thompson, P.: Smart gateways for terabit/s satellite. Int. J. Satell. Commun. Netw. **32**(5), 93–106 (2014)

A Novel Rough Sets Positive Region Based Parallel Multi-reduction Algorithm

Guangyao Dai[1], Tongbang Jiang[1], Yonglin Mu[1(✉)], Nanxun Zhang[1], Hongbo Liu[1(✉)], and Aboul Ella Hassanien[2(✉)]

[1] School of Information Science and Technology, Dalian Maritime University, Dalian, China
yonglinmu@foxmail.com, lhb@dlmu.edu.cn
[2] Information Technology Department, Cairo University, Giza, Egypt
aboitcairo@gmail.com

Abstract. How to extract more knowledge from a complex information system is a hot issue in the big data era. Attribute reduction, as an effective method in rough set theory, is widely used for knowledge acquirement. In this paper, a novel positive region based parallel multi-reduction algorithm (POSMR) is proposed. Two strategies that for simplifying decision tables and computing the attribute importance are introduced based on MapReduce mechanism at first. And a non-core attribute replacement strategy based on positive region is employed to obtain multi-reduction. The experimental results conducted on UCI Datasets show the algorithm proposed in this paper can obtain multi-reduction correctly, and it is superior to the the binary discernibility matrix based parallel multi-reduction algorithm (BDMR) and traditional single-reduction algorithm.

Keywords: Knowledge acquirement · Rough sets · Multi-reduction Positive region · MapReduce

1 Introduction

The rough set theory [12,16], as one of the most important tool in the filed of granular computing [3,14], is a mathematical technique for the identification and recognition in the uncertain and incomplete data. Attribute reduction [4,15], which can be considered as a selection process that attempts to explore a minimal subset of the original attribute set with minimum information loss, is a key idea in rough set theory for knowledge acquisition [13]. As a result of knowledge acquired by using attribute reduction is more concise and easier to understand, there are widely research on different attribute reduction algorithms [8,18].

However, because of the fast-growing data in scientific and industrial areas, traditional attribute reduction algorithms are facing the challenges on massive data in big data era in the data storage and computational complexity [10]. Besides, limited to higher time and space complexity, traditional attribute

A. E. Hassanien et al. (Eds.): AISI 2018, AISC 845, pp. 515–524, 2019.
https://doi.org/10.1007/978-3-319-99010-1_47

reduction algorithms are usually paid more attention on how to obtain a single minimal reduction. However, in practice, multiple reductions usually exist in information systems and can be thought of providing insights from different perspectives [2,19]. MapReduce [5,7], as a programming model and an associated implementation for parallel computing, is commonly used in the field of data mining and knowledge acquisition in the big data era, and the efficiency of traditional algorithms is greatly improved after parallel [9,11]. As many computations can be carried out simultaneously in the task of positive region based attribute reduction, MapReduce may be a good solution to improve the computational efficiency of positive region based attribute reduction on massive data.

In this paper, we propose a novel positive region based parallel multi-reduction algorithm (POSMR). By constructing the proper $<key, value>$ pairs and implementing MapReduce functions, the decision tables are simplified and the attribute importance is computed in parallel. Then, a non-core attribute replacement strategy based on positive region is employed to obtain multi-reduction. According to the POSMR algorithm, the knowledge contained in massive data can be interpreted from multiple perspectives efficiently.

2 Methodology

2.1 Rough Sets

In rough set theory, the information system (IS) can be defined as a quadruple: $IS = <U, AT, V, f>$, where U and AT are two non-empty finite sets containing all the objects and attributes, respectively. V is a set containing all the values of the attributes, and for each $a \in AT$, the value of attribute a is denoted as V_a. f is the information function, and $f(x, a)$ represents the value of object x in attribute a. If AT is denoted as $C \bigcup D$, where C is the set of condition attributes and D is the set of decision attributes, the quintuple $<U, C, D, V, f>$ is called a Decision Table (DT).

Definition 1. Given a DT, for $\forall X \in U$ and $A \in AT$, we can characterize X by a pair of lower and upper approximations in the condition of the attribute A as shown in Eqs. (1) and (2):

$$\underline{A}(X) = \{x \in U | [x]_A \subseteq X\}, \tag{1}$$

$$\overline{A}(X) = \{x \in U | [x]_A \cap X \neq \varnothing\}. \tag{2}$$

The positive region of A on X can be denoted by the $\underline{A}(X)$ as $POS_A(X) = \underline{A}(X)$.

According to the positive region on the equivalence relation of decision attribute D, the attribute reduction can be defined as shown in Definition 2.

Definition 2. For an attribute subset B of C in a decision table, if $POS_B(D) = POS_C(D)$ and for each attribute $b \in B$, $POS_{B-\{b\}}(D) \neq POS_C(D)$, B is a reduction of C. And the set including many reductions like B is called multi-reduction (MR).

In this paper, the multi-reduction is obtained by a non-core attribute replacement strategy based on positive region. After obtaining an initial reduction set based on attribute importance, the non-core attributes in the initial reduction set are replaced by the rest attributes that not in the set of initial reduction in order to obtain multi-reduction. The attribute importance and core attribute are defined as shown in Definitions 3 and 4.

Definition 3. Given a decision table DT, B is an attribute subset of condition attribute set C, for $\forall c \in C - B$, the attribute importance of c to B on decision attribute D can be defined as shown in Eq. (3).

$$Sig_{POS}(c, B, D) = |POS_{B \bigcup c}(D)| - |POS_B(D)|, \tag{3}$$

where $|POS_B(D)|$ represents the number of objects in the positive region of B on D.

Definition 4. For an attribute c_i in C in a decision table, if $POS_{C-c_i}(D) = POS_C(D)$, c_i is an indispensable attribute of C on D. The set contained all indispensable attribute c_i is called the core attribute.

2.2 MapReduce

MapReduce is one of the most popular parallel computing framework. Unlike parallel programming models such as OpenMP [1] and MPI [6], MapReduce does not increase the details of distributed computing, and it provides a set of distributed computing strategies (e.g., executing monitoring and checking pointing). In addition, MapReduce provides a distributed file system that provides an extensible mechanism for storing large amounts of data.

Two functions, Map and Reduce (as shown in Eqs. (4) and (5)), are used to express the computation in the MapReduce framework.

$$Map: \ <k_1, v_1> \rightarrow list \ <k_2, v_2>; \tag{4}$$

$$Reduce: \ <k_2, list(v_2)> \ \rightarrow list \ <k_3, v_3>. \tag{5}$$

All of the data can be represented as a $<key, value>$ form in the MapReduce framework. In the Map step, $<k_1, v_1>$ is assigned as the input to each processor, and an intermediate $key-value$ pair $<k_2, v_2>$ is generated. All the intermediate values v_2 associated with k_2 are combined by MapReduce, and these values will be passed to the *Reduce* function. After that, the *Reduce* function accepts these values and produces a smaller set $<k_3, v_3>$ as the final output in the *Reduce* step. The framework of MapReduce can be seen as shown in Fig. 1.

3 A Positive Region Based Parallel Multi-reduction Algorithm

In this section, a novel positive region based parallel multi-reduction algorithm (POSMR) is proposed. The POSMR algorithm is composed of 7 sub-algorithms as shown in Algorithms 1–5.

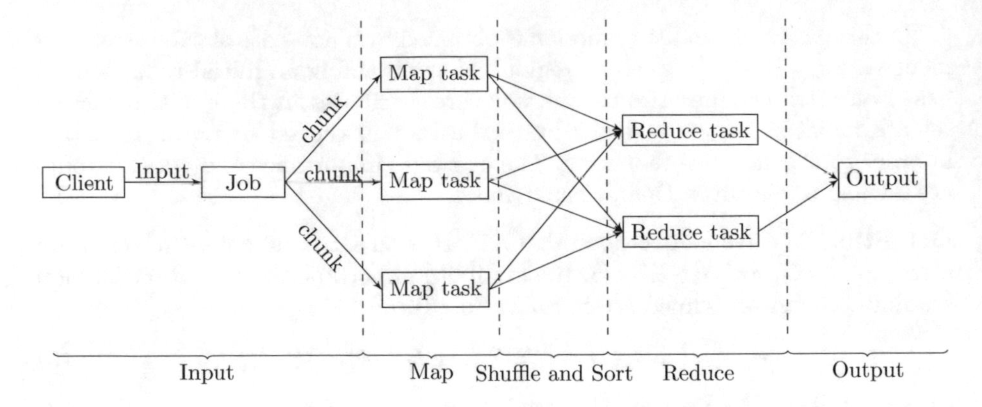

Fig. 1. Framework of MapReduce

In this paper, we adopt a parallel strategy for both data and tasks. The data is divided into several smaller pieces of data slices at first. And for each independent data slice, the equivalence class is computed based on different candidate attributes and a simplified table can be obtained at the same time. The process of simplifying the decision table can be seen in Algorithms 1 and 2.

Algorithm 1. Simplification_Mapper

Input:

A slice S_i of a decision table $DT =< U, C, D, V, f >$.

Output:

$< key', value' >$, where key' is the set of condition attributes of the equivalence class partitioned by U/C, $value'$ is the values of decision attributes corresponding to the equivalence class.

1: **for** each object $x \in S_i$ **do**
2:     **for** each $c \in C$ **do**
3:         $key' \leftarrow \{f(x, c_1), f(x, c_2), \cdots, f(x, c_n)\}$, where $f(x, c_i)$ represents the value of object x in attribute c_i;
4:         $value' \leftarrow \{f(x, d)\}$, where $f(x, d)$ represents the value of object x in decision attribute d;
5:         Output $< key', value' >$.
6:     **end for**
7: **end for**

According to the simplified table obtained by Algorithms 1 and 2, the attribute importance can be computed in parallel as shown in Algorithms 3 and 4. After that, the final multi-reduction can be obtained as shown in Algorithm 5.

Algorithm 2. Simplification_Reducer

Input:

$< key, value >$, where key is $C_n = \{f(x, c_1), f(x, c_2), \cdots, f(x, c_n)\}$, $value$ is $D_n = \{f(x_1, d), f(x_2, d), \cdots, f(x_m, d)\}$

Output:

$< key', value' >$, where key' is the serial number y of a equivalence class, $value'$ is the set of $\{C_n, D_n\}$.

1: $y \leftarrow 1$;
2: **for** each $i \in n$ **do**
3:   **for** each $j \in m$ **do**
4:     **if** $f(x_1, d) = f(x_2, d), \cdots, f(x_m, d)$ **then**
5:       $D_n = f(x_1, d)$;
6:     **else**
7:       $D_n = -1$;
8:     **end if**
9:   **end for**
10:   $key' \leftarrow y$;
11:   $value' \leftarrow \{C_n, D_n\}$;
12:   $y++$;
13:   Output $< key', value' >$.
14: **end for**

Algorithm 3. Importance_Mapper

Input:

$< key, value >$, where key is the file offset, $value$ is a set of $\{c_i, c_i(x), d(x), sum\}$ (sum is the occurrences of attribute c_i).

Output:

$< key', value' >$, where key' is the $c_i(x)$ in $value$, $value'$ is a set of $\{d(x), sum\}$.

1: **for** each object $< key, value >$ **do**
2:   $key' \leftarrow c_i(x)$
3:   $value' \leftarrow \{d(x), sum\}$;
4:   Output $< key', value' >$.
5: **end for**

4 Experimental Results

In this section, we compare the reduction results obtained by the proposed parallel algorithms in this paper (POSMR) with the results obtained in only one node and the binary discernibility matrix based parallel multi-reduction algorithm (BDMR) [17] to show the correctness and superiority of the POSMR algorithm. We conducted these experiments on 11 datasets from UCI as shown in Table 1. In Table 1 dataset KDDCPU is a large dataset with 4898432 objects and 42 attributes.

Datasets ILPD, Seeds, Iris, Glass, and Breast cancer are used to show the multi-reduction results obtained by the proposed parallel algorithms in this paper (POSMR) and the serial program in only one node as shown in Table 2.

Algorithm 4. Importance_Reducer

Input:

$< key, value >$, where key is $c_i(x)$, $value$ is a set of $\{d(x), sum\}$.

Output:

$< key^{'}, value^{'} >$, where $key^{'}$ is $c_i(x)$, $value$ is num .

1: **for** each $< key, value >$ **do**
2: **if** $|value| = 1 \bigwedge d(x) \neq -1$ **then**
3: $num = sum$;
4: **else**
5: $num = 0$;
6: **end if**
7: $key^{'} \leftarrow c_i(x)$
8: $value^{'} \leftarrow num$;
9: **end for**
10: **for** each $c_i(x) \in V_{c_i}$ **do**
11: **for** each $< key^{'}, value^{'} >$ **do**
12: $Sig_{c_i} = \Sigma num$
13: **end for**
14: **end for**

We can see form Table 2 that the results of multi-reduction obtained by serial program and POSMR are completely consistent. So, POSMR can obtain multi-reduction correctly.

The binary discernibility matrix based multi-reduction algorithm is also a commonly used algorithm for obtaining multi-reduction. A binary discernibility matrix model must be established before compute the attribute importance, so this method usually wastes a lot of time and space. We compare our algorithm with BDMR in the number of multi-reduction results on small datasets (ID:1-10 in Table 1). As getting more reductions means getting more knowledge, the multi-reduction algorithms focus on a larger number of multi-reduction results. The comparison of multi-reduction results can be seen in Fig. 2. The numbers of multi-reduction results obtained by POSMR are larger on datasets Breast Cancer (ID:1), EEG Eye (ID:3), Yeast (ID:5) and Ionosphere (ID:5) than BDMR as shown in Fig. 2, while in other datasets their numbers are almost equal. The running time on big datasets of parallel algorithms is also a important algorithm evaluation indicator. We compare our POSMR algorithm and BDMR algorithm on datasets KDDCUP (ID:11) at running tim. The dataset KDDCUP is divided into five parts, named $DS1$, $DS2$, $DS3$,$DS4$ and $DS5$, respectively. The comparison of running time can be seen in Fig. 3, the running time of POSMR algorithm is shorter than BDMR algorithm on the five data sets. So the POSMR algorithm proposed in this paper is superior to BDMR algorithm in the number of multi-reduction results and running time.

As one of a goal of attribute reduction is to improve the classification accuracy of the raw data, we compare the classification accuracy under different datasets of raw data with different multi-reduction results in Table 3. As shown in Table 3, the classification accuracy is improved after attribute reduction in the majority

Algorithm 5. Multi-Reduction Algorithm

Input:
A simplified decision table $DT' = (U', C, D, V, f)$.
Output:
The set of multi-reduction $Red = \{Red_1, Red_2, \cdots, Red_n\}$, where Red_p is a reduct.

1: $Red = Red_1 = c_{max} = Core = \phi, p = 2$;
2: $Sig_c = \{Sig_{c_1}, Sig_{c_2}, \cdots, Sig_{c_n}\}$;
3: **for** each $i \in n$ **do**
4:   **if** $POS_C(D) \neq POS_{C-c_i}(D)$ **then**
5:     $Core = Core \bigcup \{c_i\}$;
6:   **end if**
7: **end for**
8: **while** $POS_C(D) \neq POS_{Red_1}(D)$ **do**
9:   $Sig_c = Sig_c - \{Sig_{c_{max}}\}$;
10:   $c_{max} = \{\arg\max_c Sig_c\}$;
11:   $Red_1 = Red_1 \bigcup c_{max}$.
12: **end while**
13: $Red = Red \bigcup Red_1$;
14: **for** each $c' \in Red_1 - Core$ **do**
15:   $Red' = Red_1 - \{c'\}$;
16:   **for** each $c \in C - Red_1$ **do**
17:     $Red_p = Red' \bigcup \{c\}$;
18:     **if** $POS_C(D) = POS_{Red_p}(D)$ **then**
19:       $Red = Red \bigcup Red_p$;
20:       p=p+1;
21:       break;
22:     **end if**
23:   **end for**
24: **end for**
25: Output $Red = \{Red_1, Red_2, \cdots, Red_n\}$.

Table 1. UCI datasets

ID	Datasets	Objects	Attributes	Classes
1	Breast cancer	699	10	2
2	Heart	270	14	2
3	EEG eye	14980	15	2
4	German	1000	21	4
5	Yeast	1484	9	10
6	ILPD	538	11	2
7	Seeds	210	8	3
8	Iris	150	5	3
9	Ionosphere	351	35	2
10	Glass	214	10	7
11	KDDCUP	4898432	42	-

Table 2. The results of multi-reduction

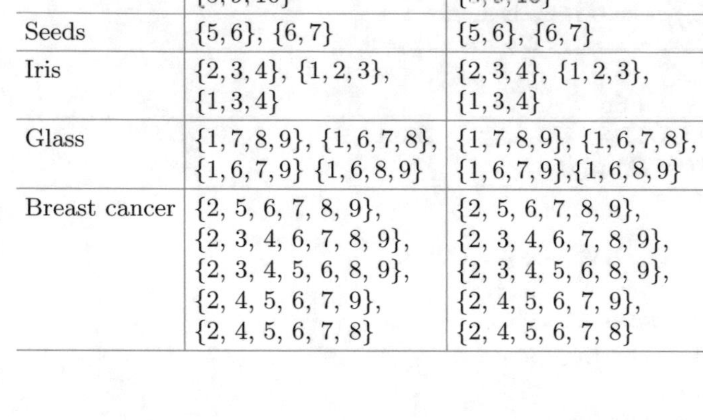

Datasets	Serial program	POSMR
ILPD	{5, 8, 9}, {5, 8, 10}, {5, 9, 10}	{5, 8, 9}, {5, 8, 10}, {5, 9, 10}
Seeds	{5, 6}, {6, 7}	{5, 6}, {6, 7}
Iris	{2, 3, 4}, {1, 2, 3}, {1, 3, 4}	{2, 3, 4}, {1, 2, 3}, {1, 3, 4}
Glass	{1, 7, 8, 9}, {1, 6, 7, 8}, {1, 6, 7, 9} {1, 6, 8, 9}	{1, 7, 8, 9}, {1, 6, 7, 8}, {1, 6, 7, 9},{1, 6, 8, 9}
Breast cancer	{2, 5, 6, 7, 8, 9}, {2, 3, 4, 6, 7, 8, 9}, {2, 3, 4, 5, 6, 8, 9}, {2, 4, 5, 6, 7, 9}, {2, 4, 5, 6, 7, 8}	{2, 5, 6, 7, 8, 9}, {2, 3, 4, 6, 7, 8, 9}, {2, 3, 4, 5, 6, 8, 9}, {2, 4, 5, 6, 7, 9}, {2, 4, 5, 6, 7, 8}

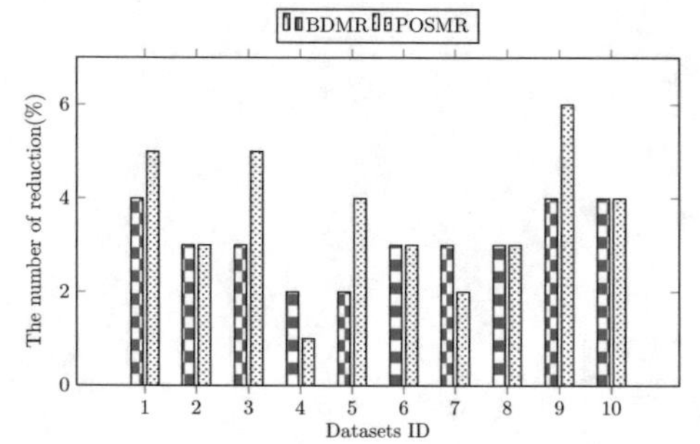

Fig. 2. The comparison of multi-reduction results

Fig. 3. The comparison of running time

Table 3. The comparison of classification accuracy

Datasets	Raw	Re_1	Re_2	Re_3	Re_4	Re_5	Re_6
Breast cancer	94.31	94.76	94.31	94.83	94.31	93.84	-
Heart	82.93	81.66	83.23	82.93	-	-	-
EEG eye	92.91	93.33	92.24	92.11	93.33	89.24	-
German	80.40	82.12	-	-	-	-	-
Yeast	52.69	52.62	51.45	50.32	53.31	-	-
ILPD	71.02	69.32	69.89	71.32	-	-	-
Seeds	92.19	92.81	93.75	-	-	-	-
Iris	91.30	95.65	91.30	92.30	-	-	-
Ionosphere	91.51	91.51	92.69	90.43	94.88	91.20	90.43
Glass	69.23	76.92	72.31	70.77	69.23	-	-

of cases, and a best classification accuracy is always can be obtained in one of the multi-reduction results. However, as far as a single reduction is concerned, the classification accuracy may be equal to or even less than the raw data. So the POSMR algorithm proposed in this paper is superior to the single reduction algorithm as a reduction result of a best classification accuracy can always be obtained.

5 Conclusion

In order to extract more knowledge from massive data effectively, we present a novel positive region based parallel multi-reduction algorithm (POSMR) using MapReduce mechanism. As the introduction of three strategies that for simplifying decision tables, computing the attribute importance, and replacing non-core attributes, respectively, the multi-reduction can be obtained successfully and effectively. The experiments show the correctness and superiority of the proposed POSMR algorithm. However, our algorithm is not guaranteed to get all the available reductions. A complete reduction set can be used to deduce a comprehensive knowledge system that reveals the all the intangible insights hidden in an information system. So, it is necessary to explore a parallel algorithm that can obtain a complete reduction set on massive data. And some detailed underlying mechanism is also worth investigating in the future works.

Acknowledgment. The authors would like to thank Zhi Wu for his scientific collaboration in this research work. This work is partly supported by the National Natural Science Foundation of China (Grant Nos. 61472058, 61602086, 61772102 and 61751205).

References

1. Aldea, S., Estebanez, A., Llanos, D.R., Gonzalez-Escribano, A.: An OpenMP extension that supports thread-level speculation. IEEE Trans. Parallel Distrib. Syst. **27**(1), 78–91 (2016)
2. Chao, Y., Liu, H., Mcloone, S., Chen, C.L.P., Wu, X.: A novel variable precision reduction approach to comprehensive knowledge systems. IEEE Trans. Cybern. **48**(2), 661–674 (2018)
3. Dai, G., Wang, Z., Yang, C., Liu, H.: A multi-granularity rough set algorithm for attribute reduction through particles particle swarm optimization. In: International Computer Engineering Conference, pp. 303–307 (2015)
4. Dai, J., Hu, Q., Hu, H., Huang, D.: Neighbor inconsistent pair selection for attribute reduction by rough set approach. IEEE Trans. Fuzzy Syst. **26**(2), 937–950 (2018)
5. Dean, J., Ghemawat, S.: Mapreduce: a flexible data processing tool. Commun. Acm **53**(1), 72–77 (2010)
6. Degomme, A., Legrand, A., Markomanolis, G., Quinson, M., Stillwell, M., Suter, F.: Simulating MPI applications: the SMPI approach. IEEE Trans. Parallel Distrib. Syst. **28**(8), 2387–2400 (2018)
7. Esposito, C., Ficco, M.: Recent developments on security and reliability in large-scale data processing with mapreduce. IGI Global **12**(1), 49–68 (2016)

8. Fan, J., Jiang, Y., Liu, Y.: Quick attribute reduction with generalized indiscernibility models. Inf. Sci. **397**(C), 15–36 (2017)
9. Igual, L., Segui, S.: Parallel Comput. Springer International Publishing, New York City (2017)
10. Qian, J., Miao, D., Zhang, Z., Yue, X.: Parallel attribute reduction algorithms using mapreduce. Inf. Sci. **279**(1), 671–690 (2014)
11. Salah, S., Akbarinia, R., Masseglia, F.: A highly scalable parallel algorithm for maximally informative k-itemset mining. Knowl. Inf. Syst. **50**(1), 1–26 (2017)
12. Swiniarski, R.W., Skowron, A.: Rough set methods in feature selection and recognition. Pattern Recogn. Lett. **24**(6), 833–849 (2003)
13. Vashist, R.: An algorithm for finding the reduct and core of the consistent dataset. In: International Conference on Computational Intelligence and Communication Networks, pp. 741–745 (2016)
14. Wang, G., Yang, J., Xu, J.: Granular computing: from granularity optimization to multi-granularity joint problem solving. Granular Comput. **2**(3), 1–16 (2017)
15. Yao, Y., Zhao, Y.: Attribute reduction in decision-theoretic rough set models. Inf. Sci. **178**(17), 3356–3373 (2008)
16. Zhang, H., Shu, L., Liao, S.: Hesitant fuzzy rough set over two universes and its application in decision making. Soft Comput. **21**(7), 1803–1816 (2017)
17. Zhang Yan, X.U.: Guilin: quick attribution reduction algorithm based on simple binary discernibility matrix. Comput. Sci. **33**(4), 155–158 (2006)
18. Zhu, X.Z., Zhu, W., Fan, X.N.: Rough set methods in feature selection via submodular function. Soft Comput. **21**(13), 3699–3711 (2017)
19. Zhu, Z., Li, H., Dai, G., Abraham, A., Yang, W.: A rough set multi-knowledge extraction algorithm and its formal concept analysis. In: International Conference on Intelligent Systems Design and Applications, pp. 25–29 (2015)

Automatic Counting and Visual Multi-tracking System for Human Sperm in Microscopic Video Frames

Nour Eldeen M. Khalifa[1,2](✉), Mohamed Hamed N. Taha[1], and Aboul Ella Hassanien[1,2]

[1] Information Technology Department, Faculty of Computers and Information, Cairo University, Giza, Egypt
{nourmahmoud, mnasrtaha, aboitcairo}@cu.edu.eg
[2] Scientific Research Group in Egypt (SRGE), Giza, Egypt
http://www.egyptscience.net

Abstract. In this paper, a proposed system for automatic counting and visual multi-tracking for human sperm in microscopic video frames is presented. It can be easily turned into a commercial computer-assisted sperm analysis (CASA) system. CASA systems help in detecting infertility in human sperm according to clinical parameters. The proposed system consists of nine phases and it counts sperm in every single frame of video in real time and calculates the average sperm count through the whole video with accuracy 94.3% if it is compared to the manual counting. Also, it tracks all identified sperm in video frames in real time. It works with different frame rates above 15 frame/s to track visually the movements of the sperm. The dataset consists of three high-quality 1080p videos with different frame rates and durations. Finally, the open challenging research points are addressed.

Keywords: Sperm counting · Sperm tracking
Computer-assisted sperm analysis (CASA)

1 Introduction

Human sperm motility is of great interest to biologists studying sperm function and to medical physician for male infertility evaluating and treating [1]. Approximately the infertility rate in Egypt is 9% according to the Egyptian IVF registry report [2]. Most of the married couples seek diagnostic semen analysis to help assess the cause.

Semen examination is divided into two groups, namely examination in macroscopic and microscopic. The macroscopic examinations are general semen inspections without the need for complicated tools. The inspections include the volume, semen color, viscosity, and the pH of the semen. The microscopic examinations are performed for more detail conditions of the semen where a sophisticated tool needed in the process. These include sperm mass movements, motility, sperm concentration, viability, and sperm morphology [3].

A. E. Hassanien et al. (Eds.): AISI 2018, AISC 845, pp. 525–531, 2019.
https://doi.org/10.1007/978-3-319-99010-1_48

Today, the general method for analyzing sperm at fertility clinics and research laboratories is strenuous and subjective [4]. Typically, lab technicians use microscopes for manually counting the number of sperm cells and for visually evaluating the quality of sperm movement according to standard protocols. The common parameters evaluated during a typical semen analysis can be divided into nine different parameters [5]. In some modern clinics, more objective Computer Assisted Semen Analysis (CASA) instruments are used to trace the swimming paths of sperm automatically in time-lapse microscopy image sequences [4]. Digital image processing algorithms could help in performing the semen analysis according to the previous four parameters. Katz and Davis were the first authors who proposed an automatic sperm tracking in the mid-1980s [6], they proposed a software contains over 60 commands. System commands enable the user to capture and manipulate data. The program automatically performs a sequence of data capture, reduction, analysis, and reporting steps with a minimum of user input.

Beresford-Smith et al. [7] applied radar tracking algorithms on sperm tracking. They adapted the probabilistic data association filter to track a single sperm in clutter. However, they reported no experimental data. Tomlinson et al. in [8] described a CASA system based on multi-target tracking algorithms. They tracked multiple sperms and graded their motility using 1-s video clips. However, the authors did not address the problem of tracking through collisions or over long durations.

Hidayatullah et al. in [9] survey on multi-sperm tracking for sperm motility measurement and illustrate the advantages and disadvantages for a group of the proposed CASA system until 2016.

The rest of the paper is organized as follows. Section 2 is a brief description of the CASA systems. Section 3 presents the proposed system for visual multi-tracking for human semen. Experimental results are discussed in Sect. 4. Finally, Sect. 5 conclude the paper and list the possible future work.

2 Computer Assisted Semen Analysis (CASA)

The term CASA stands for “computer-aided sperm analysis” or “computer-assisted sperm analysis”. Over approximately 30 years the Computer-assisted sperm analysis (CASA) systems have improved. This evolution happened through amelioration in devices to capture the image from a microscope, huge increases in computational power, new computer programming languages, and advanced software algorithms. The evolution grants great benefits for animal and human semen analysis. However, the methods for identification of sperm and their motion patterns are little changed. Outdated and slower systems stay in use [10].

In general, the CASA system consists of microscope that is given an additional digital camera. This camera will replace the function of the veterinarian or doctor eye to record video of the sperm being analyzed. This digital video is then sent to the computer for processing.

The recording results are stored in the form of a digital video file. To be processed, this video must be extracted into image frames. The analysis is done on each frame to identify sperm. After the sperm can be identified, the comparison between sequential frames is carried out to obtain sperm motility parameters [9].

Majority of fertility laboratories and semen processing facilities have a CASA system, but the extent of dependence thereon ranges exceedingly. Each system is different. Modern CASA systems can automatically assess several fields in a shallow specimen. Images of 500 to >2000 sperm can be captured, at 50 or 60 frames per second, in clear or complex extenders, and in <2 min, store information for 30 frames and deliver summary data for each spermatozoon and the population [10].

The accurate prediction fertility, from a semen sample, cannot be calculated by CASA. However, current CASA systems provide significant information for quality assurance of semen planned for understanding the responses of sperm responses to changes in the microenvironment in research, and for marketing [10].

3 The Proposed Counting and Visual Multitracking for Human Sperm System

The proposed system consists of nine phases. The first phase is reading the video file frame by frame, while the second phase is converting the video frame from RGB domain to HSV domain according to Eq. (1). In phase number 3, a thresholding value will be applied to video frame according to specific values of Hue, Saturation and Value to segment the semen from the background according to Eq. (2).

$$
\begin{aligned}
h &= \begin{cases} 0^{\circ} & \textit{if max} = \textit{min} \\ 60^{\circ} \times \frac{g-b}{max-min} + 0^{\circ} & \textit{if max} = r \textit{ and } g \geq b \\ 60^{\circ} \times \frac{g-b}{max-min} + 360^{\circ} & \textit{if max} = r \textit{ and } g < b \\ 60^{\circ} \times \frac{b-r}{max-min} + 120^{\circ} & \textit{if max} = g \\ 60^{\circ} \times \frac{r-g}{max-min} + 240^{\circ} & \textit{if max} = b \end{cases} \\
s &= \begin{cases} 0, & \textit{if max} = 0 \\ \frac{max-min}{max} = 1 - \frac{min}{max}, & \textit{otherwise} \end{cases} \\
v &= max
\end{aligned}
\tag{1}
$$

$$
\text{Thresholded}_{\text{HSV}_{\text{frame}}} = \begin{cases} \begin{cases} \mathrm{h} = 1, & 0.973 \leq \mathrm{h} \leq 0.966 \\ \mathrm{h} = 0, & \text{otherwise} \end{cases} \\ \begin{cases} \mathrm{s} = 1, & 0.000 \leq \mathrm{s} \leq 0.351 \\ \mathrm{s} = 0, & \text{otherwise} \end{cases} \\ \begin{cases} \mathrm{v} = 1, & 0.407 \leq v \leq 0.658 \\ \mathrm{v} = 0, & \text{otherwise} \end{cases} \end{cases}
\tag{2}
$$

A binarization of the image is required in phase number 3, while in phase number 5 a pixel grouping is a necessary step for phase number 6 which accept and reject pixel groups according to a pixel group value to remove noise and an unwanted group of pixels as illustrated in Eq. (3).

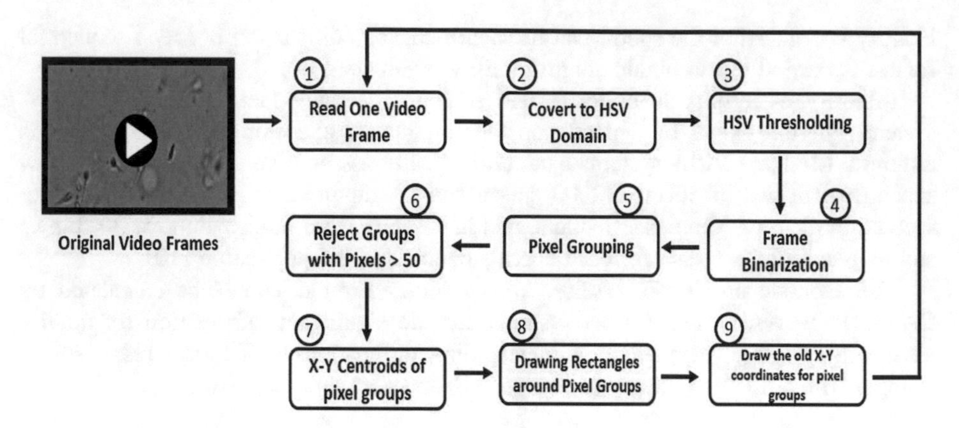

Fig. 1. Proposed system block diagram for multi-tracking and counting human sperm.

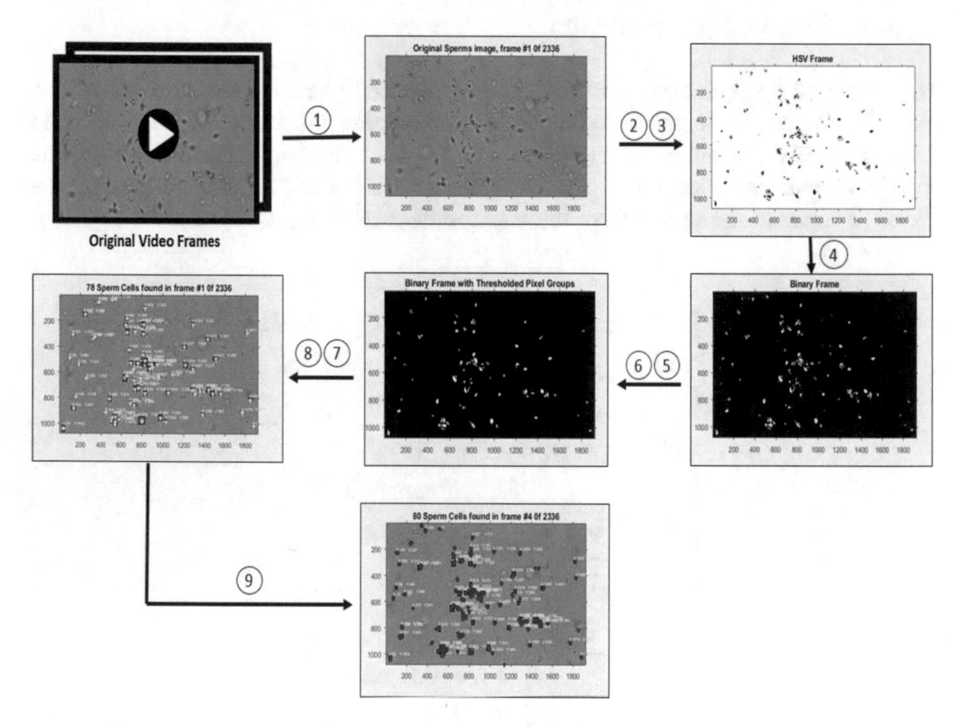

Fig. 2. The output of video frame after every phase of the proposed system

$$Threshold_{pixel\,group} = \begin{cases} 1 & number\,of\,pixels\,in\,a\,group \leq 50 \\ 0 & number\,of\,pixel\,in\,a\,group > 50 \end{cases} \quad (3)$$

In phase number 7, the proposed system calculated the center of the pixel groups according to x–y coordinates and append them to a global x–y parameter and count all pixel groups which indicated the number of sperm. Drawing a rectangle around the pixel groups is done in phase number 8 for tracking purposes. In the final phase, the system draws all the old x–y coordinates saved in the global x–y parameter. Figure 2 illustrates all the required phases in a block diagram for the proposed system.

One of the main functions of the proposed system is visually tracking all sperm in the video frame. Figure 1 presents the output video frame after every phase of the 9 phases the proposed system had.

As illustrated in Fig. 2, the resulted frame after phase 5, 6 contains only the actual visual sperm without noise while the final frame after phase 9 contains the original video frame with yellow rectangle tracking the sperms and blue dots for the previous x–y coordinates for all sperms spotted in the previous frames.

4 Experimental Results

The proposed system was implemented using a commercial software package (MATLAB). All experiments were performed on a server with Intel Xeon E5-2620 processor and 96 GB Ram. To evaluate the proposed system, three microscopic videos for human sperm sample were selected. The video's description is presented in Table 1.

Table 1. Videos description used by the proposed system

Video name	Duration (sec)	# of frames/sec	Quality
Video 1	93	25	1080p
Video 2	60	27	1080p
Video 3	28	30	1080p

The proposed system is designed to perform 2 main functions. The first function is to count the pixel groups (Human Sperm) in every frame and display the count number above the frame in real time as illustrated in Fig. 3. The average number of sperms is calculated at the end of the video. The average number of sperm for the whole video is 94.3% accuracy with +2, −2 as the standard deviation for error if it is compared to the manual counting.

The second function of the proposed system is to track every sperm and its movements through the video by tracking it is previous x–y coordinates in the previous video frames. Figure 4 illustrated a zoomed tracking for different sperm movement in video frames.

The proposed system analyze the video frames in different frame rates. Figure 4 illustrated the movements of sperms in 25 frame/s. It can operate in 1 frame/s, but the movements can't be noticed and that will affect the whole performance of the system. So, the preferable frame rate is above 15 frame/s to notice the movement of the sperms and track them visually.

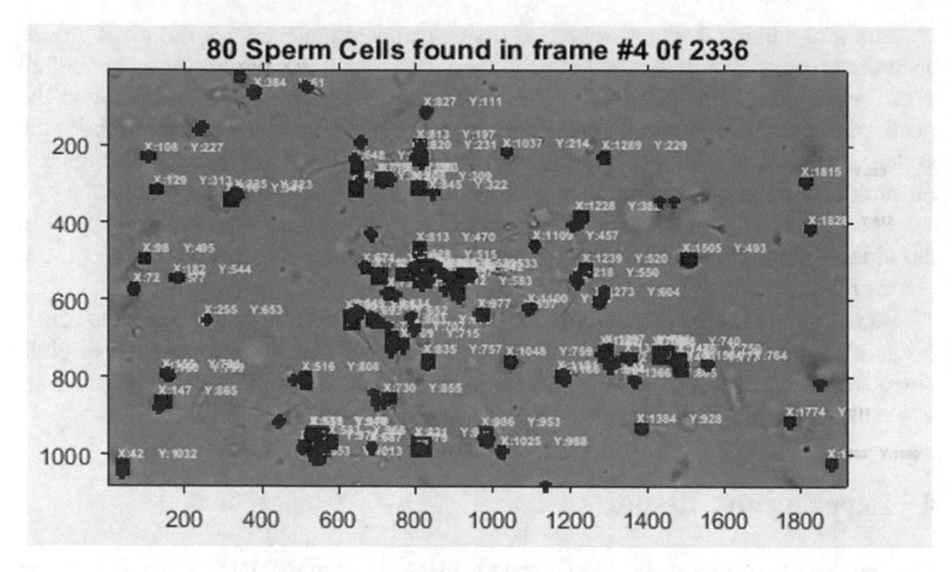

Fig. 3. Number of sperms in every frame

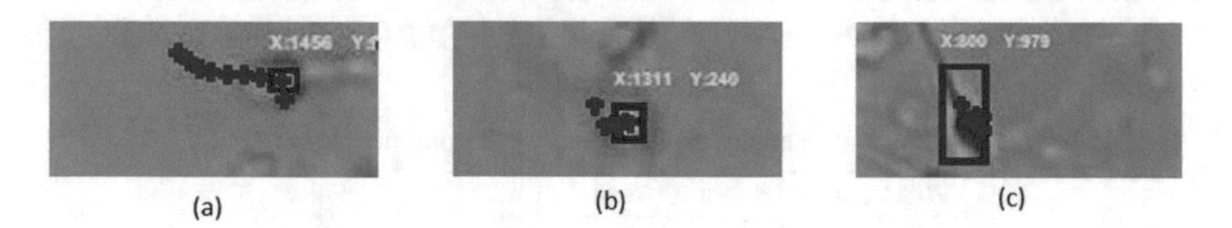

Fig. 4. (a) Sperm with fast movement, (b) sperm with medium movement and (c) sperm with slow movement

The proposed system achieved its goals to count the sperms in every frame and calculated the average the sperm on the whole video. It visually tracks every sperm in the video highlighting its path through the video sample.

5 Conclusions and Future Work

The counting and visual multitracking for human sperm is considered one of the motivating topics of interest to researchers over the past 30 decades. The development of computer-assisted semen analysis (CASA) has also attracted more researchers into this field as those systems are applied in most of the laboratories across Egypt and even worldwide. In this paper, a proposed system was introduced to count human sperm in video frames. It counts sperm in every single frame in real time and calculates the average count of sperm in the whole video with accuracy 94.3% and error +2 or −2 percent if it is compared to the manual counting. Also, the proposed system tracks visually all identified sperm across the video and visually mark the sperm with a yellow rectangle and draw blue dots for the previous x–y coordinates of sperm movement across the previous video frames.

There is still a room for improvements as a future work for the proposed system. The following future points are under open research and they are: (1) The intersections of sperm movements, it will be visually hard to track; (2) Analyzing x–y coordinates to detect the quality of the sperm by its speed, and (3) Improving the system to work with low-quality videos.

References

1. Hallak, J.: A call for more responsible use of assisted reproductive technologies (ARTs) in male infertility: the hidden consequences of abuse, lack of andrological investigation and inaction. Transl. Androl. Urol. **6**, 997–1004 (2017). https://doi.org/10.21037/tau.2017.08.03
2. Mansour, R., El-Faissal, Y., Kamal, O.: The Egyptian IVF registry report: assisted reproductive technology in Egypt 2005. Middle East Fertil. Soc. J. **19**, 16–21 (2014). https://doi.org/10.1016/j.mefs.2014.01.001
3. Gómez Montoto, L., Magaña, C., Tourmente, M., Martín-Coello, J., Crespo, C., Luque-Larena, J.J., Gomendio, M., Roldan, E.R.S.: Sperm competition, sperm numbers and sperm quality in muroid rodents. PLoS ONE **6**, e18173 (2011). https://doi.org/10.1371/journal.ponc.0018173
4. Urbano, L.F., Masson, P., VerMilyea, M., Kam, M.: Automatic tracking and motility analysis of human sperm in time-lapse images. IEEE Trans. Med. Imaging **36**, 792–801 (2017). https://doi.org/10.1109/TMI.2016.2630720
5. World Health Organization: WHO Laboratory Manual for the Examination and Processing of Human Semen. World Health Organization, Geneva (2010)
6. Katz, D.F., Davis, R.O., Delandmeter, B.A., Overstreet, J.W.: Real-time analysis of sperm motion using automatic video image digitization. Comput. Methods Programs Biomed. **21**, 173–182 (1985). https://doi.org/10.1016/0169-2607(85)90002-1
7. Beresford-Smith, B., Van Helden, D.F.: Applications of radar tracking algorithms to motion analysis in biomedical images. In: Proceedings of 1st International Conference on Image Processing. IEEE Computer Society Press, pp. 411–415 (1994)
8. Tomlinson, M.J., Pooley, K., Simpson, T., Newton, T., Hopkisson, J., Jayaprakasan, K., Jayaprakasan, R., Naeem, A., Pridmore, T.: Validation of a novel computer-assisted sperm analysis (CASA) system using multitarget-tracking algorithms. Fertil. Steril. **93**, 1911–1920 (2010). https://doi.org/10.1016/j.fertnstert.2008.12.064
9. Hidayatullah, P., Mengko, T.L.E.R., Munir, R.: A survey on multisperm tracking for sperm motility measurement. Int. J. Mach. Learn. Comput. **7**, 144–151 (2017). https://doi.org/10.18178/ijmlc.2017.7.5.637
10. Amann, R.P., Waberski, D.: Computer-assisted sperm analysis (CASA): capabilities and potential developments. Theriogenology. **81**, 5.e3–17.e3 (2014). https://doi.org/10.1016/j.theriogenology.2013.09.004

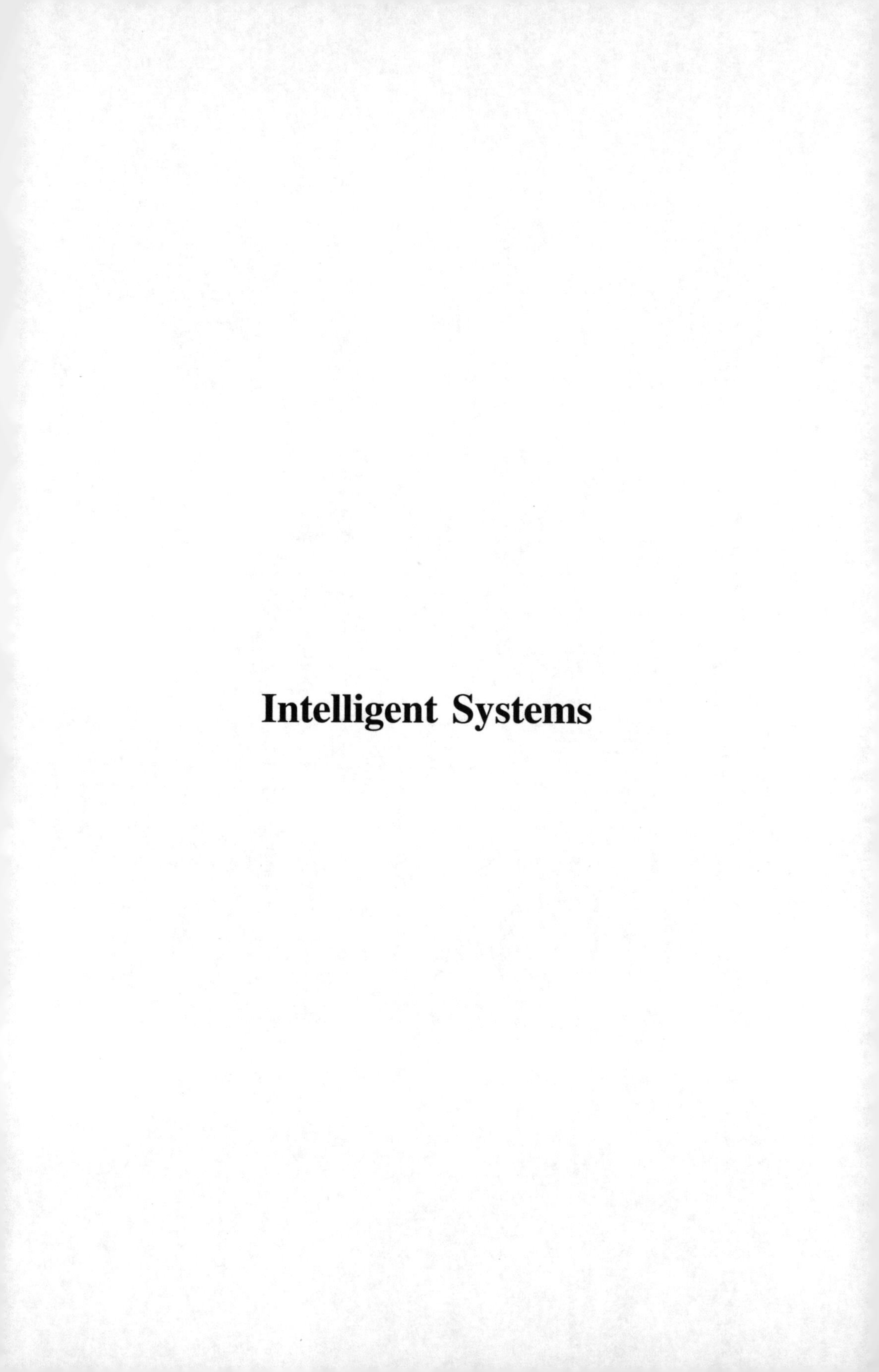

Intelligent Systems

Arabian Horse Identification System Based on Support Vector Machines

Ayat Taha[1,4](✉), Ashraf Darwish[2,4], and Aboul Ella Hassanien[3,4]

[1] Faculty of Science, Al-Azhar University, Cairo, Egypt
Ayat_taha@ymail.com
[2] Faculty of Science, Helwan University, Cairo, Egypt
ashraf.darwish.eg@ieee.org
[3] Faculty of Computers and Information, Cairo University, Cairo, Egypt
aboitcairo@gmail.com
[4] Scientific Research Group in Egypt (SRGE), Cairo, Egypt
http://www.egyptscience.net

Abstract. In this paper, a new approach for Arabian horse identification is represented base on bag of feature algorithm. This approach is based on three phases, the first is the extract bag of feature process which use speed up robust feature (SURF) to extract the features of muzzle print images then using K-mean cluster and histogram to extract features. The second phase is train support vector machine (SVM) classification which trains the features with its labels. Finally the SVM testing phase, the bag of feature is extracted from the input images and test SVM model to match the images with its labels. Arabian horse is correct identified if it matched to its label and if it matched to one of the other horse's labels, the horse not identified. We represent results for the approach for different SVM kernels at different cluster number. The results demonstrate that SVM with polynomial kernel achieve height accuracy 95.4% compare to other kernels.

Keywords: Animal biometric · Arabian horse identification · Bag of feature Speed up robust feature · Support vector machine

1 Introduction

Arabian horse is the most famous breed of all horses. The breed's had been developed in Arabia by the 7th century. The valued of the Arabian horse for its speed, stamina, beauty, intelligence, and gentleness. The Arabian horse identification is critical to keep its management of vaccination, controlling the disease outbreak, management of production, and tracking cattle with assigning the owner [1]. Traditional animal identification approaches like ear notching, tattooing, and hot or freeze branding are insufficient as they are susceptible to fraud, theft, and duplication [2]. In addition, the technology of Radio Frequency Identification (RFID) which is used currently in tracking has some problems like insulation, security, and operational problems [3]. So it is important to find new ways based on designing new and robust animal identification approaches.

A. E. Hassanien et al. (Eds.): AISI 2018, AISC 845, pp. 535–545, 2019.
https://doi.org/10.1007/978-3-319-99010-1_49

In this context, biometrics approach can be deployed instead of traditional approaches for Arabian horse identification. Biometric approach for animal identification is a promising trend which depends on the animal itself. In this trend, there is no need to insert or attach an external device. In the literature, there are many animal biometric applications such as cattle classification, cattle tracking in all its daily life phases, and understanding animal disease trajectories [4]. The muzzle print of animal is considered as an important biometric and is similar to the human's fingerprint. The muzzle print approach is preferred than traditional methods as there is no need to insert or attach any external parts within the animals and it complies with legal rules of most countries. For different animals with the same species, we can find that the muzzle print is unique [5, 6]. However, collecting inked muzzles is traditional method and time consuming and need special skills. In addition, the inked muzzle print images have no enough quality [7]. Currently, the muzzle-based method is new and prominent for identification process of Arabian horse. This approach can achieve a high accuracy to Arabian horse identification based on live captured muzzle print images as in [8] with level of accuracy is 93% [9].

Bag of features or bags of key points is used to maximize the identification accuracy and minimize computational efforts [10]. To get bag of features SURF and K-mean algorithm. SURF [11] algorithm is used for features detection and descriptions which it used to overcome as noise, illumination, transformation, rotation, invariant to variation, and occlusion [12]. K-mean also used for assign the feature descriptor to predefined clusters. K-means algorithm is a simple algorithm and mostly used in pattern recognition which based on Square-error partitioning algorithms that minimizes the within-cluster scatter or maximizes between-cluster scatter. So SURF algorithm and K-mean cluster are chosen to get bag of features.

The rest of part of this paper is organized as follows: Sect. 2 presents the basics and background of SURF, K-means, and SVM classifiers. The proposed Arabian horse identification approach is represents in Sect. 3. Section 4 describes the experimental results and evaluation of the approach. Conclusions and future work are presented in Sect. 5.

2 Background

2.1 Speeded Up Robust Feature

The SURF introduced by Bay et al. [11] is a speeded-up version of SIFT. SURF has two stages to have the descriptor. The first stage depends on detecting SURF point and then extracting the descriptor at the SURF point in the second phase. For the detection stage, a hessian based a scale invariant blob detector is used to find interest points. In an image I, the Hessian matrix calculated at point $X = (x, y)$ and scale of σ as defined by Eq. (1).

$$H(X,\sigma) = \begin{bmatrix} L_{xx}(X,\sigma) & L_{xy}(X,\sigma) \\ L_{yx}(X,\sigma) & L_{yy}(X,\sigma) \end{bmatrix} \tag{1}$$

where,

$$L_{xx}(X,\sigma) = \left(I(X), \frac{\delta^2}{\delta x^2} g(\sigma)\right) \tag{2}$$

$$L_{xy}(X,\sigma) = \left(I(X), \frac{\delta^2}{\delta x \delta y} g(\sigma)\right) \tag{3}$$

$$L_{yy}(X,\sigma) = \left(I(X), \frac{\delta^2}{\delta y^2} g(\sigma)\right) \tag{4}$$

$$g(\sigma) = \frac{1}{2\pi\sigma^2} e^{\frac{-(x^2+y^2)}{2\sigma^2}} \tag{5}$$

The convolution between the integral image and Gaussian second order derivative is represented by Eqs. (2), (3), and (4). Instead of the complicated calculation of second order Gaussian derivative a set of box filters are used to approximate it with the help of integral images [13]. The 9 × 9 box filters are approximations of a Gaussian with scale σ = 1.2. The image scale space is analyzed by changing the size of box filter and the approximated determinant of Hessian can be evaluated by using the following equation

$$\det(H_{approx}) = D_{xx}D_{yy} - (wD_{xy})^2 \tag{6}$$

Where, D_{xx}, D_{yy}, and D_{xy} are the approximated second order derivative of Gaussian. The SURF point's location and scale are obtained by applying non-maximum suppression in 3 × 3 × 3 neighborhood to find maxima.

The descriptor of SURF algorithm can be obtained by constructing a square region of size 20 s centered at the SURF point, and oriented along its main orientation, s is the scale where the interest point is detected. This region is divided into 4 × 4 sub regions and in each sub region, horizontal and vertical Haar wavelet responses denoted as *dx* and *dy* are calculated at 5 × 5 regularly placed sample points [14].

The wavelet responses and the absolute values of it are summed up for each sub-region and entered in a feature vector *v* as in Eq. (7).

$$v = \left(\sum d_x, \sum d_y, \sum |d_x|, \sum |d_y|,\right) \tag{7}$$

Where, d_x and d_y are the wavelet responses. $|d_x|$, $|d_y|$ are the absolute value of it. The feature descriptor is finally obtained by computing the feature vector *v* for all the 4 × 4 sub-regions. So the feature descriptor has length 4 × 4 × 4 = 64 dimensions.

2.2 K-Means Clustering

K-means algorithm divided the features into k cluster such that the features in each cluster are similar to each other and different from other clusters. For features of n points $X = (x_1, x_2, \ldots, x_n)$ in R^d k initial centroids are randomly chosen, where k

represent the number of clusters. Each point is assigned to the closest centroid. The closest centroid is measured by Euclidean distance as in Eq. (8) [12].

$$d^2(X, m_j) = \sum_{i=1}^{n} (x_i - m_{ij})^2 \tag{8}$$

Where, m_j are the centroid points $j = (1, 2, \ldots, k)$. The centroid of each cluster is then updated based on the points in each cluster. The assignment and update of centers and points are repeated until the centroid is converges to certain points and don't changes [15].

3 Support Vector Machine

The SVM used for classification problems of height dimensional datasets. SVM tray to find optimal separating hyperplane by maximize margin between negative and positive class [16].

For a training dataset with n sample $(X_i, Y_i) = (x_1, y_1), (x_2, y_2), \ldots, (x_n, y_n)$, X_i is the feature vector in a v-dimensional feature space, and Y_i is its labels which belong to either of two linearly separable classes C1 and C2 [17]. The optimal hyperplane which separate two classes with the maximal margin is founded by Eqs. (9) and (10).

$$\max \sum_{i=1}^{n} \alpha_i - \frac{1}{2} \sum_{i,j=1}^{n} \alpha_i \alpha_j y_i y_j . K(x_i, x_j) \tag{9}$$

$$Subject\ to: \sum_{i=1}^{n} \alpha_i y_i, 0 \leq \alpha_i \leq C \tag{10}$$

where, α_i is the weight assigned to the training sample x_i, C is the parameter of regulation which is used to trade-off the training accuracy and the model complexity so that a superior generalization capability can be achieved. K is defined as the kernel function, which is used to measure the similarity between two samples. There are many kernel functions the most popular are Radial Basis, polynomial, and linear kernel which are used in the Arabian horse identification models [18].

4 The Proposed Arabian Horse Identification Approach

The proposed Arabian horse identification approach as shown in Fig. 1 consists of three phases; creating bag of feature, training SVM classifier, and testing SVM classifier model. In creating bag of feature phase, the SURF features are extracted from the input images using grid method for more features descriptors. K-mean clustering divided and quantize the feature descriptors into k clusters each center represent bag of feature. The bag of feature uses the histogram to encode the input images into a histogram of bag of feature. The histogram length corresponds to the number of bag of feature that constructed using K-mean clustering. The histogram of bag of feature becomes a feature vector for the input images. In the training SVM classifier phase, the features are divided to training and testing features. The training features vectors of the input images are classified with different SVM kernel to train the model for identification. Finally the third step test the SVM model, the bag of feature are extracted from the test images which treated as feature vector and then test the SVM model. The horse is correct identified if matched to one of its images and if it matched to one of the other horse's images, the horse not identified. The algorithm of the proposed Arabian horse identification approach is shown in Algorithm 1.

Algorithm 1: The proposed Arabian horse identification approach
1. Input muzzle print images (Ii, $i = 1, 2, \ldots, N$), where Ii represents the ith training image and N represents the total number of training images
2. **For** *i=1 to N* ***do***
3. Extract feature from each muzzle print image using SURF algorithm at 3 different grid steps (g), where g is the number of grid steps. So each image represent as $f \times 64$, where f is feature descriptor in 64 dimensional.
4. **End for**
5. **For** m=1 to g ***do***
6. Applying k-mean clustering for all features at the same grid step to get number of cluster each cluster center represent the bag of features
7. Applying the histogram to get the feature vector for each image. The vector has length $1 \times l$, where l represent the number of cluster.
8. **End for**
9. **For** k=1:5 ***do***
10. Partition the feature vector for all images to training and testing features vector using k-fold
11. Applying SVM classifier with linear, polynomial, and RBF kernels to identify the Arabian horse.
12. **End for**

5 Experimental Results and Discussion

The experimental process is performed on 300 muzzle print Arabian horse colored image. The data set were collected from 35 female and 15 male of Arabian horses with 6 images for each horse. The dataset are divided to six K-fold to insure robustness of the approach accuracy. In each time the approach use 250 muzzle print images as training data and 50 muzzle print images as test data.

Figure 2 contains a sample of 6 muzzle print of Arabian horse and one image for its face. Class 2 also contains 35 male Arabian horse muzzle print images six images for each with minimum dimension 1000 × 1200 pixels and maximum dimension 1400 × 2100 pixels [19]. In the experimental part, special care has been given to the quality of the collected images.

The collected images have different levels of degradation factors such as the rotation of image and image partiality for simulating some real time identification conditions. In addition, some prepressing operations have been performed for these

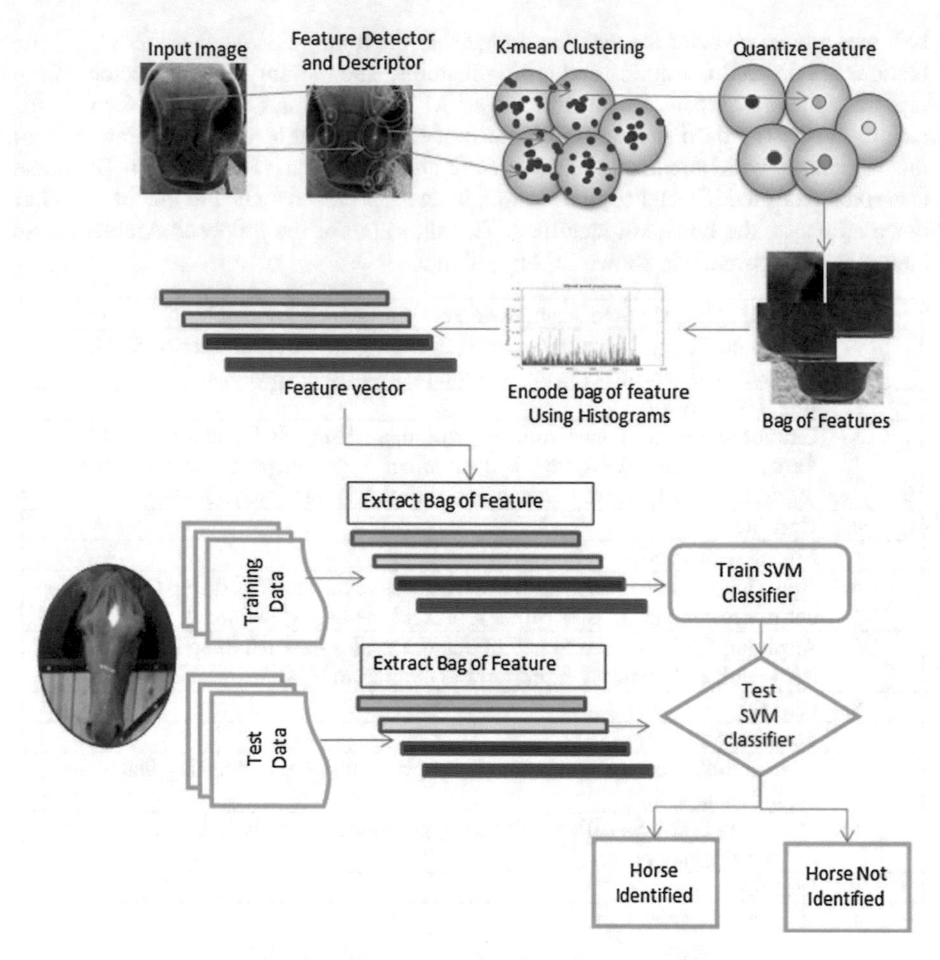

Fig. 1. Arabian horse identification approach

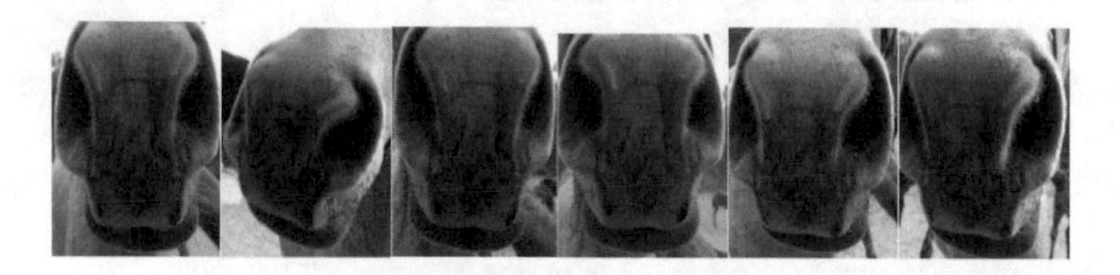

Fig. 2. Samples of six muzzle print of female Arabian horse and one image of its face

images such as, cropping, and given notation which works as FYXX, whereas XX is the cattle ID (1 to 15), and Y is the order of image (1 to 6) for female horses, and MYXX is assigned for male horse whereas XX is the cattle ID (16 to 50), and Y is the order of image (1 to 6).

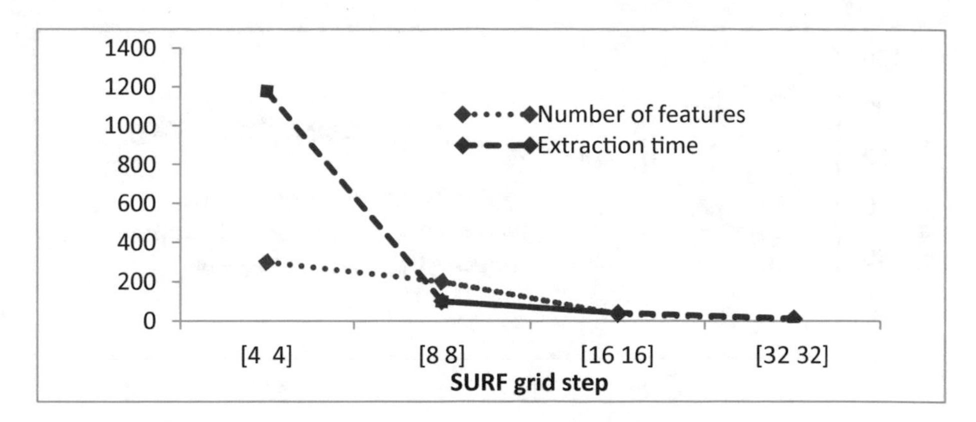

Fig. 3. The number of SURF feature extraction at different grid step and the time consuming

The Arabian horse identification approach has done on three phases creating bag of feature, training SVM model and test SVM classifier. In creating bag of feature phase the features were extracted from the input images by using SURF algorithm to get feature that invariant to scales and rotation. The grid method is used for more features extraction and then applying k-mean clustering to group the similar feature in one cluster. The number of clustering is equal to the number of bag of feature. Many scenarios are performed to determine the best number of clustering that gives best accuracy. Each cluster is quantized to determine the center. The center of each cluster is considered the bag of feature. The histogram is applied to the bag of feature to get feature vector. In SVM training phase the features vectors are divided to six K-fold. Each time 250 features vector used as training data and 50 features vector as test data to increase the robust of the accuracy. The training data are classified into 50 Arabian horses which used as training model. In last phase the model was testing using SVM with different kernels The Arabian horse was correct identified if the muzzle image matched to one of the horse images and not identified if it matched to another horse image as shown in Fig. 1.

The main processes in Arabian horse identification approach are SURF extraction; K-mean clustering, and SVM training and testing classification. In the SURF extraction process the experimental objective is directed toward setting the best SURF parameters that compromise the number of extracted features with the consumed processing time. The grid step size is effective parameter in SURF so the size of the grid step is optimized as shown in Fig. 3. The reported results are the average number (the number of feature are approximated to 0.00001) of features in feature extraction processes for 300 muzzle print images at each grid step size. The maximum number of features is achieved with (grid step size = [4 4]), however, the extraction process time is height compare with the time at other grid step size. The number of features at (grid step size = [8 8]) is reduced by 30%, and the extraction time is reduced by 1000% s. The other grid step size values achieve an acceptable number of features at reasonable time. So the Arabian horse identification approach is performed at grid step [4 4], [8 8], and [16 16] to determine the best grid step that gives best accuracy with reasonable time as shown in Figs. 4, 5, and 6.

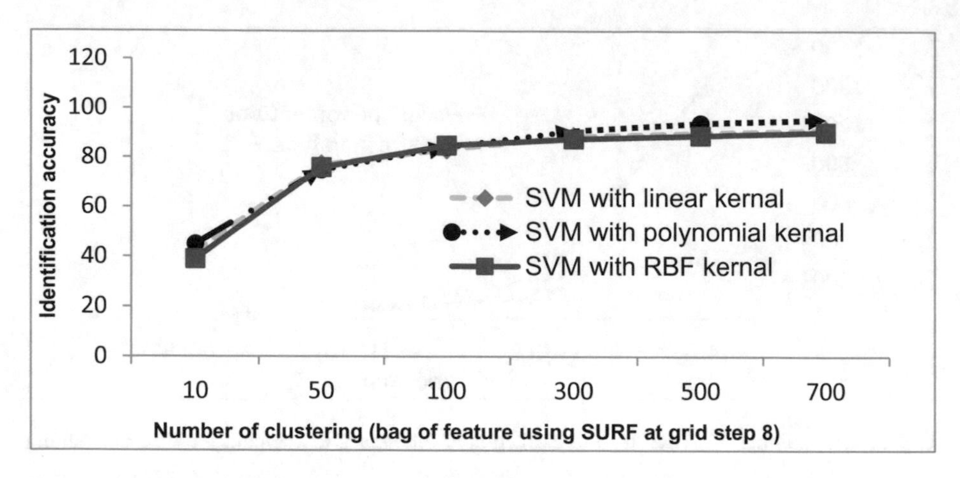

Fig. 4. The accuracy of the Arabian identification approach using SVM with different kernel and different number of cluster at grid step 8

In K-mean clustering process the features are clusters to get bag of features. So we optimize the number of clusters at grid step size [4 4], [8 8], and [16 16] as shown in Figs. 4, 5, and 6. The reported result is the Arabian horse identification approach accuracy for each number of clusters. The results show that the best accuracy is achieved with (number of clusters 700) and decrease by approximately 2% at 500 number of clustering. The other cluster number achieve unacceptable accuracy regardless the time factor. The optimum cluster number value is selected at 700 seeking for the best accuracy at reasonable time.

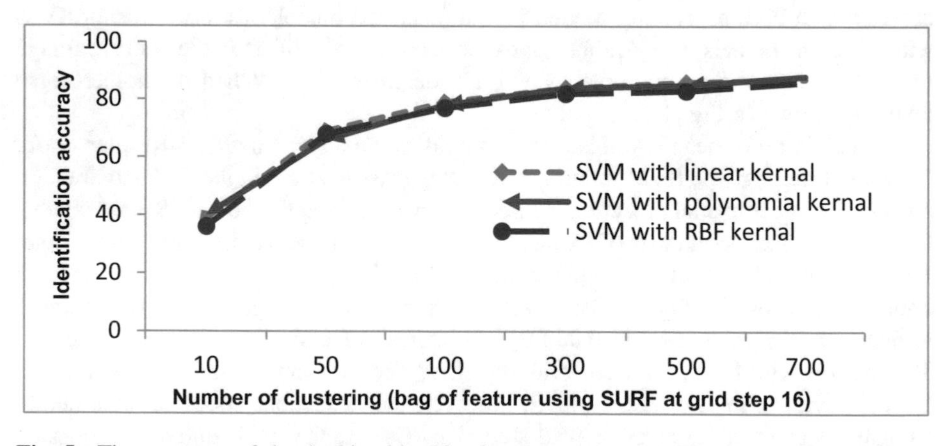

Fig. 5. The accuracy of the Arabian identification approach using SVM with different kernel and different number of cluster at grid step 16

The last process in the Arabian horse identification approach is the SVM classification. The training data are classified using SVM with RBF, linear, and polynomial kernel at three grid steps 8, 16, and 32 as show in Figs. 4, 5, and 6. The reported results of the SVM with linear, RBF, and polynomial kernel accuracy of 50 Arabian horse identification processes for each cluster value at three grid step [8 8], [16 16], and [32 32]. At grid step [8 8], the maximum accuracy is achieved with SVM polynomial kernel however, at grid step [16 16], the accuracy are reduced by 10%, and reduced by 20% at grid step [32 32]. The SVM with linear kernel achieve a good accuracy, however the time increase by 50% compare to SVM polynomial kernel at grid step [8 8] the accuracy is reduced by 5% and reduced by 20% at grid step [32 32]. SVM with radial kernel achieve the worst accuracy compared with other kernels. The best SVM classifier for Arabian horse identification approach is SVM with Polynomial kernel seeking for the best accuracy and best time for matching. The best grid step that give best accuracy is [8 8].

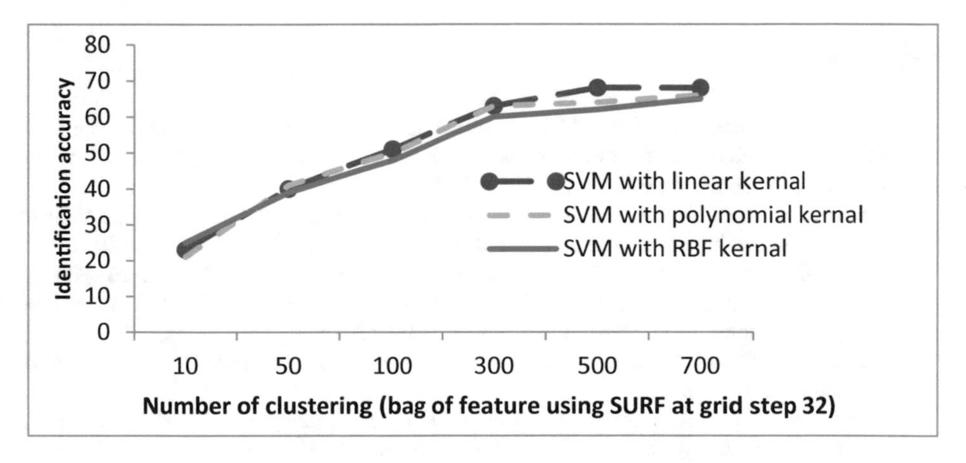

Fig. 6. The accuracy of the Arabian identification approach using SVM with different kernel and different number of cluster at grid step 32

The Arabian horse identification approach is performed under SURF extraction at grid step [8 8], 700 number of cluster on 50 Arabian horses. The accuracy of the identification approach is 95.4% for polynomial kernel and the identification time is 182 s. The accuracy for linear kernel is 91% and the identification time is 240.3 s. The accuracy for RBF kernel is 88.2% and identification time is 216.3 s.

6 Conclusions and Future Work

This paper presents a new Arabian horse identification approach based on muzzle print images and by using SURF and SVM. In the first phase, the process is performed on SURF algorithm for extract features from the muzzle print images, K-mean cluster is used to divide all features into bag of features for images, and histogram is used to assign bag of features into their images and get feature vectors for all images. The second phase is assigned to train the SVM classifier by using the trained data of features and its labels. Finally, the SVM classifier is used to determine the identity of the Arabian horse. This approach has been performed for different grid step for SURF algorithm, number of clusters for K-mean algorithm, and different SVM kernel function to optimize these parameters. At grid step [8 8], the SURF algorithm extract 2 million features from all images with reasonable time in the opposite at grid step [4 4] achieve more number of features but the time increase by 1000% s. The grid step [8 8] shows that the best results for Arabian horse identification approach. The K-mean cluster has been performed for different 700 clusters. The Arabian horse identification approach is achieved best accuracy 95.4% at grid step [8 8], 700 number of K-mean cluster and with polynomial SVM kernel. Muzzle print images database extension is future directions. The reduction of the identification time in a large database is an interesting challenge that will be tackled as a future work.

References

1. Vlad, M., Parvulet, R.A., Vlad, M.S.: A survey of livestock identification approaches. In: 13th WSEAS International Conference on Automation and Information, (ICAI12). WSEAS Press, Iasi, Romania, June 2012, pp. 165–170 (2012)
2. Klindtworth, M., Wendl, G., Klindtworth, K., Pirkelmann, H.: Electronic identification of cattle with injectable transponders. Comput. Electron. Agric. **24**(12), 65–79 (1999)
3. Roberts, C.: Radio frequency identification (RFID). Comput. Secur. **25**(1), 18–26 (2006)
4. El Hadad, H., Mahmoud, A., Mousa, F.A.: Bovines muzzle classification based on machine learning techniques. Proc. Comput. Sci. **65**, 864–871 (2015)
5. Petersen, W.: The identification of the bovine by means of nose-prints. J. Dairy Sci. **5**(3), 249–258 (1922)
6. Baranov, A., Graml, R., Pirchner, F., Schmid, D.: Breed differences and intra-breed genetic variability dermatoglyphic pattern of cattle. J. Anim. Breed. Genet. **110**(1–6), 385–392 (1993)
7. Minagawa, H., Fujimura, T., Ichiyanagi, M., Tanaka, K.: Identification of beef cattle by analyzing images of their muzzle patterns lifted on paper. In: Third Asian Conference for Information Technology in Agriculture (AFITA). Asian Agricultural Information Technology and Management, Beijing, China, October 2002, pp. 596–600 (2002)
8. Awad, A.I., Hassanien, A.E., Zawbaa, H.M.: A cattle identification approach using live captured muzzle print images. Adv. Secur. Inf. Commun. Netw. **381**, 143–152 (2013)
9. Noviyanto, A., Arymurthy, A.M.: Beef cattle identification based on muzzle pattern using a matching refinement technique in the sift method. Comput. Electron. Agric. **99**(1), 77–84 (2013)
10. Csurka, G., Dance, C.R., Fan, L., Willamowski, J., Bray, C.: Visual categorization with bags of keypoints. In: Workshop on Statistical Learning in Computer Vision (ECCV), pp. 1–22 (2004)
11. Bay, H., Ess, A., Tuytelaars, T., Gool, L.V.: Speeded-up robust features. Comput. Vis. Image Underst. **110**, 346–359 (2008)
12. Oyelade, O.J., Oladipupo, O.O., Obagbuwa, I.C.: Application of k-means clustering algorithm for prediction of students' academic performance. Int. J. Comput. Sci. Inf. Secur. **7** (2010)

13. Verma, N.K., Goyal, A., Vardhan, H., Sevakula, R.K., Salour, A.: Object matching using speeded up robust features. In: 19th Asia Pacific Symposium on Intelligent and Evolutionary Approaches (IES) (2015)
14. Hassaballah, M., Abdelmgeid, A.A., Alshazly, H.A.: Image features detection, description and matching. In: Awad, A., Hassaballah, M. (eds.) Image Feature Detectors and Descriptors. Studies in Computational Intelligence, vol. 630. Springer, Cham (2016)
15. Jin, X., Han, J.: *K*-Means Clustering. In: Sammut, C., Webb, G.I. (eds.) Encyclopedia of Machine Learning and Data Mining. Springer, Boston (2017)
16. Zawbaa, H.M., El-Bendary, N., Hassanien, A.E., Abraham, A.: SVM based soccer video summarization approach. In: Proceedings of The Third IEEE World Congress on Nature and Biologically Inspired Computing (NaBIC2011), Salamanca, Spain, October 19–21, pp. 7–11 (2011)
17. Fouad, M., Zawbaa, H.M., El-Bendary, N., Hassanien, A.E.: Automatic Nile Tilapia fish classification approach using machine learning techniques. In: 13th International Conference on Hybrid Intelligent Approaches (2013)
18. Tzotsos, A., Argialas, D.: A support vector machine approach for object based image analysis. In: Proceedings of International Conference on Object-Based Image Analysis (OBIA 2006), Salzburg, Austria (2006)
19. Taha, A., Darwish, A., Hassanien, A.E.: Arabian horse identification benchmark dataset. arXiv:1706.04870v1 (2017)

Automatic Sheep Weight Estimation Based on K-Means Clustering and Multiple Linear Regression

Aya Salama Abdelhady[1,3](✉), Aboul Ella Hassanien[1,3], Yasser Mahmoud Awad[2,3], Moataz El-Gayar[2], and Aly Fahmy[1]

[1] Faculty of Computers and Information, Cairo University, Giza, Egypt
Aya.salama@aucegypt.edu
[2] Faculty of Agriculture, Suez Canal University, Ismailia, Egypt
[3] Scientific Research Group in Egypt, Giza, Egypt

Abstract. Using a balance to estimate sheep's weight is inefficient and time consuming. Sheep's weight also fluctuates with many factors such as pregnancy, lactation, and gut fill. However, linear measurements are not highly affected by such type of factors. Therefore, in this paper, sheep weight was determined by calculating linear measurements from sheep images using visual analysis techniques. The system starts, followed by applying the K-means clustering for sheep segmentation. Then, biggest blob detection along with morphological analysis take place. After that breadth and width of sheep are extracted. Weight is then estimated from the linear dimensions using a regression function learned from the dataset. In the experiments, sheep weight estimation was tested on data set of 104 side images for 52 sheep. For performance evaluation, R-squared was measured and it reached 0.99. High accuracy of 98.75% was also achieved.

Keywords: Weight estimation · Regression · K-means clustering Segmentation · Visual analysis · Morphological image processing

1 Introduction

There has always been a huge demand for sheep products as a source of food. Weight is the main factor affecting sheep selection [1]. The weight of sheep is a vital feature that needs to be known. Heavy weighted sheep usually produce more meat than lighter ones. Therefore, most of the consumers are victims of sellers manipulation in sheep weight. Moreover, producers and consumers have no accurate measurement for sheep weight. Most of the producers can estimate weight by eye, while consumers has poor judgment. Besides, producers are also prune to mistakes. For this reason, in some of the farms and markets, sheep weight is determined by a scale balance. Using a scale balance requires significant time and labor to carry frequently sheep from their place to the balance. Furthermore, consumers cannot trust the weight given by sheep sellers. As a result, an

A. E. Hassanien et al. (Eds.): AISI 2018, AISC 845, pp. 546–555, 2019.
https://doi.org/10.1007/978-3-319-99010-1_50

easy real time application is needed for weight estimation. Since visual analysis is a powerful technique, then it can be used to measure sheep weight. Visual analysis is carried remotely without disturbing sheep and it is already available and feasible [2]. Linear dimensions can also be measured using image processing techniques [3]. Then, body weight can be estimated from the linear measurements generated by visual analysis. Research with such indirect techniques for weight estimation of many types of animals has already shown accurate results. Most of these techniques required special cameras, or multiple ones or a long video recording captured to be able to determine the body measurements. The goal of this conducted research was to have an easy application that can be used by any user in real time. For this reason, only one image of sheep was used for weight estimation. There are also many linear dimensions that can be estimated from images to determine animal's weight. However, heart girth was favoured in the reviewed studies because it was considered as the best feature to estimate the weight. Heart girth is near the chest, behind the front legs and withers. For this reason, heart girth was an important feature in this study.

In [4], Menesatti et al. used two high resolution web cameras to estimate weight and size of live sheep. The dual camera system was connected with a laptop. Authors experimented their algorithm on 27 sheep of Alpagota breed. Body height, length and chest depth were measured to extract the size and weight of sheep. This system acquired two images to determine linear dimensions by distance between their centers. Then partial least square regression was used to specify the weight by the distances between points. The efficiency was evaluated by the mean size error. Error was around 3.5% and 5.0% for body length. These were considered as high errors.

In [5], Kashiha et al. automatically identified pigs by using pattern recognition techniques on video images. Experiments were conducted on 40 pigs. The captured videos were processed offline. Pigs were localized in video frames by ellipse fitting algorithm. Therefore, the area that pig was occupying in the ellipse was calculated. Then dynamic programming obtained pigs weight. In this study, pig's weight estimation algorithm achieved 96.2% accuracy.

Mianagawa et al. in [1] also analyzed video using mesh image slide projector for pig's weight determination. Authors took use of shadow of pigs to calculate the distance between pig and floor to calculate the height. They established a geometric relationship between area and the weight of the body. Error was 2.1%.

In [6], Pradana et al. used side picture and front side images of cattle for weight estimation. Largest area was detected and localized using region based active contour. Number of pixels of this extracted area were considered as the cattle's body to estimate the weight. From the front view image, chest dimensions were used as additional characteristics to estimate cattle's weight. Linear regression modeled the relationship among these features and the weight. The accuracy of the system is 73%.

Moreover, in [7], Khojastehkey cropped the image as preprocessing to get the complete side view of new born lambs to estimate their size. Then images were cropped again to get the lamb without head, neck and limbs. Authors depended on that all the lambs are black with white background. Therefore, body was easily recognized by binarizing the image, so cattle size was measured by counting the white pixels. 89% accuracy was achieved in this research.

The goal of this paper is to have a real time application that is available for all users. Therefore, non of the previous approaches can be used as they depended on long video recordings and special camera systems. In this work, only one image is needed to estimate sheep's weight. Moreover, 40 was the maximum number of sheep in the data set used at the reviewed work. However, in this paper, data set of 52 sheep of different physiological conditions and ages was used. This research also achieves higher accuracy than that of previous approaches of sheep's weight estimation. The remainder of this paper is organized as the following. In Sect. 2, discusses the camera setting. Theoretical background is explained in Sect. 3. Section 4 then gives the details of the proposed approach for sheep weight estimation. Further, in Sect. 5, experimental results are presented and evaluated. At the end, conclusion of the paper and future work directions are provided in Sect. 6.

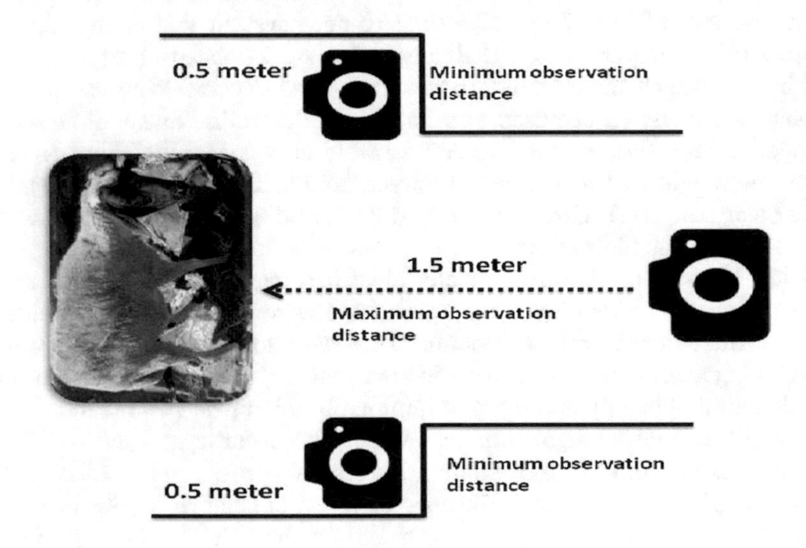

Fig. 1. Camera model setting

2 Camera Setting

All sheep in this study are from Barqi breed in the 6th of October farm in Ismailia in Egypt. The main origin of this breed is Libya.

Sheep were selected in different physiology and ages; Pregnant, non pregnant, lactating, adults and still growing. Images were taken at a distance of average 1.5 m from the sheep. As illustrated in Fig. 1, images were taken at slightly different distances and for different postures of sheep. Side images of the sheep are the best to work on because it has more information to be extracted than a top image. For each sheep, images for both sides of the sheep were captured. As a result, the used data set consists of 104 images to work on and test the feasibility of the proposed approach. High resolution mobile camera was used in the data collection to test the feasibility of the proposed approach.

3 Theory and Background

3.1 K-Means Clustering

K-means clustering assigns members of any data set into a specific number of clusters. K centers are given to each clusters. These centers have to be far away from each other [8]. Each member is then classified to the cluster with the nearest center. After each point is classified, new centroids as barycenter are created for clusters as a result of the previous step. Given points are again re-associated to the clusters with nearest new centers. This process continues until centers do not change any more. K-means clustering usually aims to minimize the following objective function:

$$\underset{S}{\operatorname{argmin}} \sum_{i=1}^{k} \sum_{x=S} ||x - \mu_i||^2 = \underset{S}{\operatorname{argmin}} \sum_{i=1}^{k} |S_i| \times varS_i \tag{1}$$

where $(x_1, x_2, \cdots, x_n)$ are the set of given points, S is the clusters and K is the number of the clusters ($\leq$n) for sets $S = S_1, S_2, \cdots S_k$ and μ_i is the mean of points in S.

3.2 Multiple Linear Regression

Linear regression models the relation among variables by a linear equation. The regression equation should define the association between these variables. Therefore, in simple linear regression, a single predictor variable is used to model the response variable [9]. For more than two variables, multiple linear regression can model the relationship between multiple variables and any response variable [10]. Every independent variable x is associated with another dependent variable y. The population regression line for r variables $x_1, x_2, \cdots, x_r$ is defined as follows:

$$y_i = B_0 + B_1 * X_{i1} + B_2 * X_{i2} + \cdots + B_r * X_{ir} + E; \, i = 1, 2, \ldots, n \tag{2}$$

Regression line defines how y changes with the r variables. y values have similar standard deviation and different means. $B_0, B_1, B_2, \cdots, B_p$ are often computed by statistical software. The coefficient of determination is called R-squared or R_2. Where R_2 is a statistical metric which measures the variation in the output that corresponds to that in the independent variables. Therefore, R2 always increases with the increase in the number of the predictors. R_2 has values between 0 and 1. If $R_2 = 0$, then it means that outcome cannot be predicted from the given variables. But if $R_2 = 1$, then the outcome can be predicted efficiently from the given variables. The correlation coefficient measures the strength of this relation of the observed data and for both variables.

4 The Proposed Automatic Sheep Weight Estimation System

The goal of this paper is to have a real time application that enables the buyer or the stockman to estimate sheep weight. For this reason, dual web cam system or video recording cannot be used. Weight has to be estimated from just a single image captured by a mobile camera. Image analysis technique was used to measure some linear dimensions to estimate weight. The most important step here was data collection. Images were collected by mobile camera for sheep in different postures to develop a robust system. Multiple images were captured for 52 sheep. The proposed weight estimation system consists of five phases: (1) Preprocessing phase, (2) Segmentation phase, (3) Post processing phase, (4) Weight estimation, and (5) Validation phase. These phases are described in details in the following subsections along with the involved steps with the characteristics for each phase. The overall architecture of the proposed system is also illustrated in Fig. 2.

4.1 Pre-processing Phase

All images were processed using Matlab processing tools to improve quality of the taken images. According to literature, legs and neck were considered to be noise. For this reason, these additional features were cropped out of the image. There were also some problems in brightness. Therefore, brightness was adjusted before working on the images. Sample input images are shown in Fig. 3 with the output of preprocessing in Fig. 4. Grey level image of the image was also saved to segment the region of interest. Binary version of the image was also established to calculate the sheep's linear dimensions needed in weight estimation.

4.2 Segmentation Phase

Segmentation is the key of success in any visual analysis algorithm. It is the process of partitioning the image into multiple objects of the same texture and color. Since unsupervised algorithm was required for segmentation, k-means clustering was used. In this study, sheep body was segmented based on color features in the images. Segmented images corresponding to sheep in Fig. 4 are shown in Fig. 5.

In the proposed system, clustering was used for color based segmentation. K-means determined the natural spectral groupings in the data set. Then clusters were determined based on the colours [11]. Therefore, Each image pixel had to belong to each of these clusters. This decision was taken based on the distance to the mean vector. Then k new centroids are re-calculated as centers of the clusters resulting from the previous step. This procedure took place until each pixel was assigned to the closest cluster so that no more change was needed.

4.3 Post-processing Phase

To remove noise from images and to have a robust algorithm to illumination conditions, morphological operations were applied. Small contours were neglected and all holes were filled [12]. Then biggest blob was extracted as it was considered to be the region of interest. In Fig. 5, the result of segmentation by k-means clustering is shown. Then in Fig. 6, the effect of morphological operators on optimizing the output is obvious.

4.4 Weight Estimation Phase with Regression Function

Then longest width of the extracted contour is considered as the sheep breadth. As mentioned before, heart girth is considered to be the most effective feature to be measured. Heart girth is a near the chest, behind the front legs and withers. For this reason, longest height in the extracted blob was considered as the sheep breadth as it is the closest to the heart girth. Even if the stomach has a bigger height, it won't make a huge difference in our calculations. Length and breadth were calculated in pixels as illustrated in Fig. 7. As a result, pixels had to be converted to millimeters for accurate calculation of linear dimensions. For this reason, we had to define an equation that converts from pixels to millimeters. We found that one pixel is equal to 2 mm relative to the average distance between the camera and the target. Therefore, to convert the calculated dimensions from pixels to millimeter, number of pixels was divided by two.

Linear measurements was determined for each sheep side with their corresponding weight. What was left was to regress the weights from these measurements. In this application, both breadth and length were influencing the response. Therefore, multiple regression was used. The output regression equation is as follows:

$$W_e = c * breadth + d * length \tag{3}$$

Where W_e is the weight of sheep, c and d are the slopes/coefficients. In this paper, c is set to 0.064443, and d is set to 0.010059. So the final equation here is: $\mathrm{W}_e = 0.064443 * breadth + 0.010059 * length$.

The regression statistics is also illustrated in Table 1. Where Multiple R is the correlation coefficient which measures the strength of the linear relationship. R Square tells how close the data are to the regression line. Adjusted R-squared represents explanatory power of the regression equation and Standard Error is the standard deviation of the error.

Table 1. Regression statistics.

Multiple R	0.99673
R square	0.993471
Adjusted R square	0.983793
Standard error	3.586779

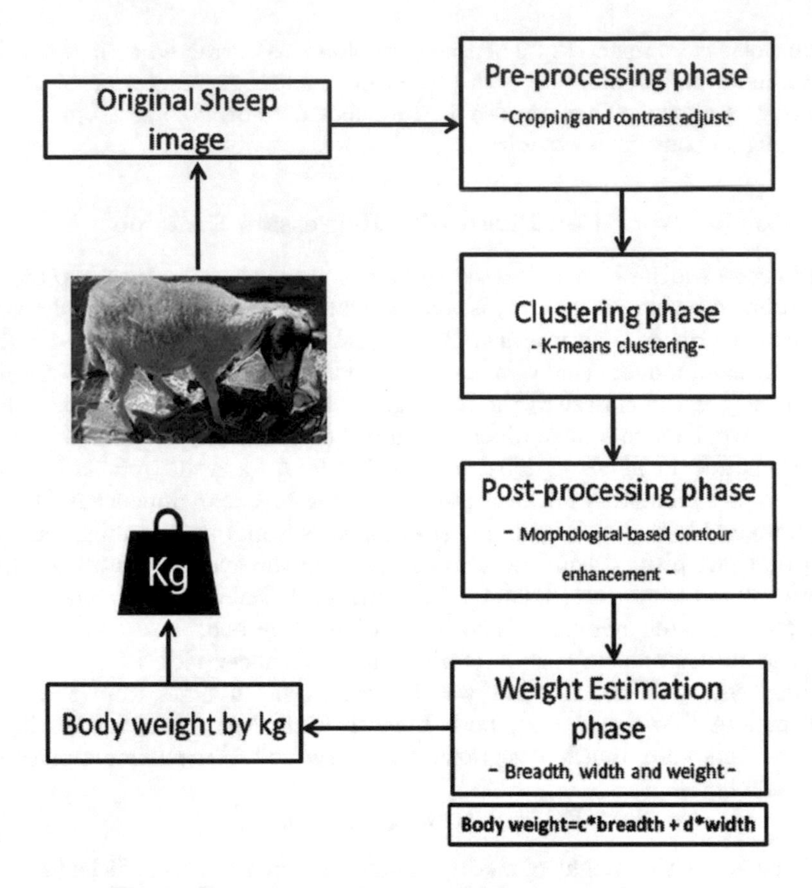

Fig. 2. The general architecture of the proposed system

Fig. 3. Sample of the input images

Fig. 4. The output of preprocessing phase

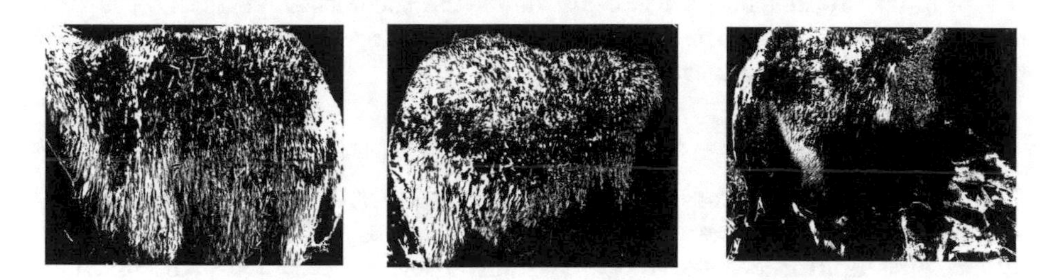

Fig. 5. The output of segmentation phase

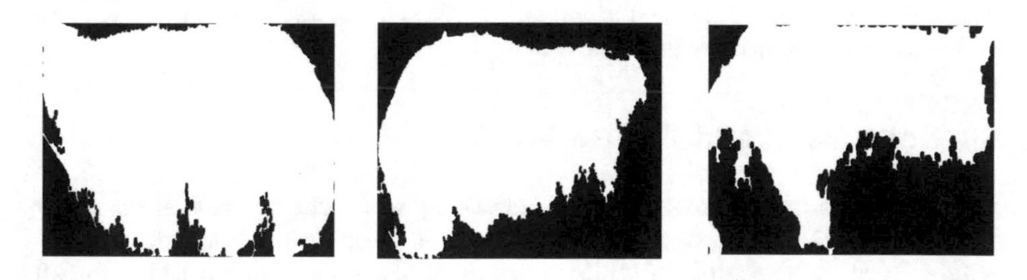

Fig. 6. The output of postprocessing phase after applying the morphological operations to fill the gaps and remove noise

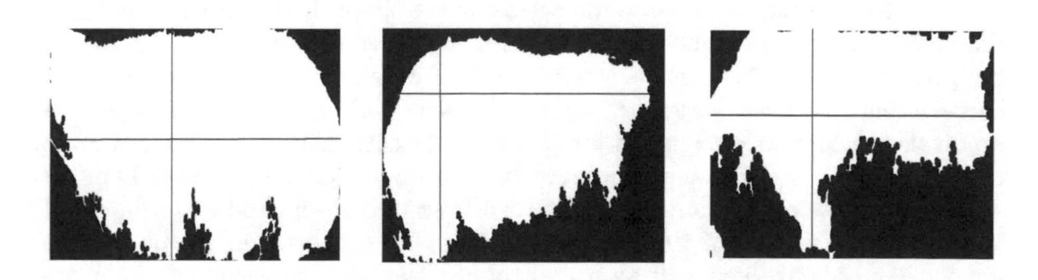

Fig. 7. The final results after measuring the linear dimensions

5 Experimental Results and Discussion

Linear regression was used to determine the relationship between the linear measurements and the weight. After training, accuracy of the proposed approach had to be calculated for validation. Accuracy was measured by the equation used before to test the accuracy of similar approach for automatic weight estimation of beef cattle as follows [5]:

$$\%error = \frac{|W_{ex} - W_{th}|}{W_t h} \times 100 \tag{4}$$

Experimental weight W_{ex} is the one that is automatically calculated using the proposed approach. On the other hand, the theoretical weight W_{th} is that measured manually with a scale balance. The system accuracy was measured using the following equation:

$$Accuracy = 1 - \%error. \tag{5}$$

The accuracy of proposed approach in this study is 98.75% with a high R-square of 0.993. These results prove that the proposed technique outperforms all previous approaches on the same topic. Previous techniques discussed in literature were also tested to compare their results with the proposed approach in this paper. Linear regression was tested to obtain the relationship between the weight and the number of pixels representing the sheep area. Regression resulted in very low R squared of value 0.21702 which resembles a dramatically lower accuracy than work in this paper.

6 Conclusion and Future Work

Visual analysis has already proved its efficiency in weight estimation for many other animals. However, previous approaches for sheep weight estimation didn't achieve very high results. In addition, most of the previous algorithms for all types of animal weight estimation required human intervention. In this work, sheep weight estimation was completely automated. The proposed approach is based on K-mean based clustering to segment the sheep. Then linear dimensions were measured to estimate the weight of the sheep. Relation between these dimensions and the weight was specified using linear regression. A significant high accuracy of 98.75% and R squared of 0.99 were achieved. For future work, this system will be a part of a bigger one that will be released as a mobile application which is responsible for sheep identification. Part of this system is to estimate the sheep's weight because it is an important feature that helps in sheep's evaluation. In addition to, this work was on sheep that are originated in Libya. In spite of that this breed of sheep is the dominant in Egypt, images for other breeds of sheep will be tested too. So the breed of sheep can be automatically identified first and then apply its corresponding regression function for weigh estimation.

References

1. Burke, J., Nuthall, P., McKinnon, A.: An Analysis of the Feasibility of Using Image Processing to Estimate the Live Weight of Sheep. Farm and Horticultural Management Group Applied Management and Computing Division, Lincoln University (2004). http://hdl.handle.net
2. Begaz, S., Awgichew, K.: Estimation of weight and age of sheep and goats. Ethiopia Sheep and Goat Productivity Improvement Program (ESGPIP). Ethiopia, No. 23 (2009). http://www.esgpip.org/PDF/Technical%20bulletin%20No.23.pdf
3. Chen, W., Wang, C.: The human-height measurement scheme using image processing techniques. Int. J. Comput. Consum. Control (IJ3C) **4**(3), 186–189 (2015)
4. Menesatti, P., Costa, C., Antonucci, F., Steri, R., Pallottino, F., Catillo, G.: A low-cost stereovision system to estimate size and weight of live sheep. Comput. Electron. Agric. **103**, 33–38 (2014)
5. Kashiha, H., et al.: Automatic weight estimation of individual pigs using image analysis. Comput. Electron. Agric. **107**, 38–44 (2014)
6. Pradana, Z., Hidayat, B., Darana, S.: Beef Cattle Weight Determine by Using Digital Image Processing. Control Electronics (2016). https://ieeexplore.ieee.org/document/7814955/
7. Khojastehkey, M., Aslaminejad, A.A., Shariati, M.M., et al.: Body size estimation of new born lambs using image processing and its effect on the genetic gain of a simulated population. J. Appl. Anim. Res. **44**(1), 326–330 (2016)
8. Chen, K.: K-means Clustering. COMP24111 Machine Learning Course (2016). https://studentnet.cs.manchester.ac.uk/ugt/COMP24111/
9. Gogtay, N., Deshpande, S., Thatte, U.: Principles of regression analysis. J. Assoc. Physicians India **65**, 48 (2017). http://www.japi.org/april_2017/08_sfr_principles_of_regression_analysis.html
10. Mouhaffel, A.G., Domìnguez, C.M., Arcones, B., Redonda, F.M., Martín, R.D.: Using multiple regression analysis lineal to predict occupation market work in occupational hazard prevention services. Int. J. Appl. Eng. Res. **12**(3), 283–288 (2017)
11. Hassan, R., Rahman Ema, R., Islam, T.: Color image segmentation using automated K-means clustering with RGB and HSV color spaces. Glob. J. Comput. Sci. Technol. **17**(3), 0975–4350 (2017)
12. Jaikla, C., Rasmequan, S.: Segmentation of optic disc and cup in fundus images using maximally stable extremal regions. In: International Workshop on Advanced Image Technology 2018 (IWAIT 2018), 7–9 January 2018, Chiang Mai, Thailand. IEEE (2018)

A Simulation-Based Optimization Approach for Assessing Sustainability in a Multi-construction Projects Environment

Areej M. Zaki(✉), Hisham M. Abdelsalam, and Ihab A. El-Khodary

Department of Operations Research and Decision Support, Cairo University, Giza, Egypt
{areej.m.zaki,h.abdelsalam,e.elkhodary}@fci-cu.edu.eg

Abstract. Sustainability development has become widely applied specially when considering construction projects as a result of the huge leap in the global urbanization rates. Researchers are continuously attempting to find accurate and reliable sustainability construction performance measures. However, the majority of these researches focus on studying the sustainability in a single construction project, although it might be worthwhile studying sustainability in multiple construction projects in order to maximize the overall sustainability. In this paper, the authors developed a simulation optimization model using simulated annealing for scheduling the different phases in their right sequence of multi-construction projects while maximizing the overall sustainability. The previous is done while assessing the sustainability development and simulating the effect of external factors such as technological advancement and change in people's perception on sustainability measures. Results showed that considering scheduling in multi-construction projects increased the overall sustainability performance compared to unscheduled ones.

Keywords: Sustainability · Construction · Optimization · Scheduling
Simulated annealing

1 Introduction

Nowadays there is a huge leap in the global urbanization rates; from around 33.5% in 1960 to 54% in 2016 [1] and it is expected to reach 66% by 2050 [2]. This is due to the huge growth in population −3 to 7.4 billion from 1960 to 2016 and industrialization activities [3]. Given this enormous increase, the demand for construction projects and economic activities is also increasing, especially in the developing countries [4]. Although this increase in construction and economic activities enhances the overall economy and standard of livings for the population, it threatens the sustainable development and increases the associated environmental losses. Therefore it is important to integrate the three pillars of sustainable development: economic development, social development and environmental protection with construction projects [5, 6].

Sustainability was first defined as "development which meets the needs of the present without compromising the ability of future generations to meet their own needs" [7]. The IUCN [8] defined the process of sustainability development as "to

A. E. Hassanien et al. (Eds.): AISI 2018, AISC 845, pp. 556–566, 2019.
https://doi.org/10.1007/978-3-319-99010-1_51

improve the quality of life while living within the carrying capacity of living ecosystems" and since that time sustainability has become widely spread around the world [9–11].

Kibert [12] was the first to introduce the term "Sustainable construction" in which he was referring to building a healthy environment without wasting any resources such that all the resources are used efficiently and also not harming the community and the environment surrounding. After that Sustainable Construction or Green Construction was described by Hill and Bowen [13] as "a process which starts well before construction per se in the planning and design stages and continues after the construction team leaves the site".

Construction sustainability is the responsibility of contractors to attain sustainable development to people by providing them with basic services, safe and healthy environment, chosen employment and ensuring social, cultural, environmental and economic development [14].

While considering sustainability measures in any construction project certain corners must be addressed for project evaluation. Hill and Bowen [13] introduced a framework that contains the main key principles that enable construction projects to contribute to sustainability development. In the framework, they considered four main corners for sustainable construction projects which are: (a) social sustainability, (b) biophysical sustainability, (c) economic sustainability, and (d) technical sustainability. After that OECD [15] identified different variables of economics, environmental and social aspects that affect the transportation infrastructure. Prasad and Hall [16] revised the four main corners for sustainable construction projects introduced earlier by Hill and Bowen [13] and came up with the following three main aspects (a) social sustainability, (b) economical sustainability and (c) environmental sustainability. These three aspects are known to be used while applying sustainability in any construction project.

A number of studies have applied sustainability rules to construction projects. Some used applications to justify their theories, while others made hard literature reviews and used expertise background to come up with the main checklist; this list helps project managers and engineers understand the main factors that must be considered in a sustainable construction project within its life cycle. While reviewing the researches it was clear that each study viewed the problem from a different perspective Zimmermann et al. [17] enable construction activities to be applied with sustainability concepts by providing benchmarks for sustainability construction. While Shen et al. [18] introduced a scoring method to measure the environmental performance of contractors by calculating the Environmental Performance Score (EPS) for each contractor. Šaparauskas and Turskis [19] introduced a multi-criteria method for evaluating construction sustainability. Shen et al. [20] explained the steps needed to analyze the ability of a construction project to attain sustainability development and developed a quantitative model to asses it's feasibility. Ochieng et al. [21] explained how sustainability principles can be integrated to construction projects and introduced a framework to do so.

While reviewing all these previous researches and another to follow, a major drawback was found which is that all these studies focused on the construction project from a single perspective, which means it examined the performance of the sustainability measures of only one single construction project during its life cycle. However if multiple construction projects are running at the same time, it might be a good idea to consider scheduling the activities of the different projects so as to improve the sustainability performance. The main contribution of this paper is to study the sustainability performance measures but in a multi-project construction environment through developing an optimization scheduling multi-project model that simulates several construction projects considering all five stages of construction from inception till demolition.

The proposed methodology was built based on Zhang et al.'s [22] model. The developed model is a simulation optimization framework to find the optimal schedule for the different projects phases that maximizes the Sustainability Development Ability value considering multi-projects instead of a single project. In the proposed framework Simulated Annealing was used as the optimization technique and Zhang et al.'s System Dynamic model for the simulation part.

Following the introduction, the rest of this paper is organized as follows. Section 2 provides a brief literature on the sustainable construction models that were developed before. Followed by the solution algorithm and methodologies used in Sect. 3. Section 4 covers the different experiments conducted and their results. Finally, conclusions are drawn in Sect. 5.

2 Sustainable Construction Models

Several studies have used Discrete Event Simulation (DES) to model the sustainable construction projects and understand the interdependencies in the operational level variables. For example, Ahn et al. [23] developed a discrete event simulation model that is integrated to an emission estimation model in construction projects in order to measure the sustainability of the construction project. Also, González and Echaveguren [24] used discrete event simulation to build a model that integrates traffic and environment models together in a road construction project.

On the other hand, many studies have used System Dynamics (SD) to model the sustainability in construction projects for studying their feedback and causal loops between the different variable and understand the inter related structure within the context level variables. Hao et al. [25] developed a model by system dynamics that simulates the construction and demolition (C&D) waste management in Shenzhen, Mainland China by integrating three different subsystems together. Shen et al. [26] used system dynamics to access the sustainability of construction projects in their life cycle, three major factors which are sustainability of environmental development, sustainability of economic development and sustainability of social development were used to examine the project sustainability performance. Shen et al. [27] then developed a checklist for assessing the sustainability performance measure of construction projects across their five stages which are Inception, Design, Construction, Operation and Demolition in which in each stage economic, social and environmental aspects where examined to assure the sustainability of projects.

Shen et al. [28] conducted a case study in a Chinese construction industry to measure its sustainability performance, by collecting 87 feasibility study reports from different projects and measuring the economic, social and environmental attributes on it. Yao et al. [4] used system dynamics to evaluate sustainability performance in highway infrastructure projects within the construction and the operation stages. In this research, indicators for measuring the sustainability performance were used and how the dynamic factors affect these indicators.

One of the major significant researches was by Zhang et al. [22] where they developed an improved Sustainability Development Ability (SDA) prototype model that captures the change of dynamical factors on sustainable projects by introducing technological advancement and change in people's perception as two major variables that will affect the sustainability performance in any construction project. SDA is recommended to be one of the most significant measures that indicates the feasibility of a construction project and it measures how well a construction project is attaining sustainability development.

In their paper, Zhang et al. used System Dynamics to calculate the SDA values. The model proposed by them was summed as follows below (1)

$$SD(t) = \int_0^t W_s(t)I_s(t)dt + \int_0^t W_E(t)I_E(t)dt + \int_0^t W_{En}(t)I_{En}(t)dt$$
$$W_s + W_E + W_{En} = 1 \quad (1)$$
$$I_s, I_E, I_{En}, V_s, V_E, V_{En} \in [-100, 100]$$

Where I_s, I_E and I_{En} are said to be the social, economic and environmental impacts that affect the SDA measures of the project and they are calculated from the V_s, V_E and V_{En} which are the Sustainability Development Values (SDV) which are given to the models as inputs in the beginning. The SDVs are conducted from experts, project managers, researchers and from excessive surveys. Both the SDV and the impact factors ranges from −100 to 100. The W_s(t), $W_E(t)$ and W_{En}(t) represent the weights of the social, economic and environmental factors and they are considered to be constant which is not logical as they are applied on construction projects which are characterized with their dynamics nature in their processes. Accordingly Zhang et al. introduced two major factors that affect the SDA values measures which are people's perception and technological advancement, where they were introduced in the model by adjustment factors for each aspect whether the social, economic or the environmental aspect A4V2S, A4V2E and A4V2En respectively, which means that when the SDA value is under a certain number which is −100 in their case, further technology will be introduced in the form of an adjustment value to increase the SDA value with 50%. When the SDA values is greater than a maximum number which is 100 in their case, people perceives a higher level of sustainability and this is reflected on the weights of each aspect in the model which will eventually reflect the overall SDA value of the project. In Zhang et al.'s paper three different scenarios were addressed (1) considering the technological advancement and the effect of the change in people's perception, (2) considering the effect of technological advancement only and (3) not considering any external changes on the sustainability measures. The results and conclusions of the addressed paper showed that the technological advancement alone is more useful and effective than the other two scenarios on the SDA values while addressing a single construction project.

In this research, an adaption was made on the model introduced by Zhang et al. [22] to calculate the SDA values for multi-construction projects, taking into consideration the scheduling of the different project's activities which is explained in the next section.

3 Methodology

The basis of the proposed methodology is to construct an optimization multi-construction projects model taking into account the sustainability measures. To do so the system dynamics model introduced Zhang et al. earlier was used to mimic the behavior of the construction project and calculate the SDA values of a construction project taking into account the technological advancement and people's perception impact on the sustainability measures. Modifications were done to the model in order to address a construction firm with multi construction projects, and to schedule the different phases of the construction projects while maintaining the right sequence of the phases of each project alone. The life cycle of each construction project consists of five phases which are the inception (A), construction (B), commission (C), operation (D) and demolition (E) phase, each phase having its own time interval. The proposed model optimized the SDA values by scheduling these construction phases of each project where the objective was to maximize the SDA end value which represents the Sustainability value of the whole project as each SDA value depends on the previous one. The scheduling optimization was performed using simulated annealing algorithm.

Simulated annealing (SA) was selected given its probabilistic heuristic nature and its ability to find a solution for the global optimization problem in a large search space without getting trapped in a local optima especially in discrete search spaces.

Also given that SA can be used easily in solving many problems due to the flexibility in adjusting its cooling schedule, and since the problem being discussed is a discrete non-linear problem, simulated annealing algorithm was selected as an optimization algorithm to schedule the phases of the construction projects taking into account its right sequence.

Figure 1 represents a flow chart of how the model works. The model starts with introducing an initial solution which consists of an initial order of the phases in the n construction projects in series following each other. For example, considering three projects the phases will be as follows:

$$A_1, B_1, C_1, D_1, E_1, A_2, B_2, C_2, D_2, E_2, A_3, B_3, C_3, D_3, E_3$$

For each phase in each project, the Sustainability Development Values (SDV) pertaining to each aspect social (V_s), economic (V_E) and environmental (V_{En}) are inserted as inputs. Given the SDV and initial schedule solution, the modified improved SDA prototype model by Zhang et al. calculates the overall SDA values for the n projects.

Afterwards the simulated annealing algorithm is activated where initialization for the temperature, the minimum temperature and the cooling rate takes place. As the temperature is greater than the minimum temperature then a new scheduled feasible solution is generated by randomly swapping two phases for the existing n projects

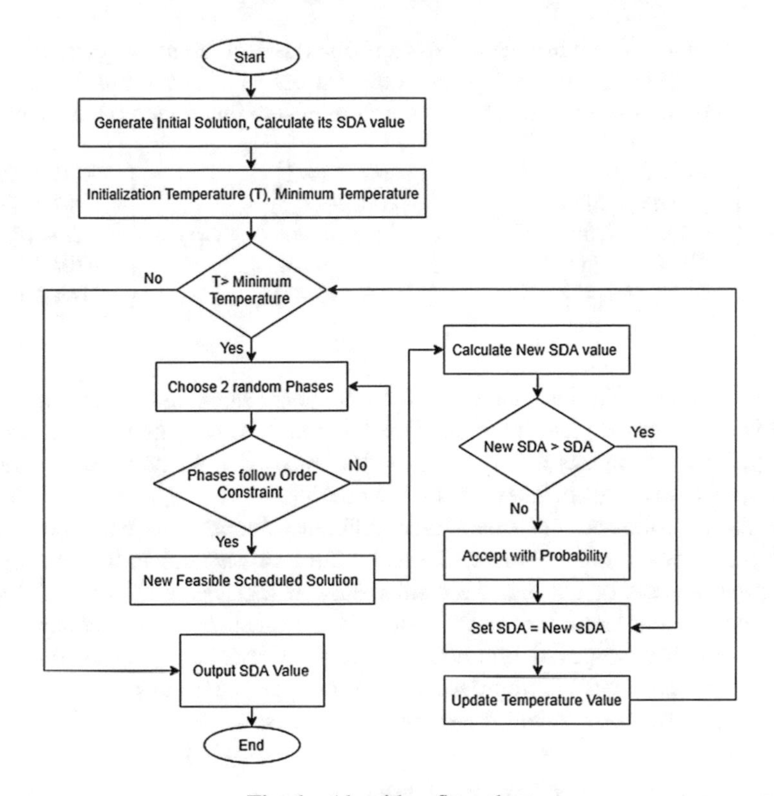

Fig. 1. Algorithm flow chart

taking into account the constraint that was mentioned earlier that each project must perform its phases in the right sequence. For example, considering three construction projects, an example for a new feasible scheduled solution will be as follows:

$$A_1, B_1, C_1, D_1, \boldsymbol{A_2}, \boldsymbol{E_1}, B_2, C_2, D_2, E_2, A_3, B_3, C_3, D_3, E_3$$

Given this new feasible scheduled solution, the overall SDA value is calculated using the modified improved SDA prototype model by Zhang et al. If the new SDA value is better than the previous one, then it is accepted as the best SDA value for this stage, if not it could be accepted with a certain probability, otherwise it will be rejected. The simulated annealing temperature is then updated. The process is repeated until no new feasible scheduling solution could be reached that results in a higher SDA value and thus a temperature greater than the minimum temperature.

4 Experiments and Results

Results of this model was based on data that were collected by Zhang et al., where they illustrated a real case which was the FD NaCN innovation project located in Chongqing, China. As the proposed model needed the data for sustainability values of economic, social and environmental aspects for three different projects, the data of Zhang et al.'s model had to be duplicated three times as a start to test the model. The initial values of the model were given in the form of step functions as follows below (2)

$$V_s(\mathrm{t}) = \begin{cases} -60t \in (0,1] \\ 50t \in (1,5] \\ -20t \in (5,6] \\ 30t \in (6,46] \\ -50t \in (46,47] \end{cases} \quad V_E(\mathrm{t}) = \begin{cases} -10t \in (0,1] \\ -100t \in (1,5] \\ 0t \in (5,6] \\ 60t \in (6,46] \\ 10t \in (46,47] \end{cases} \quad V_{En}(\mathrm{t}) = \begin{cases} -50t \subset (0,1] \\ -80t \in (1,5] \\ 0t \in (5,6] \\ -70t \in (6,46] \\ -100t \in (46,47] \end{cases} \tag{2}$$

Each time interval in this data represents a 1/4 year (3 months), for example the first interval is the inception phase with duration 1/4 year. The weights for each aspect was given equally which means that $W_s = W_E = W_{En} = 1/3$ and the adjustment values was set as constants given from Zhang et al.'s model.

The results followed the same three different scenarios to check whether the technological advancement alone will also be the best scenario in the case of multi-construction projects or not and the results showed that yes considering the technological advancement alone is the best option. In the three scenarios a comparison was made between the proposed optimization multi-construction projects model using simulated annealing and the improved SDA prototype model that was conducted by Zhang et al., considering three projects in the two cases.

4.1 Not Addressing Any External Factors

The first scenario is not considering any external factors while calculating the SDA values. The phases of the construction projects were entered in order in the improved SDA prototype model by Zhang et al. as shown in the following representation.

$$A_1, B_1, C_1, D_1, E_1, A_2, B_2, C_2, D_2, E_2, A_3, B_3, C_3, D_3, E_3$$

The same representation was entered on the optimization multi-construction projects model and the schedule of the phases resulted in obtaining the following order. As shown below the schedule of the phases is following the constraint that for each project its own phases are happening in order.

$$A_2, A_1, A_3, B_1, B_3, C_1, C_3, D_3, E_3, B_2, C_2, D_1, D_2, E_2, E_1$$

This scenario is shown to be the worst case with very low SDA values, Fig. 2 shows the difference between the two models and that in the proposed model the SDA values were lower at the beginning this is due to the operation of the construction phases of project one and three and then it rises with a steady state as the operation phase of project three started and a sudden decrease happens as the demolition of project three and the construction of project two takes place and then a long rise steady state occurs when the operation phase of project one and two occurs and finally a decreases is shown again due to the demolition of projects one and two. But as reaching the end of the construction's projects life cycle the SDA values increases slightly than their model, and this is because the proposed model optimizes the end SDA value of the projects which is considered the overall SDA value of the whole project as each SDA value is depending on the previous one.

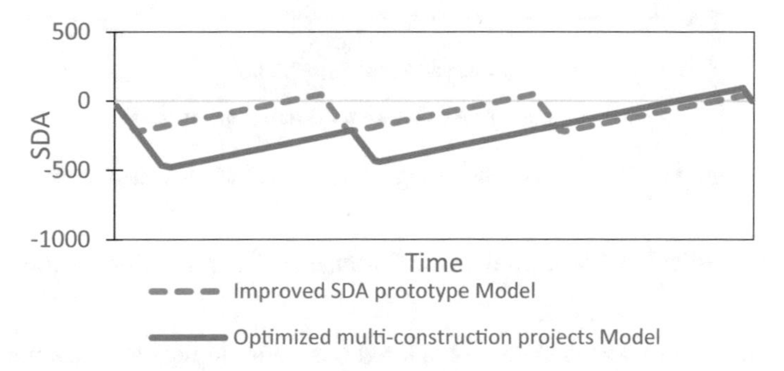

Fig. 2. SDA values with no external factors

4.2 Technological Advancement Factor

The second scenario has considered only the technological advancement while calculating the SDA values, as mentioned earlier in the improved SDA prototype model this was the best scenario from the three proposed scenarios. The order of the construction phases of the projects for the improved prototype model was entered as the last scenario, while the proposed model obtained the following order for the construction phases in this case.

$$A_3, A_1, A_2, B_1, B_2, B_3, C_1, C_3, C_2, D_3, D_1, E_3, E_1, D_2, E_2$$

Figure 3 shows the line chart for the two models and also it shows that the proposed optimization model has a very noticeable higher performance which proves what Zhang et al. concluded and validates the proposed model, at the beginning of the run the construction phases of the projects took place that's why the SDA values dropped down noticeably, then technological advancement took place so the SDA values increased again and a long steady state took place due to the operation phases of project three and one, demolition phases for both the same later projects took place and decreased the SDA value a little bit but not with a huge value due to the effect of the technological advancement, last but not least comes the operation phase of the second project followed by its demolition phase. In this scenario the effect of the technological advancement is very high as with right and proper scheduling it rises the SDA values with a great percentage and prevents it from falling again to its lower stages.

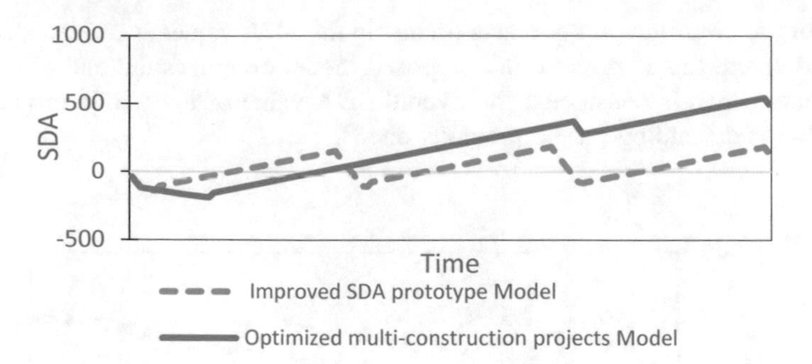

Fig. 3. SDA values considering technological advancement factor

4.3 Technological Advancement and Change in People's Perceptions Factors

The third and the last scenario has considered the technological advancement and how change in people's perception affects the overall SDA value. In this scenario the order of the phases of the construction projects in the improved SDA model were also entered in order and in the optimization model the resulted scheduling phases were as follows:

$$A_2, B_2, A_1, C_2, D_2, A_3, E_2, B_3, B_1, C_3, C_1, D_1, E_1, D_3, E_3$$

As shown previously with no factors included in the model the SDA values where very low and after introducing the technological advancement factor the SDA values were very high, now after introducing both factors together to the model the SDA values ranged between both as whenever the technological advancement increased the SDA values came people's perception to decrease it again so the results were as follows Fig. 4 shows the SDA values with respect to time in both the improved SDA prototype model and in the proposed optimization multi-construction projects model; both considering three projects. And as shown the proposed model with scheduling achieved higher SDA values at the end, explaining the behavior of the model can be summed up

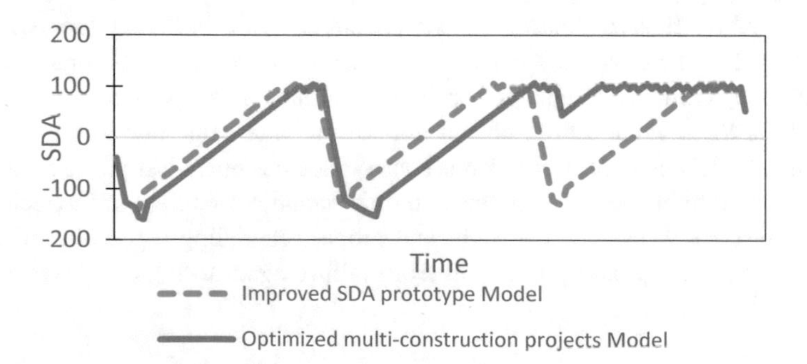

Fig. 4. SDA values considering technological advancement and change in people's perception factors

that the first decrease in the values where because of the construction phase of the second project then its operation phase took place so the values rises again, after that both project one and three experiences the construction phase so another decrease in the sustainability values takes place after that the operation phase for project one increases the SDA values for a long period of time and then comes the demolition phase of project one which decreases the values a little bit and then a steady state takes place with small variations.

5 Conclusion and Future Work

Sustainability development has been addressed many times in the field of construction projects due to its huge importance in the feasibility and the success of any project. In this paper an optimization multi-construction projects model was proposed where scheduling the phases of n projects is held by simulated annealing algorithm, taking into account the constraint that the order of the phases in each project must take place in its right sequence. Modifications on Zhang et al.'s system dynamics model has been made to adapt the multi-construction environment and be able to calculate the SDA values for the overall projects. Taking into account the three aspects of sustainability which are the social, economic and environmental aspect and also simulating the effect of external factors like the technological advancement and how the change in people's perception can affect the sustainability performance. The SDA values in the proposed optimization model were much higher than unscheduled multi-construction projects, also the technological advancement factor had higher influence on sustainability measures than people's perception and this was the conclusion in Zhang et al.'s model which validates the proposed model and proves the same point.

A real life case study is in progress at this stage where sustainability attributes from the three different perspectives social, economic and environmental are being collected and mapped with the LEED attributes which is an acronym developed by the US Green Building Council to measure the sustainability and performance of a building.

References

1. Urban population, 28 March 2018. https://data.worldbank.org/indicator/SP.URB.TOTL.IN.ZS. Accessed 28 Mar 2018
2. D. o. E. a. S. A. P. D. United Nations: World Urbanization Prospects: The 2014 Revision, Highlights (ST/ESA/SER.A/352) (2014)
3. Population Total. https://data.worldbank.org/indicator/SP.POP.TOTL. Accessed 28 Mar 2018
4. Yao, H., Shen, L., Tan, Y., Hao, J.: Simulating the impacts of policy scenarios on the sustainability performance of infrastructure projects. Autom. Constr. **20**(8), 1060–1069 (2011)
5. Desta, M.: Sustainability and sustainable development: historical and conceptual review. Environ. Impact Assess. Rev. **18**(6), 493–520 (1998)
6. Martíne, P., González, V., Da Fonseca, E.: Green-Lean conceptual integration in the project design, planning and construction. Revista Ingenieria de de Construccion **24**(1), 5–32 (2009)
7. Butlin, J.: Our common future: the world commission on environment and development. J. Int. Dev. **1**(2), 284–287 (1987)

8. Iucn, U.W.: Caring for the earth: a strategy for sustainable living. J. Int. Dev. **5**(3), 228 (1991)
9. Alexandra, W.: Efficiency in sustainability—the efficient life styles of Kerala. In: International System Dynamics Conference, Cambridge, Massachusetts (1996)
10. Choucri, N.D., Berry, R.: Sustainability and diversity of development: toward a generic model. In: International System Dynamics Conference, Tokyo (1995)
11. Hannon, B., Ruth, M., Delucia, E.: A physical view of sustainability. Ecol. Econ. **8**(3), 253–286 (1993)
12. Kibert, C.J.: Sustainable construction. In: Proceedings of the First International Conference of CIB TG 16, Tampa, Florida (1994)
13. Hill, R., Bowen, P.: Sustainable construction: principles and a framework for attainment. Constr. Manag. Econ. **15**(3), 223–239 (1997)
14. Agenda, H.: The habitat agenda: goals, principles, commitments and global plan of action. In: UN Conference on Human Settlements (Habitat II), Istanbul (1996)
15. OECD: Sustainable Development: Critical Issues. Organization for Economic Co-operation and Development, Paris (2011)
16. Prasad, D.K., Hall, M.: The construction challenge: sustainability in developing countries (2005)
17. Zimmermann, M., Althaus, H.J., Haas, A.: Benchmarks for sustainable construction: a contribution to develop a standard. Energy Build. **37**(11), 1147–1157 (2005)
18. Shen, L.-Y., Lu, W.-S., Yao, H., Wu, D.-H.: A computer-based scoring method for measuring the environmental performance of construction activities. Autom. Constr. **14**(3), 297–309 (2005)
19. Šaparauskas, J., Turskis, Z.: Evaluation of construction sustainability by multiple criteria methods. Technol. Econ. Dev. Econ. **12**(4), 321–326 (2006)
20. Shen, L.Y., Wu, M., Wang, J.Y.: A model for assessing the feasibility of construction project in contributing to the attainment of sustainable development. J. Constr. Res. **3**(2), 255–269 (2002)
21. Ochieng, E.G., Wynn, T.S., Zuofa, T., Ruan, X., Price, A.D.F., Okafor, C.: Integration of sustainability principles into construction project delivery. J. Archit. Eng. Technol. **3**(1), 1–5 (2014). https://doi.org/10.4172/2168-9717.1000116
22. Zhang, X., Wu, Y., Shen, L., Skitmore, M.: A prototype system dynamic model for assessing the sustainability of construction projects. Int. J. Proj. Manag. **32**(1), 66–76 (2013)
23. Ahn, C., Wenjia Pan, S.L., Peña-Mora, F.: Enhanced estimation of air emissions from construction operations based on discrete-event simulation. In: International Conference on Computing in Civil and Building Engineering, Nottingham, UK (2010)
24. González, V., Echaveguren, T.: Exploring the environmental modeling of road construction operations using discrete-event simulation. Autom. Constr. **24**, 100–110 (2012)
25. Hao, J.L., Tam, V.W., Yuan, H.P., Wang, J.Y., Li, J.R.: Dynamic modeling of construction and demolition waste management processes: an empirical study in Shenzhen, China. Eng. Constr. Archit. Manag. **17**(5), 476–492 (2010)
26. Shen, L., Wu, Y., Chan, E., Hao, J.: Application of system dynamics for assessment of sustainable performance of construction projects. J. Zhejiang Univ. Sci. A **6**(4), 339–349 (2005)
27. Shen, L.-Y., Hao, J.L., Tam, V.W.-Y., Yao, H.: A checklist for assessing sustainability performance of construction projects. J. Civ. Eng. Manag. **13**(4), 273–281 (2007)
28. Shen, L., Tam, V., Tam, L., Ji, Y.: Project feasibility study: the key to successful implementation of sustainable and socially responsible construction management practice. J. Clean. Prod. **18**(3), 254–259 (2010)

Detection of Water Safety Conditions in Distribution Systems Based on Artificial Neural Network and Support Vector Machine

Hadi Mohammed[1(✉)], Ibrahim A. Hameed[2], and Razak Seidu[1]

[1] Water and Environmental Engineering Group,
Department of Marine Operations and Civil Engineering, Faculty of Engineering,
Norwegian University of Science and Technology (NTNU),
Postboks 1517, 6025 Ålesund, Norway
{hadi.mohammed, rase}@ntnu.no

[2] Department of ICT and Natural Sciences,
Faculty of Information Technology and Electrical Engineering,
Norwegian University of Science and Technology (NTNU),
Postboks 1517, 6025 Ålesund, Norway
ibib@ntnu.no

Abstract. This study presents the development of artificial neural network (ANN) and support vector machine (SVM) classification models for predicting the safety conditions of water in distribution pipes. The study was based on 504 monthly records of water quality parameters; pH, turbidity, color and bacteria counts taken from nine different locations across the water distribution network in the city of Ålesund, Norway. The models predicted the safety conditions of the water samples in the pipes with 98% accuracy and 94% respectively during testing. The high accuracy achieved in the model results indicate that contamination events in distribution systems that result in unsafe values of the water quality parameters can be detected using these classification models. This can provide water utility managers with real time information about the safety conditions of treated water at different locations of distribution pipes before water reaches consumers.

Keywords: Water safety · Distribution network · Contaminant detection
Machine learning

1 Introduction

Worldwide, drinking water utilities strive to meet tight water safety regulations to lower the risks of waterborne infections from drinking water. After water is treated, the distribution network are still exposed to risks of contamination, which can occur either accidentally or intentionally. Contamination of treated through both means can have serious health consequences, particularly if carried out intentionally. Accidental contamination can occur as a result of breaches in the physical integrity of the pipe networks, such as breakages, leakages, and wastewater intrusions. For instance, backflow from cross-contaminations, cracks and leaks in the pipes can cause pathogens

A. E. Hassanien et al. (Eds.): AISI 2018, AISC 845, pp. 567–576, 2019.
https://doi.org/10.1007/978-3-319-99010-1_52

to enter treated water in the pipes [1]. Intentional contamination may however involve the introduction of chemicals, biotoxins, or biological agents such as viruses and bacteria. While drinking water utilities thrive to meet increasingly tighter water quality regulations during treatment, little is often done in terms of monitoring the quality of the treated water in distribution networks before water reaches consumers. As a result of malfunctioning in water distribution networks, increased gastrointestinal illnesses (GII) can occur from consumption of treated water, potentially causing a health risk [2].

The deployment of early warning system (EWS) at various locations across water distribution networks enables real-time estimation of the quality of water in the pipes. In these EWS, data from various sensor locations are used to simulate mathematical models. Among the essential components of EWS, the detecting algorithm plays a critical role in determining the accuracy of the detections [3, 4]. In recent years, many researchers have applied different approaches to classify water quality data from online sensors in distribution pipes as normal or abnormal. Methods already applied include multivariate Euclidean distance (MED), linear prediction filters (LPF), and time series increments [5, 6]. Other methods that have been applied include artificial intelligence models such as artificial neural network (ANN) and support vector machine (SVM) [7–9]. Many studies were based either on simulations of contamination events in the water distribution pipes or involved spiking of the measured water quality parameters to mimic contamination events. To achieve more robust that and reliable models that can efficiently predict the occurrence of contamination events in this critical infrastructure, it is essential to build the entire model based on data that reflect the actual situation in the distribution pipes. This study explores the applicability of ANN and SVM classification models for use in predicting the safety condition of water in distribution pipes. In this study, actual records of water quality parameters taken at different locations across distribution pipes were initially classified as "safe" and "unsafe". The data was subsequently used to train and test the ANN and SVM models.

2 Methodology

2.1 Study Area and Data Collection

Figure 1 shows the map of study area and the locations of the various sampling points across the water distribution network of Ålesund, Norway. The main raw water source of the Ålesund water utility is the Brusdalsvatnet Lake, which is located on the inland of Oksenøya in the Møre and Romsdal County between the Ålesund and Skodje municipalities in the West Coast of Norway. The drinking water utility draws 55,000 m^3 of water daily from the Lake at a depth of 35 m, and serves approximately 50 000 inhabitants of the city. The data used in this study consist of monthly measurements of pH, turbidity (NTU), color (mg Pt/l) and counts of total bacteria (counts/ml) in the treated water. The data composed of 504 samples were taken between January 2013 and June 2017

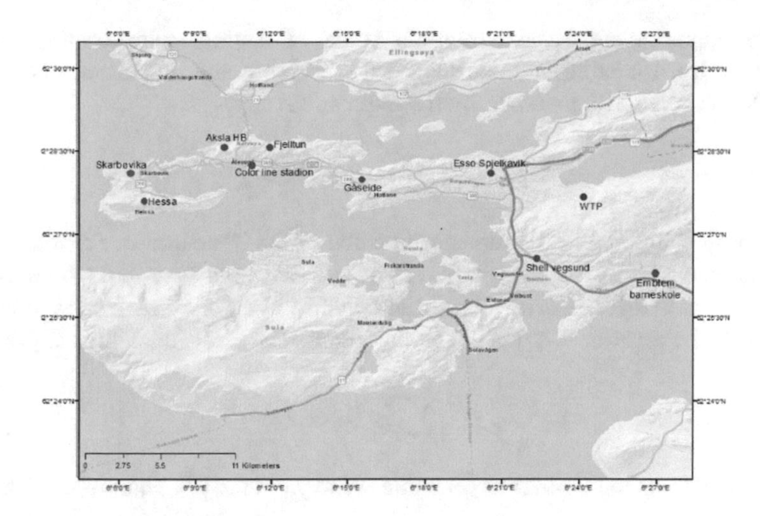

Fig. 1. Map of study site showing the locations of the distribution network where water quality parameters were measured.

2.2 Statistical Analysis of Data

Basic statistical features of the data set were evaluated using Pearson correlation analysis of the measured parameters. The correlation coefficient r_{XY} was calculated as follows:

$$r_{XY} = \frac{\sum_{i=1}^{n} (x_i - \bar{X})(y_i - \bar{Y})}{\sqrt{\sum_{i=1}^{n} (x_i - \bar{X})^2 * \sqrt{\sum_{i=1}^{n} (y_i - \bar{Y})^2}}} \quad (1)$$

where X and Y are the measured values for two different water quality parameters in the treatment plant effluent and the distribution pipe respectively x_i and y_i are the ith values in the time series, n is the number of data points used to calculate the correlations, and $\bar{X}$ and $\bar{Y}$ refer to the mathematical expectation. Further, principal component analysis (PCA) was used to determine the importance of the parameters in defining the safety conditions of the water samples. The input data were normalized (between 0 and 1) to account for the effect of dimensionality on the models. This results in scaling down the data variables to the same range, so that any bias can be avoided. In this study, the min-max normalization method was applied.

2.3 Artificial Neural Network (ANN) Classification Model

ANN is an information processing technology system that stimulates the human brain nervous system [10]. The technique offers an efficient means of modeling complex relationships between input and output variables in a system. One of the most commonly used network architectures of ANN is the feed-forward neural network. In this architecture, signals are allowed to travel in only the forward direction, from input nodes to output nodes. It uses multiple layers of neurons with nonlinear transfer functions to learn complex relationships between input and output vectors. Feedforward ANN is generally governed by the expression:

$$y_j = F_j\left(\sum_{i=1}^{m} w_{j,i}.y_i + b_i\right) \quad (2)$$

where y_j is the output, F_j is the transfer function of the jth neuron in a layer, $w_{j,i}$ is the weight that connects the output y_i of the ith neuron from one layer to the input of the jth neuron in the next layer, b_i is the bias weight on the jth neuron of each layer. It is mostly solved using the error back-propagation algorithm. The error function is expressed as:

$$E_p = \frac{1}{2}\sum_{j=1}^{m} (t_{pj} - y_{pj})^2 \quad (3)$$

where t_{pj} represents the desired target value p and y_{pj} is the jth output of the final layer.

2.4 Support Vector Machine (SVM) Classification Model

Support Vector Machine, developed by Vapnik and his collaborators [11] is a machine learning technique that is originally meant for binary classification. The principal idea of the method is to obtain a hyper-plane that separates different classes of data points. To enable a linear separation of data, SVMs employ kennel functions to map data from the input space into a higher dimensional space, creating two parallel hyper-planes to separate the data. Therefore, in classification, the geometric margin between the two hyper-planes are maximized, while minimizing the classification error [12, 13]. For a given training dataset $\{(X_i, y_i)\}_{i=1}^{N}$, the SVM optimizes the following problem.

$$\begin{cases} w^T\emptyset(x_i) + b \geq +1, \ if \ y_i = +1 \\ w^T\emptyset(x_i) + b \leq -1, \ if \ y_i = -1 \end{cases} \quad (4)$$

where $x_i \in \mathbb{R}^n$ and $y_i \in \{-1, +1\}$ are the inputs and binary class labels respectively. $\emptyset(.)$ is the non-linear function which maps the input space into some higher-dimensional Hilbert space H and $w \in$ H is orthogonal to the separating hyperplane in that space. Thus, the hyperplane $w^T\emptyset(X) + b = 0$ is constructed to discrimination the various classes in the dataset. In this study, the radial basis function (RBF) was used as the kernel function. The RBF can be expressed as:

$$K(x_i, x_j) = \exp\left(-\gamma\|x_i - x_j\|^2\right), \ \gamma > 0 \quad (5)$$

where γ is a kernel parameter.

2.5 Model Development and Performance Indices

Water quality standards described in the Norwegian Guide to Drinking Water Regulations [14] were used to initially classify the data set into "safe" and "unsafe" water samples. The acceptable concentration of total bacteria count in the clean water is 100 CFU/ml. Above this number, the water is regarded as unsafe for drinking. Moreover, there should not be irregular variations in the various water quality parameter measured in the distribution pipes. For water pH, the acceptable range is 6.5–9.5. The regulation however does not indicate specific ranges of values for turbidity and color, although it recommends that these parameters should not exceed 1 NTU and 20 mg Pt/l respectively in the clean water. The entire data set contained 13 unsafe samples as per the recommendations of the regulations. However, we tightened the recommended thresholds for the various parameters by reducing the count of bacteria from 100 CFU/ml to 80 CFU/ml, and the pH from a range of 6.5–9.5 to a range of 6.5–9. This enabled us to obtain just enough data samples to represent both classes such that the models could be efficiently trained. Using the tighter regulations we imposed on the parameters, 89 samples were found to be unsafe out of the 504 samples. For the remaining 415 safe samples, we applied only 89 samples taken randomly for building the model. This resulted in 178 samples; half safe and half unsafe. The resulting data set was normalized between 0 and 1 using the minimum-maximum method, and partitioned; 70% for model training and 30% for testing. The ANN classifier model was built with two hidden layer, the first layer composed of 10 neurons and the second layer consisting of 4 neurons. In the SVM model, the radial basis function was used as the kernel function. Both the ANN and the SVM models were was coded in Matlab R2017b. The accuracies of the two models were assessed using confusion matrix (error matrix). The matrix was used to summarize the number of correct and incorrect predictions of the models during both training and testing stages. Based on the matrix, we calculated the overall accuracies and precisions.

3 Results and Discussion

3.1 Descriptive Statistics

Figure 2 shows the variation of the normalized water quality variables taken between January 2013 and June 2017. The black lines indicate the parts of the data used for training the ANN model (70%), while the blue lines show the testing data (30%). Overall, 125 samples were used for the training and 53 for testing. Over the study period, total bacteria counts in the pipes ranged between 0 and 300 CFU/ml, with average count of 12.37 CFU/ml, and a standard deviation of 45.28 CFU/ml. The ranges of pH, turbidity and color were respectively 6.5–9.1, 0.1–1.4 NTU, and 5 to 58 mg Pt/l. Their respective average values were 7.9, 0.2, and 10.8. Thus, compared to the maximum threshold values acceptable in the safety standards, the average concentrations of the water quality parameters are lower.

The regulations also indicated that there should not be rapid variations in the water quality parameters over a given period. However, although the variances of the parameters are mostly below the thresholds, there are large variances in the data, particularly, bacteria count, turbidity and color. The figure also shows the initial classification of the safety condition of the water samples in the pipes based on the

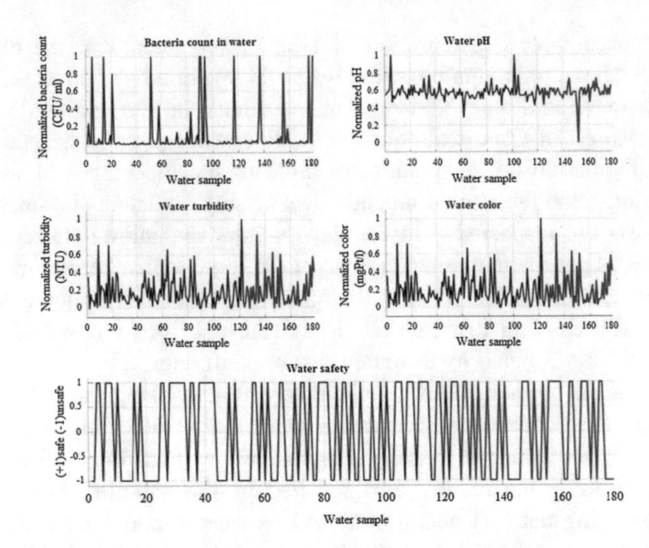

Fig. 2. Temporal variation of water quality variables measured in the distribution network between January 2013 and June 2017.

Norwegian Guide to Drinking Water Regulations. Figure 3 (b1–b4) show 3D scatter plots of the relationships among the four input variables and their effects the safety conditions of the water samples. In this figure, the two classes of the water samples to be predicted by the ANN model are shown in green colors (safe samples) and blue color (unsafe samples). The figure indicates that water samples are more safe if they have low turbidity and low bacteria count (Fig. 3b3). The effect of pH on the safety conditions of the water samples is marginal, as majority of the samples are accumulated within a short pH window. The data consists of outliers, some of which are distinct in Fig. 3b1, b2 and b4. These may pose some challenges to the outputs of the ANN and SVM model classifications. Results of the Pearson's correlation analysis for the four input parameters and the safety conditions of the water samples are shown in Fig. 3(a). Total bacteria count was the key parameter used in the initial classification of the safety condition, since the parameter had the largest variance (2050 CFU/ml) with values closer to the threshold (100 CFU/ml). This parameter had a weak negative correlation with turbidity ($r = -0.06$) and color ($r = -0.03$), but showed a weak positive correlation with water pH. With respect to the safety conditions of the water samples, all the four input parameters were negatively correlated, and the magnitudes of these negative correlations were larger with bacteria count ($r = -0.30$) and turbidity ($r = -0.58$). This indicate that these two parameters are more important determinants of the safety conditions of water in the distribution pipes than pH and color. Further, results of the PCA analysis showed that the four principal components respectively explain 54.60%, 29.82%, 9.73%, and 5.91% of the variances in the data set.

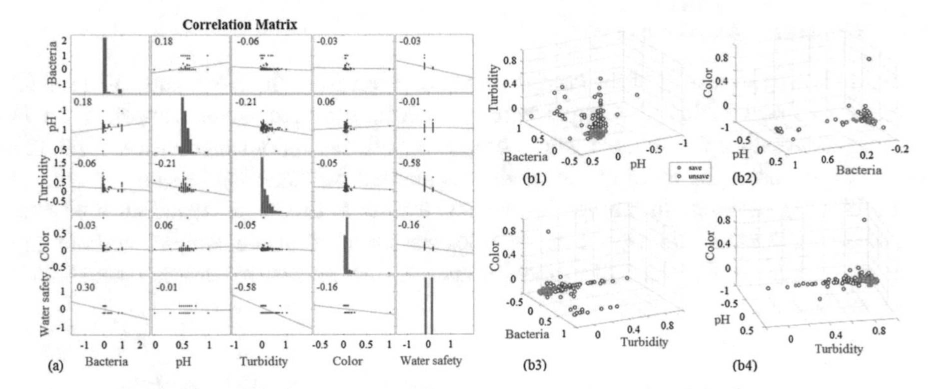

Fig. 3. Correlation matrix (a) and scatter plots (b) of normalized inputs and their effects on the safety condition of water in pipes.

3.2 Results of the ANN and SVM Models

Figure 4(a1 and a2) and (b1 and b2) respectively show the results of the performances of the ANN and SVM models in predicting the safety conditions of the water samples in the distribution pipes. In the figure, the top sections (a1 and b1) show the samples classified as safe while the bottom parts (b1 and b2) illustrate the unsafe samples. The errors associated with the model classifications are also shown in the figures. Moreover, the combined outputs of the model for both training and testing datasets are plotted. The model training stage reproduced the two different classes in the water samples with very good accuracy. However, some marginal classification errors occurred in the testing stages of both models. We further calculated the performance rates of the model during both training and testing using confusion matrix.

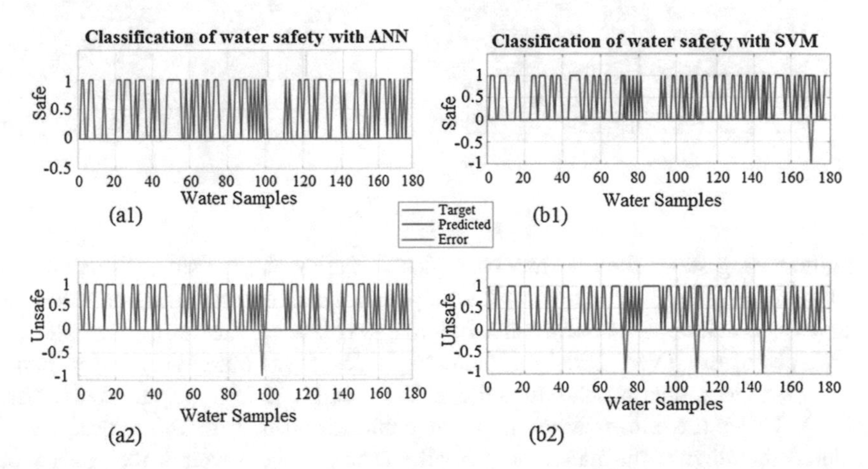

Fig. 4. Results of prediction of water safety condition with ANN model (a1, a2) and SVM model (b1, b2)

3.3 Classification Accuracy

Table 1 shows the results of the confusion matrix built from the ANN and SVM model outputs. In these tables, the actual safety conditions of the water samples used in training and testing the models were compared with the model predictions. The 125 data samples used for training the models contained 56 "safe" (+1) samples and 69 "unsafe" (−1) samples. In the training stages, the ANN model predicted all the water samples classified as safe and all the samples classified as unsafe samples without any false classification. For the SVM model however, one water sample was misclassified for each class.

Table 1. Confusion Matrix of ANN model (a) and the SVM model (b)

a. ANN

Training

n=125	Predicted +1	Predicted -1	
+1	TP=56 (45%)	FN=0 (0%)	56 (100%)
-1	FP=0 (0%)	TN=69 (62%)	69 (100%)
	56 (100%)	69 (100%)	125 (100%)

Testing

n=53	Predicted +1	Predicted -1	
+1	TP=32 (60.4%)	FN=0 (0%)	32 (100%)
-1	FP=1 (2%)	TN=20 (38%)	21 (95%)
	33 (97%)	20 (100%)	53 (100%)

b. SVM

Training

n=125	Predicted +1	Predicted -1	
+1	TP=56 (45%)	FN=1 (2%)	57 (98%)
-1	FP=1 (4%)	TN=69 (62%)	70 (98%)
	57 (98%)	70 (98%)	125 (100%)

Testing

n=53	Predicted +1	Predicted -1	
+1	TP=32 (60%)	FN=2 (4%)	34 (97%)
-1	FP=1 (2%)	TN=18 (34%)	19 (95%)
	33 (97%)	20 (95%)	53 (100%)

In the testing stage, there were 53 samples, 32 safe and 21 unsafe. While the ANN model predicted all the safe samples, it wrongly classified one sample as safe, although it was unsafe. Thus, the misclassification rate was 2% of the testing samples. The performance of the SVM model on the testing dataset was marginally lower than the ANN. The SVM model misclassified 6% of the samples (2 false negatives and 1 false positive). Using the information from the confusion matrix in the tables, we also calculated the other performance rates for the testing stage. Overall, the ANN model achieved a 98% accuracy when applied to the testing data set, while the SVM model achieved 97% classification accuracy. The calculated precisions of the two models were the same nonetheless (97%). The high accuracies for ANN and SVM obtained in this study suggests that the these model have very good potential for real time prediction of the safety condition of water in the distribution pipes. Although no time component was included in these models, the technique can be adopted and the models integrated with water quality sensors in the distribution pipes such that real time safety conditions of treated water can be established each time measurements are recorded. This is possible when all model input variables can be measured in real time.

4 Conclusion

Efficient ANN and SVM – based classification models were developed for predicting the safety conditions of water in distribution pipes. The models were built with water quality data taken from nine different locations across the water distribution network within the city of Ålesund, Norway. The results show that the ANN classification model can detect contamination events in the pipes with 98% accuracy while the SVM model achieved 94% accuracy. The low error rates achieved in the model results indicate that, signals from water quality monitoring sensors in the pipes can be integrated with ANN and SVM classification models such that water utility managers can obtain real time information about the safety conditions of treated water in the pipes before it reaches consumers. This can be especially useful for areas where online monitoring of contamination events in water distribution networks are not available.

Acknowledgement. We thank the managers of the water utility of Ålesund, Norway for providing the data used in this study.

References

1. Besner, M.C., Prevost, M., Regli, S.: Assessing the public health risk of microbial intrusion events in distribution systems: conceptual model, available data, and challenges. Water Res. **45**(3), 961–979 (2011)
2. Ercumen, A., Gruber, J.S., Colford Jr., J.M.: Water distribution system deficiencies and gastrointestinal illness: a systematic review and meta-analysis. Environ. Health Perspect. **122** (7), 651–660 (2014)
3. Liu, S., Smith, K., Che, H.: A multivariate based event detection method and performance comparison with two baseline methods. Water Res. **80**, 109–118 (2015)
4. Lambertini, E., Borchardt, M.A., Kieke Jr., B.A., Spencer, S.K., Loge, F.J.: Risk of viral acute gastrointestinal illness from nondisinfected drinking water distribution systems. Environ. Sci. Technol. **46**(17), 9299–9307 (2012)
5. Klise, K.A., McKenna, S.A.: Water quality change detection: multivariate algorithms. In: Optics and Photonics in Global Homeland Security II, vol. 6203, p. 62030J. International Society for Optics and Photonics, May 2006
6. McKenna, S.A., Wilson, M., Klise, K.A.: Detecting changes in water quality data. Am. Water Works Assoc. J. **100**(1), 74 (2008)
7. Perelman, L., Arad, J., Housh, M., Ostfeld, A.: Event detection in water distribution systems from multivariate water quality time series. Environ. Sci. Technol. **46**(15), 8212–8219 (2012)

8. Raciti, M., Cucurull, J., Nadjm-Tehrani, S.: Anomaly detection in water management systems. In: Critical Infrastructure Protection, pp. 98–119. Springer, Berlin (2012)
9. Oliker, N., Ostfeld, A.: A coupled classification–evolutionary optimization model for contamination event detection in water distribution systems. Water Res. **51**, 234–245 (2014)
10. Negnevitsky, M.: Artificial Intelligence: A Guide to Intelligent Systems. Pearson Education, New York (2005)
11. Vapnik, V., Cortes, C.: Support vector networks. Mach. Learn. **20**, 273–297 (1995)
12. Cristianine, N., Taylor, J.S.: An Introduction to Support Vector Machine and other Kernel based Learning Methods. Cambridge University Press, Cambridge (2000)
13. Singh, K.P., Basant, N., Gupta, S.: Support vector machines in water quality management. Anal. Chim. Acta **703**(2), 152–162 (2011)
14. Norwegian Guide to Drinking Water Regulations (In Norwegian): Mattilsynet statens tilsyn for planter, fisk, dyr og næringsmidler. https://www.mattilsynet.no/om_mattilsynet/gjeldende_regelverk/veiledere/veiledning_til_drikkevannsforskriften.25091. Accessed 21 Mar 2018

Clustering Stock Markets for Balanced Portfolio Construction

Omar Alqaryouti(✉), Tarek Farouk, and Nur Siyam

The British University in Dubai, Dubai, UAE
omar.alqaryouti@gmail.com, tafarouk@me.com, nur.siyam@gmail.com

Abstract. Nowadays, in the diversifying economic environment, investors need various ways to help them in taking the right decisions and maximize their returns within certain acceptable risk. This research paper incorporates data mining clustering techniques to assist investors in constructing a balanced portfolio in Abu Dhabi Securities Exchange (ADX). The study examines and analyses ADX trade market history for the year of 2015. An extensive analysis has been done for various combinations of pre-processing-clustering algorithm techniques. An unexpected conclusion was revealed from this research. It is believed that small markets such as Abu Dhabi's require different "clustering" treatment than the commonly applied ones for established markets such as New York Stock Exchange.

Keywords: Data mining in stock markets · Clustering stock markets
Portfolio construction · Cluster performance evaluation · K-means
K-medoids

1 Introduction

In the contemporary world which is computer-driven, databases comprise immense volumes of information. Therefore, the abundance and accessibility of this knowledge has made Data Mining (DM) an essential and significant aspect. DM denotes a science and technology of examining data to expose previously unrecognized patterns. It is a fundamental part of the entire practice of knowledge discovery in databases [1]. The financial entities like stock markets generate massive datasets that create a need for utilizing dynamic and enormously multifaceted DM tools. Studies concerning DM have gained high attention attributed to the significance of their applications as well as the increasing production of information. DM in finance field is important for many reasons: it helps in forecasting multidimensional time-series with high levels of noise; it accommodates particular efficiency criteria as well as forecasting accuracy; it allows for coordinated multi-resolution prediction; it helps in explaining the prediction and prediction models; and it incorporates stream of textual signals as inputs for predicting models [1].

Portfolio construction is one of the oldest investment managers' tasks and is still one of the most important activities. In its simplest form, the manager will try to construct and maintain a portfolio targeting certain return objectives with minimum risk

A. E. Hassanien et al. (Eds.): AISI 2018, AISC 845, pp. 577–587, 2019.
https://doi.org/10.1007/978-3-319-99010-1_53

levels or maximizing the returns without deviating from a predefined risk levels usually set by the customer.

Classical methods adopted statistical variance of stocks past returns as a measure of stocks volatility and risk. This technique is widely accepted and still dominant today in evaluating funds and comparing investment strategies. However, under the use of variance and covariance lies a strong assumption that stock return series have normal distribution, which is not the case. Clustering techniques, on the other hand, do not require this assumption [2]. Clustering can help investment managers in identifying stocks that are "not similar" to each other, lowering in the process the portfolio risk. As explained in the "Related Work" section, using clustering in stock markets was the subject of many papers in the past two decades [3].

In this paper, we conducted an empirical study using data from Abu Dhabi Securities and Bonds market. This study focused on the data mining side rather than the underlying investment rational. The overall structure of the research takes the form of nine sections, including this introductory section.

2 Related Work

Until recently, many researchers have shown increasing interest in the field of data mining in stock markets. For the past two decades, researchers' attention has been directed towards using clustering data mining techniques in stock markets.

Researchers across the globe have studied their local markets using various clustering techniques. For example, [4] conducted a study in Bombay stock exchange companies that involved portfolio construction and management using various clustering techniques such as k-means, Self-Organizing Maps (SOM), Fuzzy C-means and Markowitz model. The researchers concluded that using clustering techniques assists the investors in constructing portfolios with minimized amount of time. The study results indicated that using k-means clustering technique was the best among other experimented clustering techniques according to the performance evaluation measures. Similarly, [5] investigated the co-movement and relation between Taiwan and China stock markets by applying data mining association rules, clustering techniques and analysis. The study has pointed out the significance of using k-means clustering algorithm in exploring the possible clusters of stock categories between the two markets for having larger investment portfolio.

An important study by Gavrilov et al. [6] analyzed various techniques to enhance the effectiveness of stocks clustering. The authors highlighted the importance of using the appropriate similarity measures that are used for stock market data according to a given context. In another study, Dutt et al. [7] employed a technique to construct monthly stock indices pairwise correlations of returns. The authors illustrated these correlations with risk-adjusted variations in the industry. The findings indicated that countries with similar industries show higher co-movements of their stock market.

Aghabozorgi and Teh [8] proposed a novel three-phase clustering approach to categorize companies according to the similarity of their stock markets shape. This approach comprised three phases. First, low-resolution time series data are utilized to categorize companies. After that, pre-clustered companies are split into sub-clusters.

Finally, these sub-clusters are merged in the last phase. This approach outperformed existing conventional clustering algorithms.

A recent study by Tekin and Gümüş [9] classified stocks based on financial indicators extracted from the financial companies' statements. Their approach aimed to group most similar stocks in the same cluster with respect to related variables. The result of their analysis concluded 88 stocks in Borsa Istanbul (BIST 100) Index divided into 12 clusters. The stocks that were most appropriate to construct the portfolio have been chosen based on financial indicators and last years' stock performances.

3 Methodology

The goal of this study is to apply clustering algorithms on ADX and to analyze the results. The figure below (Fig. 1) summarizes the steps of the methodology used in this study.

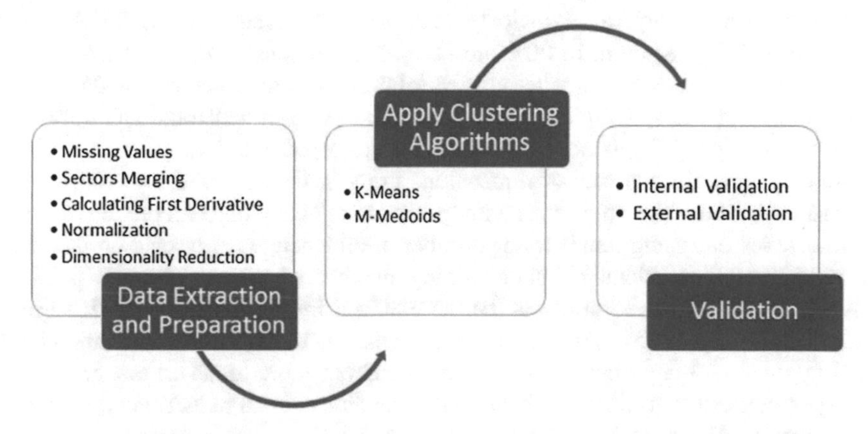

Fig. 1. Methodology

First, data was acquired through Bloomberg Terminal. ADX data was extracted for year 2015. ADX consists of 61 stocks. For each stock, 251 data points representing daily prices of year 2015 were extracted. These stocks are associated with industry sectors such as communication and industry.

Upon examining the extracted data series, one stock series was found to have many missing values and had to be excluded from the dataset. Otherwise, there were relatively a small number of missing values and never more than one per series. In such cases, the missing value was filled with the price from the previous day price. There were other options such as using the average which has the advantage of being statistically neutral, however, using previous price is more realistic in the case of stock prices.

The 61 stocks were found to be associated to seven sectors. However, three of these sectors were associated with one stock only. So, these three sectors have been merged into one sector. Another dataset was prepared by calculating the daily returns. The daily returns are calculated by dividing the difference between current price and previous price by the previous price of the stock.

Normalization in general attempts to make a dataset more comparable by addressing the varying scale of various attributes. In our case, different stocks might vary in their prices greatly. Highly priced stocks could dominate similarity calculation. Normalization can dampen this effect. Normalization is done by subtracting a stock sequence from its mean. Then dividing the resulting sequence by its Euclidean norm. In this way the resulting sequence will have a mean of zero. Both normalized and non-normalized sequences were tested and results were compared.

Dimensionality reduction is an important step. Not only it improves calculation time but, in many cases, it improves the results due to dimensionality curse. Bellman [10] explained the curse of dimensionality by the exponential increase of space volume as number of dimensions' increase which leads to data sparsity. Every stock daily prices series constitute of 251 points. So, each one of the 61 stocks can be represented by a point in a 251-dimensional space.

There are various techniques to reduce the number of dimensions used by stock daily prices series such as Principal Component Analysis (PCA), SOM, Fourier Transform and Aggregation. In PCA, number of dimensions is reduced by substituting the stock price series with a smaller number of principal components, e.g. 25. Through the projection of the 251-dimensional space into the 25-dimensional space, principal component can preserve, if not improve, the series representation.

SOM reduces the number of dimensions using artificial neural networks. Fourier transform substitutes the time series with group of frequency that capture the essence of the time series but using much lesser number of dimensions. Aggregation can reduce the number of dimensions by summarizing number of points into one point. For example, the daily prices series can be summarized into a monthly one that uses 12 dimensions instead of 251. This might improve the clustering results by eliminating the meaningless day-to-day fluctuations of stocks and put more focus on longer trends. In this paper, clustering results that did not have any dimensional reduction is presented as well as results after applying PCA to produce 5, 10, 25 and 50 principal components.

4 Applying Clustering Algorithms

This study focused on centroid based clustering algorithms, namely, k-means and k-Medoids. Clustering is responsible for producing groups of similar stocks. Sectors have been used to produce manual clustering of stocks as "ground truth".

Euclidean distance and Dynamic Time Wrapping (DTW) were used in this study. On the other hand, dynamic time wrapping is very suitable for sequences like lime series. Euclidean distance is the root square of the sum of squares of dimensional values. Euclidean distance has some weaknesses since it compares pairs of instances from the two series. Thus, it is very sensitive when there is a shift difference between the two series such as two stocks that are very similar to each other but one of them lags the other with number of days. Conversely, DTW compares each point in a series with the other, hence the time wrapping name. This makes it very suitable to sequences like stocks daily prices series. DTW is immune to transposition and scaling effects where sequences might be shifted or scaled but are essentially similar to each other. However, this advantage comes at cost. It is an expensive algorithm and requires much more time to compute, giving more importance to dimensionality reduction techniques.

5 Validation

We have employed two types of validations; internal and external evaluation methods. The internal evaluation indexes measure the quality of the resulting clustering through detecting the tightness of the clusters and how distant they are from each other. The average distance from centroids and Davies-Bouldin index have been used. The average centroid distance is calculated by measuring the distance of each series to the centroid of the clusters it belongs to and then getting the average. The lower the average the tighter the clusters. Distance between centroids themselves indicate the sparseness of the cluster set. Davies-Bouldin index is the ratio between cluster sparsity and cluster separation. Low value means tight clusters and good separation in between. However, in External Evaluation Index, each stock is associated with an industry sector. This association is usually done by professional organizations such as GIGS and are considered the DEFACTO standard. Sectors usually represent industries such as Oil and Gas, Communication, Financial and Consumers. Stocks belong to companies that are usually associated to specific industries. A company is not allowed to be associated to more than one sector at one time. GIGS can create and remove sectors as well as change the sector that a company belongs based on experts' analysis of the market and company activities. Given that the relation between the sector and stock is dependent on the company activities, stocks belonging to the sectors exhibit very similar price change behavior and therefore have similar price series. This nature is exploited to test the quality of clusters set.

Cluster similarity measure is used to compare the resultant clusters to sectors. For the purpose of this study it has been chosen to use k = number of sectors. Two values for k were used, 4 and 7. As mentioned earlier, the 61 stocks were associated to 7 sectors. We noticed that 3 of these sectors have only one stock so those were merged resulting in 4 sectors. This merge of sectors and reduction in the number of clusters improved the results. The similarity is calculated by first counting the number of stocks shared between each pair of cluster and sector. This count is doubled and divided by the sum of the number of companies in the cluster and the sector pair. Using the results, the summation of the maximum similarity for each cluster is calculated. C_i in the following formula represent clusters and C_j represent sectors. k is the number of the clusters.

$$Sim\left(C_i, C_j^{'}\right) = 2\frac{|C_i \cap C_j^{'}|}{|C_i| + |C_j^{'}|} \tag{1}$$

$$Sim(C, C^{'}) = \left(\sum\nolimits_i max_j\ Sim(C_i, C_j^{'})\right)/k \tag{2}$$

6 Experiment Results

The experiment has been run using the following configurations: (1) extracted daily price data and the derived daily return series, (2) normalized and not normalized, (3) full series or reduced to 50; 25; 10 or 5 principal components, (4) using k-means or k-Medoids using k = 4 and k = 7 for each, (5) using Euclidean distance or dynamic time wrapping, (6) using 7 sectors grouping or 4 sectors grouping for validation.

This resulted in a total of 160 runs. In each run, we produced: (1) cluster set; (2) parallel chart (each dimension is represented by a coordinate). For example, some of the parallel charts with different pre-processing and clustering parameters are shown in (Figs. 2 and 3) below; (3) internal performance vector that constitute of average centroid distance for the set and for each cluster as well as Davies-Bouldin index; (4) cluster-sector sets similarity measure.

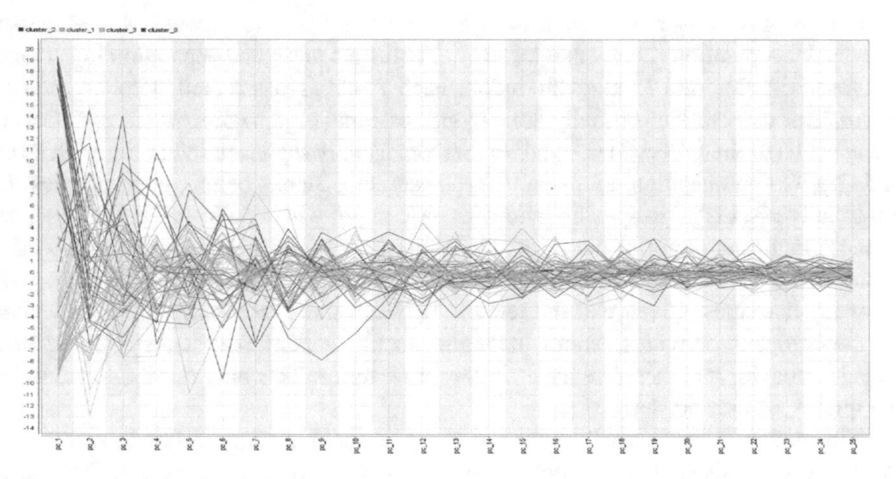

Fig. 2. K-means with 7 clusters, with raw prices, with normalization, using Euclidean and PCA(25)

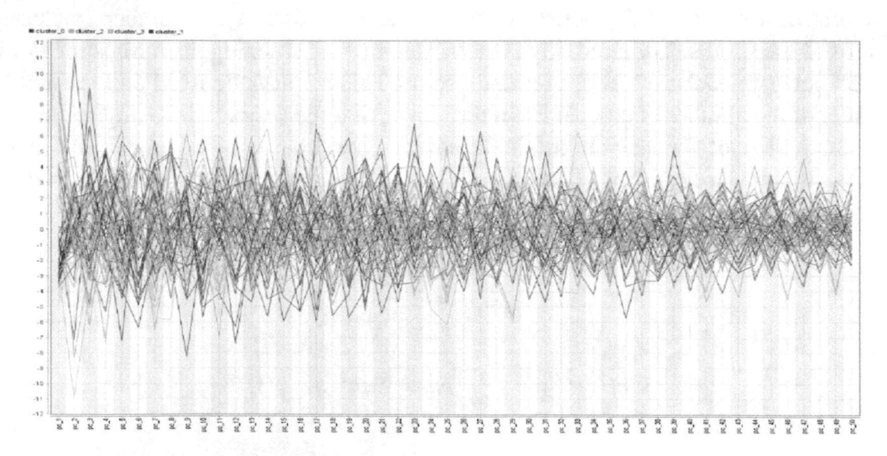

Fig. 3. K-means with 4 clusters, with returns, with normalization, using Euclidean and PCA(50)

The results of 160 runs were then compared to each other. This resulted in the following list of observation:

1. Prices vs Returns: As shown in Fig. 4, higher cluster similarity cannot be attributed to the use of prices or returns. Normalization had different effect on returns than prices. Best results from returns were mostly for the not normalized ones. On the other hand, normalization produced best prices results.
2. Using un-normalized prices: As shown in Fig. 5 the average centroid distance becomes insensitive to the number of dimensions. This could be related to the fact that the highly priced stocks dominate the clustering algorithm from the first point of the series and it becomes irrelevant to add more dimensions to the series.
3. Number of clusters: As shown in Fig. 6, cluster similarity index is higher for k = 4. There could be multiple reasons for that. k-Means/k-Medoids are sensitive to the selection of initial centroids. The probability of selecting centroids within each of the sectors decreases with the increase of k. It can be calculated using the following formula [11]: For 4 clusters, P = 9.4% while for 7 clusters, P = 0.6%.

$$P = \frac{number\ of\ ways\ to\ select\ one\ centroid\ from\ each\ cluster}{number\ of\ ways\ to\ select\ k\ centroids} = \frac{k!n^k}{(kn)^k} = \frac{k!}{k^k} \quad (3)$$

4. Variation of the results: As shown in Fig. 7, choice of similarity measure (i.e. Euclidean or DTW) and the number of dimensions or clustering algorithms seems not to have an influence on the external and internal performance indices.
5. Inconsistency between internal (Davies-Bouldin and average centroid distance) and external performance index (cluster similarity): As shown in Fig. 7, a smaller average centroid or low Davies Bouldin index, which indicates good clustering, does not mean higher cluster similarity.

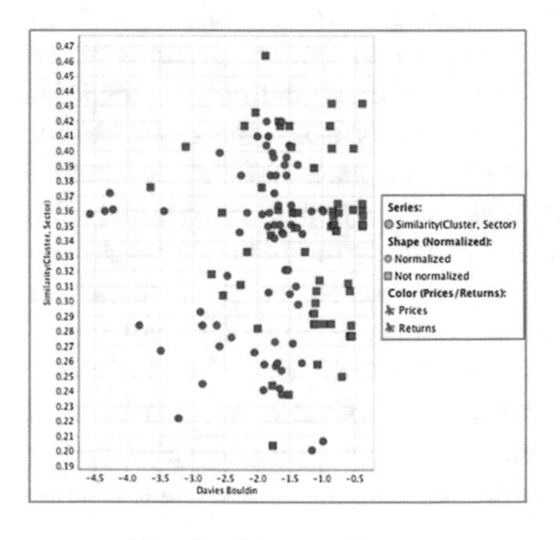

Fig. 4. Prices vs Returns

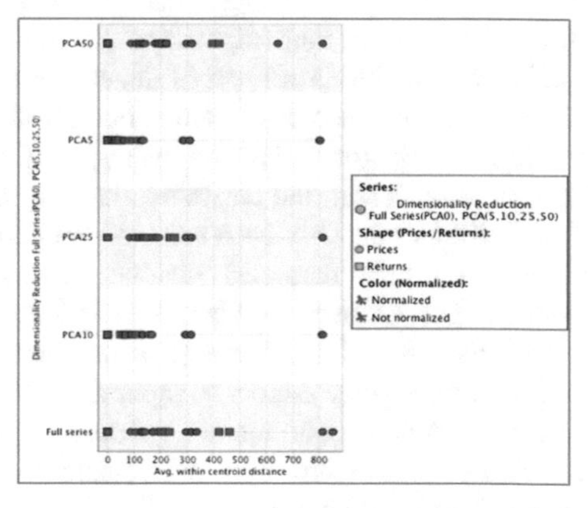

Fig. 5. Using un-normalized prices

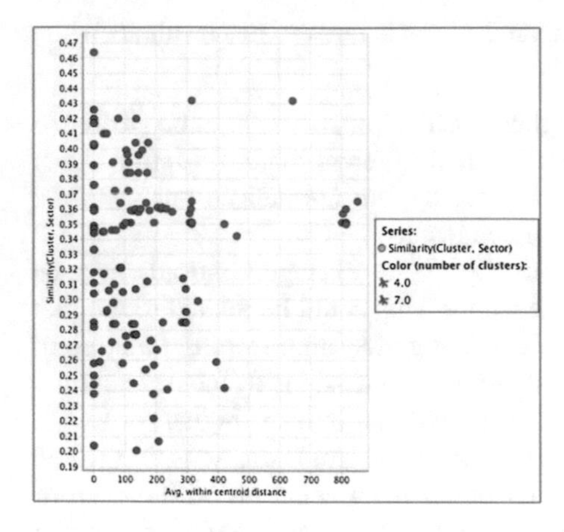

Fig. 6. Number of clusters

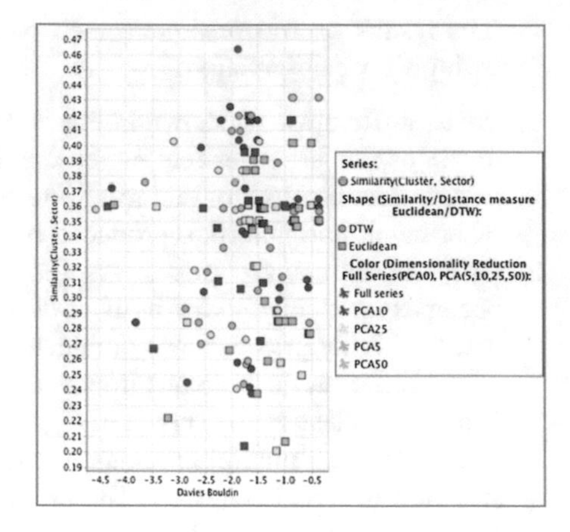

Fig. 7. Variation of the results

7 Framework Setup

We used Microsoft Excel and RapidMiner to prepare, process and evaluate the data acquired from Bloomberg. The process involves extracting the required attributes and values that include the company, sector, transaction date, and price data. It also involves examining the data, finding missing values, filling it using the previous price, and merging the sectors that are used as input to the process for comparison with the cluster output. In RapidMiner, we applied DM techniques that comprises pre-processing, normalization, dimensionality reduction, clustering and performance evaluation. In this process, we used the clustering to produce groups of similar stocks. Two clustering algorithms have been used, namely, k-means and k-medoids. The main difference between both algorithms is that the centroid in k-means represents an imaginary point that is not part of the clustering points, while the centroid in k-medoids represents one of the cluster points. Further, the internal performance was evaluated for the produced clusters using the average within centroid distance and Davies Bouldin main criterion. Then, the attributes that are required for the joining are selected. The subset attributes that has been selected are company and cluster attributes. The cluster-company outcome data and sector-company data were joined using the company attribute. The external validation is calculated using cluster-sector similarity. The process uses the cluster examples outcome joined (cluster-sector) with the actual mapping of company sectors as input data to produce the PivotTable using the clusters as rows, sectors as columns and companies count as value as shown in Table 1. The PivotTable values were used to calculate the cluster-sector and sector-cluster similarities as shown in Table 2.

The mixed results shown in the previous section could be attributed to the unbalanced market sectors. Some sectors such as the financial sector has 34 stocks, while sectors like Utilities, Communication and Energy has one stock each. It is known that centroid clustering has weakness to variation in cluster size and density [11].

Table 1. PivotTable of K-means clustering algorithm with no returns, DTW numerical measure, normalization and PCA(25)

Sectors	Communications, energy and utilities	Consumer	Financial	Industrial	Total
Cluster 1			4	3	7
Cluster 2	1	3	20	3	27
Cluster 3	2	5	4	2	13
Cluster 4		2	6	5	13
Total	3	10	34	13	60

Table 2. Similarity external calculation for K-means clustering algorithm with no returns, DTW numerical measure, normalization and PCA(25) using data in Table 2

Sectors	Communications, energy and utilities	Consumer	Financial	Industrial	Max row
Cluster 1	0.000	0.000	0.195	0.300	0.300
Cluster 2	0.067	0.162	0.656	0.150	0.656
Cluster 3	0.250	0.435	0.170	0.154	0.435
Cluster 4	0.000	0.174	0.255	0.385	0.385
Max column	0.250	0.435	0.656	0.385	
Sector-cluster similarity = (∑_Max Row Sim)/4					0.444
Cluster-sector similarity = (∑_Max Column Sim)/4					0.431

This might mean that small markets like Abu Dhabi Securities and Bond Market need different ‘clustering’ treatment than bigger markets like New York Exchange (See Table 3).

Table 3. Comparison between S&P and ADX

Sector	S&P	ADX
Consumer discretionary	86	8
Consumer staples	37	2
Energy	38	1
Financials	90	34
Health Care	58	0
Industrials	67	13
Information technology	67	0
Materials	27	0
Telecommunications services	5	1
Utilities	29	1

8 Conclusion and Future Prospects

Clustering can be a powerful technique to enhance risk level of the portfolio by choosing stocks from different clusters and avoid concentrating the portfolio in one or few clusters. However, this experiment showed that clustering results can be strongly variable-based on pre-processing tasks, similarity measures and k value. It could be a daunting task to find the best cluster set. Depending solely on internal and external measures can be misleading. Domain knowledge could offer much help in this area. We noticed that there are no individual settings that can claim best results. Instead, it is a combination of settings that produce best results. For example, while Normalization produced better results when applied to price data, this was not the case when applied to return data. The main conclusion of this study is that the use of clustering in a relatively smaller markets like Abu Dhabi and Dubai requires its own knowledge. Using the same techniques used for large indexes such as S&P would not guarantee similar success. This research can be considered as one step in a long journey rather than a conclusive one. This is can be a call for a line of research that focuses on small markets in the future. One of the areas that can be further explored is Dimensionality Reduction. As explained in Sect. 3, SOM, Fourier Transform and Aggregation are all other techniques that can be used. Furthermore, a monthly returns series can be used. Each stock will be represented by 12 points. This will eliminate the meaningless fluctuations of the daily price and put more emphasis on inherent stock characteristics. In terms of validation, an empirical study can be carried out by simulating actual investment of varying capital sizes in a portfolio constructed using clustering and then compare its performance with the index or known funds in the market. Moreover, it would be interesting to explore how clustering can be used in situations where the investment manager would like to hit certain risk mark rather than get the most dissimilar stocks. In such situation, expected returns need to be taken in consideration while comparing alternative cluster sets. Additionally, pre-processing and post-processing can be further utilized. It can enhance detecting and eliminating outliers as well as merge, split and eliminate clusters based on size.

References

1. Hajizadeh, E., Ardakani, H.D., Shahrabi, J.: Application of data mining techniques in stock markets: a survey. J. Econ. **2**(July), 109–118 (2010)
2. Durante, F., Foscolo, E.: An analysis of the dependence among financial markets by spatial contagion. Int. J. Intell. Syst. **28**(4), 319–331 (2013)
3. Musmeci, N., Aste, T., Di Matteo, T.: Relation between financial market structure and the real economy: comparison between clustering methods. PLoS ONE **10**(3), e0116201 (2015)
4. Nanda, S.R., Mahanty, B., Tiwari, M.K.: Clustering indian stock market data for portfolio management. Expert Syst. Appl. **37**(12), 8793–8798 (2010)
5. Liao, S.H., Chou, S.Y.: Data mining investigation of co-movements on the Taiwan and China stock markets for future investment portfolio. Expert Syst. Appl. **40**(5), 1542–1554 (2013)

6. Gavrilov, M., Anguelov, D., Indyk, P., Motwani, R.: Mining the stock market: which measure is best. In: ACM SIGKDD, pp. 487–496 (2000)
7. Dutt, P., Mihov, I.: Stock market comovements and industrial structure stock market comovements and industrial structure. In: INSEAD Working Paper Series, vol. 45, no. 5 (2013)
8. Aghabozorgi, S., Teh, Y.W.: Stock market co-movement assessment using a three-phase clustering method. Expert Syst. Appl. **41**(4), 1301–1314 (2014)
9. Tekin, B., Gümüş, F.B.: The classification of stocks with basic financial indicators: an application of cluster analysis on the BIST 100 index. SSRN Electron. **7**(5), 104–131 (2017)
10. Bellman, R.E.: Dynamic Programming. Princeton University Pres, Princeton (1957)
11. Tan, P.N., Steinbach, M., Kumar, V.: Introduction to Data Mining, 1st edn. Addison-Wesley Longman Publishing Co., Inc., Boston (2005)

An Improved Cache Invalidation Policy in Wireless Environment Cooperate with Cache Replacement Policy Based on Genetic Programming

Adel El-Zoghabi and Amro G. El Shenawy(✉)

Department of Information Technology, Institute of Graduate Studies and Research, Alexandria University, P.O. Box 832 163 Horreya Avenue, El Shatby, Alexandria 21526, Egypt

adel@gmail.com, amro_gme@yahoo.com

Abstract. Communication between mobile clients and database servers in wireless environment suffers from; the user's movement, disconnected modes, lots of data updates, low battery power, cache size limitation, and bandwidth limitation. Caching is used in wireless environment to overcome these challenges. The aim of this effort is to present enhanced invalidation policy that cooperates with a new cache replacement technique by using genetic programming to select the items that will be removed from the cache for improving data access in the wireless environment. Cooperation between servers and mobile clients to enhance data availability. Each mobile client Collects data like access probability, size, and next validation time and uses these parameters in a genetic programming method for selecting cached items to be removed when the cache is full. The experiments were carried using NS2 software to evaluate the efficiency of the suggested policy, and the results are compared with existing cache policies algorithms. The experiments have shown that the proposed policy outperfomed the LRU by 24% in byte hit ratio, and 11% in cache hit ratio. It is concluded that the presented policy achieves well than other policies.

Keywords: Mobile database · Cache replacement · Cache invalidation Genetic programming

1 Introduction

Different techniques for enhancing the capability and quality of wireless environments have become a hot research area. The development of mobile information systems has met some disadvantages. Mobile devices often have small storage capacity and weak power. Furthermore, mobile client users suffer from frequent disconnections from the server. Therefore, researchers have explored caching methods for effective data management in the mobile platform [1–3]. However, a major challenge is how to keep the cached items in the cache up to date with the source data at the server. The main topics in cache management are [1, 4–9]: (1) cache replacement policy removes cached items

A. E. Hassanien et al. (Eds.): AISI 2018, AISC 845, pp. 588–602, 2019.
https://doi.org/10.1007/978-3-319-99010-1_54

when the cache is full, (2) cache consistency policy tries to store data item in the cache up to date, (3) cache admission policy chooses what data items should be cached, (4) cache discovery policy collaborates between clients nodes for caching and selecting data items from nearest clients nodes. Wireless environment researchers have developed new techniques that taking into consideration the obstacles of wireless technology when dealing with the mobile database such as lack of energy, repeated disconnecting from the server. The aim of this work is to introduce an optimal cache replacement policy using genetic programming by taking into consideration invalidation policy which can greatly improve the system performance and provide high data availability.

A lot of variables [1, 10–13] have been used to remove data items from the cache when the cache is full such as data size, last time accessed, access account, distance from the node that access the document, time when replica stored in cache, mobility of the clients, update frequency and retrieve delay, etc. Therefore, the selection between the previous parameters to be used in cache replacement algorithm is great challenges. Since genetic programming is used to select the best possible solutions among the range of solutions available; it is possible to reach near the optimal solution. Therefore, the objective is to select a group of most influential variables that do not require more complex calculations to obtain their values, and use these variables in the fitness function which has been used in selecting the good solution. GP is an evolutionary computation technique that automatically solves problems without requiring the user to know or specify the form or structure of the solution in advance [14]. GP does this by randomly generating a population of computer programs (represented by tree structures) and then mutating and crossing over the best performing trees to create a new population. The main difference between GP and other evolutionary algorithm is that in GP candidate solutions are mathematical expressions built from a set of primitive functions for each branch of the to-be-evolved program, and a set of terminals [15].

This paper is an extended version of work published in [29]. We extend our previous work by introducing a cache invalidation to enhance the data availability at the mobile clients and introduce 2 new algorithms in both server side and client side in more realistic environment so, The goal is to develop not only a new algorithm for cache replacement policy but also a new cache invalidation policy to introduce an optimal solution for cache management by taking into consideration invalidation time to enhance data availability in the mobile environment using GP. Since there is no one solution for evicting a cached item from the cache when the cache is full; so GP is suitable for selecting the best solution according to a fitness function. The most difficult part when using GP is to determine the fitness function, therefore variables that will be used in the fitness function for selecting the best solution will be discussed. The next sections are planned as followings: Sect. 2 explains state-of-the-art cache replacement techniques, Sect. 3 introduces the proposed cache replacement model, Sect. 4 defines the criteria to evaluate the proposed genetic programming technique in cache replacement and shows the experimental results and finally, Sect. 5 presents conclusions and future work.

2 State-of-the-Art Cache Replacment Algorithm

Data caching in mobile systems was discussed by G. Kaur et al. [1] after that a lot of research has been done in caching mobile systems. These previous research concentrate on cache invalidation, with little attention to cache replacement techniques [1, 7]. The simple cache replacement techniques based on facilitating the assumptions, such as fixed data size, with no updates and no disconnections are discussed first. The Least Recently Used (LRU) caching approach is based on the assumption that the recently queried data are expected to be queried again; this method suffers from decision making with very limited data. The second technique is called Least Frequently Used (LFU) the least requested data items are the first out of cache [16]. In general, one recognized problem with cache replacement is that when a relatively large data item is requested, it may end up replacing many small cached items in the cache [10–13, 17, 18]. In order to avoid this problem, the authors of SIZE [1] declared that cache hit ratio can be optimized when small size documents are stored, given the fixed capacity of cache. The algorithm always evicts the largest cached items when the cache is full. The drawback of this technique is not to take the relative access frequency of each cached item into account, not to evict those stale yet small documents even if they are never visited again.

Authors of Log (Size) +LRU extended the LRU algorithm to evict the cached item who has the largest and is the least recently used data item among all data items with the same log (size) [12]. LRU-Thold method [13] suggested enforcing the document size threshold, which avoids placing any large cached items into the cache. However, the cost comes with this approach is that large data items are never cached even if the cache has enough space for the data items and they are always retrieved from the server [1]. LRU-K [1, 19] technique is an approach to database disk buffering. The essential design of LRU-K is evicting the page whose *K*-th most recent access is furthest in the past. However, most of the previous techniques use unchanging data item size, there are no updates, and no disconnections, so these techniques are unrealistic for mobile systems.

Broadly speaking, caching in the mobile environment has been addressed by many researchers [1, 2, 7, 13, 18–22]. In SAIU [19] the influence of data size, data retrieval delay, access probability and update frequency is considered together for cache replacements. Xu et al. [21] introduced cache replacement method taking into consideration these variables together, for on-demand broadcasts. In their technique, a stretch criterion is used to evaluate their technique in the cache and it is defined as the ratio of access latency to its service time for queried item, where service time is described as the ratio of the requested item's size to the broadcast bandwidth. With Min-SAUD, J. Xu et al. improved their cache replacement algorithm with by considering cache validation cost in addition to access probability, update frequency, data size, and retrieval delay. Furthest Away Replacement (FAR) cache replacement algorithm takes into consideration the current location of the client and also its movement direction. The limitation of FAR is it won't consider the access patterns of client and is not very useful for random movement of the client. MARS [23] is a gain-based cache replacement policy which takes into account parameters including the location of client movement direction and access probability. The data item with lowest value for cost function is removed first. The authors also introduced another algorithm called MARS + that enhanced MARS by taking into consideration client movement patterns for predicting the next location for the client [24].

In recent years, evolutionary computation techniques such as genetic programming and genetic algorithm has been used in to introduce optimal solutions in different domains, most of the previous work that using evolutionary have done in web caching, little research has done in the mobile database. Tirdad et al. [20] introduces how to use genetic programming in web caching. They evaluate the fitness function by the number of hits and the numbers of requests for a web page. The raw fitness function is defined as the number of runs subtracted by the number of misses. The authors did the experiment for four different cases: small caches, large caches, no constant, using the function information. The results show good performance for specific cases, but the average of performance is not optimal. As stated in [21], other authors continued Tirdad research, and uses Backus Naur Form (BNF) as a production rules to express the grammar of a language. BNF contains terminals, non-terminals, set of production rules, and start rules. They defined the same terminals and functions as K. Tirdad, but he does not define cache size terminal. None of the above genetic programming techniques have been used in cache replacement policy in the mobile database, and they ignored the cache invalidation cost in their approaches.

The innovation here is to use the time validation along with the most significant cached data item parameters available in the literature using genetic programming technique for eviction of data items from the cache when the is full. This is in contrast to the previous work, which provided the algorithm to cache replacement without taking into account the time validation and also takes into account one or two parameters when determining the element out of cache. In this regards, the proposed policy is going to extend a common used genetic programming technique to fuse cached item features like access probability, cached document size, validation time and uses these factors in a fitness function to determine the cached items that will be removed from the cache. The work shows that suitable use of cached item features can play an important role in improving this cache replacement policy, leading to higher levels of a cache hit ratio and a substantial reduction in average latency. It can be concluded from the up-to-date studies that using cached data item parameters separately cannot determine the cache replacement accurately. Cache replacement policy based on fusing multiple cached item parameters is a new trend of current development. One of the major advantages of this method is that it eliminates the need for using a lot of theoretical and complex assumptions to calculate the cached data item parameters; it uses only simple and easy calculated parameters that work efficiently together without adding burdens (calculation time) to the suggested policy. Its major limitation is that it performs a brute-force search for GP configuration variables.

3 The Proposed Model

A. Mobile System Model

In a mobile system model, the area is divided into cells, the mobile client can move fairly from one cell to another, and each cell has a base station and a number of mobile clients; each mobile client connects to the base station through wireless channel; the base station connects with the database server through wired link. To facilitate the presented model, the proposed system uses one cell only for mobile clients, and thus is considered one of the limitations of the proposed system. The mobile computing platform can be effectively described under the client/server paradigm. A data item is the basic unit for update and query. Mobile client only issues requests to read the most recent copy of a data item (see Figs. 1 and 2).

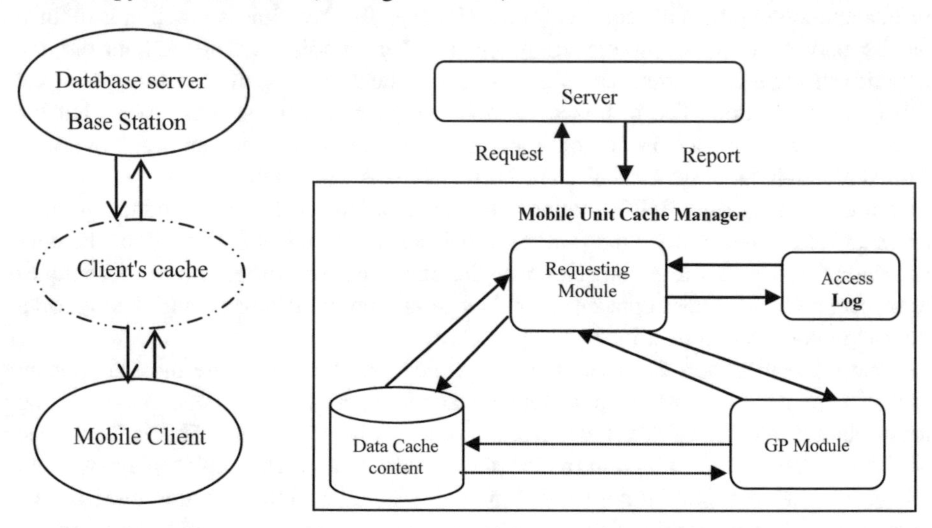

Fig. 1. The interaction between 3 nodes

Fig. 2. An overview of the proposed Cache replacement model

B. The Cache Invalidation Model

For every data item with unique identifier -, the data structure in both of database server and client's cache are as follows:

$\{d_i; v_i; t_i; s_i; f_i\}$: Characteristics for each data item in the database server. Here d_i is the data item, v_i is the next invalidation time for data item, t_i is the last update time for the data item,s_i is the size of data item at the server and f_i is the flag bit. On the other hand, the characteristics in client's cache are $\{d_i; ts_i; v_i; s_i; p_i; r_i; g_i\}$. d_i is defined above, and ts_i is the time stamp indicating thelast updated time for the cached data item (d_i), v_i the next invalidation time that will be sent by the server. While s_i is the size of cached data item, p_i is the access probability of cached data item,r_i is the retrieve delay time for the cached data item (gained from the previous request for the data item), as in SACCS [26], g_i is a two-bit flag identifying four data entry states: 0, 1, 2, and 3, indicating valid d_i, uncertain d_i, uncertain d_i with a waiting query, and ID-only, respectively. Next table shows the g_i for each data item (Table 1).

For each cached data item, Cache validation policy uses a single flag bit g_i to maintain the consistency between the database server and client's caches. It can be inferred that the flag bit in the server database reduces the size of the broadcasted IRs.

Table 1. Data items state [26]

g_i	Data item state
0	Valid
1	Uncertain
2	Uncertain with a waiting query
3	ID-only (deleted)

The proposed caching policy has communication messages used by the mobile units and communication messages sent by the base station to the clients.

- **Messages from Mobile Unit to Base Station**
  - Update (i, d_i, t_i): indicating that d_i has been updated to d_i at time t_i
  - Request (i): request for data item(i).
  - Uncertain (i, ts_i): requesting whether the data item(i)in uncertain state with the timestamp ts_i is valid.
  - Wakeup (ts): requesting for data items which have been changed since that sleep time.
- **Messages from Base Station to Mobile Unit**
  - Vdata (i, d_i, v_i, t_i): Broadcast valid data item d_i with update time at t_i and v_i is the next invalidation time.
  - IR(is): Invalidation Report containing of data items in the cache which are not valid anymore.
  - Confirmation (i, v_i, t_i): Ensure that the data items d_i is valid if $ts_i = t_i$ and v_i is the next invalidation time.
  - Wakeup Invalid (t_i, i): data items with time stamp t_i are invalid.

C. **Proposed Cache Replacement Model**

Cache content can be modeled by an information hash table consisting of a number of rows; each row associated with a specified cached item. The number of rows expresses the number of data items. Each data item is identified by its corresponding stored data item file, along with a number of related attributes. The attributes are chosen such that cache replacement could be supported and employed. the attributes in client's cache are $\{d_i.ts_i.v_i.s_i.p_i.r_i.g_i\}$. GP module deals with data cache content through requesting module, it means that requesting module sends an order to the GP module to test the cache content and determine the cached items that will be removed from the cache. One of the major differences is that the previous system uses less cooperation between invalidation policy and replacement policy. And also, our GP-CRMP differs from SACCS [26], in that SACCS depends on LRU replacement policy when improving cache invalidation policy but in our GP-CRMP we enhance both of cache invalidation policy and present a new cache replacement policy called GP-CRMP with more cooperation between them. GP-CRPMD (GP cache replacement policy for mobile database), which uses GP for selecting the cached item(s) to be removed from the cache.

- **Client-side cache management algorithm:**

```
if (going to sleep) then store the current timestamp locally;
if (waking up) then send Wakeup (ts) to the BS;
if (v_i expires for data itemi) then set the valid data item I into uncertain state (g_i =1);
if Client requests for data item i
If data item i is valid and in cache then return the data item i from the cache
  else if (d_i is in uncertain state) then send Uncertain (i, ts_i ) message to the BS; set g_i = 2; and add the ID, to query waiting
        list
  else if (i is ID-only entry in the cache) then send Query(i) message to the BS; remove the data item i in the cache; add i to
        query waiting list
  else
  send Query(i) message to the BS; add i to query waiting list;
if (it is Vdata (i, d_i , v_i , t_i ) message)
if (i is in query waiting list) then answer the request with di;remove the uncertain itemiif it exists in the cache; then add the
   valid entry i at the cache list head;
  else
        if (i is ID-only in the cache (cache miss)) then
     if enough free space in cache then store data item i in cache; Update data cache content
  Else if not enough free space in cache then
                use GP module for determining data item(s) that will be removed from the cache which has low
                fitness function
  else if (i is an uncertain entry in the cache)
if (ts_i < ts) then download di to the original entry in the cache;
        if enough free space in cache then store data item i in cache; Update data cache content
           else if not enough free space in cache then use GP module for determining data item(s) that will be
              removed from the cache which has low fitness function.
     set ts_i = t_i ; lli =v_i ; g_i= 0;
  else
  set g_i= 0;
if (it is an IR(i) message)
   if (data item i is valid or uncertain in the cache) then delete di and set g_i = 3;
if (it is Confirmation (i, v_i , t_i ) message)
if (i is an uncertain entry in the cache)
if (ts_i == t_i )
set g_i= 0 and lli = v_i ;
else
delete di and set g_i = 3;
end
```

- **Server-side cache management algorithm:**

```
For base station receiving a message
 If (it is Query(i) message)
   Fetch data item i from the database;
   Broadcast Vdata (i, d_i , v_i , t_i )to all Mobile Units;
If (fi = 0) then set fi = 1;
If (it is Uncertain (i, t_i ) message)
 Fetch data item i from the database;
If (t_i ). = ts_i )
  Broadcast Confirmation (i, v_i , t_i ).) To all Mobile Units;
  Else
 Broadcast Vdata (i,d_i ,v_i , t_i )to all Mobile Units;
If (fi == 0) then set fi = 1;
If (data item i has been changed)
  Broadcast IR(i) to all Mobile Units;
If (it is Wakeup (ts))
  Fetch all IDs of data items that has been changed after this time stamp ts;
  Broadcast Wakeup Invalid (ts,IDs);
end
```

D. **Genetic Programming Method**

Genetic programming is an evolutionary approach where instead of a solution; the population consists of programs [14]. The programs are usually a simple representation of a tree, graph or set of instructions. The genetic programming could be described as follows [12]: (1) randomly generate an initial population of random compositions of the functions and terminals (the set of data item in cache) as shown in Fig. 3. (2) Compute and save the fitness for each individual (the set of data items that will be removed from the cache) in the current population. Finally (3) Perform suitable reproduction operators and other parameters for running our GP algorithm as shown in Table 3 for generating new population. The system employs a set of variables representing the local information available about each data item (see Table 2). Each client collects this observable information that can be gathered from the data items themselves. The mobile cache contains a number of data items (*N*) which total size is *S*. A client mobile unit determines whether data item could be cache or not. In case that there is not enough space there is a need to remove one or more data items from the cache in order to free sufficient space. So a data item or more need to remove from the cache.

Table 2. Cached data items attributes

Variable	Meaning
S	The cache size at the client
N	The total number of data items in the cache of the client
S_i	The size for every cached item *i*
p_i	Access probability of cached item *i*
v	The next validation time for cached item
R	Retrieve delay time for the cached item

Herein, ITEMs = {$item_1$, $item_2$, $item_3$... $item_N$} is a set of cached items in the cache where *N* is the count of data items. To make the proposed technique more realistic, the system uses different item data size $S_s = \{S_1, S_2, S_3 \ldots S_N\}$. The total size of cached items in cache is calculated as:

$$\sum_{i=1}^{N} S_i \leq S \tag{1}$$

where *S* represents the cache size at the client. Dri = {Dr_1, Dr_2, Dr_3, ...,Dr_k} is the set of the cached items should be removed, $Sr_i = \{Sr_1, Sr_2, Sr_3, \ldots, Sr_k\}$ is the set of data items' size should be removed where *k* is the total number of cached items should be removed.

$$\sum_{i=1}^{k} Sr_i \geq S_{new} \tag{2}$$

where S_{new} is the size of data item will enter the cache and this formula means that the total size of data items that will be removed from the cache should be greater than or equal to the data item will enter the cache when the cache is full.

E. Fitness function

The fitness function incorporates the most effective variables that take into consideration the impact of invalidation policy for enhancing data availability in mobile environment.

$$F = (p_i + v + R + 1/S_i) \tag{3}$$

Where access probability of data item (*pi*) [21, 24, 25] depends on the earlier access rate of a data item at the client node. Data items will be removed from the cache if the measure of its access probability is lower. The access probability a_i for data item D_i at the client node can be defined as:

$$p_i = a_i / \sum\nolimits_{K=1}^{N} a_i \text{ For } i = 1, 2, \ldots, N \tag{4}$$

a_i signifies the mean access arrival rate of data item i, $i = 1, 2, ..., N$. The access arrival rate a_i is calculated at the client-side. A sliding average technique is employed in the implementation. Therefore a_i can be determined by using sliding window method of last K. A sliding window of K most recent access time stamps $ts_i^1; ts_i^2; ts_i^3; \ldots; ts_i^k$ for data item D is kept in the cache. The following formula is employed to update the access rate:

$$a_{i=K/}(t_{c-}ts_i^k) \tag{5}$$

t_c is the current time and ts_i^k is the timestamp of oldest access to item D_i in the sliding window. In general, K should be equal to three or two to enhance the performance. For simplifying the proposed technique, the access rate of the cached data items will not be updated during the implementation. In this case, regarding the size of the data item (s_i. For the next validation time for the cached item (***v***), the system supposes that the server sends to the client the next invalidation time for every data items. The fitness algorithm removes the data items in the cache with the smallest value. The advantages of this algorithm are that it can maintain cache validation delay. The fitness function of the suggested system is the sum of the cost of data items that tend to be removed from the cache:

$$fitness(\mathrm{i}) = \sum\nolimits_{\mathrm{i=N-k+1}}^{\mathrm{N}} F \tag{6}$$

Every data item value will be act as an f value, the data items that have smaller f value and its size is greater than or equal to the item that will be invoked to the cache will be removed from the cache. Since fitness function is crucial in determining the items that will be removed from the cache, so as mentioned in Table 3 every data item will act as F value, and the system uses only {+} function because the goal is to obtain the sum F for data items that will be removed from the cache. As an example from Fig. 3, in the first generation in (a) the genetic programming algorithm determines data item1 (D_1) to be removed from the cache.

Table 3. Genetic programming terminals and functions

1	Objective	Determining data items that will be removed from the cache which have low (*F*)
2	Terminal set	*F* value for every data item in the cache *Si* the size value for every data item in the cache
3	Function set	{+}
4	Fitness	Low *F* and $\sum_{i=1}^{k} Sr_i \geq S_{new}$
5	Parameters	Population size = 20, max tree depth = 3, generation = 5
6	Termination	Find the low sum of *F*

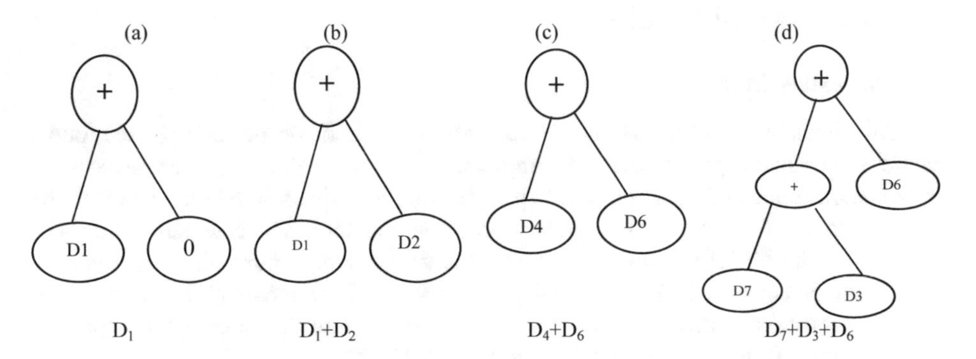

Fig. 3. Initial population of four randomly created individuals of generation 0

To make the proposed system able to deal with the fact that GPs is time-consuming and they usually converge slowly; the system is built on the following assumptions: (1) Use simple and easy parameters like data item size, next validation time, and access probability to predict which of the data item will leave the cache; unlike some other methods as in [1] that depend on complex parameters, which may require some calculations to extract them. (2) Use only one function (+) and the terminals are the data items in the cache; so the solution for GP is easy and simple. (3) Use only 5 generations to conclude the data items that will be removed by the GP module; so the time consuming is very small. (4) Specify the maximum depth for every individual to three to decline the calculations. Finally (5) Run the GP module only once to decide the data items that will be removed from the cache with sizes must be less than or equal to the data item that will enter the cache.

With regard to the complexity of the proposed system, it depends entirely on the degree of complexity of the simple GP. As stated in [14], the GP complexity could be measured in two different ways. The first is to equate complexity with tree size, based on the assumption that bloated programs are also complex. The second addresses the complexity of the program output or functional complexity. According to the first measurement, tree depth = 3 is configured to prevent the GP process from bloating, according to the second measurement, the fitness function F for each data item is calculated before running the genetic programming algorithm to determine the data items that will be removed from the cache; where the value of each data items will be represented by a value of F (terminals); and also using only one function (+) for simplifying the solution to reduce the complexity of the solution. It is commonly stated that simpler solutions to a problem should be preferred because they are less likely to be over fitted to the training examples. The GP community has tacitly assumed that simple solutions can be equated with small program trees, and that very large programs correspond to complex solutions. Therefore, bloat was assumed to be a good indicator of the program over fitting.

4 Experimental Results

A. Simulation Setup

The simulation is used to evaluate our proposed genetic programming technique in cache replacement policy. For our implementation we are using a log access file contains the requested data items and also the data items that will be removed from the cache obtained from MATLAB genetic programming tool. The NS2 simulation software [27] with CMU wireless extension is used to examine our new technique. The simulation parameters are summarized in Table 4. The movements of the mobile client node are obtained from a movement file. The next three criteria are used to compare the proposed technique with famous LRU algorithm [1, 22]:

Table 4. Simulation parameters

Channel type	Channel/wireless channel
MAC type	Mac/802_11
Radio-propagation model	Propagation/Two Ray Ground
Network interface type	Phy/Wireless Phy
Interface queue type	Queue/Drop Tail/PriQueue
Antenna model	Antenna/Omni Antenna
Link layer type	LL
Routing protocol	DSDV
Coordinate of topology	670 m × 670 m
Simulation time	300 s

- **Cache hit rate:** the total percentage of all cache hits, the aim is to maximize its value.
- **Byte hit rate:** the total percentage of all bytes that transfer to the user (Fig. 4).

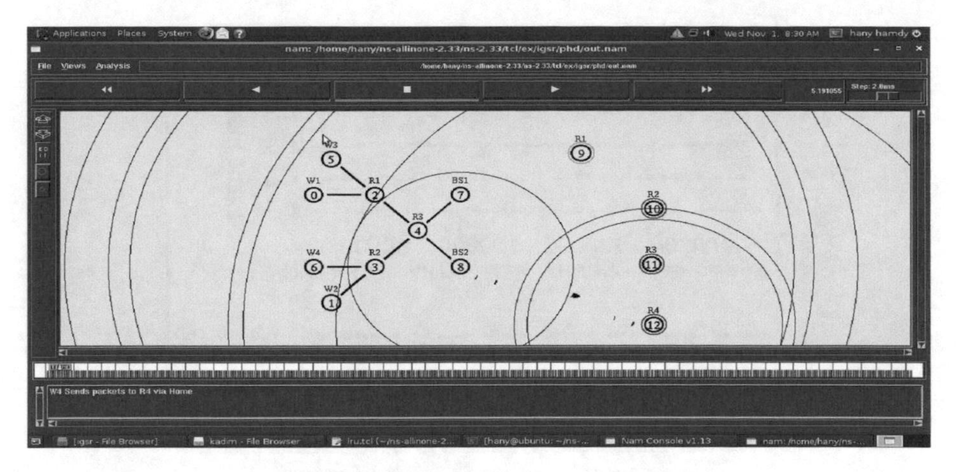

Fig. 4. Snapshot for simulation in NS2

B. Simulation results

The byte hit ratio will be used to evaluate the performance of the proposed policy as this ratio is the total percentage of all bytes that transfer to the client from the cache. Therefore, the higher the ratio, the higher the efficiency of the proposed policy. In addition, cache hit ratio will be used as this ratio is the total percentage of all cache hits; the aim is to maximize its value. The higher the cache hit ratio, the more powerful the proposed policy will gain data item from client's cache. To check the performance of the proposed GP- based cache replacement policy, the results of the current algorithm were compared with the traditional LRU replacement policy and also with state-of-the-art rule-based least profit value (R-LPV) cache replacement strategy [28] in which the relationship between data items are considered by using association rule-based data mining. This algorithm chooses the cached data item with least profit and evicts it during each replacement. Furthermore, it takes into account only average update and size of the data item in its profit function and no interest is given to the cooperative between invalidation policy and replacement policy. To facilitate the comparison between the two algorithms, a one-cell with 100 Mobile units with cache size 300 KB is used for thousands of data at different rates of update (in seconds), different data sizes and different access probability. The simulation was applied 4 times of units with an interval of 50000 ms of simulation time.

As is clear from Figs. 5 and 6 respectively, the on-hand algorithm achieves better results in the percentage of the cache hit ratio, as well as the percentage of byte, hit ratio. One explanation for these results is based on the reliability of the proposed algorithm on different data item characteristics like access probability, retrieval delay and next validation time. These characteristics have a great impact to determine the cache replacement policy, which clearly shows improvement in the cache hit rate and byte hit ratio performance.

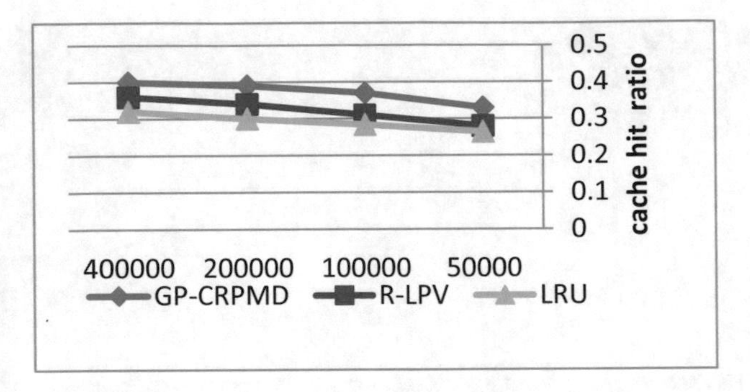

Fig. 5. The cache hit under different time intervals

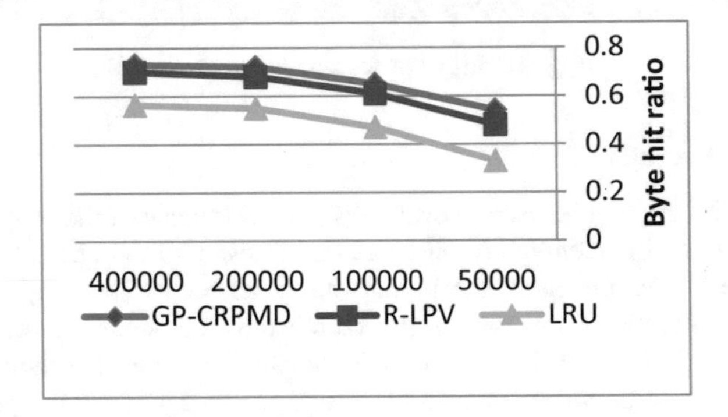

Fig. 6. Cache byte ratio under different time intervals.

5 Conclusions

This paper proposes the genetic programming cache replacement technique for a mobile database (GP-CRPMD) by taking into consideration more than one parameters such as access probability, validation time, retrieve delay and the size of the data item. A genetic programming algorithm is designed to dynamically select the data item(s) to be removed from the cache. To enhance the caching performance, the GP algorithm for cache replacement which has been used in this paper determines the data item(s) as a group of items to be removed from the cache by running the algorithm once, it means that the total size of data item(s) that will be removed from the cache allow another data item to enter the cache. Experiments show that GP-CRPMD increases byte hit ratio, cache hit ratio and decrease average query latency in comparing to the LRU and R-LPV. In the future work, the plan to enhance the fitness function which have used in the suggested technique by taking into consideration more parameters such as the relating between cached items, and also extend genetic programming in cache admission policy to determine which data item should be in the cache, and also using genetic programming in cache invalidation policy for predicting the next invalidation time for the data item.

References

1. Kaur, G., Saini, J.: Mobility aware cache management in wireless environment. In: International Conference on Methods and Models in Science and Technology, India, pp. 291–297 (2010)
2. Tassum, K., Syed, M., Damodaram, A.: Enhanced-location-dependent caching and replacement strategies in mobile environment. Int. J. Comput. Sci. Issues **8**(2), 160–167 (2011)
3. Lee, D., Chin, S., Min, J.: Efficient resource management and task migration in mobile grid environments. J. Commun. Comput. Inf. Sci. **78**(1), 384–393 (2010)
4. Shanmugarathinam, G., ViveKanandan, K.: Thread based cache consistency model for mobile environment. Int. J. Comput. Sci. Commun. **2**(2), 315–318 (2011)
5. Anandharaj, G., Anitha, R.: A Distributed cache management architecture for mobile computing environments. In: International Advance Computing Conference, India, pp. 642–648 (2009)
6. Joy, P.T., Jacob, K.P.: A comparative study of cache replacement policies in wireless mobile networks. In: Proceedings of the Second International Conference on Advances in Computing and Information Technology (ACITY) vol. 176, pp. 609–619 (2012)
7. Kottursamy, K., Raja, G., Saranya, K.: A data activity-based server-side cache replacement for mobile devices. In: Proceedings of the International Conference on Artificial Intelligence and Evolutionary Computations in Engineering Systems, pp. 579–589 (2016)
8. Tiwari, R., Kumar, N.: An adaptive cache invalidation technique for wireless environments. Telecommun. Syst. **62**(1), 149–165 (2016)
9. Chavan, H., Sane, S.: Mobile database cache replacement policies: LRU and PPRRP. In: International Conference on Computer Science and Information Technology, India, pp. 523–531 (2011)
10. Li, C., Wang, D., Zhang, X.: Enhanced adaptive insertion policy for shared caches. In: International Conference on Advanced Parallel Processing Technologies, Berlin, pp. 16–30 (2011)
11. Lee, J.H., Kwon, S.J., Chung, T.S.: ERF: efficient cache eviction strategy for e-commerce applications. In: International Conference on Mobile and Wireless Technology, pp. 295–304 (2018)
12. Chakravorty, C., Usha, J.: Cache management issues. In: Mobile Computing Environment, International Journal of Mobile Network Communications & Telematics (IJMNCT), vol. 2, No.1, February 2012
13. Gu, J., Wang, W., Huang, A., Shan, H., Zhang, Z.: Distributed cache replacement for caching-enable base stations in cellular networks. In: IEEE International Conference on Communications (ICC), pp. 2648–2653 (2014)
14. Le, N., Xuan, H., Brabazon, A., Thi, T.: Complexity measures in genetic programming learning: a brief review. In: IEEE Congress on Evolutionary Computation, pp. 2409–2416 (2016)
15. Searson, D., Leahy, D. Willis, M.: GPTIPS: an open source genetic programming toolbox for multigene symbolic regression. In: International Multi-conference of engineers and Computer Scientists pp. 77–80 (2010)
16. Shin, H., Kumar, M., Das, K.: Energy-efficient data caching and prefetching for mobile devices based on utility. J. Mob. Netw. Appl. **10**(4), 475–486 (2005)
17. Bžoch, P., Matějka, L., Pešička, L., Šafařik, J.: Towards caching algorithm applicable to mobile clients. In: Proceedings of the Federated Conference on Computer Science and Information Systems, Poland, pp. 607–614 (2012)

18. Sokolinsky, L.: LFU-K: an effective buffer management replacement algorithmIn: Lecture Notes in Computer Science (LNCS), vol. 2973, pp. 670–681 (2004)
19. Chavan, H., Sane, S., Kekre, H.: A Markov model based cache replacement policy for mobile environment. J. Commun. Comput. Inf. Sci. **145**(1), 18–26 (2011)
20. Tirdad, K., Pakzad, F., Abhari, A.: Cache replacement solutions by evolutionary computing technique. In: Spring Simulation Multi-conference, pp. 1–5 (2009)
21. Xu, J., Hu, Q., Lee, W., Lee, D.: Performance evaluation of an optimal cache replacement policy for wireless data dissemination. IEEE Trans. Knowl. Data Eng. **161**(1), 125–139 (2004)
22. Zhou, X., Zou, Z., Song, R., Yu, Z.: Cooperative caching strategies for mobile peer-to-peer networks: a survey. In: Lecture Notes in Electrical Engineering, vol. 376, pp. 279–287 (2016)
23. Lai, K., Tari, Z., Bertok, P.: Location-aware cache replacement for mobile environments. In: IEEE Global Telecommunication Conference, vol. 6, pp. 3441–3447 (2004)
24. Lai, K.Y., Tari, Z., Bertok, P.: Mobility-aware cache replacement for users of location-dependent services. In: IEEE International Conference on Local Computer Networks, pp. 50–58 (2004)
25. Jain, D.K., Sharma, S.: Cooperative caching strategy in mobile ad hoc networks for cache the replaced data item. Int. J. Wirel. Personal Commun. **84**(4), 2613–2634 (2015)
26. Wang, Z., Das, S.K., Che, H., Kumar, M.: A scalable asynchronous cache consistency scheme (SACCS) for mobile environments. IEEE Trans. Parallel Distrib. Syst. **15**(11), 983–995 (2004)
27. Qureshi, N.M.: A trace study and performance analysis of wireless network using NS2. Int. J. Adv. Res. Comput. Sci. Softw. Eng. **5**(3), 11–16 (2015)
28. Haraty, R.A.: Innovative mobile e-healthcare systems: a new rule-based cache replacement strategy using least profit values. Mob. Inf. Syst. **2016**, 1–9, Article ID 6141828 (2016)
29. Saad, M.D., Amr G.S.: An intelligent database proactive cache replacement policy for mobile communication system based on genetic programming. Int. J. Commun. Syst. e3536 (2018). https://doi.org/10.1002/dac.3536

Adopting Non-linear Programming to Select Optimum Privacy Parameters for Multi-parameters Perturbation Algorithm for Data Privacy Improvement in Recommender Systems

Reham Kamal[1(✉)], Wedad Hussein[2], and Rasha Ismail[2]

[1] MTI University, Cairo, Egypt
reham.kamal@cs.mti.edu.eg
[2] Ain Shams University, Cairo, Egypt
wedad.hussein@fcis.asu.edu.eg,
rashaismail@cis.asu.edu.eg

Abstract. Recommendation system has witnessed a significant improvement with the introduction of data mining. Data mining services require accurate input data for their results to be meaningful, but privacy concerns may influence users to provide spurious information. In order to preserve the privacy of the client in data mining process, the issue of information protection has become more urgently demanded. In this paper, an innovative system for movies recommendation is proposed. The new proposed system is fundamentally based on modified version of multi-parameters perturbation and query restriction as well as adopting non-linear programming strategy to select optimum privacy parameters. The results showed that the proposed framework is capable of providing the maximum security for the information available without decreasing the accuracy of recommendation.

Keywords: Privacy preserving · Data perturbation · Recommender system
Data mining · Data sanitization

1 Introduction

In the recent decades, the issue of sharing information, whether public or private, has become not only increasingly demanded but also a weapon of two edges. In one sense, governmental, public, and private institutions are sharing as well as receiving a massive amount of information. In another sense, those information has to be kept protected to the maximum to ensure the data privacy of respondents. The challenge lies in the fact that although the genre of the data being displayed might apparently seem anonymous, the other irrelevant data that necessarily appears with it contains other sensitive private data (race, date of birth, sex, zip code) which sets off or distinguishes specific respondents. These public data when linked with publicly available data bases, the data recipients can sort out to whom the released data belongs or at least distinguish to which group of individuals does this data belongs [1].

A. E. Hassanien et al. (Eds.): AISI 2018, AISC 845, pp. 603–613, 2019.
https://doi.org/10.1007/978-3-319-99010-1_55

Recommender systems collect behavioral data, context, social connections, or demographics to provide users with tailored recommendations. These data collection practices may pose a threat to user privacy [2]. The fundamental effectiveness of recommendation system lies in its manipulation of private data available in the users' profile. Automatically, the system classifies private data and creates a profile for the users that contains the users' preferences. Those basic inferred users' preferences are used by the system as the cornerstone for creating future prediction about the users in other perspectives. The catastrophe lies in the fact that the original data is being abused by the recommendation system for the particular interest of the system. Therefore, an accurate system that controls or restricts the amount of private data available for the recommendation system is urgently required [3].

The main target of privacy preserving data mining (PPDM) is to design proficient frameworks and algorithms that can discover relevant knowledge from large amounts of data without revealing any private information. It protects private information by providing sanitized database of original database to be mined, or a process is used in such a way that private data and private knowledge remain private even after the mining process. Thanks to PPDM the benefits of data mining be enjoyed, without compromising the privacy of concerned individuals [4].

The rest of the paper is organized as follows, related work is presented in Sect. 2. Our proposed framework is explained in Sect. 3 as well as analyzing the privacy parameters which is the main contribution of this paper. In Sect. 4, experimental data and results are shown. Conclusion and future work are discussed in Sect. 5.

2 Related Work

Several methods needs to be applied to protect the privacy of individual's data before data mining tasks are carried out. Privacy preserving data mining is the branch which includes the studies of privacy concern when mining is applied. PPDM techniques can be classified as Central/Commodity Server where the server must privatize the data prior to mining, and Distributed where the contributing parties must sanitize and privatize the data prior to publishing, here we give a brief description for PPDM techniques:

In **anonymization technique**, generalization and suppression techniques are applied to remove personally identifiable information from data sets, so that the people whom the data describe remain anonymous. To identify records uniquely it considers quasi-identifiers which can be used in conjunction with public records. Anonymization technique results in loss of information to some extent but ensures the originality of data [5].

In **data perturbation strategy**, raw data are firstly perturbed by data transformation and adding noise in data, thus sensitive information is kept private; then knowledge and rules are mined from the perturbed data. The authors in [6] have tried to add Gaussian noise to generate perturbed version of dataset for decision tree classification. In same lines, authors in [7] have proposed an individually adaptable perturbation model. A multilevel privacy can be specified by the users. This opens a new venture in field of privacy preserving – Multi-level Trust PPDM (MLT PPDM). Perturbation methods can be classified into probability distribution category and fixed data perturbation. Probability distribution allows adding noise based on some known distribution pattern like Gaussian. The data distortion techniques like addition of noise, from some known distribution, randomization and condensation are applied [8].

Randomization is a data perturbation technique where the data distortion is masked by random data. Authors in [9] have proposed a mechanism to scramble data in a manner that the central repository won't be able to judge whether the information can be classified as true or false. With large number of users, aggregate information can be estimated with accuracy. This information can be used for decision-tree classification as the latter is based on aggregate values of a dataset. Though this method causes high information loss it is a more efficient method.

Cryptography techniques are used when parties are distributed across multiple sites are legally prohibited from sharing their datasets. Authors in [6] have addressed the problem of reconstructing missing values by building a data model where the parties are distributed and data is horizontally partitioned. A cryptographic protocol based on decision-tree classification is described by them. A survey on cryptographic techniques for PPDM is studied by authors of [10]. Distributed environment where the sharing is constrained either under legal or privacy policy issues use the cryptographic techniques. Oblivious transfer is used as building block for constructing an efficient PPDM model by authors in [11]. The implementations of these protocols consist of computationally intensive operations and generally consist of hard wired circuits. The cryptography method can ensure that the transformed data is exact and secure, but it is much low efficient.

Secure-multiparty computation considered to be a subfield of cryptographic techniques where different parties jointly compute a function over their inputs while keeping these input private. These methodologies were introduced to allow different parties or data holders to mine their data together without the need to reveal their data sets to each other. However these techniques are difficult to use with huge datasets [8].

The implementation of various techniques related to **Fuzzy and Neural Networks** is still underdevelopment [9]. Authors in [12, 13] have discussed the advantage of **Rough Sets** in direction of PPDM.

Recently, **genetic algorithms** can be used to help in the problem of PPDM as they can be used in solving NP-Hard problems [5].

3 The Proposed Methodology

In our previous work, [14], we presented a methodology for data privacy preserving. The proposed architecture was divided into two main phases, first, the original data is sanitized using data perturbation and query restriction (**DPQR**) algorithm [15]. Second, comes the generating recommendation phase where modified version of **MASK** (Mining Association rules with secrecy constraints) algorithm is used to mine data privately then measuring the cosine similarity between items and choosing the top-N similar items to the user.

3.1 Data Sanitization Phase

This phase was responsible for perturbing data to ensure data privacy. It consists of two modules that are explained in the next sub sections:

Data Normalization Phase: All the transactions was converted to the format (user: item1, item2, item3, …etc.). Then the items were replaced by 1 or 0, where 1 shows the existence of item in transaction and 0 shows the non-existence.

Data Perturbation Phase: In this module we adopted data perturbation and query restriction (DPQR) algorithm to achieve data privacy [15]. We used three random parameters p1, p2 and p3 such that 0 < p1, p2, p3 < 1. Also p1 + p2 + p3 = 1. P1 indicates that information is kept the same, p2 indicates that the information is reversed, and p3 indicated that the information is hidden. The pseudocode for the algorithm is shown in Fig. 1.

```
➤ Input: the original transaction set D, random parameters p1,p2,p3
➤ Output: Sanitized data set D.
  for each transaction t ∈ D
  {
   for (k=0; k<N; k++)
      {
         Generate random number θ1;
           if (θ1≤p1)
              i=i;
           Else if (p1≤θ1≤p1+p2)
              i=1-i;
         Else if (p1+p2< θ1<p1+p2+p3)
              i=0;
      }
  }
```

Fig. 1. Data perturbation algorithm

Selection for the Optimum Privacy Parameters: Perturbation Parameters are the key of recommendation system, if the parameters are too much small, the protection level of data privacy will be too low to reach the purpose of privacy protection, while, too larger parameters will cause reducing the accuracy result of recommendation to guarantee the protection level of Privacy, so it is very important to select a perturbation parameter just right.

To the best of our knowledge choosing and selecting perturbation parameter was empirical. In this paper we present a new methodology for choosing the optimum parameters which is the main contribution of this paper.

As the purpose of PPDM is to obtain more accurate mining results in case of imprecise access to the original data, so the reconstruction probability RP of the original data can be used to evaluate the algorithm. If it has large reconstruction probability, this algorithm does not achieve better Privacy protection. Therefore the RP is given by the following equation:

$$RP = \alpha * RP1 + (1 - \alpha) * RP0 \tag{1}$$

$$RP1 = \frac{S * p_1^2}{S * p_1 + (1 - S)P_2} + \frac{(1 - p_1)^2 * S}{(1 - P_2) + (P_2 - p_1) * S} \tag{2}$$

$$RP0 = \frac{(1 - S) * p_2^2}{S * p_1 + (1 - S)P_2} + \frac{(1 - p_1)^2 * (1 - s)}{(1 - P_2) + (P_2 - p_1) * S} \tag{3}$$

Where α is the weight given to 1's over 0's. S is the average support of the item in data base item in the database.

To choose the optimum privacy parameters for perturbation we solved our equations as a **non-linear goal programming problem** as follows:

$$\text{Maximize objective function}: \quad RP = \alpha * RP1 + (1 - \alpha) * RP0$$
$$\text{Subject to these constraints}:$$

$$P1 + p2 + p3 = 1,$$

$$P1 < = 1,$$

$$P2 < = 1,$$

$$P3 < = 1,$$

$$P1, p2, p3 > = 0$$

When solving these equations together, we obtained the optimum values for p1, p2 and p3.

3.2 Generating Recommendation Phase

Generating recommendation is the main target of our work. This phase consisted of two modules: re-estimating the support of itemsets then measuring the cosine similarity between items to generate recommendations.

Re-estimating the Support of Itemsets: Since data set was normalized to Boolean format. The real dataset can be shown as a Boolean matrix T, and D is the disturbed dataset matrix by using the three parameters p1, p2 and p3. The number of '1' and '0' included in the ith column of T is defined as C_1^T, C_0^T respectively. While C_1^D and C_0^D have the similar definition. Let M be the transformation matrix,

$$C^D = M\,C^T \tag{4}$$

Where $C^D = \begin{pmatrix} C_1^D \\ C_0^D \end{pmatrix}$, $C^T = \begin{pmatrix} C_1^T \\ C_0^T \end{pmatrix}$ and $M = \begin{bmatrix} p1 & p2 \\ p2 + p3 & p1 + p3 \end{bmatrix}$.

Given the following equation:

$$C^T = M^{-1}C^D \tag{5}$$

The support of 1-item set C_1^T can be obtained by the following formula:

$$C_1^T = \frac{C_1^D - p2\left(C_1^D + C_0^D\right)}{p1 - p2} \tag{6}$$

As all items are disturbed in the same way, we can extend the formula to estimate the n-itemset support, the only difference is that M is $2^n * 2^n$ Matrix, C^T and C^D are both the $2^n * 1$ matrixes as follows:

$$C^T = \begin{pmatrix} c_{2^n-1}^T \\ . \\ . \\ . \\ C_1^T \\ C_0^T \end{pmatrix}, \quad C^D = \begin{pmatrix} c_{2^n-1}^D \\ . \\ . \\ . \\ C_1^D \\ C_0^D \end{pmatrix} \tag{7}$$

M_{2^n} = [mij] is a matrix of order 2^n, When matrix is invertible, $M_{2^n}^{-1}$ = [bij]. From Eq. (5), $c_{2^n-1}^T$ the support of n-item set can be calculated as follows:

$$c_{2^n-1}^T = b_{0,2^n-1}C_0^D + b_{0,2^n-2}C_1^D + \ldots + b_{0,0}c_{2^n-1}^D \tag{8}$$

The calculation of matrix invert is optimized using recursive relations and the counting overhead of item sets is decreased using set theory as shown in [15].

The complete algorithm of mining sanitized data using improved MASK is given in Fig. 2:

Measuring Cosine Similarity: Cosine similarity is measured between items to determine which items are most similar to each other's according to the following equation:

$$Cos_{sim(A,B)} = \frac{supp\_count(A,B)}{\sqrt{Supp\_count(A)} * \sqrt{supp\_count(B)}} \tag{9}$$

Choosing Top-N Items According to User History: Similar-items are arranged by descending order according to similarity (sim) count and support (sup) count. The result will be in the following format: Item: Item1, sim, sup; Item2, sim, sup; Item3, sim, sup…etc. Then for each user choose top-N items that are similar to his 1-items sets according to his history.

➤ Input: the sanitized database D, the parameters p1, p1, p3 and min sup.
➤ Output: the frequent itemsets after reconstruction

$C_1 = find\_candidate\_itemset(D)$;
Reconstruct (C_1, *min*_ sup); // estimate the real support of the 1-itemset
$L_1 = \{c \in C_1 \mid c.sup \geq min\_sup\}$;
For (k=2; $k_{k-1} \neq 0$; k++)
{
C_k=Apriori_gen (L_{k-1}) // generate candidate itemset
Reconstruct (C_k, min_ sup); // reconstruct real support of the itemset
If (c.sup ≥ min_ sup)
$C \in L_k$
Return $L = \cup_k L_k$
}
Reconstruct (C_k, min_ sup)
{
If (k=1)

$$M_{2^1}^{-1} = \frac{1}{p1-p2}\begin{pmatrix} 1-p2 & -p2 \\ p1-1 & p1 \end{pmatrix}$$

Else

$$M_{2^k}^{-1} = \frac{1}{p1-p2}\begin{pmatrix} 1-p2\ M_{2^{k-1}}^{-1} & -p2\ M_{2^{k-1}}^{-1} \\ p1-1\ M_{2^{k-1}}^{-1} & p1\ M_{2^{k-1}}^{-1} \end{pmatrix}$$

Foreach c_i in C_k

$C_i.supp = \sum_{j=0} M[0][j] * CtransDB$;
}

Fig. 2. Data mining algorithm using improved MASK

4 Experiments and Results

4.1 Experimental Data

We tested the recommendation accuracy of our privacy-preserving recommendation system using MovieLens dataset.

MovieLens dataset is provided by GroupLens research project team of Minnesota University for research use. In this data set there are 100234 ratings and 2213 tag applications across 8927 movies. These data were created by 718 users between March 26, 1996 and August 05, 2015. This dataset was generated on August 06, 2015. Users were selected at random for inclusion. All selected users had rated at least 20 movies. No demographic information is included.

4.2 Experimental Results

In this section we will test the effectiveness and performance of our proposed framework using two measurements.

Privacy-Preserving Degree Metrics (P): The formula for measuring privacy degree is as follows [15]:

$$P = (1 - RP) * 100 \tag{10}$$

Where RP is as (1).

In Table 1 the privacy degree measurements are shown:

Table 1. Privacy degree using different parameters

P1	P2	P3	Privacy degree %
0.7	0.15	0.15	97.4%
0.4	0.1	0.5	95%
0.4	0.2	0.4	93.5%
0.3	0.6	0.1	81.75%
0.93	0.0	0.07	99.6%

As can be seen from the above table, for different parameters values the privacy degree is above 81%. We compared our results with another work in 2014, [16] where they used only data perturbation strategy and the maximum privacy degree reached was 80% but here in our framework after using DPQR strategy and empirical perturbation parameters privacy reached 97.4% whereas by using the linear programming to choose the optimum parameters we reached 99.6%.

By analyzing the privacy parameters we found the following results as Fig. 3. It shows the relation between the privacy degree and average support of the item. As can be seen that it is a bell shape distribution where the privacy degree is maximum when the 1's or 0's is dominant in the data set, while it's above 60% when 1's and 0's are nearly equal.

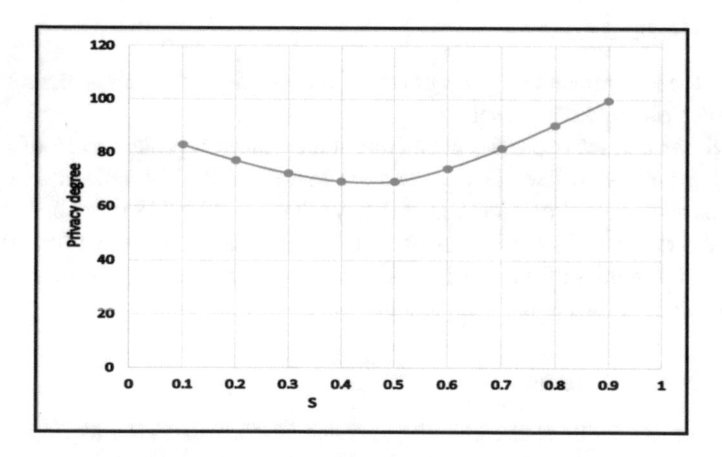

Fig. 3. Privacy degree with respect to average support using optimum parameters

Also by analyzing the nature of the data set it was proven that the privacy degree increases as the average support of items tends to 1 as shown in Fig. 4.

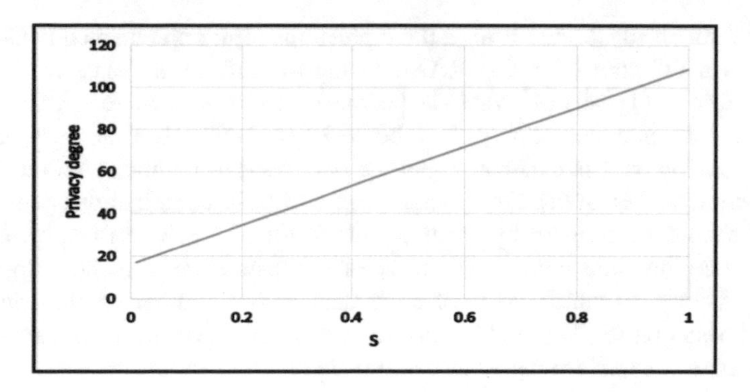

Fig. 4. Privacy degree with respect to average support using optimum parameters

Mean Absolute Error (MAE): We adopt Mean Absolute error (MAE) as one of the methods of statistical accuracy for measuring recommendations precision. For each user, statistics the counts of same itemsets of recommendation system before and after the distortion process. Therefore, the larger the value of MAE is the better. Measuring formula for MAE is as [16], where A, B respectively represents privacy preserving recommendation system and traditional recommendation system.

$$MAE = \frac{same\_num(A, B)}{Total\,Number\ of\ recommended\ items}\, x\,100 \tag{11}$$

The values of MAE compared over different values for the number of users and for different parameters selection methods are shown in Fig. 5:

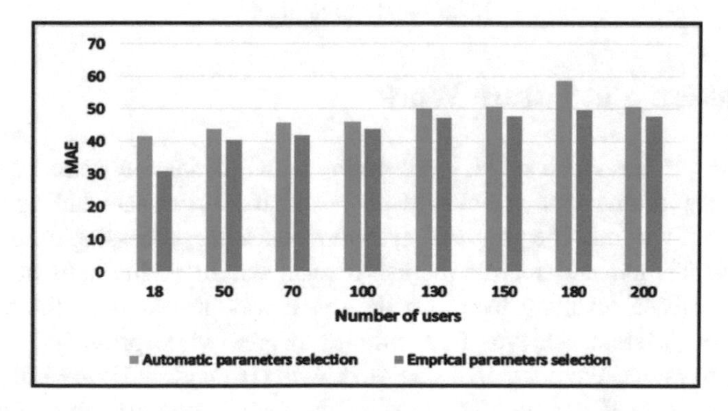

Fig. 5. MAE

As can be seen from (Fig. 5) the accuracy of recommendation increased when adopting linear programming for automatic parameters selection.

Running Time of the Algorithm: All the programs were implemented in C# language and run on a PC core i7, 6 GB RAM. In the MASK, to calculate M^{-1} should be calculated first as $(1/|M|)\ M^{*}$ where |M| is the determinant of M and M* is the adjoint matrix. So the time complexity will be $O(2^{3k}) = O(8^{k})$. But in this paper using recursive relation M^{-1} is calculated and we only need to compute. So the time complexity getting M^{-1} is O(2k). Furthermore, In the MASK, combinations generated from n-itemset should be counted by scanning the distorted database directly. But in this paper the running time complexity decreased, because we only need to query the amount of itemset in which value of each element is '1', other combinations can be calculated based on set theory [15]. So the new algorithm demonstrates excellent time performance and space efficiency as seen in Fig. 6.

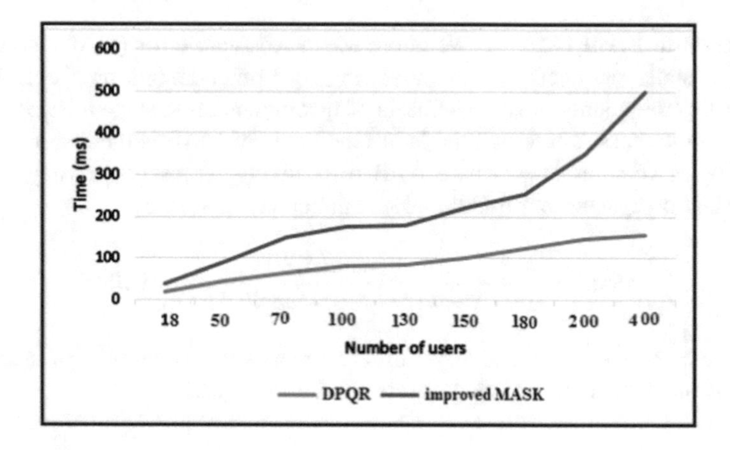

Fig. 6. Running time

5 Conclusion and Future Work

In this paper, we developed privacy preserving recommendation system using MASK and DPQR algorithms that achieves data perturbation and data hiding using multi parameters perturbation. The proposed system reduces the processing time by adopting matrix block method and matrix recursive relations. In addition to decreasing the number of database scanning using set theory. It was proven from the experimental results that our system achieves high privacy degree while providing accurate recommendation results. We compared our work with [16] and it achieved higher privacy level. Also we solved our problem as non-linear goal programming to analyze and select our privacy parameters values that increased the system accuracy. As for the system limitation, the system deals only with boolean data. And the time complexity still had room for improvement which will be addressed in our future work.

References

1. Chen, K., Liu, L.: Privacy preserving data classification with rotation perturbation. In: Proceedings of ICDM, pp. 589–592 (2005)
2. Jeckmans, A.J.P., Beye, M., Zekeriya, E., Pieter, H., Reginald, L.L., Quiang, T.: Privacy in recommender systems. In: Ramzan, N., van Zwol, R., Lee, J.S., Clüver, K., Hua, X.-S. (eds.) Social Media Retrieval. Springer, London (2013)
3. Liu, K., Kargupta, H., Ryan, J.: Random projection –based multiplicative data perturbation for privacy preserving distributed data mining. IEEE Trans. Knowl. Data Eng. **18**, 92–106 (2006)
4. Chen, K., Sun, G., Liu, L.: Towards attack-resilient geometric data perturbation. In: Proceedings of the SIAM International Conference on Data Mining, April 2007
5. Li, T., Li, N.: Towards optimal k-anonymization. Data Knowl. Eng. **303** (2008)
6. Majid, B.M., Asger Gahzi, M., Ali, R.: Privacy preserving data mining techniques: current scenario and future prospects. In: IEEE Third International Conference on Computer and Communication Technology (2012)
7. Agrawal, R., Srikant, R.: Privacy preserving data mining. In: Proceedings of the ACM SIGMOD International Conference on Management of data (2000)
8. Liu, L., Kantarcioglu, M., Thuraisingham, B.: The applicability of the perturbation based privacy preserving data mining for real-world data. Data Knowl. Eng. **65**, 5–21 (2008)
9. Shah, A., Gulati, R.: Privacy preserving data mining: techniques, classification and implications - a survey. Int. J. Comput. Appl. (0975 – 8887) **137**(12), 40–46 (2016)
10. Jagannathan, G., Wright, R.N.: Privacy-preserving imputation of missing data. Data Knowl. Eng. (2008)
11. Yang, W., Qiao, S.: A novel anonymization algorithm: Privacy protection and knowledge preservation. Expert Syst. Appl. **37**, 756–766 (2007)
12. Mukkamala, R., Ashok, V.G: Fuzzy-based methods for privacy-preserving data mining. In: Eighth International Conference on Information Technology: New Generations (ITNG), pp. 348–353 (2011)
13. Poovammal, E., Ponnavaikko, M.: An improved method for privacy preserving data mining. In: IEEE International Advance Computing Conference (IACC) Patiala, India, 6–7 March 2009
14. Kamal, R., Hussein, W., Ismail, R.: Privacy preserving recommender system based on improved MASK and query restriction. In: IEEE Eighth International Conference on Intelligent Computing and Information Systems (ICICIS) (2017)
15. Lou, H., Ma, Y., Zhang, F., Liu, M., Shen, W.: Data mining for privacy preserving association rules based on improved MASK algorithm. In: Proceedings of the IEEE 18th International Conference on Computer Supported Cooperative Work in Design (2014)
16. Xie, Y., Wulamu, A., Hu, X.: Design and implementation of privacy-preserving recommendation system based on MASK. J. Softw. **9**(10), 2607–2613 (2014)

Content Based Image Retrieval Using Local Feature Descriptors on Hadoop for Indoor Navigation

Heba Gaber[(✉)], Mohammed Marey, Safaa Amin, Howida Shedeed, and Mohamed F. Tolba

Computer and Information Sciences, Ain Shams University, Cairo, Egypt
eng.heba.gaber@gmail.com, mohammedmarey@hotmail.com, safaa_amin007@hotmail.com, dr_howida@yahoo.com, fahmytolba@gmail.com

Abstract. This paper demonstrates Content Based Image Retrieval (CBIR) algorithms implementation on a huge image set. Such implementation will be used to match query images to previously stored geotagged image database for the purpose of vision based indoor navigation. Feature extraction and matching are demonstrated using the two famous key-point detection CBIR algorithms: Scale Invariant Feature Transformation (SIFT) and Speeded Up Robust Features (SURF). The key-points matching results using Brute Force and FLANN (Fast Library for Approximate Nearest Neighbors) on various levels for both SIFT and SURF algorithms are compared herein. The algorithms are implemented on Hadoop MapReduce framework integrated with Hadoop Image Processing Interface (HIPI) and Open Computer Vision Library (OpenCV). As a result, the experiments shown that using SIFT with KNN (4, 5, and 6) levels give the highest matching accuracy in comparison to the other methods.

Keywords: Big data · HDFS · MapReduce · Computer vision tools OpenCV · SIFT · SURF · Indoor navigation · Hadoop · HIPI

1 Introduction

Several approaches have been proposed for estimating the users' indoor location, for instance, using RFID, Wi-Fi and Bluetooth. These approaches can estimate the users' location with acceptable accuracy. Most of these approaches require a dedicated infrastructure (e.g. access points, array of sensors, Bluetooth Low Energy (BLE) or cameras). Thus the system scalability can be expensive as it requires adding devices. In our survey [1], the advantages of vision based navigation from the perspective of cost and scalability were demonstrated. Vision based navigation depends on building the environment model through collecting image dataset for a specific indoor location and using query images from the navigation application. Moreover, for a dynamic environment the navigation application will continuously crowdsource the images captured during navigation to the geotagged image database. This operation shall produce tremendous amount of data that needs to be stored and processed using a reliable Big Data platform. Hadoop [2] has become defacto framework for handling large scale data

A. E. Hassanien et al. (Eds.): AISI 2018, AISC 845, pp. 614–623, 2019.
https://doi.org/10.1007/978-3-319-99010-1_56

and processing in Big Data technology. There had been a considerable amount of work done towards processing several types of textual data, for example weblogs. However, a little amount of work has been conducted on binary data like images.

In this work, the processing of images on Hadoop HDFS using MapReduce integrated with standard computer vision libraries (OpenCV) [3] and HIPI [4] framework was presented. Moreover, image matching with Brute Force algorithm and k-d tree using FLANN where compared on various levels. This comparison was drawn between SIFT [5] and SURF [6] algorithms in terms of accuracy of matching and the percent ratio of matching good features.

The paper is organized as follows: Sect. 2 discusses the related work. Section 3 elaborates MapReduce framework. Section 4 explains CBIR with OpenCV. In Sect. 5, the proposed architecture, the implementation steps and results are presented. Finally, Sect. 6 concludes the presented work and shows the future research direction.

2 Related Work

There are some recent implementations for CBIR over big data platforms applied on many domains, e.g. biomedical, social media, surveillance, satellite and geographical images processing applications. In [7] a machine learning tool, "Haar-Filter" is applied for detecting human faces by using the Haar Cascading technique. It was implemented using HIPI and OpenCV on Hadoop platform. Compressing social media images, using HIPI as image processing tool is shown in [8]. The image processing cloud with the OpenCV support was built at the Prairie View A&M University (PVAMU) [9]. The PVAMU cloud provides the platform as a service for distributed image processing and storage over Hadoop cluster. In [10], feature extraction using Hadoop and SIFT is shown. Image processing using HIPI for healthcare domain in India is shown in [11]. Using HIPI and Avro for processing surveillance images is shown in [12]. In [13], analyzing terabytes of microscopic medical images on Hadoop using Tera-soft library is demonstrated. Processing satellite and geospatial huge images databases on Hadoop is shown in [14, 15]. In [5] MapReduce Image Processing framework (MIPr) for distributed image processing using Hadoop is presented over Microsoft Azure platform. An extension for MIPr for using OpenCV is shown in [16]. Searching the images using MapReduce on Hadoop framework for processing huge volumes of data is illustrated in [17]. Hadoop image processing framework which can be used to build application containing large set of images is described in [18]. In our work we demonstrate feature extraction and matching over Hadoop platform for the purpose of indoor navigation using HIPI and OpenCV libraries. In our previous publication [19], SIFT based CBIR over Hadoop was demonstrated.

3 Hadoop MapReduce

Hadoop is an open source Java-based programming framework that supports the processing and the storage of extremely large data sets in a distributed computing environment. The two main components of Hadoop are: Hadoop Distributed File System (HDFS) and MapReduce.

- **HDFS:** is a distributed file system designed to run on a commodity hardware. It is highly fault-tolerant and designed to be deployed on low-cost hardware. Moreover, it provides high throughput access to application data and is suitable for applications that have large data sets.
- **MapReduce:** is a paradigm composed of three phases. (1) The Mapper: Each Map task operates on a single HDFS block and runs on the node where the block is stored. (2) Shuffle and Sort: Sorts and consolidates the intermediate data from all mappers, after the Map tasks are completed and before the beginning of the Reduce tasks. (3) The Reducer: Operates on shuffled/sorted intermediate data (Mapping output) to produce the final output.

In this work, the integration between HIPI and OpenCV is introduced. The images folder will be uploaded to the HDFS as a HIB bundle using HIPI. Afterwards, in the feature extraction job, OpenCV library is used to extract SIFT/SURF features descriptors from the HIB. The result of the feature extraction job will be saved on Hadoop. Another job for matching query image to the descriptors database is implemented to find the matched images with either Brute Force or FLANN matching techniques.

4 Content Based Image Retrieval (SIFT and SURF)

A typical architecture of CBIR systems, illustrated in Fig. 1, consists of four components: (1) image collection, (2) image feature extraction that mainly affects the quality of searching, (3) indexing and (4) searching that mainly affects the efficiency. The below sections demonstrate SIFT and SURF descriptor extraction algorithms.

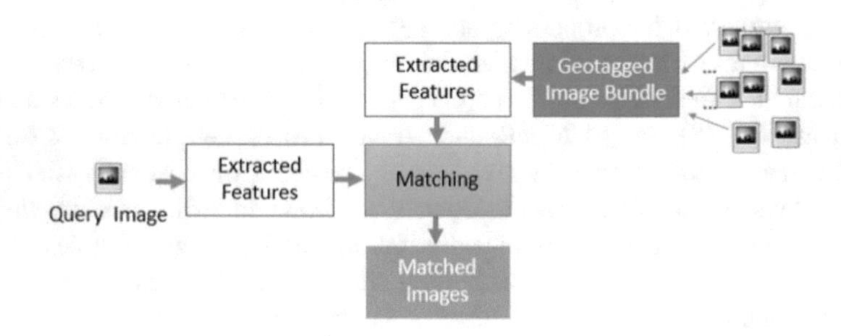

Fig. 1. Content based image retrieval paradigm

4.1 Scale Invariant Feature Transform (SIFT)

SIFT extracts blob like feature points and describes them with a scale, illumination, and rotational invariant descriptor. It generates a large number of features that densely cover the image over the full range of scales and locations as presented in Fig. 2. The correct matches can be filtered from the full set of matches by identifying subsets of the keypoints, that agree on the object and its location, scale, and orientation in the new image.

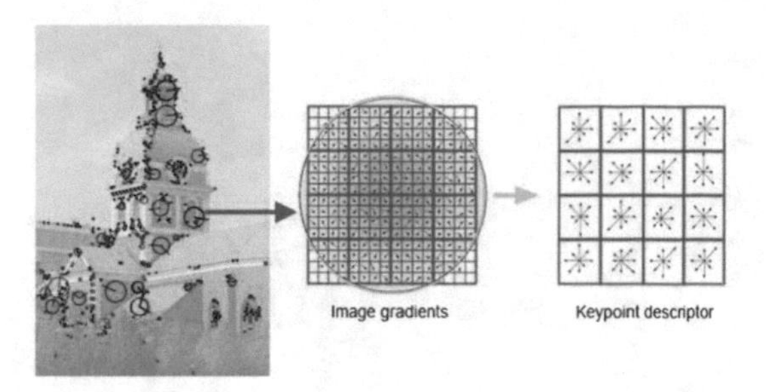

Fig. 2. SIFT key-point descriptor

The major stages of computation used to generate the set of image features are:

- **Scale-space extrema detection:** The first stage of computation searches over all scales and image locations. It is implemented efficiently by using a difference-of-Gaussian function to identify potential interest points that are invariant to scale and orientation. The scale space of an image is defined as a function, L(x, y, σ), that is produced from the convolution of a variable-scale Gaussian, G(x, y, σ), with an input image, I(x, y):

$$L(x, y, \sigma) = G(x, y, \sigma) * I(x, y) \tag{1}$$

where * is the convolution operation in x and y, and

$$G(x, y, \sigma) = \frac{1}{2\pi\sigma^2} e^{-(x^2 + y^2)/2\sigma^2} \tag{2}$$

- **Key-point localization:** At each candidate location, a detailed model is fit to determine the location and scale. Key-points are selected based on the measures of their stability.
- **Orientation assignment:** One or more orientation is assigned to each key-point location based on local image gradient directions.
- **Key-point descriptor:** The local image gradients are measured at the selected scale in the region around each key-point. These are transformed into a representation that allows for detecting significant levels of local shape distortion and changing in illumination.

4.2 Speeded Up Robust Features (SURF)

Bay et al. [20] introduced the SURF algorithm as a scale and rotation-invariant interest point detector and descriptor. SURF algorithm depends on the Hessian Matrix to detect the key-points. It uses a distribution of Haar wavelet responses at the key-point's neighborhood. The final descriptor is obtained by concatenating the feature vectors of all the sub-regions and represented with 64 elements as shown in Fig. 3. The major computational steps of SURF algorithm are as follows:

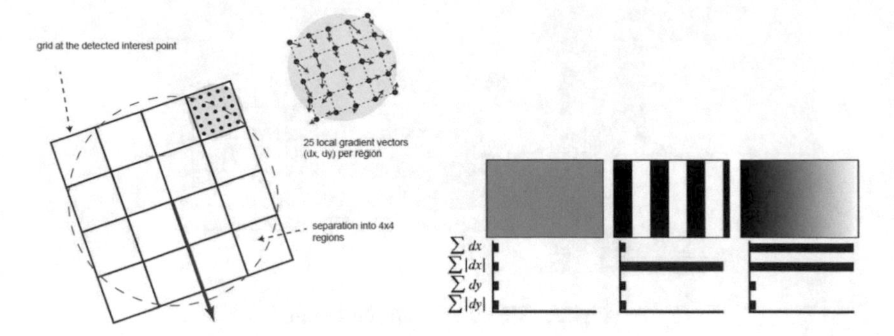

Fig. 3. SURF descriptor

- **Fast Interest Point Detection:** The Hessian matrix is defined as H(x, σ) for a given point ***x*** = (*x*, *y*) in an image as follows:

$$H(x,\sigma) = \begin{bmatrix} L_{xx}(x,\sigma) & L_{xy}(x,\sigma) \\ L_{xy}(x,\sigma) & L_{yy}(x,\sigma) \end{bmatrix} \quad (3)$$

where $L_{xx}(x,\sigma)$ is the convolution of the Gaussian second order derivative $\frac{\partial^2}{\partial x^2} g(\sigma)$ with the image I in point x and similarly for $L_{xy}(x,\sigma)$ and $L_{yy}(x,\sigma)$. The SURF approximates second order derivatives of the Gaussian with box filters. Image convolutions with these box filters can be computed rapidly by using integral images. The determinant of the Hessian matrix is written as:

$$Det(H_{approx}) = D_{xx}D_{yy} - (0.9\,D_{xy})^2 \quad (4)$$

In order to localize interest points in the image and over scales, a non-maximum suppression in a 3 × 3 × 3 neighborhood is applied. Finally, the maxima of the determinant of the Hessian matrix are then interpolated in scale and image space.

- **Interest Point Descriptor:** SURF descriptor is extracted in two steps:
  - Using Haar-wavelet responses in both x and y direction, SURF will assign an orientation based on the information of a circular region around the detected interest points.
  - The region is split up regularly into smaller square sub-regions and a few simple features at regularly spaced sample points are computed for each sub-region.

The responses of the Haar-wavelets are weighted with a Gaussian centered at the interest point in order to increase robustness to geometric deformations. Afterwards, the wavelet responses in horizontal d_x and vertical Directions d_y are summed up over each sub-region. Furthermore, the absolute values $|d_x|$ and $|d_y|$ are summed in order to obtain information about the polarity of the image intensity changes. Therefore, each sub-region has a four-dimensional descriptor vector

$$V = \left(\sum d_x, \sum |d_x|, \sum |d_y|\right) \tag{5}$$

where d_x denotes the horizontal wavelet response and d_y the vertical response. The resulting descriptor vector for all 4 by 4 sub-regions is of length 64.

5 CBIR with OpenCV over Hadoop Implementation

In this work, the implementation of CBIR using SIFT and SURF integrated with Hadoop Big Data platform with the help of Java, OpenCV and HIPI is introduced. The system configuration used for the development and experimental studies are: Cloudera Virtual Machine Version: CDH 5.12 [21], Operating System – Centos 6.9, 64 bit, Number of VCores is 4, RAM: 16 GB, Storage Disk – 250 GB, Apache Hadoop Version – 2.6. The tools needed to setup the framework were: Java Eclipse Luna Service Release 2 for MapReduce job compilation, OpenCV v3.3, Python 3.6.2 [22, 23], QT v4.8.5 [23], Gradle v4.2.1 [24] and HIPI. In this implementation the Technical University of Munich (TumIndoor) image dataset used for the indoor navigation application is selected to test the proposed approach. This dataset is constructed of 9,437 images for one floor, total size of the images is 1,395,864,371 bytes [25]. Two MapReduce jobs were executed, the first one for feature extraction and the second one for matching query image against the features dataset. Hadoop Big Data architecture was utilized to handle the large image set and the descriptors dataset with no failure due to its recovery mechanism.

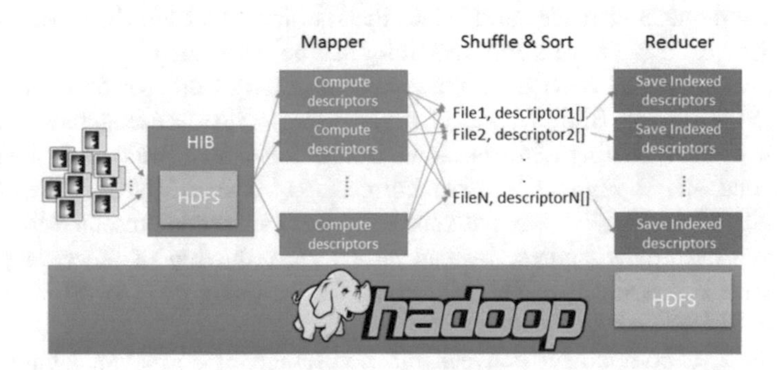

Fig. 4. Feature extraction paradigm.

Feature extraction process demonstrated in Fig. 4 consists of the below steps:

- Collecting geotagged images to construct the environment model.
- Images will be uploaded using HIPI. Afterwards, HIB will be constructed, then the initial filtration phase will be done.
- Features will be extracted and stored in features dataset on HDFS using OpenCV.

In Table 1, feature extraction job output is demonstrated. For each image file, key-point descriptor (SIFT/SURF) can be represented as a dense fixed-dimension vector of a basic type. Most descriptors follow this pattern, as it simplifies computing distances between descriptors. Therefore, a collection of descriptors is represented as "Mat", where each row is a key-point descriptor. SIFT image descriptors dataset extraction process took longer time than SURF extraction process. However, SURF descriptor dataset size is three multiples of SIFT descriptor dataset.

Table 1. Descriptor extraction

KPI	SIFT extractor	SURF extractor
HDFS: bytes read	1,449,992,560	1,449,992,560
HDFS: bytes written (bytes)	8,856,407,884	27,115,433,566
Time spent by all maps in all slots (ms)	95,196,370	66,953,905
Time spent by all map tasks (ms)	19,039,274	13,390,781
Total committed heap usage (bytes)	6,403,653,632	10,173,284,352

For the matching process, query image is compared against the descriptors dataset. In this work, the approximate k-nearest neighbor (k-NN) with k-d tree data structure was compared to the Brute Force algorithm on various levels. The purpose of k-NN is to calculate the nearest neighbor on the basis of the value of k, which specifies how many nearest neighbors used to define the class of a sample data point. The Brute Force matching implementation consists of three stages. The first stage is to calculate all the distances from each query point to every reference point in the dataset. The second stage is to sort these distances and select the k objects that are the closest from the matching image. In the third stage, matching can be performed.

The experiment had been done on sample of 55 files on two descriptor datasets SIFT and SURF using Brute Force, FLANN 2, 3, 4, 5, 6 tree matching levels. The comparison of the matching ratios is shown in Fig. 5 between Brute force matching for both SIFT and SURF. Figure 6 shows a comparison between KNN and two levels for SIFT and SURF. In Fig. 7, KNN with six levels comparison is presented for SIFT and SURF. From the below figures, we can notice the reduction of the wrong matches between Brute Force and FLANN. Moreover, the matching ratio for SIFT are always higher than SURF.

In Table 2, a comparison between the percentage of correct matching (CM) in terms of minimum number of wrong matches (MWM) and maximum matching ratio (MMR) is presented. It had been noticed that although Brute Force methods induce higher ratio in matching the training files, the possibility of wrong matching is very high. It was concluded that using SIFT matching with KNN (4, 5, and 6) gives higher matching accuracy although the matching ratio can be relatively less than the other methods.

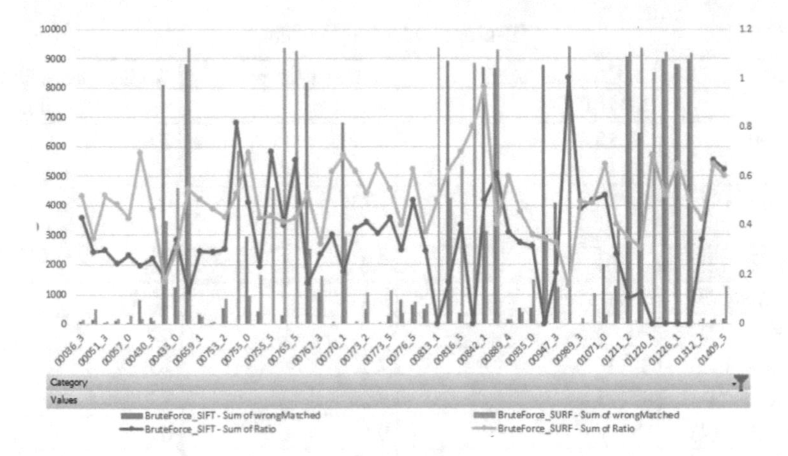

Fig. 5. Brute force matching

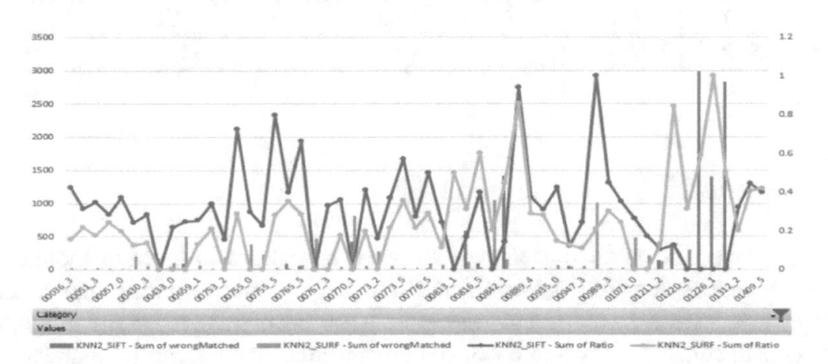

Fig. 6. KNN (2) matching

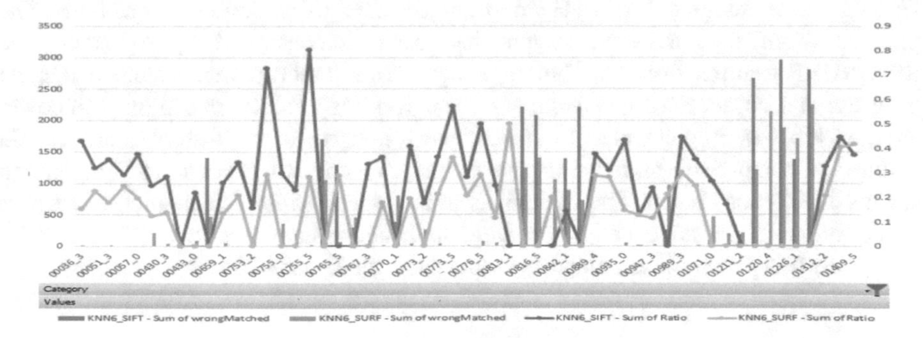

Fig. 7. KNN (6) matching results

Table 2. Matching results.

Matching method	% CM files with MWM files	% CM files with MMR	% CM files with MMR and MWM files
SIFT–brute force	14.55	12.7	3.63
SURF–brute force	0	70.1	0
SIFT–KNN (2)	43.64	3.63	1.8
SURF–KNN (2)	3.63	9.1	3.63
SIFT–KNN (3)	41.82	0	0
SURF–KNN (3)	1.82	7.27	1.8
SIFT–KNN (4)	45.45	9.09	7.27
SURF–KNN (4)	0	0	0
SIFT–KNN (5)	58.82	7.27	7.27
SURF–KNN (5)	9.1	0	0
SIFT–KNN (6)	69.1	7.27	7.27
SURF–KNN (6)	7.27	0	0

6 Conclusion

In this research, the main objective is to reach a scalable and cheap mobile based system for indoor location detection and tracking without requiring any additional infrastructure. Vision based indoor navigation has been chosen for the environment model construction. In this approach, the environment model can be extracted from previously geotagged images. Hadoop MapReduce is used as a Big Data platform for processing and storage of the data. HIPI was used to upload and filter data to HDFS. OpenCV was integrated with Hadoop for features extraction and matching. The practicality of the proposed system has been demonstrated by extracting the SIFT/SURF features from TumIndoor images. This confirms that feature extraction progress can be accelerated effectively through a MapReduce paradigm. Moreover, another job was implemented to match query images to the features database. In addition, a comparison on matching query images using Brute Force and FLANN on several levels was shown. The limitations in the presented implementation that it was tested on a single node and depending on handcrafted descriptors (SIFT/SURF). However, it is proposed to extend the work on a multi-node cluster.

References

1. Gaber, H., Amin, S., Marey, M., Tolba, F.: Localization and mapping for indoor navigation-survey. In: Handbook of Research on Machine Learning Innovations & Trends, pp. 136–160. IGI Global (2017)
2. White, T.: Hadoop: The Definitive Guide, 2nd edn. O'Reilly Media, Sebastopol (2011)

3. http://opencv.org
4. http://hipi.cs.virginia.edu/
5. Almeida, J., Torres, R.D.S., Goldenstein, S.: SIFT applied to CBIR. Revista de Sistemas de Informacao da FSMA **4**, 41–48 (2009)
6. Velmurugan, K., Baboo, S.S.: Content-based image retrieval using SURF and colour moments. Glob. J. Comput. Sci. Technol. **11**, 1–4 (2011)
7. Nausheen, K.M., Ram, M.S.: Haarfilter: a machine learning tool for image processing in Hadoop. Int. J. Technol. Res. Eng. **3** (2015)
8. Barapatre, M.H., Nirgun, M.V., Jagtap, M.H., Ginde, M.S.: Image processing using mapreduce with performance analysis. Int. J. Emerg. Technol. Innov. Eng. **I**(4) (2015)
9. Yan, Y., Huang, L.: Large-scale image processing research cloud. In: CLOUD COMPUTING 2014: The Fifth International Conference on Cloud Computing, GRIDs, and Virtualization, pp. 88–93 (2014)
10. Cheng, E., et al.: Efficient feature extraction from a wide area motion imagery by MapReduce in Hadoop. In: SPIE Defense+ Security. International Society for Optics and Photonics (2014)
11. Augustine, D.P.: Leveraging big data analytics and hadoop in developing India's healthcare services. Int. J. Comput. Appl. **89**(16), 44–50 (2014)
12. Gawde, A.U., Shah, M., Ukaye, I., Nanavati, M.: Object detection in hadoop using HIPI. Int. J. Adv. Res. Eng. Technol. (2013)
13. Bajcsy, P., et al.: Terabyte-sized image computations on Hadoop cluster platforms. In: IEEE International Conference on Big Data, pp. 729–737. IEEE (2013)
14. Han, W., Kang, Y., Chen, Y., Zhang, X.: A mapreduce approach for SIFT feature extraction. In: International Conference on Cloud Computing and Big Data, pp. 465–469 (2013)
15. Moise, D., Shestakov, D., Thor, G., Amsaleg, L.: Indexing and searching 100M images with map-reduce. In: ACM International Conference on Multimedia Retrieval, pp. 17–24 (2013)
16. Epanchintsev, T., Sozykin, A.: Processing large amounts of images on hadoop with OpenCV. In: Proceedings of the 1st Ural Workshop on Parallel, Distributed, and Cloud Computing for Young Scientists (Ural-PDC 2015), pp. 137–143. Yekaterinburg (2015)
17. Banaei, S.M., Moghaddam, H.K.: Hadoop and its role in modern image processing. Open J. Mar. Sci. **4**(4), 239–245 (2014)
18. Sweeney, C., Liu, L., Arietta, S., Lawrence, J.: HIPI: a hadoop image processing interface for image-based mapreduce tasks. University of Virginia (2011)
19. Gaber, H., Amin, S., Marey, M., Tolba, F.: Content based image retrieval with hadoop. In: Proceedings of the 2nd International Conference on Advances in Intelligent Systems & Informatics, pp. 257–265. AISI, Egypt (2016)
20. Bay, H., Tuytelaars, T., Van Gool, L.: SURF: speeded up robust features. In: European Conference on Computer Vision. Springer, Berlin (2006)
21. https://www.cloudera.com/downloads/quickstart_vms/5-12.html
22. https://www.python.org/downloads/release/python-360/
23. https://www1.qt.io/qt5-7/
24. https://gradle.org/install/
25. Huitl, R., Schroth, G., Hilsenbeck, S., Schweiger, F., Steinbach, E.: TUMindoor: an extensive image and point cloud dataset for visual indoor localization and mapping. In: 19th IEEE International Conference on Image Processing (ICIP), Orlando, FL, pp. 1773–1776. IEEE (2012)

Design and Implementation of Embedded System for Nuclear Materials Cask in Nuclear Newcomers

M. I. Youssef[1], M. Zorkany[2], G. F. Sultan[3], and Hassan F. Morsi[3](✉)

[1] Faculty of Engineering, Al Azhar University, Cairo, Egypt
[2] National Telecommunication Institute, Cairo, Egypt
[3] Egyptian Nuclear and Radiological Regulatory Authority, Cairo, Egypt
hassanmfahmy@yahoo.com

Abstract. Nuclear newcomer countries face a number of key challenges in infrastructure development, e.g. they have not Intelligent Transportation Systems. Therefore, one of challenges is the safety and security of nuclear materials during transporting, storing and disposing. Where, nuclear and radiological terrorism continues to be a worldwide concern as the nature of security threats evolves. This paper tries to solve that challenge by design and implement of an embedded system for nuclear materials cask. This system is suitable to developing countries, where it is cost effective and it uses the existing infrastructure. By using GPS, GSM/GPRS and microcontroller, the embedded system will enables the responsible bodies to remotely and continuously; tracking, monitoring and inspection of nuclear materials casks; during transporting, storing and disposing. The ORIGEN code is used to calculate the thermal and radioactivity loads of the cask. The application of this system allows the rapid intervention of the concerned bodies, which will prevent many accidents, in particular those caused by terrorists, like stealing or dispersing of nuclear materials.

Keywords: Embedded system · Monitoring · Tracking · GPS/GPRS
Nuclear materials cask

1 Introduction

Recent advancements in nuclear fission technology have led to predict that the number of Nuclear Power Plants will increase rapidly in developing countries. Therefore, Nuclear Materials (NM) inventories are predicted to increase rapidly in developing countries. Knowing that, the threat of nuclear terrorism remains one of the greatest challenges to international security, beside the weak infrastructure of developing countries, The NM will be mostly vulnerable to terrorism, especially in transportations. Therefore, additional measures are required to militate against this risk. One of these measures is the continuous monitoring of nuclear materials casks/packages; during transporting, storing and disposing. Advancements in microelectronics, wireless technology and encryption can be achieve that continuous surveillance, by integration the modern microcontrollers with sensors and wireless communication techniques. The continues monitoring of the NM

A. E. Hassanien et al. (Eds.): AISI 2018, AISC 845, pp. 624–633, 2019.
https://doi.org/10.1007/978-3-319-99010-1_57

can counter the terrorism threats by informing the first responders (e.g. police and fire fighter) to not only know the position of the incident but also the nature and severity of the accident before approaching the scene of the event, which allowing the speed response. Also, continues monitoring can be enhanced the safety and security of NM. Therefore, this paper used the existing infrastructure in these countries to builds an Embedded System (ES) that can be used for continuous monitoring and surveillance of NM, which will help nuclear newcomers to counter nuclear terrorism.

2 Related Work

For nuclear terrorisms countering, the continuous monitoring system like as Argonne and Remote Area Modular Monitoring (RAMM) systems technology [1] can be used as in advanced countries. Regarding to the lone wolves threats, where recently the world is suffering from. The most lone wolves harmful attacks were by trucks. Admittedly, the destruction will be severely increased if the truck was loaded with NM. Therefore, the NM is Vulnerable to the lone wolves threats especially during transportations. The Argonne and RAMM systems are vulnerable to counter this kind of terrorism. Therefore, to overcome this vulnerability the works in [2, 3] proposed a new design approach for Intelligent Transportation Systems (ITS) based internet of things to counter the lone wolf, just before the attacks done by trucks. For developing countries like the case of most nuclear newcomers, where they have not an ITS nor privet satellite communication like Transcom or Iridium (for two way satellite communications) [4], the system in [5] can be used. In this system, a customized Global System for Mobile communication (GSM) module is designed for wireless radiation monitoring through Short Messaging Service (SMS). This module is able to receive serial data from radiation monitoring devices such as survey meter or area monitor and transmit the data as text SMS to a host server. It provides two-way communication for data transmission, status query, and configuration setup. Integration of this module with radiation monitoring device will create mobile and wireless radiation monitoring system with prompt emergency alert at high level radiation. But, this system absent the tracking of the NM. Therefore in this paper, the proposed system used the global satellite communication for NM tracking as shown in the next sections. The ES can be attached to the NM casks.

3 Proposed Embedded System Design and Operation

The proposed system is an ES consists of microcontroller with onboard GPS and GSM modules, sensors, application software, a database server and web page, Fig. 1. The ES monitors critical parameters, including the status of seals, movement of object, and environmental conditions of the NM cask in real time. Also, it provides an instant warnings or alarms messages (i.e. SMS), when preset thresholds for the sensors are exceeded. The information collected by the system is transmitted to a dedicated central database server that can be accessed by authorized users across the responsible bodies via a secured network. The ES allows the tracking and inspecting of the casks throughout their life cycles in storage, transportation, and disposal. The software provides easy-to-use graphical interfaces that allow access to all vital information once the security and privilege requirements are met.

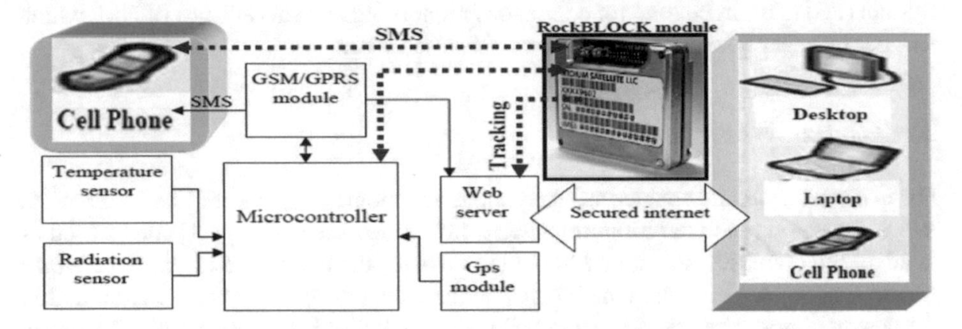

Fig. 1. Embedded system block diagram

3.1 Sensor Modules

As a prototype, the ES sensors includes; safety sensors (e.g. radiation and temperature), security sensor (e.g. the status of seals), and driver violation detector (e.g. the speed of the truck). In this paper, the Evolutionary/European Power Reactor (EPR) Spent Fuel (SF) is selected as a NM.

- Safety sensors are used to indicate the radiation and temperatures levels statues of the NM cask. The ORIGEN [6] computer code is used to calculate the thermal and radioactivity loads of the cask which will be used to determine the sensors threshold level. The ORIGEN computer code flowchart is shown in Fig. 2. The preparation details of the ORIGEN input file based EPR fuel are stated in [7]. The radiation and temperatures sensors threshold level are determined as follows.

1. Radiation: As proven next, if EPR SF (5% enriched) is placed in a real cask system, the dose rate on the external surface of the cask will be lower than 1,000 mrem/h. Therefore, the ES prototype will be used the PIN diodes to detect the increasing in gamma level, where the 1,000 mrem/h is sufficient to excite the PIN diodes. Any gamma detector PIN diode circuit is consists of low noise amplifier and comparator. The photodiode circuit stated in [8] was used for the gamma rays detection. The advantage of using a photodiode is its small sensitive area; therefore, it is suitable to the high dose rate of the cask and it is not affected by the low background rate due to cosmic rays. Therefore, this will limit the ES radiation detection sensitivity to be in milli-scale range values and it do not work in micron-scale.
2. Temperature: EPR SF can be loaded to MPC-24 basket. Using the stated equation [7] of Peak Cladding Temperature (PCT) given Decay Heat (DH), when the DH is 1.050 kw/assembly, the error free PCT is 307.12 °C in normal condition operations. The 24 PWR SF assemblies storage cask system with a burnup of 55 GW Day/Metric Ton Uranium (GWD/MTU) and 25.2 kW DH load, the normal

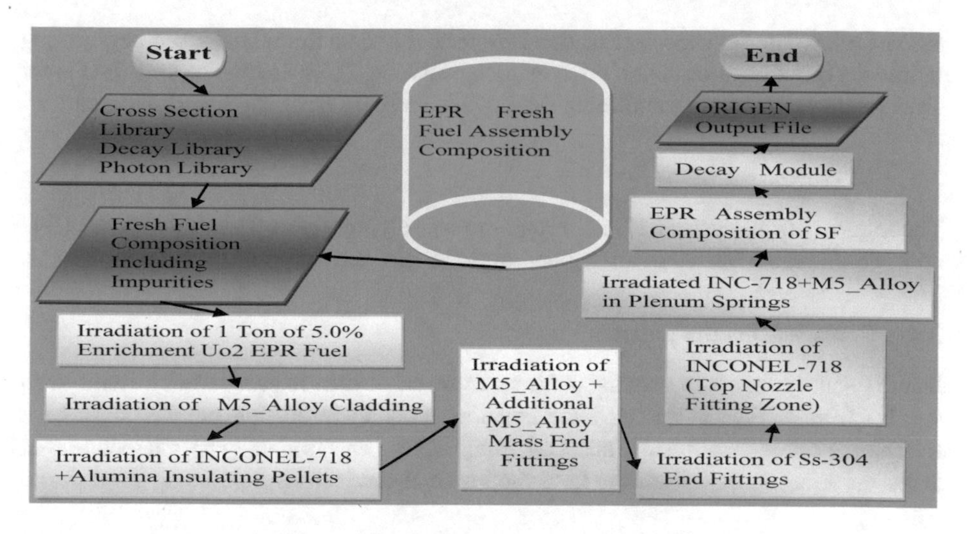

Fig. 2. ORIGEN computer code flowchart

temperature for long-term events (e.g. onsite and offsite transportations, and storage) are 302 °C, 64 °C and 67 °C for PCT, overpack outer surface and air outlet; respectively [9]. Therefore, for the EPR SF, the normal temperature are 307 °C, 69 °C and 72 °C for PCT, overpack outer surface and air outlet; respectively. The normal temperature limits for overpack outer surface and air outlet are 98 °C and 82 °C; respectively. For prototype, the circuit used two digital temperature sensors, where their positions are in overpack outer surface (near the top air outlet) and air outlet, the temperature alarm SMS will delivered to the control unit (or to an emergency specified telephone number) if the temperature exceeds 98 °C and 82 °C for overpack outer surface and air outlet; respectively.

- Status of Seals: The seal sensor can located under one or two of the seal bolts of the cask overpack. The seal sensor is a short circuit wire warped around the bolt of the cask overpack. When the bolt is loosened, the short circuit wire will opened circuit. Therefore, the microcontroller trigger an alarm, the alarm is broadcasted by SMS to the responsible bodies.

3.2 Online Cask Monitoring and Tracking

The designed and implemented ES is used for receiving locations data from satellites (via GPS module) and monitoring data (via sensors), then transmitting the received data to the desired web servers using a General Packet Radio Services (GPRS) connection (via GSM module). The ES used the recommended minimum specific GPS/Transit data ($GPRMC) frame. This frame contains information about the locations of the cask and the cask speed over ground. The speed can be used as a driver violation, if it exceeds a predefined value (e.g. 80 km/h).The ES used GSM and GPRS to provide wireless communications capabilities. Sending of the SMS messages are the functions of the GSM module. The connection of the ES to the internet is through the mobile operators

GSM/GPRS. The web servers functions are receiving data from the ES, securely storing it, and serving this information on demand to the user. There are two servers, first is for secret data e.g. the cask monitoring data, while second server is for tracking data.

3.3 Microcontroller

The microcontroller used in ES is a Programmable Cypress chip. The chip includes CPU core, configurable blocks of analog and digital logic, and programmable interconnects.

3.4 Proposed Frame Format

Data is sent to the main servers as frame format. All data is grouped in a frame with special format, Fig. 3. Frame fields contains; cask identification number (ID), cask tracking location, seal status, and cask monitoring sensors data. The microcontroller takes the location data from the GPS module and put it in its field in the frame.

Start of frame	Cask ID	Cask location N,E	Seal status	Cask radiation status	Cask overpack outer surface temperature	Cask air outlet temperature	Check sum	End of frame

Fig. 3. Embedded system frame construction

3.5 Proposed Embedded System Operation

The ES operation methodology is shown in Fig. 4. When the ES start, it read the sensors statues and sends theses data to monitoring and tracking servers by GPRS. In addition, if any one of the sensor values exceeds the limit, the ES send instantaneous

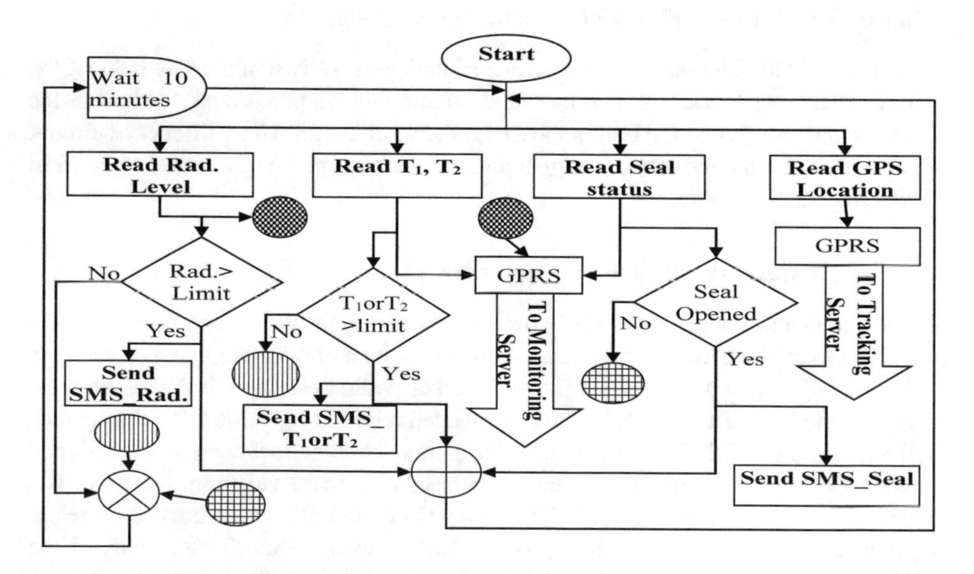

Fig. 4. Embedded system operation framework

SMS to the predefined telephone number; and the monitoring and tracking are instantaneous. For power saving, in normal operations, i.e. radiation level (Rad.), T_1 (cask overpack outer surface temperature), and T_2 (cask air outlet temperature) lower than limits; and the seal is not opened), the system is programmable to wait a time between each reading process (e.g. in casks storage site, the waiting time will be about 10 min).

4 Results

The ORIGEN computer code simulation results of the EPR SF radiation source terms and the practical results of the ES operation will be stated in the next subsections.

4.1 Gamma Source Terms Calculation

Radiation source terms of SF are photons and neutrons. In this paper, the photons source of the EPR SF is calculated using the ORIGEN code based on the EPR parameters, where the photons are the source term of gamma. The EPR SF photon source decay of the activation products, actinides and daughters, and fission products are calculated, Fig. 5(a). As shown, the main gamma source term is the fission products photons.

The radioactive characteristic of the EPR SF has previously been calculated, but for a burnup and enrichment of 60 GWD/MTU and 4% [10], respectively. To make sure that our calculations of the gamma sources (i.e. 5% enriched EPR) correspond to that calculations (i.e. 4% enriched EPR), we compared our results with the reference results, Fig. 5(b) shows the photon sources decay comparison of the results. From Fig. 5(b), beyond five years of SF cooling, the differences between the two curves are small. For example, at the cooling value of 20 year, it is found that the percentage difference is about (0.078393%), which is very small regarding to the flux-to-dose conversion coefficient deference's for photons which is about 20%. Therefore, the cask shielding

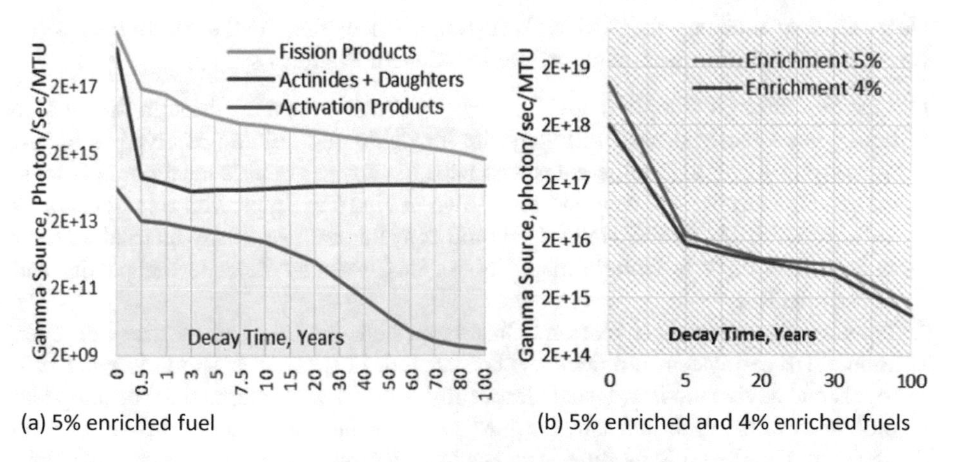

Fig. 5. EPR spent fuel gamma source

calculations in [11] can be applied to our work. This means that, when EPR SF (5% enriched) is placed in a real cask system, the dose rate on the external surface of the cask will be lower than 1,000 mrem/h, which satisfied the U.S. Nuclear Regulatory Commission requirements [12].

4.2 Online Cask Monitoring

The online NM cask monitoring data are given by accessing the server of responsible body. The cask monitoring data are cask ID, seal status, location (north and east), radiation status, overpack outer surface temperature, and air outlet temperature. As a prototype, the cask seal status are Ok or Opened, while, cask radiation and temperatures status are Ok (i.e. lower than thresholds limit) or Over Limit (i.e. larger than thresholds limit), Fig. 6.

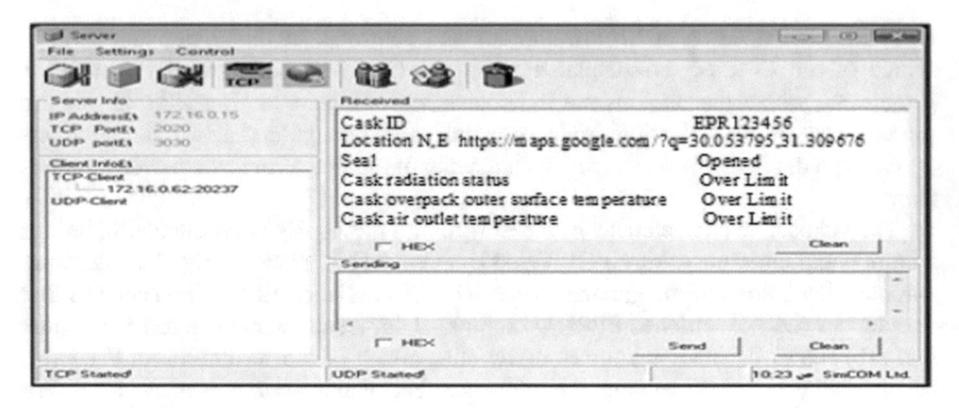

Fig. 6. Cask monitoring server screenshot

4.3 Online Cask Tracking

There are two methods for NM cask tracking, where the truck's location is given through; Google maps web page or Traccar Modern Platform system.

1. Tracking through Google maps: To track the truck's location through the Google maps, the authorized user will copy the longitude and latitude received in a cask monitoring server to Google maps web page to view the truck's location on Google maps. For example, the web address shown in Fig. 6 is (https://maps.google.com/?q=30.053795, 31.309676); the user should copy this address to any internet browser to locate the truck in Google map. This method reduces the code complexity and cost of the ES.
2. Tracking through traccar platform: In this method, the ES used the free and open source Traccar system provided by Traccar Ltd. [13]. Traccar supports more protocols and device models. It includes a fully featured web interface for desktop and mobile layouts. With Traccar, the NM cask can be viewed in real-time with no delay by the ES GPS module. Traccar have various mapping options, including

road maps and satellite imagery, Fig. 7. The cost of a single user account on a shared traccar server (for 5 devices + address information in status and reports) is $20.00/month. While the cost of the own tracking server (for 50 devices + address information in status and reports) is $100.00/month. These subscriptions include all features provided by Traccar platform like Geofencing, except SMS alerting which need a supplement subscription.

Fig. 7. Online cask tracking, tracking through Traccar modern platform

4.4 SMS Warnings and Alerts

The warning SMS about driver driving violations (e.g. speeding) can be sent to a telephone number without any delay by Traccar system. Alerts about danger states of the cask (e.g. temperature level exceeding the limit values) can be sent directly to a telephone numbers (e.g. police and/or nuclear safety staff), to insure the rapid intervention.

5 Discussions

Problems that will face the ES applications are mentioned and the solutions are stated.

5.1 Ionizing Radiation Effects on Electronic Circuits

In normal operations of the NM cask, the dose rate on the external surface of the package must be lower than 1,000 mrem/h (=1000 mrad/h, =0.01 gray/h). Therefore, the electronic components of the ES will be not affected. The ionizing radiation effect on GPS and microcontrollers modules will be shown below.

1. Renaudie et al. [14] deduced that the commercial of – the - shelf GPS receivers operate faultless up to accumulated ionizing doses of more than 9 krad (air). This is acceptable for cask in normal operation.
2. The experimental results of a collected irradiation environment of gamma rays and neutron on MCS96 microcontroller was presented by Xiao-Ming et al. [15]. The

results shown that, the microcontroller does not fail until the total ionizing dose exposure accumulates up to 11.3 krad (Si), and performs normally even when the neutron fluence is up to 3.01013 n/cm^2. Therefore, the microcontroller performs normally in the ES radiation environment, where the dose must be lower than 1000 mrad/h.

5.2 GSM Network Losing

The physical protection of stored SF and the geologic repository requirements insure the continuous surveillance of the storage and the repository sites [12]. These mean, the electrical and the communication systems in the site must be maintained. Depending on these requirements, the store and the repository sits must have more than one communication networks (e.g. two to three GSM networks, which can be used by the ES). Therefore, the unavailability of the communication networks is too rarely. Finally, the using of the Iridium satellite (e.g. RockBLOCK 9602) transceiver [16] models is another option for the ES. Where, RockBLOCK 9602 allows sending and receiving short messages from anywhere on earth, providing a clear view of the sky. The interface of the RockBLOCK to the ES board is easy, Fig. 1, where it serial interface and can be operated with a three-wire connection, which are used to transmit, receive and ground signals. The main drawback of the RockBLOCK is its cost, where costs are about 300$/module.

6 Conclusion

Advancements in microelectronics, wireless technology and encryption have opened opportunities that previously were not available to the nuclear sector. The ES tracking and monitoring system is enhancing the safety and security; reducing the need for manned surveillance; providing real-time access to status and event data; and providing overall cost effectiveness. The ES precise monitoring and tracking of the nuclear materials can perform the terrorists countering, provided that additional terrorism countermeasure like, the speed response and rapid intervention of the security bodies. The ES is suitable to nuclear newcomers, where most of them are developing countries.

Acknowledgement. Authors thank Prof. Ezzat A. Eisawy for his support. They are thankful to Eng. Nagdy Abd El Hamed for helping in laboratory works, and Eng. Emile Rushdie for his precious discussion.

References

1. Liu, Y., Sanders, K., Shuler, J.: Advances in tracking and monitoring transport and storage of nuclear material. IAEA-CN-244-186 (2016)
2. Morsi, H.F., Youssef, M.I., Sultan, G.F.: Novel design based internet of things to counter lone wolf, part-A: nice attack. In: Proceedings of the International Conference on Advanced Intelligent Systems and Informatics, Egypt, 2017, Advances in Intelligent Systems and Computing, vol. 639, pp. 875–884, Springer. https://doi.org/10.1007/978-3-319-64861-3_82

3. Morsi, H.F., Youssef, M.I., Sultan, G.F.: Novel design based internet of things to counter lone wolf, part-B: Berlin attack. In: Japan-Africa Conference on Electronics, Communications and Computers (JAC-ECC), Egypt, 2017, pp. 164–169. IEEE. https://doi.org/10.1109/jec-ecc.2017.8305802
4. U.S. Department of Energy Office of Environmental Management: TRANSCOM fact sheet: transportation tracking and communication system (2009)
5. Abd Rahman, N., Ibrahim, N., Lombigit, L., Azman, A., Jaafar, Z., Abdullah, N., Mohamad G.: GSM module for wireless radiation monitoring system via SMS. In: IOP Conference Series: Materials Science and Engineering, vol. 298, paper no. 012040 (2018). https://doi.org/10.1088/1757-899x/298/1/012040
6. INVAP, ORNL/ORIGEN Version 2.1 (2004)
7. Youssef, M.I., Sultan, G.F., Morsi, H.F.: Cooling period calculation of evolutionary power reactor spent fuel for dry management safety. Nucl. Radiat. Saf. **2**(70), 17–21 (2016). UDC 621.039.58:621.039.7
8. Kainka, B.: Improved radiation meter counter for alpha, beta and gamma radiation. Elektor **11**, 20–25 (2011)
9. Lee, J., Bang, K., Seo, K., Kim, H.: Thermal analysis of a storage cask for 24 spent PWR fuel assemblies. In: 14th International Symposium on the Packaging and Transportation of Radioactive Materials (PATRAM 2004), Germany (2004)
10. Anttila, M.: Radioactive characteristics of the spent fuel of the finish nuclear power plants. Posiva working report 2005-71 (2005)
11. Ranta-aho, A.: Review of the radiation protection calculations for the encapsulation plant. Posiva working report 2008-63 (2008)
12. Nuclear Regulatory Commission: U.S. nuclear regulatory commission regulations: title 10, code of federal regulations, Parts 71, 63, 72, and 73 (2010)
13. Traccar Ltd.: GPS tracking software - free and open source system – Traccar. https://www.traccar.org/
14. Renaudie, C., Markgraf, M., Montenbruck, O., Garcia, M.: Radiation testing of commercial-off-the-shelf GPS technology for use on low earth orbit satellites. In: 9th European Conference on Radiation and Its Effects on Components and Systems, France, 2007, pp. 1–8. IEEE. https://doi.org/10.1109/radecs.2007.5205561
15. Xiao-Ming, J., Ru-Yu, F., Wei, C., Dong-Sheng, L., Shan-Chao, Y., Xiao-Yan, B., Yan, L., Xiao-Qiang, G., Gui-Zhen, W.: Synergistic effects of neutron and gamma ray irradiation of a commercial CHMOS microcontroller. Chinese Physics B, paper no. 066104, vol. 19, no. 6 (2010)
16. Proprietary and Confidential Information: Iridium 9602 SBD transceiver developer's guide, revision 6.0, Iridium Communications Inc. (2010)

Combined Features for Content Based Image Retrieval: A Comparative Study

Nora Youssef[1(✉)], Alsayed Algergawy[2], Ibrahim F. Moawad[1], and EL-Sayed M. EL-Horbaty[1]

[1] Faculty of Computer and Information Sciences, Ain Shams University, Cairo, Egypt
nora.youssef@cis.asu.edu.eg

[2] Institute of Computer Science, Friedrich Schiller University of Jena, Jena, Germany

Abstract. Multimedia resources are rapidly growing with a huge increase of visual contents. Thus, searching these images accurately and efficiently for all types of datasets becomes one of the most challenging tasks. Content-based image retrieval (CBIR) is the technique that retrieves images based on their visual contents. So that, selecting appropriate features that describe an image sufficiently is a clue for a successful retrieval system. To this end, in this paper, a comparative study to investigate the effect of using a single and a combined set of features in the context of a CBIR is presented. To achieve this goal, several features including, edge histogram (EHD), color layout (CLD) and fuzzy color texture histogram (FCTH) as well as different combinations of these features such as, all edges (local, global and semi-global edges), all edges with CLD and finally, all edges with FCTH have been exploited. To demonstrate the effectiveness of the proposed method, a set of experiments utilizing different images datasets have been carried out. The results in terms of precision, recall, F-measure and mean average precision show a higher retrieval accuracy while using a set of combined features compared to exploiting only single features for the same retrieval task.

Keywords: CBIR · Combined features · EHD · CLD · FCTH

1 Introduction

Different forms of multimedia resources are rapidly growing with a huge development of visual contents and technologies like web content and telecommunication [1]. Thus, images play an essential role in a wide range of applications and fields, such as web, social media, and advertising [1]. Therefore, searching images rapidly, accurately, and efficiently for all types of image datasets becomes one of the most challenging tasks [1]. The traditional image retrieval techniques are text-based image retrieval (TBIR) where, users can retrieve images based on keywords or textual descriptions, which are annotated by a human [1]. TBIR

A. E. Hassanien et al. (Eds.): AISI 2018, AISC 845, pp. 634–643, 2019.
https://doi.org/10.1007/978-3-319-99010-1_58

systems have main drawbacks: (i) huge human labor for the manual annotation, (ii) the subjectivity of the annotation, (iii) the search is limited to the image text metadata, which could produce false negative result, (iv) different meanings for a word like "Jaguar" it may be an animal or a car [1,8,15,22].

Term content-based image retrieval (CBIR) appears to have been first used in the literature by Kato to describe his experiments in the automatic retrieval of images from a database by color and shape [8]. CBIR introduced as a solution for TBIR limitations, it is considered as, the technique of effective retrieval of images from an oversized data store [1,8,15,22]. It performs three major tasks: (i) feature extraction, where a set of features are extracted to describe the content of each image in the database, (ii) building an index with the extracted features, (iii) similarity measurement between the query image and each image in the database. There are various examples of CBIR systems: LIRe [12], VisualSeek [14] and SIMPLicity [20].

Image descriptors are descriptions of the visual features of an image's contents. They describe elementary characteristics, such as color, texture, shape and spatial location [8]. Selecting the appropriate features that describe an image sufficiently and accurately is the key of a successful retrieval system. Thus, it is very difficult to achieve satisfactory retrieval results using one image feature like only color or texture [9].

The evolution of multimedia, the exitance of huge digital archives have guided the research to provide effective CBIR systems [3]. CBIR should be an application based, it could vary from art galleries to biodiversity [3]. Biodiversity consists of various levels, starting from genes, then species, then communities of creatures and finally whole ecosystems, such as forests or coral reefs [16].

The proposed work is a CBIR for biodiversity domain. It has two main objectives, in other words, two main modifications of LIRe system, which is an open source CBIR Java library. Firstly, crawler modification: Flickr dataset has been replaced by an ImageNet synset that suits the biodiversity domain. The dataset is fed into the system via a user input. In addition, enhance the retrieval quality by building the image descriptor using a mixture of different image features at the same time.

This the paper is organized as follows: Sect. 2 highlights the main components of a CBIR in terms of feature extraction, indexing and similarity measurements. Section 3 presents different image descriptors and the idea of using combined features in retrieval. Section 4 explains the dataset and queries used and shows the results. Finally, Sect. 5 concludes the paper and suggests a future direction for this work.

2 Low Level CBIR Framework

This section gives an overview of a typical low-level CBIR framework with a concrete example LIRe system. The general framework consists of 3 main components: (i) feature extraction, (ii) indexing and (iii) similarity measurements.

2.1 Low-Level Feature Extraction

Low-level image feature extraction is the core of CBIR systems. Thus, these features can be either extracted from the entire image or from regions [11]. Image features can be classified into 2 main categories. On one hand, the global features, which are commonly extracted features in a CBIR. For example, color, texture, shape and spatial location. On the other hand, the local features, which describe the interesting parts in images such as region, object or edge. Such as, SIFT and SURF [1,17,19].

2.2 Dimensionality Reduction and Indexing

Indexing is the technique to efficiently retrieve images from the database based on the extracted features [19]. As feature vectors of images have the curse of dimensionality, indexing and dimensionality reduction (DR) become a challenging task in the context of the CBIR [1,19]. The commonly used techniques in DR are principal component analysis (PCA) and independent component analysis (ICA). They are statistical transformation methods. But, they are sensitive to the scaling of the original data. While, the commonly used methods in indexing are: (i) tree-based, which is fileable different shapes after partitioning the data space. (ii) hash-based techniques, which is flexible for both data-independent and data-dependent hashing [1].

2.3 Similarity Measurements

Similarity measurements are used to determine how close the query image to each image in the database. Thus, the structure of the feature vector determines the type of the measurement that shall be used [1]. The most popular measurements that are used in the context of CBIR are: (i) *Manhattan distance* which is invariant to noise in high dimensional data, (ii) *Euclidian distance* which allows normalized and weighted vectors [1,11].

2.4 Lucene Image Retrieval System (LIRe)

LIRe (Lucene Image Retrieval) is an open source Java library for CBIR. Besides providing multiple common and state-of-the-art retrieval mechanisms, LIRE allows for easy use on multiple platforms [12]. The main components of the LIRe system are given by Fig. 1 as a block diagram. These components are crawler, indexer, analyzer, and searcher.

The crawler craps a random n number of images from the Flickr dataset and saves them into a local directory. The analyzer extracts only one image feature at a time, i.e. local edge histogram. The third component is the indexer, which takes the extracted image features and build a Lucene based index for them. The last component is searcher or matcher, that takes the query image and calculates the similarity between the query and the images that are stored in that index. The application then displays the most k relevant images, where k is a number smaller than n.

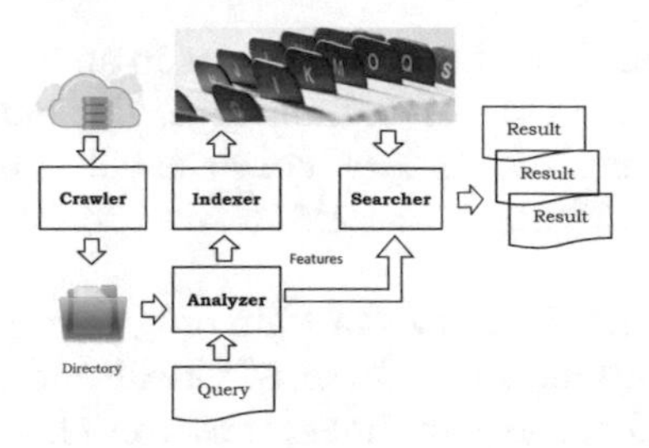

Fig. 1. LIRe block diagram.

The implementation presents two different modifications on LIRe system. On one hand, the crawler was modified, such that, it craps an n random images from an ImageNet [6] synset, given a synset Id and n from the settings page in the system. As a result, the downloaded images changed from Fig. 2(a) to (b). On the other hand, the indexer was extended to include three more image features. The newly added features are: (1) all edges histogram, (2) all edges histogram + CLD, (3) all edges histogram + FCTH.

Fig. 2. Crawler results (a) Random images from flickr database and (b) Random images from ImageNet synset n00017222 - Planet flora, planet life.

3 Image Descriptors

Visual image descriptors or image low-level features describe elementary characteristics such as shape, color, texture, edge or spatial location [7]. Feature extraction and representation are two important steps for multimedia processing [17]. The most commonly used structure to represent the global features of an image is the histogram because it is invariant to image translation and rotation [2,13]. In this paper, each of edge, color layout and fuzzy color texture histograms and the combination among them will be used the image retrieval task.

3.1 Edge Histogram Descriptor (EHD)

An image edge is an important low-level feature where it is used to describe both shape and texture features, which are essential elements for content-based image analysis [2,21]. The EHD describes 5 edge types in an image: horizontal, vertical, 45° and 135° diagonals and non-directional edge. The image space is divided into 16 (4×4) non-overlapping sub-images. For each sub-image, a histogram with 5 edge bins is generated, yielding a histogram with 80 (16×5) bins for the entire image [10,13,21,22].

However, using the local histogram bins only may not be sufficient to represent the edge distribution. Thus, both of global and semi-global edge distributions that were used for better image description and for higher retrieval accuracy. Global edge histogram is a 5 bins histogram. However, the semi-global histogram is a 65 (12×5) bins histogram. By using the combination of the three edge types, a 150 ($80+5+65$) bins edge histogram would be generated [10,13,21,22]. The distance between 2 combined edge histograms would be given by Eq. 1, such that, $L_{A,B}$, $G_{A,B}$ and $S_{A,B}$ are the histograms of local, global and semi-global edge respectively.

$$D_{Edge}(A,B) = \sum_{i=0}^{79} |L_A[i] - L_B[i]| + 5\sum_{i=0}^{4} |G_A[i] - G_B[i]| + \sum_{i=0}^{64} |S_A[i] - S_B[i]| \quad (1)$$

3.2 Color Layout Descriptor (CLD)

Color is one of the most important features in the image; it is defined subject to a particular color space [7]. For instance, the CLD is a compact descriptor that consists of representative colors. It is encoded using a non-overlay 8×8 sliding window. It could be generated by following 3 steps: (1) Extracting 3 color vectors with 84 bin. (2) Performing a Discrete Cosine Transform (DCT) on these vectors. (3) Applying a zigzag scanning [2,18]. Equation 2 gives the distance between 2 images using the CLD, such that, $Y_{A,B}$, $Cb_{A,B}$ and $Cr_{A,B}$ the 3 vectors. They are luminance, blue and red chrominance coefficients respectively.

$$\begin{aligned} D_{CLD}(A,B) &= \sqrt{\sum_{i=0}^{63} (Y_A[i] - Y_B[i])^2} \\ &+ \sqrt{\sum_{i=0}^{63} (Cb_A[i] - Cb_B[i])^2} + \sqrt{\sum_{i=0}^{63} (Cr_A[i] - Cr_B[i])^2} \end{aligned} \quad (2)$$

The CLD could be combined with other features for better retrieval accuracy: (1) Gabor filter in Wavelet transform domain for texture description [8]. (2) Dominant color (DCD) and scalable color (SCD) [18]. (3) EHD [2]. In this paper, CLD has been used with the combined edges histogram (Local + Global + Semi-Global). The similarity is calculated as in Eq. 3, such that, $D_{Edge}(A,B)$ and $D_{CLD}(A,B)$ are the distances shown in Eqs. 1 and 2 respectively.

$$D(A,B) = D_{Edge}(A,B) + D_{CLD}(A,B) \quad (3)$$

3.3 Fuzzy Color Texture Histogram (FCTH)

FCTH is in nature a combination of color and texture, that is used for accurate image retrieval systems. It represents an image with 192 bin histogram [5,9]. Distance between 2 images is given by Tanimoto coefficient in Eq. 4, where, A^T is the transpose vector of A [5].

$$D_{FCTH}(A,B) = \frac{A^T B}{A^T A + B^T B + A^T B} \tag{4}$$

Color and edge directivity descriptor (CEDD) is a combination between color and edge features [4]. It could be combined with FCTH for better retrieval accuracy. Kumar et al. [9] reached the lowest ANMRR (Averaged Normalized Modified Retrieval Rank) for that combination on Wang's database. In this paper, FCTH is used with combined edges (Local + Global + Semi-Global) histogram. Distance between the query image and the dataset is measured by Eq. 5, such that, $D_{Edge}(A,B)$ and $D_{FCTH}(A,B)$ are the distances at Eqs. 1 and 4 respectively.

$$D(A,B) = D_{Edge}(A,B) + D_{FCTH}(A,B) \tag{5}$$

4 Experimental Evaluation

This section gives an overview of the dataset and the selected queries used in experiments. as well as, the experimental results.

Table 1. Selected ImageNet synsets details.

Synset	Id	#Images	Sample
Animal Planet flora,	n00015388	400	
planet life	n00017222	400	
Owl	n01621127	100	
Elephant	n02503517	100	

4.1 Dataset and Queries

ImageNet is an image database organized according to the WordNet hierarchy, in which each node of the hierarchy has hundreds or thousands of images [6]. In this paper, experiments were done on four different synsets, that are given by Table 1. Synsets were downloaded from the public API[1], where the synset id or wnid is a user input. Queries for experiments are randomly selected that covers all the fours synsets Fig. 3 shows a sample of the selected queries.

[1] http://www.image-net.org/api/text/imagenet.synset.geturls?wnid=.

Fig. 3. A sample of the selected queries.

4.2 Experimental Results

The experiments include 13 different queries, each of them was tested against 6 algorithms. On one hand, single feature algorithms: CLD, EHD, and FCTH. On the other hand, combined features algorithms: All edges histogram, all edge histogram and CLD (Edge + Color), and all edges histogram and FCTH (Edge + FCTH). 10 users evaluated the results from their perspective. Results are shown in terms of precision Pr, recall R, F-measure and mean average precision mAP. Figures 4 and 5 show the average Pr and R for each query respectively. Moreover, Fig. 6 shows the average balanced F-measure for the selected queries.

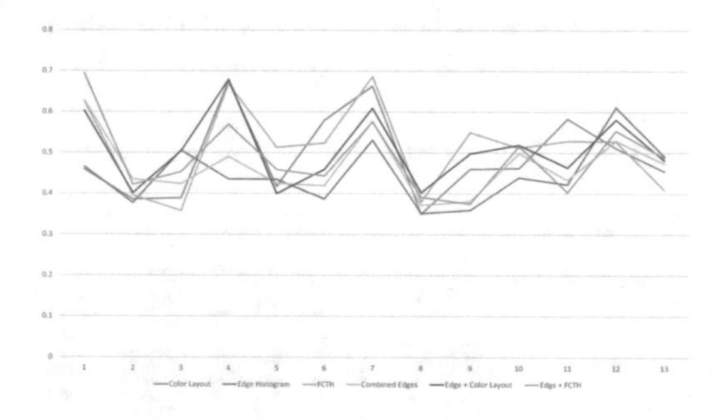

Fig. 4. Precision (pr) graph for 6 algorithms and 13 queries.

Query 1, 4 and 7 belong to Herbs and Elephant Synsets. There are many images that match the exact color and with duplicates, that's why the color feature exceeded either in Color Layout or (Edge + Color) combined. However, FCTH is secondly ranked, because it considers the texture as well. Thus, adding something extra feature like edge was not the best option while we have sufficient amount of images that could be retrieved by a single feature. Query 2 and 8 each of which is for a cat and giraffe respectively. As noticed the 6 algorithms gave their minimum Pr value, this is due to the lack of images for both of them, only 1 available match for giraffe and a 2 for a cat in the downloaded synsets. This means most of the retrieved results are irrelevant.

Nevertheless, Fig. 7. It gives the mAP for each one of the implemented algorithms. It is obvious that using multiple features like combined edges exceeded

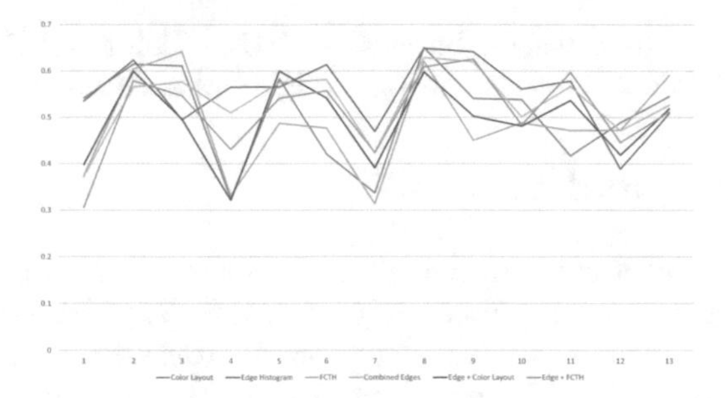

Fig. 5. Recall (R) graph for 6 algorithms and 13 queries.

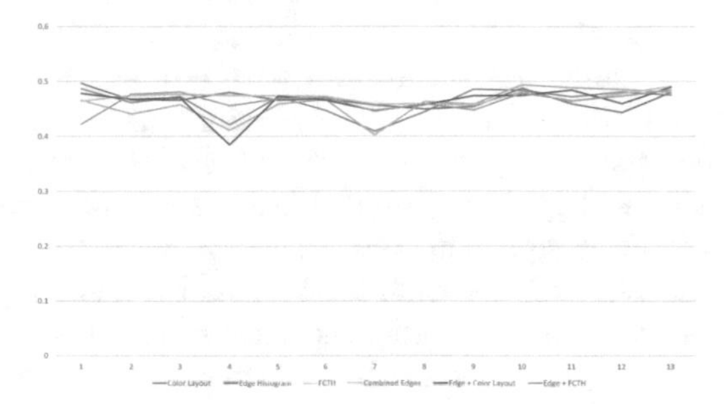

Fig. 6. F measure.

the local edge histogram. Moreover, edge + color layout exceeded both of color layout and local edge as a separate feature. However, the combination of edge and FCTH exceeded local edge histogram but retreated the FCTH.

Discussing the limitations of the implemented algorithms: Firstly, the single feature one are all fixed bins histogram (i.e. 150 bin EHD). This may not suit all types of images. Moreover, being a single low-level feature is totally insufficient for human prospective retrieval. Finally, the combined features have been

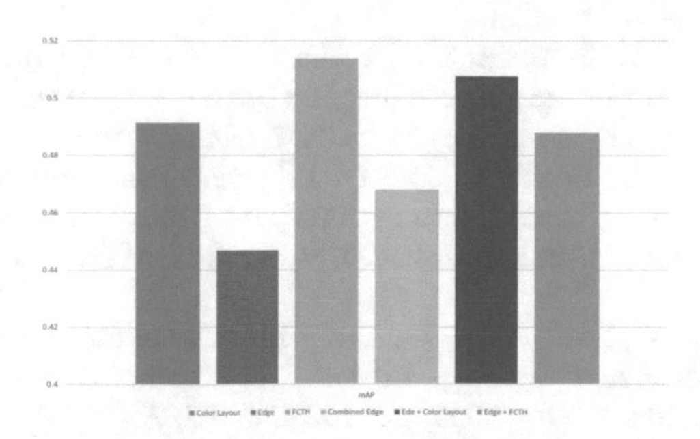

Fig. 7. mAP chart.

implemented on an equal weight basis, (i.e. EHD + Color = 50% and 50% Color layout). This didn't consider the various weights of each descriptor.

5 Conclusions and Future Work

This paper presented a comparative study between using single and combined features in image retrieval. Three single feature: *Edge histogram*, *CLD*, and *FCTH*, as well as, three combined features: *All edges histogram*, *All Edges + CLD* and *All edges + FCTH* have been implemented on LIRe system. The implementation also included a crawler modification, it is a replacement for Flickr dataset with ImageNet synsets, this modification aims at having a dataset that contains logical groups with a semantic meaning. Experiments for retrieval 13 different queries were done and the results were evaluated by 10 users. Moreover, retrieval metrics in terms of Pr, R, F-measure and mAP were calculated. The results showed that using combined features exceeded the single ones in image retrieval. As a result, using combined features in image retrieval task may give better results than using single ones. However, a semantic gap between the low-level image features and the high-level semantic meaning still exists. So, adding some sort of artificial intelligence like clustering the image dataset before retrieval then do the search on a particular cluster may lead to better results.

Acknowledgments. This work was done at the SWEP course as a part of the BioDialog project which is funded by DAAD at Friedrich Schiller University of Jena. Special thanks Prof. Birgitta König-Ries for her support.

References

1. Alzubi, A., Amira, A., Ramzan, N.: Semantic content-based image retrieval: a comprehensive study. J. Vis. Commun. Image Represent. **32**, 20–54 (2015)
2. Balasubramani, R., Kannan, D.V.: Efficient use of MPEG-7 color layout and edge histogram descriptors in CBIR systems. Glob. J. Comput. Sci. Technol. **9**(4) (2009)
3. Castelli, V., Bergman, L.D.: Image Databases: Search and Retrieval of Digital Imagery. Wiley, Hoboken (2004)
4. Chatzichristofis, S.A., Boutalis, Y.S.: CEDD: color and edge directivity descriptor: a compact descriptor for image indexing and retrieval. In: International Conference on Computer Vision Systems, pp. 312–322. Springer (2008)
5. Chatzichristofis, S.A., Boutalis, Y.S.: FCTH: fuzzy color and texture histogram-a low level feature for accurate image retrieval. In: Ninth International Workshop on Image Analysis for Multimedia Interactive Services, WIAMIS 2008, pp. 191–196. IEEE (2008)
6. Deng, J., Dong, W., Socher, R., Li, L.-J., Li, K., Fei-Fei, L.: ImageNet: a large-scale hierarchical image database. In: IEEE Conference on Computer Vision and Pattern Recognition, CVPR 2009, pp. 248–255. IEEE (2009)
7. Goshtasby, A.A.: Image registration methods. In: Image Registration, pp. 415–434. Springer (2012)
8. Jalab, H.A.: Image retrieval system based on color layout descriptor and Gabor filters. In: 2011 IEEE Conference on Open Systems (ICOS), pp. 32–36. IEEE (2011)

9. Kumar, P.P., Aparna, D.K., Rao, K.V.: Compact descriptors for accurate image indexing and retrieval: FCTH and CEDD. Int. J. Eng. Res. Technol. (UERT) **1**(8) (2012)
10. Lisin, D.A., Mattar, M.A., Blaschko, M.B., Learned-Miller, E.G., Benfield, M.C.: Combining local and global image features for object class recognition. In: IEEE Computer Society Conference on Computer Vision and Pattern Recognition-Workshops, CVPR Workshops, p. 47. IEEE (2005)
11. Liu, Y., Zhang, D., Lu, G., Ma, W.-Y.: A survey of content-based image retrieval with high-level semantics. Pattern Recognit. **40**(1), 262–282 (2007)
12. Lux, M.: Content based image retrieval with lire. In: Proceedings of the 19th ACM International Conference on Multimedia, pp. 735–738. ACM (2011)
13. Park, D.K., Jeon, Y.S., Won, C.S.: Efficient use of local edge histogram descriptor. In: Proceedings of the 2000 ACM Workshops on Multimedia, pp. 51–54. ACM (2000)
14. Smith, J.R., Chang, S.-F.: VisualSEEk: a fully automated content-based image query system. In: Proceedings of the Fourth ACM International Conference on Multimedia, pp. 87–98. ACM (1997)
15. Srinivas, M., Naidu, R.R., Sastry, C.S., Mohan, C.K.: Content based medical image retrieval using dictionary learning. Neurocomputing **168**, 880–895 (2015)
16. Takacs, D.: The idea of biodiversity: philosophies of paradise (1996)
17. Tian, D.: A review on image feature extraction and representation techniques. Int. J. Multimed. Ubiquitous Eng. **8**(4), 385–396 (2013)
18. Tsai, H.-H., Chang, B.-M., Lo, P.-S., Peng, J.-Y.: On the design of a color image retrieval method based on combined color descriptors and features. In: 2016 IEEE International Conference on Computer Communication and the Internet (ICCCI), pp. 392–395. IEEE (2016)
19. Vikhar, P., Karde, P., Thakare, V.: Comprehensive analysis of some recent competitive CBIR techniques. ICTACT J. Image Video Process. **7**(3) (2017)
20. Wang, J.Z., Li, J., Wiederhold, G.: Simplicity: semantics-sensitive integrated matching for picture libraries. IEEE Trans. Pattern Anal. Mach. Intell. **23**(9), 947–963 (2001)
21. Won, C.S.: Feature extraction and evaluation using edge histogram descriptor in MPEG-7. In: Pacific-Rim Conference on Multimedia, pp. 583–590. Springer (2004)
22. Won, C.S., Park, D.K., Park, S.-J.: Efficient use of MPEG-7 edge histogram descriptor. ETRI J. **24**(1), 23–30 (2002)

PLS-SEM in Information Systems Research: A Comprehensive Methodological Reference

Mostafa Al-Emran(✉), Vitaliy Mezhuyev, and Adzhar Kamaludin

Faculty of Computer Systems and Software Engineering, Universiti Malaysia Pahang, Gambang, Malaysia
malemran@buc.edu.om, {vitaliy,adzhar}@ump.edu.my

Abstract. The partial least squares-structural equation modelling (PLS-SEM) has become a key approach for validating the conceptual models across many disciplines in general, and the Information Systems (IS) in specific. This is guided through the assessment of the measurement and structural models. Several research articles were carried out to provide an extensive coverage of the usage and application of PLS-SEM. These articles were mainly concentrated on providing guidelines of how to use PLS-SEM in terms of reflective and formative measures, measurement and structural models, and the steps for analysing a particular conceptual model. Nevertheless, there are several steps and procedures that precede the evaluation of the measurement and structural models. The understanding of these steps and procedures is very important for many IS scholars, Ph.D. and Master students who are always struggling to find a comprehensive reference that could guide them through their research journey. Hence, the main contribution of this study is to a build a comprehensive methodological guideline of how the PLS-SEM approach can be employed in the context of IS adoption and acceptance, starting from the research design stage till the assessment of the measurement and structural models. This study may serve as a comprehensive reference for formulating the methodology in the IS adoption and acceptance related studies in the case of PLS-SEM employment.

Keywords: PLS-SEM · Information systems · Methodology · Guidelines

1 Introduction

Structural equation modelling (SEM) is a family of statistical techniques that has become very common in research studies. There are two types of SEM: the Covariance-Based SEM (CB-SEM) and the Partial Least Squares-SEM (PLS-SEM) [1]. The primary focus of this study is the PLS-SEM. As indicated by [2], PLS-SEM is considered the "most fully developed and general system" and it was called a "silver bullet" [3]. In PLS-SEM, there are two main components: the measurement model and the structural model. The former is basically referred to the "outer model" where it measures the relationships between the latent construct (i.e., variable or factor) and its relevant indicators (i.e., items or measures). The latter refers to the "inner model" where it measures the relationships between the constructs themselves [3].

A. E. Hassanien et al. (Eds.): AISI 2018, AISC 845, pp. 644–653, 2019.
https://doi.org/10.1007/978-3-319-99010-1_59

Various studies were contributed to the existing body of literature with respect to the application of PLS-SEM approach in the Information Systems (IS) research. A study conducted by [1] discusses and reviews the PLS-SEM development and provides a better understanding of the usage of PLS-SEM in a new technology research. In addition, it also provides a guideline of how to report and interpret PLS-SEM results in research studies. Another study, conducted by [4] reviews the PLS-SEM applications in IS research studies that were published between 2010 and 2014. A study carried out by [5] discusses and examines the application of PLS-SEM in all the empirical studies that were published between 1992 and 2011 in MIS Quarterly. It is apparent that the main reason for employing the PLS-SEM in these articles is the applicability of this technique in analysing small sample sizes, non-normal data distribution, and its ability to work efficiently in exploratory research studies.

Although research studies conducted by [1, 4, 5] have provided substantial understanding about the usage and application of PLS-SEM in IS research, there are some other issues that need to be considered. For instance, these studies were focused on providing guidelines of how to use PLS-SEM in terms of reflective and formative measures, measurement and structural models, and the steps for analysing a particular conceptual model. However, there are several steps and procedures that precede the

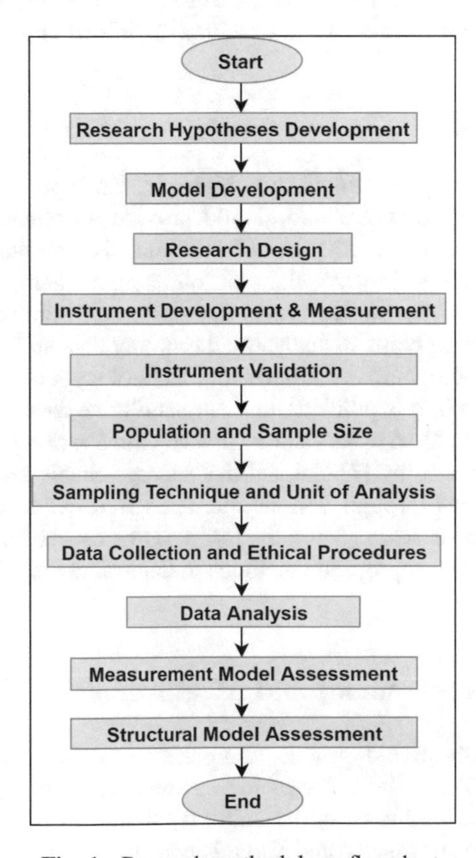

Fig. 1. Research methodology flowchart.

evaluation of the measurement and structural models. The understanding of these steps and procedures is regarded as an obstacle for many IS researchers, PhD and Master students who are always struggling with finding a particular reference that could guide them through their research journey. Based on the existing literature, there is no specific study that was dedicated to handle this issue. Thus, the main contribution of this study is to a build a guideline of how the PLS-SEM approach will be employed in the context of IS adoption and acceptance starting from the research design stage till the assessment of the measurement and structural models. In order to study the adoption and acceptance of a particular technology through the employment of PLS-SEM approach, researchers could refer to the flowchart presented in Fig. 1, which serves as a general guideline for conducting the research methodology in such type of studies.

The rest of the paper will provide a comprehensive explanation of each component in Fig. 1 starting from the research design till the assessment of the measurement and structural models. In that, the study will tackle the research design, instrument development and measurement, and instrument validation (pre-testing and pilot study). Additionally, the identification of the population, sample size, sampling technique, and the units of analysis will be demonstrated. Moreover, data collection and ethical procedures will be described. Besides, data analysis, descriptive analysis, and reliability analysis will be addressed. Furthermore, the SEM and the PLS-SEM will be described. The assessment of the measurement and structural models will be demonstrated in detail.

2 Research Design

In this section, the study should clearly highlight the type of research used (i.e., quantitative, qualitative, or both) and should provide a comprehensive justification about the used method. Since the focus of this study is quantitative, in general, and using questionnaire surveys in particular, the following are some examples about how to justify the usage of a quantitative survey approach. According to [6], it is suggested that quantitative approach helps in analysing the survey data and examining the factors that affect the users' acceptance of a particular technology. As compared to the qualitative research approach, it is pointed out that quantitative data is more efficient as the research hypotheses could be tested and evaluated using various statistical techniques. Besides, a study conducted by [7] claimed that surveys enable scholars to measure the respondents' perceptions through various questions in order to indicate their assessment for a measurement scale. Additionally, a study carried out by [8] stated that surveys are considered as appropriate methods to determine the relationship among the study factors.

3 Instrument Development and Measurement

In this section, the study should provide comprehensive details about the questionnaire survey components. Typically, most of the surveys start with questions that collect demographic data, and in this case, the author(s) should illustrate those data such as gender, age among many others, and should indicate the type of measurement used (i.e., ordinal, nominal, interval, and ratio). In terms of model/theory factors, the study should highlight the factors along with the sources that these factors were adopted from. Similarly, the author(s) should indicate the type of measurement scale used for those factors (e.g., a five-point Likert scale ranging from "1 = strongly disagree" to "5 = strongly agree").

4 Instrument Validation

An instrument should be reliable before it can be considered valid. However, reliability should not rely on validity [9]. Validity refers to the notion of how thoroughly an instrument measures what it is assumed to measure. Reliability refers to the consistency of a measure used in the study [10]. There are two consecutive stages of instrument validations; these are pre-testing and pilot study.

4.1 Pre-testing

In this stage, the researcher(s) should select some academic experts as reviewers to evaluate the developed instrument. The experts will be requested to evaluate the corresponding items to each factor in terms of easiness in understanding, clarity, and the relevance to the study objectives. Based on the experts' comments and feedback, the instrument will be modified and improved.

4.2 Pilot Study

Pilot study represents the second stage of instrument validation. The pilot study is an essential stage before conducting the final study [10]. In this section, the study should provide brief details about where the study will be conducted, who are the participants of the study, and how the survey will reach the participants (i.e., self-administration, by email, or both). Based on [11], the minimum number of the pilot study participants is 30. In this stage, the author(s) need(s) to measure the internal consistency of the items through the usage of Cronbach's Alpha in order to ensure reliability. According to [12], the Cronbach's Alpha (a) values should be above 0.7 in order to be accepted. Thus, if the scale reliability is confirmed, this will lead to the final version of the instrument that would be ready for distribution. Otherwise, the author(s) should manipulate the instrument and repeat these stages again.

5 Population and Sample Size

The population is a set that takes into account a comprehensive group of individuals, certain events, or a particular interest that needs further examination based on a particular problem [10]. In this section, the author(s) should highlight the exact population of the study (e.g., "IT undergraduate students enrolled at Universiti Malaysia Pahang") and should provide reasonable arguments about the selection of this population. In addition, the study should clearly indicate the total number of the targeted population, which could be provided by the statistics department at the targeted population (if possible) in order to determine the sample size.

It is not possible to collect data from all the individuals in the population, sampling is used instead. Sampling involves gathering data from a sufficient number of individuals in the population for the purpose of popularising the results to the whole population [12]. Besides, the usage of a small sample size will lead to lower accurate results specifically when statistical data analysis such as SEM will be employed [12]. In order to determine the sample size, it is suggested to follow the guidelines provided by [13] which specify the sample size for any given defined population.

6 Sampling Technique and Definition of the Units of Analysis

In this section, the study should clearly determine the type of sampling technique and should provide sufficient justification for the selection of a particular technique. To the best of our knowledge, there are two types of sampling techniques: probability and non-probability sampling techniques. The probability sampling includes four techniques: simple random sampling, stratified sampling, cluster sampling, and systematic sampling. Additionally, the non-probability sampling includes four techniques as well: convenience sampling, quota sampling, judgment or purposive sampling, and snowball sampling. Typically, the sampling technique type is determined based on the availability of the participants and the study nature. With respect to the unit of analysis, the study should indicate the units of analysis used in order to be obvious to the reader. Basically, there are two types of units: individual-level and organizational-level.

7 Data Collection and Ethical Procedures

In this section, the study should provide an adequate amount of information about the data collection issues and ethical procedures that need to be followed. That is, the author(s) should clearly indicate the method used for data collection (e.g., "questionnaire survey"), the period of data collection (e.g., "first semester of the academic year 2017–2018"), the duration of data collection (e.g., "four months"). In addition, the researcher(s) should be provided with a consent letter from the targeted party as a prior step to the data collection. Finally, the researcher(s) will proceed in the data collection process.

8 Data Analysis

In this section, the author(s) should determine how the data will be analysed. For instance, in order to draw a complete picture about the respondents' demographics, descriptive analysis of the data should be carried out using SPSS. This includes the frequency, mean, and standard deviations of the variables. In addition, preceding to the data analysis stage, the data screening process should be fulfilled. This includes missing data, outliers, normality, and multicollinearity. Moreover, the relationships among the factors in the conceptual model will be examined and tested using SmartPLS [14]. In terms of reliability, the internal consistency among the items within the same construct should be measured using the Cronbach's Alpha coefficient (a). As stated by [12], the Cronbach's Alpha (a) values should be greater than 0.7 in order to be accepted.

9 Structural Equation Modelling (SEM)

Structural equation modelling (SEM) is a family of statistical techniques that has become very common in research studies. SEM has become very useful for a large number of research studies due to its ability to model the latent variables, to take into consideration different types of measurement errors, and to test the entire theories [1]. In addition, SEM is an example of multivariate analysis and a statistical technique for testing and examining a set of hypotheses relationships between various independent and dependent factors [15]. Besides, SEM allows scholars to answer a set of interconnected research questions in a single, systematic, and comprehensive analysis by examining the relationships between the independent and dependent factors [15]. Furthermore, SEM enables the combined analysis of both (measurement and structural models). Hence, measurement errors of the detected constructs will be analysed as an integral part of the model. Besides, factor analysis is integrated into one process with the hypotheses testing [15]. Based on the above and according to the wide usage of SEM in the IS literature [15], this study suggests employing the SEM to test and examine the hypothesized relationships between the factors in any particular conceptual model.

According to [16, 17], there are two types of SEM: the CB-SEM and the PLS-SEM. The former is mainly used to confirm or reject theories (i.e., a set of systematic relationships among various factors that could be examined using empirical procedures). On the other side, PLS-SEM (i.e., PLS path modelling) is mainly used to develop theories in exploratory research. This is usually performed by concentrating on understanding the variance in the dependent variables when examining the model. In order to answer the question of when to use CB-SEM or PLS-SEM, scholars should concentrate on the objectives that differentiate the two techniques.

10 Partial Least Squares (PLS)

The PLS-SEM is considered as a variance-based technique to SEM. PLS-SEM is typically used when the research objective is theory development and prediction of constructs [16, 17]. PLS-SEM works efficiently with complex models and makes no assumptions about the data distribution (i.e., assumes that the data is non-normally distributed). Moreover, PLS-SEM can simply handle reflective and formative measurement models, as well as constructs with single and multi-item measures, without any identification problems. PLS-SEM has a higher statistical power than CB-SEM, in which PLS-SEM is more capable to render a particular relationship as positive when it is, in fact, positive in the population.

PLS-SEM is an appealing alternative to CB-SEM specifically when there is a slight prior knowledge on the structural model relationships or the measurement of the constructs or when the emphasis is more on exploration than confirmation [16, 17]. PLS-SEM is capable to model both factors and composites. This is appreciated by various scholars in various contexts, which makes the PLS-SEM a promising technique specifically for new technology and information systems research [1]. In the same vein, a study carried out by [18] provided an evidence through a systematic literature review that there is an increasing usage of PLS in IS research studies. Similarly, a study conducted by [4] indicated that PLS-SEM acceptance and continued usage in the IS research render the PLS-SEM as an accepted method in the IS context, specifically when applied for exploratory research, model complexity, and theory development. Furthermore, PLS-SEM is regarded as the appropriate approach if the research objective is an extension of an existing structural theory [3].

11 Assessment of Measurement and Structural Models

Model assessment delivers empirical estimations of the measurement model (i.e., the relationships between the indicators and the constructs) and the structural model (i.e., the relationships between constructs). This enables users to compare the theoretically formed measurement and structural models with reality, as characterized by the sample data. In that, we can detect how well the theory fits the data [16, 17]. A two-step approach is suggested to assess any particular conceptual model. The first one related to the assessment of the measurement model, while the second one concerned with the assessment of the structural model. The techniques used to assess each step are demonstrated in the following sub-sections.

11.1 Measurement Model Assessment

In this stage, the measurement model assessment includes two main techniques: the convergent validity and discriminant validity [16, 17].

11.1.1 Convergent Validity

With reference to [16, 17], the first criterion to be assessed is basically the internal consistency reliability. The common criterion for the internal consistency is the Cronbach's Alpha, in which, it provides a measure of the reliability according to the intercorrelations of the observed indicator variables. Due to the limitations of Cronbach's Alpha, it is more suitable to use another measure of internal consistency reliability, which is known as composite reliability. Composite reliability takes into consideration the different outer loadings of the indicator variables. Generally, the Cronbach's Alpha and composite reliability values are ranged between "0 and 1", in which higher values referred to higher reliability levels. More specifically, in exploratory research, values of 0.60 to 0.70 can be regarded as acceptable, while in more advanced research, values of 0.70 to 0.90 are considered as satisfactory.

Convergent validity refers to "the extent to which a measure correlates positively with alternative measures of the same construct" [16, 17]. In order to assess the convergent validity of reflective constructs, research studies should consider the outer loadings (i.e., indicator reliability) of the measures (i.e., indicators) and the Average Variance Extracted (AVE). Higher outer loadings on a construct reveal that the related measures have much in common, which is represented by the construct. The outer loadings values should be ≥ 0.70 in order to be acceptable [16, 17]. The AVE refers to "the grand mean value of the squared loadings of the indicators associated with the construct (i.e., the sum of the squared loadings divided by the number of indicators)" [16, 17]. The AVE values should be ≥ 0.50 in order to be accepted.

11.1.2 Discriminant Validity

Discriminant validity refers to "the extent to which a construct is truly distinct from other constructs by empirical standards" [16, 17]. In that, confirming the discriminant validity indicates that the variable is unique and represents phenomena that is not captured by other variables in the model. According to [16, 17], research studies should depend on two measures in order to assess the discriminant validity. Those measures are cross-loadings and the Fornell-Larcker criterion.

In terms of cross-loadings, an indicator's outer loading on the related construct should be greater than all its cross-loadings (i.e., its correlation) on other constructs. With regard to the Fornell-Larcker criterion, it compares the square root of the AVE values with the correlations of the latent variable. More precisely, the square root of each construct's AVE must be greater than its highest correlation with other constructs [16, 17]. Recently, a study conducted by [19] has proposed a new criterion for assessing the discriminant validity which is so-called Heterotrait–Monotrait ratio (HTMT) of correlations. In simple words, HTMT is the mean of all correlations of indicators across constructs measuring different constructs. According to the suggested threshold level of HTMT by [19], when the constructs in the path model are conceptually distinct, a value of <0.85 is regarded as warranted.

11.2 Structural Model Assessment

Once we confirm the assessment results of the measurement model, the next step is to assess the structural model. In that, the model's predictive capabilities and the relationships among the constructs will be examined. According to [16, 17], the main criteria for assessing the structural model in PLS-SEM involve the significance of the path coefficients, the level of R^2 values, the f^2 effect size, and the predictive relevance Q^2.

The path coefficients refer to the hypothesized relationships between the constructs. Values of path coefficients range between -1 and $+1$, where values close to $+1$ indicate strong positive relationships. Typically, most scholars use p values to assess the significance level. The p-value should be <0.05 when the assumption of the significance level is 5% in order to reveal that the relationship under observation is significant. Thus, the interpretation of the path model results is based on the assessment of the significance of all the structural model relationships using t and p values and the confidence intervals [16, 17].

The coefficient of determination (R^2) is regarded as the most commonly utilized measure to assess the structural model. This coefficient is a measure of the model's predictive power. Values of R^2 range between 0 and 1, where higher values represent higher accuracy levels of prediction [16, 17]. With respect to [20], values of 0.67, 0.33, and 0.19 are regarded as substantial, moderate, and weak, respectively. The effect size (f^2) refers to the change in the R^2 value when a particular exogenous construct is removed from the model could be used to assess whether the removed construct has a significant effect on the endogenous constructs [16, 17]. In terms of effect size (f^2) assessment, as indicated by [21], values of 0.02, 0.15, and 0.35 indicate small, medium, and large effects, respectively. The predictive relevance (Q^2) measure is an indicator of the model's out-of-sample predictive power. Values of $Q^2 > 0$ indicate that the exogenous constructs have predictive relevance for the endogenous construct under consideration. There are two different techniques for calculating the Q^2 value. The first one refers to the cross-validated redundancy technique, while the second one refers to the cross-validated communality. The technique that perfectly fits the PLS-SEM approach is the cross-validated redundancy as it builds on the path model measures of both (the structural and measurement models) of data prediction [16, 17].

12 Conclusion

The PLS-SEM has become a common approach for validating the conceptual models across many disciplines. In the IS context, several research studies were carried out to provide an extensive coverage about the usage and application of PLS-SEM [1, 4, 5]. These studies were mainly concentrated on providing guidelines of how to use PLS-SEM in terms of reflective and formative measures, measurement and structural models, and the steps for analysing a particular conceptual model. However, there are several steps and procedures that precede the evaluation of the measurement and structural models. The understanding of these steps and procedures is regarded as an obstacle for many IS scholars, Ph.D. and Master students who are always struggling with finding a particular reference that could guide them through their research journey. To the best of our knowledge, there is no specific study that was dedicated to handle this issue. This study has contributed to the existing literature by introducing a guideline of how PLS-SEM approach can be employed in the context of IS adoption and acceptance starting from the research design stage till the assessment of the measurement and structural models. It is important that scholars should understand the purpose of their study and apply each approach illustrated accordingly. In simple words, this study would serve as a comprehensive reference for formulating the methodology in the IS adoption and acceptance studies in the case of PLS-SEM employment. As a future work, this study might be expanded to include a case study in order to illustrate each component presented in the developed framework.

Acknowledgement. This research study was made under the support of the Faculty of Computer Systems and Software Engineering, Universiti Malaysia Pahang, and it was partially supported by the RDU170301 grant of Universiti Malaysia Pahang, Malaysia.

References

1. Henseler, J., Hubona, G., Ray, P.A.: Using PLS path modeling in new technology research: updated guidelines. Ind. Manag. Data Syst. **116**(1), 2–20 (2016)
2. McDonald, R.P.: Path analysis with composite variables. Multivar. Behav. Res. **31**(2), 239–270 (1996)
3. Hair, J.F., Ringle, C.M., Sarstedt, M.: PLS-SEM: indeed a silver bullet. J. Mark. Theory Pract. **19**(2), 139–152 (2011)
4. Hair, J., Hollingsworth, C.L., Randolph, A.B., Chong, A.Y.L.: An updated and expanded assessment of PLS-SEM in information systems research. Ind. Manag. Data Syst. **117**(3), 442–458 (2017)
5. Ringle, C.M., Sarstedt, M., Straub, D.: A critical look at the use of PLS-SEM. MIS Q. **36**(1), iii–xiv (2012)
6. Huang, Y.: Empirical analysis on factors impacting mobile learning acceptance in higher engineering education. Doctoral dissertation, University of Tennessee (2014)
7. Punch, K.F.: Introduction to Social Research: Quantitative and Qualitative Approaches, vol. 43, no. 3 (1998)
8. Malhotra, M.K., Grover, V.: An assessment of survey research in POM: from constructs to theory. J. Oper. Manag. **16**(4), 407–425 (1998)
9. Kimberlin, C.L., Winterstein, A.G.: Validity and reliability of measurement instruments used in research. Am. J. Health Syst. Pharm. **65**(23), 2276–2284 (2008)
10. Sekaran, U., Bougie, R.: Research Methods for Business: A Skill Building Approach, vol. 26 (2010)
11. Hunt, S.D., Sparkman Jr., R.D., Wilcox, J.B.: The pretest in survey research: issues and preliminary findings. J. Mark. Res. **19**(2), 269–273 (1982)
12. Hair, J.F., Black, W.C., Babin, B.J., Anderson, R.E., Tatham, R.L.: Multivariate Data Analysis (2010)
13. Krejcie, R.V., Morgan, D.W.: Determining sample size for research activities. Educ. Psychol. Meas. **30**(3), 607–610 (1970)
14. Ringle, C.M., Wende, S., Becker, J.-M.: SmartPLS 3. Bönningstedt: SmartPLS (2015)
15. Gefen, D., Straub, D., Boudreau, M.-C.: Structural equation modeling and regression: Guidelines for research practice. Commun. Assoc. Inf. Syst. **4**, 1–77 (2000)
16. Hair, J.F., Hult, G.T.M., Ringle, C., Sarstedt, M.: A Primer on Partial Least Squares Structural Equation Modeling (PLS-SEM). Sage Publications, Thousand Oaks (2014)
17. Hair Jr., J.F., Hult, G.T.M., Ringle, C., Sarstedt, M.: A Primer on Partial Least Squares Structural Equation Modeling (PLS-SEM). Sage Publications, Thousand Oaks (2016)
18. Urbach, N., Ahlemann, F.: Structural equation modeling in information systems research using partial least squares. J. Clean. Prod. Inf. Technol. Theory Appl. **11**(2), 5–40 (2010)
19. Henseler, J., Ringle, C.M., Sarstedt, M.: A new criterion for assessing discriminant validity in variance-based structural equation modeling. J. Acad. Mark. Sci. **43**(1), 115–135 (2015)
20. Chin, W.W.: The partial least squares approach to structural equation modeling. Mod. Methods Bus. Res. **295**(2), 295–336 (1998)
21. Cohen, J.: Statistical power analysis for the behavioral sciences. Stat. Power Anal. Behav. Sci. **2**, 567 (1988)

Automatic Segmentation of Chromosome Cells

Reem Bashmail[1(✉)], Lamiaa A. Elrefaei[1,2], and Wadee Alhalabi[1,3]

[1] Computer Science Department, Faculty of Computing and Information Technology, King Abdulaziz University, Jeddah, Saudi Arabia
{rbashmail,laelrefaei,wsalhalabi}@kau.edu.sa

[2] Electrical Engineering Department, Faculty of Engineering at Shoubra, Benha University, Cairo, Egypt
lamia.alrefaai@feng.bu.edu.eg

[3] Computer Science Department, College of Engineering, Effat University, Jeddah, Saudi Arabia
walhalabi@effatuniversity.edu.sa

Abstract. Chromosome's segmentation is an essential step in the automated chromosome classification system. It is important for chromosomes to be separated from noise or background before the identification and classification. Chromosomes image (Metaphase) is generated in the third phase of mitosis. During metaphase, the cell's chromosomes arrange themselves in the middle of the cell through a cellular. The analysis of metaphase chromosomes is one of the essential tools of cancer studies and cytogenetics. The Chromosomes are thickened and highly twisted in metaphase which make them very appropriate for visual analysis to determine the kind of each chromosome within the 24 classes (Chromosome karyotyping). This paper represents a chromosome segmentation method of high-resolution digitized metaphase images. Segmentation is done using Difference of Gaussian (DoG) as a sharpening filter before the classic technique (Otsu's thresholding followed by morphological operations). The proposed method is tested using 130 metaphase images (6011 chromosomes) provided by The Diagnostic Genomic Medicine Unit (DGMU) laboratory at King Abdulaziz University. The experimental results show that the proposed method can successfully segment the metaphase chromosome images with 99.8% segmentation accuracy.

Keywords: Genetic · Chromosome image · Chromosome segmentation Metaphase image

1 Introduction

Human chromosome's karyotyping is a standard process to detect fatal diseases and genetic abnormalities. Chromosome's segmentation is the first step of karyotyping; this step is done by selecting chromosomes (22 pairs, X, and Y) in the metaphase stage from the nucleus of the cell picture. Then, the segmented chromosomes are classified into 24 types. Figure 1 shows a chromosome spread image (metaphase), and its Karyotype [1].

Although there are computer software and laboratory devices to fully automate this process, it is still done manually by laboratory technicians. An automated system usually consists of four main steps: (1) enhancing the metaphase image, (2) segmentation and

A. E. Hassanien et al. (Eds.): AISI 2018, AISC 845, pp. 654–663, 2019.
https://doi.org/10.1007/978-3-319-99010-1_60

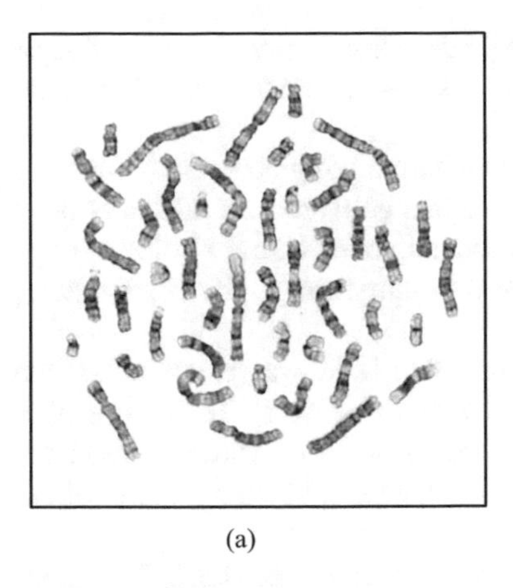

(a)

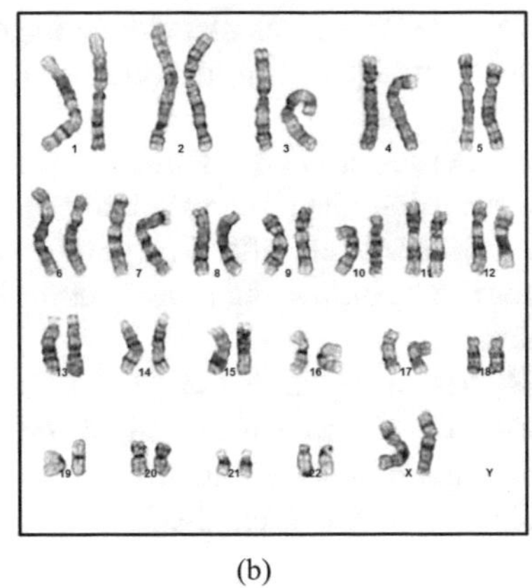

(b)

Fig. 1. (a) Metaphase image for normal female sample and (b) its Karyotype

aligning of chromosomes, (3) selecting the feature and (4) classification of chromosomes [2]. The chromosomes segmentation is an essential step because its results will affect the whole system performance.

This paper is concerned with automatic segmentation of high resolution, free overlap and separated digitized metaphase images. Mainly, the proposed algorithm is based on the classic technique, thresholding followed by a set of morphological operations. But with the addition of "Difference of Gaussian (DoG)" filtering step before thresholding.

The rest of this paper is structured as follows; Sect. 2 presents the related work. A detailed description of the proposed chromosome segmentation method is explained in Sect. 3. Section 4 illustrates experiments and results. Finally, Conclusions and future work is provided in Sect. 5.

2 Related Work

Segmenting the metaphase chromosome image into background and non-background (chromosomes) is accomplished usually by thresholding [3]. There are some limitations of this operation in its simplest form, it cannot be applied to multi-channel images and only two classes are generated. Thresholding, in addition, does not take into account the image spatial characteristics, which may cause its sensitivity to intensity and noise in homogeneities. All of these limitations corrupt the image histogram, making process of separation more difficult. For all the previous reasons, many variations of the classical thresholding algorithms have been proposed for medical image segmentation that based on local intensities and connectivity information [4].

Guimaraes et al. [5] mentioned that since their proposed method is to extract the characteristics of the chromosome from its outline, it must be working on binary image contains black chromosome on white background which will be prepared in the pre-processing step. Accordingly, the preprocessing includes filtering and binarization. The median filter was applied for filtering and the global thresholding method for the binarization.

The authors in [6], calculated the image grads using Canny edge detection operator. Then, segmented the chromosomes using a series of mathematical morphological operations.

Madian et al. in [7], presented a "Contour-Based" Segmentation; Image binarization is done using Otsu's thresholding method that converts automatically the original image to binary. Then, after image binarization, a series of morphological operations are performed by filling the holes present in the binary image, the binarization process completes.

Ji [8] has an opinion that if a very low threshold was used, holes and noisy boundaries of chromosomes might be cluster the result. To reduce the noise, rethresholding, boundary dilation or erosion could be used. These procedures will be automatically applied after the high-frequency analysis of components of boundary curvature.

Stanley et al. [9] presents three steps process for segmentation. First one is providing an under-segmentation of the image, an automatically global threshold generated and applied to the entire gray level input image. Second, an automated local thresholding procedure was used for each connected component of the globally thresholded image to facilitate object separation. Finally, using connected components analysis to label all objects. Each segmented image is the orthogonal line outer boundary that constructed from the skeleton.

In the research work by Sathyan et al. [10], segmentation is done based on algorithm of labelling 4-connectivity region. Name each pixel set of connected area to a single label, the small isolated pixels are removed, and bounding box is applied to separate each region.

Keerthi et al. [11] used the concept of connected component for segmentation. Because the 8-connected neighbors give better results than 4-connected neighbors, it was selected. Then each connected component was labeled.

In [12], the authors proposed a method of general locally adaptive thresholding using neighborhood processing. This method uses two of global thresholds, obtained from the local threshold based on neighborhood means and variances and of a global intensity distribution.

In [13], the segmentation is achieved by improved Sobel edge detection operator, with a value of global threshold which is determined by the fitness function optimization via a genetic algorithm.

Ji [14] proceeded segmentation by splitting a big cluster recursively (for each cluster) by contrast into individual chromosomes, and iteratively (for the whole cell) application of, split and evaluate hypothesis procedures. The evaluation of the result may be to confirm the segmentation or to reject either a single split or even a cell.

In this paper, we propose applying an additional step, Difference of Gaussian (DoG), before the Otsu's thresholding that followed by series of mathematical morphologies to segment the chromosome images.

3 The Proposed Chromosome Segmentation Method

The proposed chromosome segmentation method, as shown in Fig. 2, has six steps:

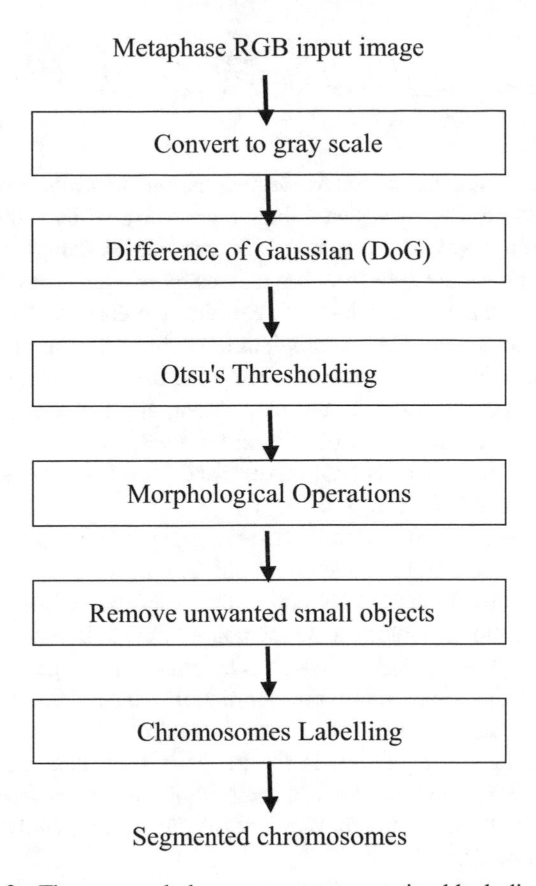

Fig. 2. The proposed chromosome segmentation block diagram

Step 1: The metaphase RGB color input image is converted to a grayscale image.

Step 2: The grayscale image $I(x, y)$ is convolved with the Difference of Gaussian (DoG) sharpening filter to produce image $g(x, y)$ using Eq. (1) [15]:

$$g(x, y) = I(x, y) * DoG \tag{1}$$

Where * is the convolution operator and *DoG* is the Difference of Gaussians filter. It is an algorithm for feature enhancement, which subtracts first blurred version of the image from a second version (less blurred). These two blurred versions of images were obtained by convolution of the original grayscale image with Gaussian kernels using different values of standard deviations. The Gaussian kernel that used for blurring the images removes the high-frequency spatial information. By subtracting one blurred image from another, the spatial information that exist between the two frequencies ranges of the blurred images is preserved. Thus, the Difference of Gaussians is a band-pass filter that keeps only the spatial frequencies that are present in the original grayscale image which are often associated to the edges. *DoG* is defined as in Eq. (2) [16]:

$$DoG = \frac{1}{\sqrt{2\pi}}\left(\frac{1}{\sigma_1}exp^{\left(-\frac{(x^2+y^2)}{2\sigma_1^2}\right)} - \frac{1}{\sigma_2}exp^{\left(-\frac{(x^2+y^2)}{2\sigma_2^2}\right)}\right) \quad (2)$$

Where σ_1 and σ_2 are the standard deviations and controls how "fat" the kernel function is going to be; higher sigma values blur over a wider radius.

Step 3: The image $g(x, y)$ from step 2 is thresholded using Otsu's method [17], where the image's pixels are split into two classes by using the best threshold (t) value through the variance maximum value between the two classes [18, 19].

Step 4: Apply some basic Morphological Operations on the binarized image from step 3. Here dilation, hole filling and erosion were used to enlarge the areas of foreground pixels (reduce the gaps in outline). Then, fill the small holes and reduce boundaries thickens for foreground objects in the image.

Step 5: Remove small objects (<50 pixels) and label the segmented chromosomes using 8-connected neighbors analysis.

Step 6: Apply the final labeled chromosomes image as a mask to the original image and store the segmented chromosomes in a cell array.

Figure 3 shows the output of each step of the proposed chromosome segmentation method. Figure 3(a) shows the grayscale metaphase image. In Fig. 3(b), the Difference of Gaussian (DoG) filtered image is shown. The binarized image using Otsu's method is shown in Fig. 3(c), where there are small holes inside the outlined regions of chromosomes. The output of dilation, filling holes followed by erosion to avoid the possibility of touching chromosomes in the image is shown in Fig. 3(d). The labeled chromosomes image is used as a mask to the original image, as shown in Fig. 3(e), to store the segmented chromosomes in a cell array. Figure 3(f) shows some samples of the segmented chromosomes.

4 Experiments and Results

4.1 Tools and Chromosomes Digital Images Dataset

The proposed chromosome segmentation method is implemented using MATLAB 2018a on a laptop having a processor Intel (R) Core™ i7-4770S CPU @ 3.10 GHz 3.09 GHz and 12 GB RAM.

A peripheral blood samples is obtained from patients by lab technicians in the Diagnostic Genomic Medicine Unit (DGMU) laboratory [20] at King Abdulaziz University based on a standard protocol. Recently, a large number of results obtained from the clinical diagnosis were stored in the DGMU database and all samples have consent to be used for public research. For this study, 130 metaphase images (6011 chromosomes) provided by the DGMU laboratory technicians that obtained from 109 patients are used. These images are selected to be free of overlap and have severe bending.

Chromosomes digital images are recorded by Olympus DP-10 Digital Camera, which has a unique 1.8-in. liquid crystal (LCD) color monitor. It provides the true color of 24-bit reproduction and sharp 1280 × 1024-pixel quality resolution in optical

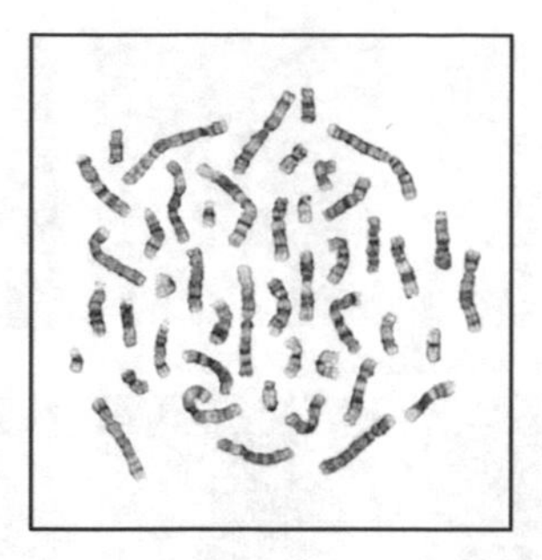

(a) The grayscale metaphase image

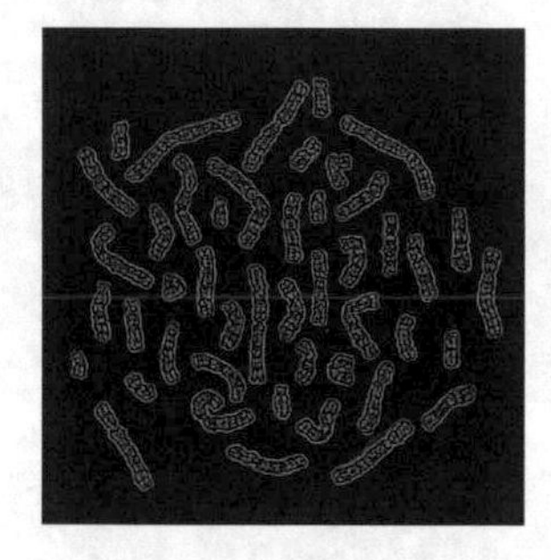

(b) Differnce of Gaussian (DoG) filtered image

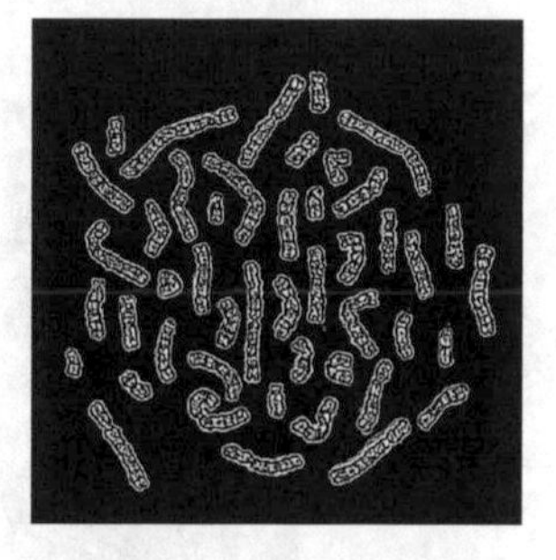

(C) Theresholded image using Otsu's method

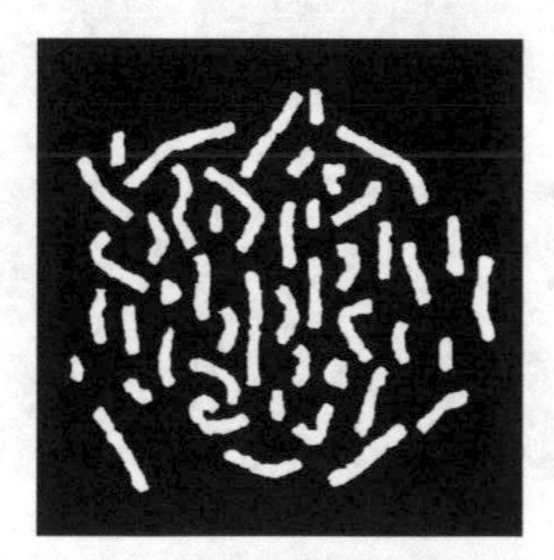

(d) Image after morphological operations (dilation, hole filling and erosion)

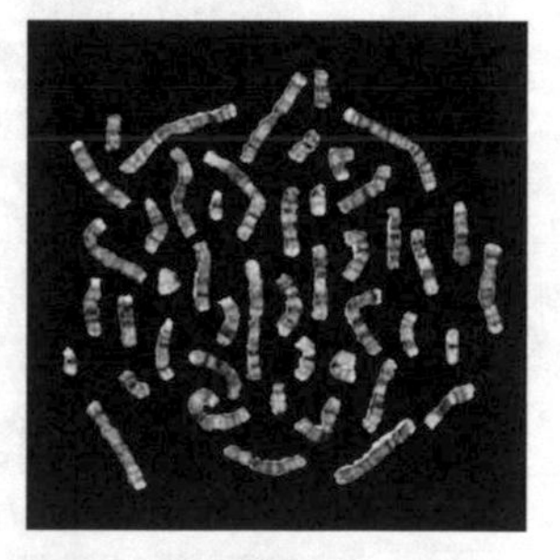

(e) Masked labeled segmented chromosomes image

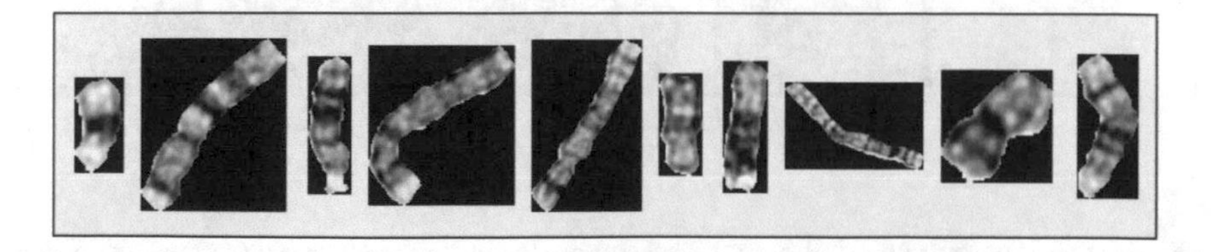

(f) Sample of the segmented chromosomes

Fig. 3. The proposed chromosome segmentation steps

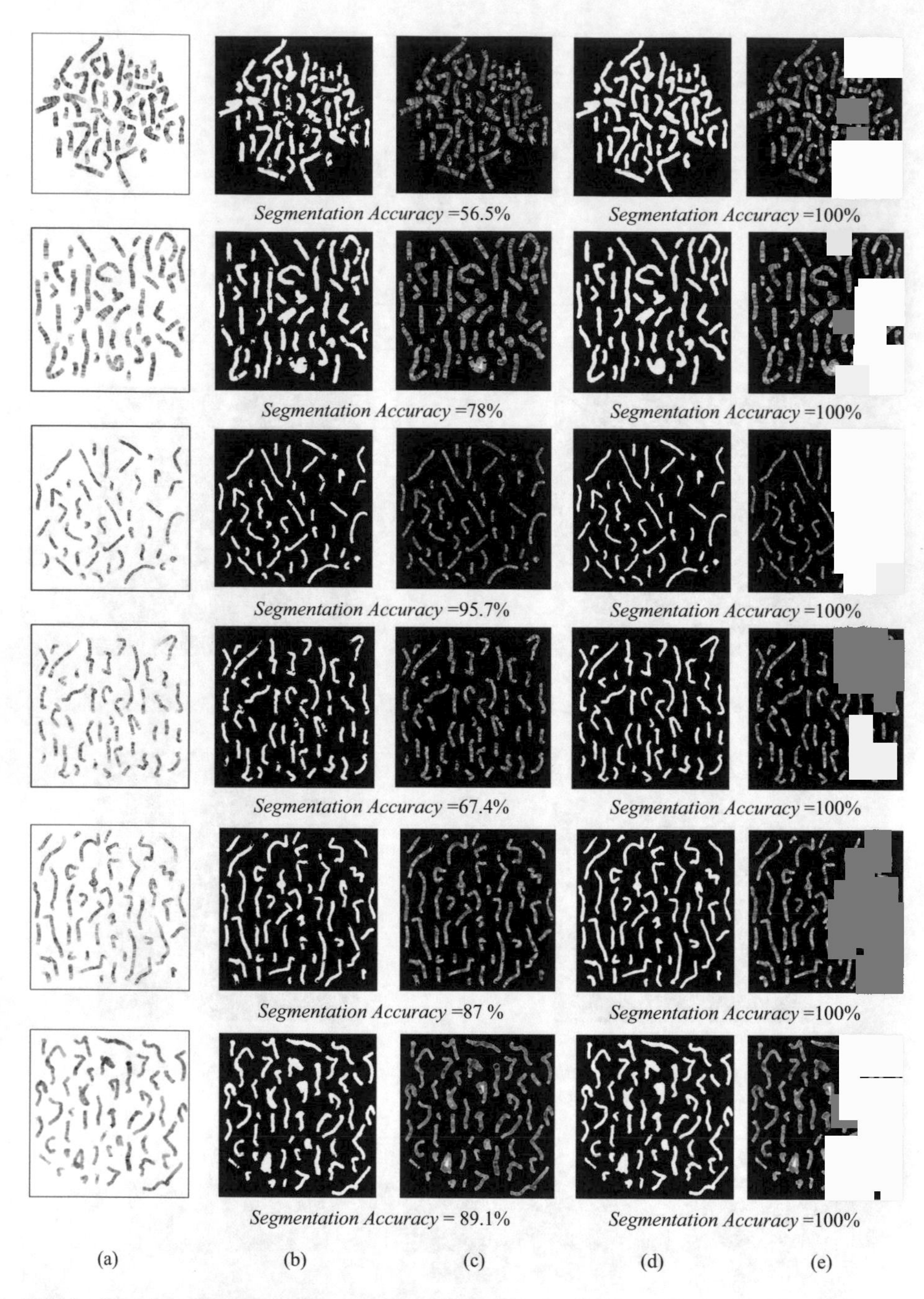

Fig. 4. Samples output of the proposed chromosome images segmentation. (a) Original chromosome images, (b) segmented chromosome images without DoG, (c) masked chromosome images without DoG, (d) segmented chromosome images with DoG, (e) masked chromosome images with DoG

photomicrography. Digital images JPEG (Joint Photographic Experts Group) are recorded and compressed at a resolution of 72 or 144 dpi [21]. Figure 4(a) shows six samples of the used chromosomes digital metaphase images.

4.2 Experiments

In this paper, a method for chromosome images segmentation based on sharpening the image by applying the Difference of Gaussian (DoG) before the classic segmentation method (thresholding then series of morphological operation) is presented. Hence, the performance of the proposed segmentation method is investigated and compared with and without the application of the DoG, step 2 in Sect. 3. The proposed chromosome segmentation performance is evaluated according to the accuracy of the segmentation as in Eq. (3).

$$Segmentation\,Accuracy = \frac{C}{N} * 100\% \tag{3}$$

Where N is the total number of chromosomes, and C is the number of the correctly segmented chromosomes in an image.

Figure 4(b) and (c) show the segmented and masked chromosome images respectively along with the Segmentation accuracy using the proposed chromosome segmentation method without the application of DoG step. Figure 4(d) and (e) show the same with the application of DoG step. It can be shown from Fig. 4 that the DoG step enhanced the chromosome segmentation process.

The proposed method is tested on 130 chromosome images (contain 6011 chromosomes). The average segmentation accuracy without and with DoG step are 74.4% and 99.8% respectively.

5 Conclusions and Future Work

This paper presents a proposed chromosome segmentation of digitized metaphase images. Segmentation is done using Difference of Gaussian (DoG) as a sharpening filter before a classic technique of using Otsu's thresholding method followed by dilation, hole filling and erosion morphological operations. The experiment proved that this algorithm can segment the chromosome images successfully. The proposed method is tested on 130 metaphase images (6011 chromosomes) that are free of overlap and have severe bending. These images are provided by The DGMU laboratory technicians at King Abdulaziz University, that are obtained from 109 patients. The average segmentation accuracy without and with DoG step are 74.4% and 99.8% respectively. The results show that the DoG step enhanced the chromosome segmentation process.

As a future work, the proposed method will be improved to handle metaphase images with overlapped and touched chromosomes. Also, an automated chromosome classification system will be implemented based on the proposed segmentation method.

Acknowledgment. This work was supported by King Abdulaziz City for Science and Technology (KACST) under Grant Number (PGP – 37 – 1165). Therefore, the authors wish to thank, KACST technical and financial support. The authors also like to thank Diagnostic Genomic Medicine Unit (DGMU), King Abdulaziz University for providing the chromosome images dataset.

References

1. Jorde, L.B., Carey, J.C., Bamshad, M.J.: Medical Genetics e-Book. Elsevier Health Sciences, New York (2015)
2. Wang, X., Zheng, B., Wood, M., Li, S., Chen, W., Liu, H.: Development and evaluation of automated systems for detection and classification of banded chromosomes: current status and future perspectives. J. Phys. D Appl. Phys. **38**, 2536 (2005)
3. Munot, M.V., Joshi, M.A., Mandhawkar, P.: Semi automated segmentation of chromosomes in metaphase cells. In: IET Conference on Image Processing (IPR 2012), pp. 1–6 (2012)
4. Pham, D., Xu, C., Prince, J.: A survey of current methods in medical image segmentation. Technical report, Johns Hopkins University, Baltimore (1998)
5. Guimaraes, L., Schuck, A., Elbern, A.: Chromosome classification for karyotype composing applying shape representation on wavelet packet transform. In: The 25th Annual International Conference of the IEEE, pp. 941–943 (2003)
6. Wenzhong, Y., Dongming, L.: Segmentation of chromosome images by mathematical morphology. In: 2013 3rd International Conference on Computer Science and Network Technology (ICCSNT), pp. 1030–1033 (2013)
7. Madian, N., Jayanthi, K., Suresh, S.: Contour based segmentation of chromosomes in G-band metaphase images. In: 2015 IEEE Global Conference on Signal and Information Processing (GlobalSIP), pp. 943–947 (2015)
8. Ji, L.: Intelligent splitting in the chromosome domain. Pattern Recognit. **22**, 519–532 (1989)
9. Stanley, R.J., Keller, J.M., Gader, P., Caldwell, C.W.: Data-driven homologue matching for chromosome identification. IEEE Trans. Med. Imaging **17**, 451–462 (1998)
10. Neethu Sathyan, M., Remya, R.S., Sabeena, K.: Automated karyotyping of metaphase chromosome images based on texture features. In: 2016 International Conference on Information Science (ICIS), pp. 103–106 (2016)
11. Keerthi, V., Remya, R.S., Sabeena, K.: Automated detection of centromere in G banded chromosomes. In: 2016 International Conference on Information Science (ICIS), pp. 83–86 (2016)
12. Yan, F., Zhang, H., Kube, C.R.: A multistage adaptive thresholding method. Pattern Recognit. Lett. **26**, 1183–1191 (2005)
13. Jin-Yu, Z., Yan, C., Xian-Xiang, H.: Edge detection of images based on improved Sobel operator and genetic algorithms. In: 2009 International Conference on Image Analysis and Signal Processing, pp. 31–35 (2009)
14. Ji, L.: Fully automatic chromosome segmentation. Cytometry **17**, 196–208 (1994)
15. Huang, M., Mu, Z., Zeng, H., Huang, H.: A novel approach for interest point detection via Laplacian-of-bilateral filter. J. Sens. **2015**, 9 (2015)
16. Mu, K., Hui, F., Zhao, X., Prehofer, C.: Multiscale edge fusion for vehicle detection based on difference of Gaussian. Optik Int. J. Light Electron Opt. **127**, 4794–4798 (2016)
17. Zhang, J., Hu, J.: Image segmentation based on 2D Otsu method with histogram analysis. In: 2008 International Conference on Computer Science and Software Engineering, pp. 105–108 (2008)

18. Fang, M., Yue, G., Yu, Q.: The study on an application of Otsu method in Canny operator. In: International Symposium on Information Processing (ISIP), pp. 109–112 (2009)
19. Neto, J.F., Braga, A.M., de Medeiros, F.N., Marques, R.C.: Level-set formulation based on otsu method with morphological regularization. In: 2017 IEEE International Conference on Image Processing (ICIP), pp. 2144–2148 (2017)
20. C. o. E. I. G. M. Research, Center of Excellence in Genomic Medicine Research. http://cegmr.kau.edu.sa/
21. Abramowitz, M., Davidson, M.W.: Digital imaging in optical microscopy. https://www.olympus-lifescience.com/en/microscope-resource/primer/digitalimaging/olympusdp10/

Cube Satellite Failure Detection and Recovery Using Optimized Support Vector Machine

Sara Abdelghafar[1,4](✉), Ashraf Darwish[2,4], and Aboul Ella Hassanien[3,4]

[1] Computer Science Department, Faculty of Science, Al Azhar University, Cairo, Egypt
Sara.abdelghafar@yahoo.com
[2] Faculty of Science, Helwan University, Cairo, Egypt
[3] IT Department, Faculty of Computers and Information, Cairo University, Giza, Egypt
[4] Scientific Research Group in Egypt (SRGE), Giza, Egypt
http://www.egyptscience.net/

Abstract. Failure detection and recovery is one of the important operations of the health monitoring management system which plays a relevant role for keeping the reliability and availability of the failed sensor over an on-orbit system whole lifetime mission especially where the maintenance may be impossible. In this paper we have implemented Grey Wolf Optimization (GWO) for optimizing Support Vector Machines (SVM) in terms of recovering the detected failure of the failed sensor. The performance of the proposed model with GWO is compared against to four swarm algorithms; Ant Lion Optimizer (ALO), Dragonfly Algorithm (DA), Moth Flame Optimizer (MFO) and Whale Optimizer Algorithm (WOA), four different evaluation aspects are used in this comparison; failure recovering accuracy, stability, convergence and computational time. The experiment is implemented using cube satellite telemetry data, the experimental results demonstrate that the optimization of SVM using GWO (SVM-GWO) model can be regarded as a promising success for satellite failure detection and recovery.

Keywords: Failure detection and recovery · Grey Wolf Optimization
Telemetry data mining · Support Vector Machine · Satellite health monitoring

1 Introduction

The health and performance of artificial satellites are monitored by analyzing the stream of telemetry data received at the ground control station during its entire mission life. Telemetry is non-stationary time series data contains the wealth information related to the health and status of the satellite and actual operating state of each subsystem, collected via a thousands of sensors distributed on all subsystems. As a satellite is operating in a very special, harsh, and challenging environment so it is difficult to eliminate the possibility of failures which resulting from a totally losing or failures to retrieve the expected response. Therefore, the failure and lose the measurements is one of the most important challenge facing the health monitoring

A. E. Hassanien et al. (Eds.): AISI 2018, AISC 845, pp. 664–674, 2019.
https://doi.org/10.1007/978-3-319-99010-1_61

operations. Hence, failure detection and recovery is addressed for space system making full use of the historical data to detect the failures or malfunctions in sensor or subsystem then automatically initiate recovery actions to manage and enhance the life time of subsystems sensors by keeping the reliability and availability of their reading and measurements over the whole satellite mission life.

This paper presents a hybrid model for an optimal failure detection and recovery of cube satellite missions, this model developed based on SVM for failure detection, then integrated with swarm algorithm for obtaining the optimal recovery solution. This model have been implemented using GWO, which is compared against four swarm algorithms; ALO, DA, MFO and WOA, with using three kernels of SVM; linear, quadratic and cubic. The comparison of performance is built using four different aspects; recovering accuracy, stability, convergence and computational time. The rest of the paper is organized as follows: Sect. 2 presents the core concepts of SVM and GWO. Section 3 introduces the proposed model. Section 4 defines the experimental dataset and the evaluation metrics, then discuss the obtained experimental results. Finally, Sect. 5 presents conclusions and future work.

2 Preliminaries

2.1 Support Vector Machine (SVM)

SVM is a supervised machine learning algorithm which can be used for both classification and regression analysis for high dimensional datasets with great results [1]. The classification problem consists of assigning a class label to data points in input space, all data points in the same class express the same pattern to be learned. SVM solves the classification problem via trying to find an optimal separating hyperplane between classes [2]. Any classification problem can be defined as follows: Given a training dataset with n samples $\{(x_1, y_1), (x_2, y_2), \ldots, (x_n, y_n)\}$, where x_i is a feature a vector in a v-dimensional feature space and with labels $yi \in \{-1, 1\}$ belonging to either of two linearly separable classes' C_1 and C_2. Geometrically, finding an optimal hyperplane with the maximal margin to separate two classes requires for solving the optimization problem, as shown in Eqs. (1) and (2).

$$\boldsymbol{maximize} \sum_{i=1}^{n} \alpha_i - \frac{1}{2} \sum_{i,j=1}^{n} \alpha_i \, \alpha_j \, y_i \, y_j . K(x_i, x_j) \quad (1)$$

$$Subject - to\!: \sum_{i,j=1}^{n} \alpha_i \, y_i, 0 \leq \alpha_i \leq C \quad (2)$$

Where, α_i is the weight assigned to the training sample x_i. If $\alpha_i > 0$, x_i is called a support vector. C is a regulation parameter used to trade-off the training accuracy and the model complexity. K is a kernel function, which is used to measure the similarity between two samples [3, 4]. SVM is being widely used for failure detection based on the correlation coefficient which is used to detect the faulty of the received data from the sensors by checking the variation of correlation coefficient against the training data set, this variation can be classified to normal class if lies in the normal areas and meets constraint conditions or faulty class if it does not meet the limit or constraint. Assuming two correlated nodes as y_i and y_j, the correlation coefficient $\left(\sigma_{y_i y_j}\right)$ of observed data among y_i and y_j is estimated based on the mean, variance (V) and covariance (CV) of the data sample at time t, it is given using the following equation [5].

$$\sigma_{y_i y_j} = \frac{CV\left(y_i y_j\right)}{V(y_i)V\left(y_j\right)} \tag{3}$$

The value of σ varies in the range between −1 and +1, the sign quantifies the direction (directly or inversely proportional). For any point that lies on defined normal boundary, we can write

$$(w.\sigma) + b = 0 \tag{4}$$

Where w a vector that defines the boundary, and b is a scalar threshold value and can be calculated as a following:

$$b = \frac{1}{N_{sv}} \sum\nolimits_{i=1}^{N_{sv}} (w.\sigma_i - y_i) \tag{5}$$

A decision function based on the variation of correlation coefficient can be created to determine whether a given data point belongs to normal or fault class, can be defined as following [6]

$$Y = sign(w.\sigma) + b \tag{6}$$

2.2 Grey Wolf Optimization (GWO)

GWO is one of the new meta-heuristic optimization algorithms, which was introduced by Mirjalili et al. [7]. GWO is an innovative algorithm based on population that stimulates mechanism of grey wolves hunting in nature. Grey wolves live in a pack and the size of group is between 5 and 12. The leader is called alpha and is responsible for making decisions for their pack such as sleep place, hunting, and wake-up time. Mostly, other individuals of the pack must obey the decision made by alpha. The second is called beta and helps the alpha in decision making. Beta is the consultant of alpha and responsible for disciplining the pack. The beta reinforces the orders of alpha and gives alpha the feedbacks. The weakest level in a pack of grey wolves is omega that have to obey other individuals' orders, and submits the information to all the others dominant wolves. The rest of grey wolves are called delta that obeys the alpha and beta wolves and dominates the omega wolves.

The GWO algorithm utilizes the simulated social leadership and encircling mechanism in order to find the optimal solution for optimization problems. This algorithm saves the first three best solutions obtained so far and obliges other search agents to update their positions with respect to them. In order to mathematically model the social hierarchy of grey wolves in the GWO algorithm, the best solution is considered as alpha (α). Therefore, the second and third best solutions are respectively considered as beta (β) and delta (δ), and other solution is assumed to be omega (ω). In the GWO algorithm, hunting (optimization) is guided by α, β and δ, and ω wolves follow them. In the process of hunting, the grey wolves' encircling behavior to hunt for a prey can be expressed as following:

$$\vec{D} = \left|\vec{C}.\vec{X}_p(t) - \bar{X}(t)\right| \tag{7}$$

$$\vec{X}(t+1) = \overrightarrow{X_p}(t) - \vec{A}.\vec{D} \tag{8}$$

Where t = iteration number; $\overrightarrow{A}$ and $\overrightarrow{C}$ = coefficient vectors; $\vec{X}_p$ = vector of the prey's position; $\vec{X}$ = vector of the wolf's positions; and $\overrightarrow{D}$ calculated vector used to specify a new position of the grey wolf. $\overrightarrow{A}$ and $\overrightarrow{C}$ can be calculated by (9) and (10) respectively:

$$\overrightarrow{A} = 2\overrightarrow{a}.\overrightarrow{r_1} - \overrightarrow{a} \tag{9}$$

$$\vec{C} = 2.\overrightarrow{r_2} \tag{10}$$

Where a is linearly decreased from 2 to 0 over the course of iterations,r_1, r_2 are random values in $[0, 1]$. In the same way, in n dimensions the search agents can move to any nodes of a hypercube around the best solution (position of the prey). The optimization process of GWO starts with creating random population of grey wolves. Assuming that, α is optimum solution, β and δ are candidate solutions. Over the iterations, α, β and δ wolves estimate the probable position of the prey (optimum solution), grey wolves update their positions based on their distances from the prey, as following.

$$\overrightarrow{D_\alpha} = \left|\overrightarrow{C_1}.\vec{X}_\alpha - \vec{X}\right|; \overrightarrow{D_\beta} = \left|\overrightarrow{C_2}.\vec{X}_\beta - \vec{X}\right|; \overrightarrow{D_\delta} = \left|\overrightarrow{C_3}.\vec{X}_\delta - \vec{X}\right| \tag{11}$$

$$\vec{X}_1 = \overrightarrow{X_\alpha} - A_1.\left(\overrightarrow{D_\alpha}\right); \vec{X}_2 = \overrightarrow{X_\beta} - A_2.\left(\overrightarrow{D_\beta}\right); \vec{X}_3 = \overrightarrow{X_\delta} - A_3.\left(\overrightarrow{D_\delta}\right) \tag{12}$$

In order to emphasize exploration and exploitation during the search process, parameter a should decrease from 2 to 0. If $\left|\vec{A}\right| > 1$, the candidate solutions diverge from the prey; and if $|A| < 1$, the candidate solutions converge to the prey. Algorithm saves three best solutions achieved so far and forces others (ω) to update their positions to achieve the best place in the decision space. This process continues and the GWO algorithm is terminated if the stopping criteria are satisfied [8, 9].

3 The Proposed Model

In the context of this paper, SVM performs two functions, namely failure detection and failure recovery. First, SVM based on the correlation coefficient is used to classify the received data from the sensors by comparing the variation of correlation coefficient against the training data set to check if it lies in normal class or faulty class, and then the type of fault data is detected. The faulty data in the satellite telemetry data may be as a complete or partial loss of data, or unexpected readings which is defined as a constant value that is either higher or lower value when compared to normal measurements for successive training sets, as shown in Fig. 1.

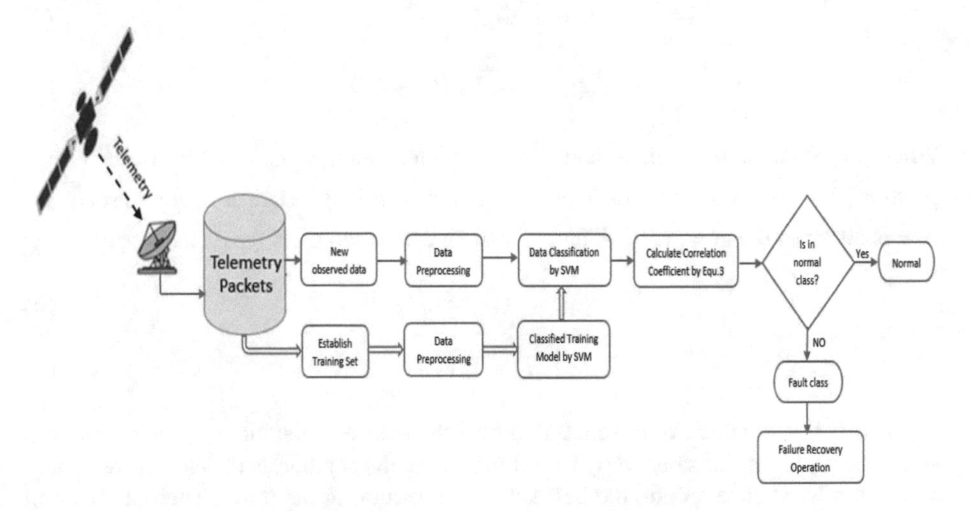

Fig. 1. The failure detection architecture

After detecting the faulty data, the proposed optimized SVM model as shown in Fig. 2, is used to execute loss recovery operation. The proposed recovery mechanism is built on predicting the lost readings of the failed sensor using the classification process, which starts with learning classifier from training data, then using the training model to predict the labels on test data and calculate the kernel function that represents the scalar of classification process for controlling the tradeoff between maximizing the margin and minimizing the training error. The kernel function is used as a fitness function in the optimization process through optimizing its parameters, which are considered as the input of swarm algorithm, whereas the output is the optimal SVMs parameters that optimizing the classification output and hence the prediction results. The proposed approach is divided to three phases; training, optimization and recovering failure on the independent test set.

Phase 1: Training phase
The first phase of approach starts with the classification process by SVM on the training set, where begin with building classifier model with training data and tested for the prediction, the output of the classification training phase is used to calculate the fitness values through finding the optimal SVMs parameters which will be as an input for the swarm algorithm to start the optimization phase at this stage as a second phase of the approach.

Phase 2: Optimization phase
The optimization phase starts with swarm algorithm which is firstly used to generate the initial positions of population then update the position of the search based on the initial parameters of SVM resulted from training phase, whereas the classification kernel function is used as the cost function of the optimization process to update the position of the search until reaches to the stopping criteria, the resulted optimal parameters return again for the classification process to start the third phase.

Phase 3: Recovering failure on the independent test set phase
The resulted optimal parameters from the optimization phase is the input of this phase which will be used to optimize the classifier model then will be tested for recovering the failure on the independent set by predicting the lost readings of the failed sensor. Stopping criteria is based on the prediction accuracy, resulting the optimal prediction result on the independent test set.

4 Experimental Results

The telemetry packets have been used for the implementation in this paper that are gathered from more than 30 sensors on the Cube Satellite [10], provide information about temperature, voltages and currents. The experiment was implemented in MATLAB-R2015a on a computer with Intel Core i3 and 4 GB memory. We firstly used SVM to detect the failure then evaluated the effectiveness of the using hybrid model of swarm and SVM for recovery mechanism. Five swarms ALO, DA, GWO, MFO and WOA have been used with SVM, and they have been compared with using 3 kernel methods of SVMs linear, quadratic and cubic. The proposed algorithms are compared in four aspects: failure recovering accuracy, stability, convergence and computational time. The proposed algorithms were tested on 10 telemetry packets (P_{1-10}) were taken at successive intervals with different modes.

4.1 Comparison of the Accuracy

Root mean square error (RMSE) is used to evaluate the accuracy performance of the proposed approach, RMSE is one of the most common metrics used to measure predicative accuracy for time series data [11]. Equation 13 defines how RMSE is calculated, where $\hat{y}_i$ is observed values and y_i is predicated values at time i, then (1-RMSE) % is used to express the accuracy.

$$RMSE = \sqrt{\frac{\sum_{i=1}^{n} (\hat{y}_i - y_i)^2}{n}} \quad (13)$$

The Table 1 lists the detailed results for the predication accuracy are obtained by the five proposed models with 10 telemetry packets taken sequentially have one failed sensor, from the table we can see that all proposed models have achieved excellent prediction accuracy for the first two packets then the accuracy for the other packets began to decrease over the time. Based on the average of the accuracy for the proposed model with different kernel of SVM, we can see the height accuracy with all different kernels of SVM has been achieved using GWO. As can be seen, the accuracy for the proposed model using all swarms have been decreased for both quadratic and cubic kernels, while GWO has the lowest decreasing for both these kernels.

4.2 Comparison of Stability

Standard deviation is used to evaluate the stability of the proposed model with different swarms. It is defined as following:

$$Std = \sqrt{\frac{\sum_{i=1}^{n_r}(X_i - \mu)^2}{n_r}} \quad (14)$$

$$\mu = \frac{\sum_{i=1}^{n_r} X_i}{n_r} \quad (15)$$

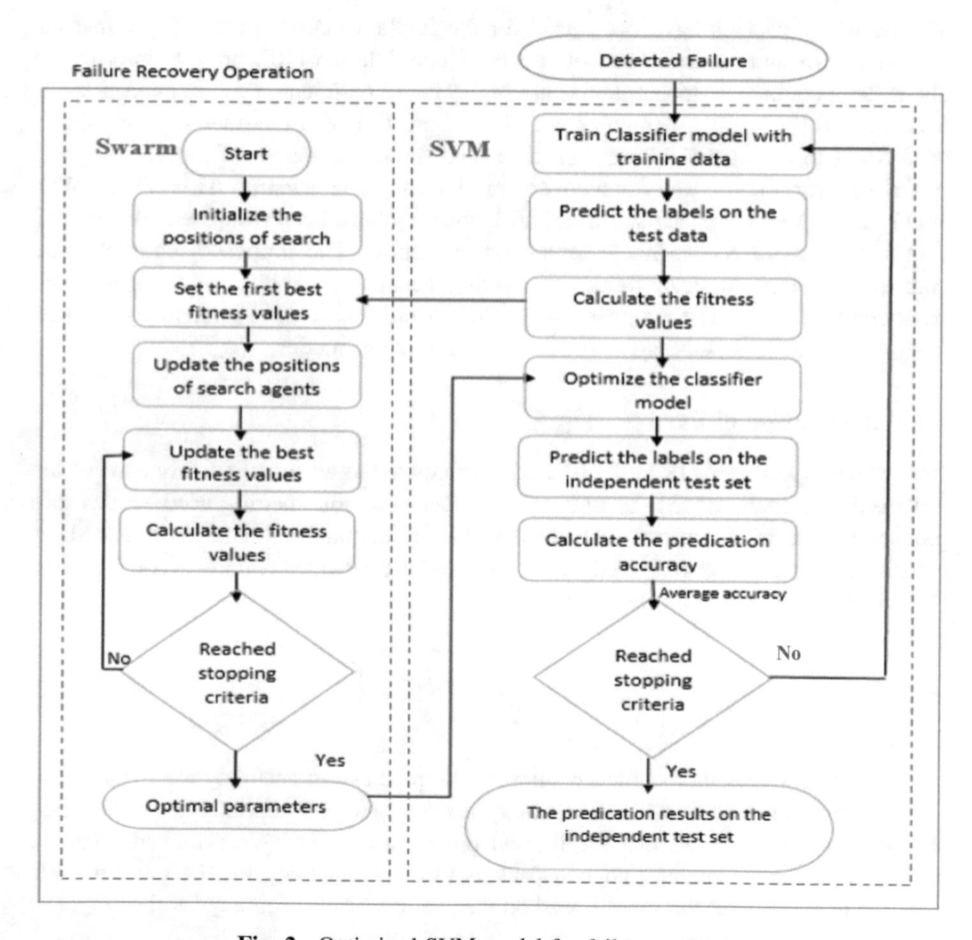

Fig. 2. Optimized SVM model for failure recovery

Where μ is the mean, and is defined in Eq. 15, n_r is the number of iterations and X is the best score obtained each time [12]. The standard deviation of the obtained fitness values are calculated for 10 runs and outlined in Table 2. As can be observed GWO owns the best stability quality as has minimum standard deviation with quadratic and cubic kernels and was placed in the third place with linear kernel while highest stability for linear was obtained with WOA.

Table 1. The obtained accuracy of proposed model with different swarms

	SVM kernel functions	P1	P2	P3	P4	P5	P6	P7	P8	P9	P10	Average
GWO	Linear	99.96	99.71	95.1	97.58	94.28	96.1	90.5	94.88	81.1	62.65	91.186
	Quadratic	99.97	99.8	95.1	96.28	90.9	96	90.6	94	81.1	58.2	90.195
	Cubic	99.97	99.8	96.02	96.5	91	96.5	89.5	94.79	81.1	60	90.518
AWO	Linear	99.96	99.71	95.1	96.2	94.57	79.54	67.99	75.71	52.35	62.66	82.379
	Quadratic	99.98	99.72	93.1	97.58	87.25	80.34	65.84	78.23	51.1	51.1	80.424
	Cubic	99.98	99.72	95.1	76.2	82.66	70	53.76	71.66	47.95	56.44	75.347
ALO	Linear	99.96	99.7	95.1	94.9	94.3	79.5	68	75.97	52.34	62.63	82.24
	Quadratic	99.98	99.72	93.1	97.6	87.25	80.34	65.84	78.23	51.1	51.1	80.426
	Cubic	99.99	99.7	95.1	98.5	82.6	71.2	58	71.7	55.2	57.4	78.939
DA	Linear	99.96	99.71	95.1	96.2	94.57	79.55	67.9	75.7	52.4	62.66	82.375
	Quadratic	99.98	99.7	93.1	97.58	87.22	80.47	67.46	80.5	64.25	51	82.126
	Cubic	99.99	99.73	95.1	76.17	82.57	69.88	53.75	71.63	49.5	56.43	75.475
MFO	Linear	99.96	99.7	95.1	96.2	94.57	79.58	67.94	75.7	52.4	62.66	82.381
	Quadratic	99.98	99.72	93.1	97.58	87.25	80.34	65.84	78.23	51.1	52	80.514
	Cubic	99.98	99.72	95.1	76.2	82.57	69.99	53.76	71.68	55.18	56.4	76.058

Table 2. Stability of the proposed model with different swarms

SVM kernel functions	Swarms	Standard deviation
Linear	WOA	1.12E−05
	MFO	1.77E−04
	GWO	1.00E−04
	ALO	2.76E−04
	DA	3.69E−04
Quadratic	GWO	0
	ALO	1.61E−05
	DA	8.50E−05
	WOA	2.70E−03
	MFO	6.70E−03
Cubic	GWO	1.05E−07
	WOA	2.93E−06
	MFO	2.08E−05
	DA	2.00E−04
	ALO	2.77E−04

4.3 Comparison of Convergence Speed

In order to compare the convergence performance of the proposed model with different swarms, convergence curves are drawn in this section for linear, quadratic and cubic kernels in Fig. 3(A), (B) and (C) respectively. As can be seen in the following figures, the proposed model with GWO has a faster convergence speed than the other swarms with quadratic and cubic kernels while the WOA has the highest fast convergence speed with linear kernel.

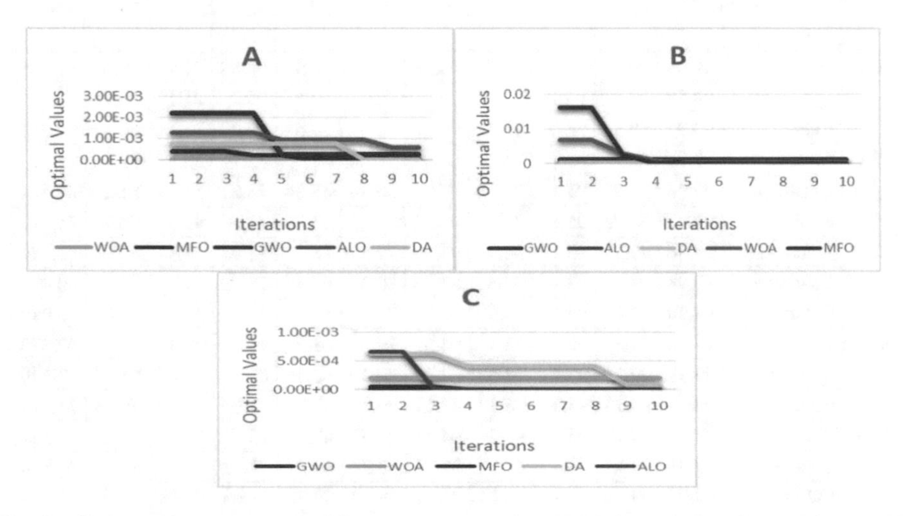

Fig. 3. Convergence curve of different swarms with SVM kernel functions; Linear (A), Quadratic (B) and Cubic (C)

4.4 Comparison of the Computational Time

The mean computational time for the proposed model with each SVM kernel was calculated and represented in Fig. 4. As can be seen, the GWO has the lowest computational time while MFO is the highest.

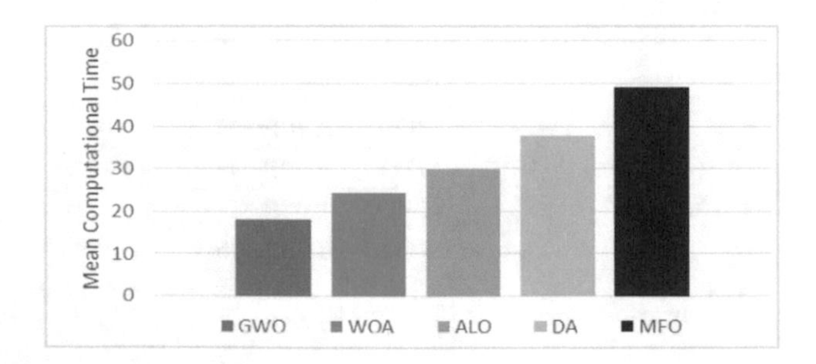

Fig. 4. Mean computational time for the proposed model with different swarms

5 Conclusion and Future Work

Failure detection and recovery is an essential requirement for any space mission, as they work and fulfill their mission in very harsh environment, so keeping the reliability and availability of all sensors over the whole mission lifetime is a big challenge and needs for intelligence approaches to detect the failure and loss recovery, which maintaining the survival of control and health monitoring operations between space system and the ground station, which has been basically based on the analyzing the stream of telemetry data.

In this paper, hybrid model for an optimal failure detection and recovery for small satellite missions like the cube satellite is proposed and developed based on SVM, which is used to perform failure detection by classifying the received telemetry data to normal and faulty classes based on the variation of the correlation coefficient. After faulty is detected, the second role of SVM is coming to combine with GWO for recovering function. The proposed model with GWO is compared against four swarm algorithms; ALO, DA, MFO and WOA, with using three kernels of SVM; linear, quadratic and cubic. From the experimental results, it can be concluded that the highest accuracy for recovering is achieved with GWO especially with SVM linear kernel function. As well as, GWO has the lowest computational time with all SVM kernels. In addition to, it has the most stable algorithm with quadratic and cubic kernels and has been placed in the third place with linear kernel. Also, it has a faster convergence speed than the other swarms with quadratic and cubic kernels. Future work includes the development of the proposed model to maintain the recovering with high accuracy even with long time periods. Also, the modes and the real time aspects of the telemetry data shall be addressed.

Acknowledgement. This work is supported by Egypt Knowledge and Technology Alliance (E-KTA), which is funded by The Academy of Scientific Research & Technology (ASRT), and coordinated by National Authority for Remote Sensing & Space Sciences (NARSS).

References

1. Hearst, M.A., Dumais, S.T., Osuna, E., Platt, J., Scholkopf, B.: Support vector machines. IEEE Intell. Syst. Their Appl. **13**(4), 18–28 (1998)
2. Chen, J., Licheng, J.: Classification mechanism of support vector machines. In: Proceedings of 5th International Conference on Signal Processing Proceedings. 16th World Computer Congress 2000, Beijing, China. IEEE (2000)
3. Qiang, W., Xuan, D.: Analysis of support vector machine classification. Comput. Anal. Appl. **8**(2), 99–119 (2006)
4. Elhariri, E., El-Bendary, N., Mostafa, M., Fouad, M., Platos, J., Hassanien, A.E., Hussein, M.M.: Multi-class SVM based classification approach for tomato ripeness. In: Proceedings of 4th International Conference on Innovations in Bio-Inspired Computing and Applications, IBICA, Ostrava, Czech Republic (2013)
5. Zai, W.Y., Guo, W.X.: The fault detection and diagnosis for fractionating tower based on correlation coefficient. In: Proceedings of International Symposium on Computer, Consumer and Control, China, pp. 268–274. IEEE Computer Society (2016)

6. Kamalesh, S., Ganesh Kumar, P.: Data aggregation in wireless sensor network using SVM-based failure detection and loss recovery. J. Exp. Theor. Artif. Intell. **29**, 1362–3079 (2016)
7. Mirjalili, S., Lewis, A.: Grey wolf optimizer. Adv. Eng. Softw. **69**, 46–61 (2014)
8. Rezaei, H., Bozorg-Haddad, O., Chu, X.: Grey Wolf Optimization (GWO) algorithm. In: Bozorg-Haddad, O. (ed.) Advanced Optimization by Nature-Inspired Algorithms. Studies in Computational Intelligence, Karaj, Iran, vol. 720, pp. 81–91. Springer (2018). https://link.springer.com/content/pdf/10.1007%2F978-981-10-5221-7_9.pdf. Accessed 1 Apr 2018
9. Elhariri, E., El-Bendary, N., Hassanien, A.E., Abraham, A.: Grey wolf optimization for one-against-one multi-class support vector machines. In: Proceedings of Soft Computing and Pattern Recognition (SoCPaR), Fukuoka, Japan, pp. 7–12. IEEE (2015)
10. FunCube Real Time Data. http://warehouse.funcube.org.uk/. Accessed 1 Nov 2017
11. Jifri, M.H., Hassan, E.E., Miswan, N.H.: Forecasting performance of time series and regression in modeling electricity load demand. In: Proceedings of 7th IEEE International Conference on System Engineering and Technology (ICSET), Shah Alam, Malaysia. IEEE (2017)
12. Bonyadi, M.R., Michalewicz, Z.: Analysis of stability, local convergence, and transformation sensitivity of a variant of the particle swarm optimization algorithm. IEEE Trans. Evol. Comput. **20**(3), 370–385 (2016)

Author Index

A. E. Hassanien et al. (Eds.): AISI 2018, AISC 845, pp. 675–677, 2019.
https://doi.org/10.1007/978-3-319-99010-1

T

Y

Z

Printed in the United States
By Bookmasters